Atlas of Hematopathology
Morphology, Immunophenotype, Cytogenetics, and Molecular Approaches

ELSEVIER *science & technology books*

Companion Web Site:

http://booksite.elsevier.com/9780123851833

Atlas of Hematopathology: Morphology, Immunophenotype, Cytogenetics, and Molecular Approaches
Faramarz Naeim, P. Nagesh Rao, Sophie X. Song, Wayne W. Grody

Resources

All figures from the book available as both PowerPoint slides and JPEG files.

Please note: Figures are password protected. Upon loggging into the site, you will be prompted for a unique password from the text of the book.

TOOLS FOR ALL YOUR TEACHING NEEDS
textbooks.elsevier.com

ACADEMIC PRESS

To adopt this book for course use, visit http://textbooks.elsevier.com.

Atlas of Hematopathology
Morphology, Immunophenotype, Cytogenetics, and Molecular Approaches

Faramarz Naeim M.D., FASCP
*Professor Emeritus, David Geffen School of Medicine, UCLA
Director of Hematopathology
Department of Pathology and Laboratory Medicine
VA Greater Los Angeles Healthcare System
Los Angeles, California*

P. Nagesh Rao Ph.D., FACMG
*Professor, Department of Pathology and Lab Medicine and Pediatrics
Director of Clinical and Molecular Cytogenetics Laboratories
David Geffen School of Medicine, UCLA
Los Angeles, California*

Sophie X. Song M.D., Ph.D.
*Associate Clinical Professor, Department of Pathology and Lab Medicine
Director of the Flow Cytometry and Bone Marrow Laboratories
David Geffen School of Medicine, UCLA
Los Angeles, California*

Wayne W. Grody M.D., Ph.D., FACMG, FCAP, FASCP
*Professor, Departments of Pathology and Laboratory Medicine,
Pediatrics, and Human Genetics
Director of Molecular Pathology Laboratories
David Geffen School of Medicine, UCLA
Los Angeles, California*

AMSTERDAM • BOSTON • HEIDELBERG • LONDON • NEW YORK • OXFORD • PARIS
SAN DIEGO • SAN FRANCISCO • SINGAPORE • SYDNEY • TOKYO

Academic Press is an imprint of Elsevier

Academic Press is an imprint of Elsevier
32 Jamestown Road, London NW1 7BY, UK
225 Wyman Street, Waltham, MA 02451, USA
525 B Street, Suite 1800, San Diego, CA 92101-4495, USA

First edition 2013

Copyright © 2013 Elsevier Inc. All rights reserved

Medicine is an ever-changing field. Standard safety precautions must be followed, but as new research and clinical experience broaden our knowledge, changes in treatment and drug therapy may become necessary or appropriate. Readers are advised to check the most current product information provided by the manufacturer of each drug to be administered to verify the recommended dose, the method and duration of administrations, and contraindications. It is the responsibility of the treating physician, relying on experience and knowledge of the patient, to determine dosages and the best treatment for each individual patient. Neither the publisher nor the authors assume any liability for any injury and/or damage to persons or property arising from this publication.

No part of this publication may be reproduced, stored in a retrieval system or transmitted in any form or by any means electronic, mechanical, photocopying, recording or otherwise without the prior written permission of the publisher Permissions may be sought directly from Elsevier's Science & Technology Rights Department in Oxford, UK: phone (+44) (0) 1865 843830; fax (+44) (0) 1865 853333; email: permissions@elsevier.com. Alternatively, visit the Science and Technology Books website at www.elsevierdirect.com/rights for further information

Notice
No responsibility is assumed by the publisher for any injury and/or damage to persons or property as a matter of products liability, negligence or otherwise, or from any use or operation of any methods, products, instructions or ideas contained in the material herein. Because of rapid advances in the medical sciences, in particular, independent verification of diagnoses and drug dosages should be made

British Library Cataloguing-in-Publication Data
A catalogue record for this book is available from the British Library

Library of Congress Cataloging-in-Publication Data
A catalog record for this book is available from the Library of Congress

ISBN: 978-0-12-385183-3

For information on all Academic Press publications
visit our website at elsevierdirect.com

Typeset by MPS Limited, Chennai, India
www.adi-mps.com

Printed and bound in United States of America

13 14 15 16   10 9 8 7 6 5 4 3 2 1

Working together to grow
libraries in developing countries

www.elsevier.com | www.bookaid.org | www.sabre.org

ELSEVIER   BOOK AID International   Sabre Foundation

Preface

When the time approached for the second edition of the *Hematopathology* book, the editors decided for a major revision in order to make the book more useful to the readers. They decided to create an *Atlas of Hematopathology* by increasing the number of the images, whilst keeping the text concise and to the point. The most important and unique feature of this book is that it is comprehensive and multidisciplinary:

1. Comprehensive, because it covers both non-neoplastic hematopathology and neoplastic hematopathology in all organs and tissues.
2. Multidisciplinary, because in addition to demonstration of conventional histpathology and cytopathology, it demonstrates results of other important complementary diagnostic tests, such as immunophenotyping (immunohistochemical stains and flow cytometry), karyotyping, FISH and DNA/molecular studies.

The explosion of new molecular, cytogenetic, and proteomic techniques applicable to pathology has not rendered histologic examination obsolete, but rather offers powerful ancillary information to facilitate differential diagnosis, predict prognostic behavior, and help in the selection of targeted molecular therapies. With this in mind, we have constructed most chapters along the general format for each disease category to demonstrate clinical aspects, morphology, laboratory findings, immunohistochemical and flow cytometric results, cytogenetics and molecular studies, and differential diagnosis.

This book offers important information to practicing physicians and those in pathology and hematology training, to help them to improve their diagnostic skills. It also functions as a valuable referral book for researchers who work in hematology-related areas. The book consists of 63 chapters. The first 6 chapters are devoted to normal structure and function of hematopoietic tissues, principles of immunophenotyping, cytogenetics and molecular genetics, an overview of abnormal bone marrow morphology, and reactive lymphadenopathies. The remaining chapters deal with various types of neoplastic and non-neoplastic hematopoietic disorders. The hematopoietic neoplasms are classified according to the World Health Organization (WHO) Classification published in 2008, with minor modifications.

Faramarz Naeim, M.D.
P. Nagesh Rao, Ph.D.
Sophie X. Song, M.D. Ph.D.
Wayne W. Grody, M.D., Ph.D.

Acknowledgments

The authors are thankful to Sunita Bhuta, M.D., Jennifer Hsiao. MD, Gholamhossein Pezeshkpour, M.D., Ryan Phan, Ph.D., Diana Tanaka-Mukai, CLS, and Jimin Xu, Ph.D. for their contribution to the collection of the images in this book.

We acknowledge all the residents, fellows, and staff members at VA Greater Los Angeles Healthcare System, UCLA Medical Center and USC Medical Center for providing the opportunity to teach, to learn, and to exchange ideas that are reflected in this book.

The authors would like to thank Mara Conner, Megan Wickline, and Caroline Johnson of Academic Press for their assistance in the production of this book.

Structure and Function of Hematopoietic Tissues

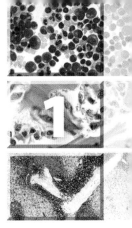

Bone marrow is a mesenchymal-derived complex structure consisting of hematopoietic precursors and a complex microenvironment that facilitates the maintenance of hematopoietic stem cells (HSCs) and supports the differentiation and maturation of the progenitors (Figure 1.1). All differentiated hematopoietic cells including lymphocytes, erythrocytes, granulocytes, macrophages, and platelets are derived from HSCs (Figure 1.2).

The most primitive (pluripotent) HSCs express CD34 and are negative for CD38 and HLA-DR. These primitive cells, which include long-term repopulating stem cells, are also characterized by low-level expression of c-kit receptor (CD117) and absence of lineage specific maturation markers. There is a spectrum of heterogeneity in the bone marrow stem cell pool: a continuum of cells with decreasing capacity for self-renewal and increasing potential for differentiation. This trend is also associated with changes in immunophenotypic features. For example, the more mature committed stem cells, in addition to CD34, appear to express CD38 and/or HLA-DR. The pluripotent HSCs comprise about 1 per 20,000 of bone marrow cells, and only a small fraction of them are active, whereas the remaining majority are in a "resting" phase, on call for action when it is necessary. The HSCs reside in microenvironmental niches. These niches—which are composed of stromal cells, accessory cells (such as T lymphocytes and macrophages), components of extracellular matrix (Table 1.1), and various regulatory cytokines (Table 1.2) (Figure 1.1)—play an important role in the regulation of hematopoiesis and proliferation of the committed stem cells, leading to the production of huge numbers of progenitor cells and differentiated mature blood cells (Figure 1.2). Every day, an estimated 2.5 billion red cells, 2.5 billion platelets, and 1.0 billion granulocytes are produced per kilogram body weight in normal conditions.

Bone marrow stromal cells are derived from pluripotent stromal stem cells. In other words, two separate and distinct pluripotent stem cells are simultaneously at work in bone marrow: hematopoietic and stromal. These two

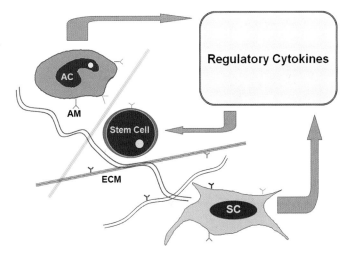

FIGURE 1.1 Cytokines released from accessory cells (AC) (e.g., macrophages, T-cells) and stromal cells have a regulatory effect on stem cells. The extracellular matrix (ECM) and adhesion molecules (AM) support cell–cell, cell–matrix, and cell–cytokine interactions.

systems not only coexist but closely interact with each other. Stromal cells are composed of a heterogeneous cell population including adipocytes, fibroblast-like cells, endothelial cells, and osteoblasts. They produce a number of cytokines and a group of proteins that are involved in facilitating cell–cell interactions and presenting the cytokines and growth factors to the hematopoietic progenitor cells. Stromal cells with their extracellular matrix make a fibrovascular mesh environment to home and support the hematopoietic precursors. The thin-walled venous sinuses are the most prominent vascular spaces in the bone marrow. They consist of an inner layer of endothelial cells supported by an outer layer of fibroblast-like (parasinal, adventitial) stromal cells. They receive blood from the branches of the nutrient artery and periosteal capillary network. The nutrient artery penetrates the bony shaft, branches into the bone marrow cavity, and forms capillary–venous sinus junctions. The periosteal capillary

Atlas of Hematopathology. DOI: http://dx.doi.org/10.1016/B978-0-12-385183-3.00001-2
© 2013 Elsevier Inc. All rights reserved.

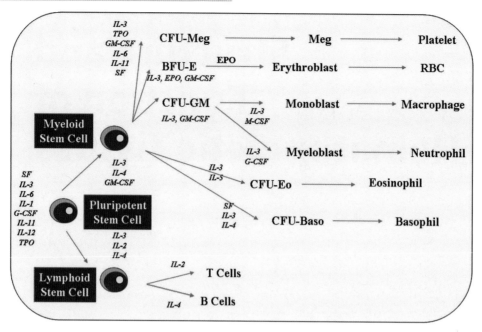

FIGURE 1.2 Current scheme of hematopoiesis demonstrating the differentiation of the multipotent stem cell to hematopoietic precursors and various levels of cytokine interaction.

Table 1.1	
Major Components of Extracellular Matrix in Bone Marrow	
Type	Comments
Collagen (reticulin)	Consisting of various subtypes. Erythroid and myeloid precursors adhere to collagen types I and VI
Fibronectin	Attaches to early erythroid precursors and other hematopoietic and stromal cells
Hemonectin	Myeloid precursors adhere to laminin. Regulates leukocyte chemotaxis.
Proteoglycans	Components containing heparin sulfate, chondritin sulfate, and hyaluronic acid. Interact with laminin and type IV collagen and play a role in cytokine presentation and cell differentiation
Thrombospondin	Interacts with collagen, fibronectin, and CD36

network connects with the sinuses at the bone marrow junction through the Haversian canals. The smaller venous sinuses drain into larger centrally located sinuses, which connect together to form the comitant vein. The comitant vein and the nutrient artery run through the bone marrow cavity adjacent to one another in the same vascular canal (Figure 1.3).

Erythropoiesis takes place in distinct anatomical foci referred to as "erythroblastic islands." These islands are rich in fibronectin and consist of erythroid precursors surrounding a central macrophage. They are usually away from the bone trabeculae and are located subjacent to the vascular structures. Erythropoiesis begins with the commitment of a small pool of pluripotent stem cells to a primitive cell committed to non-lymphoid lineages, referred to as CFU-GEMM. GEMM stands for granulocytes, erythrocytes, macrophages, and megakaryocytes. The most primitive committed erythroid progenitor in humans is the erythroid burst-forming unit (BFU-E), which divides and forms subpopulations of erythroid colonies known as colony-forming units (CFU-E). This process requires a combination of cytokines, such as erythropoietin (EPO), Steel factor (SF; c-kit ligand), and interleukin-3 (IL-3) (Figure 1.2). Proliferation and maturation of CFU-E leads to the formation of erythroblasts, more mature erythroid precursors, and eventually enucleated reticulocytes and erythrocytes. The entire process requires approximately 2 weeks. Except for newborns, only erythrocytes and polychromatic erythrocytes (reticulocytes) are released into the blood circulation.

Myelopoiesis begins with the differentiation of a small population of pluripotent stem cells to CFU-GEMM and then to the committed primitive myeloid precursors, granulocyte/macrophage colony-forming units (CFU-GM). This process requires GM-CSF, SF, IL-3, and IL-6 (Figure 1.2). CFU-GM give rise to more mature colony-forming units CFU-G, CFU-M, CFU-Eo, and CFU-Baso which in turn differentiate into neutrophils, macrophages, eosinophils, and basophils, respectively (Figure 1.2). The neutrophilic precursors in bone marrow consist of two major compartments: mitotic and postmitotic. The mitotic compartment consists of cells that are able to proliferate, such as myeloblasts, promyelocytes, and myelocytes. The postmitotic compartment includes metamyelocytes, bands, and segmented neutrophils, representing more differentiated cells with no proliferating capacity. The granulocytes released from the bone marrow into the circulation consist of two components: the marginating pool and the

Table 1.2
Regulatory Cytokines

Cytokine	Primary Effect
GM-CSF[1]	Granulocyte and macrophage colony formation, functional enhancement of mature forms
G-CSF[2]	Granulocyte colony formation, functional enhancement of granulocytes
M-CSF (CSF-1)[3]	Macrophage colony formation, functional enhancement of monocytes and macrophages
Erythropoietin (EPO)	Erythropoiesis, possible enhancement of megakaryocyte proliferation
Thrombopoietin (TPO)	Megakaryocyte proliferation, platelet production
Steel factor (c-kit ligand)	Stem cell and mast cell proliferation
Interleukin (IL)-1	Promoter of hematopoiesis, inducer of other factors, B- and T-cell regulator, endogenous pyogen
IL-2	T-cell growth factor, may inhibit G/M colony formation and erythropoiesis
IL-3 (Multi-CSF)	G/M colony formation, syngeneic effects on EPO, eosinophil, mast cell, and megakaryocyte colony formation
IL-4	B-cell proliferation, IgE production
IL-5	Eosinophil growth and B-cell differentiation
IL-6	B-cell differentiation, synergistic effects on IL-1
IL-7	Development of B and T cell precursors
IL-8	Granulocyte chemotactic factor
IL-9	Growth of mast cells and T cells
IL-10	Inhibitor of inflammatory and immune responses
IL-11	Synergistic effects on growth of stem cells and megakaryocytes
IL-12	Promoter of T_h1 and suppressor of T_h2 functions
IL-13	B-cell proliferation, IgE production
IL-14	High molecular weight B-cell growth factor
IL-15	Activates T cells, neutrophils and macrophages
IL-16	Chemotactic factor for helper T cells
IL-17	Promotes T cell proliferation, pro-inflammatory activities
IL-18	Activates T cells, neutrophils, and fibroblasts
IL-19	Member of IL-10 family; transcriptional activator of IL-10
IL-20	Member of IL-10 family with epidermal function
IL-21	Improves proliferation of T cells and B cells, and enhances natural killer (NK) cytotoxic activities
IL-22	Member of IL-10 family; induces inflammatory responses
IL-23	Activates autoimmune responses
IL-24	Member of IL-10 family; tumor suppressor molecule
IL-25	Capable of amplifying allergic inflammation
IL-26	Member of IL-10 family; plays a role in mucosal and cutaneous immunity
TGF[4]-β	Suppresses BFU-E, CFU-S, and HPP-CFC
Interferons	Suppress BFU-E, CFU-GEMM, and CFU-GM
TNF[5]-α and -β	Suppress BFU-E, CFU-GEMM, and CFU-GM
PGE[6]-1 and -2	Suppress GFU-GM, GFU-G, and GFU-M
Lactoferrin	Suppresses release of IL-1

[1] Granulocyte and macrophage colony-stimulating factor
[2] Granulocyte colony-stimulating factor
[3] Macrophage colony-stimulating factor
[4] Transforming growth factor
[5] Tumor necrosis factor
[6] Prostaglandin E

circulating pool. These two components are in equilibrium. Granulocytes reside in the blood for an average of 10 hours and leave the circulation toward the inflammation sites in the various tissues. Eosinophil and basophil maturation appear to be analogous to neutrophil maturation, though the involved regulatory cytokines are different. GM-CSF and IL-5 play a major role in the development of the eosinophilic lineage, while SF, IL-3, IL-4, IL-9, and IL-10 regulate the basophilic development (Figure 1.2).

Monocytic maturation, assisted by GM-CSF and M-CSF, begins with the formation of monoblasts followed by promonocytes and monocytes. Monocytes are released into

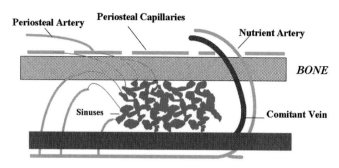

FIGURE 1.3 Schematic of microvascular circulation in the bone marrow.
Adapted from De Bruyn PPH. Structural substrates of bone marrow function. Semin Hematol 1981; 18: 179.

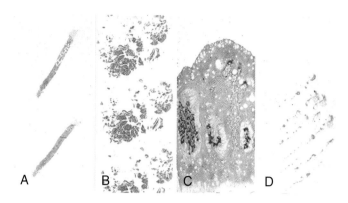

FIGURE 1.4 Representative examples of glass slide preparations of (A) bone marrow biopsy, (B) clot sections, (C) bone marrow aspirate smear, and (D) touch preparation.

the circulation, and from there they travel into the various tissues and become different types of tissue macrophages (histiocytes) and dendritic cells. Pulmonary alveolar macrophages, hepatic Kupffer cells, pleural and peritoneal macrophages, osteoclasts, Langerhans and interdigitating dendritic cells in various tissues, and perhaps microglial cells in the central nervous system are all examples of cells derived from a CFU-M.

Thrombopoiesis begins with maturation of CFU-GEMM into a colony-forming unit with a high proliferating response to cytokines (e.g., IL-1, IL-3, and IL-6) referred to as the high proliferative potential–colony-forming unit–megakaryocyte (HPP-CFU-MK). The next step is formation of a burst-forming unit (BFU-MK) which is capable of producing numerous megakaryocytic colony-forming units (CFU-Meg). This process is regulated by SF, IL-3, GM-CSF plus thrombopoietin (TPO) and IL-11 (Figure 1.2). Maturation of CFU-Meg leads to the formation of megakaryoblasts, megakaryocytes, and eventually, platelets. Megakaryocytes are mostly located in proximity of the sinuses. In this location a portion of their cytoplasm enters into the sinusoidal space to release platelets.

Lymphopoiesis begins in the bone marrow with the committed lymphoid stem cells. The first step of differentiation of the committed lymphoid stem cells appears to be the separation of B progenitor cells from non-B progenitor cells (T-cells and natural killer cells). This step is considered an "antigen-independent" process. The development of B progenitor cells is influenced by a number of regulatory cytokines including IL-1, IL-2, IL-4, IL-10, and interferon gamma (Figure 1.2). The B-cell precursors in bone marrow are known as hematogones. These cells are found in small numbers in normal bone marrow (usually 5–10% in young children and 5% in adults), but may increase in regenerating marrows.

T-lymphocytes derive from the precursor lymphoid cells in the marrow (pre-thymic phase) under the influence of several cytokines, such as IL-1, IL-2, and IL-9, and then migrate to the thymus for further maturation. A subclass of large granular lymphocytes, natural killer (NK) cells, appears to share a common progenitor cell with T-cells in the marrow. The NK-cells demonstrate HLA-non-restricted cytotoxicity and release a variety of regulatory cytokines, such as IL-1, IL-2, IL-4, and interferons.

Bone Marrow Examination

Bone marrow samples are obtained and prepared for pathologic evaluation in different ways, such as biopsies, clotted aspirated marrow particles, marrow smears, and touch preparations (Figures 1.4 and 1.5).

Bone marrow biopsy sections are evaluated for the estimation of bone marrow cellularity and for the identification of pathological processes, such as primary hematologic disorders, granulomatosis, amyloidosis, fibrosis, osteosclerosis, and metastasis. Biopsy sections are routinely stained with hematoxylin and eosin (H&E stain) (Figures 1.5 and 1.6). In addition, in some laboratories sections are stained by the periodic acid Schiff (PAS) technique. Bone marrow cellularity is defined as the percentage of the bone marrow areas occupied by cells (% cellularity). Cellularity of the bone marrow varies depending on the location of the marrow sample and the age of the individual. For example, bone marrow cellularity is higher in the vertebrae than in the pelvic bone, and higher in children than in the elderly. Bone marrow cellularity approaches 100% at birth, and continues to decline approximately 10% for each decade of life. In a 50-year-old healthy person, the average bone marrow cellularity is about 50% (Figure 1.6).

Bone marrow clot sections (particle sections) are prepared from aspirated bone marrow, and therefore are devoid of bone trabeculae. They only represent cells and lesions that are released by aspiration (Figures 1.4 and 1.5). Clot sections are routinely stained with H&E. Some laboratories may also use PAS stain.

Bone marrow smears are prepared by smearing the aspirated marrow over the glass slides. Marrow smears are usually stained with Wright's (or in some laboratories with Giemsa) stains (Figures 1.5–1.7). They are used primarily for cytological evaluations, cellular details, maturation

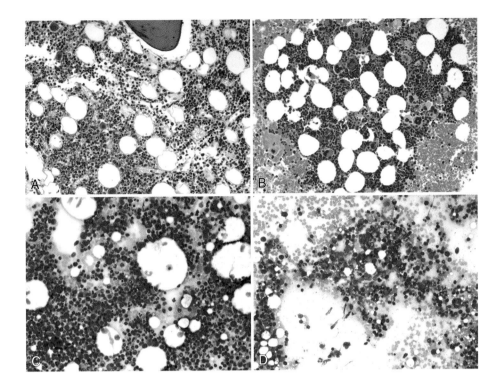

FIGURE 1.5 Bone marrow preparations: (A) biopsy section, (B) clot section, (C) aspirate smear, and (D) touch preparation.

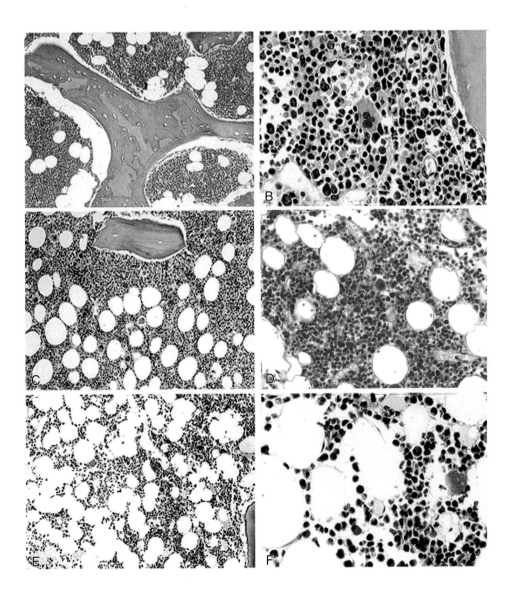

FIGURE 1.6 Bone marrow cellularity declines with age. (A) and (B), (C) and (D), and (E) and (F) are biopsy sections from 2-year, 55-year, and 75-year-old individuals, respectively.

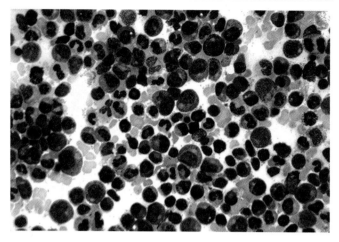

FIGURE 1.7 Bone marrow smear demonstrating myeloid cells at various stages of maturation. Scattered erythroid precursors and lymphocytes are also present.

steps, differential counts, and assessment of the myeloid : erythroid ratio (normal range: 2–3). Differential counts reflect the percent of different hematopoietic cells in bone marrow smears. At least 200 cells are counted by randomly selected areas of a properly stained and adequately cellular marrow smear to calculate the differential count (Table 1.3). Marrow smears are also useful for special cytochemical stains and evaluation of the bone marrow iron stores.

Bone marrow touch (imprint) preparations are made by gently touching (pressing) the glass slides over the biopsy sample, and are routinely stained with Wright's and/or Giemsa stains (Figures 1.4 and 1.5). Touch preparations most often are not optimal for morphologic evaluations, because their preparation creates significant artifacts. However, they are the only source of cytologic evaluation when bone marrow aspiration fails to yield (dry tap).

Morphologic Characteristics of Bone Marrow Cells

GRANULOCYTIC SERIES

The granulocytic series includes neutrophilic, eosinophilic, basophilic, and mast cell lineages. The morphologic steps in the maturation process of the granulocytic series consist of myeloblast, promyelocyte, metamyelocyte, band, and segmented cell (Figures 1.7–1.9). During this process, the ratio

Table 1.3
Approximate Ranges of Bone Marrow Differential Count (%) in Healthy Persons

Cell Type	Age	
	18 months	Adult
Myeloid		
Myeloblast	N/A	1–5
Promyelocyte	1–2	1–8
Neutrophilic Series:		
Myelocytes	2–4	5–19
Metamyelocyte	8–16	13–22
Band/Segs	14–25	21–40
Eosinophilic Series	1.5–3.5	0.5–3
Basophilic Series	<1	<1
Monocytic Series	1–3	1–4
Erythroid		
Rubriblast	<1	0.5–2
Prorubricyte	0.5–1	1.5–6
Rubricyte	4–10	5–25
Metarubricyte	<1	2–20
Other		
Megakaryocyte	<1	0.5–2
Lymphocyte	40–42	3–20
Plasma Cell	<1	0.5–2
M:E Ratio		
	4–5:1	3–3.5:1

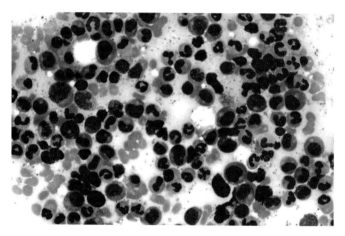

FIGURE 1.9 Bone marrow smear demonstrating myeloid cells at various stages of maturation.

   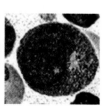

FIGURE 1.8 Left to right: myleoblast types I, II, and III, and a promyelocyte.

of cytoplasm to nucleus increases, and cytoplasm accumulates lyzosomal granules that are non-specific at first (primary granules, azurophilic granules) and become specific (secondary granules) later. The nuclear chromatin becomes coarser and denser, and the nucleoli appear less prominent and indistinct. The nuclear shape gradually changes from round/oval to kidney-shaped and segmented forms.

Myeloblasts are the earliest granulocytic precursors identified by morphologic evaluations. They range in size from 10 to 20 μm and are characterized by a high nuclear to cytoplasmic (N:C) ratio, a centrally located round or oval nucleus, finely dispersed chromatin, and several nucleoli. Myeloblasts may show no cytoplasmic granules or variable numbers of azurophilic cytoplasmic granules (Figure 1.8). Myeloblasts are positive for CD13, CD33, CD117, and HLA-DR and may express CD34 and myeloperoxidase (MPO).

Promyelocytes are overall larger than myeloblasts, ranging from 13 to 25 μm in diameter. They carry more cytoplasm and contain larger quantities of azurophilic granules than myeloblasts. They depict a perinuclear pale area (a well-developed Golgi system) and a round or oval nucleus, which is often eccentric. Highly granular myeloblasts and promyelocytes share overlapping morphologic features, and therefore their distinction at times is difficult (Figures 1.7 and 1.8). In contrast to myeloblasts, promyelocytes are negative for HLA-DR and CD34. Promyelocytes are positive for CD13, CD15, CD33, and MPO, plus variable expression of CD117.

Myelocytes are smaller than promyelocytes and are characterized by a reduced N:C ratio with ample granular cytoplasm containing both primary and secondary granules. At the myelocytic stage the production of primary granules stops and the synthesis of specific granules enhances. Myelocytes depict a round or oval nucleus with coarse chromatin and often lack distinct nucleoli (Figures 1.7 and 1.9). Myelocytes are positive for CD13, CD15, CD33, and MPO. They are negative for CD34, CD117, and HLA-DR.

Metamyelocytes are slightly smaller than myelocytes and are characterized by abundant granular cytoplasm with predominance of specific granules, kidney-shaped or indented nucleus, coarser chromatin, and lack of distinct nucleoli (Figure 1.9). Metamyelocytes express CD11b, CD13, CD15, CD16, CD33, and MPO.

Bands and **segmented cells** are the end stage cells in the granulocytic series and are distinguished by abundant cytoplasm with specific granules, lack of or sparse primary granules, condensed nuclear chromatin with indistinct nucleolus, and nuclear lobulation or segmentation. Neutrophilic bands (stabs) are cells with bilobed nuclei with no filament formation, and neutrophilic segmented cells (Segs) demonstrate up to five distinct nuclear lobules (segments) connected to one another by filaments (Figure 1.9). Neutophilic bands and segmented cells demonstrate alkaline phosphatase activity and are positive for CD11b, CD13, CD15, CD16, CD33, and MPO.

Other granulocytic lineages, such as *eosinophils* and *basophils*, undergo more or less similar differentiation steps. Mature eosinophils, unlike segmented neutrophils, usually have bilobed nuclei and are loaded with eosinophilc granules. Eosinophilic granules are larger than the neutrophilic granules (Figure 1.10). Compared with neutrophils, eosinophils express brighter CD45, but dimmer CD11b, CD16, and CD13. Mature basophils contain a large number of coarse basophilic granules and show less nuclear segmentation than the neutrophils (Figure 1.10). Therefore, compared with neutrophils, basophils reveal much lower side scatter. Basophils are positive for CD13, and uniquely express CD22 while being negative for the other B-cell markers.

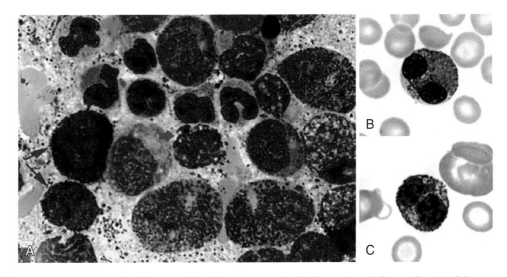

FIGURE 1.10 (A) A bone marrow smear showing granulocytic precursors including eosinophilic myelocytes (blue arrows) and a basophil (red arrow). (B) An eosinophil and (C) a basophil are demonstrated in blood smears.

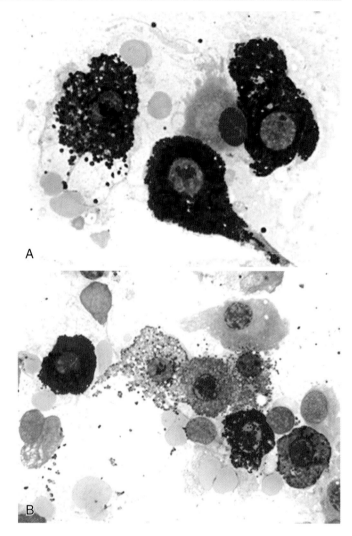

FIGURE 1.11 Bone marrow smears demonstrating mast cells with various amounts of cytoplasmic granules.

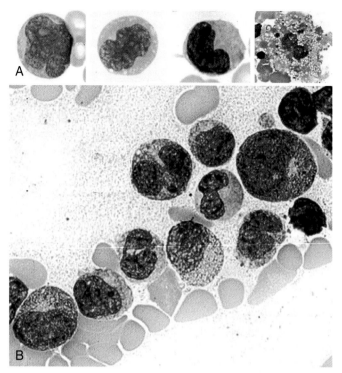

FIGURE 1.12 (A) Monocytic maturation, from left to right: monoblast, promonocyte, monocyte, and macrophages. (B) Bone marrow smear showing several monocytic cells.

Mast cells appear to be closely related to the basophils by sharing certain characteristics, such as basophilic granules, IgE receptor, and histamine content. However, mast cells live longer, are larger, have more abundant cytoplasm than basophils, and their nucleus is round, oval, or spindle-shaped without segmentation. Mast cell cytoplasmic granules are MPO negative and are more numerous and more variable in appearance than the granules in basophils (Figure 1.11). Mast cells express CD117 and tryptase.

Monocytes and macrophages are derived from the same committed stem cells (CFU-GM) as the granulocytic cells. The maturation process in this lineage starts from *monoblast*, and then goes through *promonocyte, monocyte, macrophage (histiocyte)*, and *multinucleated giant cell* (such as osteoclasts or giant cells in granulomas) (Figures 1.12 and 1.13). During the maturation process from monoblast to monocyte, nuclei become folded, nuclear chromatin becomes more condensed, nucleoli disappear, and cytoplasm acquires lysosomal granules (Figure 1.12). Monocytes are positive for dim CD4, CD11b, CD13, CD14, CD15, CD33, CD36, CD64, HLA-DR, and non-specific esterase, and variably positive for lysozyme. Monocytes are released from bone marrow into the blood circulation, and from there they migrate out into various tissues, and finally transform to soft tissue histiocytes (macrophages), which are positive for CD68. Iron is stored in bone marrow macrophages as hemosiderin (insoluble aggregates) or less abundantly as ferritin (soluble). Prussian blue (potassium ferrocyanide) stains hemosiderin as dark blue cytoplasmic granules (Figure 1.13A).

Dendritic cells (DC) are considered a subclass of histiocytic lineage and are primarily involved in antigen presentation to lymphocytes. Dendritic cells, except for the follicular dendritic cells (FDCs), are derived from bone marrow stem cells. These cells are divided into two groups: Langerhans cells (LCs) (Figure 1.13B) and interdigitating dendritic cells (IDCs). Based upon their immunophenotype, five subsets of DC have been reported, with variable expression of CD1b/c, CD16, CD123, CD141 (BDCA-3), and CD34. LCs are primarily located in the skin and are characterized by the ultrastructural Birbeck granules and expression of CD1a, CD4, S100, HLADR, and Langerin (CD207) (Figure 1.14). Unlike macrophages, LCs are negative for CD68, non-specific esterase, and lysozyme. IDCs are found in the lymphoid tissues and show immunophenotypic features similar to those of LCs, except for lack of CD1a and CD4 expression. The FDCs are derived from the mesenchymal cells in the follicular structures in the lymph nodes and express CD21, CD23, and CD35 (Figure 1.14).

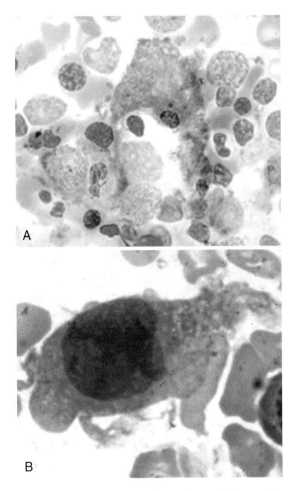

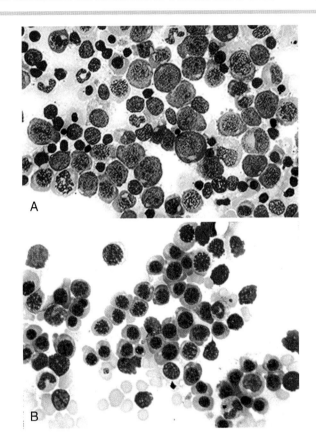

FIGURE 1.13 (A) A bone marrow clot section with iron stain showing iron-laden macrophages. (B) Bone marrow smear showing a Langerhans cell with abundant cytoplasm and irregular cytoplasmic projections.

FIGURE 1.15 Bone marrow smear showing erythroid precursors of (A) early and (B) intermediate stages of maturation.

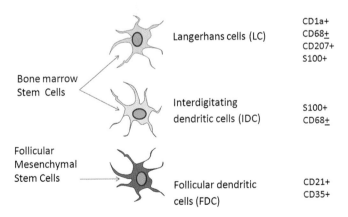

FIGURE 1.14 Subtypes of dendritic cells and their origin.

ERYTHROID PRECURSORS

During the maturation process in erythropoiesis, cells gradually become smaller, the N:C ratio decreases, cytoplasm accumulates hemoglobin, the nuclear chromatin becomes denser and pyknotic, and the nucleoli appear less prominent and indistinct. The nucleus is eventually extruded from the cell, resulting in the development of polychromatophilic erythrocytes, and then mature red blood cells (RBCs) (Figures 1.15 and 1.16). Erythroid precursors express several membrane-associated molecules, such as CD36, CD71, CD235 (glycophorin), CD238 (Kell blood group), CD240 (Rh blood group), and CD242 (LW blood group).

Rubriblasts (erythroblast, pronormoblast) are the earliest morphologically distinguished erythroid precursors. They express CD71 and CD117. They measure 15–30 μm in diameter and have a high N:C ratio with a deep blue non-granular cytoplasm and a perinuclear pale area (the Golgi system). The nuclear chromatin is fine and one to two nucleoli are present. *Prorubricytes* (basophilic erythroblasts, basophilic normoblasts) are smaller and have more condensed nuclear chromatin than rubriblasts. They depict dark blue cytoplasm and indistinct nucleoli. Prorubricytes undergo three cell divisions and continue maturation to form *rubricytes* (polychromatophilic normoblasts), and subsequently *metarubricytes* (orthochromic normoblasts) which are not able to divide but continue hemoglobin synthesis (Figure 1.16). Metarubricytes lose their nuclei and become polychromatophilic RBCs (reticulocytes). Reticulocytes gradually lose their ribosomes (in 1–2 days) and become mature RBCs. RBCs live for about 100–120 days.

PLATELET PRECURSORS

Megakaryoblasts (promegakaryoblasts, group 1 megakaryocytes) are the earliest morphologically identifiable platelet precursors (Figure 1.17). Megakaryoblasts undergo endomitosis (nuclear division without cytoplasmic division) once or twice and become promegakaryocytes (group II megakaryocytes). Endomitosis continues, cells become larger, the nuclear lobulation and volume increase, and the end result is the formation of granular megakaryocytes (group III megakaryocytes) which are able to release platelets (Figure 1.17). Granular megakaryocytes are the largest hematopoietic cells in the bone marrow. The duration from formation of megakaryoblasts to platelet production is about 1 week. Platelets are released into the bloodstream with a proportion (approximately one-third) pooled in the spleen. Their average life span is about 8–10 days. Cells from the megakaryocytic lineage express CD41, CD42, CD31, CD61, and factor VIII.

LYMPHOID LINEAGE

Lymphocytes, as is the case for the other hematopoietic cells, are derived from the multipotent hematopoietic stem cells. Lymphoblasts, the earliest morphologically identifiable lymphoid cells, have a high N:C ratio with a narrow rim of dark blue non-granular cytoplasm, a round or oval nucleus with fine chromatin and one to two nucleoli. Mature lymphocytes are slightly larger than erythrocytes and are characterized by scanty blue cytoplasm, round nucleus, coarse chromatin, and inconspicuous nucleolus (Figure 1.18A). They may be of B- or T-cell origin. A variable proportion of lymphocytes are larger with abundant cytoplasm and cytoplasmic azurophilic granules. These *large granular lymphocytes* (LGL) are more frequently identified in normal blood smears than bone marrow smears and often express CD16, CD56, and/or CD57 (Figure 1.18b). They are of two types: NK-cells and cytotoxic T-cells. NK-cells are negative for surface CD3 and show no T-cell receptor (TCR) gene rearrangement but they are positive for CD2 and CD7, and may express CD8. Cytotoxic T-cells express CD2, CD3, CD5, CD7, CD8 and show *TCR* gene rearrangement.

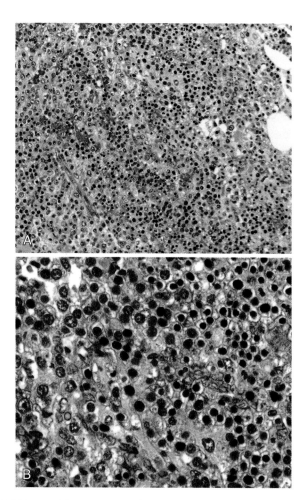

FIGURE 1.16 A bone marrow clot section (A, low power; B, high power) showing erythroid precursors at various stages of maturation.

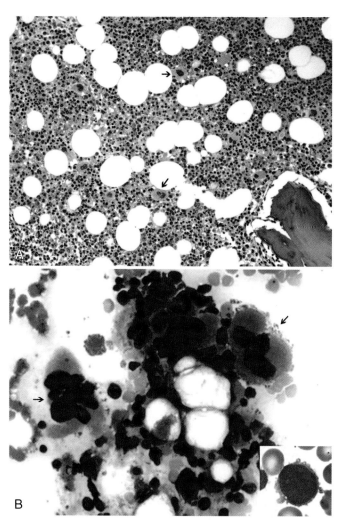

FIGURE 1.17 The megakaryocytic lineage: (A) bone marrow biopsy section demonstrating megakaryocytes (arrows), several eosinophils, and numerous erythroid and myeloid precursors; (B) bone marrow smear demonstrating megakaryocytes. A megakaryoblast with cytoplasmic budding is demonstrated in the inset.

Prolymphocytes are larger than lymphocytes (more cytoplasm and a larger nucleus), display a coarse chromatin, and often show a prominent nucleolus. They are either of B- or T-cell origin (Figure 1.18c).

Activated lymphocytes are transformed B-, T-, or NK-cells. These are large cells with abundant cytoplasm and a highly polymorphic nuclear morphology (Figure 1.18d). They are more frequently identified in blood smears than marrow smears.

Hematogones represent normal bone marrow precursor B-cells. These cells consist of a heterogeneous population of B-cell precursors at various stages of maturation. The earlier forms (pro- and pre-B-cells) often express TdT, CD34, CD10, CD19, and intracellular CD22. As these precursors mature, they lose CD34 and TdT, while beginning expression of CD20 and ultimately surface light chains.

Early forms of hematogones may morphologically resemble lymphoblasts but always coexist with more mature forms of lymphocytes. Heterogeneity is the key for distinguishing hematogones from lymphoblasts of acute leukemias by morphology as well as immunophenotype (Figures 1.19 and 1.20). Hematogones display a distinctive pattern on CD45 gating by flow cytometry (Figure 1.21).

Plasma cells are the end product of the B-cell lineage and are characterized by abundant dark blue cytoplasm, a perinuclear pale area (Golgi system), and an eccentric nucleus with coarse chromatin (cartwheel appearance) (Figure 1.22). Plasma cells may show small cytoplasmic vacuoles (Mott or morula cells), or large eosinophilic cytoplasmic inclusions (Russell bodies) or nuclear inclusions (Dutcher bodies) (Figure 1.23). Russell bodies and Dutcher bodies are more often seen in plasma cell disorders than in normal plasma cells. The vacuoles and inclusions contain immunoglobulin. Rarely, ovoid-, angular-, or rod-shaped immunoglobulin crystals are found in plasma cells. Plasma cells express CD38, CD138, and CD79, and sometimes are positive for CD19 and/or CD20.

Lymphoid aggregates are relatively common findings in bone marrow sections, particularly in the elderly. They are well-defined round or oval structures that are randomly distributed in the marrow, usually in close association with small blood vessels and apart from bone trabeculae. They primarily consist of small mature lymphocytes comprising a mixture of B- and T-cells (Figure 1.24). Scattered macrophages, eosinophils, and plasma cells may also be present within or around the lymphoid aggregates.

OTHER BONE MARROW CELLS

Osteoblasts are derived from a multipotent mesenchymal stem cell, a lineage different from that of the HSC. These cells are elongated or oval cells that contain an

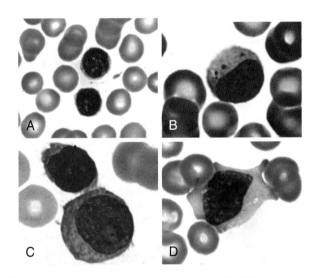

FIGURE 1.18 Examples of various lymphoid cells: (A) mature lymphocytes, (B) a large granular lymphocyte, (C) prolymphocytes, and (D) a reactive lymphocyte.

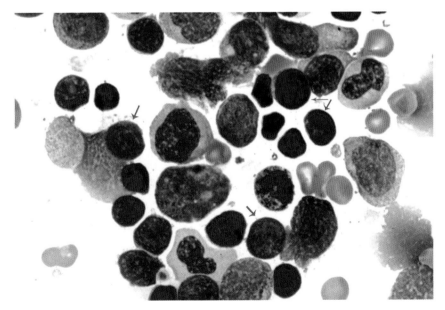

FIGURE 1.19 Several hematogones (arrows) are present in this bone marrow smear.

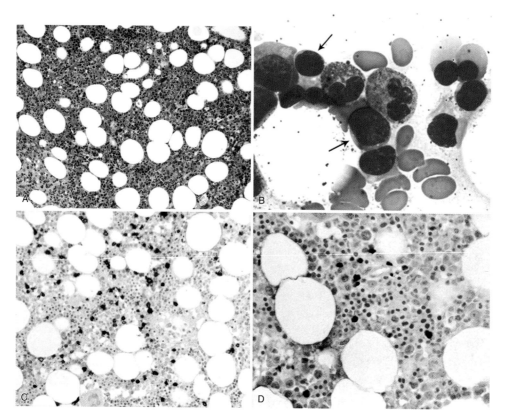

FIGURE 1.20 (A) An unremarkable bone marrow biopsy section showing progressive multilineage maturation. (B) A bone marrow smear demonstrating hematogones (arrows) and an eosinophil. (C) Scattered CD20-positive B-cells. (D) Rare TdT-positive cells in a dispersed distribution pattern which may represent early hematogones.

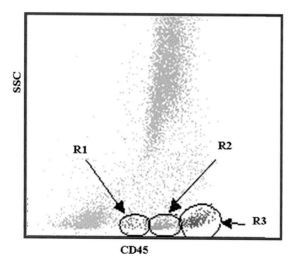

FIGURE 1.21 Flow cytometry. The CD45 gating display of a bone marrow sample demonstrates aggregates of early (R1) and intermediate (R2) and late hematogones plus mature lymphocytes (R3).

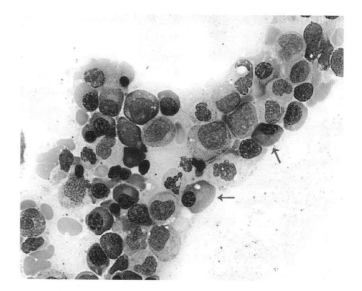

FIGURE 1.22 Bone marrow smear demonstrating several plasma cells (arrows).

eccentric round or oval nucleus and one or more nucleoli. Osteoblasts may resemble plasma cells, except that they are larger, their Golgi are not as close to the nucleus, and their nuclear chromatin is finer than that of plasma cells. Osteoblasts in biopsy sections are located along the bone trabeculae, and in bone marrow smears usually appear as individual or small clusters of cells (Figure 1.25). Osteoblasts release a number of matrix molecules and cytokines, such as collagen type 1, proteoglycans, osteoid, GM-CSF, M-CSF, and IL-6.

Osteoclasts are derived from CFU-M. They are multinucleated giant cells involved in bone resorption and

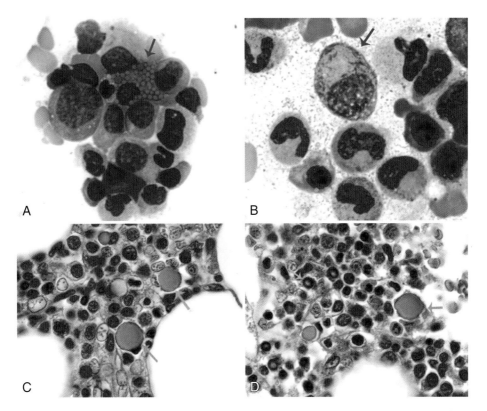

FIGURE 1.23 Plasma cells with inclusions: (A) a grape-like plasma cell (Mott cell), (B) plasma cells with cytoplasmic rod-like Ig crystals, (C) cytoplasmic Ig inclusions (Russell bodies), and (D) Russell bodies (green arrow) and nuclear Ig inclusions (Dutcher bodies) (blue arrows).

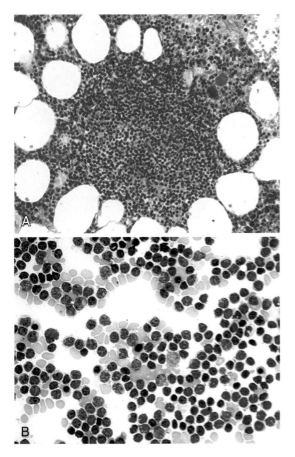

FIGURE 1.24 (A) A well-defined lymphoid aggregate is shown from a bone marrow biopsy section. (B) Numerous small mature lymphocytes are present in a bone marrow smear.

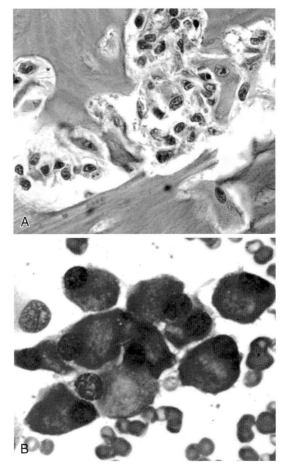

FIGURE 1.25 Numerous osteoblasts are demonstrated in (A) a bone marrow biopsy section and (B) a bone marrow smear.

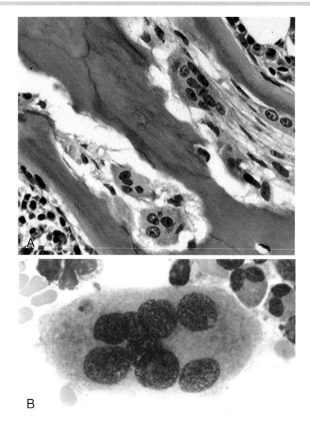

FIGURE 1.26 (A) Numerous osteoclasts are demonstrated in a bone marrow biopsy section. (B) An osteoclast with multiple separated nuclei and finely granular cytoplasm is shown.

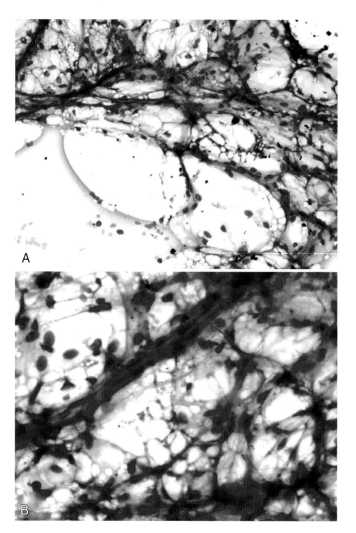

FIGURE 1.27 Bone marrow smears showing adipose tissue and stromal cells. A collapsed capillary (arrow), lined by endothelial cells, is present in (B).

remodeling (Figure 1.26). They have abundant cytoplasm which contains numerous azurophilic granules. They are found along the bone trabeculae and may resemble megakaryocytes, except they have multiple separated nuclei that are uniform in size. Osteoclasts are frequently observed in bone marrow of patients with hyperparathyroidism, chronic renal failure, and Paget's disease.

Adipocytes (fat cells, lipocytes) are mesenchymal derived cells with abundant fat-laden cytoplasm and a small nucleus often pushed toward the cell membrane (Figure 1.27). Bone marrow adipocytes can be rapidly replaced by hematopoietic tissue when there is a need for increased hematopoiesis.

Fibroblast-like cells and **endothelial cells** support the wall of the bone marrow sinuses and build the framework of the marrow stroma that supports the hematopoietic cells. These are usually elongated or polygonal cells (15–30 μm) with variable amount of pale cytoplasm and round, oval, or folded nuclei. Their nuclear chromatin is fine and they may depict one or more nucleoli. Fibroblast-like cells support proliferation of myeloid and lymphoid progenitor cells. They are negative for CD33 and CD34 but may express CD10. Endothelial cells are involved in the regulation of homing and trafficking of the hematopoietic cells, as well as proliferation and differentiation of hematopoietic precursors (Figure 1.27). They express CD31, CD34, and CD146 and carry various receptors, such as receptors for IL-3, EPO, and SF.

Blood Smear Examination

Morphologic evaluation of blood smear is important in routine hematology work-up, because unremarkable CBC results by automated instruments may not necessarily reflect normal hematopoiesis. For example, in hereditary spherocytosis, lead poisoning, or malaria, the CBC may be within normal limits, but the peripheral blood smears show spherocytes, basophilic stippling, or RBC-containing parasites, respectively. Blood smears should be thin, evenly distributed over the glass slides and quickly air-dried and stained (Wright's stain is the most popular stain).

RED BLOOD CELL MORPHOLOGY

In normal conditions, red cells are relatively uniform in shape and size and contain no inclusions. They are normocytic (an average of 7–8 μm in diameter) and normochromic (the pale central area less than half of the RBC diameter) (Figures 1.28 and 1.29). One to two percent of

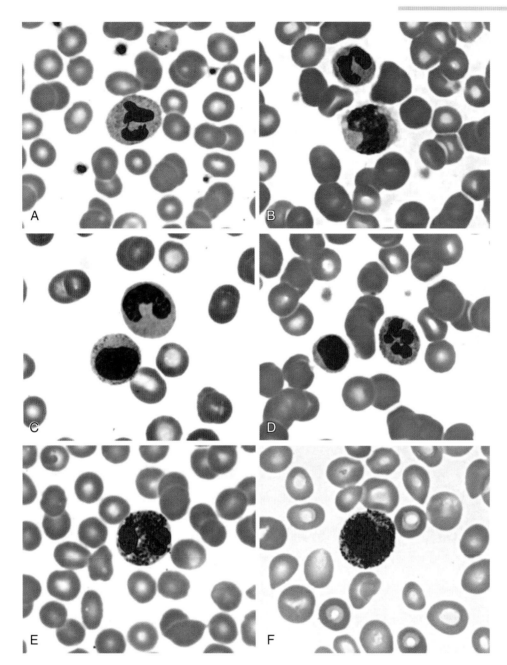

FIGURE 1.28 Blood smears demonstrating segmented neutrophils (A, B, and D), monocytes (B and C), a large granular lymphocyte (C), a lymphocyte (D), an eosinophil (E), and a basophil (F). Platelets are present in (A), (B), and (D).

erythrocytes are larger and polychromatophilic (bluish-red) (Figure 1.29). These represent reticulocytes. Except in newborns, nucleated red cells are not normally found in peripheral blood.

LEUKOCYTE MORPHOLOGY

In normal conditions, peripheral blood smears show various proportions of neutrophilic segmented cells (Segs) and bands (stabs), lymphocytes, monocytes, eosinophils, and basophils (Figure 1.28). The white blood cell (WBC) count is in the range $3-10\times10^3$ cells/µL with a differential count shown in Table 1.4. Certain conditions such as exercise, emotional distress, menstruation, anesthesia, convulsive seizures, and electric shock may be associated with a transient neutrophilic granulocytosis. This is due to the demargination of the neutrophilic granulocytes and their release into the circulating pool. The presence of immature leukocytes in the peripheral blood should be considered abnormal.

PLATELET MORPHOLOGY

Platelets are the end products of the megakaryocytic lineage and are released into the circulation as cytoplasmic fragments of granular megakaryocytes. They are the

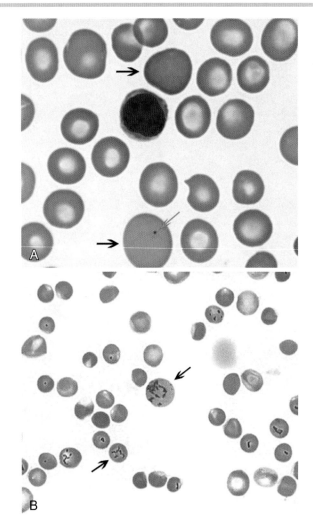

FIGURE 1.29 (A) Blood smear showing polychromatophilic red cells (black arrows) and a Howel-Jolly body (green arrow). (B) Reticulocytes are demonstrated by a supravital stain (arrows).

Table 1.4

The Range of White Blood Cell Differential Counts in Normal Adults

Cell Type	Range (%)
Granulocytes	
Segmented cells	33–72
Bands	0–13
Eosinophils	0–6
Basophils	0–3
Lymphocytes	16–48
Monocytes	1–13

smallest hematopoietic elements (measuring 2–4 μm in diameter), with a count ranging from 150,000 to 400,000/μL (Figure 1.28). A rough estimate of the platelet count is calculated in wedge smear preparations by the number of platelets per oil-immersion field×20,000. Approximately 7–21 platelets are found per 100× oil-immersion field in an evenly distributed normal blood smear. Anticoagulants or agglutinins (IgM or IgG) which are found in patients with autoimmune disorders, chronic liver disease, or malignancy may cause platelet aggregation.

Structure and Function of the Spleen

The spleen represents the largest filter of the blood circulation in our body. In normal conditions, it weighs between 75 and 200 grams and has a deep indentation (the hilum), where blood vessels enter and leave. The spleen is surrounded by a fibrous capsule with many trabeculae radiating from the internal surface of the capsule into the splenic parenchyma. The splenic artery branches into the trabecular arteries, and these branches in turn give off smaller branches that leave the trabeculae and are called central arteries. Central arteries run through the splenic lymphoid tissue (*white pulp*) and extend to the *marginal zone* and the *red pulp*. Therefore, the splenic parenchyma consists of three distinct components: the *white pulp*, the *marginal zone*, and the *red pulp* (Figures 1.30–1.32).

THE WHITE PULP

The white pulp consists of lymphoid structures organized in B- and T-cell zones (Figures 1.30–1.32). The T-cell zone is represented by the periarteriolar lymphoid sheath, primarily consisting of tightly packed lymphocytes and the presence of interdigitating dendritic cells. The B-cell zone consists of follicles, which are structurally similar to the follicular structures in the lymph nodes (see lymph node structure later). The follicles are separated from the marginal zone by a densely packed mantle zone and frequently contain germinal centers consisting of large blast-like lymphocytes (centroblasts), smaller lymphocytes (centrocytes), follicular dendritic cells, and scattered macrophages.

THE MARGINAL ZONE

The marginal zone is the transit area for cells that are leaving the bloodstream and entering the white pulp (Figures 1.30–1.32). However, a large number of cells, such as macrophages, B-cells, and dendritic cells, reside in the marginal zones in order to regulate and facilitate the back and forth transit flow of cells between the blood and the white pulp. Marginal zone B-cells are medium- to large-sized cells with pale cytoplasm and irregular nuclei, resembling monocytes. That is why they were originally called monocytoid B lymphocytes. Because of the presence of variable amounts of cytoplasm, they show nuclear spacing in sections and appear lighter in color and less dense than the cells present in the mantle zone. The marginal zone

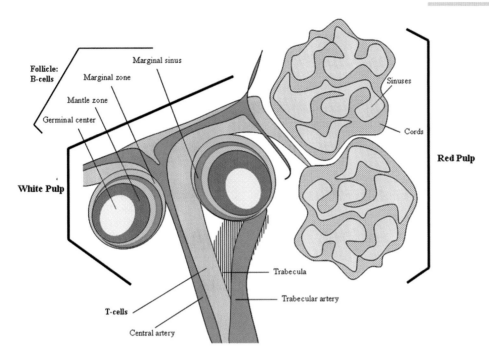

FIGURE 1.30 Schematic of a spleen demonstrating the white pulp, the red pulp, and the marginal zone.
Adapted from Greer JP, Foerster J, Lukens JN, et al. Wintrob's Clinical Hematology, 11th ed. Williams & Wilkins Lippincott, 2004.

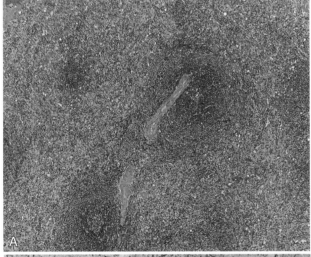

FIGURE 1.31 (A) White pulp and red pulp regions of the spleen are demonstrated in an H&E section. (B) Dual immunohistochemical staining for CD3 (brown) and CD20 (red) demonstrates T- and B-cell areas, respectively.

B-cells, unlike mantle cells, do not express CD5. They are negative for CD10 and CD23, have mutated IgV genes and express surface IgM and IgD.

The marginal zone is a place where the blood-borne pathogens are challenged by the adaptive immune system. Macrophages with their specific pattern-recognition receptors can effectively take up the pathogens and also activate the marginal zone B-cells and dendritic cells. Entry of activated marginal zone B lymphocytes and dendritic cells into the white pulp initiates an adaptive immune response against the blood-borne pathogens. The marginal zone is devoid of sinuses.

THE RED PULP

The red pulp consists of splenic cords and the sinusoidal system (Figures 1.30–1.32). Cords are composed of a meshwork of fibroblast-like cells supported by extracelluar matrix and reticulin fibers. They form cavernous spaces with no endothelial lining and directly receive arterial blood from terminal arterioles and arterial capillaries. Numerous macrophages are present in the cords, which are able to remove the damaged, abnormal, or aged blood cells, while the blood passes through into the venous sinuses. Unlike the cords, sinuses are lined by endothelial cells. There are slit-like gaps between the endothelial cells which allow blood cells to penetrate from the cordal space into the sinusoidal lumen. Abnormal RBCs, such as sickle cells, or cells with inclusions, such as Heinz bodies, might not be able to pass through these gaps. The sinus basement membrane consists of a network of contractile thick and thin reticular fibers (stress fibers) running circumferentially and longitudinally, respectively. The network is connected to the extracellular matrix of the splenic cord, and its contraction helps the blood to pass through

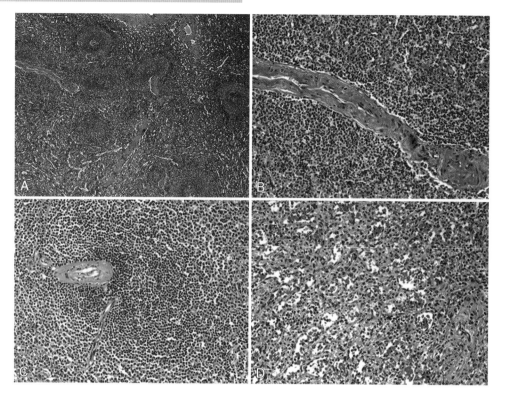

FIGURE 1.32 (A) Low power microscopic view of the spleen demonstrates white and red pulp. (B) and (C) show higher power views of the T- and B-cell regions, respectively. (D) Higher power view of the red pulp with numerous blood-containing sinuses.

the cords into the sinuses. The spleen has three major functions:

1. Micro-organisms and pathogens are removed by macrophages in the spleen. Also, abnormal, damaged, and dysfunctional blood cells are filtered and removed by macrophages when blood passes through the spleen.
2. The splenic white pulp is an important component of the cell-mediated and humoral immune systems.
3. The splenic sinusoidal system serves as a big reservoir for blood cells.

Structure and Function of the Lymph Nodes

Lymph nodes are the major components of the lymphatic system and consist of round or oval structures located along the major blood vessels, in peritoneum and mediastinum, and at the base of the extremities. They measure from several millimeters to around 1 cm in diameter and are surrounded by a fibrous capsule. Incoming lymphatic vessels penetrate the capsule and release their content into the subcapsular sinuses. Blood vessels enter and leave the lymph nodes through the hilum. Several fibrous trabeculae extend from the inner part of the capsule into the lymph node parenchyma, forming a supporting meshwork and dividing the lymph node into many subsections. The lymph node parenchyma is divided into a peripheral zone—the *cortex*—and a deeper, centrally located zone, the *medulla* (Figure 1.33). The cortex consists of a superficial part, immediately located under the capsule, and a deeper part or *paracortex*.

The following anatomical structures are recognized in a lymph node section (Figures 1.33 and 1.34).

FOLLICULAR STRUCTURES

Follicular structures are the primary home of the B lymphocytes. The ones that are not yet exposed to antigens are called *primary follicles* and consist of packed, uniform-looking small mature lymphocytes. Secondary follicles have been already exposed to antigenic stimulation. They have a pale central area, called *germinal center*, consisting of a mixture of large and small cells. The smaller uniform-looking lymphocytes surrounding the germinal center are packed as a darkly stained crescent known as *mantle zone* (Figure 1.34). Lymphocytes of the mantle zones (mantle cells) are B-cells consisting of a mixture of bone marrow-derived naive cells and lymphocytes previously exposed to antigens. Mantle zone B-cells are positive for CD5, CD23, and IgM.

Germinal centers are generated by the clonal expansion of antigen-activated B-cells with the support of follicular dendritic cells and helper T-cells (Figures 1.34 and 1.35). In these centers, B-cells undergo somatic hypermutation and Ig isotype switch and eventually differentiate into memory cells or plasma cells. The clonal expansion and somatic hypermutation occurs at the base of the germinal

STRUCTURE AND FUNCTION OF THE LYMPH NODES   19

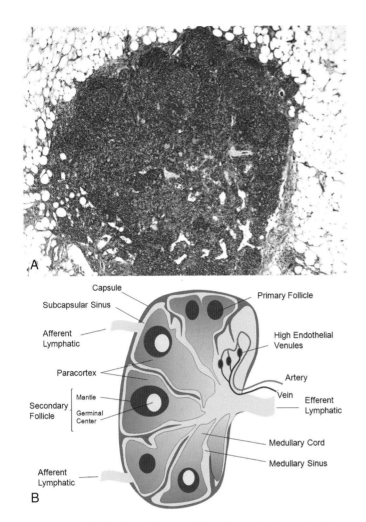

FIGURE 1.33 (A) Section of a normal lymph node demonstrating primary and secondary follicles, paracortex and medullary region. (B) Schematic of a normal human lymph node.
(B) is adapted from Fajardo LF. Lymph nodes and cancer: A review. Front Radiat Ther Oncol 1994; 28: 1–10.

center with the proliferation of the cells called *centroblasts*. Centroblasts are large, non-cleaved cells with a vesicular nuclear chromatin and multiple distinct nucleoli usually located close to the nuclear membrane. They show frequent mitotic figures. The centroblast-concentrated area of the germinal center is known as the *dark zone*. Scattered tangible body macrophages may be seen in this zone.

The dividing B-cells continue to differentiate to the *centrocytes*. These cells, which are not in cell cycle anymore, are concentrated toward the apex of the germinal center in an area called *light zone*. Centrocytes consist of small lymphoid cells with scant cytoplasm, cleaved nuclei, dispersed nuclear chromatin, and inconspicuous nucleoli. The centrocytes may eventually mature into plasma cells or memory cells. Numerous apoptotic cells and tingible body macrophages are present in the light zone (Figure 1.34).

Centroblasts and certrocytes express B-cell-associated antigens, such as CD19, CD20, CD22, and CD79a. They are positive for CD10 and bcl-6, but negative for CD5 and BCL-2 (Figure 1.36).

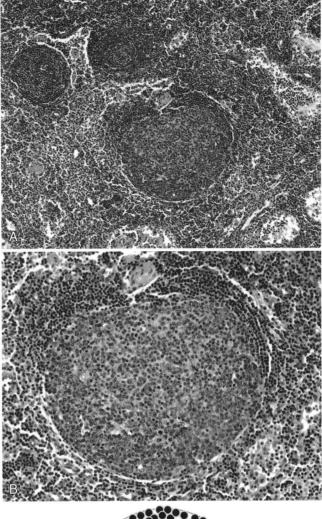

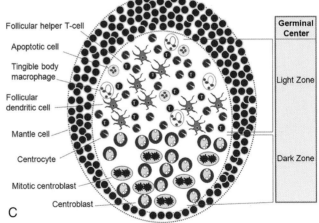

FIGURE 1.34 (A) Section of a normal lymph node demonstrating primary and secondary follicles. (B) A secondary follicle with dark and light polarized zones. (C) Schematic of a secondary follicle demonstrating cellular composition in the germinal center.
(C) is adapted from Liu YJ et al. Follicular dendritic cells and germinal centers. Int Rev Cytol 1996; 166: 139–179.

The light zone of the germinal center contains a rich network of follicular dendritic cells (FDCs). FDCs are derived from the mesenchymal cells in the follicular structures (not originated from bone marrow stem cells).

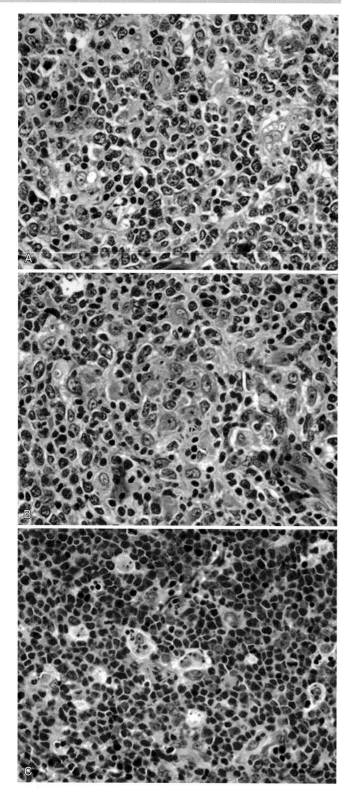

FIGURE 1.35 High power views of a germinal center in a lymph node section demonstrating centroblasts and centrocytes with the presence of mitotic figures (A, B, and C), follicular dendritic cells (B, arrows), and tingible body macrophages (C, arrows).

They often appear in pairs, have round or irregular nuclei with dispersed chromatin, and often one small, centrally located nucleolus (Figure 1.35). FDCs are characterized by expressing CD21, CD23, and CD35 (Figure 1.36).

The light zone also contains follicular helper T-cells (T_{FH}), which are positive for CD3, CD4, CD10, CD25, CD57, CD69, CD95, OX40 (CD134), and CD40L (CD154) (Figure 1.36). In recent studies, expression of PD1 and CXCL13 has been identified on T_{FH}.

THE PARACORTEX

The paracortical area is the primary home of T-cells. These cells slowly flow in the spaces provided by the paracortical cords. The cords consist of a centrally located venule lined by tall, cuboidal endothelial cells (high endothelial venules) surrounded by narrow corridors outlined by reticular fibers (Figure 1.37A). In these corridors T-cells interact with antigens presented to them by the stationary IDCs. IDCs, unlike FDCs, are derived from bone marrow stem cells and express HLA-DR and S-100 protein (Figure 1.37B). The T-cells have passed through the thymic developmental processes (post-thymic T-cells) and are divided into *helper* and *suppressor* T-cells. Helper T-cells are CD4-positive and release regulatory cytokines to facilitate the immune responses and are divided into two major subtypes: (a) Th1 and (b) Th2 cells. The Th1 cells secrete IL-2 and interferon γ and provide help to other T-cells and macrophages. On the other hand, the Th2 cells secrete IL-4, IL-5, IL-6, and IL-10 and assist B-cells in their antibody production. Suppressor T-cells express CD8 and are primarily involved in cytotoxic reactions. There are more CD4-positive than CD8-positive T-cells in lymph nodes.

THE MEDULLA

The medulla consists of medullary cords loaded with T- and B-cells, plasma cells, and macrophages.

VASCULAR AND LYMPHATIC STRUCTURES

The main artery, after entering the lymph node through the hilum, branches and gives rise to numerous arterioles that pass through the trabeculae and reach the cortex. There, they make a capillary network. The capillaries empty into the high endothelial venules in the center of the cortical corridors. Venules join together and make larger branches, extend from the cortex to the medulla, and finally leave the hilum as veins. Afferent lymphatic vessels penetrate the lymph node capsule and empty into the subcapsular sinuses, which are connected to the cortical sinuses. The sinuses are lined by endothelial cells which have no basement membrane. The sinusoidal lumen is subdivided into smaller interconnecting spaces by fibrous septa covered by endothelial cells. Sinuses guide the lymphatic flow from the capsule into the medulla and eventually terminate to the efferent lymphatics at the hilum and leave the lymph node. Sinusoidal spaces are loaded with macrophages.

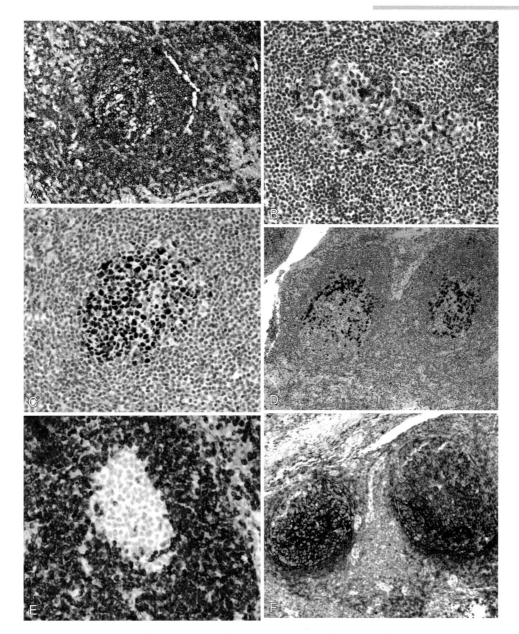

FIGURE 1.36 Immunohistochemical stains of lymph node follicular structures. Follicular B-cells express CD20 (A, brown), CD10 (B) and BCL-6 (C), and are surrounded by CD3+ T-cells (A, pink). Scattered CD3+, CD57+ and BCL-2+ cells are present within the germinal centers (A, D, and E, respectively). CD21+ cells (F) represent a mesh of follicular dendritic cells.

Extramedullary Hematopoiesis

Extramedullary hematopoiesis refers to hematpoiesis in locations other than the bone marrow medullary space. It is either physiological or pathological. Extramedullary hematopoiesis in the fetus is a physiological process which consists of two steps: (1) *primitive hematopoiesis* which develops in the yolk sack during 2.5 to 8 weeks of fetal life as a temporary red-cell-forming system, and (2) *definitive hematopoiesis* which is developed later to generate the entire blood cells and involves the fetal liver (Figure 1.38), spleen, and bone marrow.

In a variety of pathological conditions, such as myeloproliferative disorders, hemolytic anemias, bone marrow metastasis, Gaucher disease, osteopetrosis, and Paget disease, the rate of hematopoiesis in bone marrow is compromised. In these conditions, as a compensatory mechanism, liver and spleen, and sometimes other tissues, may become the source of hematopoiesis (Figures 1.39 and 1.40). Extramedullary hematopoiesis has also been observed in other tissues and organs, including lymph nodes (Figure 1.41), heart, respiratory system, gastrointestinal tract, kidney, adrenal gland, breast, prostate gland, uterus, and mediastinum.

Myelolipoma is a benign tumor of adrenal gland, usually an incidental autopsy finding. It consists of fat and hematopoietic precursors, resembling an ectopic bone marrow tissue (Figure 1.42).

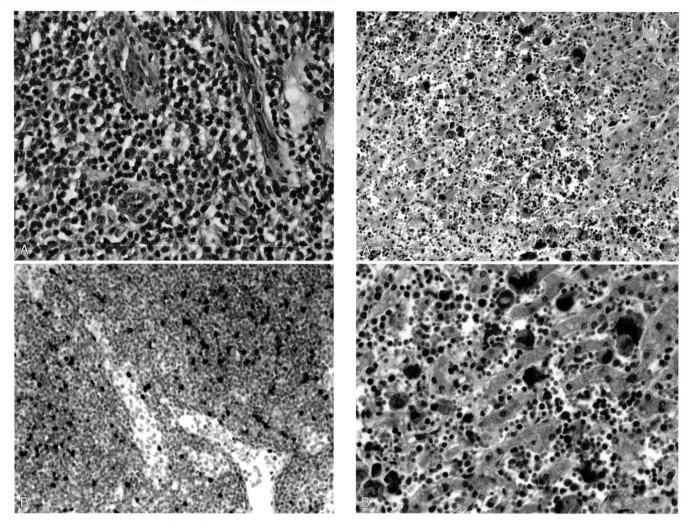

FIGURE 1.37 Lymph node paracortex: (A) high endothelial venules; (B) scattered S-100+ interdigitating dendritic cells.

FIGURE 1.39 Extramedulary hematopoiesis in the liver of an adult patient with a history of primary myelofibrosis (A, low power; B, high power).

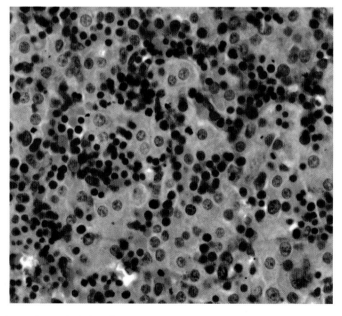

FIGURE 1.38 Extramedulary hepatic erythropoiesis in a fetus.

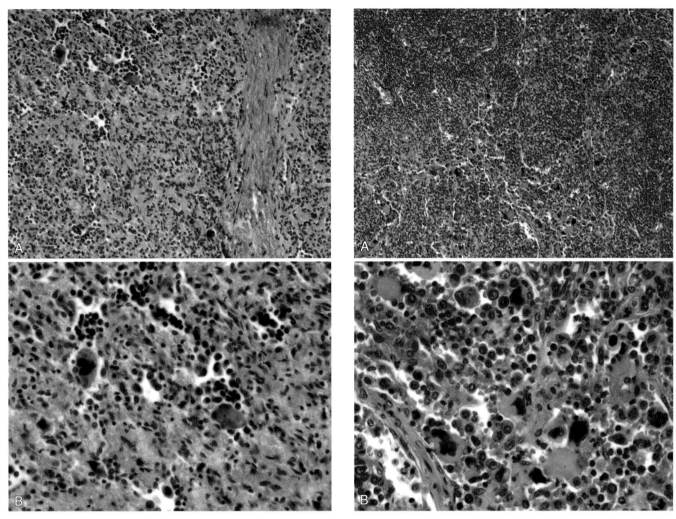

FIGURE 1.40 Extramedullary hematopoiesis in the spleen of an adult patient with a history of primary myelofibrosis (A, low power; B, high power).

FIGURE 1.41 Extramedullary hematopoiesis in the lymph node of an adult patient with a history of primary myelofibrosis (A, low power; B, high power).

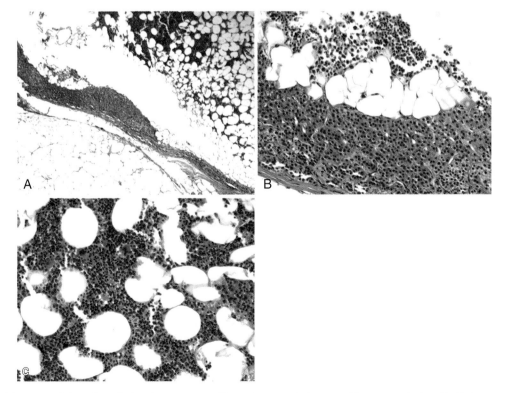

FIGURE 1.42 Myelolipoma of the adrenal gland consisting of fatty tissue and clusters of hematopoietic cells (A, low power; B, intermediate power; C, high power).

Additional Resources

Allen CD, Okada T, Cyster JG: Germinal-center organization and cellular dynamics, *Immunity* 27:190-202, 2007.

Balato A, Unutmaz D, Gaspari AA: Natural killer T cells: an unconventional T-cell subset with diverse effector and regulatory functions, *J Invest Dermatol* 129:1628-1642, 2009.

Barreda DR, Hanington PC, Belosevic M: Regulation of myeloid development and function by colony stimulating factors, *Dev Comp Immunol* 28:509-554, 2004.

Cesta MF: Normal structure, function, and histology of the spleen, *Toxicol Pathol* 34:455-465, 2006.

Choi KD, Vodyanik M, Slukvin II: Hematopoietic differentiation and production of mature myeloid cells from human pluripotent stem cells, *Nat Protoc* 6:296-313, 2011.

Foucar K: *Bone marrow pathology*, ed 2, Chicago, 2001, ASCP Press.

Gatto D, Brink R: The germinal center reaction, *J Allergy Clin Immunol* 126:898-907, 2010.

Geddis AE: Megakaryopoiesis, *Semin Hematol* 47:212-219, 2010.

Hirose J, Kouro T, Igarashi H, et al: A developing picture of lymphopoiesis in bone marrow, *Immunol Rev* 189:28-40, 2002.

Hoffbrand AV, Pettit JE, Vyas P: *Color atlas of clinical hematology*, ed 4, 2010, Mosby/Elsevier.

Kassem M: Mesenchymal stem cells: biological characteristics and potential clinical applications, *Cloning Stem Cells* 6:369-374, 2004.

Kawamoto H, Minato N: Myeloid cells, *Int J Biochem Cell Biol* 36:1374-1379, 2004.

Liu K, Nussenzweig MC: Origin and development of dendritic cells, *Immunol Rev* 234:45-54, 2010.

Maillard I, Fang T, Pear WS: Regulation of lymphoid development, differentiation, and function by the Notch pathway, *Annu Rev Immunol* 23:945-974, 2005.

Mebius RE, Kraal G: Structure and function of the spleen, *Nat Rev Immunol* 5:606-616, 2005.

Mebius RE: Organogenesis of lymphoid tissues, *Nat Rev Immunol* 3:292-303, 2003.

Metcalf D: Hematopoietic cytokines, *Blood* 111:485-491, 2008.

O'Malley DP: Benign extramedullary myeloid proliferation, *Mod Pathol* 20:405-415, 2007.

Rothenberg ME: Eosinophils in the new millennium, *J Allergy Clin Immunol* 19:1321-1322, 2007.

Seita J, Weissman IL: Hematopoietic stem cell: self-renewal versus differentiation, *Wiley Interdiscip Rev Syst Biol Med* 2:640-653, 2010.

Shiohara M, Koike K: Regulation of mast cell development, *Chem Immunol Allergy* 87:1-21, 2005.

Uston PI, Lee CM: Characterization and function of the multifaceted peripheral blood basophil, *Cell Mol Biol (Noisy-le-grand)* 49:1125-1135, 2003.

Van Lochem EG, van der Velden VHJ, Wind HK, et al: Immunophenotypic differentiation patterns of normal hematopoiesis in human bone marrow: reference patterns for age-related changes and disease-induced shifts, *Cytometry Part B* 60B:1-13, 2004.

Willard-Mack CL: Normal structure, function, and histology of lymph nodes, *Toxicol Pathol* 34:409-424, 2006.

Yona S, Jung S: Monocytes: subsets, origins, fates and functions, *Curr Opin Hematol* 17:53-59, 2010.

Principles of Immunophenotyping

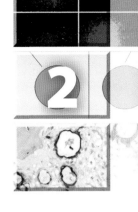

Human Cell Differentiation Molecules

Human cell differentiation molecules (HCDM) is a new terminology coined by the 8th International Workshop on Human Leukocyte Differentiation Antigens (HLDA) in 2004 to describe surface molecules associated with human cell differentiation. These molecules have been characterized in a series of international workshops studying a large number of monoclonal antibodies. The antibodies have been grouped according to their patterns of reactivity and are referred to as "clusters of differentiation" (CD). The 9th International Workshop on HLDA was held in Barcelona, Spain, in March 2010 and brought the total number of CD molecules to 363 (Table 2.1) (http://hcdm.org). These molecules characterize human leukocytes as well as other human cells such as endothelial, stromal, and epithelial cells. They can be detected on the surface and/or inside the cells.

Monoclonal antibodies are routinely used for the diagnosis and classification of hematopoietic malignancies and other hematologic disorders. However, it is important to remember the following facts.

1. If not all, then by far the vast majority of the available monoclonal antibodies raised against CD molecules are not tumor-specific and react with non-neoplastic hematopoietic cells.
2. These molecules are mostly differentiation-associated and not lineage-specific.
3. They may react with non-hematopoietic human cells.

Because of these facts, the results of immunophenotypic studies, by flow cytometry and/or immunohistochemical stains, should always be incorporated with clinical information, morphology, and other available data, such as cytogenetics and molecular studies. The following are examples of CD molecules most frequently used in diagnostic hematopathology at the present time.

B-CELL-ASSOCIATED CD MOLECULES

CD10

CD10, also known as *common acute lymphoblastic leukemia antigen* (CALLA), is a neutral endopeptidase which inactivates several peptides. It is expressed on hematogones (normal precursor B-cells in bone marrow), normal germinal center B-cells, as well as neoplastic conditions such as B-lymphoblastic leukemia/lymphoma, follicular lymphomas, Burkitt lymphoma, and some cases of diffuse large cell lymphoma, plasma cell myeloma, T-lymphoblastic leukemia/lymphomas, and myeloid leukemias. This molecule is abundantly expressed in kidney, particularly on the brush border of proximal tubules and on glomerular epithelium. It is also present in granulocytes, fibroblasts, and a variety of normal and neoplastic epithelial cells.

CD19

CD19 is a signal-transduction molecule that plays an important role in the regulation of development, activation, and differentiation of B-lymphocytes. It is one of the earliest lineage-restricted molecules expressed on B-cells, and remains expressed throughout B-cell ontogeny. Follicular dendritic cells also express CD19. This molecule is not expressed on normal T-lymphocytes, monocytic and granulocytic series, or erythroid precursors. However, CD19 is occasionally expressed in patients with AML.

CD20

CD20 is a membrane-embedded surface molecule which plays a role in the development and differentiation of B-cells into plasma cells. It appears after HLA-DR, TdT, CD19, and CD10 expression and before cytoplasmic μ chain appearance in B-cell ontogeny. Similar to CD19, CD20 is a lineage-restricted molecule and is expressed on B-cells throughout B-cell differentiation prior to terminal differentiation of

Table 2.1
The Human Cluster of Differentiation Molecules

CD	Molecule	Main Distribution
CD1a	T6/leu-6, R4, HTA1	Cortical thymocyte, LC, IDC
CD1b	R1	Cortical thymocyte, LC, IDC
CD1c	M241, R7	Cortical thymocyte, LC, IDC
CD2	T11; Tp50; sheep red blood cell (SRBC) receptor; LFA-2;	Thymocyte, T, NK, thymic B cells
CD3	CD3 complex, T3, Leu4	Precursor T, thymocyte, T
CD4	OKT4, Leu 3a, T4	Helper T, thymocyte, M
CD5	Tp67; T1, Ly1, Leu-1	Thymocytes, T, B subset
CD6	T12	Thymocyte, T, B subset
CD7	Leu 9, 3A1, gp40, T cell leukemia antigen	Precursor T, T, NK
CD8	OKT8, LeuT, LyT2, T8	Cytotoxic T, NK
CD9	Drap-27, MRP-1, p24, leucocyte antigen MIC3	Platelet, early B, Eo, Baso, endothelial
CD10	CALLA, membrane metallo-endopeptidase	Precursor B, B subset, G
CD11a	alpha L; LFA-1, gp180/95	All leukocytes
CD11b	alpha M; alpha-chain of C3bi receptor, gp155/95, Mac-1, Mo1	G, M, NK
CD11c	alpha X; α-chain of complement receptor type 4 (CR4); gp150/95	G, M, NK
CDw12	P90-120	G, M, NK
CD13	Aminopeptidase N, APN, gp150, EC 3.4.11.2	G, M, endothelial, LGL subset
CD14	LPS receptor	M, DC subset
CD15	Lewis X, CD 15u: sulphated Lewis X. CD 15s: sialyl Lewis X	G, Reed-Sternberg cells
CD16	Fc gamma R IIIa,	NK, G, M, macrophage
CDw17	LacCer, lactosylceramide	Platelet, G, M, B subset
CD18	β2-Integrin chain, macrophage antigen 1 (mac-1)	All leukocytes
CD19	Bgp95, B4	Precursor B, B
CD20	B1; membrane-spanning 4-domains, subfamily A, member 1	Precursor B subset, B
CD21	C3d receptor, CR2, gp140; EBV receptor	FDC, B subset, T subset
CD22	Bgp135; BL-CAM, Siglec2	Precursor B, B
CD23	Low affinity IgE receptor; FceRII; gp50-45; Blast-2	B, DC, M
CD24	heat stable antigen homologue (HSA), BA-1	Precursor B, B, G
CD25	Interleukin (IL)-2 receptor α-chain; Tac-antigen	Activated T, B, and M
CD26	Dipeptidylpeptidase IV; gp120; Ta1	Thymocyte, B, NK, Macrophage, activated T
CD27	T14, S152	NK, thymocyte, B subset, T subset
CD28	Tp44	Thymocyte, T, PC
CD29	Integrin β1 chain; platelet GPIIa; VLA (CD49) beta-chain	All leukocytes
CD30	Ki-1 antigen, Ber-H2 antigen	M, activated B, T, and NK
CD31	PECAM-1; platelet GPIIa'; endocam	Endothelial, platelet, leukocyte
CD32	Fc gamma receptor type II (FcgRII), gp40	M, G, Eo, Baso, B, platelet
CD33	My9, gp67, p67	Precursor G, G, M
CD34	My10, gp105-120	Hematopoietic progenitor cells, endothelium
CD35	C3b/C4b receptor; complement receptor type 1 (CR1)	Erythroid, B, Eo, M, T subset
CD36	platelet GPIV, GPIIIb, OKM-5 antigen	Platelet, M
CD37	gp40-52	Mature B
CD38	T10; gp45, ADP-ribosyl cyclase	Early and activated hematopoietic cells, PC
CD39	gp80, ectonucleoside triphosphate diphosphohydrolase 1	Leukocytes
CD40	Bp50, T NF Receptor 5	B, DC, macrophage, endothelial
CD41	platelet glycoprotein GPIIb	Platelet
CD42a	platelet glycoprotein GPIX	Platelet
CD42b	platelet glycoprotein GPIb-α	Platelet
CD42c	platelet glycoprotein GPIb-β	Platelet
CD42d	platelet glycoprotein GPV	Platelet
CD43	Leukosialin; gp95; sialophorin; leukocyte sialoglycoprotein	Leukocytes
CD44	Pgp-1; gp80-95, Hermes antigen, ECMR-III and HUTCH-I.	Leukocytes
CD45	LCA, B220, protein tyrosine phosphatase, receptor type C	Leukocytes

(Continued)

Table 2.1
The Human Cluster of Differentiation Molecules (Continued)

CD	Molecule	Main Distribution
CD45RA	Restricted T200; gp220; isoform of leukocyte common antigen	Naive T, B, M, NK
CD45RO	Restricted T200; gp180	Thymocyte, memory T, G, M
CD45RB	Restricted T200; isoform of leukocyte common antigen	T subset, B, G, M
CD46	Membrane cofactor protein (MCP)	Leukocytes
CD47	Integrin-associated protein (IAP), Ovarian carcinoma antigen OA3	Leukocytes
CD48	BLAST-1, Hulym3, OX45, BCM1	Leukocytes
CD49a	Integrin α1 chain, very late antigen, VLA 1a	Broad
CD49b	Integrin α2 chain, VLA-2-alpha chain, platelet gpIa	Broad
CD49c	Integrin α3 chain, VLA-3 alpha chain	Broad
CD49d	Integrin α4 chain, VLA-4-alpha chain	Broad
CD49e	Integrin α5 chain, VLA-5 alpha chain	Broad
CD49f	Integrin α6 chain, VLA-6 alpha chain, platelet gpIc	Broad
CD50	ICAM-3, intercellular adhesion molecule 3	Leukocytes
CD51	Integrin alpha chain, vitronectin receptor alpha chain	Platelet, endothelial cell
CD52	Campath-1, HE5	Thymocyte, B, T, NK, M
CD53	MRC OX-44	B, T, M, NK, G
CD54	ICAM-1, intercellular adhesion molecule 1	B, T, M, G, endothelial cell
CD55	DAF, Decay Accelerating Factor	Broad
CD56	NKHI, Neural cell adhesion molecule (NCAM)	NK, T subset, neuroendodermal cells
CD57	HNK1	NK, T subset, neuroendodermal cells
CD58	LFA-3, lymphocyte function associated antigen-3	Broad
CD59	MACIF, MIRL, P-18, protectin	Broad
CD60	GD3 (CD60a), 9-0-acetyl GD3 (CD60b), 7-0-acetyl GD3 (CD60c)	Platelets, T subset
CD61	Glycoprotein IIIa, beta3 integrin	Platelets
CD62E	E-selectin, LECAM-2, ELAM-1	Endothelium
CD62L	L-selectin, LAM-1, Mel-14	B, T, M, NK subset, G
CD62P	P-selectin, granule membrane protein-140 (GMP-140)	Activated platelet, endothelium
CD63	LIMP, gp55, LAMP-3 neuroglandular antigen, granulophysin	Activated platelets, G, M, endothelium
CD64	FcgR1, FcgammaR1	Precursor G, G, M, DC subset
CD65	Ceramide dodecasaccharide 4c, VIM2	G, M
CD66a	BGP, carcinoembryonic antigen-related cell adhesion molecule 1	G, epithelium
CD66b	CGM6, NCA-95	G
CD66c	nonspecific crossreaction antigen, NCA-50/90	G, epithelium
CD66d	CGM1	G
CD66e	CEA	Epithelium
CD66f	PSG, Sp-1, pregnancy specific (b1) glycoprotein	Myeloid cell lines, placenta
CD68	gp110, macrosialin	M, G, DC subset, Baso, Mast cell
CD69	AIM, activation inducer molecule, MLR3, EA1, VEA	Activated leukocytes
CD70	CD27 ligand, KI-24 antigen	Activated B and T
CD71	Transferrin receptor	Erythroid precursors, proliferating cells
CD72	Lyb-2, Ly-19.2, Ly32.2	Precursor B, B
CD73	Ecto-5′-nucleotidase	B subset, T subset
CD74	MHC Class II associated invariant chain (Ii)	B, IDC, T subset
CD75	Lactosamines	B, activated T, macrophages, activated endothelium
CD75s	Since HLDA7, CDw76 has been renamed CD75s	B, T subset
CD77	Pk blood group antigen; Burkitt's lymphoma associated antigen	Germinal center B
CD79a	MB-1; Igα	Precursor B, B, activated B
CD79b	B29; Igβ	Precursor B, B, activated B
CD80	B7-1; BB1	Macrophages, activated T and B
CD81	Target of an antiproliferative antibody (TAPA-1); M38	Broad
CD82	R2; 4F9; C33; IA4, kangai 1	Broad
CD83	HB15	IDC, LC

(Continued)

Table 2.1
The Human Cluster of Differentiation Molecules (Continued)

CD	Molecule	Main Distribution
CD84	p75, GR6	CD84
CD85	ILT5; LIR3; HL9	B, thymocytes, M, macrophages, platelets
CD86	B7-2; B70	IDC, LC, B, and M subset
CD87	Urokinase plasminogen activator-receptor (uPA-R)	Subsets of T, NK, M, and G
CD88	C5a-receptor	G, M, DC
CD89	Fcα-receptor, IgA-receptor	Precursor myeloid, G, M
CD90	Thy-1	Hematopoietic stem cell
CD91	α2-macroglobulin receptor (ALPHA2M)	Broad
CDw92	p70	G, M
CDw93	GR11	G, M, myeloid blast, endothelium
CD94	kP43, killer cell lectin-like receptor subfamily D, member 1	NK, T subset
CD95	APO-1, Fas, TNFRSF6	Thymocytes, B and T subset
CD96	TACTILE (T cell activation increased late expression)	Activated NK and T
CD97	BL-KDD/F12	DC, G, M, activated B and T
CD98	4F2, FRP-1	Activated leukocytes
CD99	MIC2, E2	Broad
CD100	SEMA4D	Leukocytes, activated T, germinal center B
CD101	V7, P126	G, M, DC, activated T
CD102	ICAM-2	M, platelet, endothelium
CD103	Integrin alpha E subunit, HML-1	Intraepithelial lymphocytes, hairy cells
CD104	Integrin beta 4 subunit, TSP-1180	Epithelium
CD105	Endoglin	Endothelium, precursor B, activated M
CD106	VCAM-1 (vascular cell adhesion molecule-1), INCAM-110	DC, activated endothelium
CD107a	Lysosomal associated membrane protein (LAMP)-1	Degranulated platelet, activated T
CD107b	Lysosomal associated membrane protein (LAMP)-2	Degranulated platelet
CD108	GPI-gp80; John-Milton-Hagen (JMH) human blood group antigen	Erythroid
CD109	Platelet activation factor; 8A3, E123	Activated platelet, endothelium
CD110	Thrombopoietin receptor; c-mpl	Hematopoietic stem cells, platelets
CD111	PRR1, Nectin 1, Hve C1, poliovirus receptor related 1 protein	34+ hematopoietic precursors
CD112	PRR2, Nectin 2, Hve B, poliovirus receptor related 2 protein	34+ hematopoietic precursors
CDw113	PVRL3, Nectin3	Epithelium
CD114	G-CSFR, HG-CSFR, CSFR3	M, platelets
CD115	M-CSFR, CSF-1, C-fms	M, macrophages
CD116	GMCSF R alpha subunit	Myeloid cells
CD117	SCFR, c-kit, stem cell factor receptor	Hematopoietic stem cells, mast cells, plasma cells, AML blasts
CD118	LIFR	Broad
CD119	IFN gamma receptor alpha chain	Broad
CD120a	TNFRI; TNFRp55	Broad
CD120b	TNFRII; TNFRp75	Broad
CD121a	Type I IL-1 receptor	Broad
CD121b	Type II IL-1 receptor	Broad
CD122	IL-2 receptor beta chain, p75	B, T, NK, M
CD123	IL-3 receptor alpha chain (IL-3Ra)	Hematopoietic precursors
CD124	IL-4 R alpha chain	Broad
CDw125	IL-5 receptor alpha chain	Baso, Eo, activated B
CD126	IL-6 receptor alpha chain	T, M, activated B
CD127	IL-7 receptor alpha chain, p90	Precursor B, B, T
CD129	IL-9 receptor alpha chain	Hematopoietic cells
CD130	gp130	Broad
CD131	Common β chain, low-affinity (granulocyte-macrophage)	Precursor myeloid, precursor B, M, G, Eo
CD132	Common gamma chain, interleukin 2 receptor, gamma	B, T, M, G, NK

(Continued)

Table 2.1

The Human Cluster of Differentiation Molecules (Continued)

CD	Molecule	Main Distribution
CD133	AC133, PROML1, prominin 1	CD34+ hematopoietic precursor
CD134	OX 40, TNFRSF4	Thymocyte, T
CD135	FLT3, STK-1, flk-2	Precursor B, precursor myleomonocytic
CDw136	Macrophage stimulating protein receptor, MSP-R, RON	Epithelium, M
CDw137	4-1BB, Induced by lymphocyte activation (ILA)	T, activated T
CD138	Syndecan-1, B-B4	Plasma cells, B subset, epithelium
CD139		B, M, G
CD140a	alpha-platelet derived growth factor (PDGF) receptor	Mesenchymal cells
CD140b	beta-PDGF receptor	Mesenchymal cells, M, G
CD141	Thrombomodulin (TM), fetomodulin	Broad
CD142	Tissue factor, thromboplastin, coagulation factor III	Epithelium, M, endothelium
CD143	Angiotensin-converting enzyme (ACE), peptidyl dipeptidase A	Broad
CD144	VE-cadherin, cadherin-5	Endothelium
CDw145	None	Endothelium
CD146	Muc 18, MCAM, Mel-CAM, s-endo	Endotheium, melanoma cells, activated T
CD147	Basigin, M6, extracellular metalloproteinase inducer (EMMPRIN)	Leukocyte, erythroid, platelet, endothelium
CD148	DEP-1, HPTP-n, protein tyrosine phosphatase, receptor type, J	G, M, T subset, DC, platelet
CD150	SLAM, signaling lymphocyte activation molecule, IPO-3	Thymocyte, B, DC, T subset, endothelium
CD151	Platelet-endothelial tetra-span antigen (PETA)-3	Platelet, endothelium, epithelium
CD152	Cytotoxic T lymphocyte antigen (CTLA)-4	Activated B and T
CD153	CD30 Ligand	Activated T and M
CD154	CD40 Ligand; TRAP (TNF-related activation protein)-1; T-BAM	Activated T
CD155	Polio virus receptor (PVR)	M, neurons
CD156a	ADAM-8, a disintegrin and metalloproteinase domain 8	G, M
CD156b	TACE, ADAM 17 snake venom like protease CSVP	Broad
CD157	BST-1 BP-3/IF7 Mo5	G, M, precursor B
CD158	killer cell Ig-like receptor, three domains, long cytoplasmic tail, 1	NK, T subset
CD159a	killer cell lectin-like receptor subfamily C, member 1	NK
CD160	BY55, NK1, NK28	NK, T subset
CD161	NKR-P1A, killer cell lectin-like receptor subfamily B, member 1	NK, T subset
CD162	P selectin glycoprotein ligand 1, PSGL-1	T, M, G, B subset
CD163	GHI/61, D11, RM3/1, M130	M, macrophage, activated T
CD164	MUC-24, MGC 24, multi-glycosylated core protein 24	M, epithelium, bone marrow stromal cells
CD165	AD2, gp 37	Thymocyte, T, platelet
CD166	ALCAM, KG-CAM, activated leukocyte cell adhesion molecule	Epithelium, activated T and M
CD167	Discoidin receptor DDR1 (CD 167a) and DDR2 (CD 167b)	Epithelium
CD168	RHAMM (receptor for hyaluronan involved in migration & motility)	Thymocyte
CD169	Sialodhesin, Siglec-1	Macrophage
CD170	Siglec 5 (sialic acid binding Ig-like lectin 5)	Myeloid cells
CD171	Neuronal adhesion molecule, LI	Neurons
CD172	SIRP, signal inhibitory regulatory protein family member	leukocytes
CD173	Blood Group H2	Erythrocytes
CD174	Lewis Y blood group, LeY, fucosyltransferase 3	Erythrocytes
CD175	Tn Antigen (T-antigen novelle)	Carcinomas
CD176	Thomsen-Friedenreich antigen (TF)	Carcinomas
CD177	NB 1	
CD178	FAS ligand, CD95 ligand	T, NK
CD179a	V pre beta	Precursor B
CD179b	Lambda 5	Precursor B
CD180	RP105, Bgp95	Mantle zone and marginal zone B
CD181	CXCR1, (was CDw128A)	Leukocytes
CD182	CXCR2, (was CDw128B)	Leukocytes

(Continued)

Table 2.1
The Human Cluster of Differentiation Molecules (Continued)

CD	Molecule	Main Distribution
CD183	CXCR3 chemokine receptor, G protein-coupled receptor 9	T, CD34+ hematopoietic cells, DC subset, Eo,
CD184	CXCR4 chemokine receptor, Fusin	M, T subset
CD185	CXCR5; Chemokine (C-X-C motif) Receptor 5, Burkitt lymphoma receptor 1	Broad
CDw186	CXCR6; Chemokine (C-X-C motif) Receptor 6	T, epithelium
CD191	CCR1; Chemokine (C-C motif) Receptor 1, RANTES Receptor	T, and NK subset
CD192	CCR2; Chemokine (C-C motif) Receptor 2, MCP-1 receptor	M
CD193	CCR3; Chemokine (C-C motif) Receptor 3, eosinophil eotaxin receptor	Eo, Baso, epithelium
CD195	CCR5 chemokine receptor	T, M
CD196	CCR6; Chemokine (C-C motif) Receptor 6	DC and T subset
CD197	CCR7; (was CDw197) Chemokine (C-C motif) Receptor 7	DC and T subset
CDw198	CCR8; Chemokine (C-C motif) Receptor 8	Thymocyte, macrophage
CDw199	CCR9; Chemokine (C-C motif) Receptor 9	Intestinal T cells
CD200	MRC OX 2	Broad
CD201	Endothelial protein C receptor (EPCR)	Endothelium
CD202b	TIE2, TEK	Endothelium, hematopoietic stem cell
CD203c	E-NPP3, PDNP3, PD-1beta	Mast cell, Baso
CD204	MSR, SRA, Macrophage scavenger receptor	Macrophage
CD205	DEC-205	DC
CD206	Macrophage mannose receptor (MMR)	M, macrophage, endothelium
CD207	Langerin	LC
CD208	DC-LAMP	IDC
CD209	DC-SIGN	DC subset
CD210	IL-10 receptor	B, T, NK, M, macrophage
CD212	IL-12 receptor beta chain	Activated T and NK
CD213a1	IL-13 receptor alpha 1	Broad
CD213a2	IL-13 receptor alpha 2	B, M
CDw217	IL-17 receptor	Broad
CDw218	IL-18R alpha	
CD220	Insulin Receptor	Broad
CD221	IGF I Receptor, type I IGF receptor	Broad
CD222	Mannose-6-phosphate receptor, insulin like growth factor II R	Broad
CD223	LAG-3 (Lymphocyte activation gene 3)	T and NK subset
CD224	Gamma-glutamyl transferase, GGT	Broad
CD225	Leu-13, interferon-induced transmembrane protein 1	Broad
CD226	DNAM-1, DTA-1	T, NK, M, platelet, B subset
CD227	MUC1; episialin; PUM; PEM; EMA; DF3 antigen; H23 antigen	Broad
CD228	Melanotransferrin, p97	Melanoma cells, endothelium
CD229	Ly9	T, B
CD230	Prion protein, PrP(c), PrP(sc) abnormal form	Broad
CD231	TALLA-1, TM4SF2	Precursor T, neuroblastoma
CD232	VESPR	Broad
CD233	Band 3, AE1 (anion exchanger 1), Diego blood group antigen	RBC
CD234	DARC, Fy-glycoprotein, Duffy blood group antigen	RBC
CD235a	Glycophorin A	RBC
CD235b	Glycophorin B	RBC
CD236	Glycophorin C/D	RBC, stem cell subset
CD236R	Glycophorin C	RBC, stem cell subset
CD238	Kell blood group antigen	RBC, stem cell subset
CD239	B-CAM, lutheran glycoprotein	RBC, stem cell subset
CD240CE	Rh blood group system, Rh30CE	RBC
CD240D	Rh blood group system, Rh30D	RBC
CD240DCE	Rh30D/CE crossreactive mabs	RBC

(Continued)

Table 2.1
The Human Cluster of Differentiation Molecules (Continued)

CD	Molecule	Main Distribution
CD241	RhAg, Rh50, Rh associated antigen	RBC
CD242	LW blood group, Landsteiner-Wiener blood group antigens	RBC
CD243	MDR-1, P-glycoprotein, pgp 170, multidrug resistance protein I	Hematopoietic stem cell
CD244	2B4; NAIL; p38	NK, T subset
CD245	p220/240, DY12, DY35	T subset
CD246	Anaplastic lymphoma kinase (ALK)	Anaplastic large cell lymphoma
CD247	T cell receptor zeta chain, CD3 zeta	T, NK
CD248	TEM1, Endosialin	Fibroblast, endothelium
CD249	Aminopeptidase A; APA, gp160	Epithelium
CD252	OX40L; TNF (ligand) superfamily member 4, CD134 ligand	T
CD253	TRAIL; TNF (ligand) superfamily member 10, APO2L	T
CD254	TRANCE; TNF (ligand) superfamily member 11, RANKL	T, M
CD256	APRIL; TNF (ligand) superfamily member 13, TALL2	Osteoclast, B subset
CD257	BLYS; TNF (ligand) superfamily, member 13b, TALL1, BAFF	B
CD258	LIGHT; TNF (ligand) superfamily, member 14	
CD261	TRAIL-R1; TNFR superfamily, member 10a, DR4, APO2	Broad
CD262	TRAIL-R2; TNFR superfamily, member 10b, DR5	Broad
CD263	TRAIL-R3; TNFR superfamily, member 10c, DCR1	Broad
CD264	TRAIL-R4; TNFR superfamily, member 10d, DCR2	NK, T subset
CD265	TRANCE-R; TNFR superfamily, member 11a, RANK	M, DC
CD266	TWEAK-R; TNFR superfamily, member 12A, type I transmembrane protein Fn14	Broad
CD267	TACI; TNFR superfamily, member 13B, transmembrane activator and CAML interactor	Lymphocytes
CD268	BAFFR; TNFR superfamily, member 13C, B cell-activating factor	B
CD269	BCMA; TNFR superfamily, member 17, B-cell maturation factor	B
CD271	NGFR (p75); nerve growth factor receptor (TNFR superfamily, member	Neurons
CD272	BTLA; B and T lymphocyte attenuator	B, T subset
CD273	B7DC, PDL2; programmed cell death 1 ligand 2	Activated B and T
CD274	B7H1, PDL1; programmed cell death 1 ligand 1	Broad
CD275	B7H2, ICOSL; inducible T-cell co-stimulator ligand (ICOSL)	Broad
CD276	B7H3; B7 homolog 3	N/A
CD277	BT3.1; B7 family: butyrophilin, subfamily 3, member A1	
CD278	ICOS; inducible T-cell co-stimulator	Activated T
CD279	PD1; programmed cell death 1	Broad
CD280	ENDO180; uPARAP, mannose receptor, C type 2, TEM22	Macrophages
CD281	TLR1; TOLL-like receptor 1	Lymphocytes
CD282	TLR2; TOLL-like receptor 2	Lymphocytes
CD283	TLR3; TOLL-like receptor 3	Lymphocytes
CD284	TLR4; TOLL-like receptor 4	Lymphocytes
CD289	TLR9; TOLL-like receptor 9	Lymphocytes
CD292	BMPR1A; Bone Morphogenetic Protein Receptor, type IA	Broad
CDw293	BMPR1B; Bone Morphogenetic Protein Receptor, type IB	Broad
CD294	CRTH2; PGRD2; G protein-coupled receptor 44,	T subset
CD295	LEPR; Leptin Receptor	Platelets, G
CD296	ART1; ADP-ribosyltransferase 1	G
CD297	ART4; ADP-ribosyltransferase 4; Dombrock blood group glycoprotein	RBC
CD298	ATP1B3; Na$^+$/K$^+$-ATPase beta 3 subunit	Broad
CD299	DCSIGN-related; CD209 antigen-like, DC-SIGN2, L-SIGN	DC
CD300	CMRF35 FAMILY; CMRF-35H	M, G, B and T subsets
CD301	MGL; CLECSF14, macrophage galactose-type C-type lectin	Macrophages
CD302	DCL1; Type I transmembrane C-type lectin receptor DCL-1	Hodgkin lymphoma cell line
CD303	BDCA2; C-type lectin, superfamily member 11	DC subtype

(Continued)

Table 2.1
The Human Cluster of Differentiation Molecules (Continued)

CD	Molecule	Main Distribution
CD304	BDCA4; Neuropilin 1	Broad
CD305	LAIR1; Leukocyte-Associated Ig-like Receptor 1	B, T, NK
CD306	LAIR2; Leukocyte-Associated Ig-like Receptor 2	B, T, NK
CD307	IRTA2; Immunoglobulin superfamily Receptor Translocation Associated	B
CD309	VEGFR2; KDR (a type III receptor tyrosine kinase)	Endothelium
CD312	EMR2 ; EGF-like module containing, mucin-like, hormone receptor-like	Lymphocytes
CD314	NKG2D; Killer cell lectin-like receptor subfamily K, member 1	NK
CD315	CD9P1; Prostaglandin F2 receptor negative regulator	Lymphocytes
CD316	EWI2; Immunoglobulin superfamily, member 8	Lymphocytes
CD317	BST2; Bone Marrow Stromal cell antigen 2	Bone marrow stromal cells
CD318	CDCP1; CUB domain-containing protein 1	Hematopoietic stem cell subset
CD319	CRACC; SLAM family member 7	Activated T
CD320	8D6; 8D6 Antigen; FDC	N/A
CD321	JAM1; F11 receptor	Epithelium, endothelium
CD322	JAM2; Junctional Adhesion Molecule 2	Epithelium, endothelium
CD324	E-Cadherin; cadherin 1, type 1, E-cadherin (epithelial)	Epithelium
CDw325	N-Cadherin; cadherin 2, type 1, N-cadherin (neuronal)	Neurons
CD326	Ep-CAM; tumor-associated calcium signal transducer 1	Epithelium
CDw327	siglec6; sialic acid binding Ig-like lectin 6	Cell–cell adhesion
CDw328	siglec7; sialic acid binding Ig-like lectin 7	Cell–cell adhesion
CDw329	siglec9; sialic acid binding Ig-like lectin 9	Cell–cell adhesion
CD331	FGFR1; Fibroblast Growth Factor Receptor 1	Fibroblasts
CD332	FGFR2; Fibroblast Growth Factor Receptor 2 (keratinocyte growth factor receptor)	Fibroblasts
CD333	FGFR3; Fibroblast Growth Factor Receptor 3 (achondroplasia, thanatophoric dwarfism)	Fibroblasts
CD334	FGFR4; Fibroblast Growth Factor Receptor 4	Fibroblasts
CD335	NKp46; NCR1, (Ly94); natural cytotoxicity triggering receptor 1	NK
CD336	NKp44; NCR2, (Ly95); natural cytotoxicity triggering receptor 2	NK
CD337	NKp30; NCR3	NK
CDw338	ABCG2; ATP-binding cassette, sub-family G (WHITE), member 2	Epithelium
CD339	Jagged-1; Jagged 1 (Alagille syndrome)	Broad
CD351	FCAMR; Fc receptor, IgA,IgM, high affinity	Lymphocytes
CD352	SLAMF6; SLAM family member 6	Lymphocytes
CD353	SLAMF6; SLAM family member 8	Lymphocytes
CD354	TREM1; triggering receptor	Myeloid cells T-cells
CD355	CRTAM; cytotoxic and regulatory T-cell molecule	
CD357	TNFRSF18; tumor necrosis factor receptor, member 18	Broad
CD358	TNFRSF21; tumor necrosis factor receptor, member 21	Broad
CD360	IL-21R; interleukin 21 receptor	T, NK
CD361	EVI28	Retrivirus-induced myeloid tumors
CD362	SDC2; syndecan 2	Epithelial cells
CD363	S1PR1; sphingosine-1-phosphate receptor 1	Broad

Adapted from http://hcdm.org
IDC: interdigitating dendrite cell; LC: Langerhans cell; DC: dendritic cell; follicular dendrite cell; M: monocyte/macrophage; G: granulocyte; PC: plasma cell; Eos: eosinophil; Baso: basophil; RBC: red blood cell.

B-cells to plasma cells. CD20 is expressed in a vast majority of mature B-cell neoplasms and some cases of B-lymphoblastic leukemia/lymphomas, plasma cell myelomas, Hodgkin lymphomas, T-cell neoplasms and AMLs.

CD21

CD21 is a receptor for EBV, C3 complement components. CD21 along with CD19 and CD81 make a large signal-transduction complex involved in B-cell activation. CD21 is expressed on mature B-cells and follicular dendritic cells, and is lost upon activation. Follicular dendritic cell neoplasms, some cases of mantle cell and marginal zone lymphomas, as well as a proportion of diffuse large B-cell lymphomas express CD21.

CD22

CD22 is a member of the immunoglobulin gene superfamily. It is involved in the regulation of expression of surface IgM on B-cells. Cytoplasmic CD22 is expressed at the earliest stages of B-cell differentiation, along with CD19 and is present prior to the expression of CD20. The majority of the precursor B-cells are also positive for cytoplasmic CD22. Expression of surface CD22 precedes or accompanies expression of surface IgM and/or IgD in mature B-lymphocytes, but it is lost in plasma cells. The neoplastic cells in various proportions of B-cell malignancies, including B-lymphoblastic leukemia/lymphomas and mature B-cell leukemia/lymphomas, express CD22. The expression of CD22 can be particularly strong in hairy cell leukemia and prolymphocytic leukemia. T-cells and their malignant counterparts, except for rare cases, do not express CD22.

CD23

CD23 is an integral membrane glycoprotein involved in the regulation of IgE synthesis. It is expressed by subsets of B-cells, monocytes, follicular dendritic cells, platelets, and eosinophils. CD23 is expressed in chronic lymphocytic leukemia (CLL), in some cases of follicular lymphomas, and primary mediastinal large B-cell lymphoma. Tumor cells in mantle cell and marginal zone lymphomas do not typically express CD23. Plasma cell myeloma, B-lymphoblastic leukemia/lymphomas, as well as T-cell and myeloid malignancies are negative for CD23.

CD24

CD24 is a redundant co-stimulatory molecule in lymphoid tissues. It is expressed on immature and mature B-cells except plasma cells. This molecule, however, is not lineage-restricted and is present on granulocytes and various benign and malignant epithelial cells. Because CD24 is a glycosyl phosphatidylinositol (GPI)-linked molecule, it is sometimes used for the diagnosis of paroxysmal nocturnal hemoglobinuria (PNH) in flow cytometric assays. Normal T-lymphocytes, monocytes, and erythroid precursors do not express CD24.

CD35

CD35 is a complement C3b/C4b receptor membrane-bound glycoprotein. It is expressed on follicular dendritic cells, B- and T-lymphocytes, erythrocytes, eosinophils, monocytes, and kidney podocytes. CD35 monoclonal antibodies are used for the identification of follicular dendritic cell tumors.

CD77

CD77 is a glycosphingolipid Pk blood group antigen which is also expressed on germinal center B lymphocytes. It is strongly positive in Burkitt lymphoma and is weakly expressed on some follicular center lymphomas.

CD79

CD79 in association with surface Ig constitutes the B-cell antigen receptor complex and plays a critical role in B-cell maturation and activation. The pattern of CD79 expression on B-cells is closely similar to that of CD19. CD79 consists of α and β heterodimers. CD79a is expressed initially in the cytoplasm of B-cells prior to cytoplasmic μ heavy chain expression, and later on, after the expression of surface Ig, appears on the surface membrane. CD79a is usually negative in CLL cells and plasma cells, whereas CD79b is expressed in a significant proportion of patients with CLL and some cases of plasma cell myeloma patients. CD79 is an excellent pan-B cell marker for the detection of B-cell neoplasms, but tumor cells in some cases of T-lymphoblastic leukemia/lymphoma and AML may react positively with CD79 monoclonal antibodies.

CD138

CD138 is a transmembrane sulfate proteoglycan which functions as a receptor for cell–matrix interactions. Plasma cells adhere to type 1 collagen through CD138 and are the only hematopoietic cells that express this molecule. CD138 is expressed by various mesenchymal and epithelial cells, such as fibroblasts, endothelial cells, and stratified epithelia.

CD5

CD5 is briefly described in the section of "T-Cell-Associated CD Molecules."

CD103

CD103 is a membrane receptor involved in the activation of intraepithelial lymphocytes. It is expressed in >90% of intestinal intraepithelial lymphocytes and also found in intraepithelial lymphocytes of bronchial mucosa. In addition, expression of CD103 can be identified in certain types of B- and T-cell lymphoid malignancies, such as *hairy cell leukemia* (B-cell), enteropathy-associated T-cell lymphoma, and rare cases of adult T-cell leukemia/lymphoma. Dendritic cells in the skin, lymphoid tissues, and gastrointestinal tract may express CD103.

Other B-Cell-Associated Markers

FMC7. This molecule binds to a particular conformation of the CD20 antigen. It is detected only when CD20 is expressed at moderate to high intensities. FMC7 is often negative in CLL, but is strongly positive in hairy cell leukemia

PAX5. The *PAX5* gene encodes the B-cell lineage-specific activator protein (BSAP), which is a member of the highly conserved paired box (PAX)-domain family of transcription factors. It plays a crucial role in B-cell development and commitment of the bone marrow multipotent progenitor cells to the B-lymphoid lineage. Therefore, PAX5 is a useful pan B-cell marker, and antibodies to PAX5 are commonly used for the diagnosis of B-lymphoid malignancies, particularly precursor B-ALL. T-cell malignancies may sometimes reveal aberrant PAX5, and PAX5 is positive in >90% classical Hodgkin lymphomas. Neuroendocrine neoplasms and t(8;21)-AML may also express PAX5.

ZAP-70 is a tyrosine kinase that plays a role in TCR-mediated signal transduction. Intracellular expression of this molecule is found in normal T-cells, NK cells, and some normal B-cells. It can also be found in precursor B-acute lymphoblastic leukemia cells, and CLL cells, particularly in those CLL cases with unmutated IgV_H genes. The positivity of ZAP-70 in CLL appears to be associated with a poor prognosis.

Annexin-A1 protein is highly expressed in differentiated hematopoietic cells, including neutrophils, monocytes, mast cells, and lymphocytes, particularly T-cells. In an immunohistochemical study of 500 B-cell neoplasms, using a specific anti-annexin A-1 monoclonal antibody, the expression of this protein was demonstrated in all cases of *hairy cell leukemia*.

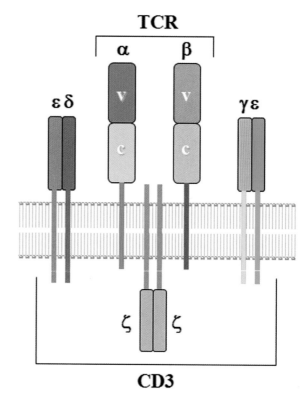

FIGURE 2.1 Schematic of TCR complex.

T-CELL-ASSOCIATED CD MOLECULES

CD1

CD1 is a member of the immunoglobulin supergene family consisting of MHC class I-like glycoproteins. So far, five distinct molecules of CD1 have been described: a, b, c, d, and e. The first three types have been extensively used in diagnostic hematopathology. CD1 molecules are expressed on thymocytes. They are absent from mature peripheral blood T-cells, but their cytoplasmic expression has been observed in activated T-lymphocytes. High levels of CD1a and, to a lesser degree, of CD1b and CD1c are present on the Langerhans cells. CD1a is an excellent marker of cortical thymocytes, which are often double positive for CD4 and CD8 (see below). CD1c is expressed by the majority of cord blood B-cells. A subset of mantle cells and follicular center B-cells also express CD1c. Follicular dendritic cells and monocytes/macrophages are CD1-negative.

CD2

CD2 is a transmembrane molecule and a member of the immunoglobulin supergene family. The existence of this molecule was originally discovered by the ability of human T-cells to spontaneously bind sheep erythrocytes (E-rosette receptor). CD2 plays an important role in T-cell activation, T- or NK-mediated cytolysis, apoptosis in activated peripheral T-cells, and the production of cytokines by T-cells. It is expressed by thymocytes, peripheral T-cells, NK-cells, and a subset of thymic B-cells. CD2 is an excellent pan-T-cell marker and one of the earliest antigens which precedes CD1 but appears after CD7 on the T lymphocytes. Some of the T-lymphoid malignancies, particularly peripheral T-cell lymphomas, may lose their CD2 expression. On the other hand, some cases of AML may aberrantly express CD2.

CD3

CD3 is a complex structure composed of three different polypeptide dimmers: $\gamma\varepsilon$, $\delta\varepsilon$, and $\zeta\zeta$. CD3 in conjunction with T-cell receptor (TCR) makes the TCR complex (Figure 2.1). TCR molecules represent two different heterodimers: $\alpha\beta$ and $\gamma\delta$. The vast majority of mature T-cells bear TCR $\alpha\beta$, whereas only about 5% of T-cells express TCR $\gamma\delta$. The $\alpha\beta$ T-cells are divided into CD4+ and CD8+ cells that are widespread and found in all hematopoietic and lymphoid tissues, whereas $\gamma\delta$ T-cells are negative for CD4 or CD8, and are primarily found in the spleen and intestinal mucosa. Surface membrane CD3 is a pan-T-cell marker and is expressed by thymocytes and all mature T-cells of peripheral blood and lymphoid tissues. NK-cells do not express TCR complex but usually show the cytoplasmic ε chain of CD3; hence, they are often positive for CD3 by immunohistochemical studies where the anti-CD3 antibody is reactive against the cytoplasmic ε chain of CD3. B-cells, granulocytic series, and monocytes/macrophages are all CD3-negative. T-lymphoblastic leukemias/lymphomas are positive for cytoplasmic CD3 expression, but often lack expression of surface CD3.

CD4

CD4 is a membrane glycoprotein and a member of the immunoglobulin supergene family and a co-receptor in MHC class II-restricted T-cell activation. CD4 is the primary receptor for HIV retroviruses. It is coexpressed with CD8 on cortical thymocytes. Over 50% of the peripheral blood T-cells and bone marrow cells are CD4+, representing the helper/inducer subtype. Monocytes/macrophages and dendritic cells express dim CD4. The majority of post-thymic T-cell neoplasms are CD4-positive.

CD5

CD5 is a signal transducing molecule interacting with TCR and B-cell receptor (BCR). It is dimly expressed on early thymocytes and at moderate to high density on all mature T-lymphocytes. It is also expressed at dim density on a small subset of mature B-lymphocytes (B1a cells), which is expanded during fetal life and in several autoimmune disorders. Certain B-cell neoplasms, such as chronic lymphocytic leukemia/small lymphocytic lymphoma and mantle cell lymphoma, express dim CD5.

CD7

CD7 is a transmembrane glycoprotein and a member of the immunoglobulin supergene family. It plays an important role in T-cell and T-cell/B-cell interactions during early lymphoid development. CD7 is the earliest T-cell-associated molecule to appear in stem cells and pre-thymic stages and extends its expression all the way to the mature T-cells. This molecule is also present on NK-cells. In addition, the pluripotent bone marrow stem cells may express CD7. A subpopulation of AML, particularly those with monocytic or megakaryocytic differentiation, may express CD7. Lack of CD7 expression can be used for the detection of T-cell lymphoproliferative disorders, such as mycosis fungoides/Sezary syndrome and adult T-cell leukemia/lymphoma. However, normal and reactive T-cells often demonstrate variable degree of CD7 loss.

CD8

CD8 is a cell surface glycoprotein and a member of the immunoglobulin supergene family. This molecule is found on cytotoxic/suppressor T-lymphocytes and a subset of medullary thymocytes. Approximately 80–90% of the thymocytes and 35–45% of the peripheral blood lymphocytes express CD8. A subpopulation of NK-cells is also CD8+.

CD45RA and CD45RO

These molecules represent two different isoforms of the CD45 cluster. CD45 is typically expressed in most hematopoietic cells (a pan-leukocyte marker). Hematopoietic blasts and plasma cells often show a dim expression of CD45, but erythroid precursors are CD45-negative. CD45RA is expressed on naive/resting T-cells and medullary thymocytes, whereas CD45RO is detected on memory/activated T-cells, cortical thymocytes, monocytes/macrophages, and granulocytes.

T-Cell Receptor Molecules

As mentioned earlier, TCR heterodimers, $\alpha\beta$ and $\gamma\delta$, in association with CD3 make the TCR complex (Figure 2.1). The vast majority of mature T-cells bear TCR $\alpha\beta$, and only about 5% of T-cells express TCR $\gamma\delta$. The $\alpha\beta$ T-cells are widespread and are found in all hematopoietic and lymphoid tissues, whereas $\gamma\delta$ T-cells are primarily found in the spleen and intestinal epithelium. NK cells do not express TCR; neither do the B-cells, monocytes/macrophages, or granulocytic cells. $\gamma\delta$ T-cells are negative for CD4 and CD8.

Other T Cell-Associated Markers

CD26 is a T-cell co-stimulatory molecule with dipeptidyl peptidase activity and is considered as a T-cell activation molecule. CD26 expression is often lost in the Sezary cells.

CD246 or anaplastic lymphoma kinase (ALK) is expressed by the neoplastic cells in a subpopulation of anaplastic large cell lymphomas. ALK can be found normally in neurons, glial cells, and endothelial cells.

CD MOLECULES ASSOCIATED WITH LARGE GRANULAR LYMPHOCYTES

CD16

CD16 is involved in antibody-dependent cell mediated cytotoxicity and is expressed on large granular lymphocytes (LGL) of both NK- and T-cell types. Approximately 15–20% of the peripheral blood lymphocytes and a much smaller fraction (5%) of bone marrow lymphocytes express dim CD16. CD16 is also expressed at moderate levels on granulocytes, tissue macrophages, and subsets of monocytes, eosinophils, and dendritic cells. CD16 expression is reduced or lost in PNH because of the structural abnormality of anchor membrane protein, GPI.

CD56

CD56 is a member of the immunoglobulin supergene family. It functions as an adhesion molecule on neural and NK cells, and a subset of T-cells. NK cells are divided into CD56 bright and CD56 dim subgroups. The CD56 dim subset represents about 90% of the NK cells, is CD16-positive, and contains higher levels of granzyme A and perforin (molecules involved in exocytosis-mediated cytotoxicity). The CD56 bright NK cells are CD16 dim or negative. CD56 is an excellent marker for the detection of NK cells and T-LGL lymphoproliferative disorders, but it is also expressed by plasmacytoid dendritic cells and neoplastic cells of some cases of plasma cell myeloma, AML, and ALL. Neuroectodermal tumors, such as small cell carcinoma of lung, neuroblastoma, medulloblastoma, and astrocytoma, are CD56-positive.

CD57

CD57 is a glycoprotein expressed on NK cells, T-cell subsets, and some cells of neuroectodermal origin. The proportion and absolute number of CD57-positive cells in peripheral blood increases with age. In adults, CD57 is expressed by 10–25% of the peripheral blood mononuclear cells. Most of the CD57+ T-cells are of cytotoxic/suppressor type. Only a small fraction of CD4+ T-cells expresses CD57, and these appear to be associated with chronic inflammatory conditions, such as tuberculosis, malaria, and AIDS. The CD57+, CD4+ T-cells constitute a major component of T-cells within the germinal centers and are known as follicular helper T-cells. Approximately 40% of the CD16-positive lymphocytes coexpress CD57. The CD16+, CD57+ subset demonstrates strong cytotoxic activities. The CD4+; CD57+ cells are increased in lymphocyte predominance Hodgkin lymphoma and chronic inflammatory conditions. CD57 is a helpful marker for the detection of LGL disorders and is also positive in a wide variety of tumors of neuroectodermal or mesenchymal origin.

Other NK-Associated Markers

The recent International Workshop on HLDA enlists several new NK-associated markers including the following.

- **CD158** or killer cell inhibitory receptor is expressed by NK cells and T-cell subsets.
- **CD161** is expressed on most NK cells and a subset of CD4+/CD8+ T-cells (thymocytes).
- **CD335** was previously known as NKp46, NCR1, (Ly94), or natural cytotoxicity triggering receptor 1.
- **CD336** was previously referred to as NKp44, NCR2, (Ly95), or natural cytotoxicity triggering receptor 2.
- **CD337** was previously known as NKp30, NCR3, or natural cytotoxicity triggering receptor 3.

GRANULOCYTIC/MONOCYTIC-ASSOCIATED CD MOLECULES

CD13

CD13 is a zinc-binding aminopeptidase expressed on the surface of about 40% of myeloid precursors, mature granulocytic/monocytic cells and basophils. This molecule is also expressed on endothelial cells, bone marrow stromal cells, osteoclasts, some early erythroid precursors, a small proportion of LGLs, and some cases of plasma cell myeloma. CD13 is commonly used as a pan-myeloid marker for the diagnosis of acute myeloid leukemia. However, about 5–15% of acute lymphoid leukemias aberrantly express CD13. The epithelia of renal proximal tubules and bile duct canaliculi may also express CD13.

CD14

CD14 is a lipopolysaccharide-binding protein, which functions as an endotoxin receptor. It is anchored to the cell surface by linkage to GPI. CD14 is strongly positive in monocytes and most tissue macrophages, but is weakly expressed or negative in monoblasts and promonocytes. Myeloblasts and other granulocytic precursors do not express CD14, but neutrophils and a small proportion of B lymphocytes may weakly express CD14. T-cells, dendritic cells, and platelets are CD14-negative. CD14 expression is reduced or lost in PNH because of the structural abnormality of GPI. Anti-CD14 monoclonal antibodies are frequently used for the identification of leukemias with monocytic differentiation.

CD15

CD15 is a carbohydrate-based molecule expressed by the granulocytic series past the myeloblast stage. A significant proportion of monocytes, a minority of macrophages/histiocytes, and a wide variety of epithelial cells and their malignant counterparts also express CD15. Erythroid precursors, B-cells, T-cells, and NK cells are CD15-negative. Reed–Sternberg cells and activated T-cells may express CD15.

CD33

CD33 is a sialoadhesin molecule and a member of the immunoglobulin supergene family. It is expressed by myeloid stem cells (CFU-GEMM, CFU-GM, CFU-G, and E-BFU), myeloblasts and monoblasts, monocytes/macrophages, granulocyte precursors (with decreasing expression with maturation), and mast cells. Mature granulocytes may show a very low level of CD33 expression. CD33 can be aberrantly expressed on some cases of plasma cell myeloma. This molecule is not expressed in erythrocytes, platelets, B-cells, T-cells, or NK cells. CD33 is an excellent myeloid marker and is commonly used for the diagnosis of AML. However, approximately 10–20% of B lymphoblastic or T lymphoblastic leukemia/lymphomas may aberrantly express CD33.

CD64

CD64 is a member of the immunoglobulin supergene family and functions as an Fc gamma-receptor of IgG. It is expressed by monocytes/macrophages, myeloid precursors, and follicular dendritic cells. CD64 and CD14 together are considered good monocyte/macrophage-associated markers and are commonly used in flow cytometric studies to identify leukemias with myelomonocytic differentiation.

CD64 alone appears to be a more sensitive but less specific monocytic marker than CD14. Langerhans cells, interdigitating dendritic cells, B-cells, T-cells, NK cells, and erythroid and megakaryocytic lineages are all CD64-negative.

CD68

CD68 is a sialomucin and a member of the scavenger receptor supergene family. It is expressed by monocytes and macrophages as well as subsets of CD34-positive hematopoietic stem cells, dendritic cells, neutrophils, basophils, and mast cells. Activated T-cells and a proportion of mature B-cells may also express CD68, which usually appears as a dot-like or finely granular cytoplasmic positivity by immunohistochemical techniques. Some non-hematopoietic cells, such as epithelium of renal tubules, may show CD68 positivity.

Other Myeloid-Associated CD Molecules

CD88 is a C5a receptor and is expressed by granulocytes, monocytes, mast cells, subsets of dendritic cells, as well as astrocytes and microglia.

CD114 is the receptor for granulocyte colony-stimulating factor (G-CSF). It is expressed by cells of the granulocytic lineage in all stages of differentiation and is found in various proportions of monocytes, platelets, endothelial cells, and trophoblastic cells.

CD115 is the receptor for macrophage colony-stimulating factor (M-CSF) and is primarily expressed on cells of the monocyte/macrophage lineage.

CD116 is α chain subunit of the GM-CSF receptor and is expressed by various myeloid cells including macrophages, neutrophils, eosinophils, and dendritic cells.

DENDRITIC CELL-ASSOCIATED CD MOLECULES

CD21

See page 32.

CD35

See page 33.

CD123

CD123 or interleukin-3 (IL-3) receptor α chain is expressed in a wide variety of hematologic malignancies such as AML, precursor B-ALL, hairy cell leukemia, a subset of CLL with strong CD11c expression, and blastic plasmacytoid dendritic cell neoplasms.

CD207

CD207 (Langerin) is a transmembrane glycoprotein necessary for formation of the Birbeck granules in the Langerhans cells. CD207 is an excellent marker for the detection of Langerhans cell disorders.

Other Dendritic Cell-Associated Markers

S100 is a calcium-binding protein which regulates protein kinase C-dependent phosphorilation. S100A is composed of an alpha and beta chain and S100B of two beta chains. Immunohistochemical stains for S100 are positive in the Langerhans cells, interdigitating dendritic cells, and their corresponding neoplasms, as well as in Schwannomas, ependymomas, astrogliomas, and malignant melanomas.

ERYTHROID-ASSOCIATED CD MOLECULES

CD71

CD71 is the transferrin receptor and is expressed on all proliferating cells. It is also expressed by erythroid precursors which need iron for the synthesis of heme molecules. CD71 in conjunction with glycophorin A (CD235A) is a helpful marker in the identification of erythroid precursors in hematologic disorders.

CD235

CD235 molecules represent glycophorins A and B, the two major sialoglycoproteins of the human erythrocyte membrane, and are restricted to erythrocytes. These molecules bear the antigenic determinants for the MN and Ss blood groups. Monoclonal antibodies against glycophorin A (GPA) are frequently used in immunophenotypic studies for the identification of erythroid precursors in hematologic disorders.

CD238

CD238 is the Kell blood group transmembrane protein.

CD240

CD240 represents the C, E, and D antigens of the Rh blood group system, the second most clinically significant, and the most polymorphic, system of the human blood group.

CD242

CD242 is an intercellular adhesion molecule (ICAM4) and represents the Landsteiner–Wiener (LW) blood group antigen(s). Expression of CD242 is increased in sickle cell disease.

Other Erythroid-Associated Markers

Anti-hemoglobin antibodies are routinely used for immunophenotypic studies of erythroid precursors.

MEGAKARYOCYTE/PLATELET-ASSOCIATED CD MOLECULES

CD36

CD36, also known as glycoprotein IV (gpIV) or glycoprotein IIIb (gpIIIb), is one of the major glycoproteins of the platelet surface and serves as a receptor for thrombospondin. Anti-CD36 monoclonal antibodies are routinely used for identification of megakaryoblasts and immature megakaryocytes in myeloproliferative disorders and myeloid leukemias.

CD42

CD42 glycoprotein complex (a, b, c, and d) is restricted to the megakaryocytic lineage and platelets. This complex facilitates adhesion of the platelets to the subendothelial matrices. Absence of the CD42 complex leads to the Bernard–Soulier syndrome. Anti-CD42 monoclonal antibodies are routinely used for identification of megakaryoblasts and immature megakaryocytes in myeloproliferative disorders and myeloid leukemias.

CD41 and CD61

CD41 (platelet glycoprotein IIb) and CD61 (platelet glycoprotein IIIa) form a calcium-dependent heterodimeric complex. This glycoprotein complex (GPIIb-IIIa) binds plasma proteins, such as fibrinogen, fibronectin, von Willebrand factor, and vitronectin, and plays a critical role in platelet aggregation. Hereditary defects of the GPIIb-IIIa receptor cause Glanzmann's thrombasthenia. Similar to CD42, anti-CD41 and -CD61 monoclonal antibodies are frequently used for identification of megakaryocytic precursors in myeloproliferative disorders and myeloid leukemias.

CD110

CD110 or thrombopoietin receptor (TPO-R) is expressed on the megakaryocytes and platelets, as well as subsets of hematopoietic stem cells, activated T-cells, and endothelial cells.

Other Megakaryocyte/Platelet-Associated Markers

Factor VIII is another useful megakaryocytic marker used for identification of megakaryocytic precursors in myeloproliferative disorders and myeloid leukemias.

PRECURSOR-ASSOCIATED CD MOLECULES

CD34

CD34 is a transmembrane glycoprotein expressed on early lymphohematopoietic stem cells, progenitor cells, and endothelial cells. Also, embryonic fibroblasts and some cells in fetal and adult nervous tissue are CD34-positive. Almost all pluripotent and committed stem cells in colony-forming assays express CD34. The uncommitted progenitor cells are CD38-negative, whereas the committed ones are CD38-positive. In normal conditions, CD34-positive cells account for about 1–2% of the total bone marrow cells. The TdT+ subset of precursor B-cells (hematogones) is also positive for CD34. Approximately 40% of AMLs and over 50% of ALLs express CD34.

CD38

CD38 is a multifunctional ectoenzyme widely expressed in hematopoietic cells. It plays a role in the regulation of cell activation and proliferation. It is expressed in committed hematopoietic stem cells and other hematopoietic precursors during early differentiation and activation. Very early erythroid and myeloid cells, precursor B-cells, thymocytes, activated T-cells, and NK cells express CD38. CD38 is also expressed at high levels on plasma cells.

CD90

The CD90 molecule is a member of the immunoglobulin supergene family and is expressed by 10–40% of CD34+ cells in bone marrow. The CD34+/CD90+ cells probably represent the most primitive hematopoietic progenitor cells. CD90 is also expressed in fibroblasts and other stromal cells.

CD99

CD99 is a transmembrane protein involved in homotypic cell adhesion, apoptosis, vesicular protein transport, and differentiation of T-cells. Its expression has been reported in acute lymphoid leukemias, Ewing's sarcoma, Granulosa cell tumor, plus other tumors of neuroectodermal origin.

CD117

CD117 (c-kit) is a tyrosine kinase receptor and a member of the immunoglobulin supergene family. It is expressed in most of the hematopoietic stem cells, CD34+ progenitor cells, and mast cells. The majority of AML cells are CD117-positive. Plasma cells may also express CD117. CD117 is an excellent marker for the detection of mast cell disorders and identification of myeloblasts in acute leukemias.

CD123

See page 37.

Other Precursor-Associated Markers

TdT (terminal deoxynucleotidyl transferase) is a DNA polymerase present in precursor T-cells and thymocytes, as well as subset of precursor B-cells. Anti-TdT antibodies are routinely used for the detection of B and T lymphoblastic leukemias/lymphomas plus lymphoid blast transformation in chronic myelogenous leukemia (CML). A small proportion of AMLs is also TdT-positive.

OTHER MARKERS ROUTINELY USED IN HEMATOPATHOLOGY

CD11

CD11a, b, and c are components of heterodimer CD11/CD18 adhesion molecules. CD11a is a panleukocyte marker and is expressed by B- and T-lymphocytes, monocytes, macrophages, neutrophils, basophils, and eosinophils. CD11b is expressed by most of the granulocytes, monocytes/macrophages, and NK cells, and subsets of B- and T-cells. CD11c is highly expressed in monocytes/macrophages, NK cells, and hairy cells.

CD30

CD30 is a member of the tumor necrosis factor (TNF) receptor family and appears to be involved in TCR-mediated cell death. It is expressed by Reed–Sternberg cells and Hodgkin cells, neoplastic cells of anaplastic large cell lymphoma (ALCL), as well as activated lymphocytes, and monocytes. Some cases of embryonal carcinoma and mixed germ cell tumors also express CD30.

CD43

CD43 is a sialomucin transmembrane molecule expressed at high levels on all leukocytes except most resting B lymphocytes. In hematopathology, CD43 is often considered as a T-cell associated marker, because it is expressed by over 95% of thymocytes and peripheral blood T-cells. But, interdigitating dendritic cells, Langerhans cells, epithelioid histiocytes, and multinucleated giant cells express CD43, whereas follicular dendritic cells and sinus histiocytes of the lymph nodes are usually CD43-negative. CD43 may be expressed in mantle cell lymphoma and other types of B-cell lymphomas, mastocytosis, some cases of plasma cell disorders and blastic plasmacytoid dendritic cell neoplasms. Loss or defect of CD43 has been reported in lymphocytes of patients with Wiskott–Aldrich syndrome.

CD55

CD55 or decay-accelerating factor (DAF) binds C3b and C4b to inhibit formation of the C3 convertases. It is anchored to the GPI in the cell membrane, and, therefore, its expression is reduced or lost in patients with paroxysmal nocturnal hemoglobinuria (PNH). It is widely expressed on cells throughout the body, including hematopoietic cells.

CD59

CD59 is also a GPI-anchored molecule and inhibits formation of membrane attack complex (MAC), thus protecting cells from complement-mediated lysis. Similar to CD55, CD59 expression is reduced or lost in patients with PNH. It is widely expressed on cells throughout the body, including hematopoietic cells.

Ki-67

Ki-67 is a proliferation-associated molecule. Its expression is upregulated during the S phase of the cell cycle and is maximized during mitosis. Anti-Ki-67 antibodies are used for the estimation of proliferating index in various malignancies.

Immunoglobin Transcription Factors

Oct1, Oct2, and BOB.1/OBF.1. Oct1 and Oct2 and their co-activator BOB.1/OBF.1 regulate immunoglobulin gene transcription. Antibodies raised against these molecules are used for the characterization of certain types of B-cell lymphoid malignancies, and also aid in distinction between classic Hodgkin lymphoma (CHL) and nodular lymphocyte predominant Hodgkin lymphoma (NLPHL). Double positivity for Oct2 and BOB.1/OBF.1 is seen in L&H cells of NLPHL but not in Hodgkin-Reed–Sternberg cells of CHL.

Principles of Flow Cytometry

Flow cytometry is now considered an integral component of diagnosis and management of hematolymphoid disorders. Availability of a vast number of antibodies against CD molecules in high quality plus numerous fluorochrome-conjugated forms, sophisticated multiparametric and user-friendly flow cytometers, advanced software, improved gating strategies, and use of "pattern recognition", which is the gold-standard in data interpretation, have made flow cytometry a powerful method of immunophenotyping.

A flow cytometer is basically a particle analyzer. It detects cell properties as a stream of single cell (or particle) suspension passes through a laser beam (Figure 2.2). The cell size, nuclear and cytoplasmic complexity (e.g., nuclear irregularity and cytoplasmic granularity), along with membrane-associated, cytoplasmic or nuclear molecules that are labeled by fluorochrome-conjugated antibodies, are measured by a set of optical detectors and analyzed. A fluorochrome is a chemical which can absorb energy from an excitation source (laser beam) and emit photons at a longer wavelength (fluorescence), which is captured by optical detectors of the flow cytometer. The current generations of flow cytometers are capable of analyzing at least six parameters simultaneously: cell size depicted by forward scatter (FSC) laser light, cell complexity represented by side scatter (SSC) laser light, plus antibodies conjugated to four different fluorochromes, defining at least six molecular characteristics of the target cells (or particles) passing through the instrument.

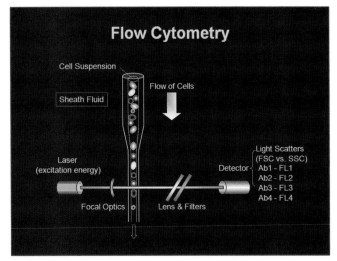

FIGURE 2.2 A schematic overview of a 4-color/6-parameter flow cytometer.

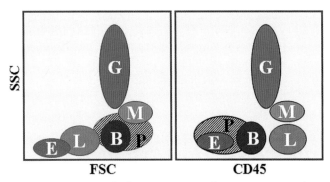

FIGURE 2.3 Patterns of cell aggregation in flow cytometry analysis of (left) FSC/SSC and (right) CD45/SSC: E, erythroid precursors; L, lymphocytes; B, blasts; M, monocytes; G, granulocytes; P, plasma cells.

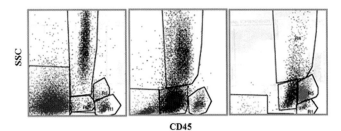

FIGURE 2.4 Open gate displays by CD45 gating of flow cytometry demonstrating variable locations of the neoplastic cells of a plasma cell myeloma (A, red cluster), an acute lymphoblastic leukemia (B, green cluster), and an acute myelomonocytic leukemia (C, green and red clusters).

GATING

Gating refers to the selection of target cell populations in a given electronic window. Cells revealing similar electronic signals tend to cluster together in the same electronic windows. For example, in flow cytometric study of peripheral blood lymphocytes that are small and non-granular tend to cluster in the lower left section of the FSC versus SSC electronic window (Figure 2.3). Currently, the recommended basic gating strategy for analyzing hematolymphoid cells includes a combination of the scatter gate (FSC versus SSC)

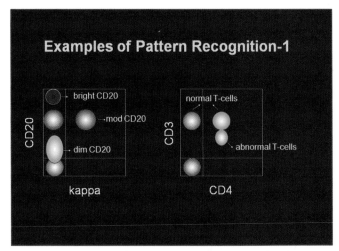

FIGURE 2.5 Examples of various antigen expression levels as determined by pattern recognition using proper internal controls. Comparing with moderate CD20 expression by normal polytypic B-cells, two subsets of abnormal B-cells are present with either dim or bright CD20 (left). Similarly, normal T-cells express moderate level of CD3, but a subset of the CD4+ T-cells is abnormal with dim CD3 (right).

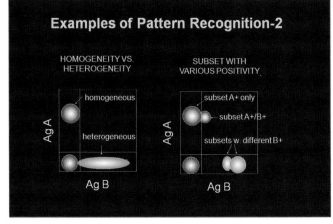

FIGURE 2.6 Examples of various antigen expression patterns. A tight cluster represents a homogeneous expression profile, whereas a widespread pattern indicates a more heterogeneous nature of the antigen expression profile (left). In addition, subsets may coexist with different patterns of positivity (right), which can often be clearly illustrated using density plot.

and CD45 gate (CD45 versus SCC) (Figure 2.3). On the CD45 gate, the hematolymphoid cells are separated into rather distinct subgroups based on their differential expression of CD45 plus side scatter characters. CD45 is strongly expressed by lymphocytes and relative strongly by monocytes, while platelets and erythroid precursors are CD45-negative. Blasts are dimly positive for CD45 in about 80% of the acute leukemias, and plasma cells are CD45-negative to CD45 dim. Hence, the CD45 gating strategy is very helpful in separating blast cells from non-blast cells, lymphoid cells from non-lymphoid cells, and monocytes from granulocytes (Figures 2.3 and 2.4). It also helps to distinguish the normal patterns of expression from the abnormal ones, especially when a combination of various gating strategies is applied (Figures 2.5 and 2.6). Proper gating is a critical

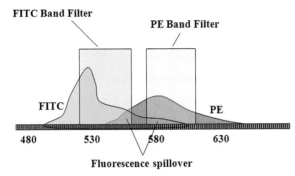

FIGURE 2.7 An example of fluorescence spillover between FITC and PE.
Adapted from Wulff S. Guide to Flow Cytometry, DakoCytomation.

step in data analysis and interpretation of the results in flow cytometry.

In order to utilize proper gating strategies and aim at the target populations, we strongly recommend correlation with microscopic review of the samples and the gathering of as much clinical information as possible prior to the selection of monoclonal antibodies and gating processes.

COMPENSATION

When multiple fluorochromes are used, there is often a fluorescence interference due to the overlapping emission spectra, which is most prominent in fluorochromes that emit at adjacent wavelengths. For example, fluorescein isothiocyanide (FITC) and phycoerythrin (PE) emit at 520 and 576 nm, respectively. Since these two emission peaks are adjacent to each other, there is an overlap or spillover between the emitted FITC and PE fluorochromes, even when the proper filters are used to limit this overlap (Figure 2.7). The process of correcting (subtracting) this fluorescence spillover is called compensation. Current models of multiparametric multicolor flow cytometers have a built-in automated compensation mechanism, which can be applied during both pre- or post-acquisition stages of analysis. In practice, both manual and automated compensation adjustments require proper antibody–fluorochrome controls

DATA ANALYSIS

Modern multiparametric flow cytometers in conjunction with powerful and user-friendly software programs offer great opportunities for hematopathologists, immunologists, and researchers to rapidly acquire data and analyze the results on large cell populations. For example, one of the most popular current models is capable of processing 10,000 cells (events) per second and detecting at least eight parameters (two light scatter and six fluorescent signals) simultaneously. Software programs provide a variety of options for the evaluation and analysis of the signals, including data collection on logarithmic or linear scales, and different options for displaying histograms (Figures 2.8 to 2.12). The logarithmic scale is the preferred scale for most immunophenotypic studies by flow cytometry.

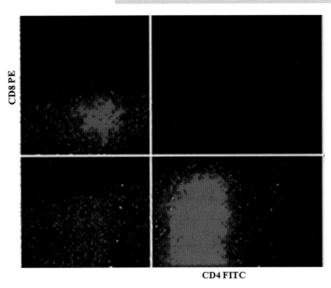

FIGURE 2.8 For monitoring compensation, in practice, two mutually exclusive markers such as CD4 and CD8 are utilized in a peripheral blood sample.

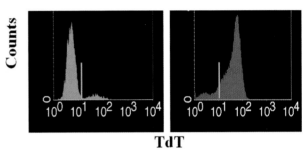

FIGURE 2.9 Histograms for a single parameter usually depict fluorescent intensity versus cell count. TdT-negative and TdT-positive samples are shown here on the left and right, respectively.

Histograms for a single parameter usually depict fluorescent intensity versus cell counts (Figure 2.9). Dot and density plots provide simultaneous information for two parameters (Figures 2.10 and 2.11). Contour histograms display the data as a series of encircling lines correlating with cellular density and distribution (Figure 2.12). Most programs also allow comparative studies of multiple samples by simultaneously overlaying their single parameter histograms.

QUALITY CONTROL AND QUALITY ASSURANCE

Similar to all other instruments in clinical laboratories, flow cytometry has its own quality control (QC) and quality assurance (QA) issues. Many steps are involved in various aspects of flow cytometry, such as the optimization of instrument function, sample processing, acquiring and analyzing data, and reporting the results. Flow cytometers should be calibrated with samples consisting of a mixture of blank and predefined fluorescence-labeled microbeads. The performance of various components, such as fluidics, optical filters, multiplier tubes, and lasers, should be checked on a regular basis. Standardized protocols for each step of the process should be implemented to ensure reliable results, including verification of accuracy of the results with known samples.

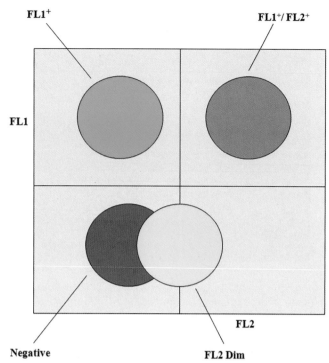

FIGURE 2.10 Two-parameter dot plot histograms depict four quadrants. The lower left quadrant represents a negative cell cluster, the upper left quadrant shows the cell population positive for one parameter, the lower right quadrant depicts cells positive for the second parameter, and the upper right quadrant represents cells that coexpress both parameters.

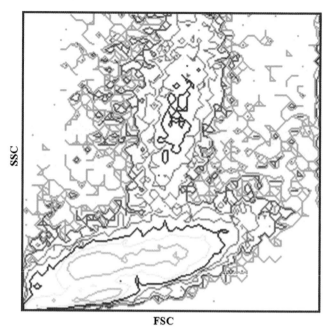

FIGURE 2.12 Counter histograms display the data as a series of encircling lines correlating with cellular density (higher in the center) and cell clusters.

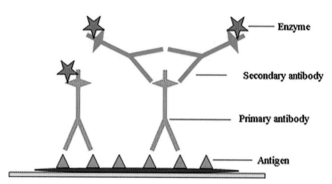

FIGURE 2.13 Schematic of immunohistochemical techniques demonstrating direct and indirect immunoenzyme methods. Adapted from Ramos-Vara JA. Technical aspects of immunohistochemistry. Vet Pathol 2005; 42: 405–426.

Principles of Immunohistochemistry

Immunohistochemistry has become a routine staining technique in most pathology laboratories. Enzyme-conjugated antibodies are used for the demonstration of antigens in tissue sections, smears, and cytospin preparations. Horse radish peroxidase and/or alkaline phosphatase are the most frequently used enzymes for signal generation. Sections from frozen or fixed tissues are used. Archival tissue blocks are sectioned and deparaffinized and then properly heated for epitope retrieval. After blocking the endogenous peroxidase, the primary antibody (1–5 µg) is applied with proper incubation time (~30 min) and then the enzyme-conjugated or biotinylated secondary antibody is added (Figure 2.13). To amplify the signals, biotin–avidin–enzyme or biotin–streptavidin–enzyme

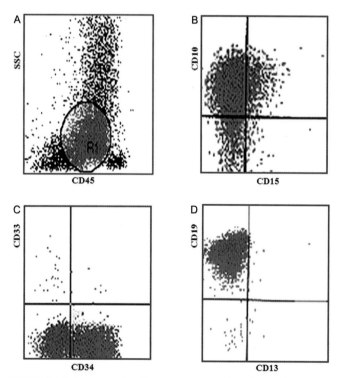

FIGURE 2.11 An example of two-parameter dot plot analysis showing (A) a population of CD45 dim blast cells expressing (B) CD10, (C) CD34, and (D) CD19.

PRINCIPLES OF IMMUNOHISTOCHEMISTRY 43

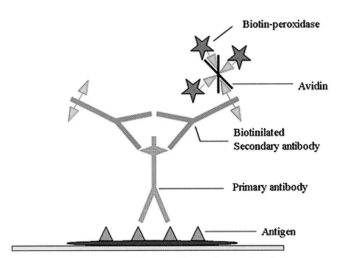

FIGURE 2.14 Schematic of immunohistochemical techniques demonstrating an indirect method using biotin–avidin–enzyme complexes.
Adapted from Ramos-Vara JA. Technical aspects of immunohistochemistry. Vet Pathol 2005; 42: 405–426.

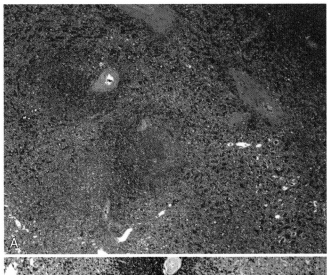

FIGURE 2.15 (A) An H&E section of spleen demonstrating white and red pulps. (B) Dual immunohistochemical stains showing T (CD3-positive, brown) and B (CD20-positive, pink) cells.

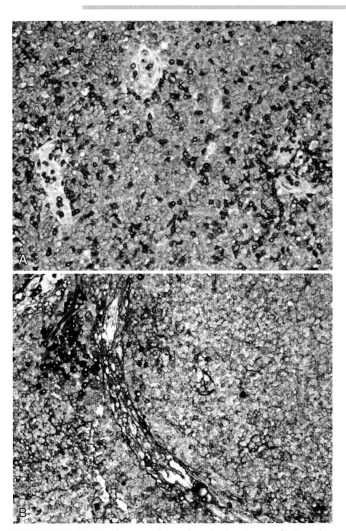

FIGURE 2.16 (A) Dual immunohistochemical staining of a lymph node section from a patient with B-cell lymphoma demonstrating large numbers of B-cells (CD20-positive, pink) and scattered T-cells (CD3-positive, brown). (B) Dual kappa (brown) and lambda (pink) staining shows a cluster of polyclonal plasma cells and sheets of kappa-positive lymphocytes.

complexes are used (Figure 2.14). Currently, there are automated machines available for performing single or dual immunohistochemical stains (Figure 2.15). Immunohistochemistry provides information regarding pattern, intensity, and location of antigen(s) in tissues and cells. It is used in hematopathology for diagnosis and classification of leukemias and lymphomas, and differential diagnosis of primary hematopoietic neoplasms from non-neoplastic hematopoietic disorders and metastatic tumors (Figures 2.16 to 2.19). For example, most of the acute lymphoid leukemias are CD10, TdT, and HLA-DR positive, whereas these markers are not expressed in neuroblastoma, rhabdomyosarcoma, or Ewing's sarcoma. Immunohistochemical techniques may help to detect occult metastatic lesions, such as metastatic breast carcinomas and neuroblastomas, or to identify their tissue of origin. For example, metastatic prostatic carcinomas are positive for prostatic acid phosphatase and prostate specific antigen (PSA), and metastatic rhabdomyosarcomas may demonstrate myosin, desmin, or myoglobin expression.

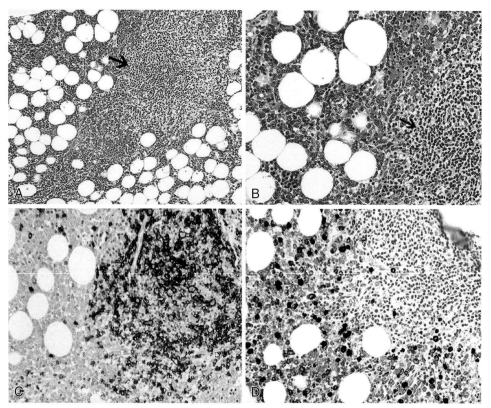

FIGURE 2.17 An H&E section of bone marrow biopsy from a patient with a residual CLL (arrow) and a secondary AML (A, low power; B, high power). (C) The residual CLL cells express CD20; (D) the AML cells express MPO.

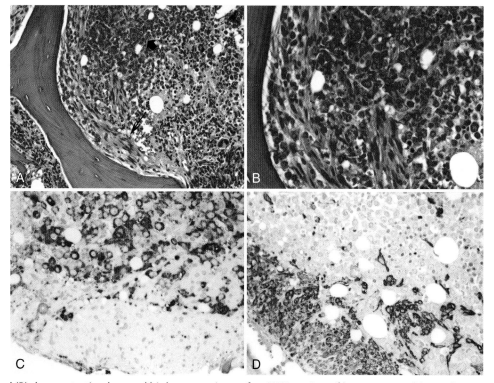

FIGURE 2.18 (A) and (B) demonstrating low and high power views of an H&E section of bone marrow biopsy from a patient with AML (short arrow) and metastatic leiomyosarcoma (long arrow). (C) AML cells are MPO+; (D) leiomyosarcoma cells express myosin.

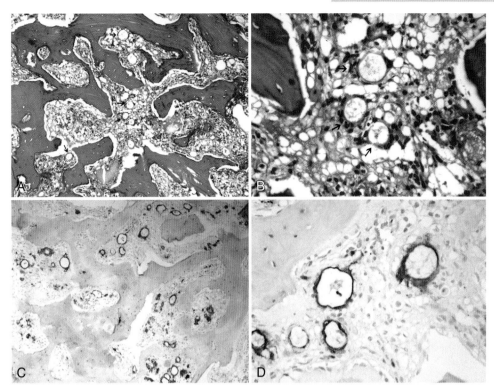

FIGURE 2.19 A bone marrow biopsy section demonstrating osteosclerosis and fibrosis with the presence of vessel-like structures (arrows) (A, low power; B, high power). These vessel-like structures are positive for cytokeratin with immunohistochemical stain representing metastatic adenocarcinoma (C, low power; D, high power).

Additional Resources

Buchwalow IB, Böcker W: Immunohistochemistry: basics and methods, Berlin, 2010, Springer.

Carey JL, McCoy JP, Keren DF: Flow cytometry in clinical diagnosis, ed 4, Chicago, IL, 2007, ASCP Press.

Craig FE, Foon KA: Flow cytometric immunophenotyping for hematologic neoplasms, *Blood* 111:3941–3967, 2008.

Human Cell Differentiation Molecules. <http://hcdm.org/>.

Matesanz-Isabel J, Sintes J, Llinàs L, et al: New B-cell CD molecules, *Immunol Lett* 134:104–112, 2011.

Nguyen D, Diamond LW, Braylan RC: Flow cytometry in hematopathology: a visual approach to data analysis and interpretation, ed 2, Totowa, NJ, 2007, Humana Press.

Novo D, Wood J: Flow cytometry histograms: transformations, resolution, and display, *Cytometry A* 73:685–692, 2008.

Ramos-Medina R, Montes-Moreno S, Maestre L, et al: Immunohistochemical analysis of HLDA9 Workshop antibodies against cell-surface molecules in reactive and neoplastic lymphoid tissues, *Immunol Lett* 134:150–156, 2011.

Ramos-Vara JA: Technical aspects of immunohistochemistry, *Vet Pathol* 42:405–426, 2005.

Stetler-Stevenson M, Davis B, Wood B, et al: 2006 Bethesda International Consensus Conference on Flow Cytometric Immunophenotyping of Hematolymphoid Neoplasia, *Cytometry B* 72B:S3, 2007.

Principles of Cytogenetics

Introduction

Cancer is a genetic disease characterized by DNA changes at either the nucleotide or chromosomal level, or both. Malignancies can develop either from a genetic predisposition followed by acquired somatic mutations, or from an accumulation of somatic mutations that develop into a cancer phenotype. At the chromosome level, these mutations include changes in chromosome number, loss of heterozygosity (LOH) brought about by whole chromosome or segmental region loss, chromosomal rearrangements (mainly balanced and unbalanced translocations, and inversions), and gene amplifications. Many of these abnormalities are cytogenetically visible, but also sometimes cryptic (submicroscopic), and are often recurrent and characteristic for a particular disease or disease subtype. Because specific chromosomal aberrations, especially in hematological malignancies, provide diagnostic, prognostic, and/or treatment information for many cancers, they are in many ways true biomarkers for human cancer.

Clear insights into the genetic basis of cancer were obtained in the 1950s and 1960s, when improved cell culture and slide preparation techniques made it possible in 1956 to accurately enumerate the number of human chromosomes as 46. Chromosome analysis is usually carried out on cells in mitosis (cell division), when the chromosomes become visible as distinct entities. After identifying each chromosome in a cell by its characteristic size, position of the centromeres, and staining properties (e.g., G-banding), a karyotype displaying the full chromosome complement of the cell can be prepared.

The first specific chromosome abnormality observed in a human tumor was seen in Philadelphia in 1960 by Nowell and Hungerford who found an unusually small chromosome—the Philadelphia (Ph) chromosome—in the leukemic cells of patients with chronic myelogenous leukemia (CML). The discovery of the Ph chromosome aroused considerable interest in cancer cytogenetics, as it gave the first direct evidence for a consistent DNA-associated change in tumor. More than 30,000 cases of hematological malignancies with chromosome aberrations have since been reported, making it the most thoroughly cytogenetically investigated group of all neoplastic disorders. This is because hematological neoplasms can be detected earlier, when the cells have fewer cytogenetic changes, which are often simple and disease specific, in contrast to solid tumors, which are often diagnosed later after multiple unbalanced and sometimes non-specific chromosomal aberrations have already taken place.

A second major breakthrough in cytogenetics was the development of microscopic staining techniques, generating a banding pattern along the length of the chromosomes. With this banding pattern, all individual chromosomes could be identified, enabling the characterization of structural changes in much greater detail. Consistent chromosome aberrations which are uncommon or extremely rare in normal tissues are found in the different cells of a tumor, and further karyotypic changes have been shown to occur during tumor progression. Modifications of culture methods to improve the yields of mitotically active (metaphase, prometaphase) cells and high resolution banding of elongated chromosomes now allow for a more precise definition of the rearrangements, as well as the identification of previously undetected abnormalities. Using these techniques, most tumor cells can be shown to have some chromosomal abnormality. For example, about 45–50% of acute leukemias have an abnormal karyotype with a recurrent chromosomal aberration, with 15% exhibiting three or more cytogenetic abnormalities (complex karyotype). Possibly the best correlation between the presence of highly specific chromosomal changes and a subtype of leukemia is the 15;17 translocation which is identified only in patients with acute promyelocytic leukemia. Since in several leukemias and lymphomas the karyotype determined at diagnosis is an independent prognostic indicator, it is imperative that the cytogenetics analyses are completed prior to clinical management. The current status of WHO classification requires not only morphology but also immunophenotyping, cytogenetics, and molecular studies for proper classification and prognosis. Because of the relative ease of obtaining bone marrow (BM) or peripheral blood specimens from leukemia patients, it is possible to do serial sampling, which enables the study of the karyotype patterns during the various stages of the clinical course, such as at diagnosis, remission, and relapse. This requires appropriate preparations and culturing of the tissue sample. A number of techniques are available to evaluate the chromosomal aberrations in hematological malignancies. It is very important to also recognize that more modern molecular techniques such as FISH and gene mutational studies may be more appropriate for certain disease conditions since standard karyotyping will provide results only from actively dividing cells which may not necessarily represent the malignant clones. Situations such as monitoring disease status following therapy or bone marrow staging in lymphomas may be better followed up by non-karyotype methods. In general karyotyping is favored for myeloid disorders, such as FISH or specific molecular methods with FISH being more sensitive for the detection of chromosomal aberrations in the lymphoproliferative disorders.

Cell Preparation

Bone marrow cells from leukemia patients are an ideal source of tissue for cytogenetics study of leukemias. The use of "direct" (same day harvest with no time in culture) preparation to avoid selection during culturing has been considered advantageous for obtaining an accurate picture of the chromosomal constitution of the leukemic cells. However, it has also been proven that short term (~24–48 hrs) unstimulated culturing of leukemic bone marrow cells *in vitro* may uncover both clonal abnormalities, and a greater number of abnormal cells than the "direct" method of preparation. Chromosomes analysis of leukemic cells is not always possible by the "direct" method, because cases are often encountered in which only a few to no mitoses are present or whose chromosomes show blurred outlines (poor morphology) with poor banding patterns. These metaphases are of inadequate quality for detailed analysis even when there are enough mitoses. It must also be noted that sometimes the 15;17 translocation characteristic of acute promyelocytic leukemia is not observed in the direct preparations but seen prominently in 24-hour cultured bone marrow cells. Thus it is important that more than one culture setup and harvest method be performed.

A prerequisite for chromosome analysis is dividing or mitotic cells. Spontaneously dividing cells suitable for chromosome preparations are found only in the rapidly proliferating tissues of the body, such as the gonads and bone marrow, or in tissues with malignancies. Bone marrow is the tissue of choice for cytogenetics studies of most hematological conditions. However, in chronic disorders where there is high white cell count, such as CML-blast crisis and chronic lymphocytic leukemia (CLL), a hypercellular peripheral blood sample is appropriate. It is also important to recognize that even a very dilute sample can overgrow and subsequently fail. Spleen tissue, or more rarely ascetic and pleural effusion, is also suitable for cytogenetics study in some hematological disorders.

Chromosome studies of malignant lymphomas are usually based on lymph node biopsies, since bone marrow may not always be involved in the early stages of the disease. But occasionally a bone marrow sample can prove informative, particularly if there is some question as to whether bone marrow involvement has occurred. The success of a cytogenetics analysis mostly depends on the quality of material investigated. Therefore the key to successful cytogenetics is to obtain cell-rich material derived from the middle part of the bone marrow. The tissue must be drawn under sterile conditions with the aspiration syringe coated with preservative-free heparin to avoid clumping of blood components. Heparin is also added to lymph node and spleen tissue after surgical removal. Where there is a likelihood of a dry tap, especially in diseases such as primary or secondary myelofibrosis, or due to faulty technique, a peripheral blood sample can be sent as an alternative. However, unless there is a sufficient number of blast cells in these samples, the abnormal clone may go undetected. Care must also taken in suspected APL cases where clotting is possible and may cause the sample to be unsuitable for culture and chromosome analysis.

A good-quality bone marrow aspirate or bore core biopsy sample of 1–2 mL is adequate in most cases, although less than this can be accepted if the marrow is very cellular. The tissue drawing must be done under sterile conditions, since the chromosome analysis is in most cases preceded by short- or long-term cell culture which mandates a high degree of sterility. It is also advisable to determine if the initial aspirate has bone spicules; and, if a previous aspirate has been obtained, aspiration of a second sample after repositioning is recommended for cytogenetics studies.

The shorter the duration of time (≤24 hours) between collection and culture setup in the cytogenetics laboratory, the greater the chances for a successful chromosome analysis with an accurate result. Every effort must be made to ensure that the bone marrow or lymph node biopsy samples are set up in cultures with minimum delay. If a delay in the transportation of the sample is expected, the sample should preferably be collected and transferred to a tube containing transportation medium, made up of preservative-free heparin in an appropriate basal medium, supplemented with serum and antibiotics. The sample must never be frozen, but can be stored at 4°C overnight or for up to 3 days. However, the cell viability is greatly reduced with time, yielding misleading or only normal karyotype results. Disorders such as acute lymphoblastic leukemia (ALL) and others with a high white cell count are particularly adversely affected by delays.

Same-day or "direct" cultures are often recommended for ALL, and sometimes for CML. Studies suggest that erythropoietic cells divide rapidly in the first few hours of culture followed by granulopoietic cell divisions. Based on these observations, short cultures in erythroleukemia are more likely to yield good results. However, in most cases a minimum of 16–24 hours of unstimulated culture is appropriate. On the other hand, CLL and some ALL, which have B- or T-cell phenotype, may need 3–5-day cultures with appropriate mitogens, as well as having some unstimulated cultures. Sometimes, in the case of poor response to the commonly used mitogens, TPA (12-O-tetradecanoylphorbol-13-acetate), AmpliB-DSP-30 (synthetic deoxyoligonucleotide of 27 bases), or EBV (Epstein–Barr virus) are used as stimulants in the cultures. Several laboratories supplement the regular media with a condition medium derived from cultures of a human urinary bladder carcinoma cell line or other commercially available media supplements, which are capable of stimulating the proliferation and growth of the leukemic cells.

The chromosomes of patients with ALL are particularly difficult, fuzzy and resistant to banding. Nevertheless, analysis of direct preparations of these marrows has shown that 50–78% of these patients have chromosomal abnormalities in their leukemic cells. But it is important to recognize that more than one technique is necessary to assess accurately the karyotypic constitution of the leukemic cells. Culturing may, but does not always, reveal more cells with abnormal chromosomes than are seen in direct preparations.

Banding Techniques

The standard cytogenetics banding method consists of culturing a suspension of cells in mitogenic media for 24–72 hours. Then, dividing cells are arrested in metaphase by the addition of an inhibitor of the mitotic spindle, such as colchicine or vinblastine. The cells are submitted to a hypotonic solution (e.g., KCl or Na citrate), followed by dropping of cells on slides, enzyme treatment, and staining with Giemsa stain (G-banding), revealing characteristic banding patterns that are specific for each chromosome. These banding patterns allow for the assignment of homologous chromosomes, and the identification of extra or of missing chromosomes as well as of structural aberrations. For G-banding patterns, pre-treatment of chromosomes by enzymes such as trypsin or pancreatin is required. The mechanism of G-banding is not yet fully understood, but the chromosomes express dark (AT sequences rich) and light (GC sequences rich) G-bands. This is the most common and traditional banding technique used in a clinical setting. Other chromosome banding techniques used in standard cytogenetics laboratories to better define the chromosomal aberrations are: a technique to produce a reverse banding pattern (R-bands); a fluorescence banding technique using quinacrine derivatives (Q-bands); and centromeric staining (C-bands). The nucleolus organizer regions (NORs) located in the short arms of acrocentric chromosomes can be visualized by staining with silver nitrate ($AgNO_3$).

Analysis

Chromosome analysis is performed on a microscope at 1000 times magnification. With the development of image analysis hardware and software, computer-aided chromosome analysis systems are in use; these have greatly reduced the turn-around times, and the quality of the karyotyped images is almost equal to that of photography. Bone marrow karyotype analysis is often concentrated on cells with poor morphology where there is a mixed population. Selection of only metaphases with good morphology can often lead to failure to detect cells from abnormal clones. Twenty or more metaphase cells are usually analyzed. If the karyotype is found to be abnormal, then analysis of fewer than 20 cells can be acceptable, but for a case to be reliably determined as cytogenetically normal, an analysis of at least 20 karyotypes from a BM sample is required. Any chromosome abnormality is considered clonal and therefore significant if it is present in a minimum of two metaphase cells in the case of an identical structural aberration, or gain of the same structurally intact chromosome (trisomy), or in at least three cells in the case of a missing chromosome (monosomy). Thus a clone is defined as at least two cells with the same structural abnormality or gain of the same chromosome, or at least three cells with the loss of the same chromosome. The karyotype results are interpreted using the International System for Human Cytogenetic Nomenclature (ISCN). For example, the normal male and female karyotypes are designated as 46,XY and 46,XX, respectively. An abnormal karyotype such as

46,XY,t(9;22)(q34;q11.2)

designates an abnormal male karyotype with a balanced translocation between chromosomes 9 and 22 at band 34 of the long arm of chromosome 9 and long arm of chromosome 22 at band 11.2. In contrast, a karyotype

47,XY,t(9;22)(q34;q11.2),+der(22)t(9;22)

delineates an abnormal male karyotype not only with a balanced 9;22 translocation as explained above, but also an additional chromosome that is derived from this translocation, i.e., an extra Ph chromosome.

There are two main types of cytogenetic aberrations in human cancer:

1. Balanced chromosomal rearrangements (reciprocal translocations, inversions) (Figures 3.1 and 3.2).
2. Gains or losses of whole chromosomes (aneuploidy), or part of a chromosome (segmental aneuploidy) and unbalanced rearrangements (Figures 3.3 and 3.4), resulting in deletions or duplications.

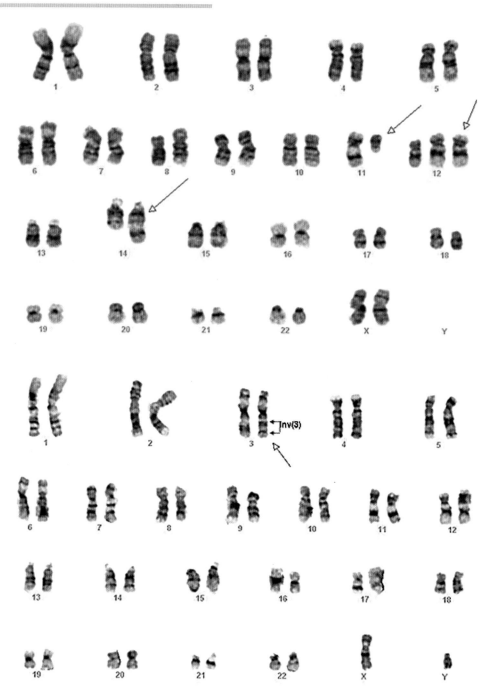

FIGURE 3.1 Abnormal female karyotype showing a balanced 11;14 translocation and trisomy 12. The karyotype is designated as 47,XX,t(11;14)(q13;q32),+12.

FIGURE 3.2 Abnormal male karyotype with a paracentric inversion in the long arm of one chromosome 3; the region of inversion is shown by arrows. The karyotype is 46,XY,inv(3)(q21q26).

These types of chromosomal aberrations typically cause cell overgrowth through over-expression/activation of an oncogene or by forming a "hybrid" gene or by deletion of a tumor suppressor gene. Identification of recurrent chromosomal aberrations has become very important in the diagnosis of soft tissue and hematologic tumors. Especially, in some hematological malignancies, the identification of recurrent chromosomal aberrations is important for diagnosis, prognosis, and therapy.

BALANCED REARRANGEMENTS

Balanced rearrangements in cancer include translocations (exchange between two or more chromosomes) and inversions (orientation change relative to the centromere, within a single chromosome). These rearrangements often result in chimeric cellular proteins that appear to disrupt the normal function of critical genes involved in normal cell growth or differentiation, resulting in an abnormal

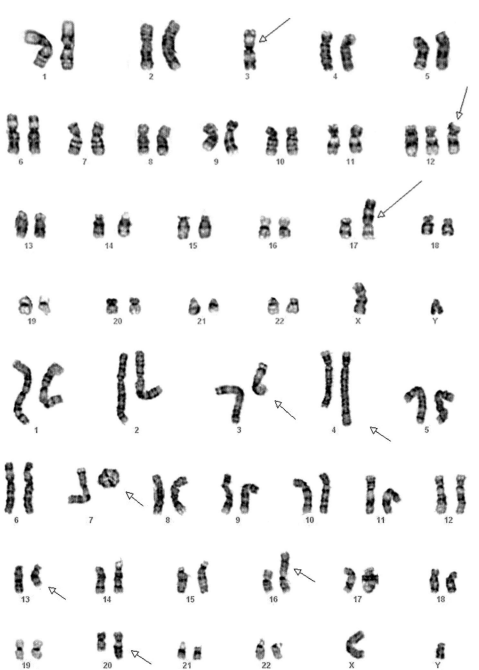

FIGURE 3.3 Abnormal male karyotype with monosomy of chromosome 3, trisomy 12, and an unbalanced translocation between the long arm of chromosome 3 and the short arm of chromosome 17 resulting in the deletions of the short arm of chromosome 3 and distal 17p segment. The karyotype is designated as 46,XY, +12,der(17)t(3;17)(q13;p13).

FIGURE 3.4 Abnormal complex male karyotype with several aberrations including a terminal deletion of 3q, an interstitial deletion of 13q, and an unbalanced translocation with unidentifiable chromosomal segments at chromosomes 4q, 16p, and 20q, respectively, and ring chromosome 7. The karyotype is written as 46,XY,del(3)(q21),add(4)(q35), r(7),del(13)(q12q14),add(16)(p13),add(20)(q13).

process. More than six hundred neoplasia-related recurrent balanced cytogenetic aberrations have been reported to date. Translocations and inversions, and in some instances cryptic insertions, usually cause cancer by fusing together two distantly removed genes, resulting in aberrant hybrid gene expression. Currently, more than two hundred chimeric fusion genes responsible for human cancers have been reported in the literature. One classic example of an important translocation is the t(9;22) in CML. The t(9;22) results in aberrant expression of an oncogene (*ABL1*) that normally functions in cellular proliferation, by coming under control of a constitutively expressed gene (breakpoint cluster region) (*BCR*). The resulting chimeric protein, *BCR-ABL1*, contains the catalytic domain of *ABL1* fused to a domain of *BCR* that mediates constitutive oligomerization of the fusion protein in the absence of physiologic activating signals, thereby promoting aberrant tyrosine kinase activity. Balanced rearrangements can also result in the deregulation of expression of normal genes. For example, in the reciprocal translocation t(8;14)(q24;q32), associated with Burkitt lymphoma, the immunoglobulin gene (*IGH@*) on band 14q32 drives the constitutive expression of the gene encoding the *MYC* transcription factor on band 8q24. Approximately 50% of hematopoietic neoplasms acquire translocations somatically; most of these are restricted to a single cell lineage (that in which the translocation originated) and are arrested in a particular stage of developmental maturation. On occasion more than one cell lineage is affected (e.g., *Mixed Lineage Leukemia* gene-related malignancies), suggesting that the involved genes were affected

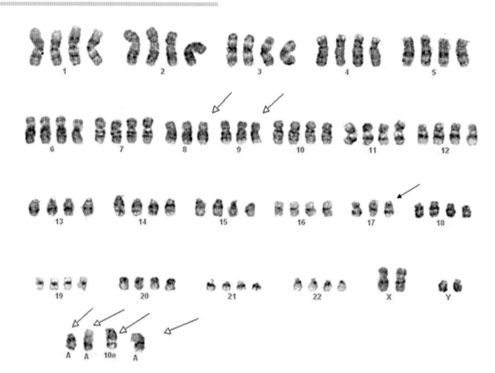

FIGURE 3.5 Abnormal male hyperdiploid (near-tetraploid) karyotype with four copies of all autosomes except for chromosomes 8, 9, and 17, and copies of small marker chromosomes of unknown origin. This karyotype is written as 93, XY, +X, +Y, −8, −9, −17, +4mar.

at the pluripotent stem cell stage. While balanced aberrations may be directly related to the etiology of the malignancy, the unbalanced translocations are often recognized as indicators of secondary tumor progression.

CHROMOSOMAL ANEUPLOIDY

Chromosomal aneuploidy is extremely common in cancers and can be either a primary or secondary event. Chromosomal gains (whole or partial) are designated in the karyotype with "+" or "add" and typically result in the over-expression of an oncogene. Despite the high frequency of aneuploidy in cancer, the exact role of aneuploidy in carcinogenesis is not very clear. Numerical aberrations including single or multiple chromosomal losses or gains as the sole karyotypic anomalies, are found in approximately 15% of all cytogenetically abnormal hematologic neoplasms (Figure 3.5). Although relatively frequent, numerical abnormalities have generally received less attention than the structural abnormalities, because the simple reciprocal translocations are easily amenable to rigorous molecular analysis. The association of numerical aberrations with hematologic disorders, although well established, also appears less disease specific. Trisomy chromosome 8, monosomy of chromosome 7, and trisomy chromosome 21 have been found in different categories of leukemias both at initial presentation or as secondary cytogenetic events. Roughly half of all numerical aberrations are trisomies. Other than trisomies for chromosomes 3, 8, 9, 11, 12, and 21, other autosomal trisomies are infrequent in hematologic disorders. Aneuploidy can be detected with the help of traditional metaphase cytogenetics (e.g., as in Figure 3.5), interphase cytogenetics (fluorescence in situ hybridization (FISH) (Figure 3.6 A,C), multicolor FISH/spectral karyotyping on metaphase cells, and comparative-array genomic hybridization techniques (CGH)), flow cytometry (FCM), and image cytometry (ICM). FCM and ICM can measure the relative DNA content of the cell with respect to reference diploid cells. Imbalances, i.e., aberrations that results in gain or loss of genetic material, are even more common than translocation and inversions in hematologic malignancies. These include amplifications, duplications, heterozygous or homozygous deletions, monosomies, and trisomies. Amplifications (several extra copies of a chromosome region), may occur in the form of supernumerary marker chromosomes (SMCs), double minutes, and homologous staining regions (HSRs) and are a result of over-expression of one or more genes (Figure 3.6B). SMCs are small additional chromosomes whose origins are not readily identifiable by banding methodologies, and are designated as "+mar" in the karyotype. Double minutes are specific types of SMCs that are characterized by a typical dumbbell shape, and represent extra-chromosomal oncogene amplification. For example, MYCN gene amplification in the form of double minutes is commonly observed in neuroblastoma. The Mixed Lineage Leukemia (MLL) gene is sometimes amplified in acute leukemia and can be easily visualized in interphase nuclei by FISH studies with MLL-specific probes (Figure 3.6D). HSRs are amplified oncogenes within the structure of a chromosome (Figure 3.7), and are designated as "hsr" in the karyotype. MYC gene amplification on chromosome 8q24 in some AMLs is an example of an HSR that is typically detected with the FISH technique. Chromosomal losses (whole or partial) are designated in the karyotype by "−" or "del" and are thought to result in deletion (or decreased activity) of tumor suppressor genes.

LOSS OF HETEROZYGOSITY

Loss of heterozygosity (LOH), defined as the loss of one parent's contribution to the cell, can be caused by deletion, gene conversion, mitotic recombination, or loss of a chromosome. LOH often occurs in cancer, where the second copy of a gene (typically a tumor suppressor gene) has been inactivated by other mechanisms, such as point mutation or hypermethylation. When a whole chromosome or a large segment of a chromosome is lost, the remaining chromosome or segment is often duplicated. With complete duplication of the remaining genetic material, the karyotype may appear normal, even though no normal genes are present. Though not easily detected by cytogenetic techniques, this duplication of the remaining chromosome or segment has been shown using molecular genetic techniques. At least in theory this type of LOH can be detected cytogenetically using chromosome heteromorphisms, though it is not often pursued.

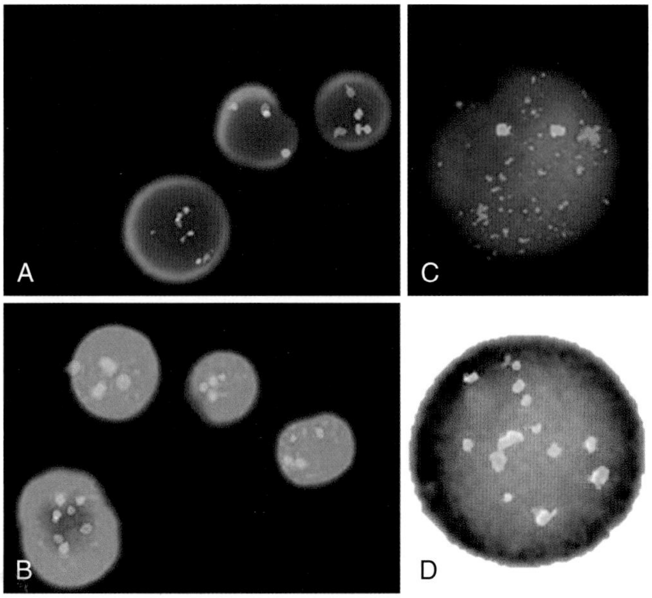

Conclusions

Cytogenetics analyses can provide valuable and extremely relevant information to establish the presence of a malignant clone, determine the cell lineages in the disease process or clarify and confirm a diagnosis, provide prognostic predictive features, and monitor response to treatment and classification of neoplasms. However, it must be noted that recent studies have recognized that in some cases where the karyotype appears "normal," particularly in AML, nucleotide-level mutations, which are undetectable by standard cytogenetics, have been discovered within a growing list of genes such as *FLT3* and can have important prognostic implications. The significance and usefulness of these will be discussed in the following chapters. A close relationship between the pathologist and the cytogeneticist is essential if maximum useful information is to be produced from the cytogenetics studies of hematological disorders.

FIGURE 3.6 Identification of various chromosomal abnormalities by FISH. (A) Four copies of 11q13 (red) and 14q32 (green) loci. (B) Amplification of the *abl* oncogene (red), and normal two copies of the *BCR* locus. (C) Four copies of 9q34 (red) and 22q11.2 (green) loci. (D) Amplification of the *MLL* locus (yellow).

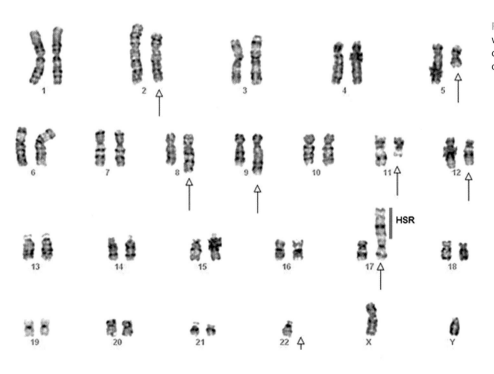

FIGURE 3.7 Abnormal male karyotype with an HSR of an unknown chromosomal region at the short arm of chromosome 17.

Additional Resources

Ferguson-Smith MA: Cytogenetics and the evolution of medical genetics, *Genet Med* 10:553–559, 2008.

Fröhling S, Döhner H: Chromosomal abnormalities in cancer, *N Engl J Med* 359:722–734, 2008.

Heim S, Mitelman F: Cancer cytogenetics – chromosomal and molecular genetics aberrations of tumor cells, edn 3, Hoboken, 2009, Wiley-Blackwell.

International Society of Chromosome Nomenclature (ISCN) Shaffer LG, Tommerup N, editors: An international system for human cytogenetic nomenclature, Basel, 2009, Karger.

Mitelman F. Cancer cytogenetics update, *Atlas Genet Cytogenet Oncol Haematol* Available at <http://atlasgeneticsoncology.org/>

Mittelman F. Database of chromosome aberrations in cancer. Cancer Genome Anatomy Project (CGAP), 2005.

Mitelman F, Johansson B, Mertens F: The impact of translocations and gene fusions on cancer causation, *Nat Rev Cancer* 7:233–245, 2007.

Principles of Molecular Techniques

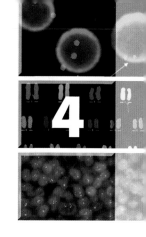

Fluorescence In Situ Hybridization

Karyotype analysis depends primarily on classical chromosomal banding techniques. It has the distinct advantage that the entire genome can be analyzed in a single experiment. In particular, it is useful for identifying whole chromosomes accurately and for identifying obvious chromosomal deletions, and translocations (low-resolution whole genome genome profile). However, karyotype studies are limited to actively dividing cells, and the resolution is limited to chromosomal aberrations greater than 3–5 Mb in size. In addition, through several decades of clinical cytogenetics analysis of cancer cells, it has become apparent that suboptimal collection, transport, and culture of clinical specimens can lead to inappropriate (e.g., normal) results. Poorly spread or contracted metaphase chromosomes, low mitotic activity, and highly rearranged karyotypes with numerous marker chromosomes, common in neoplastic cell preparations, are often difficult to interpret unambiguously. Furthermore, chromosome preparations are labor-intensive and time-consuming and the interpretation of cytogenetics findings requires extensive experience. Although automated karyotyping systems have become available, analyzing metaphase spreads remains time-consuming. Techniques such as polymerase chain reaction (PCR) analyses have the advantage of greater sensitivity, and they screen for a specific chromosome aberration without the need for dividing cells. However, such molecular analyses are limited to known genetic changes, and do not allow the screening of the whole genome for other (secondary) alterations. Thus molecular cytogenetics techniques have been developed to bridge the gap between classical cytogenetics and molecular DNA techniques. The limitations of classical chromosome studies have been overcome by the introduction of fluorescence in situ hybridization (FISH), which offers a molecular dimension to cytogenetic analysis. Different and new FISH technologies have emerged, each with their own particular advantages and applications: e.g., interphase FISH, comparative genome hybridization (CGH), fiber-FISH, and multi-color FISH. These techniques are capable of detecting aberrations of an intermediate size (~10 kb to 5 Mb), are expected to be fast and provide an accurate analysis of whole tumor genomes in a single experiment, and thus are now commonly used in cancer cytogenetics laboratories today, for both diagnostic and research applications. The FISH technologies provide increased resolution for the elucidation of structural chromosome abnormalities that cannot be resolved by more conventional cytogenetic analyses, including submicroscopic deletions, cryptic or subtle duplications and translocations, complex rearrangements involving many chromosomes, and marker chromosomes.

The FISH procedure has been developed for the tagging of DNA and RNA with labeled nucleic acid probes, and is a process whereby chromosomes or portions of chromosomes are vividly painted with fluorescent molecules that anneal to specific regions. This technique has been used widely for the identification of chromosomal abnormalities. The method enables enumeration of multiple copies of chromosomes or detection of specific regions of DNA or RNA that represent associations with certain genetic characteristics and infectious disease.

The FISH methods widely employed in clinical laboratory studies involve hybridization of a fluorochrome-labeled DNA probe to an *in situ* chromosomal target. FISH can be applied to a variety of specimen types and can be performed on non-dividing interphase cells. Recent developments have also allowed for the possibility of detection by non-fluorescence methods of the probes: CISH, or chromogenic in situ hybridization. Here the method utilizes conventional peroxidase or alkaline phosphatase reactions and the signals are observed under a bright-field microscope. This can be used for most tissues, similar to the FISH application, but has been more prominently successful in formalin-fixed, paraffin-embedded tissues. Like the FISH technique, it may be used to evaluate aneuploidy, chromosome translocation, amplification, and deletions.

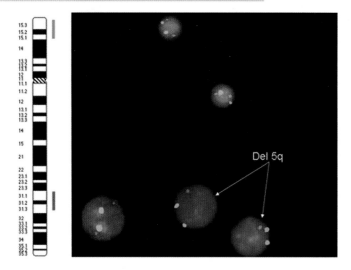

FIGURE 4.1 Deletion of 5q detected by 5p (green) and 5q (red) specific FISH probes.

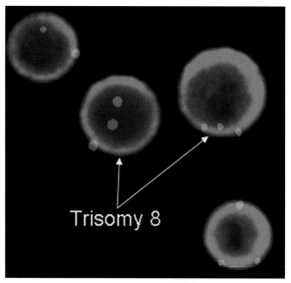

FIGURE 4.2 FISH studies reveal three copies of a chromosome 8 centromere specific probe in each cell (arrows), indicative of trisomy 8.

Interphase nucleus assessment from uncultured preparations allows rapid screening for specific chromosome rearrangement or numerical abnormalities associated with hematologic malignancies. Interphase analysis may also be performed on fixed bone marrow cell suspension, paraffin-embedded tissue sections, or disaggregated cells from paraffin blocks, bone marrow, or blood smears, and touch-preparations of cells from lymph nodes or solid tumors. It is also commonly used when rapid or direct (i.e., without culturing) results are needed, and can be performed on formalin fixed paraffin-embedded tissue sections. FISH uses fluorescently labeled DNA probes, e.g., **b**acterial **a**rtificial **c**hromosomes (BACs) hybridized to either metaphase chromosomes or interphase nuclei, depending on the application. However analytically powerful and diagnostically useful interphase FISH may be, great care should be taken in the interpretation of interphase hybridization patterns. As a rule of thumb, FISH results should be interpreted in conjunction with the neoplastic karyotype.

Four different types of probes are commonly used, each with different ranges of applications.

1. Gene-specific probes target DNA sequences (Figure 4.1) present in only one copy per chromosome. They are used to identify chromosomal translocations, inversions, deletions, duplications and gene amplifications in interphase and metaphase chromosomes. These probes are particularly useful for screening of specific gene rearrangements in metaphase spreads and interphase nuclei. For this purpose, the probes cover the chromosomal breakpoints and can specifically identify the genes involved in the chromosome alterations without the need for dividing cells. The same FISH experiments can subsequently be used to assess the efficacy of therapeutic regimens and to detect residual disease with limited sensitivity of 0.5–5%.
2. Repetitive sequence probes (Figure 4.2) (alpha-satellite sequences) bind to chromosomal regions that are represented by short repetitive base-pair sequences that are present in multiple copies (e.g., centromeric and telomeric probes).

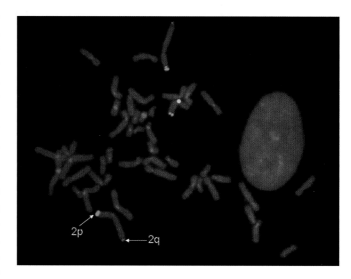

FIGURE 4.3 Subtelomere specific FISH probes are demonstrated in chromosomes 2.

Centromeres are usually A-T rich, whereas telomeres are known to have repetitive TTAGGG sequences. Centromeric probes are extremely useful for identifying marker chromosomes and for detecting copy number chromosome abnormalities in interphase nuclei.

3. Sub-telomeric probes (Figure 4.3) are frequently used to identify subtle or submicroscopic chromosomal rearrangements. The relative ease of performance and high resolution (0.5 Mb) of such unique-sequence FISH analysis has made it popular to screen for whole arm chromosomal rearrangements, subtelomeric deletions and duplications, and marker chromosomes.
4. Whole-genome painting probes (WCP) (Figure 4.4) are complex DNA probes that are generated by degenerate oligonucleotide polymerase chain reaction (DO-PCR) or through flow-sorting. WCP paints have high affinity for the whole chromosome along its entire length, with the exception of the centromeric and telomeric regions. These probes are most

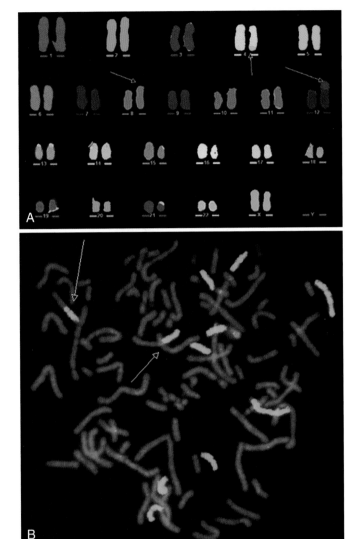

FIGURE 4.4 Chromosome painting. (A) Translocations (arrows) are demonstrated by multicolor FISH technique. (B) Chromosome 7 is highlighted with single whole-chromosome paint.

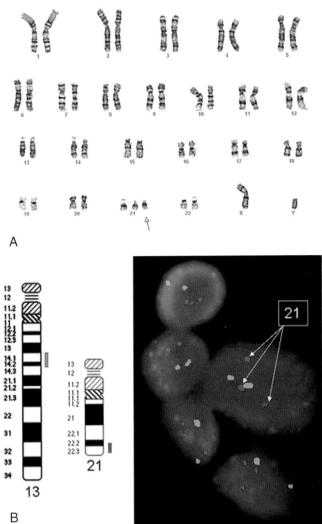

FIGURE 4.5 FISH probes (B) for chromosome 13 (green) and chromosome 21 (red) demonstrate trisomy 21 seen by standard karyotype studies (A).

suitable for identifying genomic imbalances in metaphase chromosomes, especially the complex chromosomal arrangements observed in many cancers. Two variants of the WCP have been developed: multicolor-FISH (M-FISH) (Figure 4.4A) and spectral karyotyping (SKY). WCP probes have been developed that paint all human chromosomes in different colors (48 paints). And WCP can also offer simultaneous detection of each arm of all human chromosomes in a single hybridization. This technique is usually used in conjunction with chromosomal banding techniques for a more precise identification of chromosome aberrations. The two greatest limitations of WCP are the requirement for extensive knowledge of genetic abnormality to enable the correct selection of the probes (unless the full complement of paints is to be used), and limited resolution (>5 Mb) to detect very small deletions, duplications and translocations.

Three main probe strategies are utilized in FISH: (1) enumerating probes, (2) fusion probes, and (3) "break-apart" "split" probes.

1. **Enumerating probe.** The enumerating or counting probe strategy, as its name implies, is useful for counting the number of a particular locus within the cell. Counting probes are used to detect gains or losses of whole or part chromosomes (e.g., chromosomes 5, 7, 8, and 20 in MDS (Figures 4.5, 4.6 and 4.7), or deletions and duplications of genes involved in a disease (e.g., *TP53* and *RB1* gene probes in myeloma). These probes can be either BACs containing the gene or genes of interest, or alpha-satellite repeat sequences specific for the centromeric region of each chromosome. This strategy is also useful for detecting cryptic deletions that may not be detectable by classical metaphase chromosome analysis.
2. **Fusion probe.** The fusion probe strategy is classically used to detect specific translocations or inversions, e.g., the t(9;22) in CML (Figure 4.8) and the t(15;17) in APL. BAC probes complementary to chromosomal regions involved in the rearrangements are labeled with two different fluorophores (e.g., red and green) and analyzed under the microscope for signal overlap. Normal nuclei will have two red and two green signals, corresponding to the two normal (un-rearranged) chromosomes, while nuclei with rearrangements will have one or more yellow signals, corresponding to the overlap of

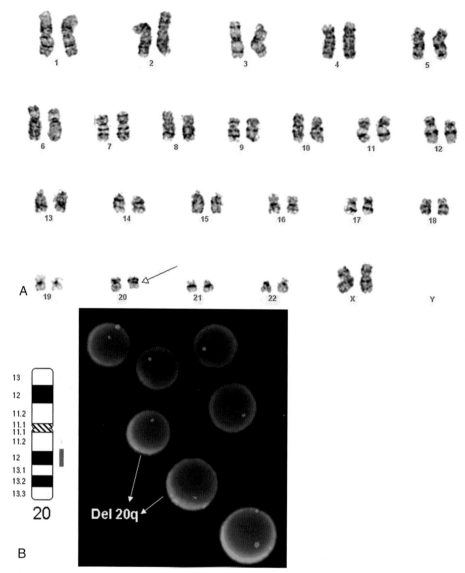

FIGURE 4.6 Deletion of chromosome 20q (46,XX,del(20)(q11.2)) is demonstrated by (A) G-banded karyotyping and (B) FISH.

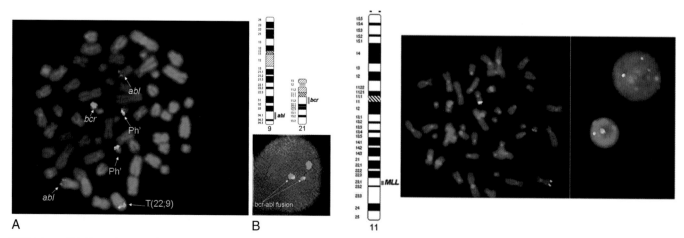

FIGURE 4.7 (A) Metaphase and (B) interphase FISH showing BCR-ABL1 fusion signals in CML.

FIGURE 4.8 Break-apart or split probe signal patterns of MLL locus (yellow) in abnormal cells.

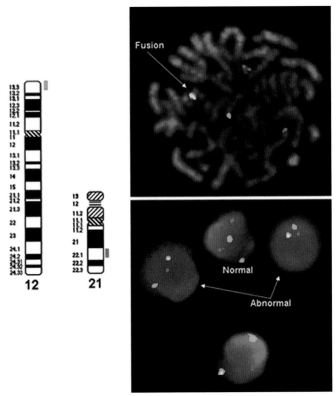

FIGURE 4.9 Demonstration of t(12;21). The fusion of *ETV6* (red) and *RUNX1* (green) genes demonstrates as a yellow spot by dual-color FISH.

the red and green signals and suggestive of rearranged chromosomes. Dual-fusion strategies are used to reduce false-positive signals produced by artifactual overlap caused by the three-dimensional structure of DNA compaction within the nucleus. Dual-fusion approaches utilize probes that overlap the two reciprocal translocation breakpoints and result in two yellow fusion signals corresponding to the two derivative chromosomes. This probe strategy is also useful to distinguish between variants, and identifying an extra Philadelphia chromosome in CML in accelerated phase.

3. **Break-apart or Split probe.** The break-apart or split probe strategy (Figure 4.9) is essentially the opposite of the fusion probe strategy, and is most useful when a single locus is involved in several different rearrangements (translocations, inversions, deletions, etc.) involving multiple partners. For example, a dual-color FISH probe has been developed with probes at either side of the *MLL* gene breakpoint, resulting in separation of the normally co-localizing signals if the *MLL* gene is rearranged. The advantage of this system is that it can detect all recurrent and possibly novel *MLL* rearrangements in a single experiment. The *MLL* gene locus is involved in >70 recurrent translocations, all of which can be detected with the break-apart strategy. Two differently labeled BAC probes (e.g., red and green) normally bind to a single locus and produce the overlapped signal color (e.g., yellow). When the locus of interest is rearranged, the colors split apart. Normal nuclei will have two yellow (overlapped) signals, while nuclei with a rearrangement will have one yellow, one red, and one green signal.

While FISH can provide rapid results, and is applicable for various sample types that are otherwise not amenable to classic cytogenetic analyses, it has limitations. FISH will only answer the particular question being asked regarding an exact probed locus. For example, cells probed with a *BCR-ABL1* fusion probe set may be positive for the t(9;22), but trisomy 8 cells within the sample (often seen in CML in transformation) would not be detected unless a chromosome 8 enumerating probe set is used in the probe mix. Similarly, an enumerating probe set consisting only of the alpha satellite repeats from the chromosome 5 centromere will not detect a deletion of the long arm of chromosome 5 (5q-). Physicians ordering tests should be mindful of the questions they are trying to address and order FISH and/or karyotypes appropriately. FISH studies may also be an integral component of the diagnostic workup if a specific genetic abnormality is suggested by histopathology, peripheral blood counts, or clinical parameters, but when cytogenetic analysis fails or provides a normal karyotypic result. For example, it is recommended that all cases of CML be studied at diagnosis by cytogenetic analysis and molecular cytogenetic methods to determine the initial clonal abnormalities and the FISH signal pattern, both for prognostic information and for follow-up studies. When questions arise regarding which FISH test to order, the laboratory should be consulted, as ordering a FISH test without specifying the probe set is inappropriate. The application of FISH further provides increased sensitivity, in that chromosomal abnormalities have been detected in samples that appeared to be normal by conventional cytogenetic analysis; for example, the chromosomally hidden translocation was evidenced for the first time in 1995 with the discovery of the t(12;21)(p13;q22), resulting in ETV6/RUNX1 gene fusion (Figure 4.9). While appearing to have a normal looking karyotype, molecular cytogenetic investigations with probes specific for this gene rearrangement revealed this translocation to be present in 25% of all pediatric B-cell ALL.

CGH is a relatively new molecular technique for identifying chromosomal gains and losses in a test sample (e.g., a patient sample) relative to a control sample. DNA is extracted from both the test and control samples and digested with restriction enzymes or sonicated to break it into short (~500bp) fragments. The test and control samples are differentially labeled with florescent dyes (e.g., red and green), denatured, and hybridized to metaphase chromosomes. Chromosomal regions that are equally represented in the test and control samples will hybridize equally to the chromosomes and produce an overlapped color (e.g., yellow). A loss (deletion) is detected when the control DNA fluoresces stronger within a region of the metaphase chromosomes, and a gain (duplication) is detected when the patient DNA fluoresces stronger. In array CGH (aCGH), the fluorescently labeled test and control samples are hybridized to an array of DNA sequences, e.g., BACs (Figure 4.10) or oligonucleotides (Figure 4.11), rather than metaphase chromosomes. aCGH has a much higher resolution than classic metaphase CGH. Although CGH and aCGH have been well established for use in

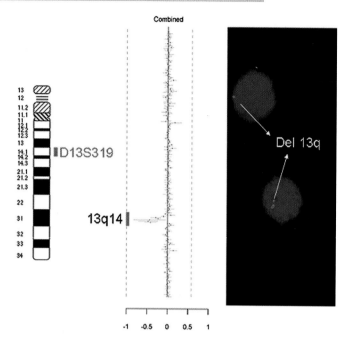

FIGURE 4.10 Whole-genome BAC-array CGH showing a deletion of 13q.

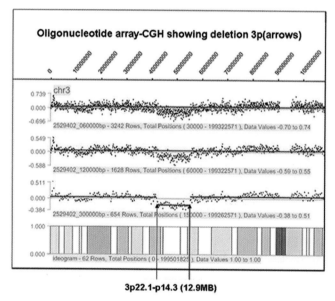

FIGURE 4.11 Shift at 42150000..55050000 bp; chromosome 3p22.1–p14.3 (12.9 Mb).

detecting copy number, submicroscopic gains and losses in constitutional (inherited) disease, neither is currently appropriately established as a stand-alone technology for diagnosis in cancer. One reason is that the CGH technique cannot detect balanced chromosomal aberrations (translocations and inversions), which are very common in cancers, especially in hematological disorders. Also, because of tumor heterogeneity and general genomic instability, i.e., several clonal populations, CGH and aCGH do not necessarily provide narrow and consistent genomic regions of interest that can be definitely implicated or identify previously unknown genomic regions of primary etiology. However, with improvements in technology, software analyses and database collection, CGH will be one of the most useful techniques adopted by diagnostic laboratories. Indeed, it would be interesting to see if aCGH enhances the efficiency of detecting subtle changes such as partial tandem duplication/amplification of the *MLL* gene observed in AML. Furthermore, in chronic diseases such as MDS or CML, where a progressively evolving karyotype relates to worsening prognosis that may be used in therapeutic decisions, array CGH would provide very useful comparative genome-wide information in sequential clinical specimens.

FISH is also useful tool to monitor remission status when clonal chromosome abnormalities have been identified at diagnosis and appropriate probes are available. For CML, sequential FISH studies are particularly useful to determine changes in clinical status in response to therapy and to assess for minimal residual disease. In patients with sex-mismatched bone marrow transplants for whom graft rejection, marrow suppression, or disease relapse is a clinical consideration, monitoring with a FISH assay that combines sex chromosome probes with or without probes to detect the patient's clonal abnormality can be valuable for graft assessment and to detect residual or recurrent disease.

FISH may be performed on formalin-fixed paraffin-embedded (FFPE) sections either by hybridizing directly to unstained thin sections (2–4 μm) of tissue that have been de-paraffinized, or thick sections of tissue from which individual cell suspensions are made, to which standard FISH techniques may then be applied. Sections must be mounted on slides that will reduce the loss of tissue during FISH-pretreatment processes. Buffered formalin (pH 7.0) is the best fixative for FISH hybridization. However, fixation in solutions that contain a heavy metal or picric acid often results in unsuccessful FISH results. It is also well established that decalcifying techniques commonly used in bone marrow core biopsy sections often do not yield good FISH signals and may be inappropriate for obtaining cytogenetic results. There are also several other limitations to FFPE FISH including overlapping and truncated cells, making assessment of individual cells difficult (Figure 4.12). Also improvements in the pretreatment, hybridization, optimal probe size, and post wash techniques will alleviate the commonly encountered problems such as background (non-specific binding) or absence of probe signals. FISH on frozen-tissue sections will yield good results if the slides are first slowly thawed at room temperature and then fixed in 10% buffered formalin.

There are only a few commercially manufactured probe kits that have been approved by the FDA for *in vitro* diagnostic testing. These FISH kits must meet the sensitivity and specificity parameters stated in package inserts provided by the manufacturer. The majority of probes used for clinical FISH testing are considered to be analyte specific reagents (ASRs) that are exempt from FDA approval. When a new ASR probe is introduced in the laboratory, extensive

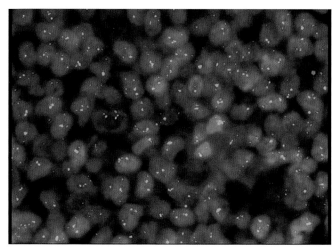

FIGURE 4.12 A typical image of FISH analysis on an FFPE section showing overlapping truncated cells as demonstrated by variable signal patterns.

validation is needed, including specific validation of the probe itself (probe validation) and validation of the procedures utilizing the probe (analytical validation). Initially, it is important to become familiar with a probe's parameters, including signal intensity and pattern, and any cross-hybridization that is likely to confound test results. Probe sensitivity, defined as the percentage of metaphases with the expected signal pattern at the correct chromosomal location, should be established. Likewise, probe specificity, the percentage of signals that hybridize to the correct locus and no other location, must also be assessed. Probes used for hematologic malignancy studies should have a high analytic sensitivity and specificity (>95%), particularly if they are to be used for minimal residual disease assessment.

FISH results should be interpreted within the broader context of probe and analytical validation. The interpretation of FISH results should include consideration of the reason for referral for testing and, when available, additional laboratory findings including conventional cytogenetic analysis, hematopathology, and immunophenotyping. When acute promyelocytic leukemia (APL) is the suspected diagnosis, FISH should be performed on a STAT basis with 24 hours turnaround to allow for timely treatment such as All-trans-retinoic acid (ATRA)-therapy.

A system for FISH nomenclature, including both metaphase and interphase analysis, has been developed. While the system may seem confusing to those not working directly with chromosomes, correct nomenclature designations are important to convey the precise nature of a result. For example, metaphase FISH ISCN nomenclature for a male patient with a 9;22 translocation resulting in fusion of the *BCR* and *ABL1* genes studied with conventional banding and with a dual-color, single-fusion BCR/ABL1 probe set would be written:

46,XY,t(9;22)(q34;q11.2).ish t(9;22)(*ABL1*−;*BCR*+,*ABL1*+)

indicating that the probe sequence from the *ABL1* locus is missing from the derivative chromosome 9 and is present on the derivative chromosome 22 distal to the *BCR* locus.

The same rearrangement expressed in interphase FISH nomenclature but using a dual-color, dual-fusion probe set would be:

nuc ish 9q34(*ABL1*×3),22q11.2(*BCR*×3)(*ABL1* con *BCR*×2)

indicating that each of the probes has been split apart and juxtaposed by the translocation. The use of such precise ISCN nomenclature is valued by laboratories in the initial diagnostic work-up and continued monitoring of patients with a specific chromosome abnormality. The report must indicate any specific limitations of the assay, some of which may be described in the probe manufacture's package insert.

While interphase FISH analysis provides information only on specific probes used and generally does not substitute for complete karyotype analysis, it may, under some disease circumstances, be the preferred means of identifying an abnormal clone, e.g., FISH with the *ATM*, CEP12, D13S319, and *TP53* probe panel in B-cell CLL, discrimination of the inversion 16 or t(12;21) in acute leukemia, or FISH for the diagnostic abnormality in post-therapy patients who have hypocellular marrows. More recently, due to the low proliferation rate of myeloma cells *in vitro*, which results in normal karyotypes, and because of the patchy nature of the disease, conventional FISH has also been hindered by normal cell contamination. However, these limitations have been overcome by enriching the plasma cells by using CD138 markers, which is followed by FISH analysis with disease specific panel of probes. This ensures that only the cells of interest are analyzed, and thus the results obtained are a more accurate reflection of the plasma cells.

FISH has now become an invaluable tool in defining and monitoring acquired chromosome abnormalities associated with hematologic and other neoplasias. The implementation of the technology into the routine diagnostic laboratory requires rigorous attention to when it is appropriate to apply the technology, a very systematic approach to the validation of probes and technical procedures involved, the training of individuals who will perform the testing, and a comprehensive but plain and simple means of reporting results. As the number of critical loci found to be involved in neoplastic chromosome rearrangements or numeric abnormalities continues to expand, the diversity of FISH probes and unique probe sets will undoubtedly improve. FISH has become an important means both for definition of the initial chromosome changes in a disease process, as well as a reliable means for the ongoing monitoring of response to therapy and disease remission.

Apart from the diagnostic approaches, FISH can be used as a research tool to refine the breakpoint regions of novel chromosome abnormalities, which is often an essential step in the identification of new (partner) genes involved in leukemogenesis. Recent progression of the Human Genome Project has facilitated the characterization of the translocation breakpoints using FISH. PAC/BAC resources, covering the entire genome are available and can be easily found using the databases of the University of California, Santa Cruz (UCSC) (http://genome.ucsc.edu/), Ensembl (http://www.ensembl.org/), and the National Center for Biotechnology Information (NCBI) (http://www.ncbi.nlm.nih.gov/genome/guide/human/). These clones can be used to determine more precisely the breakpoint regions and to search for genes involved in the translocations. The molecular cytogenetics techniques are rapidly evolving with even greater "high resolution" genome analysis being established and eventually used in a clinical setting. One such technology is whole-genome sequencing which has been proven to identify cytogenetically invisible cryptic fusion oncogenes in a clinically relevant time frame.

The genetic tools and strategies described briefly above have been applied to cancer for fewer than fifty years, but they have quickly been recognized to be invaluable in the study and diagnosis of malignancy. The importance of cytogenetics in oncology is evidenced by the reclassification of certain hematological diseases by the World Health Organization, and the application of both classical and molecular cytogenetic methods to hematologic diseases will be presented throughout this book.

Polymerase Chain Reaction and Related Techniques

It is impossible to overstate the degree to which the advent of nucleic acid amplification techniques, especially the polymerase chain reaction (PCR), has revolutionized the molecular approaches to hematopathologic diagnosis and molecular diagnostics generally. PCR and reverse transcriptase PCR (RT-PCR) not only enable robust analysis of scant or degraded specimens, but they also allow for precise quantitation of the analyte, fine dissection of a particular locus or sequence from among the over three billion nucleotides of the human genome, and access to otherwise difficult specimens such as formalin-fixed, paraffin-embedded tissue biopsies. These techniques have largely replaced the much more laborious and time-consuming Southern blot (see below) for most, but not all, applications in hematopathology. At a more basic level, they have also replaced difficult DNA cloning procedures for many purposes, including the generation of DNA probes used in a variety of downstream applications such as the Southern blot.

BASIC TECHNIQUE

PCR is so simple in its conception that it is surprising no one in molecular biology happened to think of it before the seminal paper by Saiki et al. appeared in 1988. Fundamentally it merely mimics replication of DNA *in vivo*, using essentially the same enzymes (DNA polymerases), but in a highly specific and exponential fashion. The specificity results from the use of specific *primer* sequences that are constructed to be complementary only to the target gene or region of interest. The exponential amplification results from the use of a pair of primers which flank the target region and hybridize to opposite strands of the DNA (Figure 4.13). Serving as start-sites for the polymerase reaction, they promote a bidirectional synthesis of daughter strands that then become templates for the next round of replication. The replication cycles are controlled by alternate heating and cooling of the sample, denaturing the amplification products and then enabling the primers (which are present in great excess) to re-hybridize and begin the replication again. Since each replication cycle doubles the number of template molecules, the products accumulate in exponential, rather than linear, fashion. A typical experiment encompasses about 30 cycles, performed in a programmable heating/cooling instrument called a thermal cycler, which produces amplification of many million-fold. The precise timing and temperature settings will vary depending on the target sequence and application, but generally fall in the range of about 94°C for denaturation and 50–60°C for renaturation. The elongation step, in which the actual DNA synthesis occurs, is run at 70–75°C. The reason it is not performed at physiologic temperature is because in modern PCR the DNA polymerases used are cloned from a variety of thermophilic microorganisms (e.g., *Thermus aquaticus*, source of so-called *Taq* polymerase) so that they do not degrade during each denaturation step.

PRIMER DESIGN

For PCR to deliver the specificity desired, it is most important that primer sequences be chosen carefully so they do not cross-hybridize with other regions of the genome at the renaturation temperatures used. Primers are typically about 18–25 nucleotides in length, and there are computer programs available on the internet to work out optimal base composition such that the primers do not hybridize elsewhere or to themselves. For most clinical purposes, primers are designed to hybridize a few hundred nucleotides apart in the target sequence, since replication efficiency declines as target length increases. A number of long-range PCR protocols and commercial kits are available, allowing for amplification of targets many thousands of base pairs in length, but these are mostly reserved for research purposes.

QUALITY CONTROL

Given the awesome sensitivity of PCR, theoretically down to a single target DNA molecule, extreme care must be exercised to guard against the production of spurious

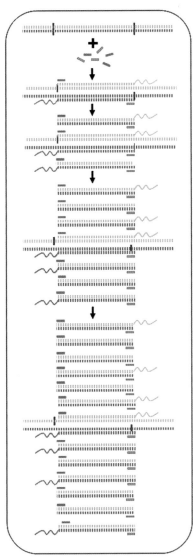

FIGURE 4.13 Polymerase chain reaction scheme. The oligonucleotide primers (rectangles) are designed to hybridize to opposite strands some distance apart on the target region of interest. When hybridized, they serve as priming sites for DNA polymerase replication of each strand. Alternate heating and cooling allows for multiple cycles of denaturation, rehybridization, and replication, increasing the amount of the target sequence exponentially.

amplification products by contaminant target molecules. In a clinical laboratory, these can come from other patient specimens, from the operator himself, or from residual amplification products of a previous assay. In practice, it is this last source that is the most concerning, since these products are present in infinitely greater excess than stray genomic contaminants from an individual. A large number of precautionary and preventative procedures have been developed to guard against it, including one-way workflow from pre- to post-amplification areas, special dedicated pipettes, re-gloving, and degradation of residual amplicons by UV light and other methods. Failing this, contamination is detected by the running of a "no DNA" or "no template" control tube in every PCR assay. This tube contains all the requisite enzymes and nucleotides for the reaction to proceed, but no target template to copy. If any amplicon is observed in this sample after PCR, it indicates a contamination problem with that run or with one of the reagents. In addition, known positive controls must be run with each assay, to demonstrate capability of the PCR conditions and reagents to detect and amplify the desired target sequence when present.

PRODUCT ANALYSIS

For most applications, PCR is not an end in itself but rather a first step in a subsequent assay. Once the desired target has been amplified, it may be analyzed by hybridization with specific DNA probes in a dot blot format, by size fractionation in agarose gel or capillary electrophoresis, by digestion with sequence-specific restriction endonucleases, or by DNA sequencing. The choice of downstream technique will depend on the nature of the disease process being tested.

REVERSE TRANSCRIPTASE PCR

Certain disease applications, for example the detection of the *BCR-ABL* translocation in chronic myelogenous leukemia, involve detection and accurate quantification of an RNA, rather than DNA, target. Since PCR uses DNA polymerases which do not replicate or transcribe RNA, amplification of such targets can only be accomplished by first converting them to DNA. Fortunately, there are viral-derived enzymes available that can do this, termed reverse transcriptases (RT). Thus, amplification and study of RNA targets utilizes the technique of RT-PCR, in which the first step is reverse transcription, followed then by conventional PCR of the resulting DNA products.

REAL-TIME PCR

As noted above, in most cases PCR amplification is but the first step in an assay that then utilizes another method to analyze or quantify the products. This is because conventional thermal cyclers are closed systems in which amplification proceeds without observation until it is finished and the reaction tubes removed. In contrast, a newer generation of thermal cycler instruments can measure the accumulation of amplicon as it occurs (i.e., in real time) by measuring the incorporation of a fluorescent label into the elongating products. These instruments can provide very precise quantification of products and, by extrapolation, the amount of starting target material, in the sample; hence the technique is also called quantitative PCR (qPCR). They are very useful for cancer applications in hematopathology designed to detect minimal residual disease after therapy.

RELATED AMPLIFICATION TECHNIQUES

A large number of innovations to the basic PCR technique have been developed over the years to address particular applications or to circumvent certain pitfalls. Included are such techniques as nested PCR, whole-genome amplification, inverse PCR, hot-start PCR, allele-specific PCR, cold PCR, and many others. For the most part they are beyond the scope of this chapter, but will be mentioned in the context of particular disease applications where relevant elsewhere in the book. In addition, a number of non-PCR amplification techniques have been developed over the years, such as Q-β replicase, ligation chain reaction, etc., but for the most part they have fallen by the wayside in favor of PCR, at least for applications relevant to hematopathology (some are used in molecular microbiology and genetics testing).

Blotting Techniques

In blotting techniques, unique segments of nucleic acid sequences (DNA probes) are used to demonstrate the presence of complementary sequence of DNA or RNA in the sample. Since the complementary target is composed of hundreds or thousands of nucleotide bases, the reaction of the DNA probe to the target (hybridization) is the tightest and most specific intermolecular interaction.

SOUTHERN BLOT

The DNA probes are labeled with a radioactive or non-radioactive signal moiety and are used as templates for DNA. The DNA target is treated with restriction enzymes (endonucleases) and the DNA fragments are separated by gel electrophoresis (usually an agarose gel). The DNA is then transferred to a membrane which is a sheet of special blotting paper. The blot is incubated with numerous copies of a single-stranded labeled DNA probe The probe hybridizes with its complementary DNA sequence of the target sample to form a labeled double-stranded DNA molecule. The radiolabeled copy of the DNA probe is then detected by autoradiography (Figure 4.14). In non-radioactive DNA labeling procedures, nucleotides (probes) are conjugated with biotin or other protein binders such as digoxigenin. Biotin binds specifically to the protein avidin with a very high affinity. Avidin is a polyvalent protein which can be linked to chromogenic enzymes, fluorescein compounds, or electron dense particles. The advantage of a radiolabeled probe over a non-radiolabeled probe is its higher sensitivity (5–10 fold), but its disadvantages are an elongated radiography step, which may extend to several days, radiation hazard, and requirement for special procedures for disposal of the radioactive contaminated wastes. In general, Southern blotting is a laborious, expensive, and time-consuming procedure, and most clinical laboratories prefer to abandon it as soon as a suitable PCR-based technique is available. Detection of clonal gene rearrangements in lymphocytic lesions is about the only remaining application for it, and even there, some laboratories have moved on to an approach using many PCR primer sets to encompass the same genetic region covered by the Southern blot.

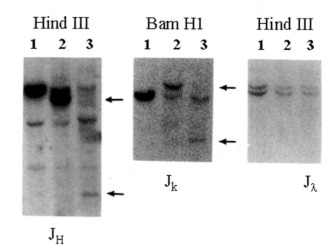

FIGURE 4.14 Southern blot analysis for Ig gene rearrangements demonstrating rearranged bands (arrows) for the joint region of the Ig heavy chain (J_H) and the joint region of the kappa light chain (J_k).

NORTHERN BLOT

Northern blot is a technique basically similar to Southern blot, except that it is used to transfer RNA from a gel to a blot, instead of DNA.

DOT BLOT

The dot blot is similar to the other blotting techniques, except it does not provide information regarding the size of the hybridized fragment. With this technique, extracted DNA or RNA from the target specimen is spotted onto the filter without the prior electrophoresis and transfer steps. In a reverse dot blot, it is the probe that is pre-bound to the filter and then hybridized with the patient's (usually PCR-amplified and colorimetrically tagged) DNA. The probe configuration may be as lines (line blot) rather than dots or circles.

Microarray Techniques

The next generation of blotting techniques is a marriage of molecular biology and information technology: the microarray. In contrast to a dot blot or Southern blot in which target DNA is hybridized to one or a few probes, microarrays enable the hybridization of hundreds or hundreds of thousands of target sequences simultaneously. Because it is constructed at nanoscale using technology similar to that used to manufacture silicon-based computer chips, the vernacular term "DNA chip" is often applied. Essentially it is a reverse dot blot on a grand (and miniaturized) scale. The individual probes, in great numbers, are bound to the solid support, while the specimen DNA (or RNA) is hybridized to them after being labeled with a fluorescent marker (Figure 4.15). A chip-reader instrument interprets these signals and generates data output detailing which sequences hybridized and which did not, as well as the intensity of hybridization (which relates to the relative amount of each particular target sequence in the material being tested).

There are two basic kinds of microarrays: DNA arrays and RNA expression arrays. The former are used to detect the presence of mutations, deletions, duplications, and other sequence variants in the tested sample, while the latter are used to assay relative expression of hundreds or thousands of genes, as for example in comparing tumor mRNA to that in a corresponding normal tissue. Depending on how many probes are placed on the array, assays of varying comprehensiveness can be developed, even to the extreme of comprising the entire human genome.

DNA Sequencing

In many ways DNA sequencing is the most definitive molecular biology technique because it gets directly at the genetic code of the specimen being analyzed, and that is what is at the core of everything else, both biologically and clinically. Other techniques, such as probe hybridization, restriction endonuclease digestion, and even PCR, are surrogate assays whose outcome depends ultimately on the inherent sequence of the target material but which do not ascertain that sequence directly. For this reason DNA sequencing is often referred to as the "gold standard" for molecular testing in the clinical setting.

Like other techniques in molecular diagnostics, DNA sequencing arose in the research setting to satisfy research needs, and resulted in Nobel Prizes for its inventors. Initially cumbersome, the techniques have been refined over the years, recently under the impetus of the Human Genome Project which demanded extremely high throughput, accurate, and inexpensive sequence analysis. The benefits of these innovations now spill over into the clinical arena where sequencing, previously considered a kind of "last resort" methodology for special cases, is now routine for a wide variety of applications.

CHEMICAL METHODS

Two basic approaches to sequencing were initially developed and used, but one has now assumed dominance for both manual and automated platforms. The chemical degradation method, developed by Maxam and Gilbert, uses various chemicals to cleave the double helix, based on their reaction with specific nucleotides. Sizing and aligning the resulting fragments by electrophoresis allows one to determine at which point along the length of the helix each nucleotide was positioned. This method has largely been abandoned for routine uses because of its technical difficulty and the noxious nature of the chemicals required.

It has been supplanted by the chain termination method, developed by Sanger at about the same time. This

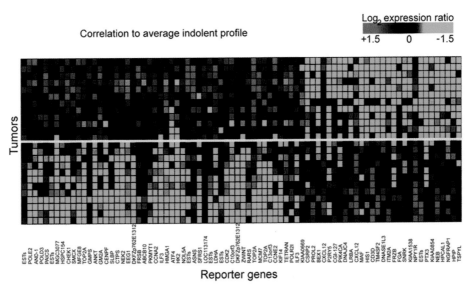

FIGURE 4.15 Gene expression profiling in follicular lymphomas using a microarray technique. Results demonstrate two populations of patients: those with aggressive disease are segregated above the solid yellow line and those with indolent clinical course are placed below the solid yellow line.
(From Glas AM, Kersten MJ, Delahaye LJ, et al. Blood 2005; 105: 301–307, with permission.)

one, too, is based on fragment sizes determining the positions of each nucleotide, but the fragments are created not by degradation but by synthesis. Four DNA polymerase reactions are set up using the target sample as the template for replication. In each reaction tube, in addition to the four regular nucleotide substrates, a dideoxynucleotide derivative is also added. When one of these is incorporated into the growing daughter strand by the polymerase, replication is halted at that point because the dideoxynucleotide lacks the 3'-hydroxyl group required for chemical linkage to the next nucleotide that would otherwise be added. Since the derivative nucleotides are present in the minority, they are not inserted at every spot but rather at a random distribution whenever that nucleotide is required. This produces a complete spread of fragments whose size is based solely on the positions of that particular nucleotide in the original target material. If the fragments from each nucleotide reaction are separated on the basis of size by electrophoresis, a "ladder" is created that reveals in order, from shortest to longest, the nucleotide sequence.

SEQUENCE DETECTION AND ANALYSIS

The electrophoresis formerly was done in a flat-bed (usually vertical) gel set-up, typically a long polyacrylamide gel, using radioactive nucleotides that can be seen by autoradiography of the dried gel after the run. But at present most clinical laboratories have moved to automated sequencing instruments which use capillary electrophoresis of fluorescently labeled oligonucleotides to generate the sequence. The reactions are carried out in a single tube instead of four, and the various reactions are discriminated because each of the four dideoxynucleotides is labeled with a different-colored fluorophore. These instruments provide more precise sizing, much higher throughput, and sophisticated software for calling out and analyzing the sequence (Figure 4.16).

LIMITATIONS OF SEQUENCING

While it may be the "gold standard," DNA sequencing, like any other technique, has its limitations and pitfalls. For many applications in which only one particular mutation or size fragment is being probed, sequencing may represent "overkill" in that it yields a tremendous amount of extraneous data not relevant to the clinical question being asked. Some of that data, moreover, can be of questionable clinical relevance. Scattered throughout the genome of every human being are countless non-pathologic nucleotide substitutions, called polymorphisms. When they occur within protein coding regions (exons), they appear as missense mutations, causing the substitution of one amino acid for another in the gene product. If a sequencing test reveals one of these changes that has not been seen or reported in the literature before, it can be very difficult or impossible to decide whether it represents a pathologic

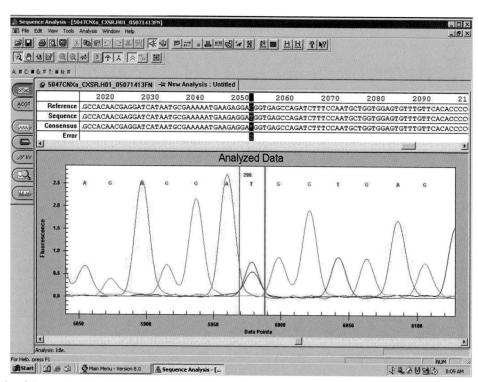

FIGURE 4.16 Example of the read-out from an automated capillary electrophoresis DNA sequencer. Each colored peak represents the capture of a fluorescently labeled DNA fragment by the laser detector of the instrument, in order of size fractionation. Each dideoxynucleotide in the reactions was labeled with a different-colored fluorophore, represented by the colors of the read-out and nucleotide calls by the instrument. Note the heterozygous single nucleotide.

missense mutation or merely a benign polymorphism. Certain physical attributes, such as its position within the protein and the biochemical nature of the amino acid substitution, may help one to deduce its impact, though there are many exceptions to these rules. Correlation with phenotype, and presence or absence of the change in other affected or unaffected family members or the population at large can also be of help. Ultimately, functional studies of the altered gene *in vitro* may be needed, but these are not applicable to the clinical laboratory.

Aside from the clinical interpretation, even the detection of nucleotide substitutions in the heterozygous state may be problematic. Ironically, this seems to be more of an issue with automated sequencers than with manual systems which depend more on the eye of the operator. A heterozygous substitution will appear on the automated read-out as two differently colored peaks superimposed at the same nucleotide position (Figure 4.16). For various technical reasons which are difficult to control, the two peaks may not be of the same intensity, even though they are supposedly present in equal amounts in the specimen. If one of the peaks is much smaller than 20–30% level, it may be ignored by the instrument's software and called out as homozygous for the other nucleotide. One way to guard against missing a heterozygous change in this way is to perform the sequencing in both directions (i.e., using opposite strands of the same fragment as template) and compare the read-outs. It is unlikely that the instrument would miss the change both times.

NEXT-GENERATION SEQUENCING

Over the last few years we have witnessed the advent of a powerful new variation on traditional Sanger sequencing, as revolutionary in its own way as was the advent of PCR in the mid-1980s. The new technology is variously referred to as highly parallel DNA sequencing or "next-generation" sequencing (NGS). Instead of targeting a single locus as sequencing template and sizing the resulting replicated fragments as in Sanger sequencing (producing about 200 nucleotides of sequence per run), NGS uses random primers to generate short "reads" across the entire genome in a "shotgun" manner but with tremendous redundancy. The addition of each successive nucleotide to the synthesized products is captured in real time by various sensitive detection methods (depending on the instrument) which distinguish the four nucleotides. A huge series of "snapshots" are taken at microsecond/nanoscale and then analyzed and aligned by the computer software to generate the completed sequence. These instruments are capable of sequencing billions of nucleotides per run, finally placing within practical reach the whole-genome sequence (3.3 billion nucleotides) of an individual. One can choose to analyze the entire genome or just the coding regions (called whole-exon sequencing), the latter accomplished by an initial exon-capture step.

Significantly, it is in the realm of hematopathology that some of the first medically relevant successes of NGS have emerged, with unexpected gene mutations and rearrangements being detected especially in acute myeloid leukemia. Of course, the downside of this approach is the acquisition of a tremendous amounts of data and novel sequence variants which then have to be assessed and interpreted (similar to what was discussed above for whole-gene sequencing but multiplied by many orders of magnitude). In addition, the detection of "off-target" mutations, such as for later-onset hereditary disorders, raises ethical challenges surrounding test reporting and informed consent.

Additional Resources

American College of Medical Genetics, Standards and Guidelines for Clinical Genetic Laboratories, Section E: Clinical Cytogenetics. Available at www.acmg.net

Cline J, Braman JC, Hogrefe HH: PCR fidelity of Pfu DNA polymerase and other thermostable DNA polymerases, *Nucl Acids Res* 24:3546–3551, 1996.

Davies JJ, Wilson IM, Lam WL: Array CGH technologies and their applications to cancer genomes, *Chromosome Res* 13:237–248, 2005.

Grody WW: Ethical issues raised by genetic testing with oligonucleotide microarrays, *Molec Biotechnol* 23:127–138, 2003.

International Society of Chromosome Nomenclature (ISCN) Shaffer LG, Tommerup N, editors: An International System for Human Cytogenetic Nomenclature Basel, 2009, Karger.

Macleod RA, Kaufmann M, Drexler HG: Cytogenetic analysis of cancer cell lines, *Methods Mol Biol* 731:57–78, 2011.

Maxam AM, Gilbert W: A new method for sequencing DNA, *Proc Natl Acad Sci USA* 74:560–564, 1977.

Nasedkina TV, Guseva NA, Gra OA, Mityaeva ON, Chudinov AV, Zasedatelev AS: Diagnostic microarrays in hematologic oncology: applications of high- and low-density arrays, *Mol Diagn Ther* 13:91–102, 2009.

Ravandi F, Kadkol SS, Ridgeway J, Bruno A, Dodge C, Lindgren V: Molecular identification of CBFbeta-MYH11 fusion transcripts in an AML M4Eo patient in the absence of inv16 or other abnormality by cytogenetic and FISH analyses—a rare occurrence, *Leukemia* 17:1907–1910, 2003.

Saiki RK, Gelfand DH, Stoffel S, et al. Primer-directed enzymatic amplification of DNA with a thermostable DNA polymerase, *Science* 239:487–491, 1988.

Sanger F, Nicklen S, Coulson AR: DNA sequencing with chain terminating inhibitors, *Proc Natl Acad Sci USA* 74:5463–5467, 1977.

Shaffer LG, Bejjani BA: Medical applications of array CGH and the transformation of clinical cytogenetics, *Cytogenet Genome Res* 115:303–309, 2005.

Shanafelt TD, Geyer SM, Kay NE: Prognosis at diagnosis: integrating molecular, biologic, insights into clinical practice for patients with CLL, *Blood* 103:1202–1210, 2004.

Singh RR, Cheung KJ, Horsman DE: Utility of array comparative genomic hybridization in cytogenetic analysis, *Methods Mol Biol* 730:219–234, 2011.

Stratton MR: Exploring the genomes of cancer cells: progress and promise, *Science* 331:1553–1558, 2011.

Ten-Bosch J, Grody WW: Keeping up with the next generation: massively parallel sequencing in clinical diagnostics, *J Molec Diagn* 10:484–492, 2008.

Welch JS, Westervelt P, Ding L, et al. Use of whole-genome sequencing to diagnose a cryptic fusion oncogene, *JAMA* 305:1577–1584, 2011.

Zordan A: Fluorescence in situ hybridization on formalin-fixed, paraffin-embedded tissue sections, *Methods Mol Biol* 730:189–202, 2011.

Additional Resources

Alter BP: Diagnosis, genetics, and management of inherited bone marrow failure syndromes, *Hematology Am Soc Hematol Educ Program*:29-39, 2007.

Bagby GC, Alter BP: Fanconi anemia, *Semin Hematol* 43:147-156, 2006.

Borowitz MJ, Craig FE, Digiuseppe JA, et al: Guidelines for the diagnosis and monitoring of paroxysmal nocturnal hemoglobinuria and related disorders by flow cytometry, *Cytometry Part B* 78B:211-230, 2010.

Galili N, Ravandi F, Palermo G, et al: Prevalence of paroxysmal nocturnal hemoglobinuria (PNH) cells in patients with myelodysplastic syndromes (MDS), aplastic anemia (AA), or other bone marrow failure (BMF) syndromes: interim results from the explore trial, *J Clin Oncol* 27:15s, 2009.

Hoffbrand AV, Pettit JE, Vyas P: *Color atlas of clinical hematology*, ed 4, Philadelphia, 2010, Mosby/Elsevier.

Ito E, Konno Y, Toki T, Terui K: Molecular pathogenesis in Diamond-Blackfan anemia, *Int J Hematol* 92:413-418, 2010.

Nakao S, Sugimori C, Yamazaki H: Clinical significance of a small population of paroxysmal nocturnal hemoglobinuria-type cells in the management of bone marrow failure, *Int J Hematol* 84:118-122, 2006.

Nishio N, Kojima S: Recent progress in dyskeratosis congenita, *Int J Hematol* 92:419-424, 2010.

Richards SJ, Whitby L, Cullen MJ, et al: Development and evaluation of a stabilized whole-blood preparation as a process control material for screening of paroxysmal nocturnal hemoglobinuria by Flow Cytometry, *Cytometry Part B* 76:47-55, 2009.

Shimamura A: Clinical approach to marrow failure, *Hematology Am Soc Hematol Educ Program*:329-337, 2009.

Sutherland DR, Kuek N, Azcona-Olivera J, et al: Use of a FLAER-based WBC assay in the primary screening of PNH clones, *Am J Clin Pathol* 132:564-572, 2009.

Young NS, Bacigalupo A, Marsh JC: Aplastic anemia: pathophysiology and treatment, *Biol Blood Marrow Transplant* 16(1 Suppl):S119-S125, 2010.

Zhang Y, Zhou X, Huang P: Fanconi anemia and ubiquitination, *J Genet Genomics* 34:573-580, 2007.

Myelodysplastic Syndromes/ Neoplasms—Overview

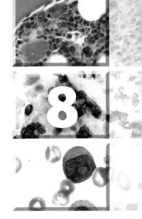

Myelodysplastic syndromes (MDS) or neoplasms are a group of hematologic disorders characterized by clonal expansion of defective hematopoietic stem cells leading to abnormal maturation and peripheral blood cytopenia. Clonality of the underlying marrow failure has been supported by various molecular and cytogenetic techniques, such as karyotyping (Table 8.1), X-chromosome inactivation studies, and fluorescence in situ hybridization (FISH) analysis.

Development of MDS appears to be a multistep process with an initial genetic insult to the multipotent stem cells, leading to the development of an abnormal clone, which in some cases may eventually lead to acute myeloid leukemia. Combined clinical and experimental data suggest that immune dysregulation mediated by expanded cytotoxic T-cell clones may result in an intrinsic stem cell defect and suppression of normal hematopoiesis. Several reports have proposed that there may be an etiologic link between MDS and T-LGL type of lymphoproliferation (Figure 8.1). The abnormal clone is the precursor of morphologically dysplastic and dysfunctional hematopoietic cells with a tendency to die prematurely. Excessive apoptosis (programmed cell death) of the hematopoietic precursors, particularly at the early stages, has been proposed as the primary mechanism for the bone marrow hypercellularity and peripheral cytopenia in patients with MDS. The overall transformation rate to acute leukemia depends on the subtype of MDS, the presence or absence of chromosomal aberrations, and types of these abnormalities.

The ineffective hematopoiesis in MDS may involve one or several hematopoietic lines, resulting in anemia, thrombocytopenia, granulocytopenia, or pancytopenia. The lymphoid lineage may occasionally be involved. In such cases, evolution of acute leukemia at the later stages may be of a lymphoblastic or biphenotypic type. The involvement of the lymphoid lineage in the myelodysplastic process may also cause immune dysfunction, such as cell-mediated suppression of bone marrow stem cells or deficient NK (natural killer) activities.

There are two major categories of MDS: primary (with no known cause), and secondary or therapy-related (usually post-chemotherapy or irradiation). The etiology and pathogenesis of the primary MDS are not clearly understood. Some familial clustering has been reported, but no causative germline mutations have been identified. Clinical features of MDS represent bone marrow failure and cytopenia. Anemia, thrombocytopenia, and/or neutropenia may lead to symptoms such as fatigue, pallor, infection, bruising, and/or bleeding. But some patients may be asymptomatic at diagnosis.

Primary or *de novo* MDS is usually a disease of the elderly and is uncommon under the age of 50 years. The median age of onset is between 60 and 70 years with an estimated annual incidence of about 3.5–10 per 100,000 in the general population. Therapy-related MDS (t-MDS) may arise at any age, usually 4–5 years after the initiation

Table 8.1

Chromosomal Abnormalities in Myelodysplastic Syndromes (MDS)

Type of Abnormality	Chromosome
Gain/loss of chromosomal material	monosomy 5,del(5)(q12 → 35) (various breakpoints within these bands),monosomy 7,del(7)(q11.2 → q36)(various breakpoints within these bands),+8,del(9q),+11,del(11q), del(12)(p11.2 → p13), del(13)(q12q14),+15, monosomy 17,del(17p),iso(17q), monosomy 20,del(20)(q11.2 → q13),Trisomy 21,Iso(Xq13),Loss of Y
Translocations/inversions	t(1;7)(q10;p10),t(3;21)(q26.2;q22), ins(3;3)(q26.2;q21q26),inv(3)(q21q26.2),t(6;9)(p23;q34), t(11;v)(q23;v);t(12;v)(p13;v)

of chemotherapy or radiation therapy. The percent of cytogenetic aberrations and risk of transformation to acute leukemia are significantly higher in t-MDS than in the primary MDS (Table 8.2). A small proportion of MDS patients, roughly 4–5%, may develop blast transformation in extramedullary sites (granulocytic sarcoma), particularly skin. Evolution of MDS to granulocytic sarcoma is associated with poor prognosis.

According to the International Prognostic Scoring System (IPSS) (Table 8.3), four distinctive risk groups are defined in the MDS patients: low risk, intermediate 1 risk, intermediate 2 risk, and high risk. The major parameters measured in the IPSS are percent blasts, nature of the chromosomal aberrations, and the extent of cytopenia (Table 8.3). These four groups show different survival rates and carry various risk levels for transformation to acute leukemia. For example, the overall survival time for the low-risk group is about 5.7 years, intermediate 1 risk group approximately 3.5 years, intermediate 2 risk group about 1.2 years, and high-risk group approximately 0.4 years.

At the present time, the only effective therapy available for MDS is hematopoietic stem cell transplantation. Other promising, newly developed therapeutic approaches include the utilization of DNA methyltransferase inhibitors, vascular endothelial growth inhibitors, and the use of thalidomide, arsenic trioxide, and anti-TNF-α.

Establishment of the diagnosis is based on a multidisciplinary approach including morphologic evaluation and utilization of the accessory laboratory tests, such as immunophenotyping, cytogenetic analysis, molecular genetic studies, and *in vitro* colony growth assays.

In addition to the clonal forms of MDS, non-clonal myelodysplastic changes have been observed in a variety of conditions, such as severe inflammatory states, viral infections, autoimmune disorders, megaloblastic anemia, exposure to arsenic, status post-chemotherapy, and endocrine dysfunctions.

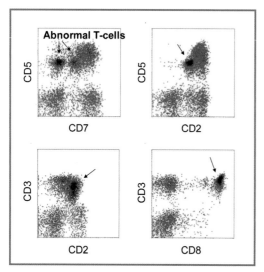

FIGURE 8.1 T-LGL-type CD8+ T-cells in MDS. Occasionally, abnormal CD8+ T-cells similar to those seen in T-LGL can be found in peripheral blood or bone marrow of MDS patients. The lymphocyte gate displays show two abnormal subsets of CD8+ T-cells expressing dim CD2, dim CD5, and dim to complete loss of CD7.

Morphologic Features—Overview

The ineffective hematopoiesis in MDS is demonstrated by mono- or pancytopenia and abnormal morphology in one or more hematopoietic lines in the bone marrow and peripheral blood.

- The bone marrow in patients with primary MDS is usually hyper- or normocellular, whereas patients with secondary MDS may show a variable marrow cellularity ranging from 5% to almost 100%. MDS bone marrow may show increased reticulin fibers with a higher frequency in the secondary MDS.
- The bone marrow biopsy sections usually show some degree of topographical alterations, such as the presence of erythroid

Table 8.2

Comparison of Chromosomal Aberrations between Primary and Secondary MDS

Abnormality	Primary MDS (%)	Secondary MDS (%)
−5 or del(5q)	10–20	40
−7 or del(7q)	10–15	30–50
Trisomy 8	15	10
Loss of 17p	3	10

Table 8.3

The International Prognostic Scoring System (IPSS) for MDS [1]

	Score				
Prognostic Variable	0	0.5	1.0	1.5	2.0
% Blasts	<5	5–10	–	11–20	21–30
Karyotype[2]	Good	Intermediate	Poor		
Cytopenia(s)[3]	0–1	2–3			

[1] Adapted from Greenberg P, Cox C, LeBeau MM, et al. International scoring system for evaluating prognosis in myelodysplastic syndromes. Blood 1997; 89: 2079–2088. Risk groups are scored as: 0, low risk; 0.5 to 2, intermediate risk; ≥2.5, high risk.
[2] Karyotype: Good = normal, −Y, del(5q),del(20q); Poor = complex (≥3 abnormalities), and chromosome 7 abnormalities; Intermediate = other abnormalities.
[3] Cytopenia: HB < 10 g/dL; Neutropenia < 1500/μL; Platelets < 100 k/μL.

clusters next to the bone trabeculae, loss of sinusoidal orientation of megakaryocytes and their placement next to bone, and centrally located aggregates of myeloid precursors (Figure 8.2).
- The morphologic appearance of aggregates of immature myeloid cells is referred to as "abnormal localization of immature precursors" (ALIP) (Figure 8.3). ALIP is defined as clusters of five or more myeloblasts and/or early immature myeloid cells, located in the marrow tissue away from bone trabeculae. More than three ALIP clusters per biopsy section are required to be diagnostically significant. Presence of ALIP, however, is not exclusive to MDS and has been observed in other hematologic conditions, such as myeloproliferative disorders and status post-bone marrow transplantation or chemotherapy.
- Signs of an inflammatory response, such as lymphoid aggregates, areas of edema, extravasation of erythrocytes, increased mast cells, plasma cells, and macrophages, disrupted sinusoids, and patchy or sometimes diffuse fibrosis are frequent findings. Occasionally, there is evidence of hemophagocytosis and/or presence of sea blue histiocytes.
- The accurate assessment of dysplastic cytologic features in blood and bone marrow smears depends on the quality of the smear preparations. Slides should be made from a fresh specimen and properly stained within preferably <2 h of exposure to anticoagulants. The recommended threshold for significant dysplasia is at least 10% for each hematopoietic lineage.

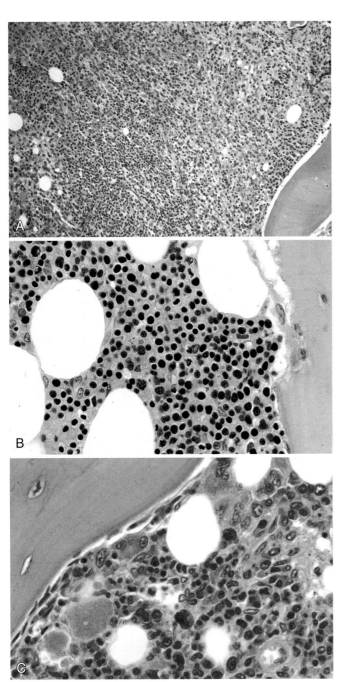

FIGURE 8.2 (A) Bone marrow section from a patient with MDS demonstrating hypercellularity with paratrabecular localization of the erythroid precursors and presence of a lymphoid aggregate. Higher power views demonstrate (B) paratrabecular localization of erythroid precursors and (C) megakaryocytes.

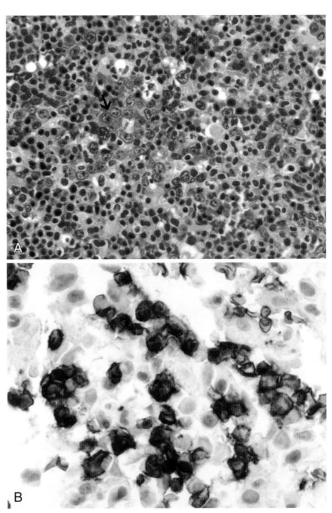

FIGURE 8.3 Hypercellular bone marrow biopsy section (A) showing several clusters of immature cells, referred to as "abnormal localization of immature precursors" (ALIP). The immunohistochemical stain (B) shows clusters of myeloperoxidase-positive immature cells.

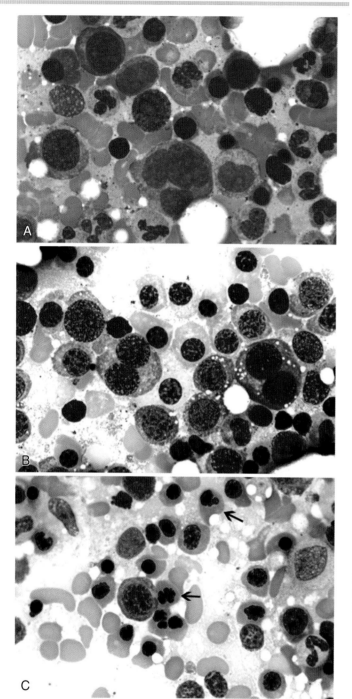

FIGURE 8.4 Bone marrow smear from a patient with refractory anemia (RA) showing dysplastic (A) multinucleated and (B) binucleated early erythroid precursors with vacuolated cytoplasm. (C) Dysplastic late erythroid precursors with irregular or multilobated nuclei (arrows).

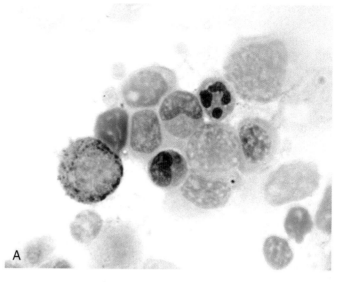

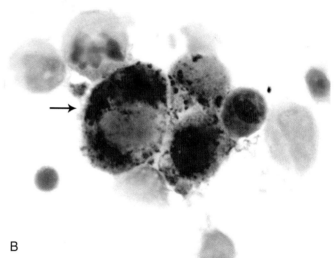

FIGURE 8.5 Bone marrow smear from a patient with RA showing coarse PAS-positve cytoplasmic granules in erythroid precursors (A and B) and a dysplastic megakaryocyte (B, arrow).

Ring sideroblasts are defined as nucleated red cells with five or more iron granules encircling at least one third of their nucleus.
- In biopsy sections, erythroid colonies may be seen next to the bone trabeculae.
- Blood smears show a wide variety of abnormal erythrocyte morphology, such as macro-ovalocytosis, microcytosis, schistocytosis, basophilic stippling, and the presence of teardrop shaped red blood cells and Howell–Jolly bodies (Figure 8.7). Occasionally, nucleated red blood cells are present.

DYSERYTHROPOIESIS

- Dysplastic features of the erythroid precursors in bone marrow smears include megaloblastic changes, irregular nuclear shape, nuclear fragmentation and budding, multinucleation, nuclear bridging, cytoplasmic vacuolization, poor hemoglobinization, presence of ring sideroblasts and periodic acid-Schiff (PAS)-positive cytoplasmic globules (Figures 8.4 to 8.6).

DYSGRANULOPOIESIS

- Granulocytic precursors may show abnormal variations in size, cytoplasmic granularity, and nuclear configuration (Figures 8.8 to 8.11). Abnormal staining of primary granules, hypergranularity or hypogranularity, is commonly observed in promyelocytes and myelocytes.

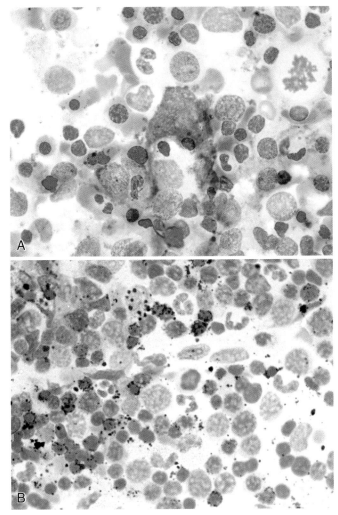

FIGURE 8.6 Bone marrow iron stains in MDS patients often show increased iron stores in histiocytes (A), or may demonstrate numerous ring sideroblasts (B).

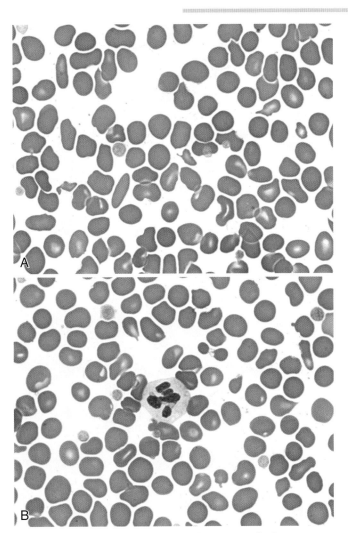

FIGURE 8.7 Peripheral blood smears showing marked anisopoykilocytosis with the presence of schistocytes and tear drop forms, and scattered large hypogranular platelets (A and B). A large hypogranular neutrophil is shown in (B).

- Irregular distribution of the cytoplasmic basophilia may be present and the cytoplasm in the perinuclear area may stain lighter than that in the periphery.
- The more mature granulocytic cells may depict a marked variation in size and decreased or absent secondary granules. There may be coarse basophilic (pseudo-Chediak–Higashi) granules.
- Nuclear hyposegmentation (pseudo-Pelger–Huet anomaly) or hypersegmentation, and other forms of abnormal nuclear morphology, such as ring (doughnut-shaped) nuclei, may be present.
- Eosinophils may be increased or show dysplastic changes, such as abnormal nuclear segmentation or abnormal granulation. Studies of bone marrow basophils on patients with MDS are very limited. However, basophilia has been observed in some MDS patients.

ABNORMAL MEGAKARYOCYTES AND PLATELETS

- Megakaryocytes may show multiple separated nuclei, hypo- or hyperlobated nuclei, vacuolated cytoplasm, and giant abnormal cytoplasmic granules. Mono- and binuclear megakaryocytes (micormegakaryocytes) are frequently seen (Figures 8.12 and 8.13). Sometimes, it is difficult to distinguish the mononuclear micromegakaryocytes from stromal cells or macrophages.
- In biopsy sections, megakaryocytes may appear in clusters or are located close to bone trabeculae.
- Blood smears show pleomorphic platelets with the presence of giant forms. They may show hypogranulation or abnormal granules. Megakaryocytic fragments, bare megakaryocytic nuclei, and sometimes micromegakaryocytes may be present.

Immunophenotype

FLOW CYTOMETRY

Although a large number of cases have been investigated in many studies, no consensus is available on evaluating MDS by flow cytometry. Several studies have proposed scoring systems according to the relative significance of

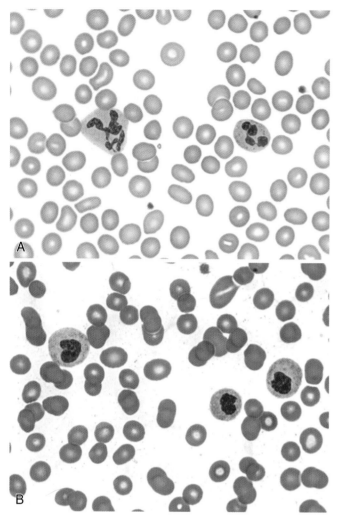

FIGURE 8.8 Peripheral blood smears showing a hypersegmented and hypogranular neutrophil (A), and several hyposegmented neutrophils (B).

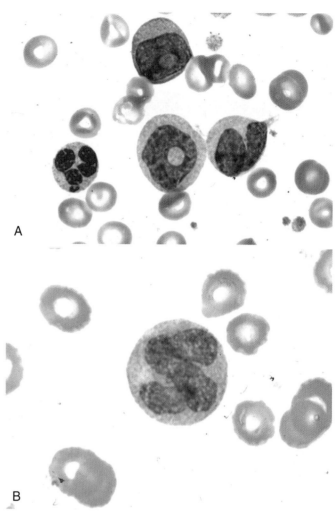

FIGURE 8.9 Peripheral blood smears showing a blast and an abnormal cell with doughnut-shaped nucleolus (A) and a large dysplastic monocyte (B).

various myeloid abnormalities, which are determined by comparison with normal myeloid cells using pattern recognition. However, there are significant variations among these scoring systems, primarily because the choice of antibodies and antibody combinations used in the MDS panel by different groups is highly variable. Until a standardized MDS panel is adopted, it is not possible to develop a consensus scoring system that can be universally applied.

Nevertheless, multiple studies have demonstrated that multiparametric immunophenotyping by flow cytometry (MIFC) plays an important role in aiding the diagnosis and classification of MDS in the context of multidisciplinary correlations. In addition, some studies have suggested that results of MIFC may have prognostic implications in certain risk groups of MDS patients.

MIFC is useful in diagnosis and classification of MDS in primarily two areas: (1) blast enumeration and (2) detecting immunophenotypic aberrancies and dysmaturation patterns of blasts, myeloid cells, monocytes, and erythroid precursors.

Blast Enumeration

Blast enumeration by MIFC should always be correlated with morphologic findings.

- Myeloblasts may not always be positive for CD34 and/or CD117.
- A combination of multiple gating strategies including back-gating is often necessary for an accurate enumeration of blasts.
- Compared with manual differential, where 500 cells are usually counted, MIFC is a much more powerful tool in estimating the number of blasts, since at least 20,000 events/cells are routinely analyzed.
- If the blast count is higher by MIFC than by manual differential, the results of flow cytometry cannot be simply dismissed and they should always be taken into account in the overall evaluation and classification of MDS.
- Flow cytometric analysis can sometimes underestimate the number of blasts due to lack of marrow particles and/or hemodilution of the sample.

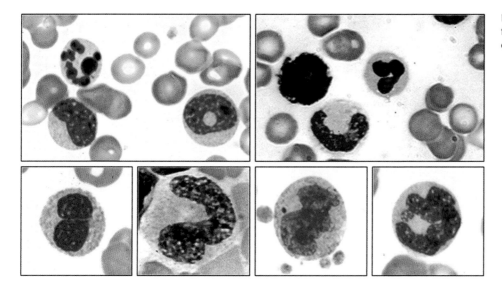

FIGURE 8.10 Peripheral blood smears from patients with MDS showing dysplastic neutrophils and monocytes.

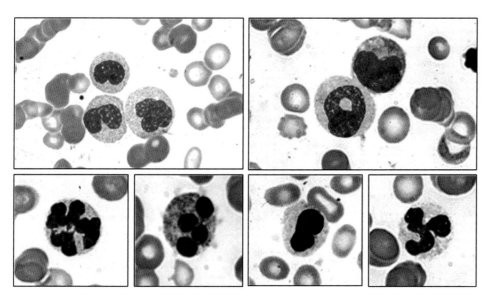

FIGURE 8.11 Peripheral blood smears from patients with MDS showing dysplastic neutrophils and monocytes.

Immunophenotypic Aberrancies and Dysmaturation Patterns

Proper recognition of immunophenotypic aberrancies and dysmaturation patterns relies on a solid understanding of the normal maturation patterns of myeloid, monocytic, and erythroid cells in a given MDS panel.

- Immunophenotypic aberrancies and dysmaturation patterns include:
  - cross-lineage aberrancies;
  - abnormal expression intensities;
  - asynchronous expressions;
  - abnormal location by CD45 gating;
  - abnormal clustering profiles.
- The single most significant finding in predicting MDS by MIFC, is the identification of abnormal myeloblasts, even though the blast count may be less than 5%. Common aberrancies of myeloblasts include abnormal location by CD45 gating, cross lineage aberrancies of CD5, CD7, CD19, and CD56, as well as asynchronous expression of CD13/CD15/CD33/CD34, plus expression of TdT (Figure 8.14).
- Dysmaturation patterns of myeloid cells are highly variable among different studies. Some of the more commonly described ones include abnormal CD45 gating, reduced side scatter, abnormal coexpression profile of CD11b/CD16/HLA-DR, loss of CD64, and loss of CD10 (Figure 8.15).
- Dysmaturation patterns of monocytes are not as well studied as those of granulocytic precursors. In general, reduced antigen expressions (e.g., CD11b, 13, CD14, CD15, CD36, and HLA-DR) may be present in addition to cross-lineage aberrancies (e.g., CD56 and CD2).
- Dysmaturation patterns of erythroid cells are rarely reported.

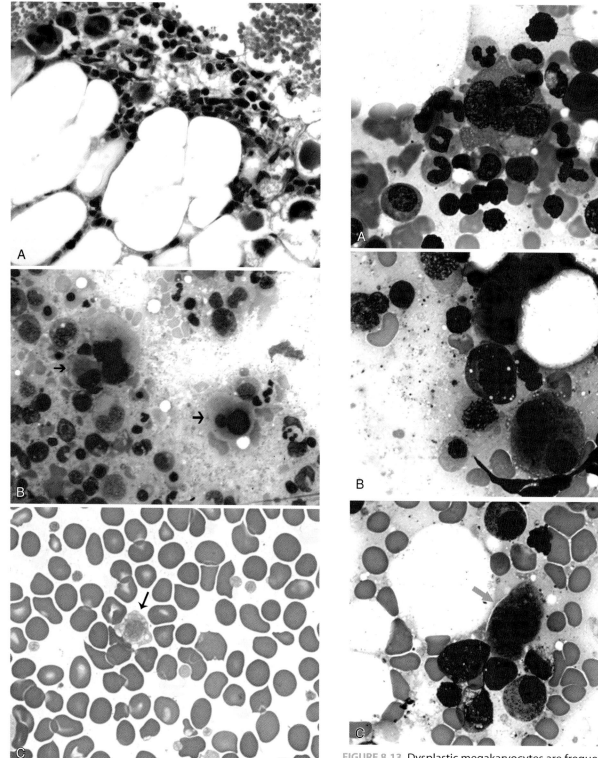

FIGURE 8.12 (A) Clot section and (B) bone marrow smear from a patient with 5q− syndrome, demonstrating numerous micromegakaryocytes. (C) Blood smear showing macrocytosis and giant platelets (arrow).

FIGURE 8.13 Dysplastic megakaryocytes are frequent bone marrow findings in patients with MDS: (A) a megakaryocyte with multiple separated nuclei; (B) a binucleated micromegakaryocyte (arrow); (C) a mononuclear micromegakaryocyte (arrow).

- In general, the higher the number of dysmaturation patterns, the more likely the findings are meaningful and specific.
- With the exception of abnormal myeloblasts, especially those with multiple aberrancies, dysmaturation patterns are not pathognomonic of MDS and can be observed in a wide variety of non-clonal conditions, such as viral infections, post-chemotherapy, heavy metal toxicity, and folate or vitamin B12 deficiencies.

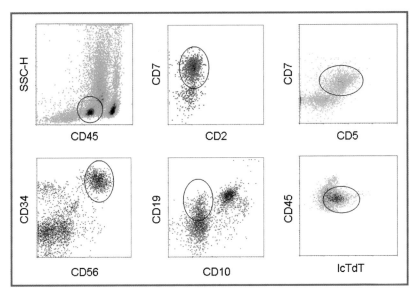

FIGURE 8.14 Common immunophenotypic aberrancies of abnormal myeloblasts in MDS. The open gate display by CD45 gating shows abnormal location of a discrete population of blasts. The blast-enriched gates demonstrate cross-lineage aberrancies of the abnormal myeloblasts, which include expression of CD7, CD5, CD56, CD19 (partial, dim), and TdT.

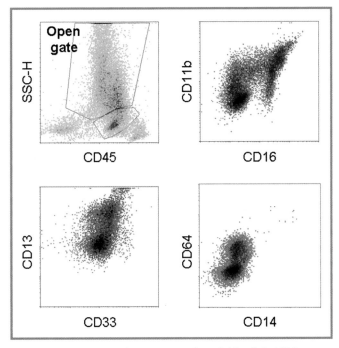

FIGURE 8.15 Dysmaturation patterns of myeloid cells in MDS. Compared with their normal counterparts, myeloid cells in MDS reveal a variable degree of dysmaturation patterns depending on the antibody panel used and disease subtypes. The open gate display by CD45 gating demonstrates an abnormal location/clustering profile of myeloid cells. The myeloid-enriched gates show abnormal coexpression profile of CD11b and CD16, with many myeloid cells lacking both antigens, abnormal clustering profile of CD13 and CD33, plus complete loss of CD64 in a major subset.

IMMUNOHISTOCHEMICAL STUDIES

Immunohistochemical stains on bone marrow biopsies of patients with suspected MDS may provide helpful information regarding the following matters.

- Evaluation of the topographical alterations and estimation of the M:E ratio by using monoclonal antibodies against hemoglobin and/or glycophorin A molecules for erythroid precursors and myeloperoxidase for myeloid precursors.
- Detection of clusters of immature cells (ALIP) and estimation of blast cell numbers by using blast-associated markers such as CD34 and CD117.
- Screening for the presence of micromegakaryocytes and topographical alterations of megakaryocytes by using monoclonal antibodies against CD31, CD61, and factor VIII.
- Evaluation of the monocytic component of the bone marrow by studying CD68 and lysozyme expressions.
- Immunohistochemical stains are also occasionally used to evaluate the nature of the lymphoid aggregates, which are frequently observed in bone marrow biopsy sections of MDS patients. Sometimes these aggregates are morphologically atypical or located adjacent to the bone trabeculae, raising the possibility of a lymphoproliferative disorder.

Molecular Studies

The diagnosis of MDS is typically based on clinical history and cell morphology in blood and bone marrow. Owing to one or more genetic defects in precursor stem cells, these are not yet sufficiently well characterized to become targets for molecular diagnostic study. The predominant molecular technique used in MDS diagnosis is FISH, often used as a supplement to the important standard karyotypic findings (see below). As already discussed, molecular techniques can be used to determine clonality of the underlying process, such as by characterization of polymorphic markers on the active and inactive X-chromosomes (applicable only in females). However, this is primarily a research tool, not directly relevant to the clinical diagnosis or management of a particular patient.

On the other hand, a number of individual point mutations, deletions, or epigenetic alterations have been observed in certain oncogenes, tumor suppressor genes, and signaling factors. The most frequent are the following.

- Point mutations in N-*ras* (found in 15–20% of MDS cases).
- Tandem duplication mutation in the *FLT3* gene (5%).
- Promoter methylation of *p15* (30–50%).
- Inactivating and deletion mutations of *p53* (5–10%). Inactivation of *p53* is of course a common occurrence in many types of cancer, but in MDS it appears to be an indicator of late stage or imminent progression to AML.
- *RUNX1/ RUNXT1* mutation screening is positive in 15% of MDS.
- Mutations of *NRAS, MLL-PTD, NPM1,* and *JAK2V617F,* and missense mutations in *RUNXT1* are also detected at low frequencies, confirming MDS diagnosis.
- Mutations in *NF1*, the causative gene of neurofibromatosis type 1, are found in some cases, which is of interest because the NF1 protein product is also involved in the RAS pathway.

Again, while these findings may have important implications for our understanding of MDS pathogenesis and progression, they are not necessary or routinely used as targets for diagnosis.

Cytogenetics

Chromosomal changes occur in about 30–50% within the diverse subtypes of MDS and are the strongest independent prognostic indicators. These changes range from balanced translocations to unbalanced karyotypes with numerical or structural gains and losses and to complex aberrant karyotypes (Table 8.1). The complex karyotypes are characterized by three or more chromosomal abnormalities (Figures 8.16 and 8.17) and show an extremely unfavorable prognosis.

Bone marrow is the tissue of choice for chromosomal studies in patients with MDS. Peripheral blood karyotyping has a higher failure rate than marrow studies and rarely adds useful information beyond that available from the marrow study.

There are no studies that suggest frequent assessment of chromosomal status in MDS for monitoring therapy. In special situations in which chromosomal information is desirable but a marrow examination is not required, such as measuring donor/host chimerism after stem cell transplantation, FISH testing (see below) may be preferable. Given the important role cytogenetics information plays in risk assessment, request for karyotype testing is an essential part of the initial evaluation.

Several recurrent and well-established cytogenetic changes have been described in MDS, and the detection of these changes can greatly facilitate diagnosis, prognosis, follow up, and treatment of patients. Although chromosomal abnormalities occur in almost half of de novo cases, aberrations are observed in up to 95% of secondary MDS. Most chromosomal defects in MDS are non-specific, and, with the exception of 5q−, none is specifically associated with any FAB or WHO subtype.

Chromosomal deletions are the most common defects observed in both de novo and secondary MDS. Deletions are generally interstitial, rather than terminal, and frequently occur in 5q, 7q, 20q, 11q, 12p, 13q, and 17p (Figures 8.18 to 8.21).

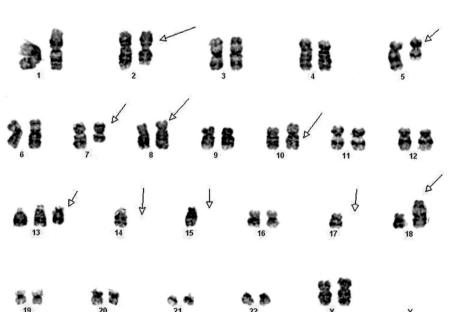

FIGURE 8.16 Complex karyotype: 44,XX,der(2)t(2;14)(q11.2;p11.2),del(5)(q13q33),del(7)(q22q34),add(8)(q24),add(10)(p13),+del(q12q14),−14,−15,−17,t(2;18)(q11.2;q21).

The most common of the chromosomal aberrations is represented by the chromosome 5q interstitial deletion (critical region reported to be between bands 5q31 and 5q33). Deletion 5q occurs either as an isolated abnormality or accompanied by additional chromosomal anomalies, and accounts for up to 28% of all cytogenetic aberrations in MDS, with an overall frequency of approximately 15%. Prognosis of patients with primary MDS and isolated del(5q) is favorable, but MDS patients with 5q deletion and presence of one or more additional

FIGURE 8.17 Complex karyotype: 43,XX,der(1;12)(q21;p11.2),−3,−5,−7,+8, −12,−15,−17,−18,+3mars.

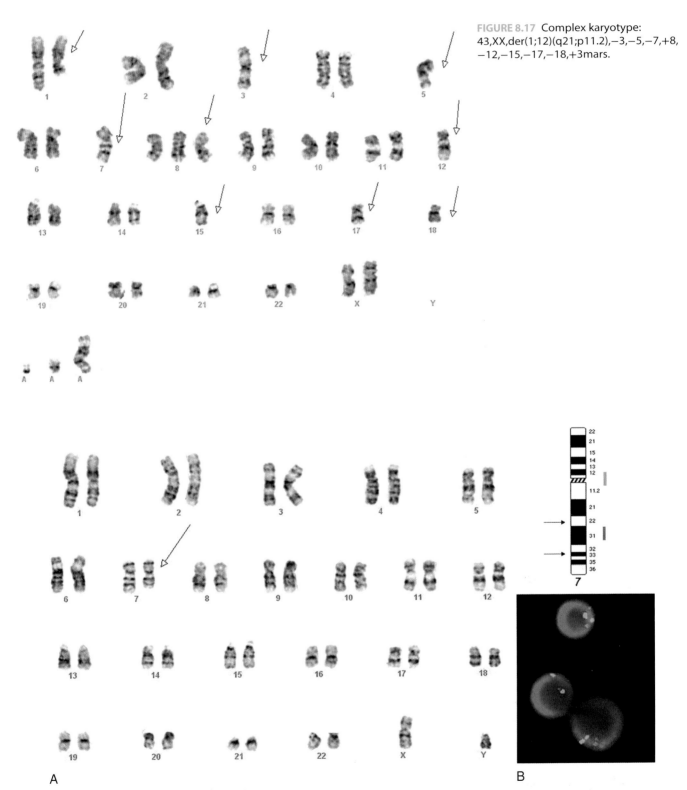

FIGURE 8.18 (A) G-banded karyotype showing a deletion of 7q. (B) An ideogram of 7q with the two critical regions of deletion (arrows), and a panel of interphase cells exhibiting 7q deletion by FISH.

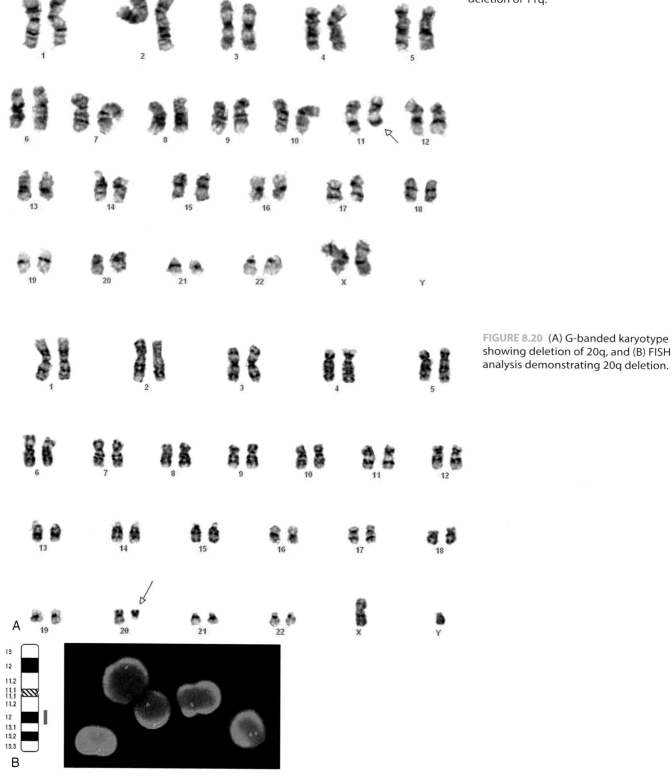

FIGURE 8.19 G-banded karyotype with a deletion of 11q.

FIGURE 8.20 (A) G-banded karyotype showing deletion of 20q, and (B) FISH analysis demonstrating 20q deletion.

chromosomal abnormalities show an aggressive clinical course.

Monosomy 7/del(7q) has been observed in all MDS subtypes, though it is much more common in advanced forms (Figures 8.18 and 8.22). 7q-/-7 occurs as a sole chromosomal abnormality in 1% of cases. it is more common in secondary MDS, seen in up to 60% of the patients, and is therefore considered a secondary event in the pathogenesis of the disease. Unlike the 5q−, there are multiple commonly deleted regions in chromosome 7q including

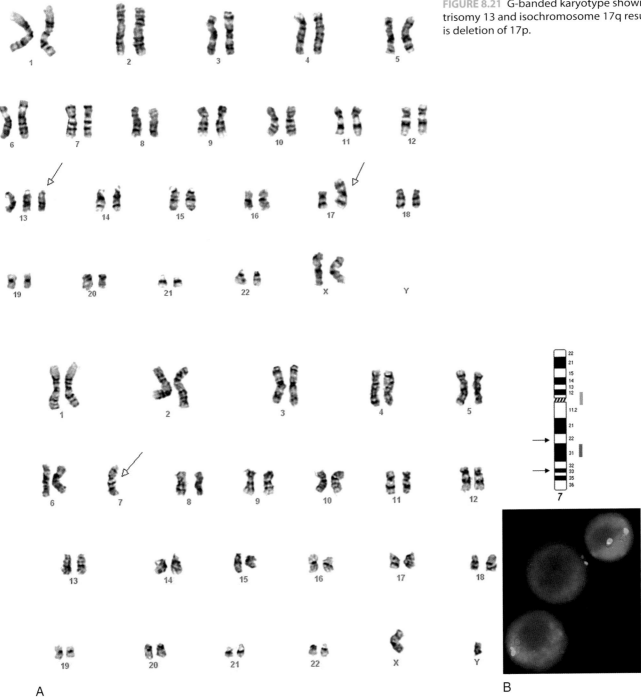

FIGURE 8.21 G-banded karyotype showing trisomy 13 and isochromosome 17q resulting is deletion of 17p.

FIGURE 8.22 (A) G-banded karyotype with monosomy 7. (B) The corresponding FISH studies.

7q22 and several in the region 7q31–7q35. Monosomy 7 is the most common chromosomal defect in bone marrow of patients with constitutional syndromes (e.g., Fanconi's anemia, type I neurofibromatosis, and severe congenital neutropenia) that predispose to myeloid disorders. Also, a recently described pediatric monosomy 7 syndrome presents with hepatosplenomegaly, leukocytosis, thrombocytopenia, male predominance, and an unfavorable outcome. Patients harboring deletions in 7q31 to 7q36 regions have an inferior response to chemotherapy and shorter survival than those with deletions in the 7q22 region.

Deletion of long arm of chromosome 20 occurs in 5% of de novo and 7% of secondary MDS. This incidence might be an underestimation, since monosomy 20 and unbalanced translocations involving chromosome 20 occur as frequently as deletions. Although the critical region seems to be 20q11.2 to 20q12, deletions are rather large and involve most of the long arm of chromosome

20. Patients with del(20q) as a sole abnormality are in the low-risk MDS categories—refractory anemia (RA) and refractory anemia with ring sideroblasts (RARS) (see Chapter 9)—whereas those presenting with this deletion as a part of a complex karyotype (three or more abnormalities in karyotypes) have a poor prognosis. Also, there is evidence that late emergence of 20q− during disease progression is an indicator of genomic instability and may be associated with poor outcome. The smallest Critically Deleted Region in the 20q deletion contains 19 genes. But none of these have been found to be mutated or otherwise directly implicated in myeloid diseases to date. Occasionally a variant of the deleted 20q is observed. Here a monosomy of chromosome 20 with small metacentric marker chromosome: 46,XX or XY,−20,+mar is seen. The marker often represents an isoderivative of chromosome 20 [ider(20)(q10)del(20)(q11q13)], a variant of del 20q (Figure 8.23) and these patients have an unfavorable prognosis.

Deletion of the short arm of chromosome 17 encompasses not only simple deletions, but also unbalanced translocations, iso17q (Figure 8.21), and (rarer) monosomy 17. Del(17p) is rare in de novo MDS (~7%), but occurs more frequently in secondary MDS. Despite its heterogeneity, all of the above-mentioned aberrations of the short arm of chromosome 17 lead to the loss of one p53 allele. Mutation or submicroscopic deletion of the other p53 allele occurs in 70% of the patients and causes inactivation of the TP53 gene.

Isolated loss of the chromosome Y is observed in about 10% of MDS patients. It also occurs in about 7% of elderly men without any hematological disorder. Therefore, MDS diagnosis cannot be based on the presence of -Y alone. However, loss of Y in the majority of cells can be a useful marker of disease burden, but likely has no bearing on the mechanism of disease. When biological and clinical parameters point to an MDS diagnosis, loss of the Y chromosome identifies patients with a favorable clinical outcome.

Interstitial deletions or balanced translocations involving band 12p13 are found in about 5% of patients with RAEB, mostly involving the ETV6 gene. These patients usually belong to an intermediate-risk cytogenetic category for MDS. However, recent studies suggest that 12p13 aberrations signify a clinical outcome similar to that of patients included within the low-risk category.

A normal karyotype, monosomies, trisomies, and unbalanced translocations are the next most common aberrations, occurring in 15% of patients. The most common monosomies in MDS involve chromosomes 5, 7, and Y. Deletions and monosomies cause loss of one allele of a tumor suppressor gene with the subsequent submicroscopic deletion of the second allele on the homologous chromosome. This recessive mechanism inactivates the cell's ability to control the cell cycle, DNA repair, and apoptosis.

Several other chromosomal aberrations are observed in MDS, but are not specific to the disease. Trisomy 8 occurs in 10% of all MDS cases, but can be found in other clonal

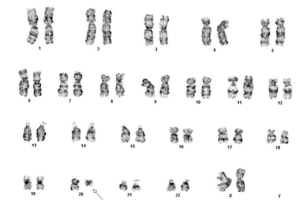

FIGURE 8.23 A MDS karyotype with an isochromosome of 20q(arrow) [46,XX,ider(20)(q10)del(20)(q11q13)]. This derivative chromosome is a variant of del(20q) but associated with a poorer prognosis.

hematological disorders. Trisomy 8 is more often associated with RARS and RAEB (Figure 8.24). Chromosome 3 rearrangements, typically translocations or inversions of the long arm, occur in 2–5% of patients with MDS (also in AML). Chromosome 3 changes are frequently associated with −7/7q and 5q−, and are associated with a short survival and a poor response to chemotherapy. Aberrations within 11q23 (the MLL gene locus) are found in 5% of MDS patients. Further karyotypic defects (e.g., rearrangements of long arm of chromosome 3) occur in 5–10% of de novo MDS cases.

Using cytogenetic abnormalities, MDS patients have been divided into three prognostic categories. Patients in the first, low-risk category exhibit a normal karyotype, deletion of long arm of chromosome 5 or 11 or deletion of 12p as a sole abnormality, or harbor an isolated deletion of the long arm of chromosome 20. Patients with trisomy 8 are categorized as an intermediate-risk group. Finally, the presence of complex karyotypes, monosomy 7, deletion of the short arm of chromosome 17, rearrangements involving chromosome 3, indicate a high-risk group of MDS patients.

About 12% of patients with MDS with normal cytogenetics progress to acute myeloid leukemia (AML), whereas 50% of those with chromosomal changes will progress to AML. Generally, therapy-related MDS (MDS-t) are much more aggressive clinically than primary MDS, and this characteristic is reflected in the karyotypes. At least 80% of patients with t-MDS have chromosomally abnormal clones in the marrow, and in a vast majority of cases these clones contain multiple abnormalities. It is has also been established that even prognostically unfavorable aberrations can be observed in cases where morphology and immunophenotyping performed by highly experienced hematologists fail to diagnose MDS, emphasizing the value of karyotyping even when MDS is not detectable with cytomorphology or immunophenotyping but is suspected from the clinical point of view, and underlines the need for cytogenetic analysis in all cases suspected of MDS.

Although balanced translocations are relatively common aberrations in acute myeloid leukemias, they are

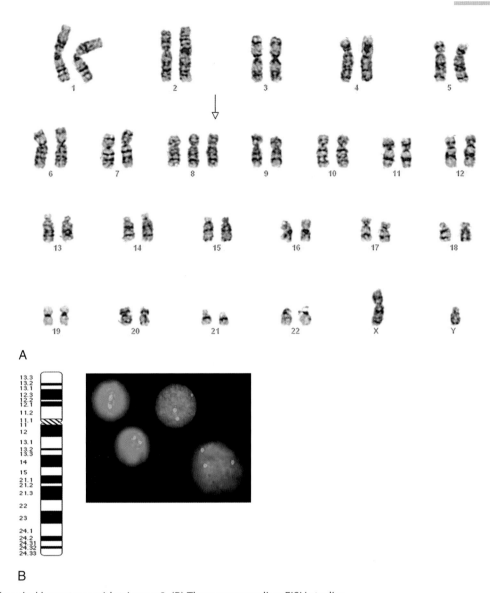

FIGURE 8.24 (A) G-banded karyotype with trisomy 8. (B) The corresponding FISH studies.

very rare in MDS, and very few such rearrangements are observed in MDS: e.g., involving 3q26 (corresponding to the zinc finger DNA binding protein, MDS1-EVI1), 3q25.1 (the p53 regulator, *MLF1*), and 1p36 (*PRDM16*, another zinc finger transcription factor). Other genes that are involved in rarer translocations include RUNX1, ETV6, MEL1, NUP98, and IER3. Many translocations in MDS occur in the setting of a more widely disturbed cytogenetic profile, making it unclear if these represent disease-modifying changes or incidental rearrangements in a highly unstable genome. A balanced translocation between chromosomes 6p and 9q is often considered a presumptive evidence of MDS in patients with unexplained anemia.

FISH AND OTHER TECHNOLOGIES

Cytogenetic data can be obtained with FISH from nondividing or terminally differentiated cells, or from poor samples (fibrotic marrow) that contain too few cells for routine karyotyping studies. Detection of chromosomal aberrations within the interphase nuclei can be achieved by hybridizing with an appropriate selection of probes (Figures 8.18B, 8.20B, 8.22B, and 8.24B). The technique permits the direct correlation of cytogenetic and cytologic features (for example, trisomy 8 in a hypogranular neutrophil), which enables cytologists to differentiate malignant from benign conditions in equivocal cases.

A typical MDS FISH panel includes probes specific for chromosomes 5/5q, 7/7q, centromere 8, and 20q. FISH analyses in MDS occasionally reveal cryptic cytogenetic abnormalities that are not recognized on metaphase analysis. However, conventional cytogenetics and FISH should be viewed as complementary because there are also abnormalities better detected by conventional cytogenetics than FISH. Reviews have shown that concurrent FISH studies confirmed the G-banding cytogenetic findings in >99% of cases, but also detected cytogenetic abnormalities in 25% of cases in which the conventional cytogenetics study failed.

An advantage of FISH is that one can combine this method with cytology and/or immunohistochemistry to examine the cytogenetic pattern of specific cell populations to monitor the effects of therapy and to detect minimal residual disease. Most of the studies using FISH for the identification of lineage involvement in MDS indicate that, in most cases, the pluripotent stem cell is not affected, since the lymphoid cells usually do not contain the chromosomal abnormality.

Some of the MDS patients may have very complex chromosomal rearrangements, which corresponds to short survival times and high rates of progression to AML. However, it is acknowledged that, because of the limitations of conventional chromosome banding techniques, the identification of the abnormalities remain unresolved. Molecular cytogenetics techniques such as multicolor FISH (M-FISH), and spectral karyotyping (SKY), may allow for the comprehensive evaluation of these karyotypes. With these methods, it is possible to analyze the origin of marker chromosomes, reveal cryptic rearrangements, and determine recurrent breakpoints and the structure of derivative chromosomes.

Single nucleotide polymorphism (SNP) array analysis showed that about 10–15% of MDS patients with apparently normal karyotypes had regions of DNA (such as on the long arm of chromosome 5) derived from only one parent, a phenomenon called Uniparental Disomy or UPD. Such high resolution genomic analysis also exhibited copy number changes in the abnormal cells such as duplications and deletions, which often go undetected by standard cytogenetics or FISH analyses. The presence of a high percentage of UPD in MDS suggests that individuals who are born with constitutional UPD may have an increased risk for developing MDS.

Additional Resources

Bejar R, Ebert BL: The genetic basis of myelodysplastic syndromes, *Hematol Oncol Clin North Am* 24:295–315, 2010.

Cazzola M, Malcovati L: Myelodysplastic syndromes – coping with ineffective hematopoiesis, *N Engl J Med* 352:536–538, 2005.

Cazzola M, Malcovati L: Prognostic classification and risk assessment in myelodysplastic syndromes, *Hematol Oncol Clin North Am* 24:459–468, 2010.

Chu SC, Wang TF, Li CC, et al: Flow cytometric scoring system as a diagnostic and prognostic tool in myelodysplastic syndromes, *Leuk Res* 35:868–873, 2011.

Della Porta MG, Lanza F, Del Vecchio L: Flow cytometry immunophenotyping for the evaluation of bone marrow dysplasia, *Cytometry B Clin Cytom* 80:201–211, 2011.

Greenberg P, Cox C, LeBeau M, et al: International Scoring System for Evaluating Prognosis in Myelodysplastic Syndromes, *Blood* 89:2079–2088, 1997.

Huh YO, Medeiros LJ, Ravandi F, et al: T-cell large granular lymphocyte leukemia associated with myelodysplastic syndrome: a clinicopathologic study of nine cases, *Am J Clin Pathol* 131:347–356, 2009.

Kern W, Haferlach C, Schnittger S, et al: Clinical utility of multiparameter flow cytometry in the diagnosis of 1013 patients with suspected myelodysplastic syndrome: Correlation to cytomorphology, cytogenetics, and clinical data, *Cancer* 116:4549–4563, 2010.

Kussick SJ, Fromm JR, Rossini A, et al: Four-color flow cytometry shows strong concordance with bone marrow morphology and cytogenetics in the evaluation for myelodysplasia, *Am J Clin Pathol* 124:170–181, 2005.

Orazi A, Czader MB: Myelodysplastic syndromes, *Am J Clin Pathol* 132:290–305, 2009.

Orfao A, Ortuno F, de Santiago M, et al: Immunophenotyping of acute leukemias and myelodysplastic syndromes, *Cytometry* 58A:62–71, 2004.

Saunthararajah Y, Molldrem JL, Rivera M, et al: Coincident myelodysplastic syndrome and T-cell large granular lymphocytic disease: clinical and pathophysiological features, *Br J Haematol* 112:195–200, 2001.

Stetler-Stevenson M, Arthur DC, Jabbour N, et al: Diagnostic utility of flow cytometric immunophenotyping in myelodysplastic syndrome, *Blood* 98:979–987, 2001.

Tormo M, Marugán I, Calabuig M: Myelodysplastic syndromes: an update on molecular pathology, *Clin Transl Oncol* 12:652–661, 2010.

van de Loosdrecht AA, Alhan C, Bene MC, et al: Standardization of flow cytometry in myelodysplastic syndromes: report from the First European LeukemiaNet Working Conference on Flow Cytometry in Myelodysplastic Syndromes, *Haematologica* 94:1124–1134, 2009.

van de Loosdrecht AA, Westers TM, Westra AH, et al: Identification of distinct prognostic subgroups in low- and intermediate-1-risk myelodysplastic syndromes by flow cytometry, *Blood* 111:1067–1077, 2008.

Vardiman JW, Thiele J, Arber DA, et al: The 2008 Revision of the WHO Classification of Myeloid Neoplasms and Acute Leukemia: Rationale and important changes, *Blood* 114:937–951, 2009.

Myelodysplastic Syndromes/Neoplasms—Classification

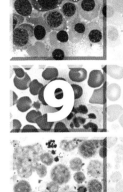

Classification of MDS by the World Health Organization (WHO) in 2008 includes the following categories (Table 9.1):

1. Refractory cytopenia with unilineage dysplasia (RCUD)
2. Refractory anemia with ring sideroblasts (RARS)
3. Refractory cytopenia with multilineage dysplasia (RCMD)
4. Refractory anemia with excess blasts-1 (RAEB-1)
5. Refractory anemia with excess blasts-2 (RAEB-2)
6. MDS associated with isolated del(5q)
7. MDS unclassifiable.

This morphologic classification correlates with clinical behavior, frequency and type of cytogenetic abnormalities, and rate of transformation to acute leukemia (Tables 9.2 and 9.3).

Refractory Cytopenia with Unilineage Dysplasia

Refractory cytopenia with unilineage dysplasia (RCUD) is a low-risk MDS which includes three subclasses: (1) refractory anemia (RA), (2) refractory neutropenia (RN), and (3) refractory thrombocytopenia (RT). The recommended levels of cytopenia are: less than 10 g/L hemoglobin, less than 1500/μL neutrophil count, and less than 100,000/μL platelet count. However, a diagnosis of RCUD could be made with levels above the recommended threshold when there is significant unilineage dysplasia and/or evidence of characteristic cytogenetic aberrations.

RCUD accounts for about 10–20% of MDS cases, with RA being the most frequent subtype. The vast majority of patients with RCUD fall into the low or intermediate risk groups according to the International Prognostic Scoring System (IPSS) and have good to intermediate cytogenetic profiles. The 5-year progression rate to AML is about 2%.

Table 9.1

WHO Classification of Myelodysplastic Syndromes [1]

Type	Blood Findings	Bone Marrow Findings
Refractory cytopenia with unilineage dysplasia (RCUD)	Cytopenia with dysplasia; No or rare blasts	Unilineage dysplasia, >10% of the cells; <5% blasts; <15% ring sideroblasts
Refractory anemia with ring sideroblasts (RARS)	Anemia, abnormal erythrocytes; No blasts	Dysplastic erythropoiesis; ≥15% ring sideroblasts; <5% blasts
Refractory cytopenia with multilineage dysplasia (RCMD)	Bicytopenia or pancytopenia; No or rare blasts; No Auer rods; No absolute monocytosis	Multilineage dysplasia; <5% blasts; No Auer rods; ±15% ring sideroblasts
Refractory anemia with excess blasts-1 (RAEB-1)	Cytopenia(s); <5% blasts; No Auer rods; No absolute monocytosis	Unilineage or multilineage dysplasia; 5–9% blasts; No Auer rods
Refractory anemia with excess blasts-2 (RAEB-2)	Cytopenia(s); 5–19% blasts; Auer rods ±; No absolute monocytosis	Unilineage or multilineage dysplasia 10–19% blasts; Auer rods±
MDS associated with isolated del(5q)	Anemia; Normal or increased platelets; No or rare blasts	Micromegakaryocytes; Isolated del(5q); <5% blasts; No Auer rods
Myelodysplastic syndrome-Unclassified (MDS-U)	Cytopenia; No or rare blasts	<10% dysplasia in one or more lineages; <5% blasts; No Auer rods; Evidence of cytogenetic abnormalities

[1] Adapted from Swerdlow SH, Campo E, Harris NL, et al. WHO Classification of Tumours of Haematopoietic and Lymphoid Tissues. IARC Press, Lyon, 2008.

Table 9.2
Rate of Chromosomal Aberrations in MDS Subclasses

Subgroup	Common Alterations	Frequency (%)
RA (or RCUD)	−5/d(5q)	30
	−7/del(7q)	10–15
	+8	20
RARS	+8	30
	−5/(5q)	25
	del(11q)	10
	del(20q)	10–15
RAEB	−5/del(5q)	35–40
	−7/)del(7q)	30–35
	+8	20

RA, refractory anemia; RARS, refractory anemia with ring sideroblasts; RAEB, refractory anemia with excess blasts; RCUD, refractory cytopenia with unilineage dysplasia.

Table 9.3
Survival and Rate of Transformation of MDS to Acute Leukemia According to the WHO Subtypes [1]

MDS Subtype	Median Survival (years)	Evolution to AML (%)	Stratification by IPSS Score (%)
RA (or RCUD)	5.7	7.5	Low (57) Intermediate 1(33) Intermediate 2 (10) High (0)
RARS	5.7	1.4	Low (96) Intermediate 1(4) Intermediate 2 (0) High (0)
5q− Syndrome	9.7	8	Low (61) Intermediate 1 (30) Intermediate 2 (9) High (0)
RCMD	2.7	10	Low (55) Intermediate 1 (40) Intermediate 2 (5) High (0)
RCMD-RS	2.6	13	Low (56) Intermediate 1 (36) Intermediate 2 (8) High (0)
RAEB-1	1.5	21	Low (0) Intermediate 1 (33) Intermediate 2 (55) High (12)
RAEB-2	0.8	34.5	Low (0) Intermediate 1 (0) Intermediate 2 (25) High (73)

RA, refractory anemia; RAEB, refractory anemia with excess blasts; RARS, refractory anemia with ring sideroblasts; RCMD, refractory cytopenia with multilineage dysplasia; RCMD-RS, refractory cytopenia with multilineage dysplasia and ring sideroblasts; RCUD, refractory cytopenia with unilineage dysplasia.

[1] Adapted from Germing U, Gattermann N, Strupp C, Aivado M, Aul C. (2000). Validation of the WHO proposals for a new classification of primary myelodysplastic syndromes: a retrospective analysis of 1600 patients. *Leuk Res* **24**, 983–92.

MORPHOLOGY

Refractory Anemia (RA)

- RA is characterized by anemia, dyserythropoiesis in >10% of erythroid precursors, and absent or very low percentage of blasts in bone marrow (<5%) and peripheral blood (≤1%) (Figure 9.1).
- Bone marrow is often hypercellular and frequently shows erythroid preponderance. The degree of dysplasia in the erythroid precursors varies and may include megaloblastic changes, multinucleation, nuclear bridging, nuclear

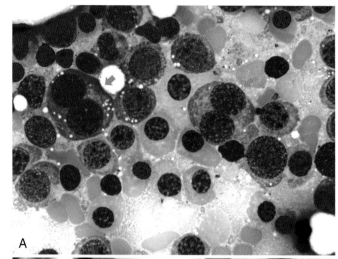
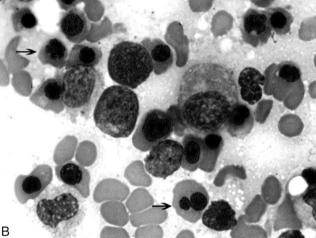

FIGURE 9.1 REFRACTORY ANEMIA. Bone marrow smears demonstrating dysplastic early binucleated erythroid precursors (A, arrows) and dysplastic late erythroid precursors with irregular or lobated nuclei (B, arrows).

fragmentation or budding, cytoplasmic vacuolization, and abnormal hemoglobinization.
- Iron stores are increased. Ring sideroblasts, if present, account for <15% of the nucleated red blood cells.
- In the peripheral blood, red blood cells often show some degree of anisopoikilocytosis with reduced polychromasia (reticulocytes).
- The granulocytic and megakaryocytic lines are either normal or show minimal dysplastic changes.

Refractory Neutropenia (RN)

- RN is characterized by ≥10% dysplastic neutrophils in peripheral blood or bone marrow (Figure 9.2).
- Dysplastic changes include hyper- or hyposegmentation of the neutrophils and/or hypogranulation.
- Lack of significant dysplasia in other lineages.

Refractory Thrombocytopenia (RT)

- RT is characterized by ≥10% dysplastic megakaryocytes in at least 30 observed megakaryocytes.
- Mononuclear, hypolobated, binucleated micromegakaryocytes and megakaryocytes with separated nuclei are frequent findings (Figure 9.3).
- Lack of significant dysplasia in other lineages.

REFRACTORY CYTOPENIA WITH UNILINEAGE DYSPLASIA **131**

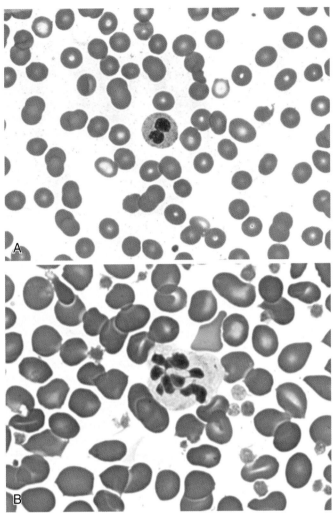

FIGURE 9.2 **REFRACTORY NEUTROPENIA.** Peripheral blood smears demonstrating a hypolobated neutrophil (A) and a large hypersegmented, hypogranular neutrophil (B).

According to the WHO recommendations, in order to establish a diagnosis of RCUD, all other causes of erythroid, neutrophilic, and megakaryocytic dysplasias should be excluded. These include drug and chemical exposure, immunologic disorders, viral infections, congenital abnormalities, and vitamin deficiencies. Also, there should be an observation period of at least 6 months if cytogenetic and molecular studies show no evidence of clonal disorder. Development of multilineage dysplasia and/or an increase in the percent of blast cells are suggestive of progression of the disease into a more aggressive type of MDS.

IMMUNOPHENOTYPE (FIGURE 9.4)

- Blast enumeration of bone marrow by flow cytometry is less than 5% and needs to be correlated with morphologic findings.
- There may be early changes with low number of dysmaturation patterns in myelomonocytic cells.
- Identification of abnormal myeloblasts (<5%) expressing discrete aberrancies is the single most significant finding in evaluating dysmaturation patterns of myeloid cells.

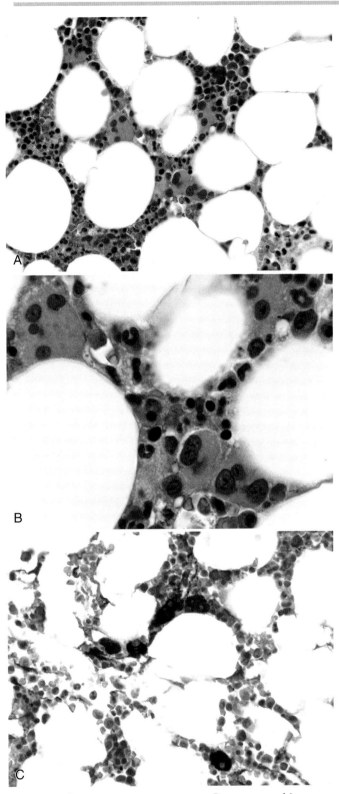

FIGURE 9.3 **REFRACTORY THROMBOCYTOPENIA.** Bone marrow biopsy section demonstrating numerous micromegakaryocytes (A, low power; B, high power). (C) The micromegakaryocytes are highlighted by immunohistochemical stain for factor VIII.

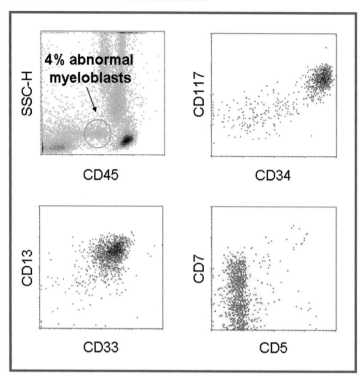

FIGURE 9.4 Abnormal myeloblasts detected at early stages of MDS by flow cytometric studies. Although myeloblasts account for less than 5% of the total, they reveal discrete immunophenotypic aberrancies with expression of very dim CD45, bright CD34, and dim CD7.

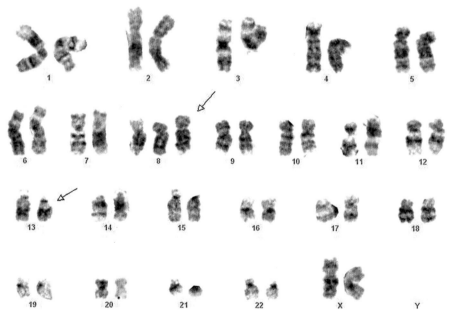

FIGURE 9.5 G-banded karyotype showing deletion 13q and trisomy 8.

MOLECULAR AND CYTOGENETIC STUDIES

For detailed molecular and cytogenetic studies in MDS, see Chapter 8. The most frequent cytogenetic abnormalities are partial deletions or loss of chromosomes 5, 7, and 20 and trisomy 8 (Figure 9.5).

Refractory Anemia with Ring Sideroblasts

Refractory anemia with ring sideroblasts (RARS) is a low risk MDS characterized by anemia, dyserythropoiesis, and

REFRACTORY ANEMIA WITH RING SIDEROBLASTS 133

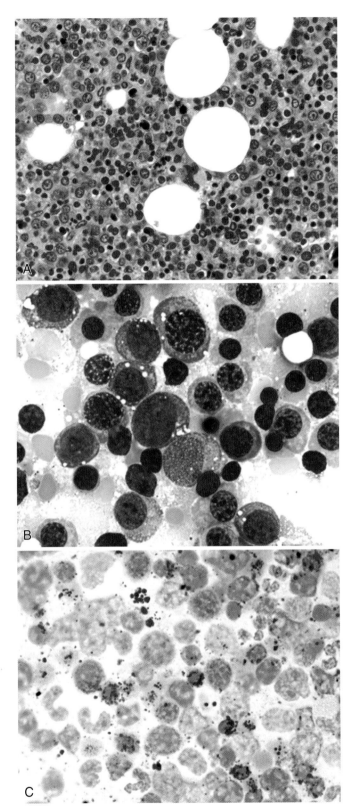

FIGURE 9.6 **REFRACTORY ANEMIA WITH RING SIDEROBLASTS.**
(A) Bone marrow biopsy reveals hypercellularity and erythroid preponderance. (B) Bone marrow smear shows erythroid preponderance and dysplastic vacuolated early erythroid precursors. (C) Numerous ring sideroblasts are present by iron stain.

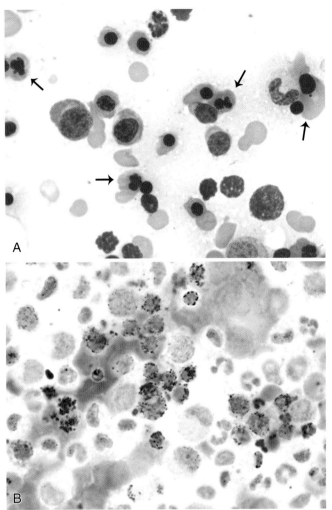

FIGURE 9.7 **REFRACTORY ANEMIA WITH RING SIDEROBLASTS.** Bone marrow smears reveal numerous dysplastic intermediate and late erythroid precursors (A, arrows) and numerous ring sideroblasts by iron stain (B).

presence of 15% or more ring sideroblasts in the bone marrow nucleated red blood cells. Anemia is usually moderate and may be associated with clinical symptoms such as fatigue and pallor. RARS accounts for about 5–10% of MDS cases, and the median age ranges from 65 to 70 years. It has an indolent clinical course with 1–2% chance of evolving to AML

MORPHOLOGY

- RARS is morphologically similar to RA except for the presence of ≥15% ring sideroblasts. Ring sideroblasts are defined as nucleated red cells with five or more iron granules encircling at least one third of their nucleus (Figure 9.6). The iron granules are precipitated in the mitochondria, which are usually located around the nucleus.
- Evidence of dyserythropoiesis in >10% of erythroid precursors, and absent or very low percentage of blasts in bone marrow (<5%) and peripheral blood (<1%) (Figures 9.6 and 9.7).

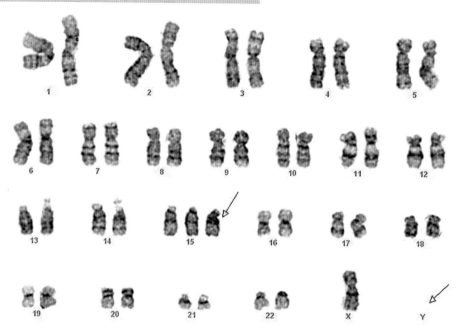

FIGURE 9.8 G-banded karyotype demonstrating trisomy 15.

- Bone marrow is often hypercellular and frequently shows erythroid preponderance. The degree of dysplasia in the erythroid precursors varies and may include megaloblastic changes, multinucleation, nuclear bridging, nuclear fragmentation or budding, cytoplasmic vacuolization, and abnormal hemoglobinization.
- Iron stores are increased with the presence of ≥15% ring sideroblasts.
- In the peripheral blood, red blood cells often show some degree of anisopoikilocytosis with reduced polychromasia (reticulocytes). A dimorphic pattern is common with a mixture of normochromic and hypochromic red cells.
- There is no evidence of absolute monocytosis (monocyte count < 1000/μL).
- Lack of significant dysplasia in other lineages.

Ring sideroblasts are observed in other conditions such as congenital sideroblastic anemia, alcohol intoxication, exposure to lead, benzene, zinc and isoniazid, and copper insufficiency.

IMMUNOPHENOTYPE

Findings are similar to those seen in RCUD.

MOLECULAR AND CYTOGENETIC STUDIES

For detailed molecular and cytogenetic studies in MDS see Chapter 9. The most frequent cytogenetic abnormalities are partial deletions or loss of chromosomes 5, 7, and 20 and trisomy 8. Other cytogenetic abnormalities are listed in Table 9.2 in Chapter 9. Figure 9.8 shows a karyotyping image with +15, −Y.

Refractory Cytopenia with Multilineage Dysplasia

Refractory cytopenia with multilineage dysplasia (RCMD) is characterized by the presence of dysplasia in two or more hematopoietic lineages and corresponding cytopenias. It accounts for close to 30% of the MDS cases. The median age is between 70–75 years. RCMD has a more aggressive clinical course than RCUD and RARS, with a 2-year rate of progression to AML of about 10%.

MORPHOLOGY (FIGURES 9.9 AND 9.10)

- At least 10% of the precursor cells in two or more hematopoietic lineages show dysplastic changes. The spectrum of dysplastic changes is discussed in Chapter 8.
- Bone marrow is usually hypercellular and contains <5% myeloblasts. Occasional blasts (<1%) may be present in the peripheral blood.
- Iron stores are often increased, and variable percentage of ring sideroblasts may be present.
- There is no evidence of absolute monocytosis (monocyte count <1000/μL).

IMMUNOPHENOTYPE

- Blast enumeration of bone marrow by flow cytometry is less than or equal to 5% and needs to be correlated with morphologic findings.
- There may be several dysmaturation patterns of myelomonocytic cells.

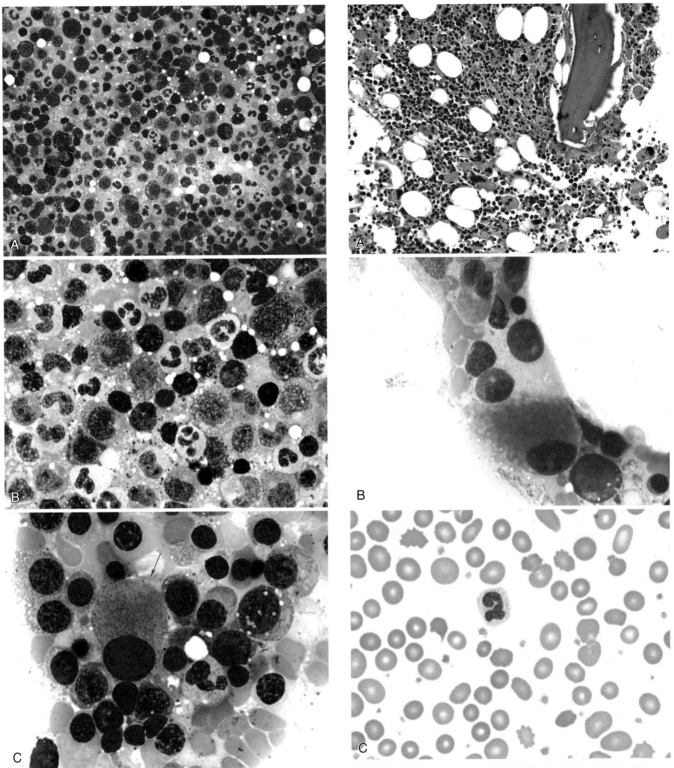

FIGURE 9.9 **REFRACTORY CYTOPENIA WITH MULTILINEAGE DYSPLASIA.**
Bone marrow smears from a patient with RCMD demonstrating erythroid dysplasia and hypogranular neutrophils (A, low power; B, high power) and a micromegakaryocyte (C, arrow).

FIGURE 9.10 **REFRACTORY CYTOPENIA WITH MULTILINEAGE DYSPLASIA.**
(A) Biopsy section shows numerous micromegakaryocytes. (B) Bone marrow smear depicts a mononuclear micromegakaryocyte. (C) Peripheral blood smear reveals anisopoykilocytosis and a hypogranular neutrophil.

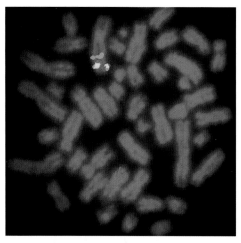

FIGURE 9.11 t(3;3) *EVI1* rearrangement demonstrated by FISH.

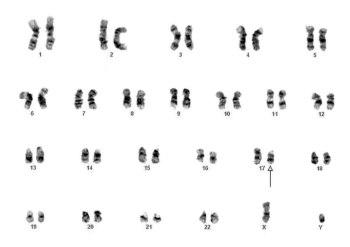

FIGURE 9.12 G-banded karyotype demonstrating deletion of 17p.

- Identification of abnormal myeloblasts expressing discrete aberrancies is the single most significant finding in evaluating dysmaturation patterns of myeloid precursors.

MOLECULAR AND CYTOGENETIC STUDIES

For detailed molecular and cytogenetic studies in MDS see Chapter 8. The most frequent cytogenetic abnormalities are partial deletions or loss of chromosomes 5, 7, and 20 and trisomy 8. Other cytogenetic abnormalities are listed in Table 9.2 in Chapter 9. Figures 9.11 and 9.12 demonstrate t(3;3) and del(17p), respectively.

Refractory Anemia with Excess Blasts

Refractory anemia with excess blasts (RAEB) is characterized by multilineage dysplasia and increased myeloblasts (5–19%) in the bone marrow and/or peripheral blood (1–19%). RAEB accounts for about 30–40% of MDS cases, and it usually affects patients older than 50 years.

Refractory anemia with excess blasts is divided into two groups: RAEB-1 and RAEB-2.

- RAEB-1 refers to the group with 5–9% blasts in the bone marrow and/or 2–4% blasts in the peripheral blood.
- RAEB-2 refers to the group with 10–19% blasts in the bone marrow and/or 5–19% blasts in the peripheral blood. The presence of Auer rods is a significant finding. According to the WHO recommendation, cases with 1% blasts or fewer in the peripheral blood and <5% blasts in bone marrow with Auer rods should be classified as RAEB-2.

RAEB is considered a high grade MDS, with a 25% and 33% chance of evolving to AML in RAEB-1 and RAEB-2, respectively.

MORPHOLOGY

- At least 10% of the cells in two or more hematopoietic lineages show dysplastic changes. The spectrum of dysplastic changes is discussed in Chapter 8.
- Bone marrow is usually hypercellular, but in <15% of the cases is hypocellular. It shows myeloid preponderance and left shift with increased blasts accounting for 5–19% of the bone marrow cells (Figures 9.13 and 9.14). Auer rods may be present.
- Centrally located small aggregates of blasts, referred to as ALIP, are often present in biopsy sections.
- Peripheral blood smears show multilineage dysplasia, myeloid left shift, and the presence of 1–19% myeloblasts. Auer rods may be present.
- There is no evidence of absolute monocytosis (monocyte count <1000/µL).

IMMUNOPHENOTYPE (FIGURES 9.15 AND 9.16)

- Blast enumeration of bone marrow by flow cytometry is between 5% and 19% and needs to be correlated with morphologic findings.
- Usually there are multiple dysmaturation patterns of myelomonocytic cells.
- Discrete phenotypic aberrancies of myeloblasts are easily seen, and cross-lineage aberrancies are common.

MYELODYSPLASTIC SYNDROME, UNCLASSIFIABLE

MOLECULAR AND CYTOGENETIC STUDIES

For detailed molecular and cytogenetic studies in MDS see Chapter 8. The most frequent cytogenetic abnormalities are partial deletions or loss of chromosomes 5, 7, and 20 and trisomy 8. Other cytogenetic abnormalities are listed in Table 9.2 in Chapter 9. Figure 9.17 demonstrates +8 and del(12p), and Figure 9.18 shows a complex karyotype abnormality.

Myelodysplastic Syndrome with Isolated Del(5q)

Myelodysplastic syndrome with isolated del(5q) (5q– syndrome) is characterized by anemia and isolated del(5q31-33) chromosome abnormality. Anemia is usually macrocytic. Platelet counts are normal or elevated, and there may be a modest leukopenia.

5q– syndrome has female preponderance with a median age of 65-70 years. Clinically, this disorder falls into low grade MDS with less than 10% chance of evolving to AML

MORPHOLOGY

- Bone marrow is usually hypercellular and shows numerous micromegakaryocytes with mono- or hypolobated nuclei (Figures 9.19 and 9.20). Erythroid precursors may show some dysplastic changes. Blasts are < 5% and show no Auer rods.
- Blood smears show macrocytosis, mild leukopenia, and sometimes, occasional blasts (<1%).

IMMUNOPHENOTYPE

Findings are similar to those seen in early stages of MDS (RCUD, RARS, and RCMD), and blast enumeration of bone marrow should be less than 5%.

MOLECULAR AND CYTOGENETIC STUDIES

Cytogenetic studies reveal del(5q31-33) (Figure 9.21).

Myelodysplastic Syndrome, Unclassifiable

This term is recommended by the WHO for those MDS cases that lack proper features to fall into one of the well-defined, above-mentioned categories, such as RCUD, RARS, RCMD, and RAEB. This category refers to the cases that: (1) are otherwise classifiable as RCUD or RCMD, but with ≤1% blasts in peripheral blood, (2) cases

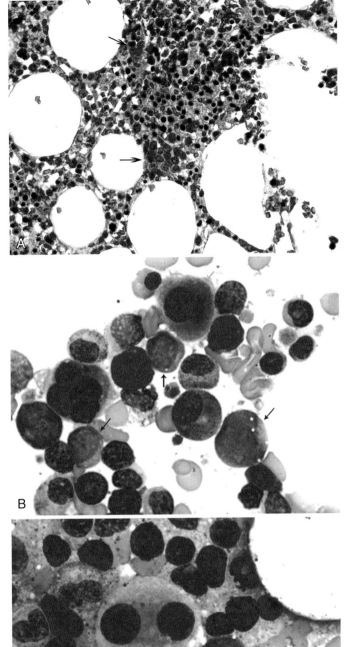

FIGURE 9.13 **REFRACTORY ANEMIA WITH EXCESS BLASTS (RAEB-1).** (A) Bone marrow biopsy section demonstrating clusters of immature myeloid cells known as ALIP (arrows). (B) Bone marrow smear showing increased blasts (black arrows) and a micromegakaryocyte (green arrow). (C) A binucleated micromegakaryocyte is noted next to a myeloblast.

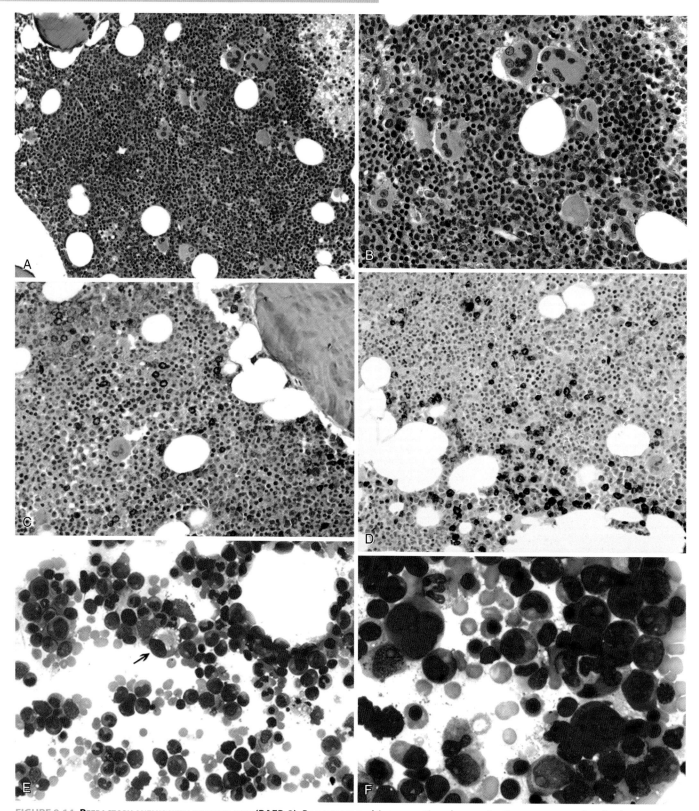

FIGURE 9.14 **REFRACTORY ANEMIA WITH EXCESS BLASTS (RAEB-2).** Bone marrow biopsy section demonstrating numerous dysplastic megakaryocytes and micromegakaryocytes (A, low power; B, high power). Immunohistochemical stains for CD117 (C) and CD34 (D) show numerous positive cells indicative of increased myeloblasts. Bone marrow smear (E, low power; F, high power) demonstrates increased blasts. A mononuclear micromegakaryocyte is present (E, arrow).

MYELODYSPLASTIC SYNDROME, UNCLASSIFIABLE

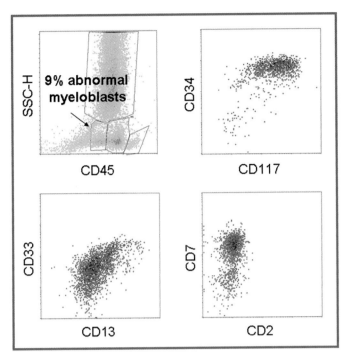

FIGURE 9.15 **FLOW CYTOMETRY OF MYELOBLASTS IN RAEB-1.** The open gate display by CD45 gating (in light blue) shows abnormal location of excess blasts, accounting for 9% of the total. The blast-enriched gate displays (in purple) demonstrate abnormal myeloblasts with asynchronous expression of CD13 and CD33, plus cross-lineage aberrancy of CD7.

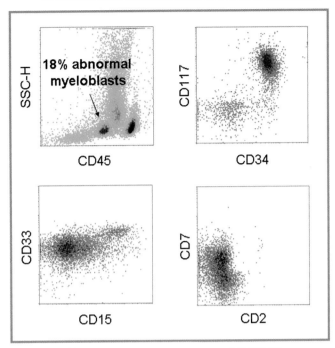

FIGURE 9.16 **FLOW CYTOMETRY OF MYELOBLASTS IN RAEB-2.** The open gate display by CD45 gating (in green) reveals excess and abnormal blasts (18% of the total) with very dim expression of CD45. The blast-enriched gate displays show abnormal myeloblasts with asynchronous expression of CD33 and partial CD15, plus cross-lineage aberrancy of CD7.

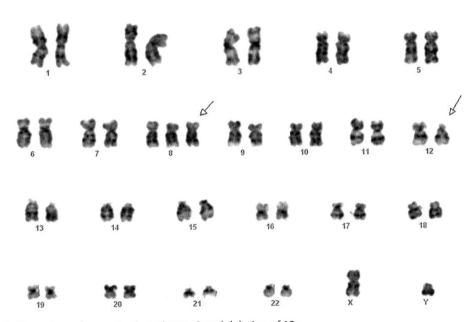

FIGURE 9.17 G-banded karyotype demonstrating trisomy 8 and deletion of 12p.

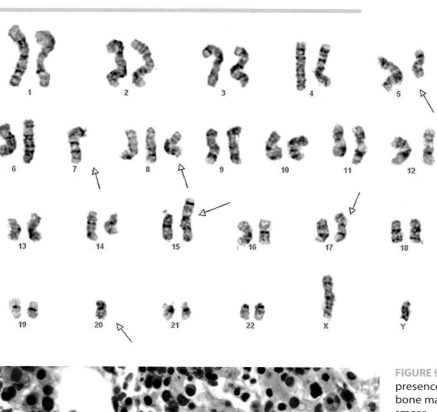

FIGURE 9.18 A complex abnormal karyotype in a case of MDS transformed to AML, showing deletion 5q, monosomy 7, trisomy 8, monosomy 20, and unbalanced translocation at 15p and 17p resulting in trisomy 1q and deletion 17p, respectively.

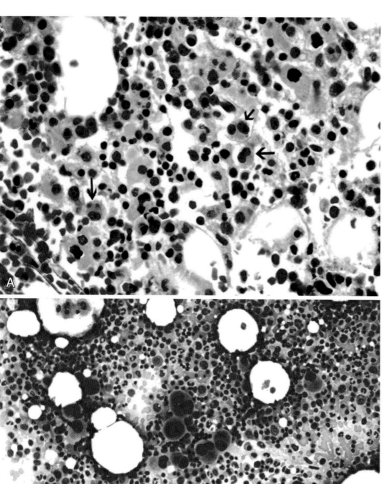

FIGURE 9.19 5Q– SYNDROME. This is characterized by the presence of numerous micromegakaryocytes in the bone marrow: (A) biopsy section and (B) bone marrow smear.

OTHER TYPES OF MYELODYSPLASTIC SYNDROMES

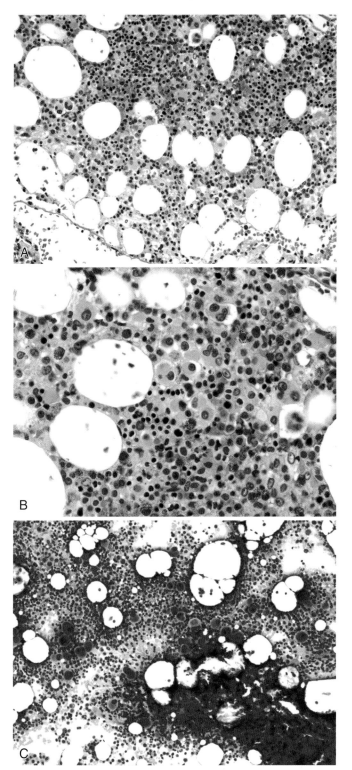

FIGURE 9.20 **5Q– SYNDROME.** Numerous micromegakaryocytes are present in the bone marrow biopsy section (A, low power; B, high power) and bone marrow smear (C).

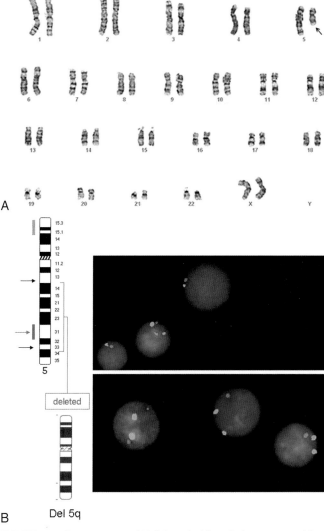

FIGURE 9.21 **5Q– SYNDROME.** (A) G-banded female karyotype with a deletion of 5q. (B) Ideogram of chromosome 5 showing the commonly deleted region in the long arm (red bracket). FISH studies on interphase cells show monosomy 5 (top right) and deleted 5q (bottom right).

otherwise classifiable as RCUD, but with pancytopenia, and (3) cases with no significant dysplasia, but evidence of cytopenia and cytogenetic abnormalities typical of MDS (Figures 9.22 and 9.23).

Other Types of Myelodysplastic Syndromes

THERAPY-RELATED (SECONDARY) MDS

The development of MDS following cytotoxic chemotherapy and/or radiotherapy has been extensively investigated and reported in the literature. The emergence of therapy-related MDS (t-MDS) is usually associated with a long latency period and is particularly seen following the use of alkylating agents. The pathologic manifestations of t-MDS are, in general, similar to those of primary MDS except for higher frequency of the following features:

1. High-risk variants
2. Bone marrow fibrosis

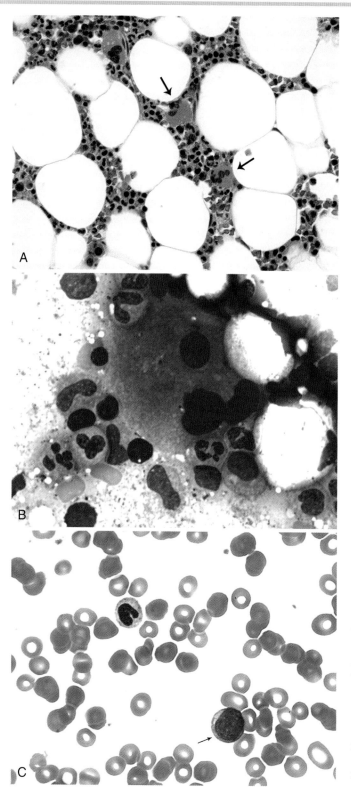

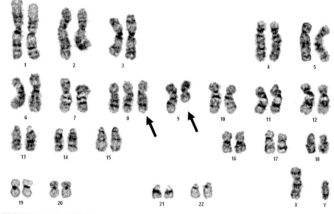

FIGURE 9.23 **MDS, UNCLASSIFIABLE.** G-banded karyotype demonstrating trisomy 8 and deletion of 9q.

3. Bone marrow hypocellularity
4. Unclassifiable forms
5. Chromosomal aberrations
6. Transformation to acute leukemia.

t-MDS represents one spectrum of a broader syndrome now designated as *therapy-related AML and myelodysplasia* (see Chapter 12). It constitutes about 10–15% of the total MDS cases. The interval between initiation of therapy and the onset of MDS varies and depends on the type, duration, and dose of the therapeutic agent(s). This period in most studies ranges from 1 to 8 years, with a mean of 5 years. The topoisomerase II inhibitors may occasionally cause MDS but more often are associated with *de novo* AML without going through dysmyelopoiesis. Prolonged environmental or occupational exposure to benzene and benzene-derivative compounds may also lead to MDS.

PEDIATRIC MDS (FIGURE 9.24)

Myelodysplastic syndromes are rare in children, accounting for approximately 3–5% of all pediatric clonal hematologic disorders. They may also present themselves differently from the adult forms, particularly the MDS categories with <5% blasts. The following observations have been reported in pediatric patients with MDS:

1. Hematopoietic dysplasia is frequently observed in a variety of conditions, such as infection, metabolic disorders, and nutritional deficiencies.
2. Neutropenia and thrombocytopenia are more frequently observed than anemia.
3. The rate of cytogenetic aberrations in low risk MDS is higher in children (about 65%) than in adults (20–30%). Monosomy 7 is one of the most frequent findings.
4. The majority of the cases fall into the MDS/myeloproliferative categories (see Chapter 10), particularly in patients younger than 5 years.

FIGURE 9.22 **MDS, UNCLASSIFIABLE.** (A) Bone marrow biopsy section shows presence of dysplastic megakaryocytes with multiple separated nuclei (arrows). (B) Bone marrow smear reveals a megakaryocyte with separated nuclei. (C) Blood smear depicts a blast cell (arrow).

HYPOCELLULAR MDS

Hypocellular MDS accounts for about 5–10% of MDS cases. It is usually therapy related and is often associated

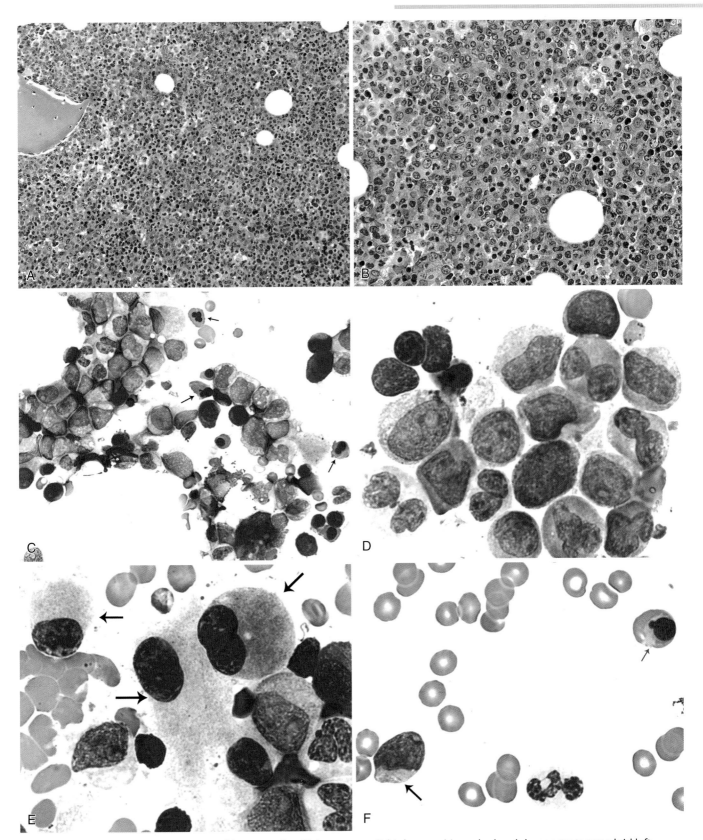

FIGURE 9.24 **PEDIATRIC MDS.** Bone marrow biopsy section (A, low power, B, high power) is packed and demonstrates myeloid left shift. Bone marrow smears (C to E) show myeloid left shift with erythroid dysplasia (C, arrows), and micromegakaryocytes (E, arrows). Peripheral blood smear (F) demonstrates a myeloblast with an Auer rod (black arrow) and a dysplastic nucleated red cell (red arrow).

with more severe pancytopenia. Most investigators consider the diagnosis of hypocellular MDS when bone marrow cellularity is ≤25% of the age-matched normal range (Figure 9.25). Aplastic bone marrow conditions (such as aplastic anemia, Fanconi anemia, and paroxysmal nocturnal hemoglobinuria) and hypocellular variants of hairy cell and acute leukemias are in the list of differential diagnosis. Dysplastic hematopoiesis distinguishes hypocellular MDS from the bone marrow aplasia group, and blast counts of <20% separate this entity from acute leukemias. Dysplastic erythroid and myeloid cells are present in the blood smears and/or the bone marrow samples, and abnormal megakaryocytes are often identified. Estimation of blast number is facilitated by immunophenotypic studies by using blast-associated markers, such as CD34 and CD117.

MDS WITH FIBROSIS

Approximately 15% of MDS patients may demonstrate a significant increase in bone marrow reticulin fibers (MDS-F). The incidence of marrow fibrosis is higher in t-MDS. Also, most MDS-F cases fall into the RAEB category. Because of significant fibrosis, bone marrow aspiration attempts may fail (dry taps) (Figure 9.26). Evidence of excess blasts could be demonstrated by immunohistochemical studies by showing increased CD34+ and/or CD117+ cells. MDS-F is often associated with increased megakaryocytes, including dysplastic forms. Figure 9.27 demonstrates monosomy 7 and del(20q) in a patient with MDS-F.

Non-Clonal Myelodysplasia

Non-clonal myelodysplastic changes have been observed in a variety of conditions, such as autoimmunity, infections, nutritional deficiencies, heavy metal intoxication, and postchemotherapy and/or radiotherapy. Dysplastic changes in these conditions are often reversible upon elimination of the causative factors, and are not associated with chromosomal aberrations. Representative examples of non-clonal MDS are briefly discussed below.

AUTOIMMUNE MYELODYSPLASIA

Myelodysplasia has been observed in a small proportion of patients with autoimmune disorders. The dysplastic changes in these patients are not associated with chromosomal aberrations, and patients respond positively to immunosuppressive therapy. These patients are usually pancytopenic, show macrocytic anemia, with a variable marrow cellularity and dysplastic morphologic changes similar to those of classic MDS. Bone marrow blasts are under 5%.

HIV-ASSOCIATED MYELODYSPLASIA

Myelodysplastic changes observed in HIV-infected patients may be related to HIV, secondary infections, or medications. Compared with classical MDS, patients with HIV-associated myelodysplasia frequently show bone marrow hypocellularity, plasmacytosis, and eosinophilia. In these patients the degree of anemia and erythroid dysplasia is less severe, micromegakaryocytes are less frequent, blasts are <5%, no ring sideroblasts are present, and cytogenetic studies are normal.

PARANEOPLASTIC MYELODYSPLASIA

Dysplastic changes similar to those of MDS have been reported in rare patients with solid tumors, such as carcinoma of colon, lung, kidney, prostate, and stomach. The cause of dysplastic changes is not clear but does not appear to be drug-related. Production and release of growth factor-like proteins by the neoplastic cells is among the possibilities. Dysplastic changes are observed in both bone marrow and peripheral blood samples.

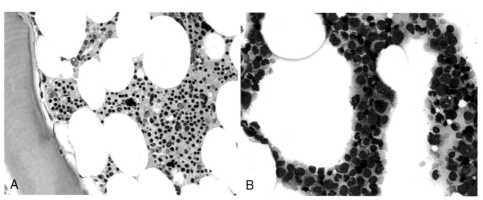

FIGURE 9.25 HYPOCELLULAR MDS. Bone marrow biopsy section (A) and bone marrow smear (B).

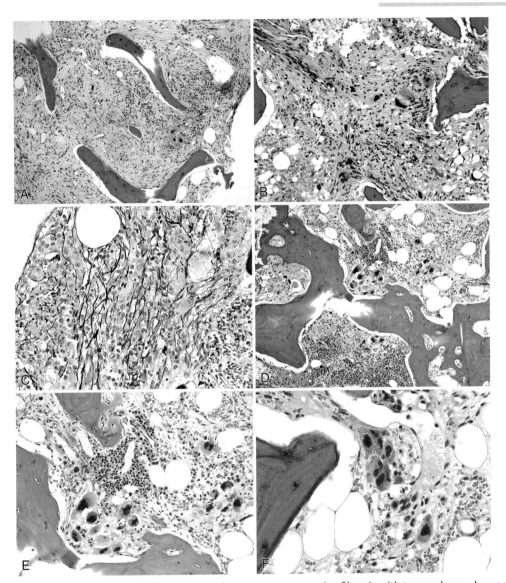

FIGURE 9.26 **MDS WITH FIBROSIS.** Bone marrow biopsy section demonstrates extensive fibrosis with trapped megakaryocytes (A and B) and increased reticulin fibers shown by reticulin stain (C). Areas of sclerotic bone, erythroid clusters and megakaryocytic aggregates are demonstrated in (D), (E), and (F).

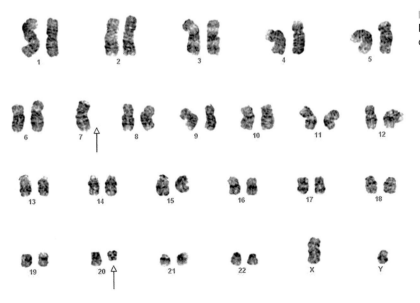

FIGURE 9.27 **MDS WITH FIBROSIS.** G-banded karyotype demonstrating monosomy 7 and deletion of 20q.

MYELODYSPLASIA ASSOCIATED WITH HEAVY METAL INTOXICATION OR DEFICIENCY

There are occasional reports of myelodysplastic changes induced by arsenic and uranium intoxication, or copper deficiency. The dysplastic changes mimic RA or RCMD, and are often associated with anemia or pancytopenia. It is also interesting to know that arsenic trioxide, which acts through pro-apoptotic and anti-angiogenesis mechanisms, has been used to treat a variety of hematologic malignancies, including RAEB.

MYELODYSPLASIA ASSOCIATED WITH CHEMOTHERAPY OR IRRADIATION

Myelodysplastic changes in bone marrow are common features of post-chemotherapy and radiotherapy. Bone marrow is hypocellular and shows multilineage dysplasia and myeloid left shift, mimicking hypocellular MDS.

Differential Diagnosis

Diagnosis of MDS is based on a multidisciplinary clinicopathologic approach. It is accomplished by obtaining adequate, pertinent clinical and environmental histories, careful pathologic review of peripheral blood and bone marrow, immunophenotyping, cytogenetic analysis, and molecular genetic studies. It should be noted that a broad spectrum of hematologic disorders may mimic MDS, and should therefore be considered in the differential diagnosis (Table 9.4).

DISORDERS WITH DYSPLASTIC ERYTHROPOIESIS

Congenital dyserythropoietic anemias are hereditary disorders with bone marrow erythroid hyperplasia and marked dyserythropoiesis, such as megaloblastic changes, and the presence of erythroid precursors with bi- and multilobular nuclei (see Chapter 61). Bone marrow morphologic features of congenital dyserythropoietic anemias may mimic those of RA and RARS. In congenital dyserythropoietic anemias, ring sideroblasts are usually absent, myeloid and megakaryocytic lineages are unremarkable, and there is no abnormal karyotype.

Megaloblastic anemia may share morphologic features, such as erythroid dysplastic and megaloblastic changes, macrocytosis and neutrophilic hypersegmentation, with MDS. Megaloblastic anemia is characterized by low levels of serum folate or vitamin B12 and lack of ring sideroblasts. A mild myeloid left shift may be seen in the bone marrow of some cases of megaloblastic anemia, but in such cases the blast cells are usually <5%. Cytogenetic and molecular studies are normal in megaloblastic anemia (see Chapter 61).

Table 9.4
Differential Diagnosis of MDS

Disorder	Characteristics
Disorders with Dysplastic Erythropoiesis	
Congenital dyserythropoietic anemias	Inherited disorders with marked erythroid dysplasia and unremarkable myeloid series and magakaryocytes. Normal karyotypes
Megaloblastic anemia	Reduced serum levels of folate or vitamin B12. Normal karyotype
Disorders with ring sideroblasts	Observed in hereditary sideroblastic anemia (rare) and patients with pyridoxine deficiency, zinc or alcohol toxicity, or as post-medication effect in some patients treated with chloramphenicol, cycloserine, or anti-tuberculosis drugs
Acute erythroleukemia	Erythroid lineage accounts for >50% of the bone marrow cells, and myeloblasts constitute ≥20% of the non-erythroid component
Hyperplastic Bone Marrows with Myeloid Preponderance	
Chronic myeloproliferative disorders	Peripheral blood cytosis, splenomegaly, no significant dyserthropoiesis, t(9;22) in CML
Transient myeloproliferative disorder	A transient neonatal condition in Down syndrome; usually disappears in 4–6 weeks
Acute myeloid leukemias	Blasts ≥20%; frequent balanced chromosomal aberrations
Hypoplastic Bone Marrows	
Aplastic anemias	No increased blasts; no significant dysplastic changes; mutated genes in FA; mutated *PIG-A* gene in PNH with loss of GPI-linked proteins (CD55, CD59)
Hypoplastic AML	Blasts ≥20%; frequent balanced chromosomal aberrations
Nonclonal myelodysplasia	Observed in viral infections, autoimmune disorders, paraneoplastic syndromes, heavy metal intoxication, and post-chemotherapy and -radiation therapy. Normal karyotype

Ring sideroblasts are seen in rare cases of hereditary sideroblastic anemia, in patients with pyridoxine deficiency, zinc or alcohol toxicity, or as post-medication effect in some patients treated with chloramphenicol, cycloserine, or anti-tuberculosis drugs.

Erythroleukemia shares many morphologic features with RAEB and at times a distinction between these two entities is difficult. However, in erythroleukemia, the erythroid lineage accounts for >50% of the bone marrow cells, and myeloblasts make up ≥20% of the non-erythroid component (see Chapter 21).

HYPERPLASTIC BONE MARROWS WITH MYELOID PREPONDERANCE

Myeloproliferative neoplasms (chronic myeloproliferative disorders) may share overlapping bone marrow findings with some variants of MDS, such as 5q− syndrome, RCMD, and RAEB-1, either because of the presence of micromegakaryocytes or myeloid left shift. The accelerated phase of chronic myelogenous leukemia (CML) may mimic RAEB-2. However, in myeloproliferative neoplasms, there is peripheral blood cytosis (thrombocytosis, granulocytosis, erythrocytosis, or pancytosis), often with evidence of leukoerythroblastosis (presence of immature myeloid and erythroid cells in the blood) and no significant dysplastic changes. Splenomegaly is a frequent clinical presentation in myeloproliferative disorders but rare in MDS.

Transient myeloproliferative disorder associated with Down syndrome is a neonatal condition which may mimic RAEB or AML with increased blasts in bone marrow and the presence of blasts in peripheral blood. This condition is transient and usually disappears within 4–6 weeks (see Chapter 22).

Acute myeloid leukemias with relatively low blast counts (20–25%) are at times difficult to distinguish from RAEB-2, particularly when the blasts are dysplastic and do not fall into the classical morphologic criteria defined for normal blasts. Another challenge in estimating percent blasts in inadequate bone marrow aspirate smears (dry tap) due to fibrosis. In such cases, immunohistochemical stains for CD34 and CD117 may help to estimate the proportion of myeloblasts in the biopsy sections. For the determination of percent blasts, the WHO recommends 500 and 200 differential counts for the bone marrow and blood smears, respectively. Also, disorders with evidence of recurrent or known AML-associated cytogenetic abnormalities, such as t(8;21) and inv (16), should be considered as AML even if the blast count is <20%. Both MDS and AML show frequent cytogenetic abnormalities. Whereas in MDS most aberrations are unbalanced chromosomal changes (deletions, monosomies, and trisomies), AMLs often show balanced (reciprocal) chromosomal changes, such as translocations or inversions.

HYPOPLASTIC BONE MARROWS

Aplastic anemias, both acquired and constitutional forms, have features overlapping with those of hypocellular MDS. Aplastic anemia usually lacks significant dysplasia or evidence of increased blasts, but may occasionally show chromosomal aberrations. Most of the acquired aplastic anemias with cytogenetic abnormalities probably represent hypocellular MDS. Fanconi anemia is associated with FA complementary gene groups, and PNH shows evidence of mutations of the *PIG-A* gene and loss of the expression of GPI-linked proteins, such as CD55 and CD59 on the hematopoietic cells (see Chapter 7).

Hypocellular AML shares many morphologic features with hypocellular MDS except for the higher percentage (≥20%) of blast cells. Multilineage dysplasia is commonly found in hypocellular MDS, but it may be present or absent in hypocellular AML. Both lesions show frequent cytogenetic abnormalities; AML often with balanced chromosomal changes and MDS with unbalanced chromosomal aberrations.

NON-CLONAL MYELODYSPLASIA

Non-clonal myelodysplastic changes are associated with conditions such as viral infections, autoimmune disorders, paraneoplastic syndromes, heavy metal intoxication, and post-chemotherapy and -radiation therapy. Dysplastic changes may be very similar to those of MDS. The clinical history and lack of chromosomal aberrations help to distinguish this broad entity from MDS.

Additional Resourses

Abel GA, Van Bennekom CM, Stone RM, et al: Classification of the myelodysplastic syndrome in a national registry of recently diagnosed patients, *Leuk Res* 34:939-941, 2010.

Boruchov AM: Thrombocytopenia in myelodysplastic syndromes and myelofibrosis, *Semin Hematol* 46(1 Suppl 2):S37-S43, 2009.

Boultwood J, Pellagatti A, McKenzie AN, et al: Advances in the 5q− syndrome, *Blood* 116:5803-5811, 2010.

Garcia-Manero G: Prognosis of myelodysplastic syndromes, *Hematology Am Soc Hematol Educ Program*:330-337, 2010.

Giagounidis AA, Aul C: The 5q− syndrome, *Cancer Treat Res* 142:133-148, 2008.

Jaffe ES, Harris NL, Vardiman JW, et al: Hematopathology, Philadelphia, 2010, Saunders/Elsevier.

Komrokji RS, Zhang L, Bennett JM: Myelodysplastic syndromes classification and risk stratification, *Hematol Oncol Clin North Am* 24:443-457, 2010.

Marinier DE, Mesa H, Rawal A, Gupta P: Refractory cytopenias with unilineage dysplasia: A retrospective analysis of refractory neutropenia and refractory thrombocytopenia, *Leuk Lymphoma* 51:1923-1926, 2010.

Niemeyer CM, Kratz CP, Hasle H: Pediatric myelodysplastic syndromes, *Curr Treat Options Oncol* 6:209-214, 2005.

Polychronopoulou S, Panagiotou JP, Kossiva L, et al: Clinical and morphological features of paediatric myelodysplastic syndromes: A review of 34 cases, *Acta Paediatr* 93:1015-1023, 2004.

Steensma DP, Hanson CA, Letendre L, et al: Myelodysplasia with fibrosis: A distinct entity?, *Leuk Res* 25:829-838, 2001.

Swerdlow SH, Campo E, Harris NL, et al: WHO classification of tumours of haematopoietic and lymphoid tissues, ed 4, International Agency for Research on Cancer Lyon, 2008.

Valent P, Orazi A, Büsche G, et al: Standards and impact of hematopathology in myelodysplastic syndromes (MDS), *Oncotarget* 1:483-496, 2010.

Yin CC, Medeiros LJ, Bueso-Ramos CE: Recent advances in the diagnosis and classification of myeloid neoplasms—comments on the 2008 WHO classification, *Int J Lab Hematol* 32:461-476, 2010.

Myeloproliferative Neoplasms—Overview

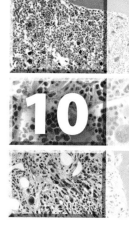

Myeloproliferative neoplasms (MPN) (or chronic myeloproliferative disorders, CMPD) are a group of hematologic disorders distinguished by clonal expansion of abnormal hematopoietic stem cells leading to a hypercellular marrow with excessive terminal proliferation of the hematopoietic cells and peripheral blood granulocytosis, erythrocytosis, and/or thrombocytosis. This hyperproliferative process, in certain conditions, is associated with bone marrow fibrosis and extramedullary hematopoiesis. The extramedullary hematopoiesis along with excess sequestration of the hematopoietic cells in the spleen often leads to massive splenomegaly, one of the clinical hallmarks of MPN. Hepatomegaly, though less frequent, may also be present.

The exact mechanisms of the myeloid proliferation in these disorders are not well understood. The hypersensitivity of the affected stem cells to certain growth factors and/or defective negative regulatory feedback mechanisms may play a role. However, recent investigations suggest abnormalities in tyrosine kinase genes as the central core to the pathogenesis of myeloproliferative neoplasms. The classical example of tyrosine kinase involvement in MPN is the fusion of *ABL1* and *BCR* genes brought about by a reciprocal chromosomal translocation—t(9;22)(q34;q11.2), resulting in a smaller chromosome 22, the Philadelphia chromosome (Ph^1)—in chronic myelogenous leukemia. Several tyrosine kinase genes, other than *ABL1*, have been identified, such as *ABL2*, *PDGFRA*, *PDGFRB*, *FGFR1*, and *JAK2*. Fusion of the *PDGFRA*, *PDGFRB*, and *FGFR1* genes and an activating V617F mutation in the *JAK2* gene have been reported in a significant proportion of non-CML myeloproliferative cases.

The WHO classification (Swerdlow et al., 2008) includes the following categories in this group:

1. Chronic myelogenous leukemia (CML)
2. Primary myelofibrosis (with extramedullary hematopoiesis) (PMF)
3. Polycythemia vera (PV)
4. Essential thrombocythemia (ET)
5. Mastocytosis
6. Chronic neutrophilic leukemia (CNL)
7. Chronic eosinophilic leukemia (CEL), not otherwise specified
8. Chronic myeloproliferative neoplasm, unclassifiable (CMPD-U)

Morphology

Morphologic features shared by various types of MPN are briefly discussed below (Figures 10.1 and 10.2):

BONE MARROW

- Hypercellular bone marrow with mono- or multilineage hyperplasia, predominance of mature cells, and no significant dyserythropoiesis or dysgranulopoiesis.
- Megakaryocytosis, often present in clusters with abnormal morphology.
- Dilated sinuses containing clusters of hematopoietic cells.
- Frequent focal or diffuse marrow fibrosis.
- Frequent osteosclerosis.
- Frequent basophilia and/or eosinophilia.

BLOOD

- Granulocytosis, erythrocytosis, and/or thrombocytosis, often with a leukoerythroblastic picture.
- Tear-drop-shaped erythrocytes.
- Giant platelets.
- Frequent basophilia and/or eosinophilia.
- Lack of significant dysgranulopoiesis.
- Lack of toxic granulation in neutrophils.

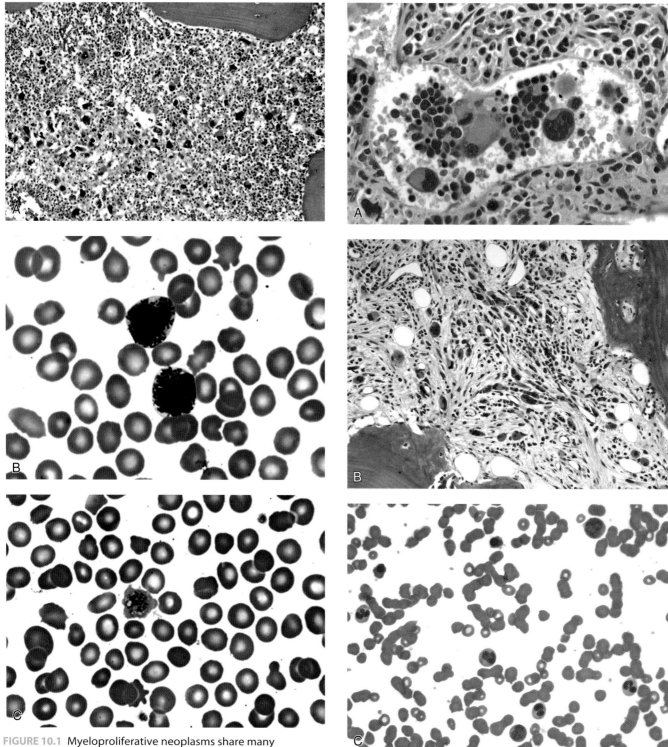

FIGURE 10.1 Myeloproliferative neoplasms share many morphologic features, such as bone marrow hypercellularity with increased megakaryocytes (A), basophilia (B), and the presence of giant platelets (C, arrow).

FIGURE 10.2 Myeloproliferative neoplasms share many morphologic features, such as dilated sinuses containing clusters of hematopoietic cells (A), marrow fibrosis (B), and leukoerythroblastosis (C).

Immunophenotype

Immunophenotypic studies are of limited value, but are sometimes helpful in monitoring of MPN patients in the following areas:

- Blast enumeration and lineage assignment;
- Identification of abnormal blasts with various aberrancies including expression of cross-lineage markers;
- Detection of dysmaturation patterns of myeloid cells, similar to those seen in MDS.

Molecular Studies

The two hallmark genetic alterations in myeloproliferative disorders are the *BCR-ABL1* fusion gene (Figure 10.4) and mutations in the *JAK2* gene (Figure 10.3). These findings are usually, but not always, mutually exclusive and therefore are frequently used to distinguish benign from malignant processes.

- Fusion of the *BCR* gene on chromosome 22 with the *ABL* oncogene on chromosome 9, which is the molecular counterpart of the microscopically visible Philadelphia chromosome, is present in some form in all cases of CML, and is essentially pathognomonic. It should *not* be present in other myeloproliferations.
- The fusion gene can be detected at extremely low levels by real-time (quantitative) PCR, the technique used for monitoring relapse in patients on anti-tyrosine kinase therapies.
- The test requires cellular RNA (not DNA) as starting material, which means that specimens must be transported to the laboratory without delay, owing to the lability of RNA.
- The non-malignant MPN are characterized by mutations in the *JAK2* kinase gene which, like the *BCR-ABL1* fusion, confer upon it heightened upregulation.
- The predominant mutation, by far, is V617F; a few other mutations have been described in the gene, but they are so infrequent that most laboratories do not routinely test for them.
- The mutation can be detected by a variety of allele-specific PCR and probe hybridization techniques, or by DNA sequencing.
- The *JAK2* mutation is most prevalent in PV (>90%), progressively less so in ET (50%), PMF (50%), CMML, and CNL.
- A small minority of these lesions will show mutations in the *MPL* gene.

Cytogenetics

The spectrum of cytogenetic aberrations in MPN is heterogeneous, ranging from numerical gains and losses to structural changes including unbalanced translocations (Table 10.1).

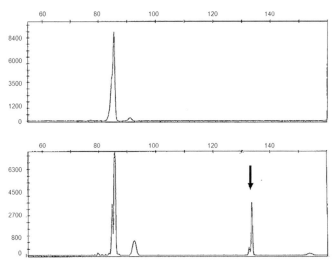

FIGURE 10.3 Detection of the *JAK2* mutation by allele-specific PCR followed by capillary electrophoresis of the amplification products. A negative control sample (top) yields only the normal PCR product peak, whereas a positive sample (bottom) yields both the mutant PCR product (arrow) and the internal control normal gene product.

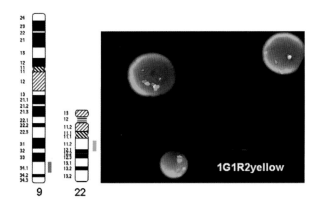

FIGURE 10.4 *BCR-ABL1* fusion in CML demonstrated by FISH.

Table 10.1

Chromosomal Aberrations in Myeloproliferative Neoplasms (MPN)

Chromosome abnormality	Frequency (%)	Prognosis
Trisomy 1q	8	
4q12 deletion (*CHIC2*)		Response to imatinib
5q32 aberrations (*PDGFB*)		Response to imatinib
Monosomy 7	5	
Trisomy 8	16	Good prognosis
8p11 rearrangements (*FGFR1*)		Poor prognosis
Trisomy 9	10	Unclear
12p aberrations	3	
13q deletion	7	Unclear to good if sole aberration
20q deletion	9	Good

Table 10.2

Incidence of Chromosomal Abnormalities in Subtypes of Myeloproliferative Neoplasms (MPN)

Subtype	Frequency (%)
CML	95
PV	34
CIMF	40
CEL	7–12
ET	<3

CIMF, chronic idiopathic myelofibrosis; CML, chronic myelogenous leukemia; ET, essential thrombocythemia; CEL, chronic eosinophilic leukemia; PV, polycythemia vera.

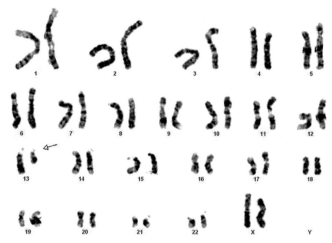

FIGURE 10.6 Karyotype with 13q deletion (arrow).

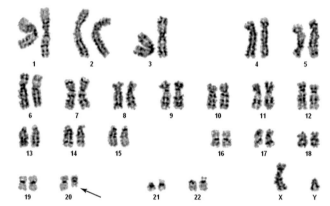

FIGURE 10.5 Karyotype with 20q deletion (arrow).

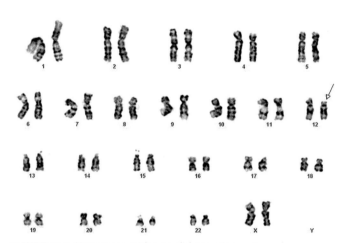

FIGURE 10.7 Karyotype with 12p deletion (arrow).

- Chromosomal gains and losses rather than balanced translocations appear to be common in MPN. Standard karyotyping along with a panel of FISH probes is important in establishing the diagnosis and may provide very useful information for disease outcome.
- Cytogenetic abnormalities in myeloproliferative neoplasms other than CML occur at different frequencies ranging from 3% to 40%, depending on the subtype (Table 10.2). Compared with CML, the other MPN subtypes are more clinically and cytogenetically heterogeneous. In fact, at least 27 different chromosomal anomalies have been associated with MPN. Unlike the "Philadelphia chromosome" in CML, there is no pathognomonic chromosomal abnormality associated with the MPN.
- Chromosomal anomalies are found most frequently in PMF (up to 50%), followed by PV, whereas anomalies in ET and CEL are so infrequent that cytogenetics can be omitted when the diagnosis is clear.
- The most common chromosomal anomalies reported in MPN are: t(9;22)(q34;q11.2), del(20)(q11q13) (Figure 10.5); del(13)(q12q14) (Figure 10.6); del(5)(q13q33), and del(12)(p12) (Figure 10.7); dup(1q) (Figure 10.8); loss of Y, + 8 (Figure 10.9); +9 (Figure 10.10); and monosomy 7 (Figures 10.3 to 10.7). Only the t(9;22) or its variants is diagnostic of CML (Figure 10.11). Relatively strong associations are observed for the del(13) in PMF and the del(20), +8, and +9 in PV.

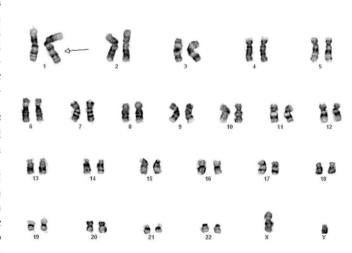

FIGURE 10.8 Karyotype showing a tandem duplication of 1q (arrow).

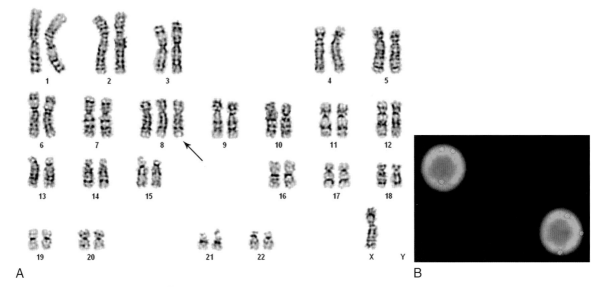

FIGURE 10.9 Karyotype (A) and FISH (B) demonstrating trisomy 8.

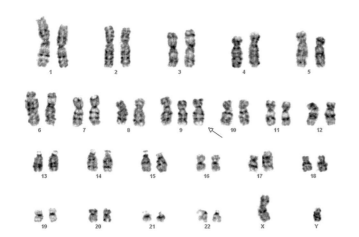

FIGURE 10.10 Karyotype showing trisomy 9 (arrow).

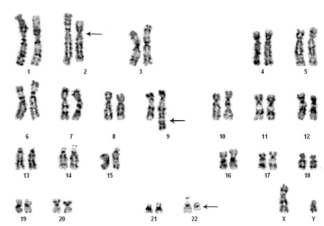

FIGURE 10.11 Karyotype from a patient with CML demonstrating three way translocation: t(2;9;22) (arrows).

Additional Resources

Burke BA, Carroll M: BCR-ABL: a multi-faceted promoter of DNA mutation in chronic myelogeneous leukemia, *Leukemia* 24:1105–1112, 2010.

Chen AT, Prchal JT: JAK2 kinase inhibitors and myeloproliferative disorders, *Curr Opin Hematol* 17:110–116, 2010.

Foroni L, Gerrard G, Nna E, et al: Technical aspects and clinical applications of measuring BCR-ABL1 transcripts number in chronic myeloid leukemia, *Am J Hematol* 84:517–522, 2009.

Kantarjian H, Schiffer C, Jones D, Cortes J: Monitoring the response and course of chronic myeloid leukemia in the modern era of BCR-ABL tyrosine kinase inhibitors: practical advice on the use and interpretation of monitoring methods, *Blood* 111:1774–1780, 2008.

Klco JM, Vij R, Kreisel FH, et al: Molecular pathology of myeloproliferative neoplasms, *Am J Clin Pathol* 133:602–615, 2010.

Kvasnicka HM, Thiele J: Prodromal myeloproliferative neoplasms: the 2008 WHO classification, *Am J Hematol* 85:62–69, 2010.

Oh ST, Gotlib J: JAK2 V617F and beyond: role of genetics and aberrant signaling in the pathogenesis of myeloproliferative neoplasms, *Expert Rev Hematol* 3:323–337, 2010.

Patnaik MM, Tefferi A: Molecular diagnosis of myeloproliferative neoplasms, *Expert Rev Mol Diagn* 9:481–492, 2009.

Rice KN, Jamieson CH: Molecular pathways to CML stem cells, *Int J Hematol* 91:748–752, 2010.

Swerdlow SH, Campo E, Harris NL, et al: WHO classification of tumours of haematopoietic and lymphoid tissues, ed 4, International Agency for Research on Cancer, 2008, Lyon.

Tefferi A, Skoda R, Vardiman JW: Myeloproliferative neoplasms: contemporary diagnosis using histology and genetics, *Nat Rev Clin Oncol* 6:627–637, 2009.

Wadleigh M, Tefferi A: Classification and diagnosis of myeloproliferative neoplasms according to the 2008 World Health Organization criteria, *Int J Hematol* 91:174–179, 2010.

Wadleigh M, Tefferi A: Classification and diagnosis of myeloproliferative neoplasms according to the 2008 World Health Organization criteria, *Int J Hematol* 91:174–179, 2010.

Chronic Myelogenous Leukemia

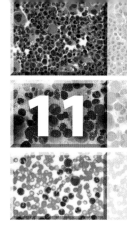

Chronic myelogenous leukemia (CML), also referred to as chronic myeloid, chronic myelocytic, or chronic granulocytic leukemia, was the first malignant disorder reported in association with a chromosomal aberration, the Philadelphia chromosome (Ph^1). CML has been at the forefront in the understanding of molecular mechanisms in hematopoietic malignancies and targeted treatment approaches. It is characterized by clonal expansion of bone marrow stem cells leading to selective granulocytic hyperplasia with or without thrombocytosis. Ph^1 is a microscopically visible short 22q resulting from an asymmetrical reciprocal t(9;22)(q34;q11.2) chromosomal translocation. CML demonstrates an evolutionary process passing through *chronic*, *accelerated*, and *acute* (blast transformation) phases.

The t(9;22)(q34;q11.2) translocation creates a *BCR-ABL1* fusion gene with three different principal protein products, based on the sites of the breakpoints on chromosome 22. All three *BCR-ABL1* protein products (p190, p210, and p230) demonstrate increased tyrosine kinase activity which is not detected in normal hematopoietic cells. The effects of BCR-ABL1 protein products include the following.

1. Insensitivity of Ph^1 positive hematopoietic progenitor cells to growth-inhibiting regulatory cytokines
2. Mitogenic activity of the *BCR-ABL1* fusion proteins on hematopoietic cells
3. Resistance of Ph^1 positive hematopoietic cultured cell lines to apoptosis
4. Decreased adherence of Ph^1 positive hematopoietic progenitors to bone marrow stroma and fibronectin, leading to increased circulation of myeloid cells.

CML represents between 15% and 20% of adult leukemias with a median age at diagnosis of about 50 years. The only known risk factor is exposure to ionizing radiation. The disease is asymptomatic in over 30% of cases and is only suspected by the elevated WBC counts on routine blood examinations. Frequent clinical symptoms include fatigue, pallor, night sweats, and weight loss.

Splenomegaly is observed in 50–75% of patients. Occasionally, the disease presents at the blast phase without prior clinical manifestation of the chronic phase. The diagnosis is established by the demonstration of t(9;22)(q34;q11.2) (Ph^1) and/or *BCR-ABL1* fusion. Transformation into accelerated or blast phase is usually associated with marked splenomegaly, severe anemia, and/or marked thrombocytopenia.

Molecular targeted therapy is now the recommended therapeutic approach for CML. The most exciting breakthrough in the treatment of CML has been the development of imatinib mesylate (IM) or Gleevec® as an oral therapeutic agent. IM binds to a cleft between the N-terminal adenosine triphosphate binding domain and the C-terminal activation loop that forms the catalytic site of the Abl tyrosine kinase, locking the protein into the inactive conformation.

Morphology and Laboratory Findings

- Bone marrow sections are hypercellular with marked myeloid preponderance and mild to moderate myeloid left shift (Figures 11.1 to 11.3). The paratrabecular myeloid regions are expanded with less mature forms next to the bone trabeculae and more mature forms closer to the center. Eosinophilia is a frequent finding.
- Megakaryocytes are usually increased and may appear in clusters. They often show dysplastic changes, including the presence of numerous micromegakaryocytes ("dwarf" forms) as well as many large, bizarre, multilobulated forms.
- Scattered or clusters of pseudo-Gaucher cells or sea blue histiocytes with abundant wrinkled cytoplasm are often present (Figure 11.2). These cells in Wright's-stained bone marrow smears may show abundant light blue cytoplasm; hence referred to as "sea-blue histiocytes" (Figure 11.2). Sea-blue histiocytes and pseudo-Gaucher cells are loaded with phagocytic

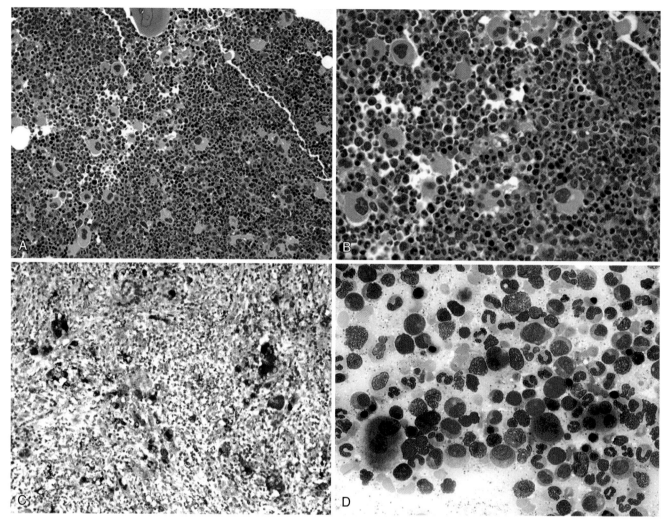

FIGURE 11.1 **CHRONIC MYELOGENOUS LEUKEMIA.** Bone marrow biopsy section showing hypercellularity with marked myeloid preponderance and increased megakaryocytes with numerous small forms (A, low power; B, high power). (C) Immunohistochemical stain for factor VIII shows numerous stained micromegakaryocytes. (D) Bone marrow smear demonstrating several micromegakaryocytes (arrow).

particles and cell membrane debris resulting from the increased turnover of the bone marrow cells.
- There may be patchy or diffuse fibrosis along with osteosclerosis. The extent of fibrosis to some degree correlates with the number of megakaryocytes.
- Bone marrow smears are highly cellular with an elevated M:E ratio of usually >10:1. Eosinophils are increased, and basophilia is a frequent feature but is usually <20%. Dysgranulopoiesis may be present but is not prominent. There may be mild to moderate myeloid left shift, but myeloblasts are usually below 10%.
- Blood smears show marked leukocytosis with a white blood cell count of often >100,000/μL (Figure 11.3). The morphologic findings of the myeloid cells mimic bone marrow with the presence of myeloid left shift and a wide spectrum of myeloid precursors, including myeloblasts and promyelocytes. Similar to the bone marrow smears, myeloblasts in blood smears are usually below 5%. Usually, myelocytes are more numerous than metamyelocytes (myelocyte bulge).

- Absolute basophilia and eosinophilia is common, and there may be absolute monocytosis, but the monocytes in differential count are usually below 5%.
- Scattered nucleated red blood cells may be present. Platelet count is normal or elevated but occasionally reduced during the chronic phase. Neutrophils and bands show reduced alkaline phosphatase activity by cytochemical stains—known as "leukocyte alkaline phosphatase (LAP) score".
- Elevated levels of serum lactic dehydrogenase, uric acid, and vitamin B12 are also frequently observed.
- Splenomegaly is common. The red pulp is diffusely expanded with infiltration of the cords and sinuses filled with mature and immature myeloid cells (Figure 11.4). The malpighian corpuscles (white pulp) are reduced in size and number or are completely absent.

The recent implementation of effective targeted therapy in CML, such as treatment with Gleevec®, (imatinib, nilotinib) has created a necessity for follow-up bone marrow and molecular studies.

MORPHOLOGY AND LABORATORY FINDINGS 157

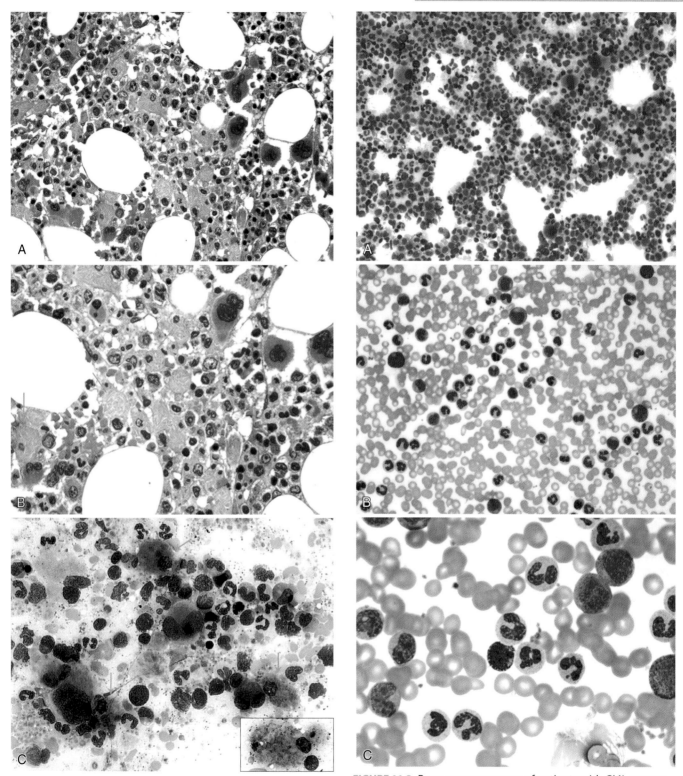

FIGURE 11.2 Bone marrow biopsy section showing aggregates of pseudo-Gaucher cells (A, low power; B, high power, arrows). Several sea-blue histiocytes are present in the bone marrow smear (C, arrows and inset).

FIGURE 11.3 Bone marrow smears of patients with CML appear cellular and show marked myeloid preponderance (A), and peripheral blood smears show marked leukocytosis with myeloid left shift and high proportion of myelocytes and metamyelocytes (B and C); basophils are often present (C).

The evaluation of post-therapy bone marrow samples shows progressive changes toward normal morphology. However, the post-therapy bone marrow samples may show certain morphologic features such as:

1. Frequent presence of non-diagnostic lymphoid aggregates, sometimes paratrabecular, consisting of a mixture of B and T lymphocytes
2. Frequent presence of histiocytic aggregates (pseudo-Gaucher cells)

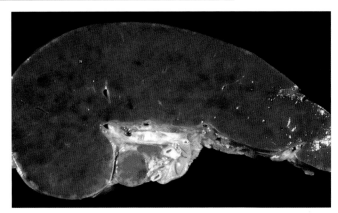

FIGURE 11.4 Splenomegaly is a frequent clinical finding in CML. Splenic involvement is usually diffuse.

Table 11.1
Evolution of Chronic Myelogenous Leukemia (CML)[1]

Stage	Characteristics
Chronic Phase	Leukocytosis (often >100,000/μL) Hypercellular marrow with marked myeloid preponderance Blasts <10% Basophilia <20% Eosinophila Megakaryocytosis with the presence of micromegakaryocytes Splenomegaly Low LAP score t(9;22)(q34;q11) and/or BCR-ABL1 fusion by FISH
Accelerated Phase	Hypercellular marrow with myeloid left shift Blasts 10–19% Basophilia ≥20% Thrombocytopenia <100,000/μL, or thrombocytosis >1000,000/μL Increasing WBC count Increasing spleen size Additional cytogenetic abnormalities
Blast Phase	Blasts ≥20% Sheets or large clusters of blasts in the bone marrow biopsy Extramedullary tissue infiltration by blast cells

[1]Adapted from Jaffe ES, Harris NL, Stein H, Vardiman JW. Pathology and Genetics: Tumors of Haematopoietic and Lymphoid Tissues. IARC Press, Lyon, 2001.

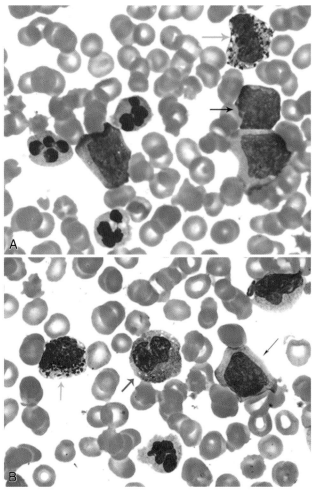

FIGURE 11.5 Peripheral blood smear of a patient with CML-AP showing myeloid left shift with blasts (A and B, black arrows), basophils (A and B, green arrows), and an eosinophil (B, red arrow).

1. The presence of 10–19% blasts in blood or bone marrow samples
2. Basophilia of >20%
3. Persistent thrombocytopenia of <100,000/μL or thrombocytosis of >1,000,000/μL
4. Progressive splenomegaly and/or increasing leukocyte count >100,000/μL
5. Cytogenetic or molecular evidence of clonal evolution.

Increased marrow fibrosis, marked megakaryocytosis with the presence of large clusters or sheets of megakaryocytes, and severe dysgranulopoiesis are also suggestive of CML-AP. CML-AP is a transient phase between the chronic phase and the blast phase.

3. Bone marrow hypocellularity, particularly in cases with a long history of treatment. The degree of hypocellularity in some instances is so severe that the bone marrow biopsy sections resemble aplastic anemia.

ACCELERATED PHASE

Accelerated phase of CML (CML-AP) is often associated with a decline in the patient's clinical condition along with certain laboratory findings. The diagnosis of CML in accelerated phase according to the WHO recommendations is based on the presence of one or more of the following (Table 11.1) (Figure 11.5).

BLAST PHASE

Blast Phase (blast crisis, blast transformation) refers to the evolution of CML into acute leukemia (CML-BP). Enhanced proliferation and differentiation arrest, the characteristic features of blast transformation in CML, seem to be dependent upon the cooperation of BCR-ABL1 with the TP53 (17p13) and RB1 (13q14) genes that appear to be directly or indirectly dysregulated in this process. According to the WHO

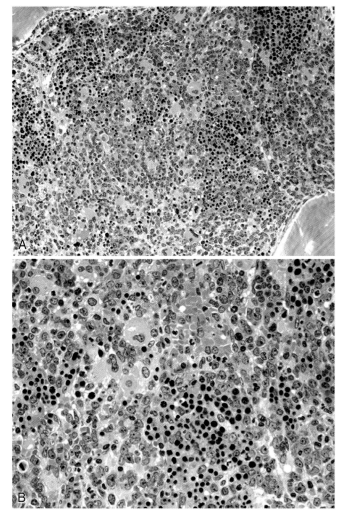

FIGURE 11.6 Bone marrow biopsy section of a patient with a history of CML showing increased blasts consistent with blast transformation (A, low power; B, high power).

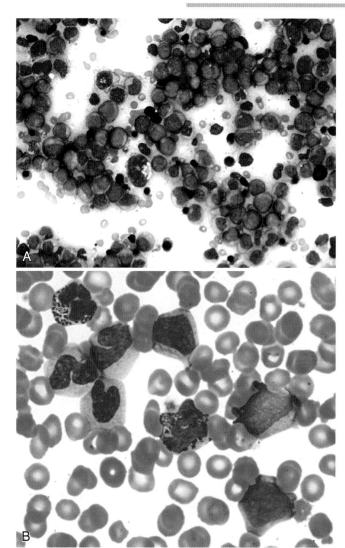

FIGURE 11.7 Bone marrow smear (A) and blood smear (B) of a patient with CML in blast transformation showing numerous blasts and a few basophils (B).

recommendations, the diagnosis of CML-BP is made when (Table 11.1) (Figures 11.6 and 11.7):

1. Blasts are ≥20% of bone marrow nucleated cells or peripheral blood differential counts
2. Large foci or clusters of blasts are present in the bone marrow biopsy sections
3. There is evidence of extramedullary tissue infiltration by blast cells.

Extramedullary CML-BP may involve any tissue, but is frequently observed in spleen, lymph node, skin, and central nervous system. Morphologic features are similar to other acute leukemic infiltrations. There is often the presence of immature eosinophils (eosinophilic myelocytes) which may provide a hint that blasts are of myeloid lineage.

The blast cells in over 70% of CML-BP cases are of myeloid (non-lymphoid) origin and express granulocytic, monocytic-, erythroid-, and/or megakaryocytic-associated CD molecules in immunophenotypic studies. In close to 30% of the CML cases, blast transformation is of lymphoid lineage or biphenotypic.

There may be some degree of dysmyelopoiesis or bone marrow fibrosis.

Immunophenotype

FLOW CYTOMETRY

Multiparametric immunophenotyping by flow cytometry (MIFC) can be useful in diagnosis and monitoring of CML in the following areas:

- Blast enumeration and lineage assignment
- Identification of abnormal blasts with various aberrancies including expression of cross-lineage markers
- Detection of dysmaturation patterns of myeloids, similar to those seen in MDS/MPN
- Enumeration of basophils. Basophils are located at a unique position by CD45 gating, which is similar to that of

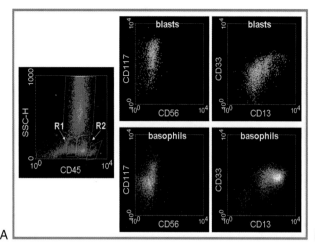
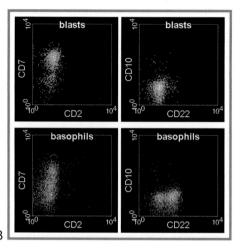

FIGURE 11.8 (A AND B) **FLOW CYTOMETRIC FINDINGS OF CML IN ACCELERATED PHASE.** The CD45 gating reveals excess blasts (R1, 11% of the total) at an abnormal location. In addition, there is a population of abnormal basophils (R2, 10% of the total) present at near-monocytic region but with lower side scatter than normal monocytes. The excess and abnormal myeloblasts (the R1 gate, in green) express dim CD13, dim CD33, CD45 (very dim), CD117, and aberrant CD7. In comparison, the abnormal basophils (the R2 gate, in magenta) express brighter CD13, brighter CD33, moderate CD45, partial dim CD117, and aberrant partial dim CD7. Those abnormal basophils are also positive for dim CD22.

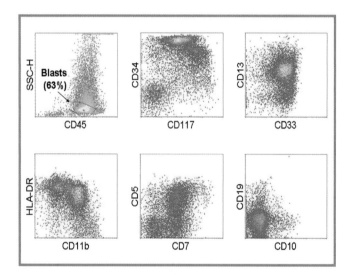

FIGURE 11.9 **FLOW CYTOMETRIC FINDINGS OF CML IN BLAST PHASE.** The patient has a 10-year history of CML, presents with a WBC of 106,000, and is under consideration for a Ponatinib trial. The open gate display by CD45 gating shows excess blasts (63% of the total) present at abnormal location. The density plots of the blast-enriched gate demonstrate abnormal myeloblasts expressing CD11b (subset), CD13, CD33, CD34 (two subsets), and CD117 (two subsets), plus cross-lineage aberrancies of dim CD5, dim CD7, and partial dim CD19.

monocytes but with slightly lower side scatter. Basophils express CD13, CD33, and dim CD22. Immature and abnormal basophils may also express CD117 and/or cross-lineage aberrancies (Figure 11.8).

Findings of CML-BP may be indistinguishable from those of other acute leukemias (Figure 11.9).

IMMUNOHISTOCHEMICAL STAINS

Immunohistochemical stains are helpful for the estimation of blast counts and identification of their lineage, particularly when there is no access to flow cytometry or there is inadequate marrow aspirate (dry tap).

The following markers are frequently used: CD34 (blasts); CD117 (myeloblasts); TdT (lymphoblasts); CD31, CD61, and factor VIII for megakaryoblasts; glycophorin A and hemoglobin A for erythroid precursors; myeloperoxidase; lysozyme and CD68 for granulocytic and monocytic lineages.

Molecular Studies

Detection of the *BCR-ABL1* fusion gene is the hallmark of CML in diagnosis, monitoring, and targeted therapy. While traditionally this has been done by cytogenetic analysis, and more recently molecular cytogenetic (FISH) testing, the most sensitive and quantitative approach is by molecular methods.

The sensitivity of classical cytogenetics is limited by the number of cells cultured to the number of metaphases counted. Moreover, about 5% of CML cases have t(9;22) translocations that may not be visible under the light microscope and will thus be missed by this approach. Molecular testing should be able to detect such "cryptic" translocations and will therefore approach 100% sensitivity for initial diagnosis.

BCR stands for "breakpoint cluster region," reflecting the fact that the point of breakage at that site can occur over a fairly broad region. The important rearrangements for clinical diagnosis are the major breakpoint (*M-BCR*), which produces the p210 gene product and is found primarily in CML, and the minor breakpoint (*m-BCR*) which produces the p190 gene product and is typical of acute lymphoblastic leukemia, more commonly seen in adult patients with that disease. However, some overlap is seen in a minority of cases, owing to alternative splicing of RNA transcripts.

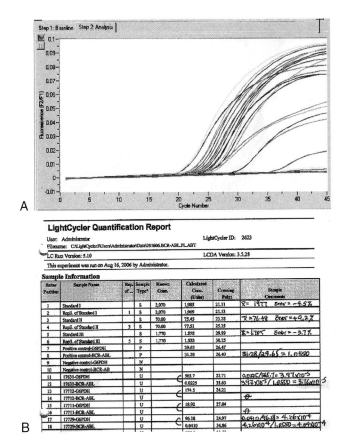

FIGURE 11.10 Data obtained from real-time PCR analysis of a series of patients with suspected, diagnosed, or treated CML using the LightCycler instrument (Roche Molecular Diagnostics, Indianapolis, IN). (A) Amplification curves produced from extracted cellular mRNA that was first reverse transcribed to create cDNA templates; the specimens producing curves that enter log phase at lower PCR cycle number had higher amounts of starting *BCR-ABL* fusion mRNA. (B) Quantitative readout of data from the LightCycler; specimens that are *BCR-ABL* negative show no amplification product from this target (but do show output from the control *GAPDH* gene target), whereas those specimens that are *BCR-ABL* positive show amplification products from both genes. Quantitation of the *BCR-ABL* fusion gene is done mathematically by comparing its amplification to that of the *GAPDH* control gene.

Most laboratories use a real-time PCR system that is both highly sensitive and highly quantitative for the *BCR-ABL* target (Figure 11.10). It is also capable, depending upon how the primer pairs are chosen, of differentiating between the *M-BCR* and *m-BCR* forms, which can be of great importance in the differential diagnosis (in a newly ascertained patient) of acute leukemia versus the blast crisis phase of CML.

The breakpoint region on chromosome 22 spans such a large area that it cannot be efficiently amplified from a genomic DNA target. Instead, messenger RNA must be isolated to serve as the template; since it has had the large intronic regions spliced out, it is of an amplifiable size.

The real-time PCR method is sensitive down to a level of 1 CML cell in 1 million normal cells. It is also highly quantitative, so that trends (upward or downward) in treated patients obtained during periodic monitoring can be used to assess minimal residual disease, relapse, or development of resistance to the anti-tyrosine kinase therapies (imatinib and its successors). (Very few treated patients, even if in apparent clinical and cytogenetic remission, actually go down to undetectable *BCR-ABL1* levels with these sensitive assays).

Most relapses are due to either amplification of the *BCR-ABL1* fusion gene or, more commonly, the development of point mutations affecting amino acid residues at the site of binding of the drug or at more distal sites causing allosteric effects. A total of about 20 such mutations have been described, which can be detected by DNA sequencing or microarray hybridization. Since certain mutations render either similar resistance or sensitivity to newer-generation drugs targeting the same protein, it will be increasingly important to accurately genotype these patients to guide management.

Cytogenetics

As described earlier, the Ph^1 chromosome was the first recurrent cytogenetic rearrangement found in a hematologic disease. Banding techniques developed during the 1970s allowed for the identification of the Ph^1 chromosome as being derived from a translocation between chromosomes 9 and 22, t(9;22)(q34;q11.2) (Figure 11.11).

The translocation was subsequently described as resulting in the fusion of the *ABL1* proto-oncogene (a homolog of the Abelson murine leukemia virus oncogene) on chromosome 9q34 with a gene called *BCR* on chromosome 22q11.2. The *ABL1* gene encodes a tyrosine kinase that phosphorylates several proteins involved in signaling for cell proliferation, and the *BCR* gene encodes a 160 kDa phosphoprotein with kinase activity.

FISH or molecular techniques can be used to establish diagnosis in cases where the t(9;22) cannot be identified by standard karyotyping. CML-BP is often predicted by cytogenetic findings prior to morphology changes; 75–80% of patients develop additional chromosome aberrations as the disease progresses.

Approximately 90–95% of patients present with the t(9;22), whereas the remaining 5–10% of patients have cryptic or complex rearrangements but all of these cases eventually have evidence of *BCR-ABL1* gene fusion. Variant translocations with deletions at the involved breakpoints signify a poorer prognosis, albeit controversial, than the more common t(9;22) (Figure 11.12).

Abnormalities of chromosomes 8, 17, 19, and 22 are most often observed in disease evolution (major route). These account for 70% of patients with evolving disease. Trisomy 8, isochromosome 17, trisomy 19, or an extra Ph^1 chromosome (derivative chromosome 22), singly or in combinations are the most frequently observed secondary changes in blast crisis (Figure 11.13). The remaining 30%

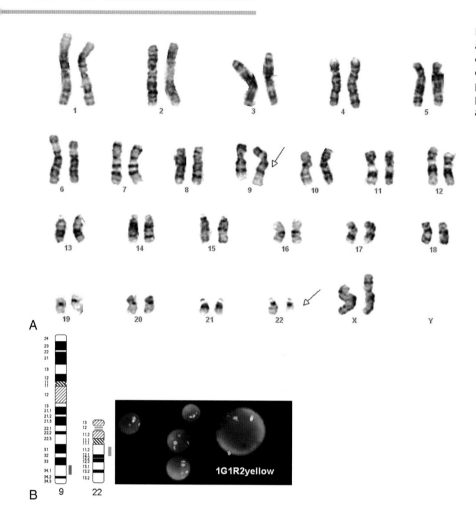

FIGURE 11.11 (A) G-banded karyotype with a classic t(9;22) showing the "Philadelphia" chromosome. (B) An ideogram of chromosomes 9 and 22 showing the FISH probes for the ABL1 and BCR loci. The FISH panel shows cells with red ABL1; green BCR, and yellow (fusion) signals.

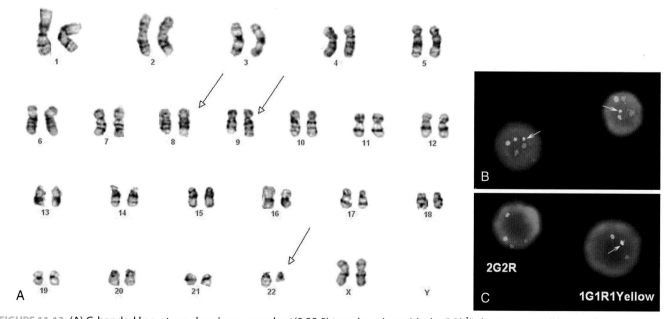

FIGURE 11.12 (A) G-banded karyotype showing a complex t(9;22;8) translocation with the "Ph[1]" chromosome. (B) Interphase FISH analyses on these cells show one fusion signal (yellow) whereas the second fusion signal got rearranged on 9q and 8q subsequent to the initial 9;22 translocation. (C) Cells with a normal signal pattern (2 red, 2 green) and a cell with a deletion of the reciprocal 22:9 fusion signal (1 red, 1 green, 1 yellow).

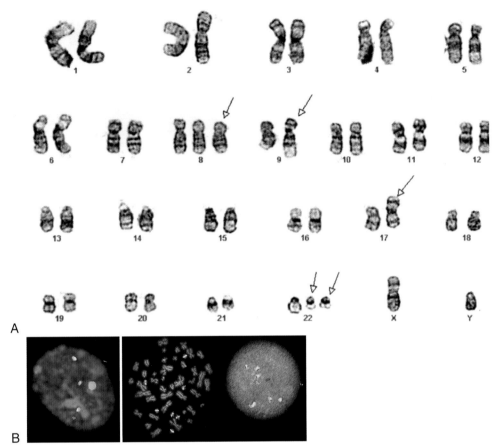

FIGURE 11.13 (A) G-banded karyotype of CML-BP, with an extra Ph^1 chromosome, Trisomy 8, and isochromosome 17q. (B) FISH shows three fusion signals (yellow) representing the additional Ph^1 chromosome. (C) Gains of additional "Ph^1" chromosome during development of imatinib resistance.

of patients with evolving disease develop various secondary aberrations that may include trisomy 21, loss of the Y, monosomy 7 or 17, rearrangements of 3q, or others. Genes known to have roles in transformation include *TP53*, *RB1*, *CDKN2A*, *INK4α*, *MINK*, *AML1*, and *EVI1*, although their role in transformation is currently unknown.

In a minority (~20%) of patients with CML, there are large deletions within the der(9q) with conflicting data regarding the prognostic implications. The size of the deletion on 9q along with the reciprocal *ABL1-BCR* fusion region is variable and the segmental loss houses not only the *ABL1-BCR* fusion, but also two micro-RNAs (miR-219-2 and miR-199b). The identification of these miRNAs presents an exciting possibility in that their loss from the chromosomal deletions could lead to increased target gene expression, promoting CML proliferation and conferring a poor prognosis.

Monitoring, by molecular methods, of response is a critical feature of the management of CML, and international consensus guidelines have been developed suggesting therapeutic end points that should be achieved after specific durations of therapy. It is generally accepted that, at a minimum, patients should attain a complete cytogenetic response (CCyR) because there is almost no disease progression in patients in CCyR after 2–3 years of treatment. Thus it is important that accurate and sensitive measurement of minimal residual disease be established that could help to identify patients in early relapse, allowing changes in therapy that may decrease the rate of disease progression. It is recommended that a CCyR be confirmed by bone marrow cytogenetics before relying exclusively on molecular monitoring. Some practices follow patients serially with peripheral blood FISH analyses, doing a marrow when the FISH becomes negative. Thereafter, patients are followed every 3–4 months with peripheral blood Q-PCR.

Differential Diagnosis

The differential diagnosis includes leukemoid reactions and other subtypes of MPN, such as primary myelofibrosis, essential thrombocythemia, and chronic neutrophilic leukemias.

Leukemoid reaction refers to leukocytosis in excess of 50,000/μL with a left shift, caused by conditions other than a leukemic process (Figure 11.14). The major distinguishing features between CML and leukemoid reaction are summarized in Table 11.2.

Major clinicopathologic differences between MN subtypes are presented in the next chapter, Table 12.2.

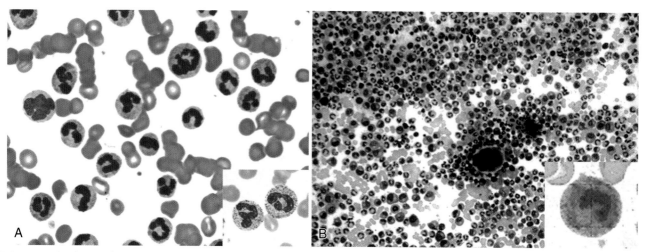

FIGURE 11.14 Leukemoid reaction. (A) Peripheral blood smear showing neutrophilia with (inset) toxic granulation. (B) Bone marrow smear demonstrating myeloid preponderance. A neutrophil strongly positive for LAP is demonstrated in the inset.

Table 11.2

Comparison of Pathologic Findings in CML and Leukemoid Reaction

		CML	Leukemoid Reaction
Peripheral Blood	WBC	Usually >100,000/μL	Usually <50,000/μL
	Granulocytes	All stages of maturation	Almost all mature
	LAP score	Reduced	Elevated
Bone Marrow	M:E ratio	>10:1	<10:1
	Pseudo-Gaucher cells	Frequent	Less frequent
	Paratrabecular fat	Less frequent	Frequent
	Mast cells	Less frequent	Frequent
Splenomegaly		Marked	Variable
Cytogenetic/DNA	t(9;22)	Present	Absent
	Ph^1 positive *BCR-ABL1* rearrangement	Present	Absent

Additional Resources

An X, Tiwari AK, Sun Y, et al: BCR-ABL tyrosine kinase inhibitors in the treatment of Philadelphia chromosome positive chronic myeloid leukemia: a review, *Leuk Res* 34:1255–1268, 2010.

Bixby D, Talpaz M: Seeking the causes and solutions to imatinib-resistance in chronic myeloid leukemia, *Leukemia* 25:7–22, 2011.

Breccia M, Efficace F, Alimena G: Imatinib treatment in chronic myelogenous leukemia: what have we learned so far? *Cancer Lett* 300:115–121, 2011.

Goldman JM: Chronic myeloid leukemia: a historical perspective, *Semin Hematol* 47:302–311, 2010.

Hochhaus A, Weisser A, La Rosée P, et al: Detection and quantification of residual disease in chronic myelogenous leukemia, *Leukemia* 14:998–1005, 2000.

Jaffe ES, Harris NL, Vardiman JW, et al: *Hematopathology*, Philadelphia, 2010, Saunders/Elsevier.

Karbasian Esfahani M, Morris EL, Dutcher JP, et al: Blastic phase of chronic myelogenous leukemia, *Curr Treat Options Oncol* 7:189–199, 2006.

La Rosée P, Deininger MW: Resistance to imatinib: mutations and beyond, *Semin Hematol* 47:335–343, 2010.

Perrotti D, Jamieson C, Goldman J, Skorski T: Chronic myeloid leukemia: mechanisms of blastic transformation, *J Clin Invest* 120:2254–2264, 2010.

Rice KN, Jamieson CH: Molecular pathways to CML stem cells, *Int J Hematol* 91:748–752, 2010.

Shah NP: Advanced CML: therapeutic options for patients in accelerated and blast phases, *J Natl Compr Canc Netw* (Suppl 2) :S31–S36, 2008.

Sullivan C, Peng C, Chen Y, Li D, Li S: Targeted therapy of chronic myeloid leukemia, *Biochem Pharmacol* 80:584–591, 2010.

Swerdlow SH, Campo E, Harris NL, et al: WHO classification of tumours of haematopoietic and lymphoid tissues, ed 4, Lyon, 2008, International Agency for Research on Cancer.

Valent P: Standard treatment of Ph+ CML in 2010: how, when and where not to use what BCR/ABL1 kinase inhibitor? *Eur J Clin Invest* 40:918–931, 2010.

Myeloproliferative Neoplasms Associated with *JAK2* Mutation

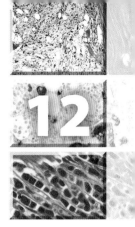

Molecular studies have shown that three subtypes of myeloproliferative neoplasms—polycythemia vera (PV), primary myelofibrosis (PMF), and essential thrombocythemia (ET)—are strongly associated with *JAK2* V617F mutation, suggesting a common pathogenetic pathway for these disorders (Figure 12.1). The close relationship between these three entities was first suggested by William Dameshek in 1951, who considered PV, PMF, and ET to be phenotypically related. He noted significant overlapping features between them. For example, patients with PV frequently show evidence of panmyelosis, and often develop bone marrow fibrosis, leukoerythroblastosis, and progressive splenomegaly. Also, a small proportion of patients with ET may eventually develop bone marrow fibrosis with myeloid metaplasia.

Polycythemia Vera

Polycythemia vera (PV) is characterized by erythrocytosis or increased red blood cell mass with clinical symptoms of hyperviscosity, increased risk of thrombosis, and varying degrees of thrombocytosis, leukocytosis, and splenomegaly. PV, similar to other myeloproliferative neoplasms, is the result of clonal expansion of a pluripotent stem cell with the involvement of the myeloid lineages and a variable proportion of B lymphocytes. However, most T and NK cells seem to be unaffected. The diagnosis of PV is established by complex clinical, laboratory, and morphologic features. The WHO (2008) recommended criteria for the diagnosis of PV are as follows.

1. Major criteria:
   a. Elevated RBC mass >25% above normal range, of Hb >18.5 g/dL in men and >16.5 g/dL in women
   b. Presence of *JAK2* V617F or other functionally similar mutations, such as *JAK2* exon 12 mutation.

2. Minor criteria:
   a. Hypercellular bone marrow for age with panmyelosis and erythroid preponderance
   b. Serum erythropoietin below normal reference range
   c. Evidence of *in vitro* endogenous (erythropoietin-independent) erythroid colony formation.

Diagnosis is established by the presence of both major criteria and one minor criterion, or the presence of the first major criterion with two minor criteria.

One of the interesting current hypotheses in the pathogenesis of PV is the presence of a defect in transcription regulation that affects cytokine receptor signaling. It has been shown that the *JAK2* gene plays an important role in the EOP–EPO receptor signaling in erythropoiesis. A unique clonal mutation in the *JAK2* gene has been reported, resulting in a valine to phenylalanine substitution at position 617, in 65–97% of PV patients.

PV is more frequent in men than in women (about 2:1) with a peak incidence around 70–80 years of age. Most common complaints are non-specific and include headache, weakness, and dizziness. Approximately 5–20% of the patients may complain of arthritis. Itching (particularly after a warm bath) and erythromelalgia (erythroderma with burning pain of extremities) are amongst the common clinical presentations. The survival time for treated patients usually exceeds 10 years. The major causes of death are thrombosis, transformation to MDS or AML, secondary non-hematologic malignancies, and hemorrhage.

Thrombosis is either venous or arterial and appears to be primarily related to the patient's blood hyperviscosity. Transformation to MDS or AML in most instances is related to chemotherapy. In one report the incidence of AML in PV patients was 1.5%, 10%, and 13% for phlebotomy, ^{32}P, and chlorambucil therapy, respectively. Approximately 3–12% of patients may develop a secondary malignancy (carcinoma, non-Hodgkin lymphoma). The current therapeutic modalities include phlebotomy, IFN-α, anagrelide (a quinazolone derivative), and allopurinol.

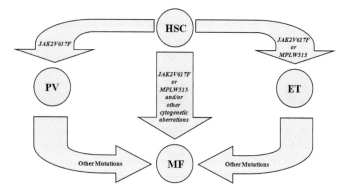

FIGURE 12.1 *JAK2* mutation in myeloproliferative neoplasms. HSC, hematopoietic stem cell; PV, polycythemia vera; ET, essential thrombocythemia; and MF, primary myelofibrosis.

MORPHOLOGY AND LABORATORY FINDINGS

Two distinct clinicopathologic phases have been described in PV: the *"polycythemic"* phase and the *"spent"* phase. The polycythemic phase describes the earlier active phase of the disease when the bone marrow shows panmyelosis and the blood displays erythrocytosis, sometimes in association with thrombocytosis and/or leukocytosis. The spent phase refers to the later stage of the disease when bone marrow is fibrotic or hypocellular, and there is evidence of extramedullary hematopoiesis, progressive splenomegaly, and anemia. In addition, a *transitional phase* has been described, referring to a process between these two phases characterized by erythrocytosis and bone marrow fibrosis. The morphologic features of the polycythemic and spent phases are described below.

Polycythemic Phase (Figure 12.2)

- Bone marrow is hypercellular and shows panmyelosis. Usually, there is marked erythroid preponderance and normoblastic erythropoiesis.
- Megakaryocytes are abundant and pleomorphic, and have a tendency to appear in clusters and/or next to bone trabeculae.
- At this phase, in the majority of PV cases, the bone marrow biopsies show no increase in reticulin fibers, though variable degrees of fibrosis are present in about 30% of cases.
- No significant dysplastic changes are noted in the granulocytic series and there is no evidence of increased myeloblasts. Basophilia is a common feature, and eosinophilia is not infrequent.
- Stainable iron is reduced or absent in the vast majority of cases.
- Peripheral blood shows increased red cell mass and elevated hemoglobin levels (Hb >18.5 g/dL in men and >16.5 g/dL in women). Erythrocytes are usually normochromic and normocytic, but sometimes are hypochromic and microcytic. The presence of deeply basophilic reticulocytes has been described.
- Serum erythropoietin is below normal reference range, the activity of red cell glycolytic enzymes is elevated, and the proportion of fetal hemoglobin is increased.
- Often, there is evidence of leukocytosis and basophilia with thrombocytosis and presence of giant, hypogranular platelets. There may be mild myeloid left shift but blasts are not usually found.
- The LAP score is often elevated.

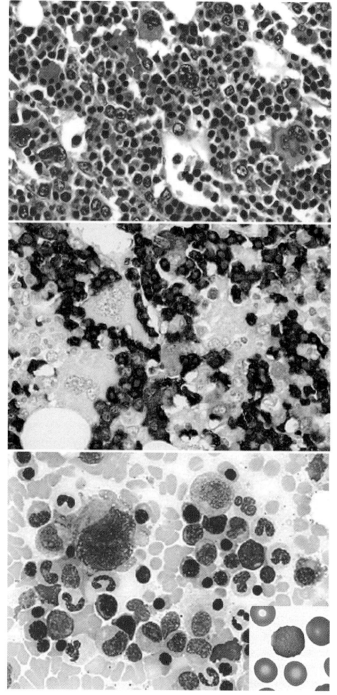

FIGURE 12.2 Bone marrow biopsy section (top) and smear (bottom) from a patient with PV demonstrating erythroid preponderance. The figure in the middle depicts hemoglobin A by immunohistochemical stains. The inset shows a deeply stained polychromatophilic erythrocyte.

Spent Phase

Spent phase is the late stage of PV. It is characterized by (Figure 12.3):

- Normalization of erythrocytosis and then progression to anemia
- Progressive splenomegaly

POLYCYTHEMIA VERA

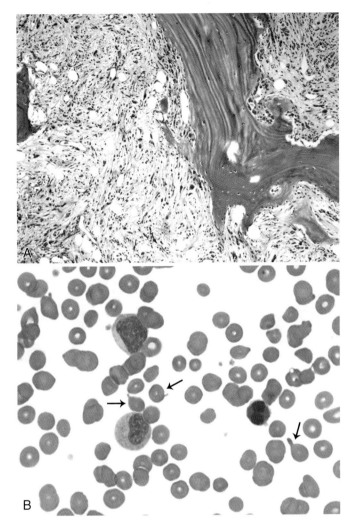

FIGURE 12.3 **POLYCYTHEMIA VERA, SPENT PHASE.** (A) Bone marrow biopsy section demonstrating sclerotic bone and marked fibrosis. (B) Blood smear showing tear drops (arrows), myelocyte, metamyelocyte, and nucleated RBC.

- Myelofibrosis and extramedullary hematopoiesis with a leuko-erythroblastic blood picture
- Myelodysplastic changes, sometimes with increased blasts.

There is some debate regarding the mechanism(s) involved in the evolution of PV to the spent phase. The original idea of the development of anemia in PV as the natural history of the disease due to bone marrow exhaustion has been challenged by the causative effects of chemotherapy, hemorrhage, and deficiencies of iron, vitamin B12, or folic acid. Similarly, the development of myelodysplasia and/or marrow fibrosis may be more secondary to therapy than a naturally occurring phenomenon. For example, the incidence of myelofibrosis is higher in PV patients exposed to chemotherapy or radiation than in those treated by phlebotomy.

IMMUNOPHENOTYPE

No pathognomonic immunophenotypic features have been described in the blood or bone marrow of PV patients.

MOLECULAR STUDIES

- A specific molecular marker for a number of MPD is a point mutation in the *JAK2* gene, V617F.
- It is most frequently seen in PV, where some series find it in >90% of affected patients.
- Testing for the mutation can be of help in the diagnosis of PV, and in its differential diagnosis from other disorders such as CML, chronic myelomonocytic leukemia, and MDS.
- Allele-specific PCR primers can be used to amplify both the mutant region of the gene (if present) and a nearby invariate portion of the gene (as an internal amplification control) (Figure 12.4).
- If more accurate quantitation is desired, one can use real-time PCR incorporating melting-curve analysis (Figure 12.5).
- The major diagnostic mutation is V617F in exon 14. Some studies have identified four additional mutations located in

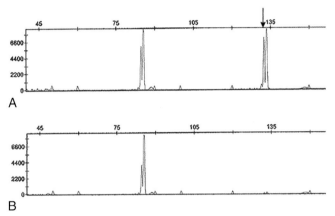

FIGURE 12.4 Detection of the *JAK2* mutation by allele-specific PCR followed by capillary electrophoresis of the amplification products. A positive sample (A) yields both the mutant PCR product (arrow) and the internal control normal gene product, whereas a negative sample (B) yields only the normal PCR product peak.

Sample	%V617F	% WT
PC-WT	0.16	99.84
PC-VF	99.97	0.03
35775	0.27	99.73
35796	0.26	99.74
35818	72.23	27.77
35827	0.24	99.76
35829	0.17	99.83
35851	0.28	99.72
35862	0.37	99.63
35873	0.32	99.68
35881	0.29	99.71
35894	0.27	99.73
35886	0.22	99.78

FIGURE 12.5 Tabular read-out of real-time PCR analysis with quantitation of target DNA templates containing either the V617F mutation or the wild type sequence (reagents from Ipsogen, Marseille, France). Mutant signals under 1% are considered to be background noise. All the samples shown here are negative for the mutation except for #35818 which shows the mutation in about 72% of the cells in the specimen.

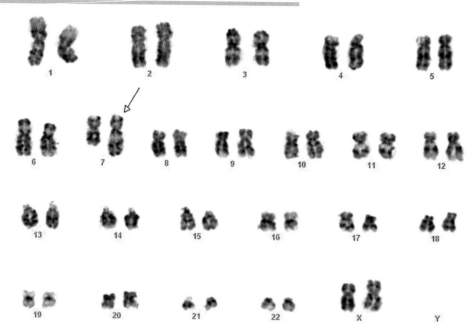

FIGURE 12.6 G-banded karyotype exhibiting an unbalanced 1;7 translocation, resulting in three copies of 1q and loss of one copy of 7q.

exon 12, but they appear to be extremely rare and of uncertain clinicopathologic value at this time.

CYTOGENETICS

- A greater proportion of patients with advanced disease (spent phase) have chromosomally abnormal clones (40–78%) than patients with early (polycythemic) stage (19%).
- The most common chromosomal anomalies are monosomy 20/del(20)(q11.2q13), +8, and +9, with trisomies of 8 and 9, often occurring together in the same clone. Additional abnormalities observed include del(1p), del(3p), t(1;6)(q11.2;p21), and t(1;7)(q10;p10) (Figure 12.6).
- In some patients with PV, a *de novo* leukemia or MDS develops; in these patients, chromosomal anomalies are more similar to the secondary disease than those associated with untreated PV.
- Other patients with PV develop a chromosomally abnormal clone as a consequence of therapy. The most common chromosomal anomaly associated with therapy-related leukemia involves abnormalities of chromosomes 5 or 7 or both and unbalanced translocations such as t(1;7)(q10;p10), resulting in duplication 1q and loss of 7q.
- Almost all the PV patients who develop acute leukemia in late disease stages have chromosomal abnormalities. Trisomy 8 or 9 may persist in PV without further clonal evolution or leukemia development for up to 15 years, whereas other chromosomal abnormalities, such as −7/7q− or −5/5q− or complex changes, may signal the terminal phase of the disease.
- Use of FISH methods does not result in the detection of a significantly increased incidence rate of the chromosomal abnormalities, nor does it reveal submicroscopic deletions in patients with a normal karyotype. However, both FISH and comparative genomic hybridization studies have revealed frequent abnormalities of chromosome 9p.

Primary Myelofibrosis

Primary myelofibrosis (PMF) (chronic idiopathic myelofibrosis, agnogenic myeloid metaplasia, or myelofibrosis with myeloid metaplasia) is a clonal stem cell disorder and a subtype of myeloproliferative neoplasms. It is characterized by myeloproliferation, atypical megakaryocytic hyperplasia, bone marrow fibrosis, extramedullary hematopoiesis, and marked splenomegaly.

The etiology of PMF is not known. However, in a small proportion of the cases, development of PMF has been linked to ionizing radiation, thorium dioxide, and petroleum derivatives such as toluene and benzene.

The incidence of PMF is approximately 1 per 100,000 with a median age of about 65 years. Men and women are equally affected. Marked splenomegaly is the hallmark, particularly in advanced stages. Non-specific symptoms such as fatigue, weight loss, night sweats, and fever are often present. Patients frequently show anemia with abnormal (low or high) white cell and/or platelet counts. Marked splenomegaly may lead to splenic infarction, portal hypertension, and thrombosis of the small portal veins.

Advanced age, anemia, and chromosomal abnormalities are considered poor prognostic indicators. Anemia in about 20% of the patients is severe (Hb <8 g/dL) and may be due to several factors, such as reduced bone marrow erythropoiesis, splenic sequestration and destruction of red cells, autoimmune hemolysis, and/or bleeding.

Therapeutic approaches include treatment with hydroxyurea, splenectomy, splenic irradiation, allogeneic or autologous stem cell transplantation, and the use of antiangiogenic drugs.

MORPHOLOGY AND LABORATORY FINDINGS

Pathologic features consist of two fundamental components: myelofibrosis and extramedullary hematopoiesis. Myelofibrosis appears to be a reactive process secondary to the activation of bone marrow stromal cells, particularly fibroblasts. The morphologic findings, such as bone marrow cellularity, extent of fibrosis, peripheral blood findings, and extramedullary hematopoiesis, vary considerably in different stages of the disease. The evolutionary process of PMF in bone marrow and blood can be divided into two major phases: prefibrotic and fibrotic.

Prefibrotic Stage

Prefibrotic stage or cellular phase is the early stage of the disease, when bone marrow fibrosis is lacking or there is only a minimal amount of reticulin fibrosis (Figures 12.7 and 12.8). At this stage:

- The bone marrow, similar to the other MPDs, is hypercellular and displays panmyelosis.
- Megakaryocytes are atypical and often appear in clusters around the sinusoids and/or bone trabeculae. Abnormal nuclear lobulation, naked nuclei, and large bizarre forms are frequent findings. Micromegakaryocytes are often present.
- Megakaryocytes may show increased rate of emperipolesis (entry of hematopoietic cells into megakaryocytic cytoplasm) (Figure 12.9). Emperipolesis, which is probably induced by abnormal localization of P-selectin on the megakaryocytes, may cause megakaryocytic damage and release of the growth factors leading to the activation and proliferation of fibroblasts.
- There may be a myeloid left shift with a higher proportion of the intermediate cells and <10% myeloblasts.
- Dilatation of bone marrow sinusoids is a common feature. The dilated sinusoids often contain aggregates of hematopoietic precursors, including dysplastic megakaryocytes.
- Lymphoid aggregates are present in up to 25% of cases.
- Osteosclerosis may be present.
- Blood examination may reveal mild to moderate anemia, leukocytosis, thrombocytosis, and presence of scattered teardrop red cells. There may be mild leukoerythroblastosis (the presence of immature myeloid and erythroid cells).

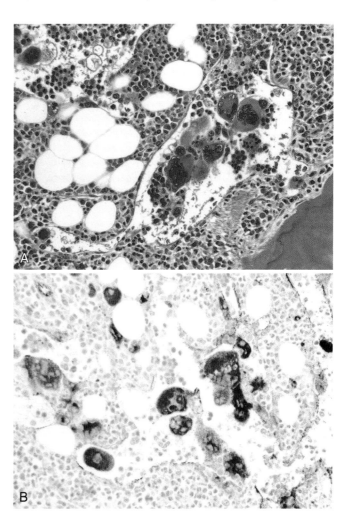

FIGURE 12.7 Bone marrow biopsy sections in cellular phase of PMF showing dilated sinuses with clusters of hematopoietic cells (A) and minimal amount of increased reticulin fibers (B, reticulin stain).

FIGURE 12.8 (A) Bone marrow biopsy section of cellular phase of PMF demonstrating a dilated sinus containing numerous megakaryocytes and erythroid and myeloid precursors. (B) Numerous megakaryocytes are demonstrated by immunohistochemical stain for factor VIII.

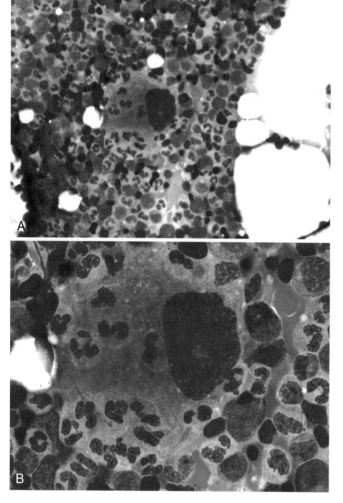

FIGURE 12.9 Bone marrow smear from a patient with PMF demonstrating a megakaryocyte with emperipolesis (A, low power; B, high power).

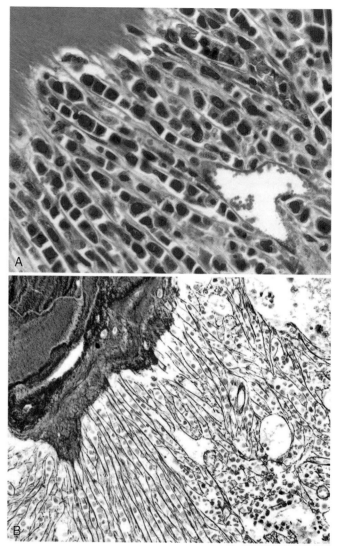

FIGURE 12.10 Bone marrow biopsy section from a patient with PMF demonstrating separation of hematopoietic cells by delicate fibers (A) which are positive with reticulin stain (B).

Fibrotic Stage

- The bone marrow in advanced stages of the disease is virtually replaced by a dense fibrous tissue with markedly reduced cellularity, scattered trapped dysplastic megakaryocytes, and small islands of erythroid and myeloid precursors (Figures 12.10 to 12.12).
- The bone marrow aspiration is usually unsuccessful and results in a "dry" tap.
- Blood examination reveals leukoerythroblastic morphology with the presence of immature myeloid and erythroid cells (Figure 12.13). Blasts may be present but are usually <5%.
- Tear-drop-shaped red cells (dacrocytes) are commonly present, and anemia is a frequent finding.
- There is often mild to moderate leukocytosis. Hypersegmented neutrophils may be present.
- The platelet count may be increased or reduced with abnormal forms present. Bare megakaryocytic nuclei and micromegakaryocytes are often detected.
- The LAP score is often increased, and serum levels of lactate dehydrogenase and uric acid may be elevated.
- Significant dysplastic changes in myeloid series and blasts >10% are suggestive of an accelerated phase. Transformation to AML has been reported in over 5% of the cases (Figure 12.14).

Extramedullary Hematopoiesis

Extramedullary hematopoiesis is often observed in the spleen and liver (Figures 12.15 and 12.16), but is also seen in other sites, such as lymph nodes (Figure 12.17), lung, serosal surfaces, urogenital system, skin, and retroperitoneal and paraspinal spaces. In the spleen, the red pulp is involved with the presence of erythroid, myeloid, and megakaryocytic cells in the sinuses. The extent of red pulp involvement and the proportion of each hematopoietic lineage vary from case to case, but megakaryocytes

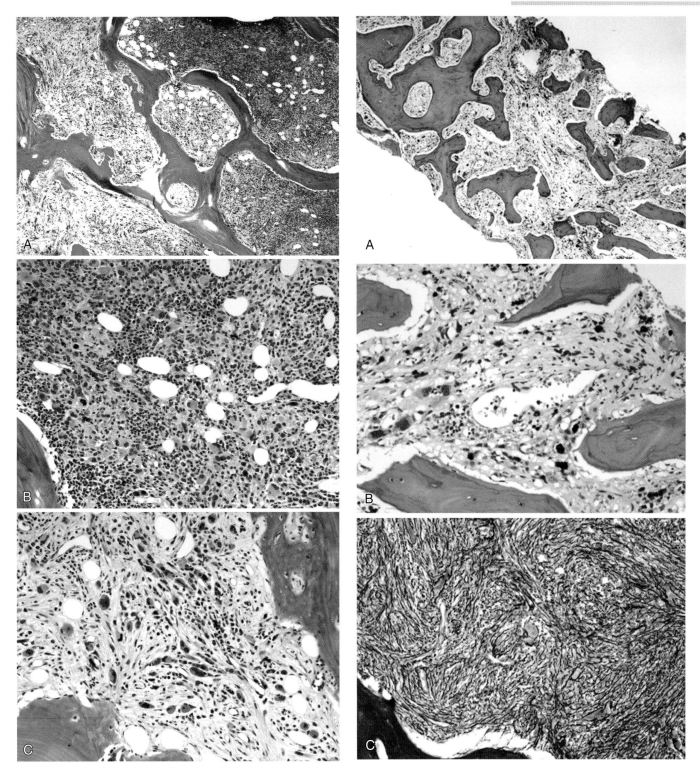

FIGURE 12.11 Bone marrow biopsy section from a patient with PMF demonstrating two distinct areas (A) adjacent to each other. One area is more cellular and less fibrotic (B) and the other area is more fibrotic and less cellular (C).

FIGURE 12.12 The advanced (fibrotic) stage of PMF is characterized by extensive fibrosis and reduced number of hematopoietic cells (A, low power; B, high power; C, reticulin stain).

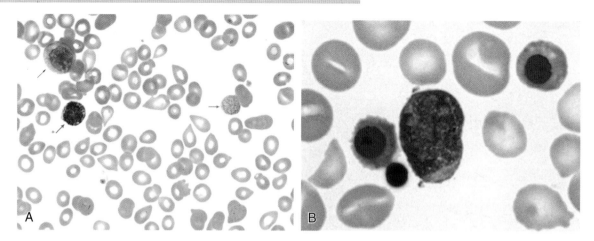

FIGURE 12.13 Peripheral blood smears from a patient with PMF. (A) Low power demonstrating numerous tear-drop-shaped erythrocytes, basophilic stippling (blue arrow), a myelocyte (black arrow), and a basophil (red arrow). (B) High power showing two nucleated red cells and a myeloblast (leukoerythroblastosis).

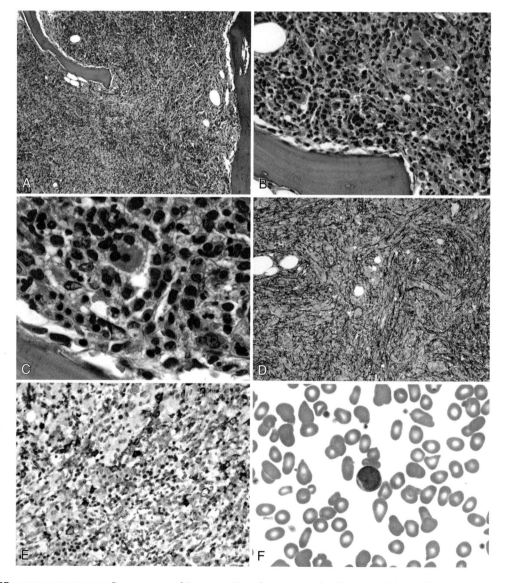

FIGURE 12.14 **PMF IN ACCELERATED PHASE.** Bone marrow biopsy section showing marked hypercellularity (A, low power) with myeloid left shift, increased blasts and micromegakaryocytes (B, intermediate and C, high power). There is an increase in reticulin fibers (D, reticulin stain), and increased CD117+ cells (E). A myeloblast is shown in blood smear (F).

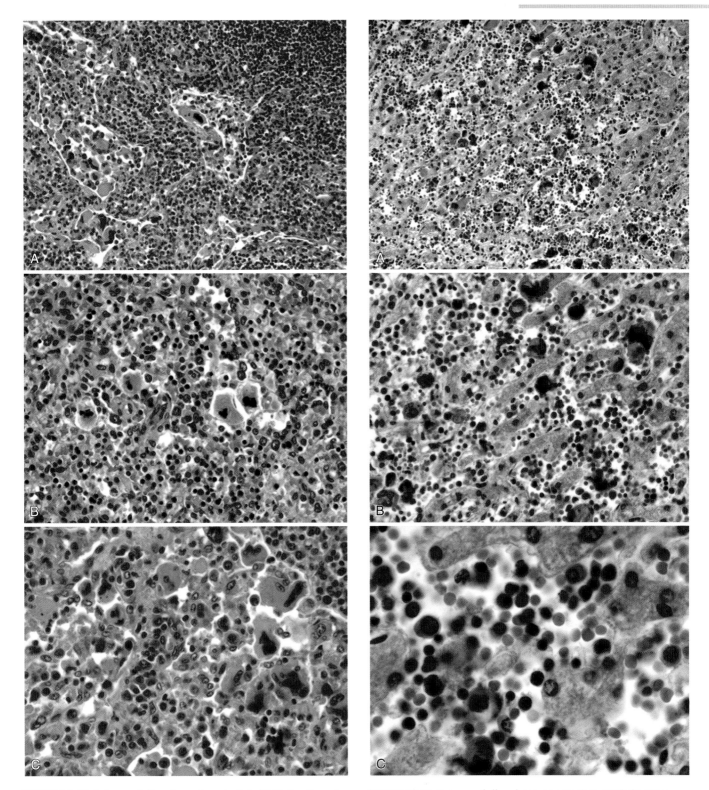

FIGURE 12.15 Extramedullary hematopoiesis in PMF. Section of spleen demonstrating numerous megakaryocytes and scattered erythroid and myeloid precursors (A, low power; B, intermediate power; C, high power).

FIGURE 12.16 Extramedullary hematopoiesis in PMF. Section of liver demonstrating numerous megakaryocytes and erythroid cells with scattered myeloid precursors (A, low power; B, intermediate power; C, high power).

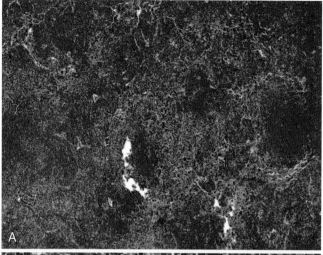
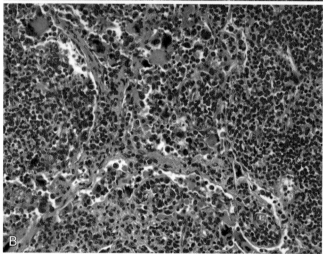
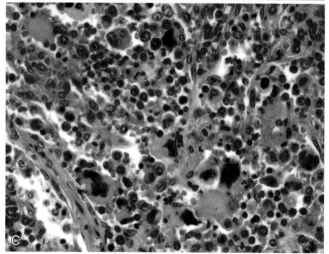

FIGURE 12.17 Extramedullary hematopoiesis in PMF. Section of a lymph node demonstrating numerous megakaryocytes and a mixture of erythroid and myeloid precursors (A, low power; B, intermediate power; C, high power).

are commonly prominent. Similarly, sinuses are the main sites of extramedullary hematopoiesis in the liver and the lymph nodes. The splenic cords and the hepatic parenchyma may show various degrees of fibrosis.

IMMUNOPHENOTYPE

No pathognomonic immunophenotypic features have been described in the blood, bone marrow, or other tissues in patients with PMF.

MOLECULAR AND CYTOGENETIC STUDIES

- There are numerous reports showing an association between *JAK2* mutation V617F and *BCR-ABL1*-negative MPN.
- The mutation has been found in about half of cases of MPF.
- The proportion of cases of PMF with abnormal karyotypes ranges from 30% to 75%, and distinct recurrent chromosomal aberrations have been reported in 40–50% of patients. This discrepancy is mostly due to difficulty in sampling adequate numbers of quality metaphases from the few cells aspirated from fibrotic marrow.
- The most common aberrations—del(13q) and translocations involving chromosome 13q14—likely interrupt the *RB1* gene, an important tumor-suppressor gene in retinoblastoma, osteosarcoma, and other solid tumors.
- Duplication of 1q, del(13q), del(20q), and trisomy 8 appear in approximately two-thirds of patients, and rarer anomalies include trisomy 9 and del(12p).
- Isolated cases with balanced translocations mostly involving chromosomes 1q and 12p with different partners have been reported.
- Specific cytogenetic abnormalities in PMF are associated with significantly different survival outcomes. Prognostically favorable aberrations include 13q and 20q-, whereas prognostically unfavorable abnormalities include 12p- and +8.

Essential Thrombocythemia

Essential thrombocythemia (ET), or primary thrombocytosis, is a clonal stem cell disorder characterized by protracted thrombocytosis in the peripheral blood and increased number of megakaryocytes with atypical features in the bone marrow. The thrombopoietic gene (*TPO*) and its receptor *c-MPL* (myeloproliferative leukemia virus oncogene) have been implicated in the pathogenesis of ET (see Figure 12.1). Several approaches based on X-linked DNA and transcript analysis have demonstrated clonal involvement in the development of ET. However, more recent studies suggest heterogeneity in this process, with the involvement of stem cells at different levels and evidence of polyclonal hematopoiesis in up to 50% of cases.

ET is diagnosed by the exclusion of reactive thrombocytosis (Table 12.1), familial thrombocytosis, and other subtypes of MPN. The WHO criteria for the diagnosis of ET are presented in Box 12.1.

The incidence of ET is approximately 2.5 per 100,000, with a median age of about 60 years. Women are affected

Table 12.1

Major Differences between Essential Thrombocytosis (ET) and Reactive Thrombocytosis (RT)[1]

Features	ET	RT
Underlying disorder (inflammation, infection, malignancy, ischemia, tissue damage)	No	Yes
Digital or cerebrovascular ischemia	Yes	No
Arterial or venous thrombosis	Increased risk	No
Bleeding complications	Increased risk	No
Splenomegaly	May be present (40%)	No
Iron deficiency	No	May be present
Platelet function	May be abnormal	Normal
Large atypical megakaryocytes	Yes	No
Cytogenetic abnormalities	May be present	No
Plasma IL-6	Low	High
Plasma C-reactive protein	Low or normal	High
Spontaneous colony formation	Yes	No

[1] Adapted from Naeim F. Pathology of Bone Marrow, 2nd edn. Williams and Wilkins, Baltimore, 1988 and Schafer AI. Thrombosis. N Engl J Med 2004; 350: 1211–1219.

Box 12.1 Criteria for the Diagnosis of Essential Thrombocythemia (ET)[1]

Positive Criteria
1. Sustained platelet count of ≥450,000/μL
2. Megakaryocytosis with the presence of enlarged forms in the bone marrow
3. No significant increase or left shift in erythroid or myeloid lineages
4. Evidence of JAK2V617F or other cytogenetic aberrations

Criteria of Exclusion
1. No evidence of other types of myeloproliferative neoplasms, such as polycythemia vera, primary myelofibrosis (lack of collagen fibrosis, minimal or lack of reticulin fibrosis), CML (no Ph^1 or *BCR-ABL1* fusion gene)
2. No evidence of myelodysplastic syndrome
3. No evidence of reactive thrombocytosis (no underlying inflammation, infection, cancer or tissue damage; no iron deficiency and no history of splenectomy)

[1] Adapted from Jaffe ES, Harris NL, Stein H, Vardiman JW. Pathology and Genetics: Tumors of Haematopoietic and Lymphoid Tissues. IARC Press, Lyon, 2001 and Swerdlow SH, Campo E, Harris NL, et al. WHO Classification of Tumours of Haematopoietic and Lymphoid Tissues, 4th edn. International Agency for Research on Cancer, Lyon, 2008.

---

more than men, with a female to male ratio of about 2. Up to 50% of patients are asymptomatic. The remaining patients may show splenomegaly (25–40%) and/or other clinical symptoms such as headache, lightheadedness, syncope, erythromelalgia, transient visual disturbances, and thrombohemorrhagic incidents (15–25%). A history of prior thrombosis has an adverse prognostic value. Some ET patients with markedly elevated platelet counts may show acquired von Willebrand deficiency with an increased tendency of bleeding. ET in approximately 2–4% of the patients may transform to PV, PMF, or AML after a median follow-up of about 10 years. The criteria for post-ET myelofibrosis are summarized in Box 12.2

Hydroxyurea, anagrelide, interferon, and pipobroman are the agents most frequently used for the treatment of ET to reduce the platelet count and the risk of thrombosis. The therapeutic use of JAK2 inhibitors is under evaluation.

MORPHOLOGY AND LABORATORY FINDINGS

- The bone marrow is usually normocellular for age or moderately hypercellular, but occasionally hypocellular.
- The morphologic hallmarks are lack or minimal reticulin fibrosis and increased number of megakaryocytes, including large or giant forms (Figures 12.18 to 12.22). These

Box 12.2 Criteria for Post-ET Myelofibrosis[1]

Required Criteria
- Documented previous diagnosis of ET
- Moderate to severe myelofibrosis

Additional Criteria (2 required)
- Development of anemia
- Leukoerythroblastosis
- Progressive splenomegaly
- Increased serum LDH levels
- Development of two to three constitutional symptoms (>10% weight loss in 6 months, night sweats, and/or unexplained fever)

[1] Adapted from Swerdlow SH, Campo E, Harris NL, et al. WHO Classification of Tumours of Haematopoietic and Lymphoid Tissues, 4th edn. International Agency for Research on Cancer, Lyon, 2008.

megakaryocytes may show hyperlobulated nuclei and/or appear in clusters or diffusely dispersed. The megakaryocytic clusters may be found around the sinusoids or close to the bone trabeculae.

- Bone marrow smears show increased number of large megakaryocytes and the presence of platelet aggregates. Emperipolesis (internalization of hematopoietic cells) is a frequent finding in megakaryocytes (Figure 12.20B).
- There is no evidence of increased myeloblasts or significant dysplastic changes in the granulocytic or erythroid lineages. Some cases may show marked erythroid preponderance mimicking PV.
- Blood smears show marked thrombocytosis with variation in platelet size and the presence of giant forms (Figure 12.21B). Megakaryocytic fragments may be present. The white blood cell count is normal or slightly elevated with normal differential counts or mild granulocytosis. Rarely, early granulocytic forms may be present. Red blood cells are normochromic and normocytic. Dacrocytes (tear-drop-shaped red cells) and leukoerythroblastosis are not characteristic features of ET.

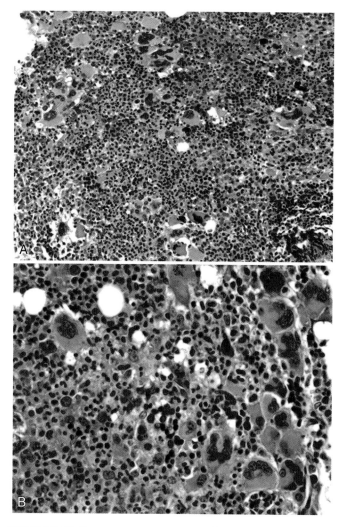

FIGURE 12.18 **ESSENTIAL THROMBOCYTHEMIA.** Bone marrow biopsy section from a patient with ET demonstrating marked hypercellularity with clusters of megakaryocytes, including large and bizarre forms (A, low power; B, high power).

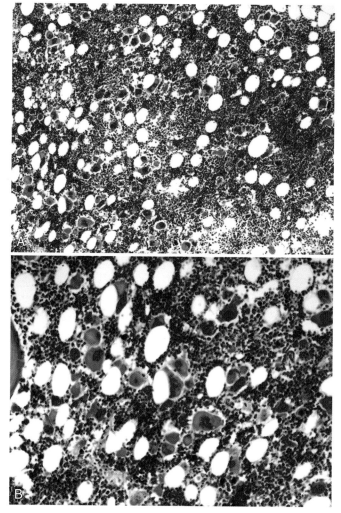

FIGURE 12.19. Bone marrow biopsy sections from a patient with ET demonstrating hypercellularity and several clusters of megakaryocytes with no evidence of marrow fibrosis (A, low power; B, high power).

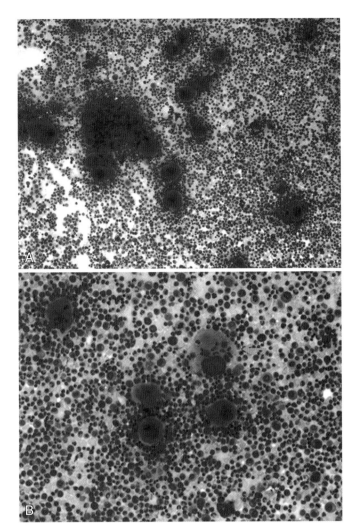

FIGURE 12.20 Numerous megakaryocytes are present in the bone marrow smear from a patient with ET (A), one showing emperipolesis (B, arrow).

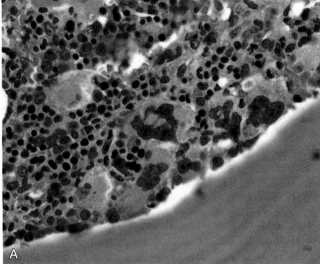

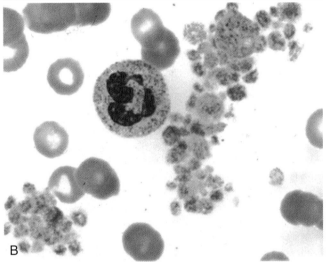

FIGURE 12.21 Bone marrow biopsy section from a patient with ET demonstrating a paratrabecular megakaryocytic aggregate (A) and platelet aggregates in blood smear (B).

In summary, morphologic findings in ET are not specific and may be seen in reactive thrombocytosis as well as other subtypes of myeloproliferative neoplasms, particularly PV and cellular phase of PMF.

IMMUNOPHENOTYPIC STUDIES

No pathognomonic immunophenotypic features have been described in the blood or bone marrow of patients with ET.

MOLECULAR AND CYTOGENETIC STUDIES

- *JAK2* V617F mutation is present in about 50% of ET patients. The presence of *JAK2* mutation in ET has been correlated with a higher frequency of transformation of ET to PV in some studies.
- While no consistent chromosomal anomalies are associated with ET, a few aberrations have been observed in about 5–7% of patients, such as der(1;15)(q10;q10), resulting in duplications of 1q,
- Chromosomal anomalies in ET may have developed as a consequence of therapy or as a *de novo* leukemic clone. However, one group detected a low percentage (<10%) of trisomy 8 and/or Trisomy 9 in about 55% of their patients by FISH.

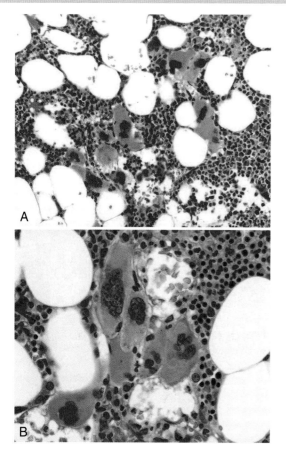

FIGURE 12.22 Bone marrow biopsy sections from a patient with ET demonstrating increased megakaryocytes, including large and bizarre forms (A, low power; B, high power).

Differential Diagnosis

At the beginning of Chapter 10 we displayed a long list of morphologic findings shared by various types of MPN. Diagnosis of a MPN subtype is based on the exclusion of a wide variety of reactive conditions as well as other subtypes of MPN. A summary of the distinctive features of CMPD subtypes is presented in Table 12.2). Following are pathologic features that may help to distinguish the four major different types of MPN: PV, CML, PMF, and ET.

BONE MARROW FINDINGS

Fibrosis is the hallmark of the advanced stage of PMF, whereas it is absent or minimal in ET. Other types of MPN may show various degrees of fibrosis. Development of marrow fibrosis in CML may indicate accelerated phase or evolving blast transformation.

Megakaryocytes are commonly increased and show abnormal morphology in various subtypes of MPN. Dwarf forms or micromegakaryocytes are frequent in CML, giant megakaryocytes with frequent emperipolesis are common in ET and PMF, and dysplastic clusters of megakaryocytes are frequently found in the dilated sinusoids in early stages of PMF.

Eosinophilia and **basophilia** are most common in CML and least frequent in ET and PV. Progressive increase in basophils in CML patients is suggestive of AP.

Table 12.2

The Major Clinicopathologic Features in Myeoplroliferative Neoplasms (MPN)[1]

Findings	PV	ET	PMF	CML
Bone Marrow				
Hypercellularity	+	±	Variable	+++
Increased myelopoiesis	+	−	+	+++
Increased erythropoiesis	+	−	±	−
Increased megakaryocytes	+	+	+	+
Fibrosis	Variable	Minimal	+ to +++	+
Osteosclerosis	−	−	+	+
Peripheral Blood				
Increased WBC	+	+	++	+++
Basophilia	±	±	+	++
Eosinophila	±	±	±	+
RBC abnormalities	±	±	+++	±
Leukoerythroblastosis	−	−	+++	+
Thrombocytosis	+	+++	±	+
Abnormal platelets	±	+++	++	++
LAP (NAP) score	Elevated	Normal	Elevated	Low
Others				
Splenomegaly	+	+	++++	+++
Cytogenetics	+8, +9, del(20q)	+8, +9	+8, +9, del(20q)	t(9;22)
Molecular	JAK2	JAK2	JAK2	BCR-ABL1

[1] Adapted from Foucar K. Bone Marrow Pathology, 2nd edn. ASCP, 2001.

PERIPHERAL BLOOD FINDINGS

Leukoerythroblastosis is considered the morphologic hallmark of PMF, but it may also be seen in CML in chronic, accelerated, or blastic phases. Leukoerythroblastosis is not a feature of PV or ET.

Thrombocytosis is frequently seen in all types of MPN, but its persistent elevation of >600,000/μL is more characteristic of ET. Giant platelets and megakaryocytic fragments have been observed more frequently in ET than in the others.

Granulocytosis is marked (>50,000/μL) and left-shifted with a significant number of myelocytes and metamyelocytes in CML. It is mild to moderate, but often left-shifted, in PMF, and less frequent in PV or ET.

Dacrocytes (tear-drop-shaped red blood cells) are the morphologic indicators of bone marrow fibrosis and, therefore, the most frequent finding in PMF. Dacrocytosis is often associated with other red blood cell morphologic abnormalities, such as anisopoikilocytosis and the presence of schistocytes (fragmented red blood cells).

MOLECULAR AND CYTOGENETIC STUDIES

At the molecular level, the finding of a *BCR-ABL1* fusion gene can help to separate CML from the other MPN subtypes. Conversely, detection of the *JAK2* mutation can distinguish PV, ET, and PMF. Molecular methods that allow specific subtyping of the three *BCR-ABL1* isoforms (p190, p210, and p230) can assist in distinguishing ALL from the blast phase of CML, as well as rare conditions such as chronic neutrophilic leukemia.

Additional Resources

Abdel-Wahab OI, Levine RL: Primary myelofibrosis: update on definition, pathogenesis, and treatment, *Annu Rev Med* 60:233–245, 2009.

Beer PA, Green AR: Pathogenesis and management of essential thrombocythemia, *Hematology Am Soc Hematol Educ Program*: 621–628, 2009.

Charafeddine KM, Mahfouz RA, Zaatari GS, et al: Essential thrombocythemia with myelofibrosis transformed into acute myeloid leukemia with der(1;15)(q10;q10): case report and literature review, *Cancer Genet Cytogenet* 200:28–33, 2010.

Fabris F, Randi ML: Essential thrombocythemia: past and present, *Intern Emerg Med* 4:381–388, 2009.

Klco JM, Vij R, Kreisel FH, Hassan A, Frater JL: Molecular pathology of myeloproliferative neoplasms, *Am J Clin Pathol* 133:602–615, 2010.

Landolfi R, Nicolazzi MA, Porfidia A, et al: Polycythemia vera, *Intern Emerg Med* 5:375–384, 2010.

McMullin MF: The classification and diagnosis of erythrocytosis, *Int J Lab Hematol* 30:447–459, 2008.

Randi ML, Ruzzon E, Tezza F, et al: JAK2V617F mutation is common in old patients with polycythemia vera and essential thrombocythemia, *Aging Clin Exp Res* 23:17–21, 2011.

Rumi E, Elena C, Passamonti F: Mutational status of myeloproliferative neoplasms, *Crit Rev Eukaryot Gene Expr* 20:61–76, 2010.

Tefferi A, Skoda R, Vardiman JW: Myeloproliferative neoplasms: contemporary diagnosis using histology and genetics, *Nat Rev Clin Oncol* 6:627–637, 2009.

Tefferi A, Vainchenker W: Myeloproliferative neoplasms: molecular pathophysiology, essential clinical understanding, and treatment strategies, *J Clin Oncol* 29:573–582, 2011.

Vannucchi AM, Guglielmelli P: Advances in understanding and management of polycythemia vera, *Curr Opin Oncol* 22:636–641, 2010.

Wadleigh M, Tefferi A: Classification and diagnosis of myeloproliferative neoplasms according to the 2008 World Health Organization criteria, *Int J Hematol* 91:174–179, 2010.

Zhan H, Spivak JL: The diagnosis and management of polycythemia vera, essential thrombocythemia, and primary myelofibrosis in the JAK2 V617F era, *Clin Adv Hematol Oncol* 7:334–342, 2009.

Chronic Neutrophilic and Chronic Eosinophilic Leukemias

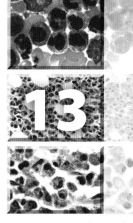

Chronic neutrophilic leukemia and chronic eosinophilic leukemia are rare subtypes of myeloproliferative neoplasms. The diagnosis of these two entities is based on the exclusion of a long list of reactive and neoplastic conditions.

Chronic Neutrophilic Leukemia

Chronic neutrophilic leukemia (CNL) is characterized by persistent peripheral blood neutrophilia, bone marrow hypercellularity, and hepatosplenomegaly. It is a rare condition which shares many morphological features with leukemoid reactions, but unlike leukemoid reactions, CNL is not associated with fever, infection, inflammatory process, or malignancy. In order to establish a diagnosis of CNL, all other myeloproliferative disorders and all causes of secondary (reactive) neutrophilia should be excluded.

Clonality of CNL has been suggested by methylation studies of the X-linked hypoxanthine phosphoribosyl transferase gene and other probes. Also, reports of the evolution of PV into CNL and the transformation of CNL to AML in certain cases support the clonal nature of this disorder.

CNL generally affects women and men over 60 years of age. In most cases, the disease behaves aggressively with a mean survival of <2 years. CNL has been frequently associated with plasma cell myeloma. The cause of death is often cerebral hemorrhage or infection. Due to the rarity of the disease, no standard therapeutic protocols are currently available. At the present time, allogeneic bone marrow transplantation appears to be the only potential cure. Certain drugs, such as hydroxyurea and IFN-α, may help to control granulocytosis and splenomegaly.

Box 13.1 Criteria for the Diagnosis of Chronic Eosinophilic Leukemia, NOS[1]

Criteria
1. Persistent peripheral blood eosinophilia ≥1500/μL
2. Bone marrow eosinophilia
3. Blasts <20% in blood or marrow.

Exclusions
1. All causes of reactive eosinophilia secondary to allergic, parasitic, infections, pulmonary, and collagen vascular diseases.
2. All neoplastic disorders with secondary, reactive eosinophilia, such as T-cell lymphoid malignancies, Hodgkin lymphoma, acute lymphoblastic leukemia/lymphoma, and mastocytosis.
3. Neoplastic disorders for which eosinophils are a part of the neoplastic clone, such as other chronic myeloproliferative neoplasms, myelodysplastic syndromes, and acute myelogenous leukemia.
4. Conditions associated with aberrant expression or abnormal cytokine production of T lymphocytes.
5. Evidence of *BCR-ABL1* fusion or t(9;22)(q34;q11.2), t(5;12)(q31–35;p13), inv(16)(p13q22), or t(16;16)(p13;q22)
6. Rearrangement of *PDGFRA*, *PDGFRB* or *FGFR1*.

[1] Jaffe ES, Harris NL, Stein H, et al. Pathology and Genetics: Tumors of Haematopoietic and Lymphoid Tissues. IARC Press, Lyon, 2001 and Swerdlow SH, Campo E, Harris NL, et al. WHO Classification of Tumours of Haematopoietic and Lymphoid Tissues, 4th edn. IARC Press, Lyon, 2008.

MORPHOLOGY (FIGURE 13.1)

- Bone marrow is hypercellular with marked granulocytic hyperplasia and an elevated M:E ratio approaching 10:1 or higher (Figure 13.1A). There is no evidence of increased blasts or promyelocytes. Bone marrow fibrosis is infrequent.
- The erythroid line is unremarkable, and megakaryocytes are either adequate or increased. No significant dysplastic changes are present.

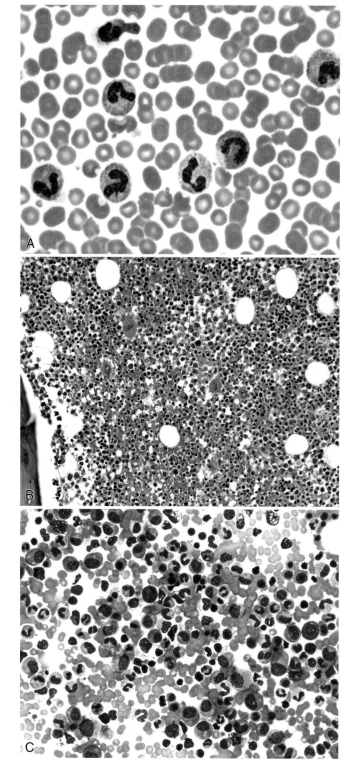

FIGURE 13.1 **CHRONIC NEUTROPHILIC LEUKEMIA.** (A) Blood smear reveals neutrophilia with toxic granulation. (B) Bone marrow biopsy section is hypercellular with myeloid preponderance, and (C) bone marrow smear shows marked myeloid preponderance with numerous middle and late stage myeloid forms.

- Blood smears show marked neutrophilia, usually >25,000/µL, with a modest myeloid left shift and presence of scattered (5–10%) myelocytes and metamyelocytes. Promyelocytes are rare, and myeloblasts are commonly absent. Neutrophils may show toxic granulation. The leukocyte alkaline phosphatase (LAP) score is often elevated. Mild anemia and/or thrombocytopenia may be present.
- Splenomegaly and hepatomegaly are due to neutrophilic infiltration in the splenic red pulp and hepatic sinusoids and/or portal areas.

IMMUNOPHENOTYPIC STUDIES

The immunophenotypic characteristics of the bands and neutrophils in CNL are similar to those of normal neutrophils and bands. So far, no aberrant expression or significant alteration of CD molecules have been reported.

MOLECULAR AND CYTOGENETIC STUDIES

- No consistent chromosomal anomaly has been associated with CNL, and the primary genetic event is likely cryptic (submicroscopic). Chromosomal anomalies reported to date may reflect secondary anomalies associated with chromosomal evolution in CNL.
- Sporadic reports of patients with trisomy 8, trisomy 9, del(20)(q11q13), del(11q), trisomy 21 (Figure 13.2), and complex karyotypes are described in the literature.
- Some cases may demonstrate the *JAK2* V617F mutation.

DIFFERENTIAL DIAGNOSIS

CNL is a diagnosis of exclusion. All the conditions listed in Box 13.1 should be excluded, such as all causes of physiologic and reactive neutrophilia, all other subtypes of chronic myeloproliferative neoplasms, myelodysplastic/myeloproliferative neoplasms, and conditions associated with rearrangement of *PDGFRA*, *PDGFRB* or *FGFR1*.

Chronic Eosinophilic Leukemia, Not Otherwise Specified

Chronic eosinophilic leukemia, not otherwise specified (CEL, NOS) is a rare subtype of chronic myeloproliferative neoplasms characterized by clonal expansion of eosinophilic precursors leading to myeloid left shift (blasts <20%) and a sustained eosinophilia in the peripheral blood (≥1500/µL) (Box 13.1). Eosinophilic disorders with no known etiology, lack of clonality, or increased blasts fall into the category of *"idiopathic hypereosinophilic syndrome"* (IHES) (Figure 13.3). Neither CEL nor IHES demonstrate Ph^1 chromosome (or *BCR-ABL1* fusion gene), or *PDGFRA*, *PDGFRB*, or *FGFR1* rearrangement.

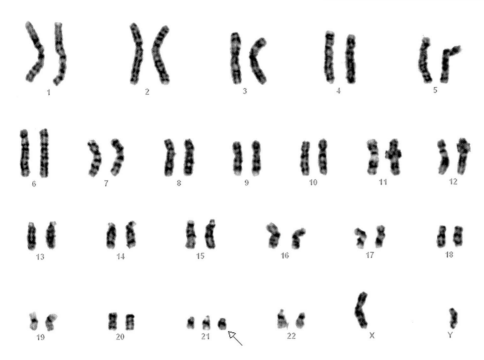

FIGURE 13.2 G-banded karyotype showing trisomy 21.

CEL and IHES involve men much more frequently than women (M:F ratio about 9:1). They are usually detected between the ages of 20 and 50 years and are rare in children. The most common clinical symptoms include fatigue, cough, dyspnea, myalgia, angioderma, rash, fever, and rhinitis. The release of eosinophilic granules may damage the endocardium and the endothelial cells and lead to thrombus formation and emboli. Endocardial thrombosis and fibrosis may cause insufficiencies of the mitral or tricuspid valves.

MORPHOLOGY

- The bone marrow is hypercellular and shows eosinophilic hyperplasia. Eosinophil counts may range from 10% to 70% of the bone marrow nucleated cells, with a mean of about 30%. The maturation of eosinophils and myeloid cells is progressive but often left-shifted with increased blasts (>5% and <20%).
- Eosinophils may show dysplastic changes such as nuclear hypersegmentation or hyposegmentation, cytoplasmic vacuolization or hypogranularity, and/or abnormal eosinophilic granules. Charcot–Leyden crystals are frequent findings. These crystals are colorless, long hexagonal, double-pointed, or needle-like lysophospholipase containing structures which are formed from the breakdown of eosinophils. However, both abnormal morphologic changes and Charcot–Leyden crystals have been observed in cases of reactive eosinophilia.
- Myelofibrosis may be present but is not common.
- The peripheral blood shows absolute eosinophilia (≥1500/μL), often with moderately elevated leukocyte count (between 20,000 and 30,000/μL), with eosinophils accounting for 30–70% of the differential counts.
- Eosinophilic infiltration may also be present in the extramedullary sites. The site of infiltration usually shows

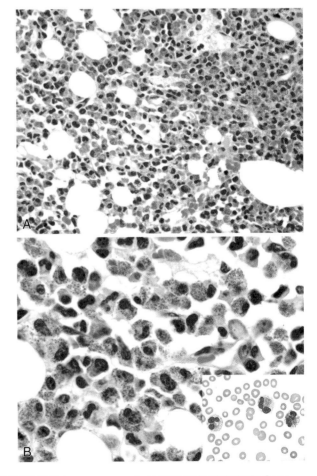

FIGURE 13.3 IDIOPATHIC HYPEREOSINOPHILIC SYNDROME. Bone marrow biopsy section (A and B) and blood smear (inset) of a patient with IHES.
(Adapted from Naeim F. Atlas of Bone Marrow and Blood Pathology. Saunders, 2001; by permission.)

some degree of fibrosis, often with the presence of Charcot–Leyden crystals.

IMMUNOPHENOTYPE

- The eosinophils show different characteristic features than the neutrophils by flow cytometry. They appear as distinct clusters in the CD45 and scatter gate displays, and their intensity of expression of myeloid-associated markers is different from neutrophilic granulocytes.
- A number of CD molecules are expressed on eosinophils, such as CD9 (leukocyte antigen MIC3), CD32 (FcgRII), CDw125 (IL-5 receptor alpha chain), and CD193 (chemokine receptor 3), but these molecules are not eosinophil-specific and are also expressed by other leukocytes.
- Blast enumeration by flow cytometry or immunohistochemical stains may be helpful in blood samples, bone marrow aspirates, or biopsy sections.

MOLECULAR AND CYTOGENETIC STUDIES

No specific molecular or cytogenetic abnormalities have been associated with CEL, NOS. All myeloid and lymphoid neoplasms with eosinophilia and abnormalities of *PDGFR A*, *PDGFR B*, and *FGFR1* are discussed in Chapter 16.

DIFFERENTIAL DIAGNOSIS

CEL, NOS is a diagnosis of exclusion. All the conditions listed in Box 13.1 should be excluded, such as all causes of reactive eosinophilia, myelodysplastic syndromes, acute myeloid leukemias, all other subtypes of chronic myeloproliferative neoplasms, and conditions associated with *BCR-ABL1* fusion or t(9;22)(q34;q11.2), (5;12)(q31–35;p13), inv(16)(p13q22), or t(16;16)(p13;q22), and rearrangement of *PDGFRA*, *PDGFRB*, or *FGFR1*.

Additional Resources

Dinçol G, Nalçaci M, Doğan O, et al: Coexistence of chronic neutrophilic leukemia with multiple myeloma, *Leuk Lymphoma* 43:649–651, 2002.

Elliott MA, Dewald GW, Tefferi A, et al: Chronic neutrophilic leukemia (CNL): A clinical, pathologic and cytogenetic study, *Leukemia* 15:35–40, 2001.

Elliott MA: Chronic neutrophilic leukemia and chronic myelomonocytic leukemia: WHO defined, *Best Pract Res Clin Haematol* 19:571–593, 2006.

Elliott MA: Chronic neutrophilic leukemia: a contemporary review, *Curr Hematol Rep* 3:210–217, 2004.

Gotlib J: Eosinophilic myeloid disorders: New classification and novel therapeutic strategies, *Curr Opin Hematol* 17:117–124, 2010.

Reilly JT: Chronic neutrophilic leukaemia: A distinct clinical entity? *Br J Haematol* 116:10–18, 2002.

Swerdlow SH, Campo E, Harris NL, et al: WHO classification of tumours of haematopoietic and lymphoid tissues, ed 4, International Agency for Research on Cancer, Lyon, 2008.

Tefferi A, Gotlib J, Pardanani A: Hypereosinophilic syndrome and clonal eosinophilia: Point-of-care diagnostic algorithm and treatment update, *Mayo Clin Proc* 85:158–164, 2010.

Wadleigh M, Tefferi A: Classification and diagnosis of myeloproliferative neoplasms according to the 2008 World Health Organization criteria, *Int J Hematol* 91:174–179, 2010.

Mastocytosis

Mastocytosis refers to an abnormal proliferation of mast cells in one or multiple organs. It covers a wide spectrum of clinicopathologic disorders from localized to systemic and from indolent to aggressive forms. Most systemic variants of mastocytosis are the result of clonal expansion of mast cells. Mast cells are derived from the hematopoietic stem cells. The committed mast cell progenitors express CD13, CD34, and CD117 (*c-kit*) and are detectable in the bone marrow and peripheral blood. Mast cells are distinguished from other granulocytic cells by their unique phenotypic and functional properties (Tables 14.1 and 14.2). Mast cells produce a substantial amount of histamine and heparin and express surface IgE receptor. In contrast to basophils and other granulocytic cells, mast cells have a significantly longer *in vivo* life span ranging from several months to years.

Mastocytosis has been associated with the somatic *c-kit* mutation at codon 816, substituting valine to aspartate (Figure 14.1). Activated mutated *c-kit* along with increased production of stem cell factor (SCF) may play a role in the pathogenesis of systemic mastocytosis. Genetic factors appear to play an important role in childhood mastocytosis, based on a report of 25% of congenital mastocytosis in pediatric cases and concordant symptoms of mastocytosis observed in monozygotic twins.

Table 14.1
Immunophenotypic Features of Mast Cells[1]

CD	Mast Cells Normal	Mast Cells Neoplastic	Basophils	Monocytes
CD2	−	+	−	−
CD13	−	±	+	+
CD14	−	−	−	+
CD15	−	−	−	+
CD25	−	+	+	±
CD33	+	+	+	+
CD34	−	−	−	−
CD45	+	+	+	+
CD117	+	+	−	−

[1]Valent P, Akin C, Sperr WR, et al. Diagnosis and treatment of systemic mastocytosis: state of the art. Br J Haematol 2003; 122: 695–717.

Table 14.2
Major Mast Cell-Derived Mediators and their Effects in Systemic Mastocytosis[1]

Mediators	Clinicopathologic Effects
Histamine	Vascular instability, urticaria, headache, edema, flushing, gastric hypersecretion, abdominal pain, bronchial constriction, diarrhea
Heparin	Coagulation abnormalities, bleeding
Tryptases	Fibrosis, angiogenesis, tissue remodeling, degradation of matrix molecules, bone resorption
tPA	Hyperfibrinolysis
VEGF	Increased angiogenesis, edema
PGD2	Edema, urticaria, flushing, bronchial constriction
Bfgf	Fibrosis, osteosclerosis, angiogenesis
TNF-α	Activation of endothelial cells, vascular instability
TNF-β	Fibrosis, abnormal bone remodeling, osteopenia
Interleukins (IL-1, -2, -3, -5, -6, -9, -10, -11, -13, GM-CSF)	Eosinophilia, bone marrow lymphocytosis, myeloid hyperplasia, activation of stromal cells, fibrosis
Chemokines (MCP-1, MIP-1α, others)	Leukocyte activation, accumulation of lymphocytes, monocytes, and eosinophils

[1]Adapted from Valent P, Akin C, Sperr WR, et al. Diagnosis and treatment of systemic mastocytosis: state of the art. Br J Haematol 2003; 122: 695–717.

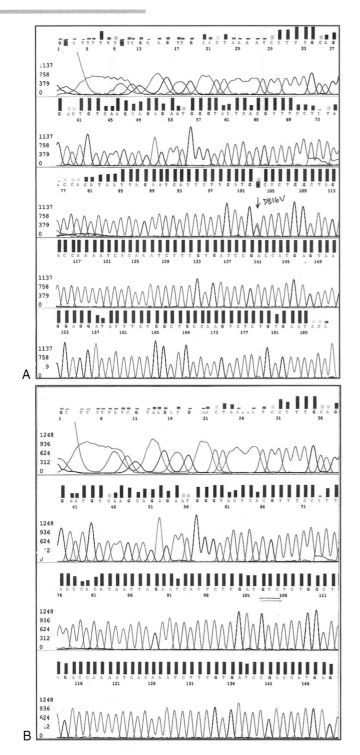

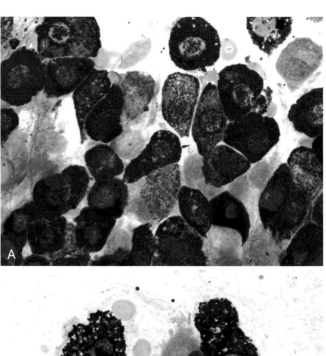

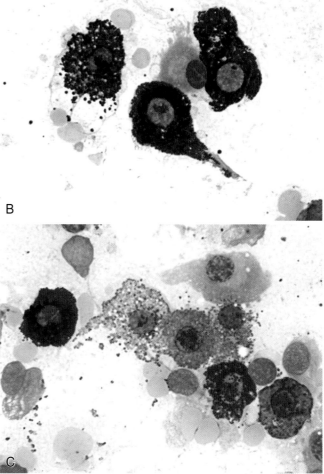

FIGURE 14.1 Detection of the D816V *c-kit* mutation in systemic mastocytosis by DNA sequencing. (A) Positive case showing heterozygous signal at nucleotide position 103 (arrow). (B) Negative control, showing only wild type sequence at same position (double line).

FIGURE 14.2 Mast cells. Wright-stained bone marrow smears demonstrating mast cells with abundant cytoplasm containing variable amounts of small deeply basophilic granules (A, B, and C).

Morphology and Laboratory Findings

Mastocytosis is demonstrated as multifocal clusters or diffuse infiltration of mast cells in the skin, bone marrow, spleen, liver, gastrointestinal tract, and other tissues.

- Mast cells in smears stained with Wright's or Giemsa stains are very distinct and appear as medium- to large-sized cells with abundant cytoplasm loaded with small deeply basophilic granules, often masking the nucleus (Figure 14.2). The nuclei are round, oval, or spindle-shaped and show a dense chromatin. Some mast cells may be hypogranular or appear immature. Binucleated or multinucleated mast cells may be present.
- Mast cells in the H&E-stained sections appear as medium-sized cells with variable amounts of granular cytoplasm and a round, oval, or spindle-shaped nucleus with condensed chromatin (Figure 14.3). The cytoplasmic granules are faintly eosinophilic in the H&E sections and variable in amount.
- The granules in some mast cells are sparse and difficult to detect. The hypogranular mast cells may demonstrate an abundant pale cytoplasm resembling histiocytes, monocytoid B-cells, or hairy cells.
- The spindle-shaped mast cells may mimic fibroblasts.
- Mast cell infiltration is often associated with various degrees of fibrosis and presence of inflammatory cells, such as lymphocytes and eosinophils (Figures 14.4 and 14.5).
- Bone marrow biopsy sections often show multiple mast cell aggregates. These aggregates may be either paratrabecular, interstitial, or both (see Figures 14.3 to 14.5). Paratrabecular infiltrates may show extensive fibrosis and osteosclerosis with scattered or aggregates of spindle-shaped mast cells which are identified by the Giemsa stain or immunohistochemical stains for tryptase or CD117 (Figures 14.6 and 14.7).
- Mastocytosis may be a part of a clonal primary hematopoietic disorder, such as myelodysplastic syndrome, chronic myeloproliferative disorder, or acute myeloid leukemia.

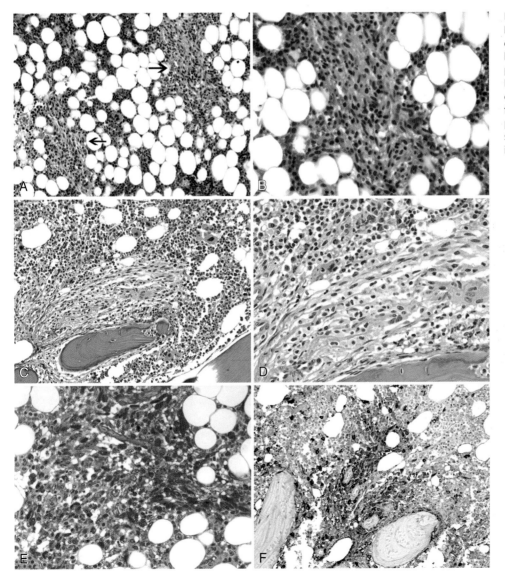

FIGURE 14.3 SYSTEMIC MASTOCYTOSIS. Bone marrow biopsy section demonstrating interstitial (A, low power; B, high power) and paratrabecular (C, low power; D, high power) clusters of mast cells resembling histiocytes or fibroblasts. These clusters are positive with Giemsa stain (E) and show expression of tryptase by immunohistochemical stain (F).

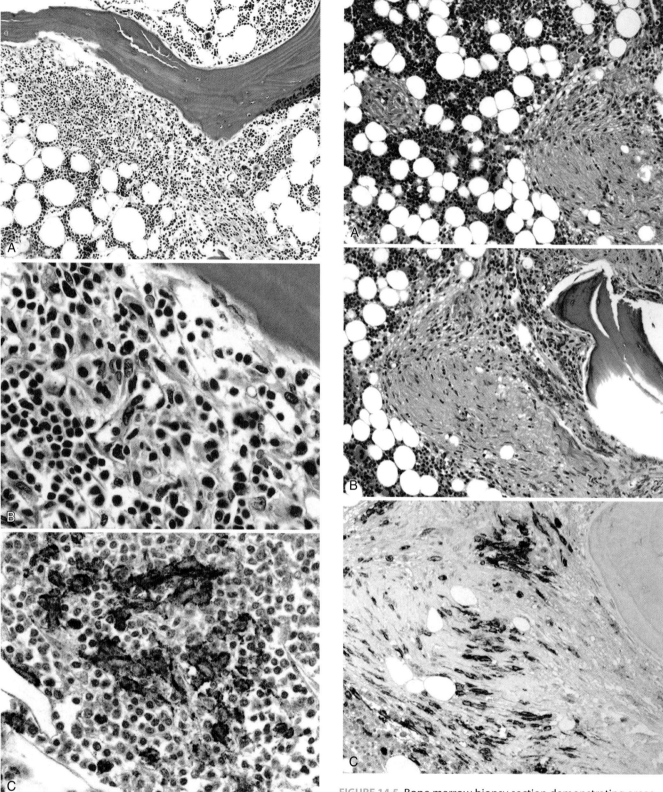

FIGURE 14.4 A cluster of mast cells mixed with lymphocytes and histiocytes is present next to bone (A, low power; B, high power). Mast cells show expression of CD117 by immunohistochemical technique (C).

FIGURE 14.5 Bone marrow biopsy section demonstrating areas with dense fibrosis (A, low power; B, high power). (C) Mast cells appear as clusters of CD117-positive elongated cells.

MORPHOLOGY AND LABORATORY FINDINGS 193

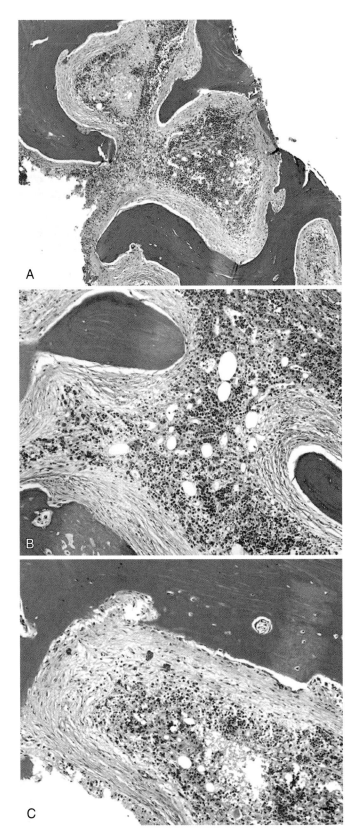

FIGURE 14.6 Bone marrow involvement in systemic mastocytosis. Extensive paratrabecular fibrosis and some osteosclerosis are present (A, low power; B, intermediate power; C, high power).

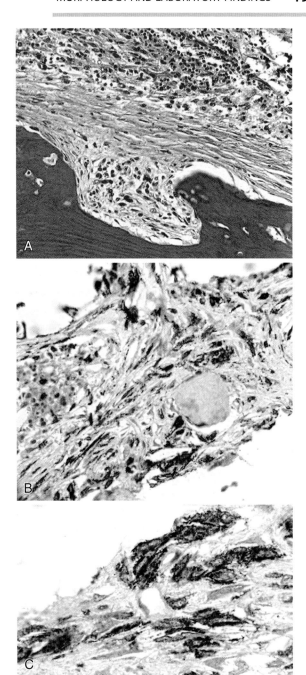

FIGURE 14.7 Systemic mastocytosis. Bone marrow biopsy section showing paratrabecluar fibrosis (A). Numerous tryptase-positive cells are present in the fibrotic area (B, low power; C, high power).

- Splenic involvement in mastocytosis usually consists of infiltrating mast cells randomly distributed in the red pulp or appearing in aggregates adjacent to the white pulps or trabeculae. The mast cell clusters are often associated with variable amounts of fibrosis and are mixed or surrounded by lymphocytes, plasma cells, and/or eosinophils.
- In the liver, mast cells are found in the sinuses and/or portal areas with focal areas of fibrosis (Figure 14.8).
- The involved skin demonstrates aggregates of atypical mast cells infiltrating into the papillary dermis and extending into the reticular dermis in compact aggregates to large sheets (Figure 14.9).

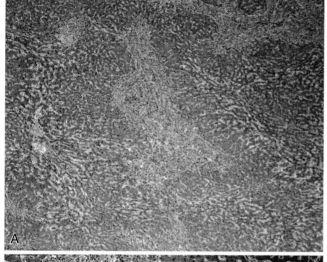
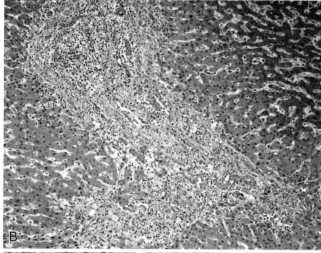
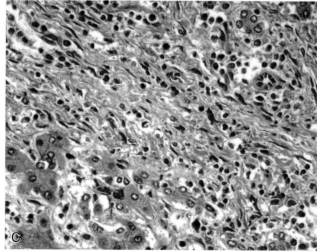

FIGURE 14.8 **MASTOCYTOSIS INVOLVING THE LIVER.** Fibrosis and infiltration of the inflammatory cells are noted in the expanded portal areas: (a) low power and (b) intermediate power views. Cells with elongated or spindle-shaped nuclei represent mast cells: (c) high power view.

Because of a wide spectrum of morphologic features, it is highly recommended that additional accessory studies be performed, such as cytochemical stains, immunophenotyping, and molecular analysis to establish the diagnosis of mastocytosis.

Immunophenotypic Studies

- Mast cells express CD13, CD33, CD45, CD68, CD117, and tryptase and are negative for CD14, CD15, CD16, and MPO (Figures 14.4, 14.5, and 14.7).
- In addition, the neoplastic mast cells may show aberrant expression of CD2, CD25, and CD30 (particularly in aggressive mastocytosis and mast cell leukemia).

Molecular and Cytogenetic Studies

The hallmark of systemic mastocytosis (SM) at the molecular level is somatic *c-kit* mutations. The following characteristics pertain.

- The most common mutation is reported at codon 816, substituting valine to aspartate and designated D816V.
- It is found in almost all cases, though is only considered one of the "minor criteria" for diagnosis.
- This variant can be detected by allele-specific or restriction endonuclease-treated PCR or by DNA sequence analysis (see Figure 14.1).
- Ongoing clinical trials are exploring small-molecule inhibitor targeting of the mutant c-kit enzyme (it is resistant to imatinib).
- Rarely, SM is seen in association with neoplasms with *FIP1L1/PDFRA* fusion and eosinophilia. However, this does not mean that *FIP1L1/PDGFRA* is a molecular marker of SM. In rare cases, both mutants (*FIP1L1/PDGFRA* and KIT D816V) are detectable.

Classification

Mast cell diseases are divided into two major clinical entities: (1) cutaneous mastocytosis and (2) systemic mastocytosis. The WHO criteria for the diagnosis of mastocytosis are presented in Box 14.1.

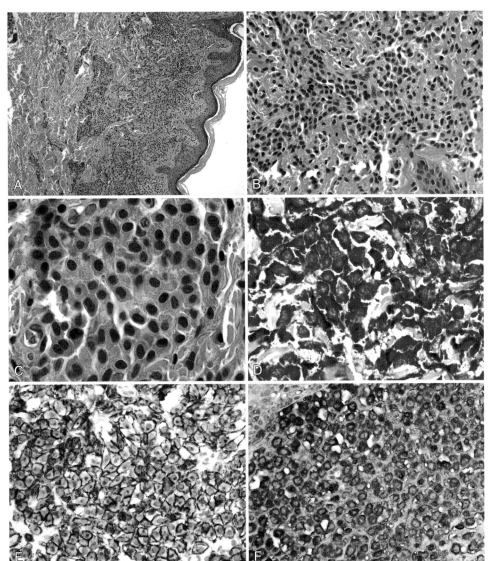

FIGURE 14.9 CUTANEOUS MASTOCYTOSIS. Wrist skin biopsy of a 3-year-old girl demonstrates aggregates of atypical mast cells involving the papillary dermis and extending into the reticular dermis in compact aggregates to large sheets (A, low power; B, intermediate power; C, high power). The mast cells contain oval, elongated, or spindle-shaped nuclei, dense chromatin, and dispersed coarse granules within the cytoplasm. Giemsa stain reveals coarse metachromic granules in the atypical mast cells (D), and immunohistochemical stains are positive for CD117(E) and tryptase (F).

Box 14.1 WHO Criteria for the Diagnosis of Cutaneous and Systemic Mastocytosis[1]

Cutaneous Mastocytosis
Mast cell infiltrates in a multifocal, solitary, or diffuse pattern in skin biopsies, with typical clinical findings.

Systemic Mastocytosis
Major criterion
Multifocal infiltrates of mast cells (≥15 mast cells in each aggregate) in one or more extracutaneous sites confirmed by tryptase immunohistochemistry or other special stains.

Minor criteria
1. More than 25% of mast cells are spindle-shaped or atypical in extracutaneous infiltrates in biopsy sections and/or smear preparations.
2. Detection of *KIT* mutation at codon 816.
3. Co-expression of CD117, CD2, and/or CD25 by the infiltrating mast cells.
4. Persistent total serum tryptase levels >20 ng/mL in cases not associated with clonal myeloid disorders.

[1] Adapted from Jaffe ES, Harris NL, Stein H, et al. Pathology and Genetics: Tumors of Haematopoietic and Lymphoid Tissues. IARC Press, Lyon, 2001 and Swerdlow SH, Campo E, Harris NL, et al. WHO Classification of Tumours of Haematopoietic and Lymphoid Tissues, 4th edn. International Agency for Research on Cancer, Lyon, 2008.

CUTANEOUS MASTOCYTOSIS

Cutaneous mastocytosis consists of mast cell disorders limited to skin without evidence of systemic involvement, such as elevated levels of serum tryptase, bone marrow infiltration, or organomegaly. There are four major clinicopathologic subtypes:

- urticaria pigmentosa
- cutaneous mastocytoma
- diffuse cutaneous mastocytosis
- telangiectasia macularis eruptiva perstans.

The diagnosis is made based on the clinical presentation and by a skin biopsy demonstrating significant increase in mast cells (usually ≥20 mast cells per high power field), which are particularly found around vascular structures.

> **Box 14.2 WHO Classification of Systemic Mastocytosis (SM)**[1]
>
> - Indolent systemic mastocytosis (ISM)
> - Systemic mastocytosis with associated clonal, hematological non-mast cell lineage disease (SM-AHNMD)
> - Aggressive systemic mastocytosis (ASM)
> - Extracutaneous mastocytoma
> - Mast cell sarcoma (MCS)
> - Mast cell leukemia (MCL)
>
> [1] Adapted from Jaffe ES, Harris NL, Stein H, et al. Pathology and Genetics: Tumors of Haematopoietic and Lymphoid Tissues. IARC Press, Lyon, 2001 and Swerdlow SH, Campo E, Harris NL, et al. WHO Classification of Tumours of Haematopoietic and Lymphoid Tissues, 4th edn. International Agency for Research on Cancer, Lyon, 2008.

Urticaria Pigmentosa

Urticaria pigmentosa (UP) is the most common mast cell disorder in children and adults. The *KIT* point mutation in pediatric and adult UP appears to be different from the mutation of codon 816 observed in SM. There are reports of mutations in codon 839 and codon 516 in pediatric and adult UP, respectively. The cutaneous lesions are usually small yellow-tan to reddish-brown macules or papules. Plaque-like lesions may also occur. The upper and lower extremities are the most frequently affected sites. The face, palms, and soles are not involved. Most of the affected children are under the age of 1 and rarely show systemic involvement. UP-associated pruritus is exacerbated by a variety of stimulants, such as change in temperature, spicy food, or local friction.

Cutaneous Mastocytoma

Cutaneous mastocytoma of skin is typically a solitary lesion occurring in early childhood, usually before 6 months of age. The trunk and wrist are frequent sites of involvement. Large clusters or sheets of mast cells are present in the papillary and reticular dermis.

Diffuse Cutaneous Mastocytosis

Diffuse cutaneous mastocytosis is a childhood disorder usually occurring before the age of 3. The skin is diffusely infiltrated but is relatively smooth. It may show increased thickness and/or a yellowish-brown color. The maculopapular lesions are usually absent.

Telangiectasia Macularis Eruptiva Perstans

Telangiectasia macularis eruptiva perstans is a rare cutaneous mast cell disorder mainly occurring in adults. It is characterized by tan-brown macules with telangiectasia but no blisters or pruritus.

SYSTEMIC MASTOCYTOSIS

Systemic mastocytosis is mast cell disease beyond skin. The increased mast cells are found in extracutaneous sites, with or without skin involvement. The frequently involved extracutaneous sites include bone marrow, liver, spleen, lymph nodes, and gastrointestinal tract.

The WHO requirements (2008) for the diagnosis and classification of SM are demonstrated in Boxes 14.1 and 14.2.

Clinical manifestations of mastocytosis are the result of two different mechanisms: (1) mediator release from mast cells and (2) growth and infiltration of the mast cells in various organs.

A wide variety of mediators are released from mast cells, resulting in clinical symptoms such as headache, flushing, pruritus, hypotension, and diarrhea (Table 14.2). The growth and infiltration of mast cells in various organs may lead to organomegaly as well as organ dysfunction, leading to ascites, cytopenia, malabsorption, and pathologic fractures. The organopathy-related clinical symptoms are referred to as *C-findings*, whereas organomegalies without any evidence of organopathy are termed *B-findings*. SM has been divided into the following categories (Box 14.2).

Indolent Systemic Mastocytosis

Indolent systemic mastocytosis (ISM) refers to cases with relatively low burden of mast cells and therefore to an indolent clinical course and good prognosis. The majority of patients with ISM have UP and show evidence of systemic involvement but lack C-findings. ISM accounts for >80% of all cases of SM. Two subtypes of ISM have been described: *smoldering systemic mastocytosis* (SSM) and *isolated bone marrow mastocytosis* (BMM). In SSM, B-findings are present, and in isolated BMM there is lack of skin involvement (Table 14.3).

Table 14.3
Comparison of B- and C-Findings in Subtypes of Systemic Mastocytosis[1]

Findings	Typical ISM	BMM	SSM	SM-AHNMD	ASM	MCL
B-Findings						
Hepatomegaly	−	−	±	±	±	±
Splenomegaly	−	−	+	±	+	±
Lymphadenopathy	−	−	±	±	±	±
Tryptase >200 ng/mL	−	−	+	±	±	+
C-Findings						
Anemia (Hb <10 g/dL)	−	−	−	±	+	+
Thrombocytopenia (<100×10^9/L)	−	−	−	±	+	+
Neutrophil count <1×10^9/L	−	−	−	±	+	+
Ascites or portal hypertension	−	−	−	−	+	+
Hypersplenism	−	−	−	±	+	±
Malabsorption with weight loss	−	−	−	−	±	+
Osteolysis	−	−	−	−	+	±
Others						
Urticaria-pigmentosa-like lesions	+	−	±	±	±	−
Elevated serum LDH	−	−	−	±	±	+
Abnormal coagulation	−	−	±	±	±	+

ASM, aggressive systemic mastocytosis; BMM, isolated bone marrow mastocytosis; ISM, indolent systemic mastocytosis; MCL, mast cell leukemia; SM-AHNMD, systemic mastocytosis with associated clonal hematological non-mast cell lineage disease; SSM, smoldering systemic mastocytosis.
[1] Adapted from Valent P, Akin C, Sperr WR, et al. Diagnosis and treatment of systemic mastocytosis: state of the art. Br J Haematol 2003; 122: 695–717.

Systemic Mastocytosis with Associated Clonal Hematological Non-Mast Cell Lineage Disease

Systemic mastocytosis with associated clonal hematological non-mast cell lineage disease (SM-AHNMD) is mastocytosis associated with acute myeloid leukemias, acute lymphoid leukemias, myelodysplastic syndromes, chronic myeloproliferative disorders, or lymphoma (Figures 14.10 and 14.11).

Aggressive Systemic Mastocytosis

Aggressive systemic mastocytosis (ASM) is characterized by the presence of organ-function impairment and C-findings, leading to an aggressive clinical course. C-findings include (1) anemia, thrombocytopenia, and/or leukopenia, (2) hepatomegaly with ascites or portal hypertension, (3) splenomegaly with hypersplenism, (4) malabsorption and weight loss, and (5) osteolysis and pathologic fractures (Table 14.3). Less than 50% of patients in this category show UP lesions. The *KIT* mutation in codon 816 is the typical molecular finding.

Mast Cell leukemia

Mast cell leukemia is a rare condition characterized by diffuse infiltration of the bone marrow by atypical and/or immature mast cells. The pattern of bone marrow infiltration is usually interstitial. Mast cells comprise ≥20% of the nucleated cells in bone marrow smears and ≥10% of the leukocyte differential counts in peripheral blood (Figure 14.12). Prognosis is extremely poor, with an estimated survival of 6–12 months.

Other Types of Systemic Mastocytosis

Extracutaneous mastocytoma is an extremely rare lesion consisting of an accumulation of mature mast cells in extracutaneous sites, such as the lung.

Mast cell sarcoma is another extremely rare lesion consisting of an infiltrating growth of atypical and/or immature mast cells with a potential of distant metastasis or progression to a leukemic phase.

In general, the therapeutic approaches depend on the clinical symptoms and extent of the disease. Mediator-related symptoms are treated with drugs that interfere with mediator production/release or mediator functions, such as histamine and leukotriene antagonists, glucocorticoids, cromylin sodium, and aspirin. In addition to the anti-mediator drugs, patients with cutaneous mastocytosis may receive psoralen and ultraviolet-A. Cytoreductive drugs such as interferon-α, cytosine arabinoside, cladribine, vincristine, and doxorubicin are preserved for patients who have clear signs of aggressive disease.

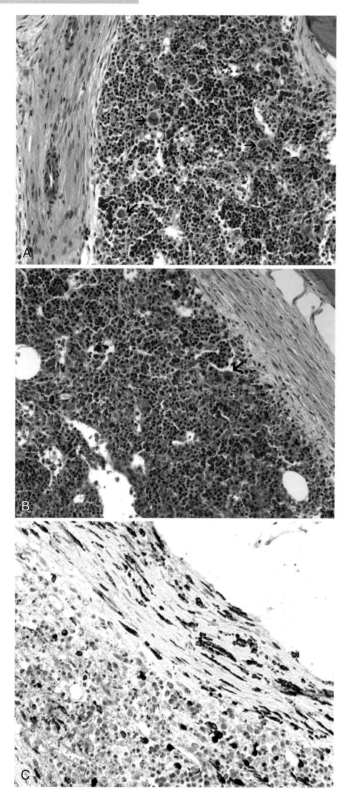

FIGURE 14.10 **MASTOCYTOSIS IN A PATIENT WITH MYELODYSPLASTIC SYNDROME.** Bone marrow biopsy section showing paratrabecular fibrosis. The bone marrow is hypercellular with the presence of numerous micromegakaryocytes (A and B, arrows). Numerous spindle cells are positive for CD117 in the fibrotic area (C).

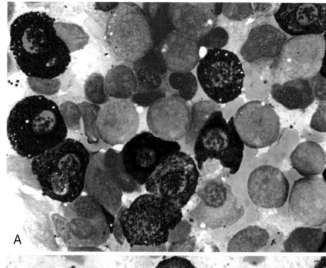
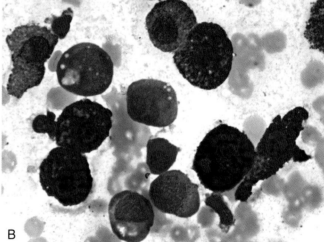

FIGURE 14.11 **MASTOCYTOSIS IN A PATIENT WITH ACUTE MYELOGENOUS LEUKEMIA.** Bone marrow smear consisting of a mixture of myeloblasts and mast cells (A and B).

Differential Diagnosis

The differential diagnosis includes two major categories: (1) disorders with similar clinical manifestation but lack of histologic evidence of cutaneous mastocytosis or SM and (2) disorders associated with increased mast cells or elevated serum tryptase (Table 14.4).

Disorders with similar clinical manifestation but lack of histologic evidence of cutaneous or systemic mastocytosis include anaphylaxis (may show elevated serum tryptase during but not in the period between the acute events), angioedema, carcinoid syndrome, pheochromocytoma, and Zollinger–Ellison syndrome. All these disorders except acute episodes of anaphylaxis lack elevated serum tryptase or urinary histamine.

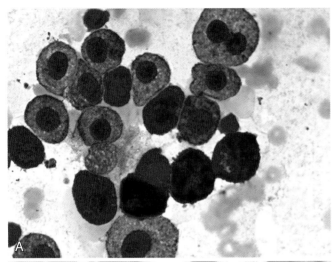

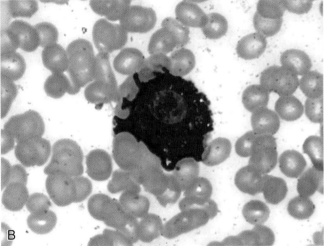

FIGURE 14.12 (A) Bone marrow smear of a patient with mast cell leukemia showing numerous mast cells. (B) Mast cell in a peripheral blood smear.

Table 14.4
Differential Diagnosis of Mastocytosis

Categories	Examples
Similar clinical manifestations but lack of histologic evidence of cutaneous or systemic mastocytosis	Anaphylaxis, angioderma, carcinoid syndrome, pheochromocytoma, Zollinger–Ellison syndrome
Associated with increased mast cells	Basal cell carcinoma, melanoma, lymphoma, helminth infection
Associated with elevated serum tryptase but lack of mastocytosis	Acute myelogenous leukemia, myelodysplastic syndrome, chronic myeloproliferative disorders
Morphologic overlap	Basophilic leukemia, histiocytic disorders, bone marrow metastasis, disorders associated with paratrabecular bone marrow fibrosis, such as chronic renal failure

Disorders associated with increased mast cells or elevated serum tryptase consist of reactive conditions associated with mastocytosis or hematopoietic malignancies associated with elevated serum tryptase but lack of mastocytosis. Reactive mast cell hyperplasia has been observed in various conditions, such as basal cell carcinoma, melanoma, and lymphomas. Elevated serum tryptase levels have been reported in a variety of clonal myeloid disorders, such as acute myeloid leukemia, myelodysplastic syndromes, and chronic myeloproliferative disorders. Also, some patients with myelodysplastic syndrome may show *c-kit* mutation.

Additional Resources

Akin C: Molecular diagnosis of mast cell disorders: a paper from the 2005 William Beaumont Hospital Symposium on Molecular Pathology, *J Mol Diagn* 8:412–419, 2006.

Amon U, Hartmann K, Horny HP, et al: Mastocytosis – an update, *J Dtsch Dermatol Ges* 8:695–711, 2010.

Arock M, Valent P: Pathogenesis, classification and treatment of mastocytosis: state of the art in 2010 and future perspectives, *Expert Rev Hematol* 3:497–516, 2010.

Arredondo AR, Gotlib J, Shier L, et al: Myelomastocytic leukemia versus mast cell leukemia versus systemic mastocytosis associated with acute myeloid leukemia: a diagnostic challenge, *Am J Hematol* 85:600–606, 2010.

Horny HP, Sotlar K, Valent P: Differential diagnoses of systemic mastocytosis in routinely processed bone marrow biopsy specimens: a review, *Pathobiology* 77:169–180, 2010.

Ozdemir D, Dagdelen S, Erbas T: Systemic Mastocytosis, *Am J Med Sci* 342:409–415, 2011.

Patnaik MM, Rindos M, Kouides PA, et al: Systemic mastocytosis: a concise clinical and laboratory review, *Arch Pathol Lab Med* 131:784–791, 2007.

Swerdlow SH, Campo E, Harris NL, et al: WHO classification of tumours of haematopoietic and lymphoid tissues, ed 4, Lyon, 2008, International Agency for Research on Cancer.

Valent P, Cerny-Reiterer S, Herrmann H, et al: Phenotypic heterogeneity, novel diagnostic markers, and target expression profiles in normal and neoplastic human mast cells, *Best Pract Res Clin Haematol* 23:369–378, 2010.

Valent P, Sperr WR, Akin C: How I treat patients with advanced systemic mastocytosis, *Blood* 116:5812–5817, 2010.

Vano-Galvan S, De la Hoz B, Nuñez R, Jaen P: Indolent systemic mastocytosis, *Isr Med Assoc J* 12:185–187, 2010.

Yamada Y, Cancelas JA: FIP1L1/PDGFR alpha-associated systemic mastocytosis, *Int Arch Allergy Immunol* 152(Suppl 1):101–105, 2010.

Myelodysplastic/Myeloproliferative Neoplasms

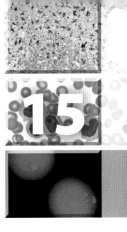

The myelodysplastic/myeloproliferative neoplasms (MDS/MPN) are a group of hematologic disorders distinguished by clonal expansion of abnormal hematopoietic stem cells that share clinicopathologic features of both MDS and MPN. They are characterized by hypercellular bone marrows with excessive terminal proliferation of one or more hematopoietic lineages as well as dysplastic changes. This combination of proliferation and dysplasia may lead to increased production of one or more lineages (cytosis) and decreased production of other lineages simultaneously. Myeloid preponderance and left shift are common bone marrow findings, but blast cells are <20% in the bone marrow and/or the peripheral blood samples. According to the WHO classification, these disorders are divided into the following major groups:

1. Chronic myelomonocytic leukemia
2. Atypical chronic myeloid leukemia
3. Juvenile myelomonocytic leukemia
4. MDS/MPD, unclassifiable

Chronic Myelomonocytic Leukemia

Chronic myelomonocytic leukemia (CMML) is a clonal hematopoietic disorder characterized by both dysplastic and proliferative features including persistent monocytosis (>1000/μL) of at least 3 months, bone marrow hypercellularity with myeloid preponderance, myelomonocytic dysplasia and left shift with <20% blasts (including promonocytes) in the peripheral blood or bone marrow. Cytogenetic and molecular studies are negative for $Ph1$ and/or BCR-$ABL1$ fusion (Box 15.1).

CMML is a disease of the elderly, with a median age of about 70 years. It affects men more than women, with a male:female ratio of 1.5–3:1 and an estimated incidence of about 4 per 100,000 per year. Childhood CMML is relatively infrequent.

Clinical manifestations are related to anemia, thrombocytopenia, and splenomegaly. Some patients may

---

Box 15.1 Diagnostic Criteria for Chronic Myelomonocytic Leukemia [1]

1. Peripheral blood monocytosis greater than 1000/μL
2. Less than 20% blasts in the peripheral blood or the bone
3. No Ph^1 or BCR-$ABL1$ fusion gene
4. No $PDGFRA$ or $PDGFRB$ rearrangement
5. Dysplastic myelopoiesis. If myelodysplasia in minimal or absent:
   a. Presence of an acquired clonal cytogenetic abnormality, or
   b. Persistent monocytosis for at least 3 months and
   c. Exclusion of all other causes of monocytosis

[1] Adapted from Jaffe ES, Harris NL, Stein H, et al. Pathology and Genetics: Tumors of Haematopoietic and Lymphoid Tissues. IARC Press, Lyon, 2001 and Swerdlow SH, Campo E, Harris NL, et al. WHO Classification of Tumours of Haematopoietic and Lymphoid Tissues, 4th edn. International Agency for Research on Cancer, Lyon, 2008.

demonstrate autoimmune disorders, such as vasculitis, pyoderma, and idiopathic thrombocytopenia. Others may develop skin infiltration and serous effusions. The reported unfavorable prognostic factors include low hemoglobin levels, low platelet counts, high percentage of marrow blasts (>10%), lymphocytosis (>2500/µL), elevated serum lactate dehydrogenase (LDH) and β2-microglobulin levels, and abnormal cytogenetics.

Approximately 15–30% of CMML patients progress to acute leukemia. Evidence of erythrophagocytosis has been suggested as an indicator of evolving blast transformation. The median survival is about 2 years. Therapeutic approaches include conventional chemotherapy, such as hydroxycarbamide (hydroxyurea) and allogeneic bone marrow transplantation (BMT). The therapeutic trial of farnesyl transferase inhibitors (inactivating RAS protein) has opened an avenue in the area of targeted biological therapy.

MORPHOLOGY AND LABORATORY FINDINGS

- Bone marrow biopsy sections are mostly (> 75%) hypercellular and display myeloid preponderance and left shift with increased number of immature myelomonocytic precursors (Figure 15.1). Bone marrow fibrosis is reported in about one-third of patients.
- Aggregates of monocytic and/or plasmacytoid dendritic cells have been observed in the bone marrow biopsy sections of patients with CMML (Figure 15.2).

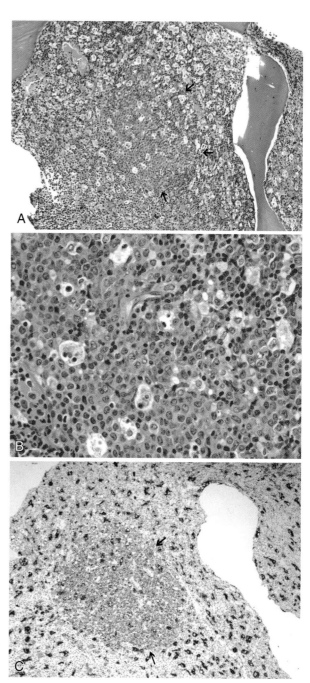

FIGURE 15.2 Monocytic nodules have been observed in bone marrow biopsy sections of patients with CMML. (A, low power; B, high power) (H&E stains). The arrows in (A) indicate the monocytic/histiocytic nodule. (C) The nodule (arrowed) is highlighted by immunohistochemical stain for CD68.

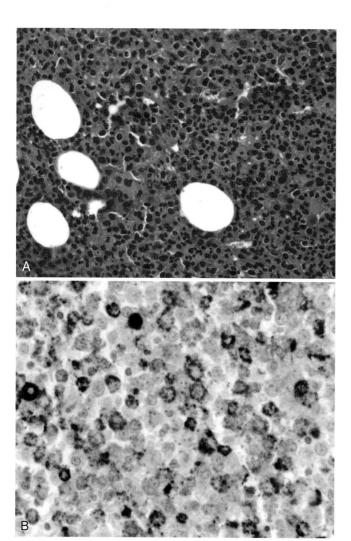

FIGURE 15.1 **Chronic myelomonocytic leukemia (CMML).** Bone marrow biopsy section from a patient with CMML demonstrating a hypercellular marrow with myeloid preponderance (A). The immunohistochemical stain for CD68 shows numerous positive monocytic/histiocytic cells (B).

- The bone marrow smears often show dysplastic changes in the monocytic and the granulocytic series, such as bizarre morphology, nuclear hyper- or hyposegmentation, and cytoplasmic hypo- or hypergranularity. The morphologic identification of monocytic precursors is much easier in bone marrow aspirate smears than in biopsy sections (Figure 15.3). The total number of myeloblasts, monoblasts, and promonocytes is < 20%. Auer rods may be detected in some cases.
- Erythroid dysplasia such as megaloblastic changes, irregular nuclei, nuclear fragments, and ring sideroblasts are frequently observed. Micromegakaryocytes are often present.
- The peripheral blood reveals monocytosis of >1000/μL. In most instances the monocyte count is between 1000 and 5000/μL, but occasionally it may exceed 50,000/μL (Figure 15.4). CMML was divided into two myeloproliferative and myelodysplastic subtypes based on the WBC count. Cases with >13,000 leukocytes/μL are considered to be of myeloproliferative subtype (Figure 15.5). However, significant clinical or biological differences between these two groups are debatable. There are often various degrees of anemia and/or thrombocytopenia. Mild eosinophilia and/or basophilia are present.
- Monocytes and granulocytes are dysplastic (Figure 15.6) and often left-shifted, with myelomonocytic blasts and promonocytes >5% and <20% of the leukocyte differential count.
- The serum lysozyme levels are elevated. Also, elevated serum levels of lactate dehydrogenase (LDH) and β2-microglobulin have been reported in some cases.
- Extramedullary involvement is observed in the spleen and sometimes in the liver and the lymph nodes. There seems to be a correlation between elevated WBC and splenomegaly. The myelomonocytic cells infiltrate into the splenic red pulp and hepatic and lymph node sinuses.

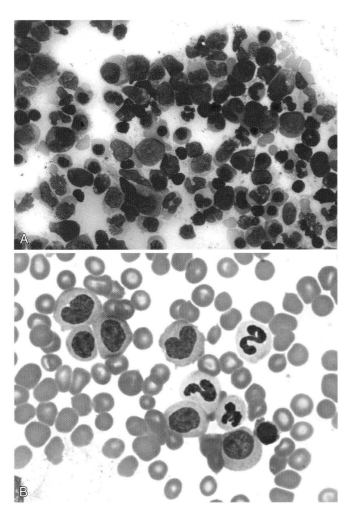

FIGURE 15.3 (A) Bone marrow smears in patients with CMML demonstrate increased number of monocytes and promonocytes with dysplastic features. (B) Blood smears reveal monocytosis with atypically featured monocytes and hypogranular neutrophils.

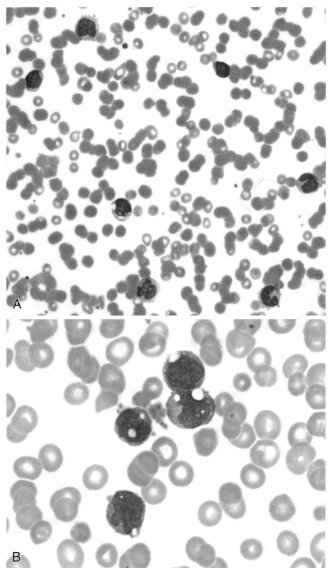

FIGURE 15.4 Blood smear from a patient with CMML demonstrating monocytes with vacuolated cytoplasm (A, low power; B, high power).

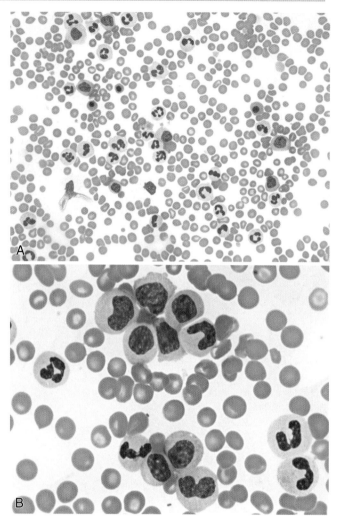

FIGURE 15.5 Dysplastic monocytes and granulocytes are often present in the blood smears of patients with CMML. Neutrophils may be hypersegmented, hyposegmented, and/or hypogranular (A, low power; B, high power).

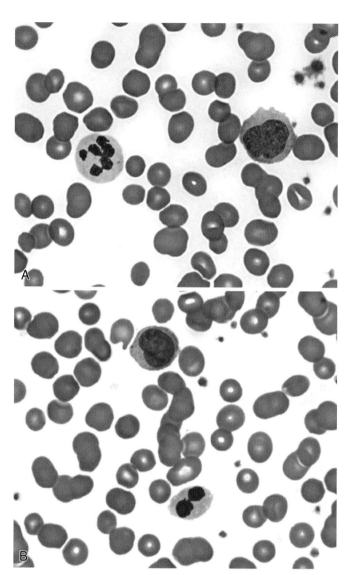

FIGURE 15.6 Blood smear from a patient with CMML demonstrating a hypogranular, hypersegmented neutrophil (A) and a hypogranular, hyposegmented neutrophil (B).

Chronic myelomonocytic leukemia has been divided into two subtypes in the WHO classification (2008):

- **CMML-1**: Blasts plus promonocytes are <5% in the peripheral blood and <10% in the bone marrow (see Figure 15.5).
- **CMML-2**:
  - Blasts plus promonocytes are between 5% and 19% in the peripheral blood and/or between 10% and 19% in the bone marrow (Figure 15.7).
  - There is presence of Auer rods irrespective of blast plus promonocyte count.

IMMUNOPHENOTYPE AND CYTOCHEMICAL STAINS

Flow Cytometry

Multiparametric immunophenotyping by flow cytometry (MIFC) plays important roles in diagnosing and monitoring CMML (Figure 15.8).

- Neoplastic monocytes in CMML are phenotypically mature monocytes with respect to coexpression of CD14 and CD64. In addition, they express other monocytic markers like dim CD4, CD11b, CD11c, and CD36.

CHRONIC MYELOMONOCYTIC LEUKEMIA 205

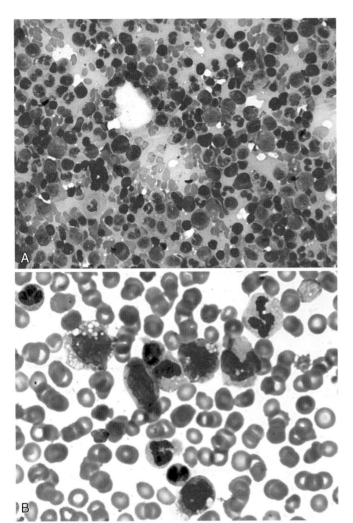

FIGURE 15.7 **CMML-2**. Bone marrow smear demonstrating myeloid left shift with increased promonocytes and blasts. (B) Blood smear showing a myeloblast and several vacuolated promonocytes.

- Neoplastic monocytes in CMML display a unique clustering profile of CD14 and CD64 coexpression, which is often broader or more heterogeneous than their normal or reactive counterparts.
- Neoplastic monocytes often show cross-lineage aberrancies of CD56 and/or CD2.
- They also demonstrate reduced expression of multiple and various antigens, including CD11b, CD13, CD15, and HLA-DR.
- Granulocytes of CMML can reveal dysmaturation patterns similar to those seen in MDS/MPN.
- Blast enumeration is helpful in distinguishing CMML-1 from CMML-2, as well as in determining transformation of CMML to AML. When the transformation occurs, AML displays variable features of monocytic differentiation (e.g., acute myelomonocytic leukemia, acute monoblastic leukemia, or acute monocytic leukemia) (Figure 15.9).

Distinction between CMML and AML with monocytic differentiation can be difficult sometimes, since blasts in this group of AMLs may be negative for both CD34 and CD117. In addition to clinical history and morphology, careful evaluation of the CD14/CD64 coexpression profile can be invaluable. In contrast to CMML, monocytes in AML with monocytic differentiation (even in acute monocytic leukemia) are phenotypically immature without complete coexpression of CD14 and CD64.

Immunohistochemistry and Cytochemical Stains (Figure 15.10)

- Immunohistochemical (IHC) studies are less sensitive in identifying monocytic differentiation than are MIFC and cytochemistry. The most helpful monocytic markers by IHC studies are CD68, CD163, and lysozyme.
- Cytochemical stains, such as MPO and non-specific esterase, are sometimes useful for estimation of the granulocytic and monocytic components and the lineage confirmation of the blast cells.

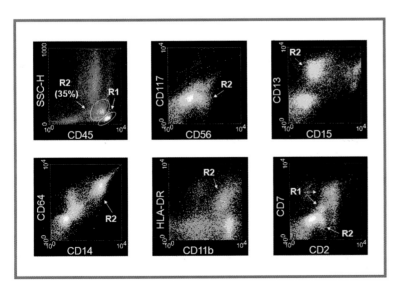

FIGURE 15.8 **IMMUNOPHENOTYPIC FEATURES OF CMML DETECTED BY MULTIPARAMETER FLOW CYTOMETRY.** A 75-year-old gentleman presents with a longstanding history of transfusion-dependent anemia and more recent pancytopenia. Open gate density plot display reveals monocytosis by CD45 gating. The monocytes (R2) account for 35% of the total. Lymphocytes are shown in R1. Compared with the internal controls of the open gate displays, the monocytes demonstrate significantly reduced CD15, reduced CD11b and HLA-DR, plus a more spread-out coexpression profile of CD14 and CD64. In addition, there is aberrant and partial expression of CD2 and CD56. Interestingly, an abnormal T-cell subset is present expressing dim CD2 and dim CD7.

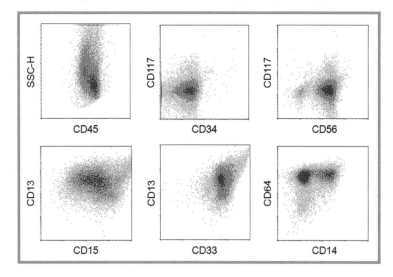

FIGURE 15.9 **FLOW CYTOMETRIC FEATURES OF TRANSFORMATION OF CMML-2 TO ACUTE MONOBLASTIC LEUKEMIA.** Compared with findings seen in Figure 15.8, transformation of CMML has occurred in this case with characteristic phenotypic features of acute monoblastic leukemia. The density plots of the blast-enriched gate demonstrate blasts that are negative for both CD34 and CD117, but positive for CD13, CD14 (one subset), CD15 (heterogeneous), CD33 (bright), CD45 (dim to moderate), CD56 (uniform), and CD64 (bright).

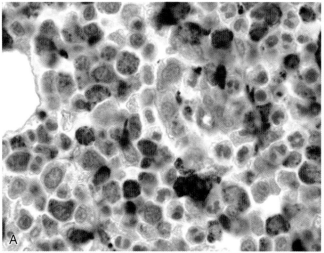

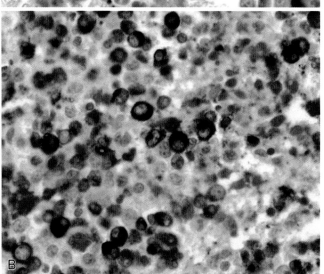

FIGURE 15.10 Immunohistochemical stains on bone marrow biopsy sections for lysozyme (A) and myeloperoxidase (B) in a patient with chronic myelomonocytic leukemia.

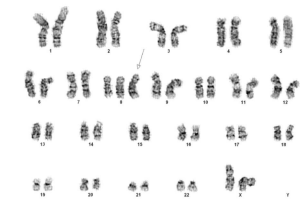

FIGURE 15.11 Trisomy 8 (arrow) in a patient with chronic myelomonocytic leukemia.

MOLECULAR AND CYTOGENETIC STUDIES

- At the molecular level, *K-RAS* and *N-RAS* mutations are reported in approximately 50% of the CMML cases. There is also a report of elevated levels of *survivin* in patients with CMML.
- Approximately 20–40% of patients with CMML show cytogenetic abnormalities including trisomy 8 (Figure 15.11), −7/del(7q), i(7q), and structural abnormalities of 12p. Four to five copies of chromosome 8, trisomy of chromosome 19, monosomy of chromosome 15, and t(1;3)(p36;q21) have been reported in occasional cases.
- Fusion of the *ETV6* and *PDGFR β* genes [t(5;12)(q33;p13)] has been demonstrated in the CMML subtype with eosinophilia. (This subtype now is classified separately, see Chapter 16).

Box 15.2 Criteria for the Diagnosis of Atypical Chronic Myeloid Leukemia[1]

- Peripheral blood granulocytosis (WBC ≥13,000/μL) with left-shift and presence of ≥10% immature forms (promyelocytes, myelocytes, metamyelocytes)
- Prominent dysgranulopoiesis
- No Ph^1 or *BCR-ABL1* fusion gene
- No rearrangement of *PDGFRA* or *PDGFRB*
- No or minimal absolute monocytosis (<10% of leukocytes)
- No or minimal absolute basophilia (<2% of leukocytes)
- Bone marrow hypercellularity with myeloid preponderance, left shift and dysgranulopoiesis, with or without other hematopoietic dysplasias
- Less than 20% blasts in the peripheral blood or in the bone marrow.

[1] Adapted from Jaffe ES, Harris NL, Stein H, et al. Pathology and Genetics: Tumors of Haematopoietic and Lymphoid Tissues. IARC Press, Lyon, 2001 and Swerdlow SH, Campo E, Harris NL, et al. WHO Classification of Tumours of Haematopoietic and Lymphoid Tissues, 4th edn. International Agency for Research on Cancer, Lyon, 2008.

Atypical Chronic Myeloid Leukemia

Atypical chronic myeloid leukemia (aCML) is a clonal hematopoietic disorder characterized by both dysplastic and proliferative features, including persistent granulocytosis with left shift, bone marrow hypercellularity with dysplastic hematopoiesis, and myeloid preponderance.

Cytogenetic and molecular studies are negative for Ph^1 and *BCR-ABL1* fusion gene (Box 15.2).

Atypical CML is a disorder of older adults with apparently no sex predominance. The incidence of aCML is not yet established. Clinical manifestations, similar to those of CMML, are related to anemia, thrombocytopenia, and splenomegaly. The median survival is <2 years with 20–40% chance of evolving to acute myeloid leukemia. Therapeutic approaches include conventional chemotherapy, such as hydroxycarbamide. Allogeneic BMT is potentially curative for eligible patients.

MORPHOLOGY (FIGURE 15.12)

- The bone marrow biopsy and clot sections are hypercellular and show myeloid preponderance and left shift.
- The bone marrow smears show an elevated M:E ratio, often >10:1, with dysgranulopoiesis and left shift. Hyposegmentation (pseudo Pelger–Huet) or hypersegmentation of the neutrophils, bizarre nuclear morphology of granulocytic precursors, and cytoplasmic hypo- or hypergranulation are frequently noted.
- Myeloblasts range from 1% to 10% but occasionally may reach up to 19%. Dyserythropoiesis with or without the presence of abnormal megakaryocytes is frequently observed. There is no evidence of monocytosis.
- Some cases may show increased reticulin fibers.
- The peripheral blood shows elevated WBC, usually ranging from 30,000 to 90,000/μL, but in occasional cases exceeding 100,000/μL. The leukocytosis is primarily due to the increased number of neutrophilic granulocytes which are also left-shifted. Granulocytic precursors account for about 10–20% or more of the leukocytes, but myeloblasts are always <10% and often range from 0% to 10%. Dysplastic granulopoiesis, as mentioned earlier, is always present.
- There is a variable degree of anemia which may be associated with abnormal morphology, such as anisopoikilocytosis and/or macrocytosis. Thrombocytopenia is a frequent feature.

The overall morphologic features mimic those of CML, except for more significant dysplasia, lack of basophilia, and no absolute monocytosis and basophilia.

IMMUNOPHENOTYPE AND CYTOCHEMICAL STAINS

- No specific immunophenotypic features have been described for aCML.
- The leukocyte alkaline phosphatase (LAP) score is variable and ranges from low to high depending on the case.

MOLECULAR AND CYTOGENETIC STUDIES

Atypical CML is negative for Philadelphia (Ph^1) chromosome and shows no evidence of *BCR-ABL1* rearrangement. A high frequency of *RAS* mutations is reported in *BCR-ABL1* negative CML.

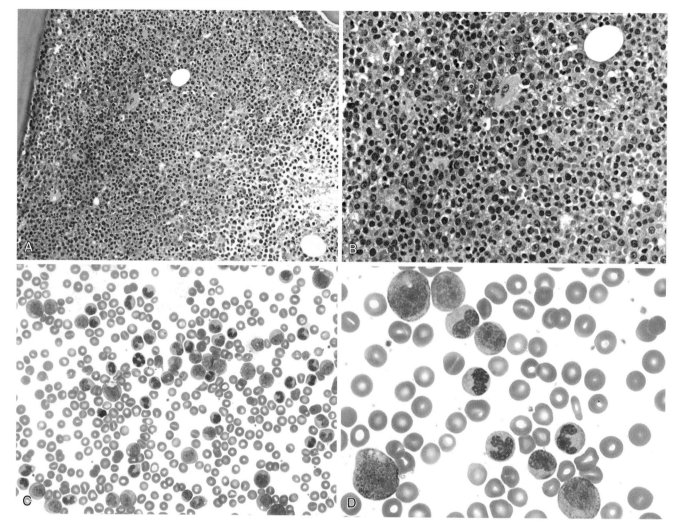

FIGURE 15.12 **ATYPICAL CML.** Bone marrow biopsy section of a 65-year-old man demonstrating marked hypercellularity with myeloid preponderance and left shift (A, low power; B, high power). Peripheral blood smear reveals marked granulocytosis with frequent immature forms (C, low power; D, high power). Dysgranulopoiesis is demonstrated by hyposegmented nuclei, abnormal chromatin pattern, and abnormal cytoplasmic granules.

Juvenile Myelomonocytic Leukemia

Juvenile myelomonocytic leukemia (JMML) is a clonal hematopoietic disorder of early childhood characterized by hepatosplenomegaly, granulocytosis, and monocytosis with left shift and dysplastic changes, elevated hemoglobin F levels, and frequent skin involvement. JMML shares considerable pathologic features with CMML. The WHO criteria for the diagnosis of JMML are presented in Box 15.3.

Juvenile myelomonocytic leukemia is a rare early childhood disorder with roughly 0.6 new cases per year per million children at risk, accounting for <2% of hematologic malignancies in children. The majority of patients are under the age of 4 years with a male:female ratio of about 2:5. Splenomegaly, hepatomegaly, lymphadenopathy, and skin rashes are noted in >90%, 80%, 70%, and 35% of the patients, respectively.

The cutaneous manifestations include neoplastic infiltration, eczema, xanthoma, and café-au-lait spots (Figure 15.13). JMML shows a high frequency (7–14%) of association with neurofibromatosis type 1. The prognosis is poor, but affected infants younger than 1 year of age appear to do better than older children. Elevated hemoglobin F levels (>15%) and low platelet counts (<33,000/μL) are amongst the unfavorable prognostic indicators.

Box 15.3 Criteria for the Diagnosis of Juvenile Myelomonocytic Leukemia[1]

- Peripheral blood monocytosis >1000/µL
- Presence of myeloid blasts and promonocytes in the peripheral blood and/or the bone marrow; less than 20% of the differential counts.
- No Ph^1 or *BCR-ABL1* fusion gene.
- Plus two or more of the following:
  - Elevated levels of hemoglobin F for age
  - Presence of immature granulocytes in the peripheral blood
  - WBC >10,000/µL
  - Clonal chromosomal aberrations
  - Hypersensitivity of myeloid precursors to GM-CSF *in vitro*

[1] Adapted from Jaffe ES, Harris NL, Stein H, et al. Pathology and Genetics: Tumors of Haematopoietic and Lymphoid Tissues. IARC Press, Lyon, 2001 and Swerdlow SH, Campo E, Harris NL, et al. WHO Classification of Tumours of Haematopoietic and Lymphoid Tissues, 4th edn. International Agency for Research on Cancer, Lyon, 2008.

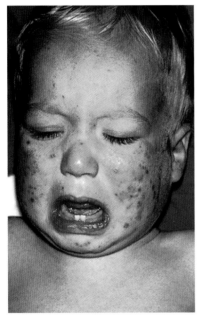

FIGURE 15.13 **JUVENILE CHRONIC MYELOID LEUKEMIA.** Eczematoid facial rash and lip bleeding in an 8-month-old infant. *(Courtesy of Professor JM Chessells).*
From Hoffbrand AV, Pettit JE, Vyas P. Color Atlas of Clinical Hematology, 4th edn. Mosby/Elsevier, Philadelphia, 2010, by permission.

Allogeneic BMT is the only available cure with an approximately 50% 5-year event-free survival rate. Recent reports on the effects of zoledronic acid (ZOL), a blocker of RAS activity, are promising.

The entity *infantile (childhood) monosomy 7 syndrome* shares most of the clinicopathologic features of JMML. Both disorders affect children at early ages (often <1 year old), show male predominance, show an association with neurofibromatosis type 1, and similar frequency of *RAS* gene mutation. Also, monosomy 7 is the most frequent chromosomal abnormality in JMML. However, children with JMML who lack monosomy 7 often display elevated levels of hemoglobin F. It seems that infantile monosomy 7 is a subtype of JMML.

MORPHOLOGY AND LABORATORY FINDINGS (FIGURE 15.14)

- The bone marrow samples are cellular and display myeloid preponderance and left shift with increased number of immature myelomonocytic precursors. However, the total number of myeloblasts, monoblasts, and promonocytes is <20%. Auer rods are not present.
- Dysplastic changes are frequently observed in the monocytic and the granulocytic series, such as bizarre morphology, nuclear hyper- or hyposegmentation, and cytoplasmic hypo- or hypergranularity.
- Erythroid dysplasia is often minimal or absent. Megakaryocytes may be reduced or show some degree of dysplastic changes including presence of micromegakaryocytes.
- Eosinophilia and basophilia are rare.
- The peripheral blood reveals monocytosis and granulocytosis. The average leukocyte count is about 30,000/µL. Monocytes and granulocytes are left-shifted and dysplastic with the presence of metamyelocytes, myelocytes, promyelocytes, and promonocytes. Myeloid blast cells and promonocytes are often <5% but never >19% of the leukocyte differential count.
- There is often some degree of anemia and/or thrombocytopenia. Anisopoikilocytosis and macrocytosis are frequent features and nucleated red blood cells are often present.

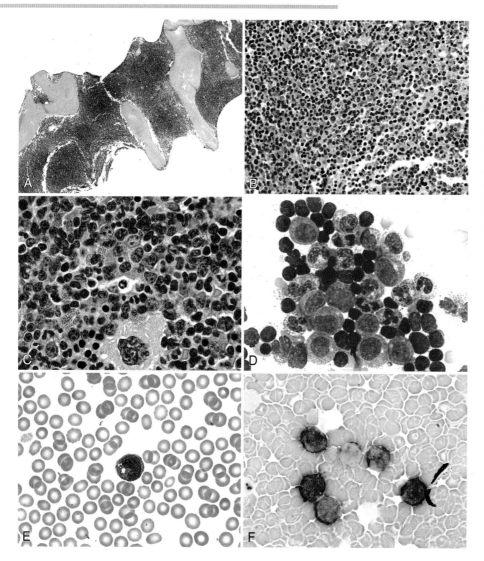

FIGURE 15.14 JUVENILE MYELOMONOCYTIC LEUKEMIA. Bone marrow core biopsy section reveals a markedly hypercellular marrow (A, low power) with preponderance of myelomonocytic precursors (B, intermediate power) and eosinophilia (C, high power). Bone marrow smear shows myeloid preponderance with moderate hematogone hyperplasia and no evidence of increased blasts (D). An abnormal, vacuolated monocyte is shown in blood smear (E), and non-specific esterase stain highlights increased number of monocytic precursors (F).

- The serum lysozyme levels are elevated. The hemoglobin F levels are elevated, the glucose-6-phosphatase activity is increased, and there is a high incidence of antinuclear (50%) and anti-IgG (40%) antibodies. There may be evidence of polyclonal hypergammaglobulinemia.
- Extramedullary involvement is a frequent feature with the infiltration of myelomonocytic cells in the dermis, the lung parenchyma, the hepatic sinusoids, and the splenic red pulp.

IMMUNOPHENOTYPE AND CYTOCHEMICAL STAINS

Immunophenotypic features by multiparametric flow cytometry are similar to those seen in CMML, as described above.

MOLECULAR AND CYTOGENETIC STUDIES

- Mutations of RAS, NF, and PTNP11 genes are frequently detected in patients with JMML. Quantitative measurements of RAS and PTPN11 have been made by an allele-specific polymerase chain reaction (PCR) assay called TaqMan, and increased levels have been correlated with relapse of JMML in transplanted patients.
- Methylation of p15, which is a frequent finding in patients with MDS (78%), is a rare event (17%) in JMML patients.
- Cytogenetic aberrations are often non-specific. Monosomy of chromosome 7 is the most frequent cytogenetic abnormality, along with deletion 7q (Figure 15.15), trisomy 8, and trisomy 21. Rare cases with t(3;12) (q21;p13) or t(3;15)(q21; q26) have also been reported.

Differential Diagnosis

The myeloproliferative/myelodysplastic neoplasms show significant overlapping morphologic features among themselves and with CML. CML patients are usually younger and show much more severe leukocytosis than patients with CMML or aCML. Basophilia is a common feature in CML but not present in CMML or aCML. Dysplastic myelopoiesis is a characteristic feature of aCML, CMML, and JMML, whereas it is insignificant in CML. Monocytosis is the hallmark of CMML and JMML and is lacking in CML and aCML. Ph^1 and/or *BCR-ABL1* fusion gene are present in CML but absent in aCML, CMML, and JMML. JMML is a disease of early childhood (usually under the age of 4) and is commonly associated with skin rashes and elevated hemoglobin F levels. There is a high frequency (7–14%) of neurofibromatosis type 1 in JMML patients. The major clinicopathologic features of CMML, aCML, and JMML are compared with one another and with those of CML in Table 15.1.

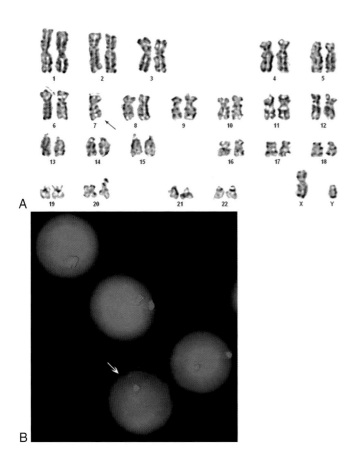

FIGURE 15.15 Monosomy 7 (arrows) in a child with JMML: (A) karyotype and (B) FISH.

Table 15.1

Clinicopathologic Features of Chronic Myelogenous Leukemia (CML), Atypical Chronic Myeloid Leukemia (aCML), Chronic Myelomonocytic Leukemia (CMML), and Juvenile Myelomonocytic Leukemia (JMML)[1]

Features	CML	aCML	CMML	JCML
Average age (years)	46	57	72	<4
Male:Female	>1	>1	>1	>1
Splenomegaly	+++	++	+	+
Blood				
Average leukocyte count	>100,000/μL	60,000/μL	35,000/μL	30,000/μL
Absolute monocytosis	Often no	Often no	Yes	Yes
Basophilia	Often yes	No	No	No
Myeloid precursors	+++	++	++	++
LAP	Reduced	Variable	Variable	Variable
Anemia	Present	Present	Present	Present
Elevated Hb F	No	No	No	Yes
Platelet count	Variable	Reduced	Reduced	Reduced
Bone Marrow				
Cellularity	Increased	Increased	Increased	Increased
Myeloid preponderance	Yes	Yes	Yes	Yes
Myeloid left shift	Yes	Yes	Yes	Yes
Monocytosis	No	No	Yes	Yes
Significant dysplasia	No	Yes	Yes	Yes

[1] Adapted from Martiat P, Michaux JL, Rodhain J. Philadelphia-negative (Ph−) chronic myeloid leukemia (CML): comparison with Ph+ CML and chronic myelomonocytic leukemia. Groupe Français de Cytogénétique Hématologique. Blood 1991; 78: 205–201 and Jaffe ES, Harris NL, Stein H, et al. Pathology and Genetics: Tumours of Haematopoietic and Lymphoid Tissues. IARC Press, Lyon, 2001.

Additional Resources

Bennett JM: The myelodysplastic/myeloproliferative disorders: The interface, *Hematol Oncol Clin North Am* 17:1095–1100, 2003.

Beran M: Chronic myelomonocytic leukemia, *Cancer Treat Res* 142:107–132, 2008.

Emanuel PD: Juvenile myelomonocytic leukemia and chronic myelomonocytic leukemia, *Leukemia* 22:1335–1342, 2008.

Foucar K: Myelodysplastic/myeloproliferative neoplasms, *Am J Clin Pathol* 132:281–289, 2009.

Germing U, Strupp C, Knipp S, et al: Chronic myelomonocytic leukemia in the light of the WHO proposals, *Haematologica* 92:974–977, 2007.

Hall J, Foucar K: Diagnosing myelodysplastic/myeloproliferative neoplasms: laboratory testing strategies to exclude other disorders, *Int J Lab Hematol* 32:559–571, 2010.

Jaffe ES, Harris NL, Vardiman JW, et al: Hematopathology, Philadelphia, 2010, Saunders/Elsevier.

Jäger R, Kralovics R: Molecular pathogenesis of Philadelphia chromosome negative chronic myeloproliferative neoplasms, *Curr Cancer Drug Targets* 11:20–30, 2011.

Loh ML: Recent advances in the pathogenesis and treatment of juvenile myelomonocytic leukaemia, *Br J Haematol* 152:677–687, 2011.

Orazi A, Germing U: The myelodysplastic/myeloproliferative neoplasms: myeloproliferative diseases with dysplastic features, *Leukemia* 22:1308–1319, 2008.

Oscier D: Atypical chronic myeloid leukemias, *Pathol Biol (Paris)* 45:587–593, 1997.

Swerdlow SH, Campo E, Harris NL, et al: WHO classification of tumours of haematopoietic and lymphoid tissues, ed 4, International Agency for Research on Cancer Lyon, 2008.

Vardiman JW: Myelodysplastic syndromes, chronic myeloproliferative diseases, and myelodysplastic/myeloproliferative diseases, *Semin Diagn Pathol* 20:154–179, 2003.

Vardiman JW: Myelodysplastic/myeloproliferative diseases, *Cancer Treat Res* 121:13–43, 2004.

Hematologic Neoplasms Associated with Eosinophilia and *PDGFRA*, *PDGFRB*, or *FGFR1* Rearrangement

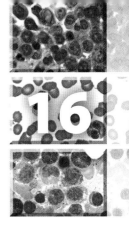

A diverse group of hematologic disorders may demonstrate eosinophilia, such as acute myeloid leukemias, acute lymphoid leukemias, chronic myeloproliferative neoplasms, myelodysplastic neoplasms, and Hodgkin's lymphoma (Box 16.1). Eosinophilia may be part of a clonal hematologic neoplasm or may represent a reactive, non-clonal process.

This chapter reviews the hematologic neoplasms that are associated with clonal eosinophilia and the rearrangement of platelet-derived growth factor receptor α/β (*PDGFRA*, *PDGFRB*) or fibroblast growth factor receptor 1 (*FGFR1*), resulting in a fusion gene encoding an aberrant tyrosine kinase. The clinical importance of these disorders is that the tyrosine kinase inhibitors, such as imatinib (Gleevec), may play an important role in their treatment.

Myeloid and Lymphoid Neoplasms Associated with PDGFRA Rearrangement

PDGFRA rearrangement is associated with a cryptic interstitial del(4)(q12) (Figure 16.1) or rearrangements of the 4q12 such as t(4;10)(q12;p11.2). This is a rare entity and may clinically present itself in the following forms:

- Chronic eosinophilic leukemia (CEL) (most common)
- Acute myeloid leukemia (AML)
- T-lymphoblastic leukemia/lymphoma (T-ALL)
- Both AML and T-ALL.

Box 16.1 Clinical Conditions Associated with Eosinophilia[1]

Primary

1. Clonal
   a. Acute myeloid leukemias
   b. Acute lymphoid leukemias
   c. Myelodysplastic neoplasms
   d. Chronic mylroproliferative neoplasms
   e. Myelodysplastic myeloproliferative neoplasms
   f. Hematologic neoplasms associated with *PDGFRA*, *PDGFRB*, or *FGFR1*
2. Non-clonal (Idiopathic): Hypereosinophilic syndrome

Secondary

1. Infectious
   a. Parasitic infections (most common)
   b. Bacterial or viral infections (rare)
2. Non-infectious
   a. Drugs and toxins (such as sulfa, toxic oil syndrome)
   b. Allergy and autoimmune disorders (such as asthma, eosinophilic fasciitis)
   c. Malignancies (such as metastatic cancer, Hodgkin lymphoma)
   d. Endocrinopathies (such as Addison disease)

[1] Adapted from Tefferi Y. Blood eosinophilia: A new paradigm in disease classification, diagnosis, and treatment. Mayo Clin Proc 2005; 80: 75–83.

FIGURE 16.1 Deletion of the 4q12 region as identified by three-color FISH.

Patients with CEL and *PDGFRA* rearrangement may demonstrate systemic symptoms such as pruritis, restrictive cardiomyopathy (due to endomyocardial fibrosis), pulmonary fibrosis, gastrointestinal distress, and/or splenomegaly. The disease may eventually transform to acute leukemia. Patients usually respond well to treatment by tyrosine kinase inhibitors, but some may develop resistance.

MORPHOLOGY AND LABORATORY FINDINGS

- Bone marrow is hypercellular and may show morphologic features consistent with CEL, AML, or T-ALL (Figure 16.2).
- There is marked eosinophilia, including some early forms. Dysplastic changes such as clumping of eosinophilic granules, presence of primary granules, and nuclear hypersegmentation may be present, and there may be areas of necrosis.

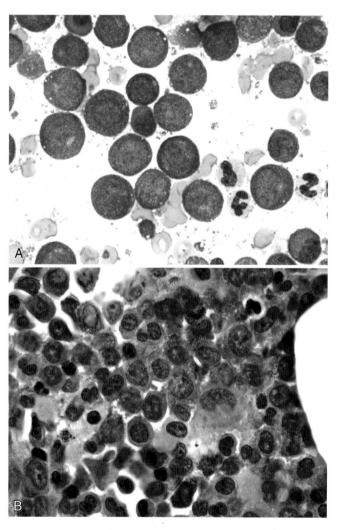

FIGURE 16.2 Bone marrow smear (A) and biopsy section (B) from a patient with AML and eosinophilia.

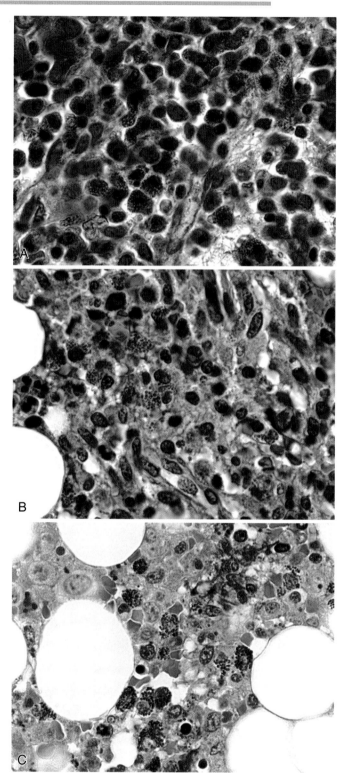

FIGURE 16.3 (A and B) Bone marrow biopsy sections from a patient with mastocytosis and eosinophilia (H&E stain). (C) Giemsa stain demonstrating the mast cells.

- Charcot–Leyden crystals may be present.
- There may be evidence of mastocytosis with or without a *KIT* mutation (Figure 16.3).
- The peripheral blood shows absolute eosinophilia (≥1500/ μL) (Figure 16.4).
- Serum tryptase and/or vitamin B_{12} levels may be elevated.

IMMUNOPHENOTYPE

There are no specific immunophenotypic aberrancies of CEL. Immunophenotypic features of the other subtypes are similar to these seen in AML or T lymphoblast leukemia.

Myeloid Neoplasms Associated with PDGFRB Rearrangement

This entity is characterized by t(5;12)(q33;p13) with formation of an *ETV6-PDGRFB* fusion gene (Figure 16.5). The clinical manifestation mimics chronic myelomonocytic leukemia (CMML), CEL, or unclassifiable myeloproliferative neoplasm with eosinophilia. This entity was considered a subtype of CMML (CMML with eosinophilia).

Patients often show splenomegaly. Hepatomegaly is rare, and skin infiltration and cardiomyopathy may be present. Patients usually respond well to treatment by tyrosine kinase inhibitors

MORPHOLOGY AND LABORATORY FINDINGS

- The bone marrow is hypercellular with myeloid preponderance and eosinophilia. Similar to CMML, the bone marrow may show myeloid left shift with increased immature granulocytic and monocytic series and dysplastic changes (Figure 16.6).
- Increased spindle-shaped mast cells and areas of fibrosis may be present.
- The peripheral blood shows an elevated WBC count with eosinophilia, and often neutrophilia and monocytosis. The eosinophils, neutrophils, and monocytes are left shifted and dysplastic.
- Serum tryptase levels may be elevated.

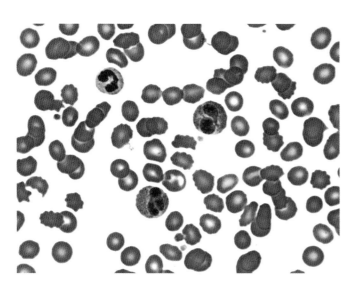

FIGURE 16.4 Blood smear demonstrating hypersegmented eosinophils.

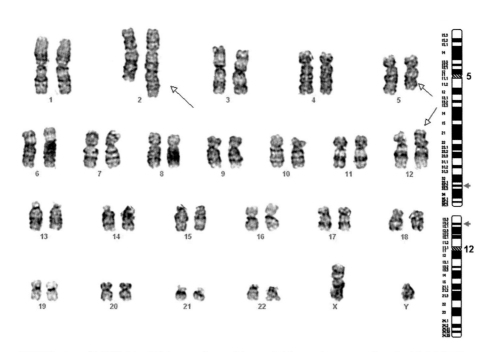

FIGURE 16.5 t(5;12)(q33;p13) in a patient with myeloid neoplasm associated with *PDGFRB* rearrangement (previously known as CMML with eosinophilia). The 2q aberration is non-specific.

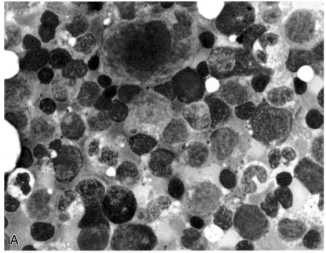
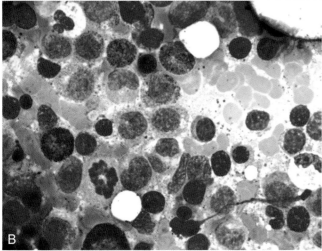

FIGURE 16.6 Bone marrow smears of a patient with myeloid neoplasm associated with *PDGFRB* rearrangement (previously known as CMML with eosinophilia).

IMMUNOPHENOTYPE

Findings of multiparametric flow cytometry (MFC) are similar to those observed in CMML, as described in Chapter 15.

Myeloid and Lymphoid Neoplasms Associated with *FGFR1* Rearrangements

This entity consists of a variety of hematologic neoplasms demonstrating 8p11.2 translocation and rearranged *FGFR1* gene. Different partner chromosomes are involved in this translocation, and therefore different diffusion genes are generated with *FGFR1*. The clinicopathologic manifestations range for CEL to AML, ALL, or bilineage acute leukemia. The prognosis is poor, and so far, no established treatment protocol with tyrosine kinase inhibitors is available.

MORPHOLOGY

- The bone marrow is hypercellular and demonstrates myeloid preponderance and eosinophilia. In the case of CEL, the bone marrow blasts are <20%. CEL may eventually transform to an acute phase demonstrating ≥20% blasts. In cases presenting as AML, ALL, or bilineage leukemia, a significant proportion of bone marrow cells (≥20%) consist of myeloblasts, lymphoblasts, or both populations, respectively.
- The peripheral blood shows eosinophilia (>1500/μL). Neutrophilia and monocytosis may be present. Leukocytes may show dysplastic changes. In the case of AML, ALL, or bilineage leukemia, blasts may be present in the peripheral blood.

IMMUNOPHENOTYPE

There are no specific immunophenotypic aberrancies of CEL. Immunophenotypic features are similar to those seen in AM, ALL, or bilineage leukemia.

MOLECULAR AND CYTOGENETIC STUDIES

The 8p11 myeloproliferative syndrome is associated with chromosomal rearrangements involving the fibroblast growth factor receptor 1 tyrosine kinase gene on chromosome 8p11.2. Translocations involving more than 10 partner loci have been identified. These abnormalities disrupt the *FGFR1* and various partner genes, and result in the creation of novel fusion genes and chimeric proteins. The latter include the N-terminal portion of the partner genes and the C-terminal portion of *FGFR1*. The most common partner is ZNF198 on chromosome 13q12. The other common sites are 9q33, 6q27, and 22q11.2, among others.

Additional Resources

Gotlib J: Molecular classification and pathogenesis of eosinophilic disorders: 2005 update, *Acta Haematol* 114:7-25, 2005.

Huang Q, Snyder DS, Chu P, et al: PDGFRA rearrangement leading to hyper-eosinophilia, T-lymphoblastic lymphoma, myeloproliferative neoplasm and precursor B-cell acute lymphoblastic leukemia, *Leukemia* 25:371-375, 2011.

Jackson CC, Medeiros LJ, Miranda RN: 8p11 myeloproliferative syndrome: A review, *Hum Pathol* 41:461-476, 2010.

Jaffe ES, Harris NL, Vardiman JW, et al: Hematopathology, Philadelphia, 2010, Saunders/Elsevier.

Rathe M, Kristensen TK, Møller MB, et al: Myeloid neoplasm with prominent eosinophilia and PDGFRA rearrangement treated with imatinib mesylate, *Pediatr Blood Cancer* 55:730-732, 2010.

Swerdlow SH, Campo E, Harris NL, et al: WHO classification of tumours of haematopoietic and lymphoid tissues, ed 4, International Agency for Research on Cancer Lyon, 2008.

Valent P: Pathogenesis, classification, and therapy of eosinophilia and eosinophil disorders, *Blood Rev* 23:157-165, 2009.

Vladareanu AM, Müller-Tidow C, Bumbea H, et al: Molecular markers guide diagnosis and treatment in Philadelphia chromosome-negative myeloproliferative disorders, *Oncol Rep* 23:595-604, 2010.

Wadleigh M, Tefferi A: Classification and diagnosis of myeloproliferative neoplasms according to the 2008 World Health Organization criteria, *Int J Hematol* 91:174-179, 2010.

Acute Myeloid Leukemia—Overview

Acute myeloid leukemia (AML) is a group of hematopoietic neoplasms derived from the bone marrow precursors of the myeloid lineage. The neoplastic process is the result of clonal proliferation of an aberrant, committed stem cell at the level of CFU-S or later stages of differentiation, leading to the accumulation of immature forms without or with limited maturation. Other terms used for AML include acute non-lymphoid leukemia (ANLL), acute myelogenous leukemia, and acute myeloblastic leukemia. The current WHO classification under "AML and related precursor neoplasms" consists of the following categories (Box 17.1).

1. AML with recurrent genetic abnormalities
2. AML with myelodysplasia-related changes
3. AML, therapy-related
4. AML not otherwise specified
5. Myeloid sarcoma
6. Myeloid proliferations related to Down syndrome
7. Blastic plasmacytoid dendritic cell neoplasm

In this book, items 1 through 6 are discussed in Chapters 18 to 22. The blastic plasmacytoid dendritic cell neoplasm is moved to Chapter 56, where dendritic cell disorders are discussed.

The etiology of AML is not clearly understood. It has been demonstrated that environmental factors and family background play important roles in the development of AML. Three major environmental insults have been implicated in the increased incidence of AML: (1) ionizing radiation, (2) chemotherapeutic agents, and (3) occupational exposure to chemicals.

Ionizing radiation induces DNA damage leading to chromosomal breaks which may cause mutations, deletions, and translocations. The extent of this damage depends on the type of radiation, the amount and rate of absorption, distribution of the absorbed energy in the tissue, and the intervals between the radiation exposures.

Alkylating agents and topoisomerase type II inhibitors are amongst the most potent chemical factors in the development of acute leukemia and make up the bulk of the subcategory of therapy-related AMLs (t-AML) in the

---

Box 17.1 WHO Classification of Acute Myeloid Leukemia (AML) and Related Precursor Neoplasms[1]

AML with recurrent genetic abnormalities
1. AML with balanced translocations/inversions
   a. AML with t(8;21)
   b. AML with abnormal eosinophils and inv(16) or t(16;16)
   c. Acute promyelocytic leukemia with t(15;17) or variants
   d. AML with t(9;11)
   e. AML with t(6;9)
   f. AML with inv(3) or t(3;3)
   g. AML with t(1;22)
2. AML with gene mutations

AML with myelodysplasia-related changes
1. Following myelodysplastic syndrome or myelodysplastic/myeloproliferative disorder
2. Without antecedent myelodysplastic syndrome

Therapy-related AML
1. Alkylating agent-related
2. Topoisomerase type II inhibitor-related
3. Ionizing radiation therapy
4. Others

AML not otherwise specified
1. AML minimally differentiated
2. AML without maturation
3. AML with maturation
4. Acute myelomonocytic leukemia
5. Acute monoblastic and monocytic leukemia
6. Acute erythroid leukemia
7. Acute megakaryoblastic leukemia
8. Acute basophilic leukemia
9. Acute panmyelosis with myelofibrosis

Myeloid sarcoma
Myeloid proliferations related to Down syndrome
Blastic plasmacytoid dendritic cell neoplasm

[1] Adapted from Swerdlow SH, Campo E, Harris NL, et al. WHO Classification of Tumours of Haematopoietic and Lymphoid Tissues, International Agency for Research on Cancer, Lyon 2008 and Jaffe ES, Harris NL, Vardiman JW, et al. Hematopathology. Saunders/Elsevier, Philadelphia, 2010.

WHO classification (see Chapter 20). Also, immunosuppressive therapy in transplant patients and in patients with immune-associated disorders may increase the risk of AML.

Occupational exposure to petroleum products (such as benzene), insecticides, and other organic solvents increases the risk of AML. Cigarette smoking, particularly in individuals over the age of 60, has shown a twofold increase in the risk of AML.

Certain familial disorders are associated with a higher risk of AML. There is a 10- to 20-fold increased chance of leukemia, particularly AML, in patients with Trisomy 21 (Down syndrome). A significant proportion of AMLs in these patients is of megakaryoblastic subtype.

The incidence of AML is also high in inherited disorders with defective DNA repair, such as Bloom's syndrome, Fanconi's anemia, Wiscott–Aldrich syndrome, neurofibromatosis, Kostmann's syndrome (infantile agranulocytosis), and Diamond–Blackfan anemia.

Leukemogenesis, similar to most other cancer developments, appears to be a multistep process involving structural and functional changes in a cascade of genes leading to the clonal expansion of defective stem cells. These genetic alterations often include mutations of oncogenes and/or loss of tumor suppressor genes. The specific genetic events in the process of leukemogenesis are not currently well understood, though it has been suggested that at least two mutations are required: one leading to a proliferative advantage and the other causing impairment of the maturation process (the "double-hit" hypothesis). The following examples represent the multistep concept of leukemogenesis in AML.

In the chronic phase of chronic myeloid leukemia (CML), leukemic cells show t(9;22) resulting in the *BCR-ABL1* fusion gene. As CML progresses to the accelerated phase and then blast transformation, additional genetic abnormalities, such as mutation of *TP53* (a tumor suppressor gene), evolve.

The high frequency of AML in patients with MDS strongly supports the double-hit hypothesis for leukemogenesis. MDS represents the first step or the first hit with frequent detectable chromosomal aberrations, including -5/del(5q), -7/del(7q), and +8. Evolution to AML, often with additional molecular and/or cytogenetic changes, depicts the final stage or the second hit.

A significant proportion of AMLs are associated with specific recurrent cytogenetic abnormalities such as t(8;21), inv(16)/t(16;16), t(15;17), t(9;11), t(6;9), inv(3q)/t(3;3) and t(1;22) (see Chapter 18).

The overall incidence of AML is about 3 per 100,000 persons per year. The median age for AML onset is 60 years, with a male to female ratio of about 1. The clinical symptoms are primarily related to cytopenias and include weakness, fatigue, recurrent infections, and hemorrhagic episodes, such as gum bleeding or ecchymoses. Bone pain is infrequent. Extramedullary infiltration (chloroma, granulocytic sarcoma) is occasionally seen, particularly in AMLs with monocytic differentiation. On rare occasions, extramedullary involvement may be the very first presenting symptom.

Cytogenetic results are the most informative indicators of prognosis. The favorable karyotypes include t(8;21), t(15;17), and inversion 16q/t(16q). Karyotypes with adverse clinical outcomes are monosomy 5/del(5q) or −7/del(7q), and abnormalities involving 3q. These aberrations are often accompanied by other anomalies and thus complex karyotypes with three or more chromosomal abnormalities fare very poorly. Resistant disease after first course of chemotherapy (>15% blasts in the bone marrow) also indicates poor prognosis. Five-year survival for the favorable prognostic category has been reported as to be 70% with a 33% chance of relapse, whereas the figures for the poor prognostic category are 15% and 78%, respectively.

Morphology

Acute myeloid leukemia refers to neoplasm of non-lymphoid hematopoietic progenitor cells. Therefore, it consists of subtypes representing various myeloid differentiations such as myeloblasts, promyelocytes, monoblasts, promonocytes, erythroblasts, and megakaryoblasts. In general, myeloblasts are the most predominant precursor cells in AML categories. The requirement for the diagnosis of AML is the presence of 20% or more blast cells in the bone marrow or blood differential counts. In addition to myeloblasts, "blast" counts in certain categories of AML include monoblasts, megakaryoblasts, promonocytes, or promyelocytes (Figures 17.1 to 17.6). Erythroblasts are excluded from the blast count, but are an important component of erythroleukemia, an AML subtype (see Chapter 21).

In certain conditions, such as in the category of AML with recurrent genetic abnormalities, the requirement for ≥20% blasts is sidestepped, and the presence of cytogenetic abnormalities even with <20% blasts is sufficient for the diagnosis of AML.

- Myeloblasts are large cells with scanty blue cytoplasm, round, oval or irregular nuclei, fine chromatin and multiple prominent nucleoli. The cytoplasm may contain no granular or variable numbers of azurophilic granules (Figure 17.1)
- Promyelocytes are overall larger and carry larger quantities of azurophilic granules than myeloblasts. They depict a well-developed Golgi system and a round or an oval nucleus, which is often eccentric. The granular myeloblasts and promyelocytes share overlapping morphologic features, and, therefore their distinction at times is difficult (Figure 17.2). Immunophenotypic studies may help to distinguish them (see below).
- Monoblasts are large cells with a variable amount of blue or grey-blue cytoplasm. The cytoplasm may be vacuolated and/or contain a few fine azurophilic granules. The nucleus is round, irregular or convoluted with fine chromatin and one or more prominent nucleoli (Figure 17.3). Morphologic distinction between myeloblasts and monoblasts may be difficult;

MORPHOLOGY

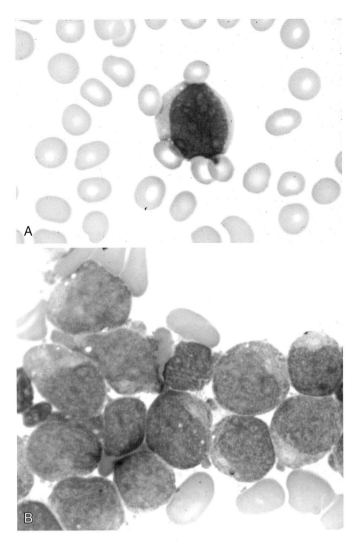

FIGURE 17.1 Blood smear (A) and bone marrow smear (B) demonstrating myeloblasts from a patient with acute myeloid leukemia, without maturation.

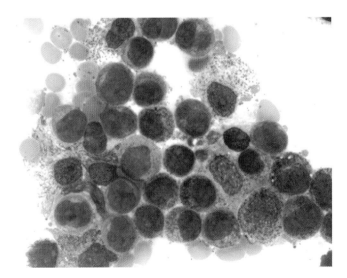

FIGURE 17.2 **ACUTE PROMYELOCYTIC LEUKEMIA.** Bone marrow smear demonstrating numerous promyelocytes with abundant cytoplasmic granules.

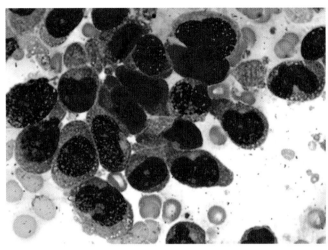

FIGURE 17.3 **ACUTE MONOBLASTIC LEUKEMIA.** Bone marrow smear demonstrating numerous monoblasts with cleaved nuclei and prominent nucleoli.

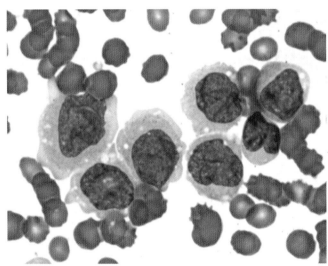

FIGURE 17.4 **ACUTE MONOCYTIC LEUKEMIA.** Blood smear demonstrating numerous promonocytes with abundant cytoplasm, convoluted nuclei and fine nuclear chromatin.

cytochemical stains and immunophenotypic studies are helpful to distinguish them (see below).

- Promonocytes are medium to large cells with variable amount of grey-blue cytoplasm. The cytoplasm may be vacuolated and/or contain a few fine azurophilic granules. The nucleus is convoluted and shows fine chromatin. Nucleoli are often inconspicuous or small (Figure 17.4).
- Erythroblasts are large cells with dark blue non-granular cytoplasm, often showing a perinuclear pale Golgi system. The nucleus is round with a thick nuclear membrane, fine nuclear chromatin, and inconspicuous or prominent nucleoli (Figure 17.5)
- Megakaryoblasts are small to large cells with a small amount of dark blue, non-granular cytoplasm, round nuclei, and fine chromatin. Cytoplasmic blebs may be present (Figure 17.6).
- Leukemic myeloblasts and promyelocytes, may contain Auer rods (Figure 17.7). Also, Auer rods may be occasionally present in the neoplastic monoblasts and promonocytes.

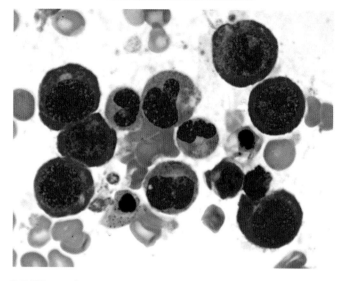

FIGURE 17.5 Bone marrow smear demonstrating numerous pronormoblasts (erythroblasts).

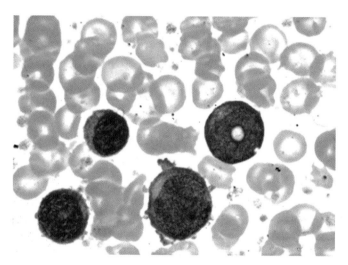

FIGURE 17.6 **ACUTE MEGAKARYOBLASTIC LEUKEMIA.** Blood smear demonstrating several megakaryoblasts.

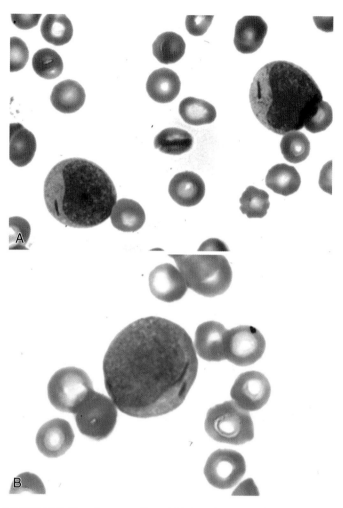

FIGURE 17.7 Blood smears (A and B) showing myelobalsts with Auer rods.

In general, the bone marrow biopsy/clot sections are hypercellular and show diffuse infiltration of the bone marrow by immature myeloid cells. Occasionally, the bone marrow is hypocellular. Blood examination may reveal anemia, leukopenia, and/or thrombocytopenia. Myeloid left shift is a common feature, and often a variable number of blasts are present. However, in some cases, at the time of bone marrow diagnosis, the peripheral blood smears may show no evidence of blast cells (aleukemic leukemia).

Cytochemical Stains

Most special cytochemical stains have been replaced by immunophenotyping in most laboratories. However, certain stains are still being used in the differential diagnosis and the classification of acute leukemia.

MYELOPEROXIDASE STAIN

Myeloperoxidase (MPO) is a lysosomal enzyme present in granulocytic and monocytic cells (Figure 17.8). MPO is expressed in neutrophilic and eosinophilic lineages in all stages of maturation, but in basophils it is more often detected in the immature forms. The mature basophils are usually negative for MPO. The intensity of MPO staining is less in monocytes than in granulocytes. Erythroid precursors and lymphocytes are MPO-negative. A peroxidase isoenzyme has been detected by electron microscopy in the dense tubular system of platelets and megakaryocytes, but, by conventional techniques, these cells are MPO-negative.

Myeloperoxidase activity declines rather rapidly. Air dried unstained smears should be stored at cool temperature, in the dark, and be used within 1–2 weeks.

CYTOCHEMICAL STAINS   223

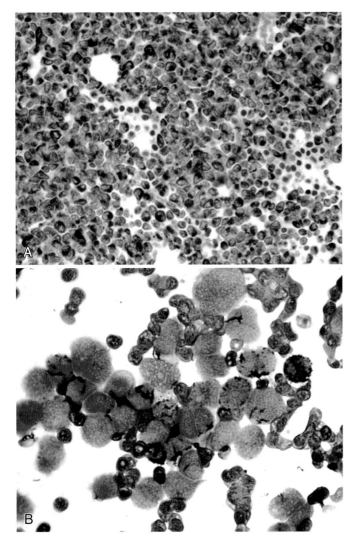

FIGURE 17.8 Immunohistochemical (A) and cytochemical (B) stains for MPO demonstrating numerous MPO-positive cells.

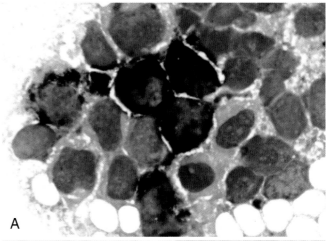

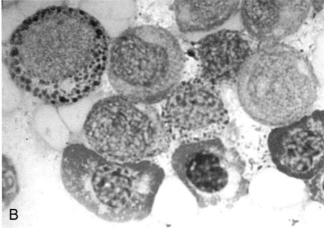

FIGURE 17.9 Bone marrow smears subject to cytochemical stains. (A) Sudan Black B stain shows dense darkly stained cytoplasmic granules in myeloid cells. (B) Dysplastic erythroid precursors show coarse PAS-positive cytoplasmic granules.

SUDAN BLACK B STAIN

Sudan Black B is a lipophilic dye that stains the granulocytic series. The pattern of reactivity of Sudan Black B in the granulocytic series is similar to that of MPO (Figure 17.9A). Monocytes are either negative or weakly positive with this stain. Lymphocytes, erythroid cells, megakaryocytes, and platelets are usually Sudan Black B-negative. Unlike MPO, Sudan Black B is stable, and therefore archival cytologic materials could be used for staining. Sudan Black B stain does not work in paraffin sections.

PERIODIC ACID–SCHIFF REACTION

Periodic acid–Schiff (PAS) reaction in hematopoietic cells is primarily due to the presence of cytoplasmic glycogen. The granulocytic lineage and plasma cells show diffuse, fine PAS-positive granules, whereas dysplastic erythroid precursors (Figure 17.9B) and sometimes blasts in acute lymphoid leukemia, monocytic leukemia, and megakaryocytic leukemia show coarse PAS-positive cytoplasmic granules.

ALPHA-NAPHTHYL BUTYRATE ESTERASE

Alpha-naphthyl butyrate esterase, also known as *non-specific esterase* (NSE), is a monocytic marker (Figure 17.10A). This stain is helpful in distinguishing acute leukemias with monocytic differentiation, as well as histiocytic lesions. However, lymphoblasts, erythroblasts, and megakaryoblasts may also show a few punctuate cytoplasmic-positive granules. Granulocytic series are negative for an NSE stain. NSE activity is fluoride-sensitive.

NAPHTHOL AS-D ACETATE ESTERASE

Naphthol AS-D acetate esterase is demonstrated in all stages of maturation in the granulocytic and monocytic series. Lymphoblasts, erythroblasts, and megakaryoblasts may also show a few punctuate cytoplasmic-positive granules. The enzyme activity is inhibited by sodium fluoride in monocytes but not in granulocytes.

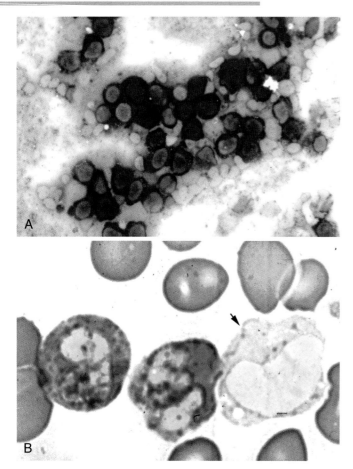

FIGURE 17.10 (A) Bone marrow smear showing numerous monocytic cells positive for an NSE stain. (B) Granulocytes stain for naphthol AS-D chloroacetate. A monocyte with a few cytoplasmic granules is present (arrow).

NAPHTHOL AS-D CHLOROACETATE

Naphthol AS-D chloroacetate is primarily expressed in the granulocytic series (Figure 17.10B) and mast cells. Other hematopoietic elements are essentially negative, though some monocytes, megakaryocytes, lymphoid and erythroid cells, and their leukemic counterparts may show a weak reaction. Naphthol AS-D chloroacetate is very stable and is demonstrated in archival cytologic materials and paraffin-embedded tissue sections.

Immunophenotypic Studies

Multiparametric immunophenotyping by flow cytometry (MIFC) is an integral component of the current diagnostic work-up in acute myeloid leukemias for the following reasons:

- Blast enumeration to confirm the presence of ≥20% blasts in bone marrow or blood at diagnosis and to evaluate post-treatment samples for residual disease

Table 17.1

Consensus Reagents for *Initial* Evaluation of Hematologic Neoplasms[1]

Lineage	Primary Reagents
B-cells	CD5, CD10, CD19, CD20, **CD45**, K/L
T- and NK-cells	CD2, CD3, CD4, CD5, CD7, CD8, **CD45**, CD56
Myelomonocytic	CD7, CD11b, CD13, CD14, CD15, CD16, CD33, CD34, **CD45**, CD56, CD117, HLA-DR
Myelomonocytic (limited)	CD13, CD33, CD34, **CD45**
Plasma cells	CD19, CD38, **CD45**, CD56

[1]Adapted from Wood BL, Arroz M, Barnett D, et al. 2006 Bethesda International Consensus recommendations on the immunophenotypic analysis of hematolymphoid neoplasia by flow cytometry: optimal reagents and reporting for the flow cytometric diagnosis of hematopoietic neoplasia. Cytometry B Clin Cytom 2007; 72(Suppl 1): S14–S22.

- Lineage assignment to distinguish AML from other types of acute leukemias, and to assign the leukemia to the proper subcategories, such as myeloblastic, monoblastic, erythroblastic, or megakaryoblastic
- Identification of phenotypic aberrancies to discriminate normal versus abnormal myeloblasts. This distinction is essential and serves as the molecular basis for the detection of residual disease and minimal residual disease (MRD). In addition, certain phenotypic aberrancies may have prognostic significance (e.g., aberrant CD56 with worse prognosis) and/or immunophenotypic-genotypic implications: e.g., concurrent aberrancies of CD56 and CD19 with AML t(8;21).

Phenotypic aberrancies include (see also Chapter 8):

- cross-lineage aberrancies
- abnormal expression intensities
- asynchronous expressions of maturation molecules
- abnormal location by CD45 gating
- abnormal clustering profiles.

ANTIBODY PANEL

No standardized antibody panel is currently available that can be used universally by different flow cytometry laboratories with various instrumentations and in diverse practice settings. Nevertheless, the 2006 Bethesda International Consensus recommendations on the immunophenotypic analysis of hematolymphoid neoplasia by flow cytometry have proposed a panel of antibodies that can be used for initial screening of hematolymphoid disorders (Table. 17.1). Additional lineage-associated markers and markers for detecting common cross-lineage aberrancies to consider are:

- Myeloid—myeloperoxidase (MPO)
- Monocytic—CD4, CD11c, CD36, CD64, or Lysozyme (by IHC)
- Erythroid—Glycophorin A, CD36 and CD71
- Megakaryocytic—CD41 and CD61, plus CD31 and factor VIII (by IHC)
- Common cross-lineage aberrancies—CD2, CD5, CD7, CD19, and CD56

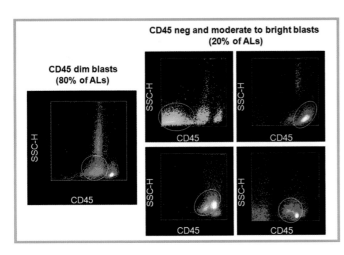

FIGURE 17.11 Variations of blasts by CD45 gating. Blasts can demonstrate various locations by CD45 gating. In about 80% of the acute leukemias, blasts express dim CD45 (left middle histogram). In the remaining 20% of cases, the blasts can be either CD45 negative (upper left), or CD45 positive at dim to moderate (lower left and right) to bright (upper right) levels.

MIFC DIAGNOSTIC PERILS AND PITFALLS

- Blast enumeration can be affected by various factors (e.g., dry tap, hemodilution, etc.) and needs to be correlated with morphology.
- Diagnosis of AML with t(8;21), inv(16), t(16;16), or t(15;17) does not require a blast count of ≥20%.
- Blasts may be present at variable locations by CD45 gating (Figure 17.11).
- Not all myeloblasts coexpress CD34 and CD117. Some myeloblasts can be negative for both CD34 and CD117.
- Phenotypic changes or switches may occur during or after therapy.
- A combined approach for immunophenotyping is essential to include MIFC, cytochemistry, and immunohistochemistry.
- Interpretation of MIFC results should always be in the context of multidisciplinary correlations.

Molecular Studies

Activating mutations in *FLT3*, a tyrosine kinase, have emerged as important changes in some cases of AML (found in about 30%), and hopeful targets of anti-kinase therapy as has worked so successfully in CML. The mutations also confer a less favorable prognosis.

- *FLT3* mutations are of two types: point mutations, usually involving codon 835, and internal tandem duplications (ITD). They are usually detected by some form of allele-specific PCR.
- Less common mutations occurring in other genes include *NPM1*, *CEBPA*, *c-KIT*, and *IDH1*. These mutations can have either synergistic or opposite prognostic effects with the *FLT3* mutations.

Cytogenetics

- A large variety of molecular genetic and cytogenetic abnormalities have been reported in AML.
- The occurring, recurrent and more specific abnormalities include [t(8;21)(q22;q22); (*RUNX1/RUNXT1*))], [t(15;17)(q24;q21.1); (*PML/RARα*)], [inv(16)(p13q22) or t(16;16)(p13;q22); (*CBFβ /MYH11*)], [t(9;11)(p21;q23); *MLLT3-MLL*], [t(6;9)(p23;q34); *DEK-NUP214*], [inv(3)(q21q26.2) or t(3;3)(q21;q26.2); *RPN1-EVI1*], and [t(1;22)(p13;q13); *RBM15-MKL1*] and 11q23 (*MLL*) abnormalities (see Chapter 18).
- The non-specific abnormalities include translocations, trisomies, monosomies, deletions, and other structural changes of all other chromosomes which are described later with each AML subtype.

Additional Resources

Bacher U, Schnittger S, Haferlach T: Molecular genetics in acute myeloid leukemia, *Curr Opin Oncol* 22:646–655, 2010.

Betz BL, Hess JL: Acute myeloid leukemia diagnosis in the 21st century, *Arch Pathol Lab Med* 134:1427–1433, 2010.

Falini B, Tiacci E, Martelli MP, et al: New classification of acute myeloid leukemia and precursor-related neoplasms: changes and unsolved issues, *Discov Med* 10:281–292, 2010.

Heerema-McKenney A, Arber DA: Acute myeloid leukemia, *Hematol Oncol Clin North Am* 23:633–654, 2009.

Jaffe ES, Harris NL, Vardiman JW, et al: Hematopathology, Philadelphia, 2010, Saunders/Elsevier.

Kern W, Bacher U, Haferlach C, et al: The role of multiparameter flow cytometry for disease monitoring in AML, *Best Pract Res Clin Haematol* 23:379–390, 2010.

Motyckova G, Stone RM: The role of molecular tests in acute myelogenous leukemia treatment decisions, *Curr Hematol Malig Rep* 5:109–117, 2010.

Orazi A: Histopathology in the diagnosis and classification of acute myeloid leukemia, myelodysplastic syndromes, and myelodysplastic/myeloproliferative diseases, *Pathobiology* 74:97–114, 2007.

Peters JM, Ansari MQ: Multiparameter flow cytometry in the diagnosis and management of acute leukemia, *Arch Pathol Lab Med* 135:44–54, 2011.

Swerdlow SH, Campo E, Harris NL, et al: WHO Classification of Tumours of Haematopoietic and Lymphoid Tissues, ed 4, Lyon, 2008, International agency for research on cancer.

Vardiman JW, Thiele J, Arber DA, et al: The 2008 revision of the World Health Organization (WHO) classification of myeloid neoplasms and acute leukemia: rationale and important changes, *Blood* 114:937–951, 2009.

Watt CD, Bagg A: Molecular diagnosis of acute myeloid leukemia, *Expert Rev Mol Diagn* 10:993–1012, 2010.

Yin CC, Medeiros LJ, Bueso-Ramos CE: Recent advances in the diagnosis and classification of myeloid neoplasms–comments on the 2008 WHO classification, *Int J Lab Hematol* 32:461–476, 2010.

Acute Myeloid Leukemias with Recurrent Genetic Abnormalities

Several specific chromosomal translocations, inversions, and gene mutations have been identified in AML patients, with significant prognostic values. Some of these genetic abnormalities are strongly associated with specific morphologic features, such as association of acute promyelocytic leukemia with t(15;17) or association of acute myelomonocytic leukemia with abnormal eosinophils with t(16;16) or inv(16). In this chapter we will discuss two major categories with recurrent genetic abnormalities: (a) AML with balanced translocations or inversions (Table 18.1) and (b) AML with gene mutations.

Table 18.1
Major Characteristics of Acute Myeloid Leukemias with Balanced Translocations/Inversions

Cytogenetic/Molecular Alteration	Predominant Morphology	Clinical Remarks
t(8;21)(q22;q22); RUNX1-RUNX1T1	AML with maturation (AML-M2)	Favorable prognosis
t(15;17)(q24;q21); PML-RARA	Acute promyelocytic leukemia	Favorable prognosis
inv(16)(p13.1q22) or t(16;16)(p13.1;q22) or del (16)(q22);CBFB-MYH11	Acute myelomonocytic leukemia with abnormal eosinophils	Favorable prognosis
t(9;11)(p22;q23); MLLT3-MLL	AML with monocytic differentiation	Intermediate prognosis
t(6;9)(p23;q34); DEK-NUP214	Variable	Poor prognosis
inv(3)(q21q26.2) or t(3;3)(q21;q26.2); RPN1-EVI1	Variable, often with multilineage dysplasia	Poor prognosis
t(1;22)(p13;q13); RBM15-MKL1	Megakaryoblastic	Guarded, responding to current therapy
t(9;22) (q34;q11.2); BCR-ABL1	Variable	Poor prognosis
t(8;16)(p11.2;p13.3); KAT6A/CREBBP	AML with monocytic differentiation and erythrophagocytosis	Variable

AML with Chromosomally Balanced Rearrangements: Translocations/Inversions

AML WITH t(8;21)(q22;q22); (RUNX1-RUNX1T1)

Acute myeloid leukemia with t(8;21)(q22;q22) is a balanced translocation in which *RUNX1* (transcription factor 1), previously called *AML1* gene, on the long arm of chromosome 21q22, fuses with the *RUNX1T1* gene (also named *MTG8*, *ETO*, or *CBFA2T1*) on the long arm of chromosome 8q22. This fusion results in an *RUNX1-RUNX1T1* chimeric product which appears to inhibit apoptosis by activating the expression of the anti-apoptosis gene *BCL-2*.

AML with t(8;21) accounts for 5–10% of AMLs and involves both children and adults. The average age for adults is about 30 years, which is significantly lower than the average age for other types of AML. It has a favorable prognosis in adults, but the clinical outcome is poor in children.

Morphology

The morphologic features in a significant proportion of AML with t(8;21) are similar to those described in the category of AML with maturation (Figure 18.1) (see Chapter 21).

- The myeloblasts are large with variable amount of blue cytoplasm, and often indented nuclei. Azurophilic cytoplasmic granules are often present, and some blasts may contain large granules mimicking cytoplasmic granules seen in the Chediak–Higashi syndrome.
- Auer rods are frequent and also may be detected in the more mature myeloid forms.
- Promyelocytes, myelocytes, metamyelocytes, bands, and segmented neutrophils are present and often show dysplastic changes.
- Eosinophilia is common, and some cases may show increased bone marrow basophils and/or mast cells.

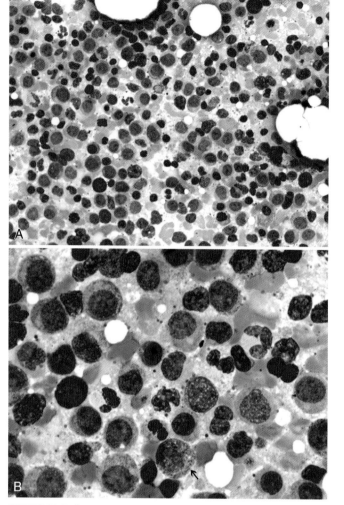

FIGURE 18.1 **ACUTE MYELOID LEUKEMIA WITH MATURATION.** Bone marrow smears (A, low power; B, high power) of a patient with AML with maturation associated with t(8;21)(q22;q22) (see Figure 18.3).

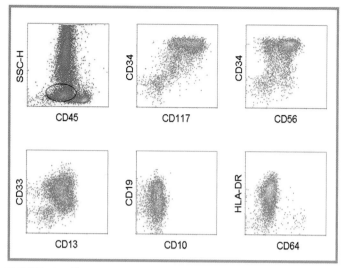

FIGURE 18.2 **FLOW CYTOMETRY RESULTS OF AML WITH t(8;21)(q22;q22).** Open gate display by CD45 gating reveals excess blasts (10% of the total, circled). Density plots of the blast-enriched gate (in blue) show abnormal myeloblasts, which are positive for CD13, CD33, CD34 (bright), CD45 (dim), CD117, and HLA-DR. In addition, there are distinct cross-lineage aberrancies of CD56 and partial dim CD19, which are characteristic of this entity. The concurrence of CD19 and CD56 aberrancies should raise a high index of suspicion for AML with t(8;21), even though the blast count may be less than 20%.

- Approximately 7% of AML cases with t(8;21) show morphologic features of acute myelomonocytic leukemia (see Chapter 21). These patients depict peripheral blood monocytosis with the presence of immature forms and increased myeloblasts, monoblasts, and promonocytes in their bone marrows.
- Rare cases of AML with t(8;21) may show blast counts of <20% in their blood and bone marrow.

Immunophenotype

Most cases demonstrate a characteristic phenotype (Figure 18.2), which may show features overlapping with those seen in AML with maturation.

- Aberrancies of CD56 and partial dim CD19 expression are common.
- Blasts often reveal bright CD34, moderate to bright CD13, dim CD33, plus CD117 and HLA-DR.
- MPO is positive; TdT is sometimes weakly positive.

Molecular and Cytogenetic Studies

- The molecular detection of *RUNX1-RUNX1T1* fusion is usually accomplished by the FISH technique or reverse transcriptase polymerase chain reaction (RT-PCR). It is important to know that some patients who have remained in continuous remission for a long period may still show *RUNX1-RUNX1T1* mRNA in their leukocytes by RT-PCR techniques. The clinical significance of the persistence of *RUNX1-RUNX1T1* is not clear. However, the detection of *RUNX1-RUNX1T1* fusion by itself may not indicate relapse or active disease.
- The characteristic cytogenetic finding is the t(8;21)(q22;q22) (*RUNX1-RUNX1T1*)(Figure 18.3A). Complex translocations, involving three or more chromosomes along with chromosomes 8 and 21, have also been reported in occasional cases. In some cases, the t(8;21) rearrangement is cryptic, and can go undetected by standard karyotyping. In such cases, FISH (Figure 18.3B) or RT-PCR methods are required to establish the fusion *RUNX1-RUNX1T1* rearrangement.

ACUTE PROMYELOCYTIC LEUKEMIA WITH t(15;17)(q24;q21)

Acute promyelocytic leukemia (APL) is one of the variants of AML associated with t(15;17)(q24;q21.2) (*PML-RARA*) or other forms of chromosomal rearrangements involving the retinoic acid receptor alpha (*RARA*) gene. This translocation leads to the production of PML-RARA fusion protein, which is less sensitive to retinoic acid. This reduced sensitivity to retinoic acid may lead to persistent

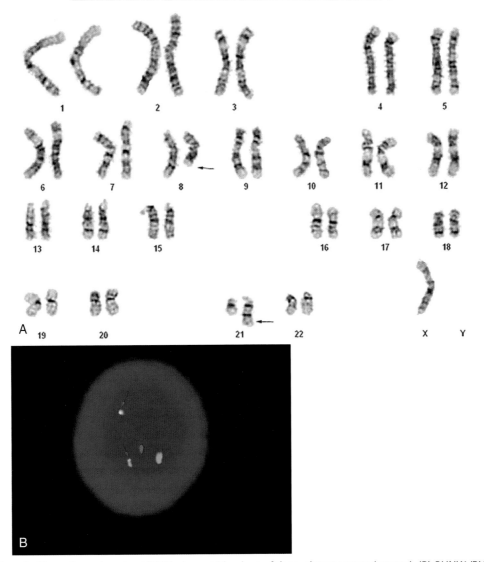

FIGURE 18.3 (A) G-banded karyotype showing t(8;21) along with a loss of the y-chromosome (arrows). (B) *RUNX1/RUNXT1* fusion demonstrated by FISH (arrows).

transcriptional repression, and therefore prevention of further differentiation of promyelocytes.

APL accounts for 5–10% of all AMLs and is primarily seen in young adults and middle-aged patients, but it may occur at any age. Clinical symptoms are related to complications of cytopenia and disseminated intravascular coagulopathy (DIC). Weakness/fatigue, infection, and hemorrhagic episodes are often complications of anemia, granulocytopenia, and thrombocytopenia, respectively. DIC is either present at diagnosis or detected soon after chemotherapy. DIC is a serious complication which may lead to cerebrovascular or pulmonary hemorrhage in up to 40% of patients. The risk is reported to be higher in the microgranular variant of APL. Three major factors may contribute to the mechanism of DIC: (1) release of tissue factor which is involved in the activation of factor X through factor VII, (2) release of cancer procoagulants which activate factor X independent of factor VII, and (3) increased expression of annexin II receptor on leukemic promyelocytes. Annexin II receptor binds plasminogen and increases plasmin formation.

Acute promyelocytic leukemia is one of the favorable types of AML. Favorable prognostic factors include age under 30 years, initial leukocyte count <10,000/µL, and platelet count >40,000/µL. There are studies suggesting that the expression of CD56 on the leukemic promyelocytes, methylation of *p15* kinase inhibitor gene, and variant translocations such as t(11;17)(q23;q21.2) (*ZBTB16-RARA*) are associated with less favorable prognosis.

All-*trans* retinoic acid (ATRA) is a highly effective therapeutic agent. It accelerates the terminal differentiation of leukemic promyelocytes and induces clinical remission. For complete molecular remission and long-term survival, a combination of ATRA and cytotoxic chemotherapy is necessary. APL patients with t(11;17) (*ZBTB16-RARA*) do not respond to ATRA.

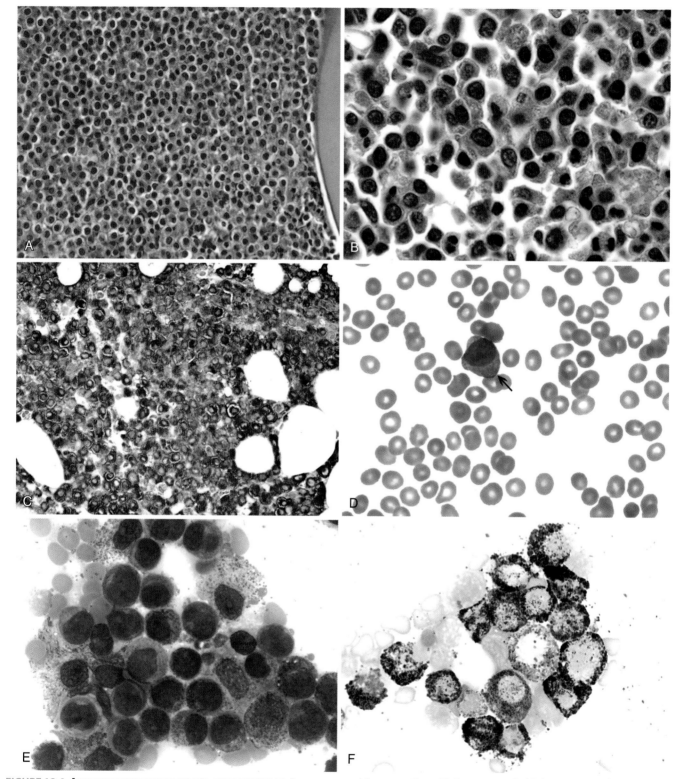

FIGURE 18.4 **ACUTE PROMYELOCYTIC LEUKEMIA, HYPERGRANULAR.** Bone marrow biopsy sections (A, low power; B, high power) are hypercellular and show sheets of promyelocytes with abundant granular cytoplasm. Immunohistochemical stain for MPO is strongly positive (C). A promyelocyte is depicted in blood smear (D), and numerous highly granular promyelocytes are demonstrated in bone marrow smear (E). These cells are positive for MPO by cytochemical stain (F).

Morphology

- Bone marrow biopsy sections are hypercellular and show clusters and/or sheets of immature myeloid cells with abundant granular cytoplasm and nuclear spacing.
- Peripheral blood smears often show leukocytosis with the presence of atypical promyelocytes.

Two morphologic variants of APL have been described: APL with hypergranular promyelocytes and APL with

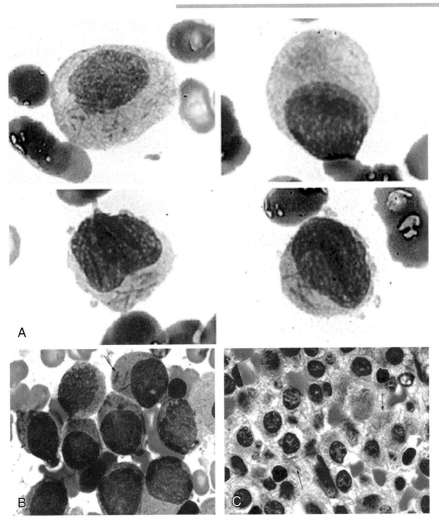

FIGURE 18.5 ACUTE PROMYELOCYTIC LEUKEMIA, FAGGOT CELLS. Blood smear (A), bone marrow smear (B) and marrow clot section (C) demonstrating several promyleocytes with bundles of Auer rods (arrows).

microgranular (hypogranular) promyelocytes. The hypergranular variant, according to the literature, accounts for about 75–80% of APLs. However, in our experience at the UCLA Medical Center and the VA Greater Los Angeles Healthcare System, we have seen more cases of the microgranular variant than the hypergranular type.

Hypergranular APL

- Hypergranular promyelocytes have a cytoplasm heavily loaded with azurophilic granules, which are often coarser and more numerous than the ones seen in normal promyelocytes (Figures 18.4 and 18.5). Auer rods are often present and in some cells appear in bundles (faggot cells) (Figure 18.5).
- Nuclei are usually round or oval but may appear irregular, folded, or dumbbell-shaped. The densely packed granules may obscure the visibility of the nuclei.
- The hypergranular promyelocytes are the predominant cells in the marrow, but smaller promyelocytes with basophilic cytoplasm and fewer azurophilic granules and microgranular promyelocytes are also present.
- Myeloblasts are fewer than promyelocytes and average around 10% of the bone marrow cells.

Hypogranular APL

- Hypogranular promyelocytes show abundant cytoplasm with lack of or sparse azurophilic granules. The azurophilic granules appear finer than the granules seen in the hypergranular variant (Figures 18.6 and 18.7).
- Auer rods and faggot cells may be present, but not so frequent as in the hypergranular subtype.
- The nuclei are predominantly bilobed, but folded and convoluted forms are often present, mimicking monocytic features. The nuclear chromatin is fine, often with prominent nucleoli.
- A small proportion of bone marrow cells may consist of myeloblasts, hypergranular promyelocytes, and small hyperbasophilic promyelocytes.
- A marked elevation of leukocyte count is seen more frequently in the microgranular variant than in hypergranular APL.

Immunophenotype

- Blasts and promyelocytes express heterogeneous CD13, bright homogenous CD33, often partial CD15 and CD64, and bright intracellular MPO (Figure 18.8).

232 ACUTE MYELOID LEUKEMIAS WITH RECURRENT GENETIC ABNORMALITIES

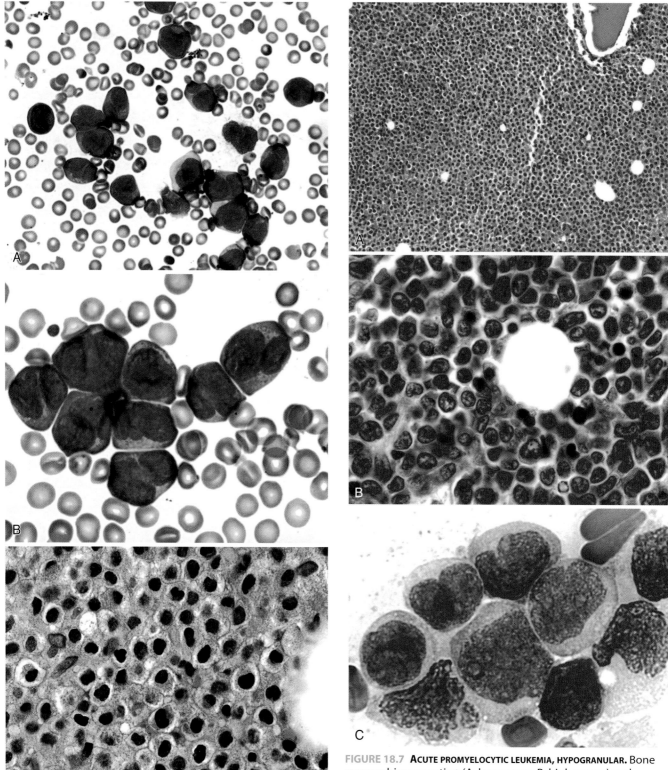

FIGURE 18.6 **ACUTE PROMYELOCYTIC LEUKEMIA, HYPOGRANULAR.** Blood smear (A, low power; B, high power) showing hypogranular promyelocytes with folded or convoluted nuclei. Bone marrow biopsy section (C) reveals a packed marrow consisting of cells with abundant finely granular cytoplasm and round, irregular, or folded nuclei.

FIGURE 18.7 **ACUTE PROMYELOCYTIC LEUKEMIA, HYPOGRANULAR.** Bone marrow biopsy section (A, low power; B, high power) and bone marrow smear (C) showing clusters of minimally granular promyelocytes with folded or convoluted nuclei.

- They often express CD117, but show absent or partial weak expression of CD34 and HLA-DR.
- In microgranular variant APL, the neoplastic cells are often positive for CD34 and aberrant CD2. Side scatter is low (Figure 18.9).
- CD56 aberrancy can sometimes be present, which may predict adverse outcome.

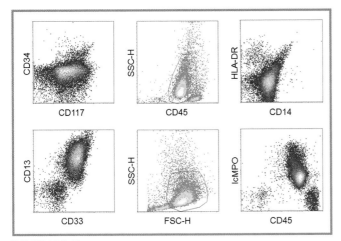

FIGURE 18.8 FLOW CYTOMETRIC FINDINGS IN ACUTE PROMYELOCYTIC LEUKEMIA. Open gate displays of density plots illustrate characteristic findings of APL, including unique positions by CD45 gating and light scatter gate (in blue, circled). The blasts (promyelocytes) are positive for CD13 (heterogeneous), CD15 (partial; not shown), CD33 (bright, tight cluster), CD45 (dim), CD117, and bright intracellular myeloperoxidase. They are negative for CD34 and HLA-DR. The blasts in APL are typically negative for HLA-DR, and generally show no uniformly double positivity for CD34 and CD117, plus typically no coexpression of CD15 and CD34.

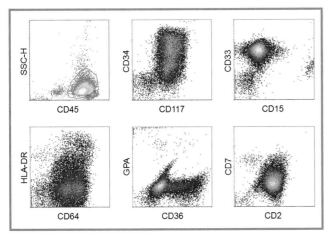

FIGURE 18.9 FLOW CYTOMETRIC FINDINGS IN HYPOGRANULAR ACUTE PROMYELOCYTIC LEUKEMIA. The immunophenotypic patterns of APL can be atypical, as often seen in cases with hypogranular variant. The open gate displays of density plots demonstrate that all blasts are positive for both CD34 and CD117, plus many are also positive for CD15. In addition, there is a partial expression of HLA-DR. Cross-lineage aberrancy CD2 and partial expression of erythrocyte/monocyte-associated marker CD36 are also noted in this case.

Molecular and Cytogenetic Studies

- The genetic hallmark of APL is a translocation involving the *RARA* gene. Four major translocations with the involvement of the *RARA* gene have been associated with APL. These include:
  - t(15;17)(q24;q21.1);(*PML;RARA*)
  - t(11;17)(q23;q21.1);(*ZBTB16;RARA*)
  - t(11;17)(q13;q21.1);(*NUMA1;RARA*)
  - t(5;17)(q35;q21.1);(*NPM1;RARA*).
- The most common translocation is t(15;17)(q24;q21.1), which is associated with the expression of PML-RARA fusion protein (Figure 18.10) and is observed in over 90% of APLs. The t(11;17)(q23;q21.1) variant represents fusion of the *RARA* gene with the ZBTB16 or *PLZF* (promyelocytic leukemia zinc finger) gene and is seen in about 1% of APLs (Figure 18.11). PLZF protein is expressed in myeloid lineages and its expression is downregulated during differentiation.
- Another rare variant, t(11;17)(q13;q21.1), involves the *NuMA1* (nuclear matrixmitotic apparatus protein) gene fusion with the *RARA* gene. Translocation of (5;17)(q35;q21.1) is also another rare variant reported, which involves the nucleophosmin (*NPM1*) gene. This gene plays a role in the regulation of ribosomal nuclear processing and its transport. Other infrequently reported genetic abnormalities include complex four-way variant rearrangements that also involve the significant 15q22 and 17q11.12 breakpoints leading eventually to the clinically relevant t(15;17).
- FISH and RT-PCR studies (Figures 18.10B and 18.12) are routinely performed for the detection of *PML-RARA* fusions either to establish the diagnosis of promyelocytic leukemia or to rule out or monitor residual disease after therapy. 97% of APL patients have a *PML-RARA* fusion on the der(17). 7% are cryptic insertions or complex rearrangements. The dual-color *RARA* "breakapart" FISH probe is designed to detect most rearrangements involving the 17q21.1 band.

AML WITH inv(16)(p13q22) or t(16;16)(p13;q22) or del(16)(q22)

Acute myeloid leukemia with inv(16)(p13q22) or t(16;16)(p13;q22) is one of the variants of AML characterized by fusion of the *CBFB* and *MYH11* genes. The fusion protein inhibits the function of the *AML1-CBFB* transcription factor leading to the repression of transcription. This leukemia displays myelomonocytic differentiation with the presence of abnormal eosinophils. It accounts for about 10% of all AMLs and mostly occurs in middle-aged patients. This leukemia is associated with a favorable prognosis with a complete remission rate of >90%. A combination of cytotoxic drugs, such as daunorubicin and cytarabine with or without mitoxantrone, has been used for induction therapy.

Morphology

- Bone marrow biopsy and clot sections usually are hypercellular with myeloid preponderance and left shift, and increased blasts. There is often evidence of eosinophilia.
- Bone marrow smears show myeloid left shift, increased number of immature myelomonocytic cells, and the presence of atypical eosinophilic precursors (Figures 18.13 and 18.14).
- Myeloblasts, monoblasts, and promonocytes usually account for ≥20% of the total marrow cells, but occasionally may be less.
- Eosinophils usually constitute >5% of marrow differential counts and appear to be a part of the leukemic clone. Some of the eosinophilic promyelocytes, myelocytes, and metamyelocytes contain large purple–violet, or basophilic granules in addition to eosinophilic granules. These atypical granules are rarely found in more mature eosinophils.
- Blood smears may show eosinophilia with the presence of blasts and promonocytes. Absolute monocytosis is a frequent finding.

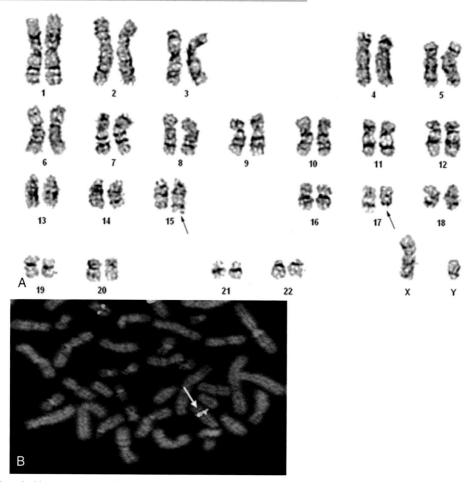

FIGURE 18.10 (A) G-banded karyotype showing t(15;17) (arrows). (B) *PML-RARA* fusion is demonstrated by FISH (arrow).

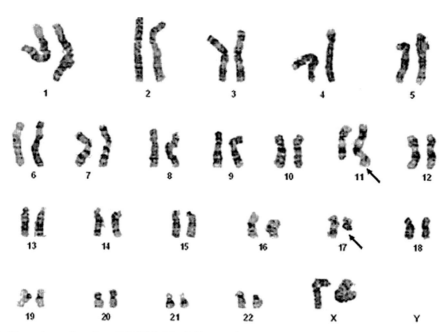

FIGURE 18.11 G-banded karyotype showing t(11;17) (q23:q21)(arrows).

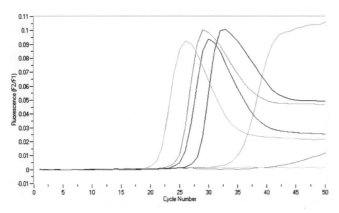

FIGURE 18.12 RT-PCR technique demonstrating a positive signal (the blue curve) in a patient with *PML-RARA* fusion (acute promyelocytic leukemia). Other curves are positive controls. The flat green and blue lines are negative controls.

In rare cases of AML with inv(16)(p13q22) or t(16;16)(p13;q22), the atypical eosinophils may not be present, or instead of both myeloid and monocytic differentiation, the acute leukemia may represent only myeloid or only monocytic features.

Immunophenotype and Cytochemical Stains (Figure 18.15)

- Blasts demonstrate complex immunophenotype with variable expression of myeloid (CD13, CD15, CD33, CD65, and MPO) as well as some monocytic markers (CD4, CD11b, CD11c, CD14, CD36, and CD64 by MIFC, plus CD68 and lysozyme by IHC).
- Subset of blasts is positive for CD34 and CD117.
- Aberrant expression of CD2 is common.
- The cytochemical stains show strong MPO positivity for the granulocytic lineage and variable degrees of diffuse cytoplasmic NSE staining for the monocytic population. The abnormal eosinophils may be weakly positive for naphthol AS-D chloroacetate esterase.

Molecular and Cytogenetic Studies

- *CBFB-MYH11* fusion with inv(16)(p13q22) or t(16;16)(p13;q22) is the characteristic genetic feature of this leukemia (Figures 18.16 and 18.17).
- The fusion transcript is detected by RT-PCR. Karyotyping and FISH studies reveal chromosome 16 pericentric inversion or translocation between the two chromosome 16s. Because of the difficulty of identifying the inversion in chromosomes with poor morphology, FISH with the CBFB probe is recommended for confirmation.
- Other associated cytogenetic abnormalities include trisomies 8, 21, and 22, as well as loss of the Y-chromosome.

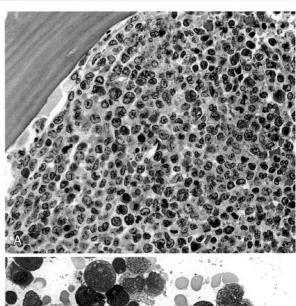

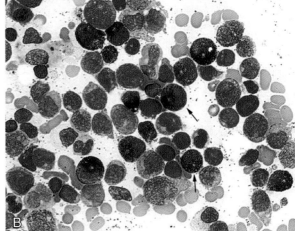

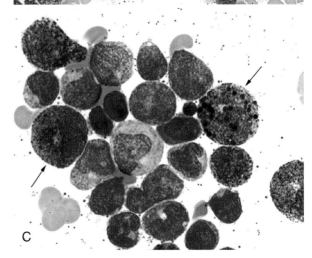

FIGURE 18.13 Bone marrow biopsy section (A) and bone marrow smear (B and C) from a patient with acute myelomonocytic leukemia with atypical eosinophilia and inv(16)(p13q22). Atypical eosinophils contain a mixture of eosinophilic and basophilic granules (arrows).

AML WITH t(9;11)(p22;11q23)

AML with t(9;11)(p21.3;q23) *MLLT3-MLL* usually represents acute leukemias with monocytic differentiation, such as acute myelomonocytic or acute monocytic leukemias.

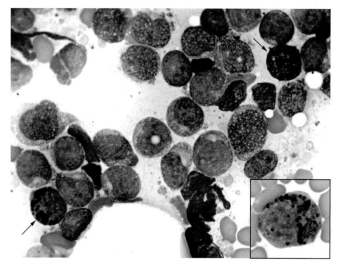

FIGURE 18.14 Bone marrow smear from a patient with acute myelomonocytic leukemia with atypical eosinophilia and inv(16)(p13q22). Atypical eosinophils contain a mixture of eosinophilic and basophilic granules (arrows and inset).

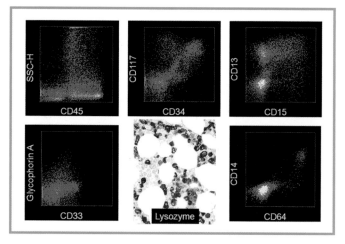

FIGURE 18.15 **FLOW CYTOMETRIC FINDINGS OF AMML WITH inv(16) (p13q22).** Open gate displays show excess blasts by CD45 gating with heterogeneous dim to moderate expression of CD45. A subset of the blasts coexpresses CD34 and CD117. The blasts are positive for CD13 (rather uniform), dim CD33, plus (not shown) CD11b, CD38, HLA-DR, and partial intracellular myeloperoxidase. The blasts are negative for CD14 and CD64, but positive for lysozyme by immunohistochemical studies.

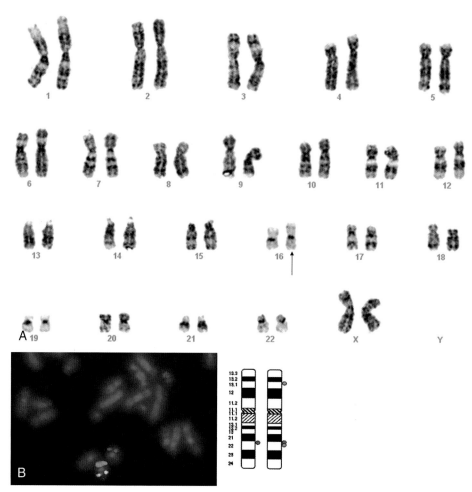

FIGURE 18.16 A t(16;16) is demonstrated by karyotyping (A) and FISH analysis (B) in a patient with acute myelomonocytic leukemia with atypical eosinophilia.

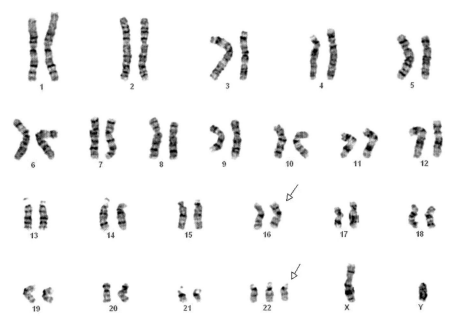

FIGURE 18.17 Karyotype of bone marrow cells in a patient with acute myelomonocytic leukemia demonstrating inv(16)(p13q22) and trisomy 22.

Myeloid leukemias with t(9;11) account for about 2% of adult and 10% of pediatric AML cases, and in general have a poor prognosis.

Morphology

- Monocytic differentiation is the morphologic hallmark of AMLs with t(9;11). Most cases fall into the category of acute myelomonocytic or acute monocytic leukemia with increased numbers of monoblasts and/or promonocytes in the bone marrow or peripheral blood (Figure 18.18) (see Chapter 21).
- Monoblasts have variable amounts of blue cytoplasm with no or few azurophilic granules or vacuoles. The nuclei are round or slightly indented or folded. The nuclear chromatin is finely dispersed, and one or more prominent nucleoli are present (Figure 18.18A).
- Promonocytes have more abundant cytoplasm which is less basophilic. They may contain few cytoplasmic azurophilic granules or vacuoles. The nuclei are irregular, folded, or convoluted with fine nuclear chromatin and often inconspicuous nucleoli (see Figure 18.18B).
- Bone marrow biopsy and clot sections are usually hypercellular with myeloid preponderance with increased blasts and myelomonocytic precursors.

Immunophenotype and Cytochemical Stains

- The monoblasts and promonocytes express CD4 (dim), CD11c, and CD14. CD34 and CD117 are usually negative.
- Myeloblasts often express CD13, CD33, and CD117 and may express CD34. CD36, CD64, and HLA-DR are usually expressed on both myeloblasts and monocytic precursors.
- Immunohistochemical stains, such as MPO, lysozyme, and CD68 are used for the evaluation of the bone marrow myelomonocytic component.

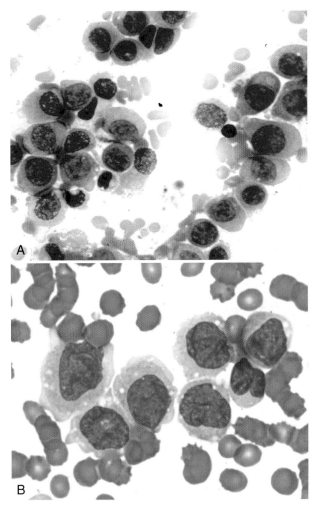

FIGURE 18.18 Acute myeloid leukemias associated with 11q abnormalities often show monocytic differentiation. (A) An aggregate of monoblasts is demonstrated in bone marrow smear. (B) Blood smear shows several promonocytes.

- The cytochemical stains show strong MPO positivity for the granulocytic lineage and various degrees of diffuse cytoplasmic NSE staining for the monocytic population.

Molecular and Cytogenetic Studies

- The gene involved in the translocation of 11q23 is the *MLL* (also known as *ALL1* or *HRX*) gene. Over 73 different translocations involving the *MLL* gene have been reported in both acute myeloid and lymphoid leukemias.
- The rearrangement of 11q23 is common in patients with acute myelomonocytic and acute monocytic leukemias, particularly in children. These leukemias are of acute lymphoblastic type or AML with monocytic differentiation.
- The translocation (9;11)(p22;q23) (*MLLT3;MLL*) (Figure 18.19) or t(10;11)(p12;q23) (*AF10;MLL*) have been reported in association with adult myeloid leukemias.

AML WITH t(6;9)(p23;q34)

Acute myeloid leukemia with t(6;9)(p23;q34) (*DEK-NUP214*) accounts for about 1% of the AML cases and is often associated with multilineage dysplasia and basophilia. Prognosis is poor, particularly in patients with elevated WBC counts and increased bone marrow blasts.

Morphology

- AML with t(6;9) may represent a variety of morphologic features except for promyelocytic and megakaryoblastic subtypes. Myelomonocytic AML (AML-M4) and AML with maturation (AML-M2) are the most frequent forms (see Figures 18.1 and 18.20).
- Bone marrow biopsies show variable cellularity with myeloid left shift and increased blasts. Occasional Auer rods may be present. Erythroid preponderance is a frequent feature.
- Multilineage dysplasia is a frequent finding, particularly dyserythropoiesis and dysmyelopoiesis. Micromegakaryocytes may be present.
- Bone marrow and/or peripheral blood basophilia (>2%) is often present.
- Ringed sideroblasts may be present.
- Peripheral blood smears may show anisopoikilocytosis, nucleated RBCs, hypogranular neutrophils, and hypogranular platelets.

Immunophenotype

- No specific immunophenotypic features.
- Blasts express myeloid-associated antigens CD13, CD15, CD33, and MPO.
- Most cases are positive for CD34 and CD117.
- TdT can be positive.

Molecular and Cytogenetic Studies

- The chromosomal translocation (6;9)(p23;q34) (Figure 18.20) is a rare translocation associated with a poor prognosis and has distinct clinical and morphologic features. This translocation t(6;9) results in a chimeric fusion gene between *DEK* (6p23) and *CAN/NUP214* (9q34).
- Prevalence of *FLT3-ITD* is as high as 70% among patients with t(6;9) AML, and patients with t(6;9) AML and *FLT3-ITD* mutations usually have higher white blood cell counts, higher bone marrow blasts, and significantly lower rates of complete remission.

AML WITH inv(3)(q21q26.2) or t(3;3)(q21;q26.2)

AML with inv(3)(q21q26.2) or t(3;3)(q21;q26.2) (*RPN1-EVI1*) is rare and accounts for 1–2% of AML cases (Figure 3.2 and Figure 18.21). It is commonly reported in adults and is frequently observed in patients with a history of MDS. Some patients may demonstrate hepatosplenomegaly. Prognosis is poor.

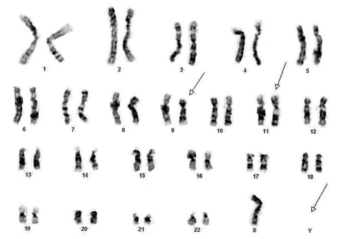

FIGURE 18.19 Karyotype of bone marrow cells in a patient with acute myelomonocytic leukemia with t(9;11)(p22;q23) and loss of chromosome Y.

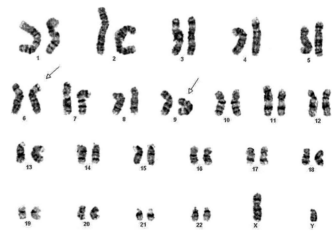

FIGURE 18.20 G-banded karyotype showing t(6;9)(p23;q34).

Morphology

- AML with inv(3)(q21q26.2) or t(3;3)(q21;q26.2) displays a variety of morphologic features except for the promyelocytic type.
- Bone marrow biopsies may show variable cellularity with myeloid left shift and increased blasts. Micromegakaryocytes are often present.
- Bone marrow may show increased eosinophils, basophils, and/or mast cells.
- Anemia is a frequent finding, but platelet count may be normal or elevated. In some cases thrombocytosis may exceed 1,000,000/µL.
- Blood smears may show anisopoikilocytosis with the presence of tear drop forms, hypersegmented and/or hypogranular neutrophils, and bare megakaryocytic nuclei.

Immunophenotype

- Blasts are generally positive for CD13, CD33, CD34, CD38, and HLA-DR.
- Aberrant expression of CD7 is sometimes detected.
- Occasional cases display megakaryocytic differentiation-associated antigens CD41 and CD61.

ACUTE MEGAKARYOBLASTIC LEUKEMIA WITH t(1;22)(p13;q13)

t(1;22)(p13;q13) (*RBM15-MKL1*) AML is of megakaryocytic lineage and reported in children less than 3 years old (most frequently in the first 6 months of life) without Down syndrome (Figure 18.22). Hepatosplenomegaly is a frequent finding, and extramedullary involvement may be present without bone marrow involvement. Originally, this was considered an aggressive disease with poor prognosis, but recent therapeutic approaches have improved the outcome.

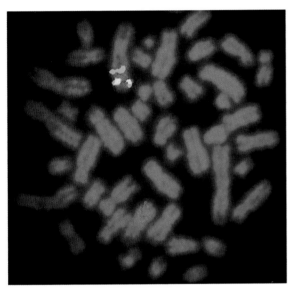

FIGURE 18.21 FISH demonstrates t(3;3)(q21;q26.2).

Morphology (Figure 18.23)

- Blast cells in the bone marrow or peripheral blood predominantly consist of megakaryoblasts with a small amount of dark blue, non-granular cytoplasm and often cytoplasmic

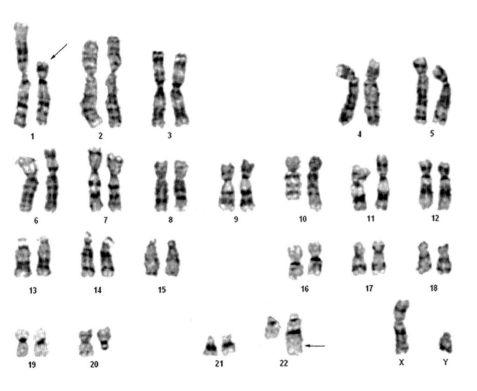

FIGURE 18.22 G-banded karyotype with t(1;22)(p13;q13) in a patient with acute megakaryoblastic leukemia.

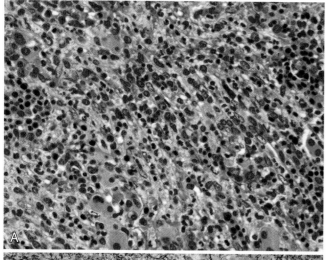

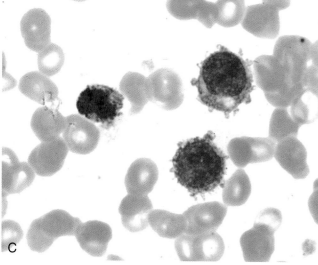

FIGURE 18.23 ACUTE MEGAKARYOBLASTIC LEUKEMIA ASSOCIATED WITH t(1;22)(p13;q13). (A) Bone marrow biopsy section demonstrates fibrosis with increased immature cells and several micromegakaryocytes. (B) Reticulin stain shows increased reticulin fibers. (C) Two megakaryoblasts with cytoplasmic blebs are present in blood smear.

blebs, round or slightly irregular nucleus, fine nuclear chromatin, and inconspicuous nucleoli.
- Bone marrow biopsy sections are usually hypercellular and may show clusters of megakaryocytes and areas of fibrosis. Micromegakaryocytes are common.
- No significant dysplastic changes are present in the erythroid and myeloid lineages.
- Some cases may exhibit extramedullary involvement (myeloid sarcoma) without evidence of bone marrow involvement.

Immunophenotype

- Blasts are positive for megakaryocytic associated antigens CD41 and CD61, as well as CD13, CD33, and CD36.
- Blasts are negative for CD34, HLA-DR, and MPO.
- Aberrant expression of CD7 is common.

OTHER TYPES OF AML WITH RECURRENT TRANSLOCATIONS

AML with t(9;22)(q34;q11.2)

AML with t(9;22)(q34;q11.2) (*BCR-ABL1*) often represents blast transformation of CML. However, a very small proportion of these leukemias (<1%) are *de novo* without clinical history of CML. These cases are different from blast transformation of CML by demonstrating less frequent basophilia and splenomegaly. All the morphologic variations described in the American-French-British (FAB) classification may apply to AML with t(9;22), except for acute promyelocytic leukemia (see Chapter 21).

AML with t(8;16)(p11.2;p13.3)

AML with t(8;16)(p11.2;p13.3) (*KAT6A-CREBBP*) is a rare leukemia characterized by myelomonocytic or monocytic features and erythrophagocytosis (Figures 18.24 and 18.25). AML with t(8;16) may be therapy-related or *de novo*. It may occur at any age or could be congenital.

AML with Gene Mutations

A significant proportion of AMLs with normal cytogenetics demonstrate gene mutations. The gene mutations are divided into two categories (Table 18.2):

1. Mutations that lead to proliferation or survival advantage without affecting differentiation, such as fms-related tyrosine kinase 3 (*FLT3*) and *KIT* mutations.
2. Mutations that impair hematopoietic cell differentiation and programmed cell death (apoptosis), such as nucleoplasmin 1 (*NPM1*) and CCAAT/enhancer binding protein-α (*CEBPA*).

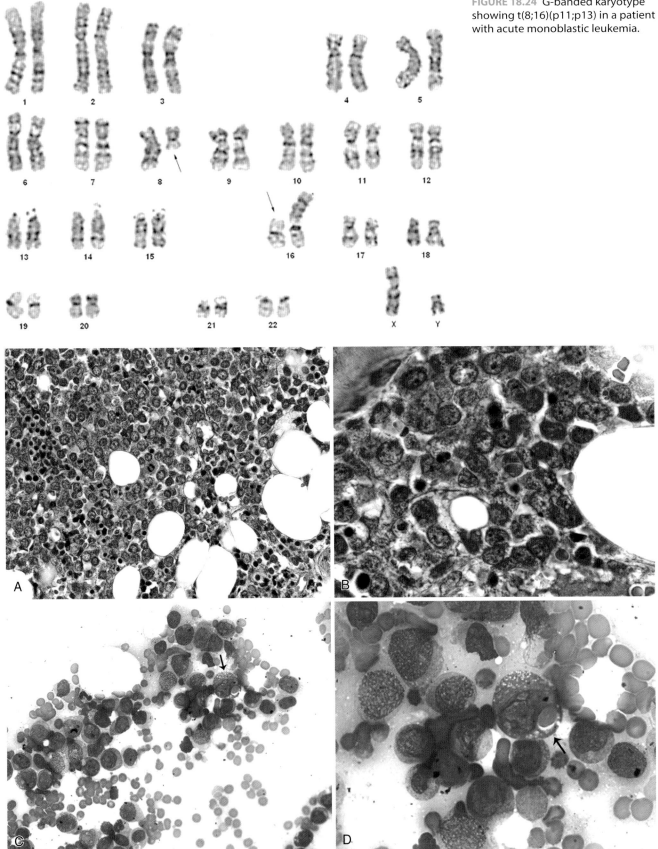

FIGURE 18.24 G-banded karyotype showing t(8;16)(p11;p13) in a patient with acute monoblastic leukemia.

FIGURE 18.25 **ACUTE MONOBLASTIC LEUKEMIA WITH t(8;16)(q11;p13).** Bone marrow biopsy section (A, low power; B, high power) is hypercellular with numerous blast cells. Bone marrow smear (C, low power; D, high power) demonstrates clusters of blasts, one of which (arrow) shows erythrophagocytosis.

Table 18.2
Mutated Genes which Occur Recurrently in Cytogenetically Normal AMLs[1]

Gene Symbol	Gene Name	Chromosome
NPM1 (NPM, B23)	Nucleophosmin (nucleolar phosphoprotein B23, numatrin)	5q35
FLT3 (STK1, FLK2, CD135)	Fms-related tyrosine kinase 3	13q12
MLL (ALL-1, HRX, HTRX1, CXXC7, TRX1, MLL1A)	Myeloid/lymphoid or mixed-lineage leukemia (trithorax homolog, *Drosophila*)	11q23
BAALC	Brain and acute leukemia gene, cytoplasmic	8q22.3
CEBPA (CEBP)	CCAAT/enhancer binding protein (C/EBP), alpha	19q13.1
ERG	v-ets erythroblastosis virus E26 oncogene-like (avian)	21q22.3

[1] Adapted from Krzysztof Mrózek, Guido Marcucci, Peter Paschka, et al. Clinical relevance of mutations and gene-expression changes in adult acute myeloid leukemia with normal cytogenetics: are we ready for a prognostically prioritized molecular classification? Blood 2007; 109: 431–448.

Table 18.3
Features Distinguishing Acute Promyelocytic Leukemia from Acute Monocytic Leukemia

Features	AML-M3	AML-M5
Azurophilic granules	More frequent	Less frequent
Auer rods	Frequent	Rare
MPO stain	Strong	Weak
NSE	Negative/weak	Strong
CD4	Negative	Often positive
CD14	Negative	Positive
HAD-DR	Negative	Positive

The *FLT3* and *KIT* mutations are the most common forms of mutations and are involved in a garden variety of morphologic subtypes of AML. Approximately 30% of AMLs with normal karyotypes show *FLT3* mutations. In general, AMLs associated with *FLT3* mutations demonstrate shorter remission duration and shorter overall survival. *KIT* mutations are frequently found in AML associated with t(8;21) and AMLs associated with 11q23 translocations. Table 18.2 shows genes whose mutations recurrently occur in cytogenetically normal AML.

AML WITH MUTATED *NPM1*

NPM1 mutation is one of the most frequently recurring genetic abnormalities in both adult and pediatric patients with AML. It is found in approximately 50% of adult AML and 20% of childhood AML with normal karyotypes. Approximately 40% of *NPM1*-mutated cases may also show *FLT3* mutation. Patients with *NPM1* mutation and lack of *FTL3* mutation have a favorable prognosis.

Morphology

- Bone marrow is markedly hypercellular with myeloid left shift and increased blasts.
- In adults, the predominant morphology associated with *NPM1* mutation is AML with monocytic differentiation (AML-M4 and AML-M5).
- In children, a wider range of morphology has been reported including myelomonocytic AML, AML without differentiation, and erythroleukemia.

Immunophenotype

- Blasts express myeloid-associated antigens CD13, CD33, and MPO.
- Occasionally express monocytic markers.
- Blasts are negative for CD34.

AML WITH MUTATED *CEBPA*

CEBPA mutations are detected in about 13% of adults and up to 20% of children with AML and normal karyotypes. *CEBPA* mutations lead to a smaller isoform that inhibits the wild-type CEBPA tumor suppressor protein. AML patients with *CEBPA* mutations and no other molecular/cytogenetic aberrations have a favorable prognosis.

Morphology

- Bone marrow is hypercellular and shows myeloid left shift with increased blasts.
- AMLs with and without differentiation are the most commonly associated morphologic subtypes. AMLs with myelomonocytic or monocytic differentiation are less common.

Immunophenotype

- Blasts express one to several myeloid-associated antigens.
- Blasts in most cases are positive for CD34 and HLA-DR.
- Blasts in more than half of the cases show aberrant expression of CD7.

Differential Diagnosis

Cytogenetic aberrations and gene mutations are the main pathognomonic markers in AMLs with recurrent genetic abnormalities. However, certain morphologic and immunophenotypic features are helpful in distinguishing these leukemias (Table 18.1). For example, a significant proportion of AMLs with t(8;21) demonstrate morphologic features of AML with maturation (AML-M2) (see Chapter 21).

Acute promyelocytic leukemia, particularly the microgranular variant, may mimic acute leukemias with monocytic differentiation. The leukemic promyelocytes, unlike monocytic cells, often show several Auer rods, are strongly MPO- and Sudan Black B-positive, and do not express CD4, CD14, or HLA-DR (Table 18.3). Acute myeloid leukemia with inv(16)(p13q22) or (16;16)(p13;q22) depicts myelomonocytic differentiation with the presence of abnormal eosinophils. Most of the acute leukemias with 11q23 (MLL) abnormalities are morphologically of myelomonocytic or monocytic types.

Additional Resources

Bacher U, Schnittger S, Haferlach T: Molecular genetics in acute myeloid leukemia, *Curr Opin Oncol* 22:646-655, 2010.

Bernstein J, Dastugue N, Haas OA, et al: Nineteen cases of the t(1;22)(p13;q13) acute megakaryblastic leukaemia of infants/children and a review of 39 cases: report from a t(1;22) study group, *Leukemia* 14:216-218, 2000.

Chandra P, Luthra R, Zuo Z, et al: Acute myeloid leukemia with t(9;11)(p21-22;q23): common properties of dysregulated ras pathway signaling and genomic progression characterize de novo and therapy-related cases, *Am J Clin Pathol* 133:686-693, 2010.

Chi Y, Lindgren V, Quigley S, et al: Acute myelogenous leukemia with t(6;9)(p23;q34) and marrow basophilia: an overview, *Arch Pathol Lab Med* 132:1835-1837, 2008.

Chou WC, Dang CV: Acute promyelocytic leukemia: recent advances in therapy and molecular basis of response to arsenic therapies, *Curr Opin Hematol* 12:1-6, 2005.

Delaunay J, Vey N, Leblanc T, Fenaux P, et al: Prognosis of inv(16)/t(16;16) acute myeloid leukemia (AML): a survey of 110 cases from the French AML Intergroup, *Blood* 102:462-469, 2003.

Gaidzik V, Döhner K: Prognostic implications of gene mutations in acute myeloid leukemia with normal cytogenetics, *Semin Oncol* 35:346-355, 2008.

Grimwade D, Mistry AR, Solomon E, et al: Acute promyelocytic leukemia: a paradigm for differentiation therapy, *Cancer Treat Res* 145:219-235, 2010.

Hernández JM, González MB, Granada I, et al: Detection of inv(16) and t(16;16) by fluorescence in situ hybridization in acute myeloid leukemia M4Eo, *Haematologica* 85:481-485, 2000.

Jaffe ES, Harris NL, Vardiman JW, et al: *Hematopathology*, Philadelphia, 2010, Saunders/Elsevier.

Mrózek K, Marcucci G, Paschka P, et al: Clinical relevance of mutations and gene-expression changes in adult acute myeloid leukemia with normal cytogenetics: are we ready for a prognostically prioritized molecular classification?, *Blood* 109:431-448, 2007.

Larson RA, Williams SF, Le Beau MM, et al: Acute myelomonocytic leukemia with abnormal eosinophils and inv(16) or t(16;16) has a favorable prognosis, *Blood* 68:1242-1249, 1986.

Lion T, Haas OA: Acute megakaryocytic leukemia with the t(1;22)(p13;q13), *Leuk Lymphoma* 11:15-20, 1993.

Lugthart S, Gröschel S, Beverloo HB, et al: Clinical, molecular, and prognostic significance of WHO type inv(3)(q21q26.2) / t(3;3)(q21;q26.2) and various other 3q abnormalities in acute myeloid leukemia, *J Clin Oncol* 28:3890-3898, 2010.

Mardis ER, Ding L, Dooling DJ, et al: Recurring mutations found by sequencing an acute myeloid leukemia genome, *N Engl J Med* 361:1058-1066, 2009.

Odenike O, Thirman MJ, Artz AS, et al: Gene mutations, epigenetic dysregulation, and personalized therapy in myeloid neoplasia: are we there yet?, *Semin Oncol* 38:196-214, 2011.

Pedersen-Bjergaard J, Philip P, Ravn V, et al: Therapy-related acute nonlymphocytic leukemia of FAB type M4 or M5 with early onset and t(9;11) (p21;q23) or a normal karyotype: a separate entity?, *J Clin Oncol* 6:395-397, 1988.

Peterson LF, Boyapati A, Ahn EY, et al: Acute myeloid leukemia with the 8q22;21q22 translocation: secondary mutational events and alternative t(8;21) transcripts, *Blood* 110:799-805, 2007.

Reiter A, Lengfelder E, Grimwade D: Pathogenesis, diagnosis and monitoring of residual disease in acute promyelocytic leukaemia, *Acta Haematol* 112:55-67, 2004.

Scholl C, Schlenk RF, Eiwen K, et al: The prognostic value of MLL-AF9 detection in patients with t(9;11)(p22;q23)-positive acute myeloid leukemia, *Haematologica* 90:1626-1634, 2005.

Sperr W, Valent P: Biology and clinical features of myeloid neoplasms with inv(3)(q21q26) or t(3;3)(q21q26), *Leuk Lymphoma* 48:2096-2097, 2007.

Swerdlow SH, Campo E, Harris NL, et al: WHO Classification of Tumours of Haematopoietic and Lymphoid Tissues, ed 4., Lyon, 2008, International Agency for Research on Cancer.

von Lindern M, Fornerod M, Soekarman N, et al: Translocation t(6;9) in acute non-lymphocytic leukaemia results in the formation of a DEK-CAN fusion gene, *Baillieres Clin Haematol* 5:857-879, 1992.

Wong KF, Siu LL: Acute myeloid leukaemia with variant t(8;21)(q22;q22) as a result of cryptic ins(8;21), *Pathology* 43:180-182, 2011.

Acute Myeloid Leukemia with Myelodysplasia-Related Changes

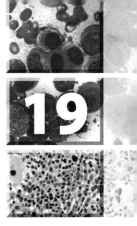

Acute myeloid leukemia with myelodysplasia-related changes is defined in the WHO classification as an acute leukemia with the presence of ≥20% myeloid blasts in bone marrow or blood and evidence of: (a) multilineage dysplasia, (b) arising from previous MDS or MDS/MPN, or (c) MDS-related cytogenetic abnormalities.

Acute myeloid leukemia with multilineage dysplasia is a disease of the elderly with a median age of 60 years. Patients often demonstrate severe pancytopenia with a poor response to chemotherapy.

Morphology (Figures 19.1 to 19.4)

- The presence of ≥20% blasts and significant multilineage dysplasia are the diagnostic hallmarks for this category.
- Biopsy and clot sections are usually hypercellular and show myeloid left shift with increased blasts.
- Dysplastic changes occur in more than one lineage with ≥50 cells in each lineage affected. Dysplastic changes may precede the development of acute leukemia but remain as a part of the picture.
- Megakaryocytic dysplasia is a dominant feature. Micromegakaryocytes with hypogranular cytoplasm, hypolobated nuclei, and/or mono- or binucleated forms are prominent. Large bizarre megakaryocytes may also be present.
- Neutrophilic series are often hypogranular and hyposegmented or may show abnormal segmentation.
- Erythropoiesis may appear megaloblastic or show nuclear budding or fragmentation. Ring sideroblasts may be present.
- Blood examination usually reveals pancytopenia. Blasts are often present in various numbers and dysplastic changes may be more obvious in the peripheral blood smears than the bone marrow smears.

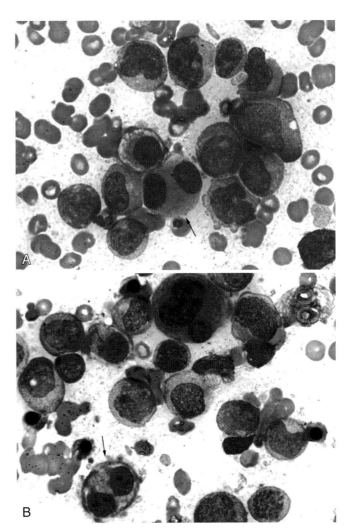

FIGURE 19.1 **Acute myeloid leukemia with myelodysplasia-related changes.** Bone marrow smears (A and B) demonstrating dysplastic immature myelomonocytic cells and binucleated micromegakaryocytes (arrows).

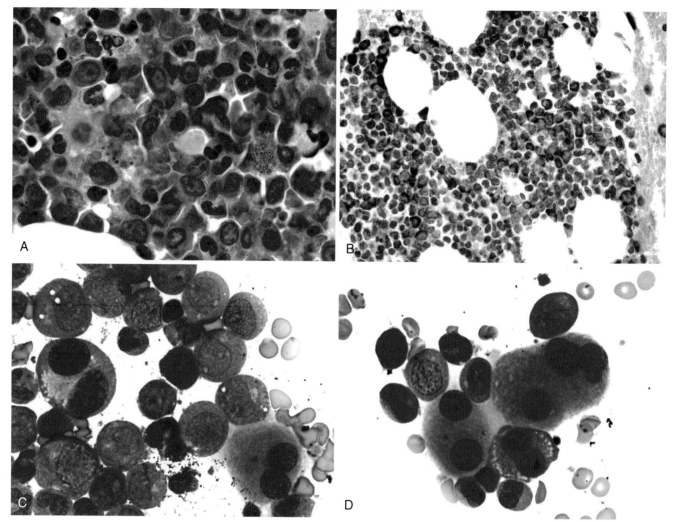

FIGURE 19.2 (A) Bone marrow biopsy section reveals excess blasts and aggregates of immature myelomonocytic precursors. (B) Immunohistochemical studies for myeloperoxidase highlight excess myeloblasts. (C and D) Bone marrow aspirate smears demonstrate myeloid left-shift, excess blasts and dysplastic megakaryocytes including disjointed nuclei and micro forms.

Immunophenotype

Multiparametric immunophenotyping by flow cytometry is useful in diagnosing and managing this type of AML in the follow areas:

- Blast enumeration
- Phenotypic characterization of blasts
- Identification of myelomonocytic dysmaturation
- Therapeutic follow-up and detection of minimal residual disease
- Immunophenotypic–genotypic correlation.

The immunophenotypic pattern of the myeloblasts varies and there are no specific findings that are diagnostic of AML with myelodysplasia-related changes.

Molecular and Cytogenetic Studies

- No recurrent cytogenetic changes have been reported in this group. Various recurrent and common aberrations normally observed in MDS and MPN such as −5/5q−, −7/7q− del(11q) (Figure 19.5), del(12p), del(20q) (Figure 19.6), and trisomy 8, 9, 11, 19, and 21 have been observed.
- Other translocations, such as t(1;7)(q10;p10), t(3;21)(q26;q21) and t(6;9)(p23;q34), have been also occasionally observed. There is a great amount of genetics heterogeneity manifested in the karyotypes of this group. Usually they exhibit numerous non-specific (monosomies, deletions) and sometimes random chromosomal aberrations, in addition to those commonly seen in myelodysplasia.

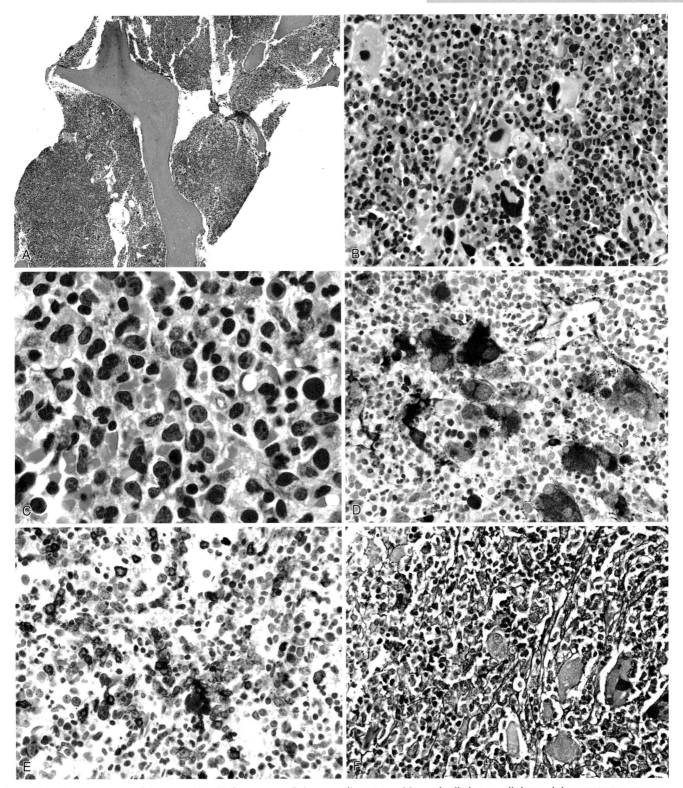

FIGURE 19.3 Bone marrow biopsy section (A, low power; B, intermediate power) is markedly hypercellular and demonstrates excess blasts and dysplastic megakaryocytes. (C) Aggregates of blast cells and scattered eosinophils are shown in a high power view. (D) Megakaryocytes are highlighted by immunohistochemical stain for Factor VIII. (E) Immunohistochemical stain for CD117 confirms excess blasts. (F) Reticulin stain highlights mild to moderate increase in reticulin fibrosis.

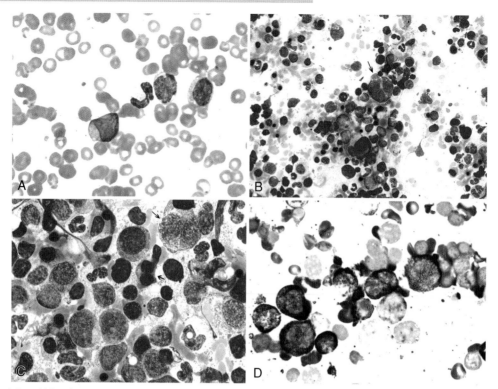

FIGURE 19.4 (A) Peripheral blood smear demonstrates thrombocytopenia and a myeloblast. (B and C) Bone marrow aspirate smears reveal erythroid dysplasia, including bi- and multinucleated forms (arrows) and dysgranulopoiesis such as hypogranulation, Pseudo Pelger–Huët anomaly, and clumped nuclear chromatin. (D) Myeloblasts are increased and are highlighted by cytochemical stain for myeloperoxidase.

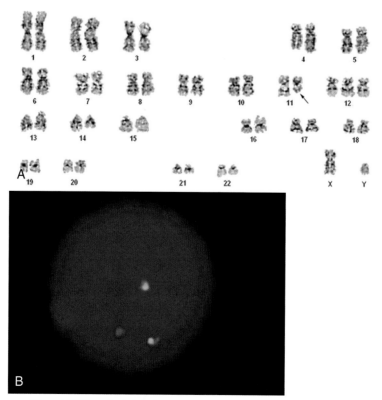

FIGURE 19.5 G-banded karyotype (A) and FISH analysis (B) showing del(11q) in a patient with AML and myelodysplasia-related changes.

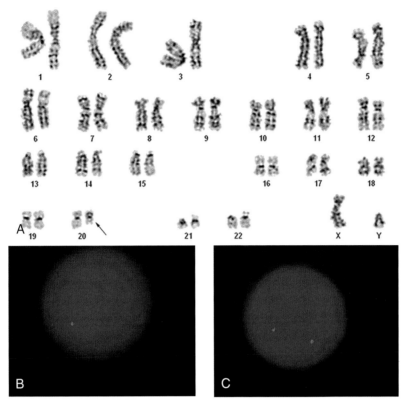

FIGURE 19.6 G-banded karyotype (A) with FISH analysis (B) showing del(20q) in a patient with AML and myelodysplasia-related changes. (C) A normal cell with two red dots representing 20q.

- Because of this genomic instability, the possibility of distinguishing AML with multilineage dysplasia from other subtypes based on gene expression profiling has been suggested.

More recently the technique of next-generation DNA sequencing has been applied to these malignancies for discovery of additional distinguishing gene mutations. Such studies have revealed:

- *AML1/RUNX1* point mutations in a subset of less differentiated cases of AML with myelodysplasia-related changes
- These mutations act by both inhibiting cellular differentiation and enhancing proliferative activity
- A considerable prevalence of *IDH1* gene mutations, particularly in those cases with normal karyotype.

Differential Diagnosis

Acute myeloid leukemia with myelodysplasia-related changes is distinguished from refractory anemia with excess blasts (RAEB) by the presence of ≥20% blasts in bone marrow and/or peripheral blood. It has morphologic overlapping features with acute erythroid leukemia (erythroid/myeloid), but unlike erythroid leukemia, in this leukemia the bone marrow erythroid component is not >50%. The requirement of significant multilineage dysplasia of ≥50% of the affected cells distinguishes this leukemia from most other types of AML.

Additional Resources

Bagby GC, Meyers G: Myelodysplasia and acute leukemia as late complications of marrow failure: future prospects for leukemia prevention, *Hematol Oncol Clin North Am* 23:361–376, 2009.

Haferlach T, Schoch C, Löffler H, et al: Morphologic dysplasia in de novo acute myeloid leukemia (AML) is related to unfavorable cytogenetics but has no independent prognostic relevance under the conditions of intensive induction therapy: results of a multiparameter analysis from the German AML Cooperative Group studies, *J Clin Oncol* 21:256–265, 2003.

Jacobs P, Wood L: Myelodysplasia and the acute myeloid leukaemias, *Hematology* 7:325–338, 2002.

Jaffe ES, Harris NL, Vardiman JW, et al: *Hematopathology*, Philadelphia, 2010, Saunders/Elsevier.

Miyazaki Y, Kuriyama K, Miyawaki S, et al: Cytogenetic heterogeneity of acute myeloid leukaemia (AML) with trilineage dysplasia: Japan Adult Leukaemia Study Group-AML 92 study, *Br J Haematol* 120:56–62, 2003.

Pedersen-Bjergaard J, Christiansen DH, Andersen MK, et al: Causality of myelodysplasia and acute myeloid leukemia and their genetic abnormalities, *Leukemia* 16:2177–2184, 2002.

Swerdlow SH, Campo E, Harris NL, et al: WHO classification of tumours of haematopoietic and lymphoid tissues, ed. 4, Lyon, 2008, International Agency for Research on Cancer.

Yin CC, Medeiros LJ, Bueso-Ramos CE: Recent advances in the diagnosis and classification of myeloid neoplasms–comments on the 2008 WHO classification, *Int J Lab Hematol* 32:461–476, 2010.

Therapy-Related Myeloid Neoplasms

Therapy-related myeloid neoplasms (t-MN) represent a spectrum of progressive clonal myelogenous disorders which are evolved following cytotoxic chemotherapy and/or irradiation. These include therapy-related AML (t-AML), therapy-related MDS (t-MDS), and therapy-related MDS/MPN (t-MDS/MPN). The reason for chemotherapy or irradiation is usually a primary malignancy. However, approximately 5% of patients with t-MPN have no prior malignancy and undergo cytotoxic chemotherapy for autoimmune disorders.

The latency period between the initiation of chemotherapy and/or irradiation and the development of t-MPN ranges from several months to several years. Overall, the latency period is shorter in patients treated with topoisomerase II inhibitors than in those treated with alkylating agents or radiation, and longer in younger patients and patients with a non-malignant primary diagnosis.

The primary feature separating t-MDS (or t-MDS/MPN) from t-AML is the percent blast count, which is <20% in t-MDS (or t-MDS/MPN) and ≥20% in t-AML. Some patients, particularly those treated with topoisomerase II inhibitors, may bypass the MDS phase (Table 20.1).

Table 20.1
Clinicopathologic Features of Therapy-Related AML

Features	Alkylating Agents	Topoisomerase II Inhibitors
Latency period	5–7 years	<5 years
Preceded by MDS	Often present	Often absent
AML subtype	Variable	Mostly monocytic; sometimes promyelocytic or other types
Cytogenetics	Deletions: often del(5) and del(7)	Translocations: t(9;11); t(6;11); t(15;17); t(8;21); t(3;21); t(6;9)
Molecular findings	AML1/RUNX1 mutations	AML1/RUNX1 mutations
Clinical outcome	Poor	Favorable

Alkylating Agent/Radiation-Related Myeloid Neoplasms

Alkylating agent / radiation-related AML has a latency period of about 5–7 years and is usually (>70%) preceded by MDS. The average time for progression from MDS to AML is about 5 months. The occurrence rate appears to be dependent on the age of the patient and the total accumulative dose of the chemotherapeutic agents and/or radiation.

The latency period between the diagnosis of the primary disease and the occurrence of t-AML appears to be longer in the younger patients and patients who have been treated with alkylating agents for non-malignant conditions, such as autoimmune disorders. The overall median latency period for the entire t-AML patient population is approximately 65 months with a median survival of about 7 months. Patients with chromosomal deletion of 5 and/or 7 have a shorter median survival time than those with chromosomal translocations.

MORPHOLOGY (FIGURE 20.1)

- The characteristic morphologic features are dysplastic hematopoiesis and increased blasts (including promonocytes). Blasts are <20% in t-MDS and ≥20% in t-AML.
- Dysplastic changes are usually multilineage and involve myeloid, erythroid, and megakaryocytic series. Hypogranulation and abnormal segmentation of the granulocytic cells, megaloblastic changes in the erythroid series, ring sideroblasts, and micromegakaryocytes are frequent findings.
- Most t-AML cases correspond to AML with maturation (AML-M2), but a minority of the cases fit into acute myelomonocytic (AML-M4), acute monocytic (AML-M5), acute erythroleukemia (AML-M6), or acute megakaryocytic leukemia (AML-M7) (See Chapter 21).
- Bone marrow is often hypercellular, but in about 25% of cases is hypocellular. Bone marrow fibrosis may be present in one-fourth of the cases. Basophilia is sometimes present.

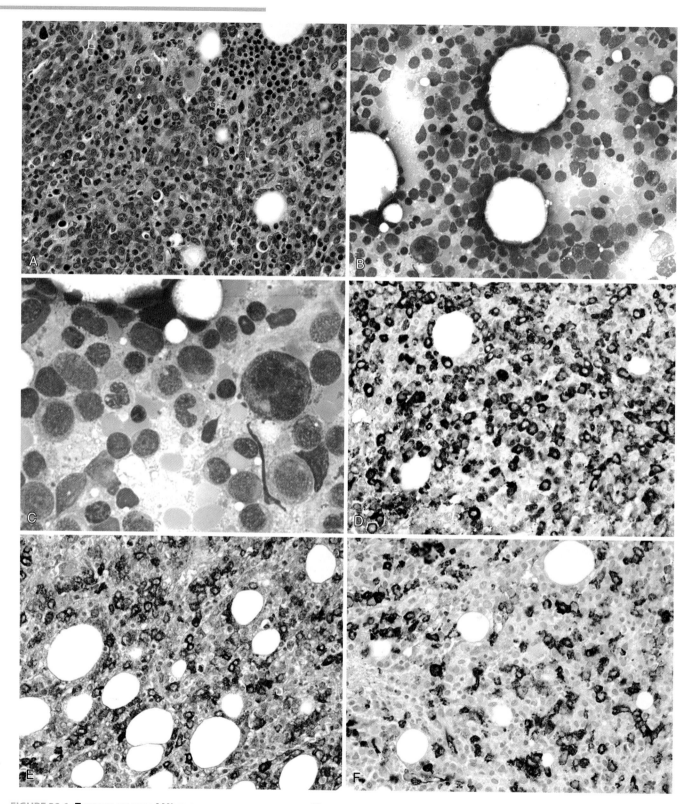

FIGURE 20.1 **THERAPY-RELATED AML IN A PATIENT WITH A HISTORY OF HODGKIN LYMPHOMA.** Bone marrow biopsy demonstrating hypercellularity with increased blasts (A). Bone marrow smear (B, low power; C, high power) showing dysplastic erythropoiesis and increased blasts. Blasts are positive for MPO (D), CD117 (E) and CD34 (F) by immunohistochemical stains.

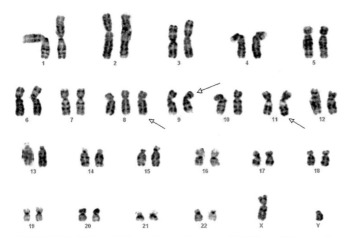

FIGURE 20.2 G-banded karyotype showing t(9;11) and trisomy 8 in a patient with therapy-related AML.

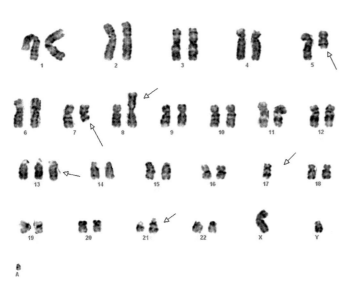

FIGURE 20.3 G-banded karyotype showing a complex karyotype involving chromosomes 5, 7, 8, 13, 17, and 21 in a patient with therapy-related AML.

- The peripheral blood may show anemia or pancytopenia with anisopoikilocytosis and leukoerythroblastic features and presence of blast cells. Dysplastic changes may be noted in the myeloid cells and/or platelets.

IMMUNOPHENOTYPE

Flow cytometry is important in enumerating and determining phenotypic patterns of the blasts, as well as detecting dysmaturation patterns of myelomonocytic cells. The immunophenotypic pattern of the blasts in t-AML is similar to that seen in various subtypes of AML, NOS (Chapter 21).

MOLECULAR AND CYTOGENETIC STUDIES

- Over 90% of alkylating agent / radiation-related AMLs show clonal chromosomal aberrations, most frequently involving loss of all or part of chromosome 5, chromosome 7, or both.
- Balanced chromosomal translocations are rare and mostly involve 11q23 or 21q22 (Figure 20.2). Some reports show an association between radiation t-AML and t(15;17) or inv(16). The karyotypes are often complex (Figure 20.3) with nonclonal or single cell aberrations and not consistently observed in consecutive follow-up studies.
- These treatment-related aberrations are frequently seen in addition to those seen in primary myeloid malignancies.
- Radiotherapy-related AML cases have been associated with acquired *AML1/RUNX1* mutations.

Topoisomerase II Inhibitor-Related AML

Topoisomerase II inhibitor-related AML generally has a shorter latency period than the alkylating agent-related neoplasms, ranging from 1 to 3 years. An antecedent myelodysplastic phase is usually lacking. Anthracyclines, doxorubicin, etoposide, epipodophyllotoxins, and teniposide are among the major drugs targeting DNA topoisomerase II. The median survival time is longer for topoisomerase II inhibitor-related AML than for alkylating agent-related AML. Balanced translocations are frequent cytogenetic abnormalities.

MORPHOLOGY (FIGURE 20.4)

- Myelodysplastic features are infrequent and in most instances the disease presents itself as a *de novo* AML.
- The most common morphologic features are those of acute myelomonocytic or acute monocytic leukemias (see Chapter 21), but some cases may present morphologic and cytogenetic features consistent with acute promyelocytic leukemia (APL).

IMMUNOPHENOTYPE

See immunophenotypic features of APL, acute myelomonocytic leukemia, and acute monocytic leukemia (Chapters 18 and 21).

MOLECULAR AND CYTOGENETIC STUDIES

- Topoisomerase II inhibitor-related AML is commonly associated with complex chromosomal aberrations and translocations, particularly involving *11q23* and the *MLL* gene (Figures 20.5 and 20.6). The 11q23-associated cytogenetic changes include del(11q23), (6;11)(q26–27;q23), t(9;11)(p22;q23), (10;11)(p12;q23), and t(11;19)(q23;p13.1).

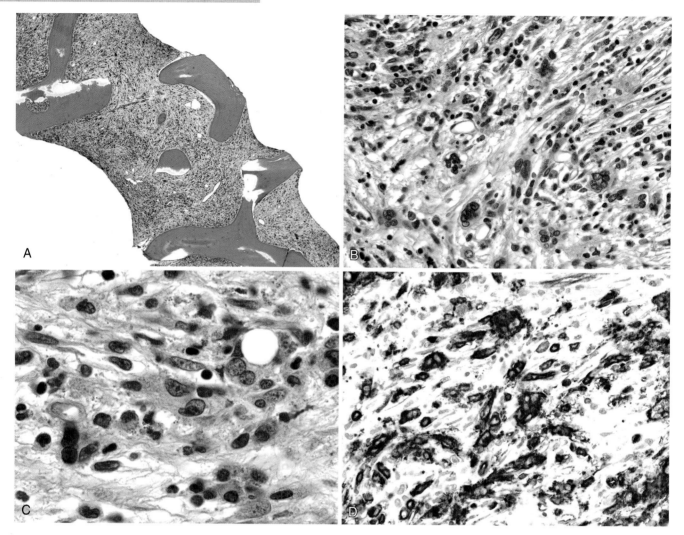

FIGURE 20.4 THERAPY-RELATED AML IN A 20-YEAR-OLD MALE WHO HAD A HISTORY OF ANAPLASTIC LARGE CELL LYMPHOMA STATUS CHEMOTHERAPY 3 YEARS PREVIOUSLY. Marrow core biopsy section shows extensive fibrosis (A) with excess blasts in small abnormal clusters (B). Many of the blasts are spindle-shaped, containing smooth to slightly irregular nuclei, dispersed chromatin, and prominent nucleoli (C). Immunohistochemical stain for CD34 highlights numerous blasts (D).

- Other cytogenetic abnormalities such as t(15;17)(q22;q11-12), (3;21)(q26;q22), t(8;21)(q22;q22), t(6;9)(p23;q34), and t(8;16)(p11;p13) have also been reported in topoisomerase II inhibitor-related AMLs.
- Occasionally, the chromosomes exhibit homogeneous staining regions (hsr) or double minutes (dm) at various genomic sites and represent amplifications of the MLL locus (Figures 20.7 and 20.8).
- As in radiation-associated AML, *RUNX1* gene mutations have been found in topoisomerase-associated AML cases.

Differential Diagnosis

The differential diagnosis of the therapy-related myeloid neoplasms includes various morphologic categories of AML (see Chapter 21) as well as AML with myelodysplasia-related changes. The most distinguished feature is a history of cytotoxic or radiation therapy.

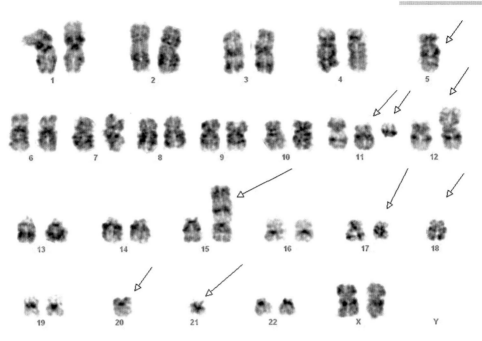

FIGURE 20.5 G-banded karyotype demonstrating complex chromosomal aberrations involving chromosomes 5, 11, 12, 15, 17, 18, and 20, and 21 in a patient with therapy-related AML.

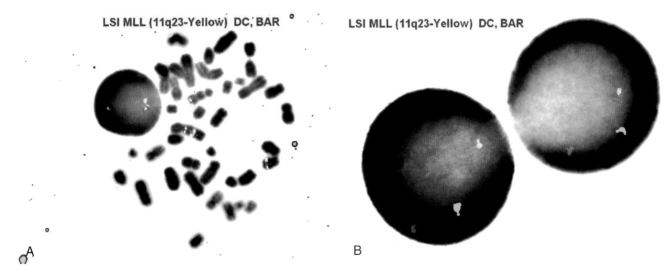

FIGURE 20.6 FISH analysis (A and B, yellow dots) showing *MLL* rearrangement in a patient with therapy-related AML.

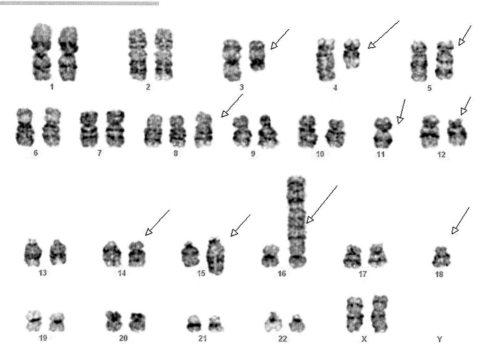

FIGURE 20.7 G-banded karyotype showing complex chromosomal aberrations with homogeneous staining regions (hsr) at 16p in a patient with therapy-related AML.

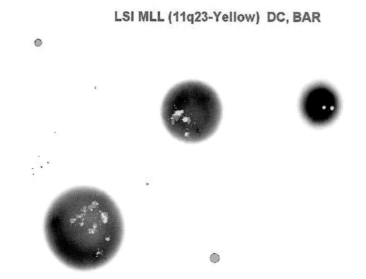

FIGURE 20.8 FISH analysis demonstrating amplification of the *MLL* locus in a patient with therapy-related AML.

Additional Resources

Borthakur G, Estey AE: Therapy-related acute myelogenous leukemia and myelodysplastic syndrome, *Curr Oncol Rep* 9:373–377, 2007.

Czader M, Orazi A: Therapy-related myeloid neoplasms, *Am J Clin Pathol* 132:410–425, 2009.

Godley LA, Larson RA: Therapy-related myeloid leukemia, *Semin Oncol* 35:418–429, 2008.

Jaffe ES, Harris NL, Vardiman JW, et al.: *Hematopathology*, Philadelphia, 2010, Saunders/Elsevier.

Joannides M, Grimwade D: Molecular biology of therapy-related leukaemias, *Clin Transl Oncol* 12:8–14, 2010.

Kwong YL: Azathioprine: association with therapy-related myelodysplastic syndrome and acute myeloid leukemia, *J Rheumatol* 37:485–490, 2010.

Qian Z, Joslin JM, Tennant TR, et al.: Cytogenetic and genetic pathways in therapy-related acute myeloid leukemia, *Chem Biol Interact* 184:50–57, 2010.

Sill H, Olipitz W, Zebisch A, Schulz E, Wölfler A: Therapy-related myeloid neoplasms: pathobiology and clinical characteristics, *Br J Pharmacol* 162:792–805, 2011.

Swerdlow SH, Campo E, Harris NL, et al.: WHO classification of tumours of haematopoietic and lymphoid tissues, ed 4, Lyon, 2008, International Agency for Research on Cancer.

Wong KF, Siu LL: Acute myeloid leukaemia with variant t(8;21)(q22;q22) as a result of cryptic ins(8;21), *Pathology* 43:180–182, 2011.

Yin CC, Medeiros LJ, Bueso-Ramos CE: Recent advances in the diagnosis and classification of myeloid neoplasms–comments on the 2008 WHO classification, *Int J Lab Hematol* 32:461–476, 2010.

Acute Myeloid Leukemia, Not Otherwise Specified

AML, not otherwise specified (AML-NOS) represents acute myeloid leukemias that are excluded from all previously described subclasses. Lineage differentiation and the extent of maturation of the leukemic cells are distinguishing features in this category. Differentiation and maturation are evaluated based on morphological, immunophenotypic, and cytochemical characteristics. This category includes most of the AML subtypes defined by the French–American–British (FAB) classification, including M0, M1, M2, M4, M5, M6, and M7 variants plus acute basophilic leukemia and acute panmyelosis with myelofibrosis (Table 21.1). AML-NOS subtypes often show characteristic flow cytometric patterns that are helpful for distinguishing them from each other (Figure 21.1).

The WHO diagnostic criterion for acute myeloid leukemia is the presence of 20% or more blast cells in the bone marrow or blood differential counts. Myeloblasts, monoblasts, promonocytes, and megakaryoblasts are included in the "blast" counts. The WHO recommends counting of 500 nucleated cells on the bone marrow smears and/or 200 on the blood smears in order to establish a diagnosis of AML.

Clinical symptoms are the result of bone marrow involvement and extramedullary infiltration by the leukemic cells. Fatigue, fever, bleeding disorders, and organomegaly, especially hepatosplenomegaly, are among frequent clinical manifestations.

AML, Minimally Differentiated

Minimally differentiated AML (AML-M0) is defined as an AML with no morphologic or cytochemical evidence of myeloid differentiation based on conventional light microscopic examinations. The myeloid lineage in this category is established by immunophenotypic characteristics and/or ultrastructural studies.

AML-M0 accounts for <5% of all AMLs in most reported studies. The affected individuals are usually older than 60 years, and the male to female ratio is about 2. The prognosis is poor with a median survival of <6 months. The coexpression of CD7 and CD56 has been associated with poorer prognosis in patients younger than 46 years in some reports.

MORPHOLOGY

- The leukemic blasts (≥20% of bone marrow or blood differential counts) lack features of morphologic differentiation. They are often medium-sized with scant non-granular basophilic cytoplasm, round or slightly irregular nuclei, fine chromatin, and one or more prominent nucleoli (Figure 21.2).
- Myeloblasts with cytoplasmic granules are absent or extremely rare (<3%), and Auer rods are not present.
- Special cytochemical stains, such as MPO, Sudan Black B, and NSE, are negative (<3% blasts show positive staining).
- Bone marrow is usually hypercellular and packed with the leukemic blast cells, but remnants of normal hematopoietic cells may be noted. The presence of residual normal maturing myeloid precursors may create a morphologic pattern mimicking AML with maturation (AML-M2).
- Blood smears often show variable numbers of blast cells. Various degree of anemia, thrombocytopenia or pancytopenia are present.
- The distinguishing features between AML with minimal differentiation and AML with maturation are lack of granular myeloblasts, absence of Auer rods, and negative cytochemical staining in the former.

IMMUNOPHENOTYPE

- Expression of antigens that are associated with early hematopoiesis including CD34, CD38, and HLA-DR;
- Expression of CD13 and/or CD117, and sometimes CD33;
- Common expression of intracellular TdT;
- Variable presence of cross-lineage aberrancies;
- Lack of mature myeloid markers or monocytic antigens;

Table 21.1

Classification of Acute Myeloid Leukemia (AML), Not Otherwise Specified

Type	FAB Classification	Definition	Immunophenotype
AML with minimal differentiation	AML-M0	Blasts lack morphologic and cytochemical evidence of myeloid lineage	CD13+, CD33±, CD38+, CD117+, CD34+, HLA-DR+, and TdT±; MPO−
AML without maturation	AML-M1	Blasts are MPO+, SBB+ by cytochemistry and ≥90% of non-erythroid cells	CD13+, CD33+, CD38+, CD117+, CD34+, HLA-DR+, and MPO+ (partial); CD15−
AML with maturation	AML-M2	Blasts are MPO+, SBB+ by cytochemistry and ≥20% of non-erythroid cells are maturing myeloid cells	CD11b±, CD13+, CD15+, CD33+, CD65±, CD38+, CD117+, CD34±, MPO+
Acute myelomonocytic leukemia	AML-M4	Blasts consist of myeloblasts, monoblasts, and promonocytes (total of ≥20%), and ≥20% to 79% of monocytic lineage	CD13+, CD15+, CD33+, CD65+, CD38+, CD117+, CD34±, HLA-DR+ (most), MPO+ (partial); Plus one or more monocytic markers
Acute monoblastic/monocytic leukemia	AML-M5	Monoblasts/promonocytes ≥20%, and ≥80% of marrow cells with monocytic features	CD13+, CD15+, CD33+ (bright), CD65+, HLA-DR+, CD117±; CD34−; Plus two or more monocytic markers
Acute erythroid leukemia	AML-M6		
Erythroid/myeloid		≥50% marrow cells are erythroid, and ≥20% of non-erythroid cells are myeloblasts	Immature erythroid cells are CD36±, CD71+, CD235 (GPA) ±, CD117±
Pure erythroid		≥80% marrow cells are pronormoblasts and/or early basophilic pronormoblasts	Myeloblasts are CD13+, CD33+, CD117+, CD34±
Acute megakaryoblastic leukemia	AML-M7	≥20% blasts; ≥50% of blasts are megakaryoblasts	CD7+ (often), CD36+, CD41+, CD61+, Factor VIII+, CD38+, CD13±, CD33±; CD117−, CD34−
Acute basophilic leukemia		≥20% of blasts show basophilic differentiation	CD11b+, CD13+, CD33+, CD123+, CD22±, CD34±; MPO−, CD117−
Acute panmyelosis with myelofibrosis		Fibrosis with increased number of immature myeloid, erythroid, and megakaryocytic cells	CD34+; one or more myeloid markers; MPO−

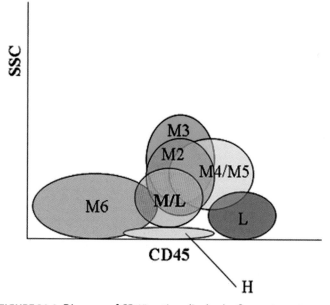

FIGURE 21.1 Diagram of CD45 gating display by flow cytometry demonstrating features of AML-NOS subtypes: M/L, myeloblast/lymphoblast; M, AML minimally differentiated (M0) or AML without maturation (M1); L, mature lymphoid malignancies; M2, acute myeloid leukemia with maturation; M3, acute promyelocytic leukemia; M4/M5, acute myelomonocytic and acute monocytic leukemia; M6, erythroleukemia; H, hematogones.

- Lack of intracellular MPO in most cases;
- Immunohistochemical stains are negative for CD68, CD3, and CD20, but may show focal positive reaction for MPO in the blast population.

MOLECULAR AND CYTOGENETIC STUDIES

- A high frequency of mutation in the *RUNX1(AML1)* gene has been reported in AML-M0 patients.
- Cytogenetic aberrations are variable and have been reported in ~50% of cases. Trisomy 4, 8, 11, 13, and 14 and monosomy 7 are among the most common reported abnormalities.
- Cases with t(11;12)(q23;q24) and del(20q) (Figure 21.3) are reported less frequently.

DIFFERENTIAL DIAGNOSIS

ALL, AML without maturation, acute monoblastic leukemia, acute megakaryoblastic leukemia, and, occasionally, large cell lymphoma are amongst the list of differential diagnoses.

The blast cells in minimally differentiated AML express myeloid-associated markers (such as CD13, CD33, and CD117) and show <3% positivity for MPO and Sudan Black B in routine cytochemical stains.

AML without Maturation

AML without maturation (AML-M1) is defined as an acute leukemia with no significant myeloid maturation and ≥90% blast cells in the non-erythroid population. The myeloid origin of blast cells is confirmed by positive (≥3%) staining for MPO and/or Sudan Black B by cytochemical techniques, as well as expression of myeloid-associated markers by immunophenotypic analysis.

AML without maturation accounts for 10–15% of all AMLs and is rare in children. The prognosis is poor, particularly in those with marked leukocytosis and increased circulating blasts.

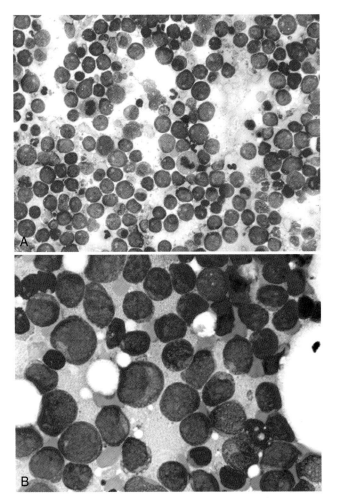

FIGURE 21.2 MINIMALLY DIFFERENTIATED AML (AML-M0). Bone marrow smear showing numerous blast cells with scant non-granular basophilic cytoplasm, round or slightly irregular nuclei, fine chromatin, and one or more prominent nucleoli (A, low power; B, high power).

MORPHOLOGY

The morphologic features of the blast cells overlap with those described in AML with minimal differentiation except for:

- The presence of blast cells with cytoplasmic azurophilic granules.
- Positive MPO and/or Sudan Black B cytochemical staining in >3% of the blast cells (Figure 21.4). Auer rods are not present.
- Similar to most other AMLs, bone marrow is hypercellular and packed with blasts. Variable degrees of marrow fibrosis may be present in a minority of the cases.
- Blood smears often show variable numbers of blast cells. Various degrees of anemia, thrombocytopenia, or pancytopenia are present.

IMMUNOPHENOTYPE

- Expression of at least partial intracellular MPO;
- Expression of one or more other myeloid-associated antigens (CD13, CD33, CD117);
- Expression of CD34 and HLA-DR;
- Variable presence of cross-lineage aberrancies;
- Lack of mature myeloid markers, e.g., CD15 or CD65;
- General lack of monocytic markers (CD14 or CD64).

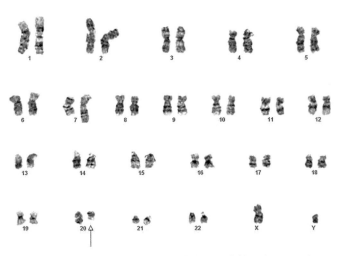

FIGURE 21.3 G-banded karyotype showing del(20q) in a patient with AML-M0.

MOLECULAR AND CYTOGENETIC STUDIES

- There are reports suggesting a reciprocal exchange between *D12S158* at 12p13.3 and the *MYH11* gene at 16p13 in AML-M1 leukemia.
- Cytogenetic aberrations are variable and include both numerical (aneuploidy) abnormalities and translocations. Trisomy 11 (Figure 21.5), trisomy 13, and trisomy 14, as well as t(11;19)(q23;p13), t(14;17)(q32;q11.2), t(12;17)(p13;q11.2), and der(16), t(16;20)(p13;p11.2) have been reported in this leukemic subtype.

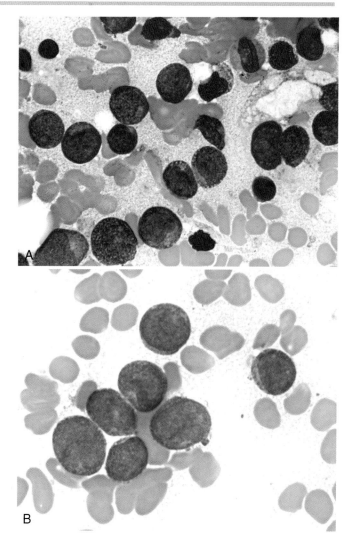

FIGURE 21.4 **AML WITHOUT MATURATION (AML-M1).** Bone marrow smear (A) and blood smear (B) showing numerous blast cells with scant non-granular basophilic cytoplasm, round or slightly irregular nuclei, fine chromatin, and one or more prominent nucleoli.

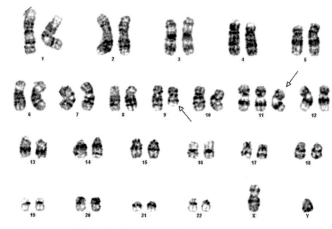

FIGURE 21.5 G-banded karyotype showing trisomy 11 and del(9q) in a patient with AML without maturation.

DIFFERENTIAL DIAGNOSIS

The differential diagnosis of AML without maturation includes ALL, minimally differentiated AML, acute monoblastic leukemia, and acute megakaryoblastic leukemia. The blast cells in AML without maturation express myeloid-associated markers (such as CD13, CD33, and CD117) and show ≥3% positivity for MPO and Sudan Black B cytochemical stains.

AML with Maturation

AML with maturation (AML-M2) is defined as an acute leukemia with ≥20% blast cells in the bone marrow and/or peripheral blood and evidence of granulocytic maturation. The maturing non-blast granulocytic cells account for ≥10% and monocytic cells ≤20% of the bone marrow cells. The myeloid origin of blast cells is confirmed by positive (≥3%) staining for MPO and/or Sudan Black B, as well as expression of myeloid-associated markers by immunophenotypic studies.

AML with maturation accounts for about 10% of all AMLs and occurs in both children and adults. It is the most frequent AML in children. The ones with t(8;21) have a more favorable prognosis in adults (see Chapter 18), but the clinical outcome for children and cases with other types of chromosomal aberrations is poor.

MORPHOLOGY

- The bone marrow biopsy sections are often hypercellular and packed with blasts. Occasionally, bone marrow may appear normocellular or hypocellular. Variable degrees of marrow fibrosis may be present in a minority of cases.
- The myeloblasts are large, often with indented nuclei and basophilic cytoplasm. Numerous myeloblasts show cytoplasmic azurophilic granules, and some contain larger granules mimicking cytoplasmic granules seen in the Chediak–Higashi syndrome (Figure 12.6). Auer rods are frequent and may also be detected in the more mature myeloid forms.
- Promyelocytes, myelocytes, metamyelocytes, bands, and segmented neutrophils are present and often show dysplastic changes.
- Eosinophilia is common, and some cases may show increased bone marrow basophils and/or mast cells.
- Blood smears often show variable numbers of blast cells. Various degrees of anemia, thrombocytopenia, or pancytopenia are present.

IMMUNOPHENOTYPE (FIGURE 21.7)

- Expression of MPO;
- Expression of one or more other myeloid-associated antigens (CD11b, CD13, CD15, CD33, and CD65);

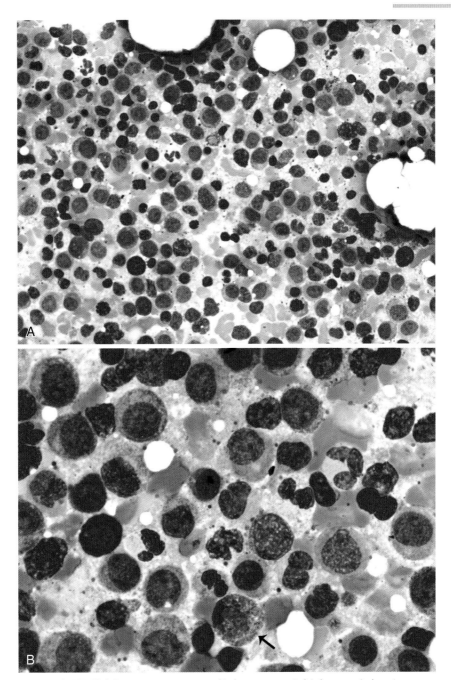

FIGURE 21.6 **AML WITH MATURATION (AML-M2).** Bone marrow smear (A, low power; B, high power) showing numerous blasts mixed with more mature myeloid cells and erythroid precursors. Arrow shows a blast with Chediak–Higashi-like granules.

- Variable expression of CD34 and/or CD117;
- Variable presence of cross-lineage aberrancies.

CYTOGENETIC STUDIES

Reported cytogenetic abnormalities include t(2;9)(q14;p12), t(5;11)(q35;q13) (Figure 21.8), t(10;11)(p13;q14), t(8;19)(q22;q13), t(8;16)(p11.2;p13), del(12)(p11–p13), and various complex translocations.

DIFFERENTIAL DIAGNOSIS

The differential diagnosis includes refractory anemia with excess blasts, acute promyelocytic leukemia, and acute myelomonocytic leukemia. Immunophenotypic and cytogenetic studies are helpful to achieve a definitive diagnosis. Blast cells in AML with maturation often express CD13, CD33, and/or CD117 and lack CD14 expression.

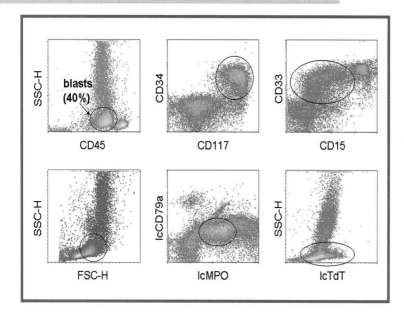

FIGURE 21.7 **FLOW CYTOMETRIC FINDINGS IN A PATIENT WITH AML WITH MATURATION (AML-M2).** Open gate displays using density plots reveal excess and abnormal myeloblasts (40% of the total; circled), which are positive for CD13 (not shown), CD15, CD33 (moderate to bright), CD34, CD45 (dim to moderate), CD64 (partial; not shown), CD117, HLA-DR (partial; not shown), plus intracellular myeloperoxidase, and TdT. Since myeloid maturation is part of the leukemic process, there is some continuity of the phenotypic patterns between the myeloblasts and maturing myeloid cells. It can be best appreciated in this case on the displays of FSC-H vs. SSC-H, CD15 vs. CD33, as well as intracellular MPO vs. CD79a.

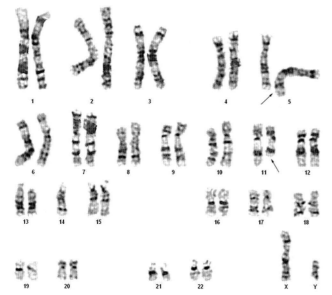

FIGURE 21.8 Translocation of 5;11 in a patient with AML with maturation.

males than in females. Similar to other acute leukemias, clinical symptoms are the result of bone marrow involvement and extramedullary infiltration by the leukemic cells. Fatigue, fever, bleeding disorders, gingival hyperplasia, lymphadenopathy, hepatosplenomegaly, and skin involvement are among frequent clinical findings. As mentioned earlier (see Chapter 20), patients with inv(16) have a favorable prognosis, and those with translocation of 11q23 fall into the category of leukemias with intermediate survival rate. Some studies show a correlation between the expression of CD56 by the leukemic cells and severe fatal hyperleukocytosis in patients with acute myelomonocytic leukemia. Successful treatment with NUP98-HOXD11 fusion transcripts and monitoring of minimal residual disease in patients with acute myelomonocytic leukemia has been reported.

MORPHOLOGY (FIGURES 21.9 TO 21.11)

- Bone marrow biopsy sections are often hypercellular and packed with immature myelomonocytic cells. Nuclear spacing is often present due to abundant cytoplasm of the monocytic cells. Variable degrees of marrow fibrosis may be present in a minority of cases. Monocytic nodules may be present.
- Bone marrow smears show myeloid left shift and increased number of immature myelomonocytic cells.
- Myeloblasts show scant-to-moderate amounts of dark blue cytoplasm, some of which contain various numbers of azurophilic granules. Auer rods may be present. The nuclei are usually round or oval, but they may be irregular. The nuclear chromatin is fine, and multiple prominent nucleoli are often present.
- Monoblasts are usually larger than myeloblasts (>40 µm), with abundant dark to light blue cytoplasm and scattered fine azurophilic granules. The nucleoli are round, oval or folded, and the nuclear chromatin is fine. There is often a single large nucleolus, but multiple prominent nucleoli may be present.

Acute Myelomonocytic Leukemia

Acute myelomonocytic leukemia (AML-M4) is defined as an acute leukemia with increased immature granulocytic and monocytic cells. Myeloblasts, monoblasts, and promonocytes account for ≥20% of the total bone marrow nucleated cells and/or peripheral blood differential counts. More than 20%, but less than 80% of the bone marrow cells are of monocytic lineage

Acute myelomonocytic leukemia accounts for about 5–10% of all AMLs. The median age is around 50 years, but it may occur at any age. The incidence is slightly more in

ACUTE MYELOMONOCYTIC LEUKEMIA

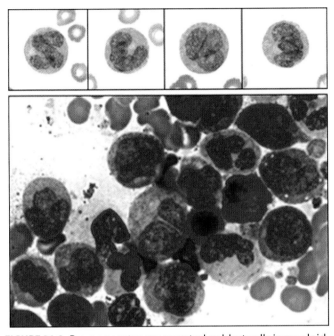

FIGURE 21.9 Promonocytes are counted as blast cells in myeloid leukemias and are prominent in acute myelomonocytic and monocytic leukemias.

- Promonocytes are larger than monocytes (~30–35 μm) and have abundant light blue to gray cytoplasm. Scattered cytoplasmic azurophilic granules and/or cytoplasmic vacuoles may be present. The nuclei are delicately folded or convoluted, often with a cerebriform pattern. The nuclear chromatin is fine, and nucleoli are present but not prominent. Bone marrow smears show an increased number (≥20%) of promonocytes, monoblasts, and myeloblasts.
- Peripheral blood smears show absolute monocytosis (often >5000/μL) with the presence of promonocytes, left shifted granulocytic series, and various numbers of circulating blasts (see Figures 21.10B and 21.11C).
- In a significant proportion of patients (5–35%), there is evidence of extramedullary leukemic infiltration, such as involvement of skin, mucosal membranes, lymph nodes, liver, and/or spleen.

IMMUNOPHENOTYPE AND CYTOCHEMICAL STAINS

- Presence of more than one blast subset, best viewed by density plot display;
- Variable expression of CD13, CD15, CD33, and CD65, plus MPO;

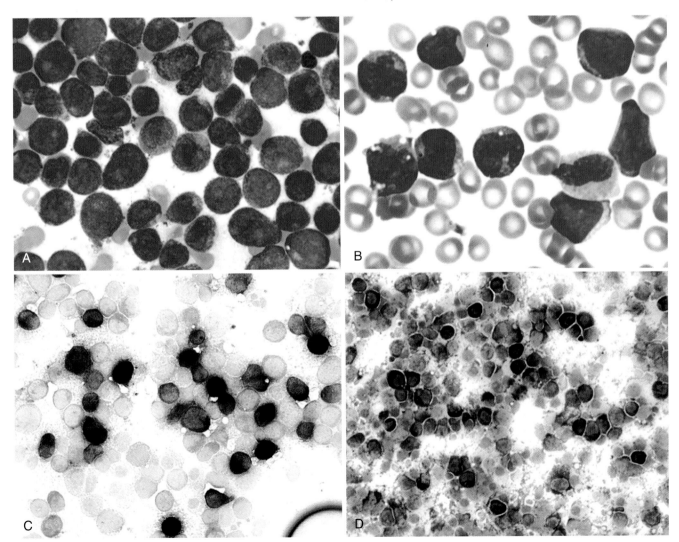

FIGURE 21.10 **ACUTE MYELOMONOCYTIC LEUKEMIA (AML-M4).** Bone marrow smear (A) and blood smear (B) demonstrating a mixture of myloblasts and monoblasts. Cytochemical stains for non-specific esterase (C) and MPO (D) show numerous positive cells.

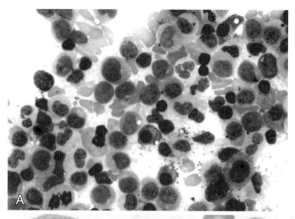
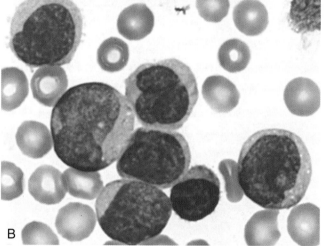

FIGURE 21.11 **ACUTE MYELOMONOCYTIC LEUKEMIA.** (A) Bone marrow smear depicts myeloid left shift with increased immature myelomonocytic cells and blasts. (B) Blood smear shows several blasts and immature monocytic cells.

- Expression of one or more monocytic markers (CD4, CD11b, CD11c, CD14, CD36, CD64, CD68, CD163, and lysozyme);
- Blast subset expression of CD34 and/or CD117;
- Variable presence of cross-lineage aberrancies;
- Lack of HLA-DR in a small number of cases, which is of particular interest in distinguishing them from APL by morphology and immunophenotype;
- Cytochemical stains show strong MPO positivity for the granulocytic lineage (see Figure 21.10D) and various degrees of diffuse cytoplasmic NSE staining for the monocytic population (see Figure 21.10C).

MOLECULAR AND CYTOGENETIC STUDIES

- Mutations in the *TET2* gene have been associated with myelomoncytic leukemias.
- A number of non-specific cytogenetic abnormalities have been reported in acute myelomonocytic leukemias. These include dup(1)(p31.2p36.2), t(1;3)(p36;q21), t(8;12)(q13;p13), t(9;21)(q13;q22), and t(6;7)(q23;q35).

DIFFERENTIAL DIAGNOSIS

The differential diagnosis of acute myelomonocytic leukemia includes chronic myelomonocytic leukemia (CMML), AML with maturation, acute promyelocytic leukemia, and acute monocytic leukemia. The diagnosis of acute myelomonocytic leukemia is established by demonstration of the sum of ≥20% myeloblasts and monoblastic/promonocytes in the bone marrow or peripheral blood. It is distinguished from the microgranular variant of acute promyelocytic leukemia by the expression of NSE, CD4, CD14, and HLA-DR and by the absence of t(15;17) (see Table 20.1).

Acute Monoblastic and Acute Monocytic Leukemias

Acute monoblastic and acute monocytic leukemias (AML-M5) are acute leukemias in which ≥80% of the leukemic cells are of monocytic lineage consisting of monoblasts, promonocytes, and monocytes. When monoblasts are the major cellular component (≥80% of the leukemic cells) the term acute monoblastic leukemia (AML-M5a) is used, and when promonocytes and monocytes account for the majority of the leukemic cells, the condition is referred to as acute monocytic leukemia (AML-M5b).

Acute leukemias of monocytic lineage account for about 3–6% of all AMLs. The median age is around 50 years, but it may occur at any age. The incidence is higher in men than in women (male to female ratio is about 1.8). Clinical symptoms are the result of bone marrow involvement and extramedullary infiltration by the leukemic cells. Fatigue, fever, bleeding disorders, gingival hyperplasia, lymphadenopathy, CNS involvement, and hepatosplenomegaly are among the frequent symptoms. In one major study, the monocytic type (M5b) had a better clinical outcome than the monoblastic type (M5a) with a reported 3-year disease-free survival of 28% and 18%, respectively. In another study, the complete remission rate and disease-free survival did not differ significantly between patients with M5a and those with M5b.

MORPHOLOGY

- The bone marrow biopsy sections are often hypercellular and packed with blasts and immature myelomonocytic cells. Nuclear spacing is present due to abundant cytoplasm of the monocytic cells. Variable degrees of marrow fibrosis are noted in a minority of cases.
- Monoblasts usually show abundant dark to light blue cytoplasm with no or a few scattered fine azurophilic granules (Figures 21.12 and 21.13). The nucleoli are round, oval, or folded, and the nuclear chromatin is fine. There is often a

ACUTE MONOBLASTIC AND ACUTE MONOCYTIC LEUKEMIAS

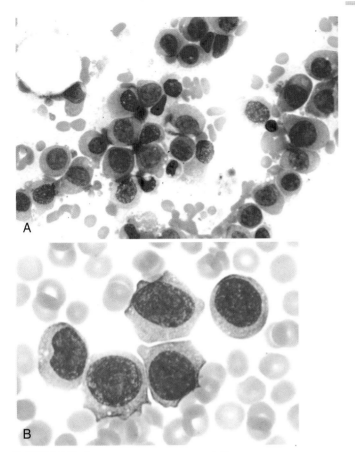

FIGURE 21.12 **ACUTE MONOBLASTIC LEUKEMIA.** Bone marrow smear demonstrates numerous blast cells with abundant blue cytoplasm, round nuclei, fine chromatin, and one or more prominent nucleoli (A). Similar blast cells are present in the peripheral blood smear (B).

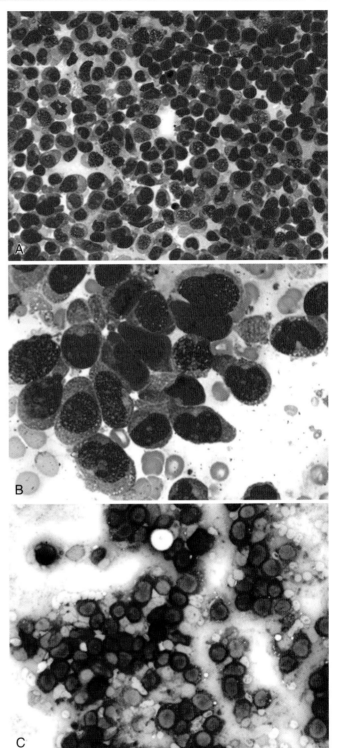

FIGURE 21.13 **ACUTE MONOBLASTIC LEUKEMIA.** Bone marrow smear demonstrates sheets of blast cells with variable amounts of blue cytoplasm, round or irregular nuclei, fine chromatin, and one or more prominent nucleoli (A, low power; B, high power). Blasts are strongly positive for non-specific esterase (C).

single large nucleolus, but multiple prominent nucleoli may be present. Auer rods are rare.
- Promonocytes have abundant light blue to gray cytoplasm (Figures 21.14 and 21.15). Scattered cytoplasmic azurophilic granules and/or cytoplasmic vacuoles may be present. The nuclei are delicately folded or convoluted, often with a cerebriform pattern. The nuclear chromatin is fine and nucleoli are present but not prominent.
- Granulocytic precursors account for <20% of the bone marrow non-erythroid nucleated cells.
- The peripheral blood smears show absolute monocytosis (often ≥5000/µL) with the presence of monoblasts, promonocytes, and monocytes (see Figures 21.12B and 21.15). The proportion of monocytes and promonocytes in blood smears may sometimes be much greater than that of blast cells, whereas in bone marrow smears the difference is not so marked.
- Extramedullary infiltration by leukemic cells is relatively common, such as involvement of gum, skin (Figure 21.16), central nervous system, lymph nodes, liver, and/or spleen.

IMMUNOPHENOTYPE AND CYTOCHEMICAL STAINS

- Expression of CD13, CD15, CD33 (often bright) and CD65;
- Expression of at least two or more monocytic markers (CD4, CD11b, CD11c, CD14, CD36, CD64, CD68, CD163, and lysozyme), often with aberrant expression of CD56 (Figures 21.17 and 21.18);

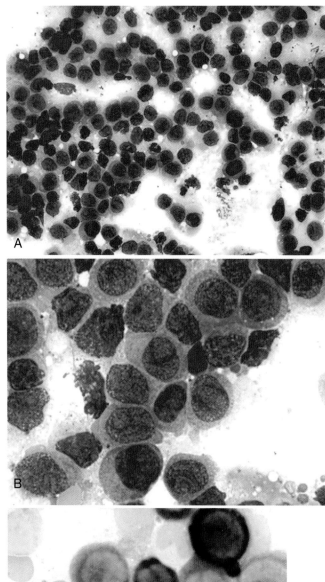

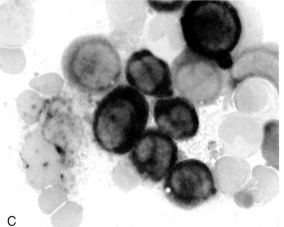

FIGURE 21.14 **ACUTE MONOCYTIC LEUKEMIA.** Bone marrow smear showing increased number of promonocytes with scattered blasts: (A, low power; B, high power). The majority of blast cells stain for NSE (C).

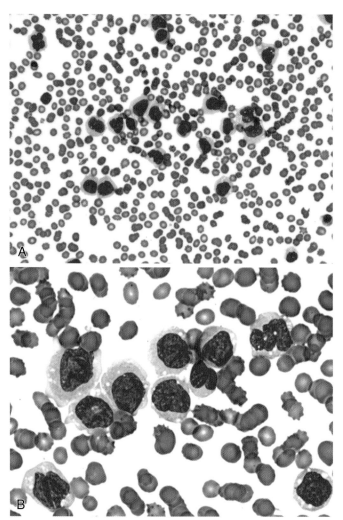

FIGURE 21.15 **ACUTE MONOCYTIC LEUKEMIA.** Blood smear showing numerous promonocytes (A, low power; B, high power).

- Common expression of CD117;
- Frequent absence of CD34;
- Expression of HLA-DR in nearly all cases;
- Variable presence of cross-lineage aberrancies;
- Cytochemical stains show strong diffuse cytoplasmic NSE staining. Monoblasts are usually MPO-negative, but promonocytes may show a weak positive reaction.

MOLECULAR AND CYTOGENETIC STUDIES

Reported cytogenetic abnormalities and affected genes include:

- t(11;17)(q23;q21) (*MLL-AF17*), t(10;11)(p11.2;q23) (*ABI-1;MLL*), t(11;20)(p15;q11.2) (*NUP98-TOP1*), ins(X;11)(q24;q23q13) (*Septing6-MLL*), (5;11)(q31;q23q23) (*MLL/GRAF*).

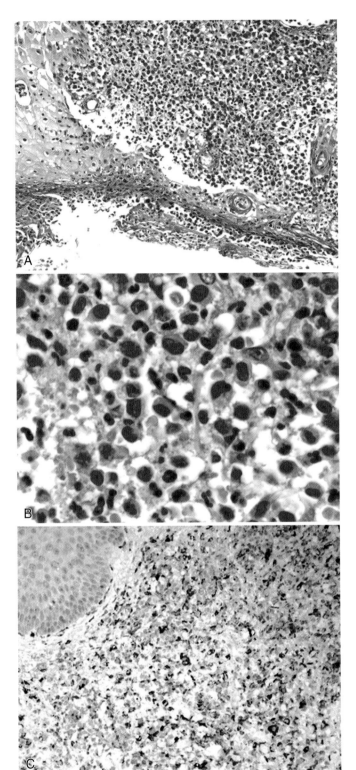

FIGURE 21.16 Dermal infiltration of leukemic cells in a patient with acute monocytic leukemia (A, low power; B, high power). Many of the leukemic cells are positive for CD68 by immunohistochemistry (C).

DIFFERENTIAL DIAGNOSIS

The differential diagnosis of acute monoblastic leukemia (AML-M5a) includes ALL, minimally differentiated AML, AML without maturation, acute megakaryoblastic leukemia, and large cell lymphoma.

Acute monocytic leukemia (AML-M5b) should be distinguished from chronic myelomonocytic leukemia (CMM), acute myelomonocytic leukemia, and microgranular variant of acute promyelocytic leukemia. Leukemic cells of monocytic/monoblastic origin may express NSE, lysozyme, CD4, CD14, CD64, and/or CD68.

Acute Erythroid Leukemias

Acute erythroid leukemia (AML-M6) is defined as a subtype of AML-NOS with predominance of erythroid precursors. 50% or more of bone marrow nucleated cells should be of erythroid origin. Acute erythroid leukemia is divided into two morphologic categories:

1. Erythroleukemia, consisting of myeloblasts and erythroid precursors (AML-M6a);
2. Pure erythroid leukemia (AML-M6b).

Acute erythroid leukemia accounts for less than 5% of all AMLs. The vast majority (≥90%) are of erythroid/myeloid (AML-M6a) subtype. The median age is around 57 years, ranging from 20 to 80 years. The incidence is higher in men than in women (male to female ratio is about 2:1). A significant proportion (up to 50%) of acute erythroid leukemia represents either therapy-related or myelodysplasia-related AMLs. Severe anemia, usually with granulocytopenia and/or thrombocytopenia, are common features. Acute erythroid leukemia is an aggressive disease, but the erythroid/myeloid type (AML-M6a) does significantly better than the pure erythroid leukemia type (AML-M6b). In one report the average survival time for the AML-M6a was 30 months compared with 3 months for the AML-M6b.

MORPHOLOGY

In general, multilineage dysplasias, particularly dyserythropoiesis, are common bone marrow features. These include megaloblastic changes, nuclear budding and fragmentation, multinuclearity, and basophilic stippling in the erythroid precursors. Hypogranulation, abnormal nuclear segmentation, and giant forms may be present in the granulocytic series. Micromegakaryocytes and megakaryocytes with separated

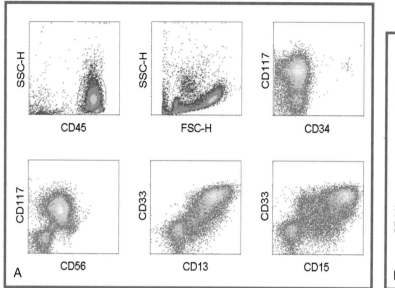
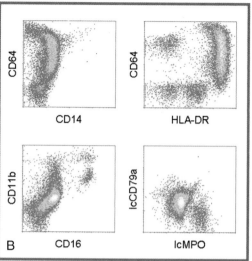

FIGURE 21.17 **FLOW CYTOMETRIC FINDINGS IN A PATIENT WITH ACUTE MONOBLASTIC LEUKEMIA.** Open gate displays by density plots show excess and abnormal myeloblasts (72% of the total, circled) within the monocytic region on CD45 and scatter gate displays. The blast-enriched displays demonstrate abnormal myeloblasts expressing CD11b (partial), CD13, CD15, CD33 (bright), CD38 (not shown), CD45 (moderate), CD64 (heterogeneous), CD117, and HLA-DR, plus cross-lineage aberrancy CD56 (partial). The abnormal myeloblasts are negative for CD34 and intracellular myeloperoxidase.

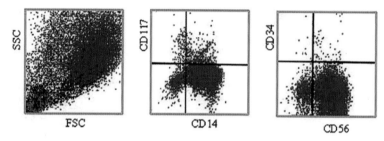

FIGURE 21.18 Flow cytometric studies reveal a large population of monocytic cells which are positive for CD14 and negative for CD34 and CD117. They also aberrantly express CD56.

nuclei are not infrequent. Erythroid leukemia is divided into two subcategories according to the WHO classification.

Erythroleukemia (Erythroid/Myeloid, AML-M6a)

This subtype is defined by at least 50% of the bone marrow nucleated cells being of erythroid origin and ≥20% myeloblasts in the non-erythroid component (Figures 21.19 and 21.20).

- Bone marrow biopsy sections are usually hypercellular with clusters or sheets of immature cells and a marked reduction in the normal hematopoietic components.
- The erythroid series are dysplastic and left shifted, but are usually found in all stages of maturation.
- Some of the myeloblasts may show cytoplasmic granules. Occasionally, Auer rods may be present.
- Bone marrow iron stores are often increased, and ring sideroblasts may be present.
- Blood smears show anisopoikilocytosis with the presence of schistocytes, tear-drops, and macrocytes. Basophilic stippling is present. Granulocytic series may show hypogranulation and hyposegmentation. Giant and/or hypogranular platelets

are often present. Various numbers of nucleated red blood cells and blasts are often present.

The WHO requirements for the diagnosis of erythroleukemia leave significant overlapping features with refractory anemia with excess blasts (RAEB). If we follow the WHO requirements, a simple calculation tells us that any bone marrow sample with dysplastic changes, >50% and <80% erythroid precursors, and 5–10% myeloblasts in the total bone marrow cells gating may be qualified for the diagnosis of acute erythroleukemia. For example, a bone marrow sample with 79% erythroid precursors and 5% myeloblasts in the total bone marrow cells, or a bone marrow sample with 51% erythroid precursors and 10% myeloblasts should be diagnosed as erythroleukemia, whereas a bone marrow sample with 49% erythroid precursors and 15% myeloblasts is called RAEB II and not an acute leukemia!

This problem with the diagnostic criteria for acute erythroleukemia has been raised in the literature. Further clinical investigations are needed for clarification of the status of erythroleukemias with low myeloblast counts.

ACUTE ERYTHROID LEUKEMIAS

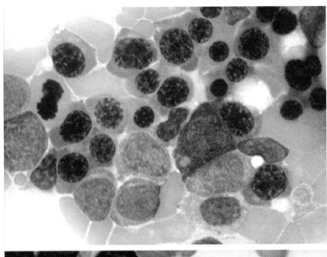

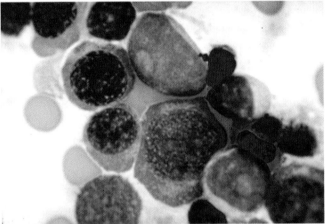

FIGURE 21.19 **ERYTHROID/MYELOID LEUKEMIA (ERYTHROLEUKEMIA, AML-M6A).** Bone marrow smears show erythroid preponderance and left shift with increased myeloblasts (top, low power; bottom, high power).

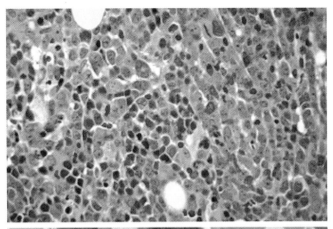

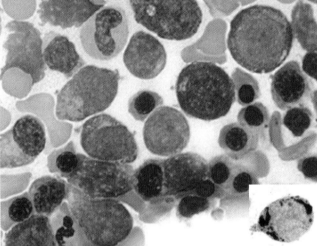

FIGURE 21.20 **ERYTHROID/MYELOID LEUKEMIA (ERYTHROLEUKEMIA, AML-M6A).** (top) Bone marrow biopsy section showing hypercellular marrow with increased immature cells. (bottom) Bone marrow smear shows erythroid preponderance and left shift with increased myeloblasts. Inset shows an erythroid precursor with coarse PAS-positive cytoplasmic granules.

Pure Erythroid Leukemia (AML-M6b)

This category is defined as a disorder with ≥80% of bone marrow nucleated cells consisting of erythroid precursors (Figures 21.21, and 21.22).

- These cells are predominantly early erythroblasts (pronormoblasts) with dark blue cytoplasm, round nuclei, fine nuclear chromatin, and one or more prominent nucleoli.
- The cytoplasm often contains poorly demarcated vacuoles. Evidence of dyserythropoiesis and/or megaloblastic changes is often present.

Pure erythroid leukemia is far less frequent than the erythroleukemia (erythroid/myeloid) type.

IMMUNOPHENOTYPIC STUDIES AND CYTOCHEMICAL STAINS

Erythroblasts:

- Asynchronous or aberrant expression intensity of CD36, CD71, and CD235a (Glycophorin A or GPA) (Figure 21.23);
- Lack of GPA in subset;
- Occasional expression of myeloid markers CD13 and/or CD117;
- Erythroblasts are negative for cytochemical MPO, Sudan black B, and NSE stains, but often show globular or coarsely granular cytoplasmic PAS positivity.

Myeloblasts:

- Variable expression of CD13, CD33, CD34, and CD117;
- Variable presence of cross-lineage aberrancies;
- Myeloblasts may show positive reactions for MPO and Sudan black B by cytochemical stains.

MOLECULAR AND CYTOGENETIC STUDIES

- Acute erythroid leukemia shares many cytogenetic features with MDS (see Chapters 8 and 9). Partial loss or monosomy of chromosomes 5 and 7 is the most frequent chromosomal aberration reported in acute erythroid leukemias, followed by abnormalities of chromosomes 8, 16, and 21.
- Some of the cases of pure erythroid leukemia are found to carry the *BCR-ABL1* fusion gene, but no specific molecular abnormalities are yet correlated with this disorder.
- Complex chromosomal abnormalities are frequent findings.

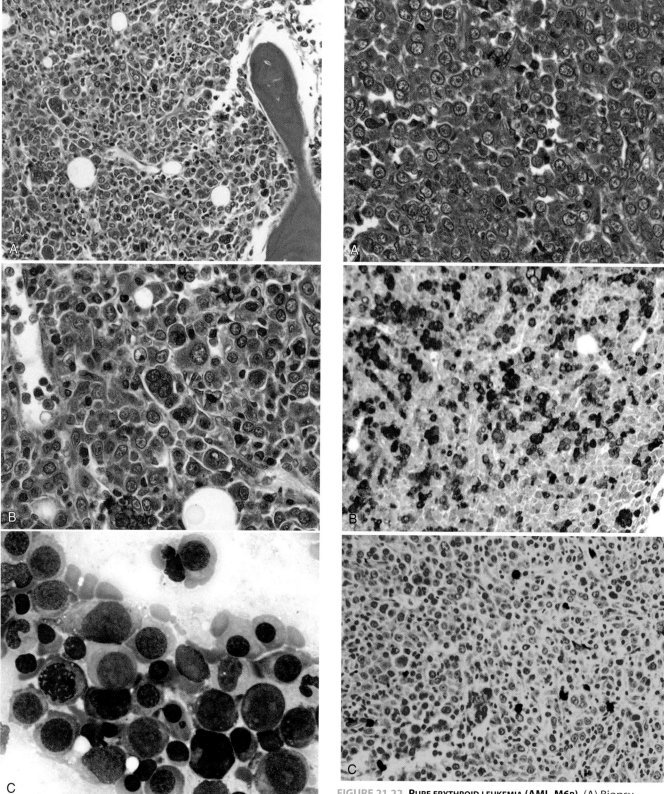

FIGURE 21.21 **PURE ERYTHROID LEUKEMIA (AML-M6B).** Biopsy section reveals a markedly hypercellular marrow with sheets of blasts and immature erythroid cells (A, low power; B, high power). Bone marrow smear shows erythroid preponderance with increased erythroblasts (C).

FIGURE 21.22 **PURE ERYTHROID LEUKEMIA (AML-M6B).** (A) Biopsy section reveals a markedly hypercellular marrow with sheets of blasts and immature cells. (B) Immunohistochemical stain for hemoglobin A shows numerous positive cells. (C) MPO stain is negative.

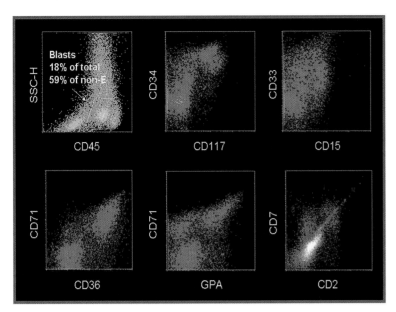

FIGURE 21.23 **FLOW CYTOMETRIC FINDINGS IN ACUTE ERYTHROLEUKEMIA (ERYTHROID/MYELOID).** Open gate displays illustrate erythroid preponderance (69% of the total) and excess myeloblasts (18% of the total or 59% of the non-erythroid cells) by CD45 gating. The myeloblasts coexpress CD34 and CD117 with cross-lineage aberrancy of partial dim CD7. The erythroids are immature, expressing CD36, CD71 (dim to moderate), and Glycophorin A (a subset).

DIFFERENTIAL DIAGNOSIS

The feature distinguishing acute erythroleukemia (AML-M6a) from RAEB is the proportion of the erythroid component in the bone marrow. In AML-M6a, 50% or more bone marrow nucleated cells are of erythroid lineage, and ≥20% of the non-erythroid population consists of myeloblasts.

AML with myelodysplasia-related changes should also be included in the differential diagnosis. According to the WHO recommendation, if >50% of the myeloid or megakaryocytic lineages show dysplasia, the case should be classified as AML with myelodysplasia-related changes.

Pure erythroid leukemia (AML-M6b) should be distinguished from megaloblastic anemia. The erythroid left shift and dysplastic changes are not so severe in megaloblastic anemia as in pure erythroid leukemia. Besides, in megaloblastic anemia, there is often evidence of vitamin B_{12} or folate deficiency, whereas serum levels of vitamin B_{12} and folate are normal or elevated in pure erythroid leukemia. The differential diagnosis of pure erythroid leukemia also includes ALL, minimally differentiated AML, AML without maturation, and acute megakaryoblastic leukemia.

Acute Megakaryoblastic Leukemia

Acute megakaryoblastic leukemia (AMKL, AML-M7) is defined as an AML in which megakaryoblasts account for ≥50% of the total blast cells. Similar to the other types of AML, total blasts comprise ≥20% of the bone marrow nucleated cells or peripheral blood differential counts.

AMKL represents 3–5% of the AMLs. It occurs in all ages with two distribution peaks: children between 1–3 years old and adults. A significant proportion of affected children have Down syndrome. In children with Down syndrome, the incidence of AMKL is 46-fold greater than in the normal age group, accounting for at least 50% of the AML cases in patients with Down syndrome. AMKL has also been associated with mediastinal germ cell tumors in young adult males. Hepatosplenomegaly is rare in adults but frequently observed in children, particularly in association with t(1;22). This translocation has distinctive clinicopathologic features including onset in infancy and extensive bone marrow fibrosis with clustering of leukemic blasts mimicking metastatic tumors (see Chapter 18).

MORPHOLOGY

- Bone marrow fibrosis, and as a consequence, dry tap (failure of bone marrow aspiration) is a common feature. Bone marrow cellularity varies depending on the extent of fibrosis, but there is evidence of increased blast cells, either in clusters or as diffuse interstitial infiltration. There is also evidence of increased megakaryocytes with dysplastic morphology.
- Megakaryoblasts are the predominant component of the blast population in the bone marrow and/or peripheral blood (Figures 21.24 to 21.26A). They are markedly pleomorphic, ranging from small, round cells with scanty cytoplasm and inconspicuous nucleoli, resembling hematogones, to large cells with abundant cytoplasm and prominent nucleoli. They often display cytoplasmic blebs or pseudopods and may appear in clusters mimicking metastatic tumors (see Figures 21.24D, and 21.26A).
- Peripheral blood smears show circulating micromegakaryocytes, megakaryocytic fragments, atypical giant platelets, and blasts. Red blood cells may show anisopoikilocytosis and scattered teardrop shapes. Granulocytes may show dysplastic changes, such as hypogranulation and abnormal segmentation of their nuclei. Leukoerythroblastosis is uncommon.

Since megakaryoblasts may resemble hematogones, lymphoblasts, non-granular myeloblasts, or metastatic tumors

274 ACUTE MYELOID LEUKEMIA, NOT OTHERWISE SPECIFIED

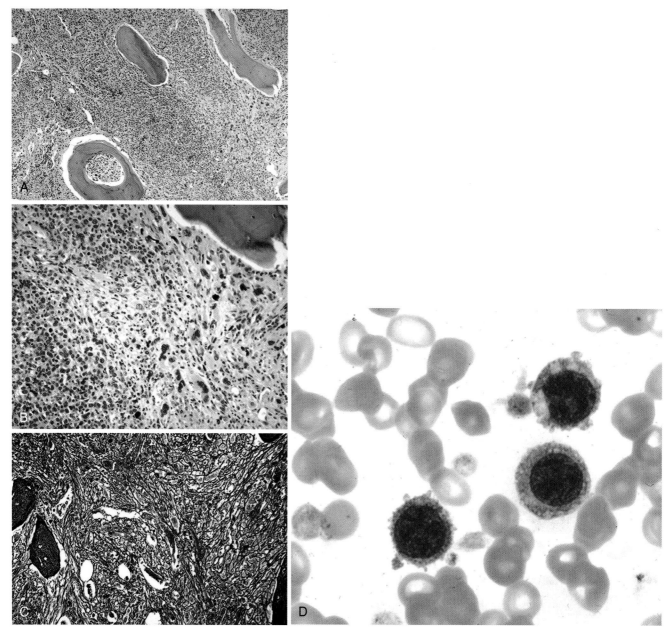

FIGURE 21.24 ACUTE MEGAKARYOBLASTIC LEUKEMIA. Bone marrow biopsy sections show large clusters of immature cells with areas of fibrosis and increased megakaryocytes (A, low power; B, high power). Reticulin stain reveals marked reticulin fibrosis (C). Blood smear demonstrates megakaryoblasts with cytoplamic blebs (D).

and do not react with a specific cytochemical stain, it is often necessary to perform immunophenotypic and/or ultrastructural studies to be able to assign their megakaryocytic lineage.

IMMUNOPHENOTYPE AND CYTOCHEMICAL STAINS

The immunophenotypic features of megakaryoblasts include (Figure 21.26B):

- Expression of CD41, CD61, Factor VIII;
- Expression of CD36 and CD38;
- Very common aberrancy of CD7;
- Variable expression of CD13 and CD33;
- Negativity of other myeloid markers, including MPO;
- Absence of CD34 and/or CD117;
- Negative or dim CD45;
- Negative HLA-DR.

No conventional cytochemical stain is specific for megakaryoblasts. However, megakaryoblasts may show:

- Diffuse and/or globular PAS reaction and punctuate NSE staining.
- MPO and Sudan Black B are negative, but ultrastructural cytochemical staining or immunostaining reveals expression of platelet peroxidase.
- Megakaryoblasts also express acid phosphatase.

ACUTE MEGAKARYOBLASTIC LEUKEMIA

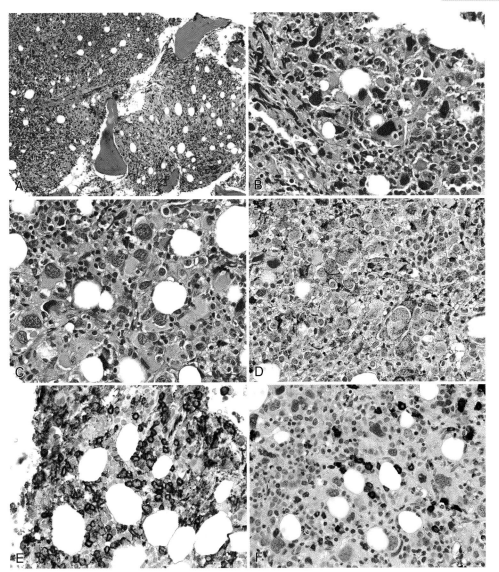

FIGURE 21.25 **ACUTE MEGAKARYOBLASTIC LEUKEMIA TRANSFORMED FROM ESSENTIAL THROMBOCYTHEMIA.** Core biopsy sections reveal a markedly hypercellular marrow with fibrosis and increased blasts, as well as many abnormal megakaryocytes in large aggregates to focal sheets (A). The abnormal megakaryocytes are prominent in some areas (B) with hyperchromatic nuclei, and are admixed with scattered blasts. Myelopoiesis and erythropoiesis are significantly reduced. In many areas, megakaryoblasts and immature megakaryocytes are abundant (C), and are admixed with clusters to aggregates of myeloblasts. Immunohistochemical studies for CD31 (D) highlight the immature megakaryocytes, including megakaryoblasts, while CD34 (E) and CD117 (not shown) reveal excess myeloblasts in clusters to aggregates. The blasts are negative for myeloperoxidase (F).

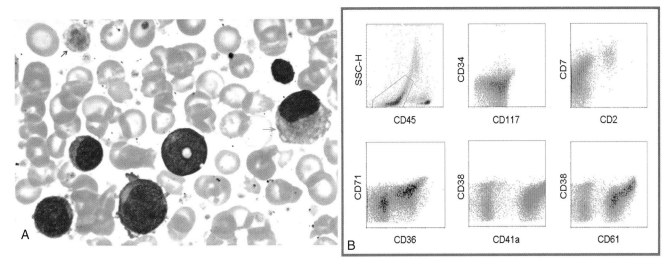

FIGURE 21.26 **ACUTE MEGAKARYOBLASTIC LEUKEMIA.** (A) Blood smear demonstrating numerous blast cells of various sizes, some with cytoplasmic blebs. A micromegakaryocyte (green arrow) and a giant platelet (red arrow) are present. (B) Flow cytometry of peripheral blood. The CD45 gating open display reveals a predominant population of megakaryoblasts (marked region; 60% of the total) that is CD45 negative to dim positive. The megakaryoblasts are positive for CD36 (major subset), CD38 (partial), CD41a, CD61, and CD71 (partial), plus cross-lineage aberrancy CD7 (dim). They are negative for CD34 and CD117.

MOLECULAR AND CYTOGENETIC STUDIES

- Mutations or altered expression of the megakaryoblastic leukemia protein-1 (*MKL1*) gene are found in some cases of megakaryoblastic leukemia and are felt to be involved in the pathogenesis.
- Mutations in the *GATA1* gene are also frequently seen.
- In addition to trisomy 21 in Down syndrome, chromosomal aberrations, such as −7/7q−, −5/5q−, and trisomy 8 have been reported.
- Other chromosomal aberrations include abnormalities of 3q21–26, 17q22, 11q14–21, and 21q21–22.
- Occasional cases of highly complex chromosomal abnormalities have been reported.

DIFFERENTIAL DIAGNOSIS

The differential diagnosis of AMKLs includes primary myelofibrosis, minimally differentiated AML, AML without maturation, ALL, acute panmyelosis with myelofibrosis, and metastatic tumors. The presence of megakaryoblastic clusters embedded in the fibrotic bone marrow may mimic metastatic carcinoma, non-Hodgkin lymphoma, neuroblastoma, or rhabdomyosarcoma.

The diagnosis of AMKL is suggested by certain clinicopathologic features, such as Down syndrome, myelofibrosis with no or minimal leukoerythroblastosis, and/or presence of megakaryoblasts in the peripheral blood. The diagnosis is confirmed when at least 50% of the total blast population is of megakaryocytic origin evidenced by the presence of platelet peroxidase (PPO) by electron microscopy and/or expression of CD41, CD42, CD61, and factor VIII.

Acute Basophilic Leukemia

Acute basophilic leukemia (ABL) is a rare type of AML with basophilic differentiation. It represents less than 1% of all AMLs. ABL is either *de novo* or the result of basophilic blast transformation in CML. It occurs at any age including infancy. Clinical findings include anemia, skin rashes, hepatosplenomegaly, and gastric ulcers

MORPHOLOGY (FIGURE 21.27)

- Bone marrow biopsy sections appear hypercellular with large clusters or sheets of blast cells and marked reduction in the normal hematopoietic components. There may be dysplastic erythropoiesis. A variable degree of reticulin fibrosis is often present, particularly in advanced stages of the disease.
- Blasts show a high nuclear to cytoplasmic ratio with variable amounts of cytoplasm containing coarse basophilic granules. The nuclei are round, oval, or bilobed with a fine chromatin and one or more prominent nucleoli. Scattered mature basophils are often present.

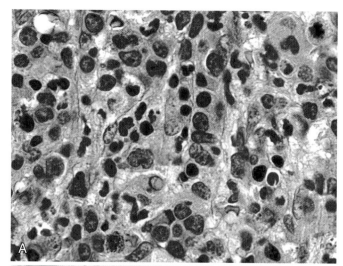
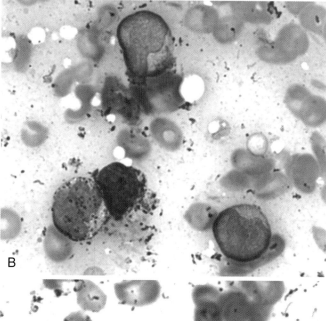
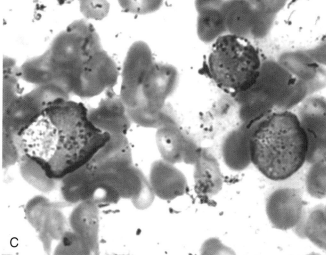

FIGURE 21.27 **ACUTE BASOPHILIC LEUKEMIA.** (A) Bone marrow biopsy section. (B and C) Bone marrow smears.

- Blood smears usually show the presence of blasts with basophilic granules and mature basophils. Anisopoikilocytosis and thrombocytopenia may be present.
- Electron microscopy reveals characteristic features of basophilic granules.

IMMUNOPHENOTYPE AND CYTOCHEMICAL STAINS

- Expression of myeloid markers CD13, and CD33; but lack of MPO;
- Common expression of CD22;
- Common expression of CD11b and CD123;
- Variable expression of CD34 and HLA-DR; but negative CD117;
- Variable cross-lineage aberrancies;
- The cytoplasmic basophilic granules may stain with toluidine blue but are usually negative for MPO and NSE.

MOLECULAR AND CYTOGENETIC STUDIES

No specific chromosomal aberration is known for ABL. There are sporadic reports of abnormalities of deletion of 12p, t(6;9)(p23;q34) and t(X;6)(p11;q23). Cases of basophilic transformation in CML show t(9;22)(q34;q11.2).

DIFFERENTIAL DIAGNOSIS

Differential diagnosis includes blast transformation of CML, certain subtypes of AML with basophilia, particularly the ones associated with t(6;9) or abnormalities of 12p, and occasional cases of ALL with coarse azurophilic granules.

Acute Panmyelosis with Myelofibrosis

Acute panmyelosis with myelofibrosis (APMF), previously referred to as *acute myelofibrosis*, *acute myelosclerosis*, or *acute myelodysplasia with myelofibrosis*, is an acute leukemia with panmyeloid proliferation, increased blasts, and bone marrow fibrosis.

This entity shares significant overlapping features with AMKL, such as bone marrow fibrosis, unaspirable marrow, dysplastic megakaryocytes, and the presence of megakaryoblasts. However, in AMKL, megakaryoblasts comprise the predominant proportion of the blast population, whereas in APMF the majority of the blasts are of non-megakaryocytic origin.

APMF is a rare type of AML often presenting with marked cytopenia. Splenomegaly is minimal or absent. It has an unfavorable prognosis with a median survival of <1 year in some reports

MORPHOLOGY (FIGURE 21.28)

- Bone marrow fibrosis is one of the morphologic hallmarks. It often leads to unaspirable bone marrow (dry tap) and inadequate marrow smears. The bone marrow cellularity varies depending on the extent of fibrosis. The megakaryocytes are increased and show dysplastic morphology, including small size and non-lobulated and/or hypolobulated nuclei.
- There is evidence of increased blast cells, either in clusters or as diffuse interstitial infiltration. The majority of the blast cells are of non-megakaryocytic origin.
- Peripheral blood smears may show absent-to-mild anisopoikilocytosis with macrocytes and occasional teardrop shapes. Granulocytes may show dysplastic changes, such as hypogranulation and abnormal segmentation of their nuclei. Abnormal platelets may be present. Circulating blasts are variable, ranging from a few to a frank leukemic picture.

IMMUNOPHENOTYPE AND CYTOCHEMICAL STAINS

- The majority of the blast cells express CD34, often with one or more myeloid-associated markers, such as CD13, CD33, and/or CD117 and cytoplasmic MPO (see Figure 21.28E and F);
- Only a minority of blast cells express platelet-associated molecules, such as CD41, CD42, or CD61.
- Cytochemical stains may show presence of MPO-positive blast cells.

MOLECULAR AND CYTOGENETIC STUDIES

Monosomy 7 and loss of 5q or 7q are the most frequent chromosomal aberrations in this condition. Also, interstitial deletion of the long arm of chromosome 11 has been reported.

DIFFERENTIAL DIAGNOSIS

As mentioned earlier, this entity shares many overlapping features with AMKL, such as bone marrow fibrosis, unaspirable marrow, dysplastic megakaryocytes, and the presence of megakaryoblasts. However, unlike AMKL, in APMF the majority of the blasts are of non-megakaryocytic origin. The differential diagnosis also includes primary myelofibrosis, minimally differentiated AML, AML without maturation, ALL, and metastatic tumors.

Other Types of Acute Myeloid Leukemia

HYPOPLASTIC ACUTE MYELOID LEUKEMIA

Hypoplastic acute myeloid leukemia or hypocellular acute myeloid leukemia is a leukemic condition characterized by marked bone marrow hypocellularity. The etiology and

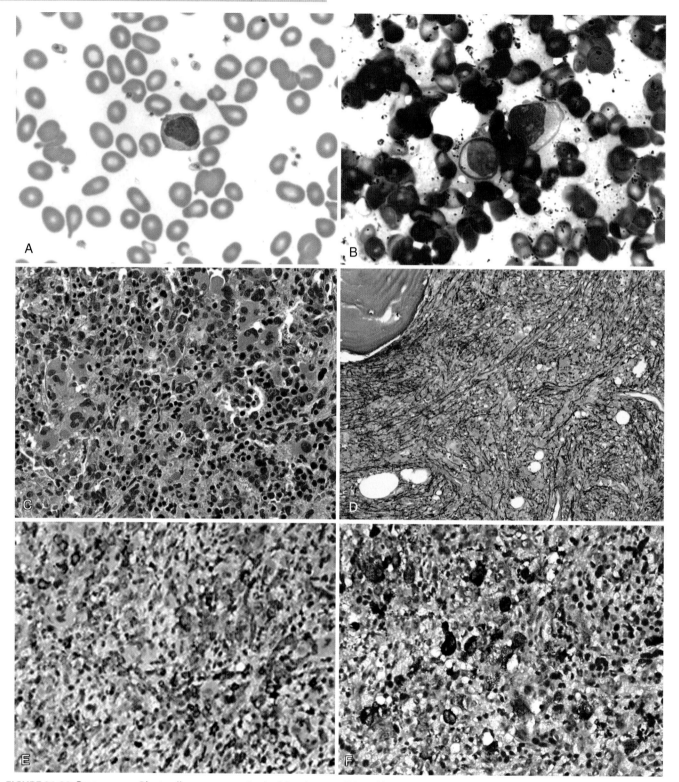

FIGURE 21.28 **PANMYELOSIS.** Blast cells are present in the blood smear (A) and bone marrow touch preparation (B). Bone marrow biopsy section (C) is hypercellular and demonstrates numerous megakaryocytes and excess blasts. Reticulin stain (D) shows diffuse marrow fibrosis, with scattered and small clusters of cells which are positive for CD117 (E) and MOP (F).

pathogenesis of this leukemia are not known, but bone marrow toxicity may play a contributing role. This type of leukemia tends to involve elderly people, primarily men. The bone marrow is hypocellular, usually <30% of the average cellularity in the normal matching age group. Blast cells are prominent and account for ≥20% of the bone marrow cells (Figure 21.29). They display scant-to-moderate amounts of cytoplasm, round or oval nuclei with fine

OTHER TYPES OF ACUTE MYELOID LEUKEMIA **279**

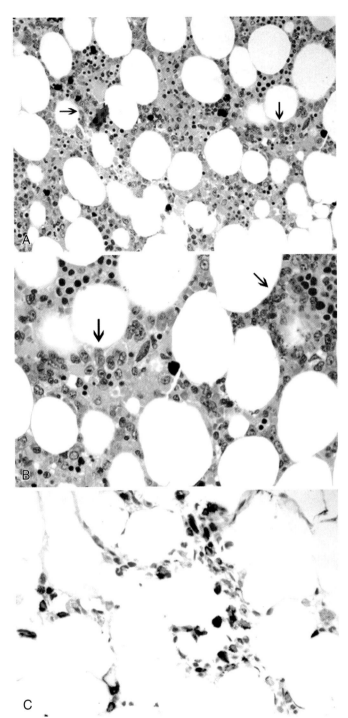

FIGURE 21.29 **HYPOPLASTIC ACUTE MYELOID LEUKEMIA.** Bone marrow biopsy section showing hypocellularity and increased blasts (A, low power; B, high power). Immunohistochemical stain for CD34 reveals numerous positive cells (C).

nuclear chromatin, and one or more prominent nucleoli. In most studies, blast cells appear to be of myeloid origin by the presence of cytoplasmic azurophilic granules, Auer rods, positive reactions for MPO, Sudan Black B, or expression of myeloid-associated molecules, such as CD13, CD33, and CD117. The differential diagnoses include aplastic anemia and hypocellular MDS. Diagnosis is made based on the presence of ≥20% blasts in a hypocellular marrow.

MYELOID SARCOMA

Myeloid sarcoma (granulocytic sarcoma, chloroma) refers to extramedullary tumors of myeloid precursors. The term "chloroma" was originally used in response to the green color of the tumor due to the presence of myeloperoxidase (MPO) in the tumor cells. Myeloid sarcoma is usually associated with CML in blast crisis, or *de novo* AML. It either develops during the active phase of the disease or represents relapse without evidence of recurrent disease in the blood or the bone marrow. The most frequent sites of involvement include skin, lymph nodes, respiratory system, gastrointestinal tract, CNS, and subperiosteal structures of the skull, ribs, vertebrae, and pelvis. The incidence of myeloid sarcoma is higher in children than in adult patients with AML.

Myeloid sarcomas are composed of various proportions of immature and mature myeloid cells (Figure 21.30). Myeloblasts, monoblasts, and immature myelomonocytic cells are the predominant cell types, though a proportion of megakaryocytic and erythroid precursors may also be present. Megakaryocytic and erythroblastic sarcomas are extremely rare.

Myeloid sarcoma may resemble lymphomas or nonhematopoietic malignancies. The presence of immature eosinophils is a distinctive feature of myeloid sarcomas. Tumor cells in these lesions express myeloid-associated molecules in the biopsy sections, such as MPO, NES, and/or lysozyme. Flow cytometric studies demonstrate a population of CD45 dim cells expressing CD13, CD33, CD34, CD64, CD117, and MPO, indicating a myeloblastic population. The presence of monoblasts/promonocytes is demonstrated by CD4, CD14, CD64, and CD68 expression.

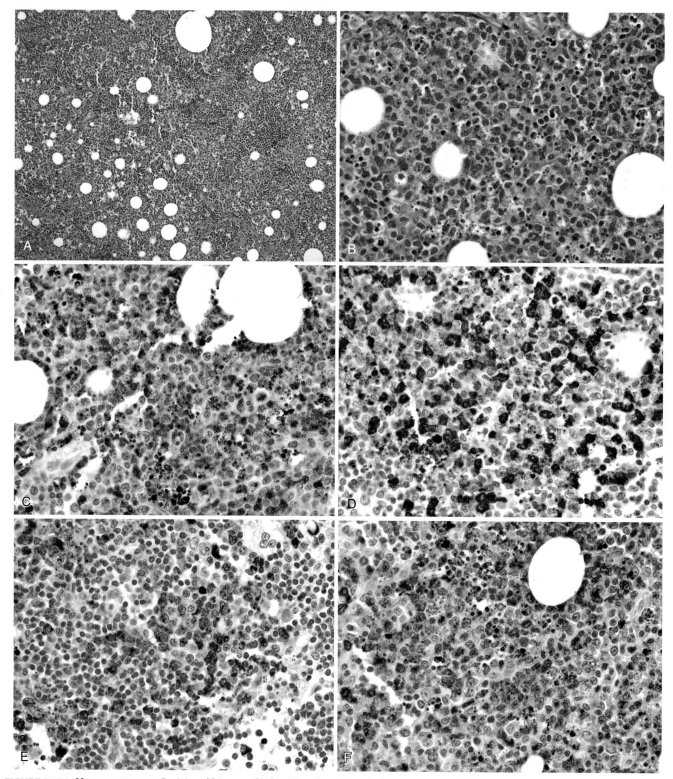

FIGURE 21.30 **MYELOID SARCOMA.** Excisional biopsy of left axillary lymph node from a 64-year-old female. Sections reveal partial effacement of the nodal architecture (A) by predominantly a sinusoidal infiltration of large neoplastic cells in sheets. These large neoplastic cells (B) contain irregular to folded nuclei, dispersed chromatin, prominent nucleoli, and moderate to abundant granular cytoplasm. Apoptotic bodies, mitoses, and tingible body macrophages are easily seen. Immunohistochemical studies demonstrate that the neoplastic cells are positive for myeloperoxidase (C), CD15 (D) and partially express CD68 (E) and lysozyme (F).

Additional Resources

Amadori S, Venditti A, Del Poeta G, et al: Minimally differentiated acute myeloid leukemia (AML-M0): a distinct clinico-biologic entity with poor prognosis, *Ann Hematol* 72:208-215, 1996.

Audouin J, Comperat E, Le Tourneau A, et al: Myeloid sarcoma: clinical and morphologic criteria useful for diagnosis, *Int J Surg Pathol* 11:271-282, 2003.

Avni B, Rund D, Levin M, et al: Clinical implications of acute myeloid leukemia presenting as myeloid sarcoma, *Hematol Oncol* 30:34-40, 2012.

Berger R, Le Coniat M, Flexor MA, Leblanc T: Translocation t(10;11) involving the MLL gene in acute myeloid leukemia: importance of fluorescence in situ hybridization (FISH) analysis, *Ann Genet* 39:147-151, 1996.

Campidelli C, Agostinelli C, Stitson R, et al: Myeloid sarcoma: extramedullary manifestation of myeloid disorders, *Am J Clin Pathol* 132:426-437, 2009.

Gougounon A, Abahssain H, Rigollet L, et al: Minimally differentiated acute myeloid leukemia (FAB AML-M0): prognostic factors and treatment effects on survival–a retrospective study of 42 adult cases, *Leuk Res* 35:1027-1031, 2011.

Jaffe ES, Harris NL, Vardiman JW, et al: Hematopathology, Philadelphia, 2010, Saunders/Elsevier.

King JA, Nye DM, O'Connor MB, et al: Acute myelogenous leukemia (FAB AML-M1) in the setting of HIV infection and G-CSF therapy: a case report and review of the literature, *Ann Hematol* 77:69-73, 1998.

Latif N, Salazar E, Khan R, Villas B, Rana F: The pure erythroleukemia: a case report and literature review, *Clin Adv Hematol Oncol* 8:283-290, 2010.

Masood A, Holkova B, Chanan-Khan A: Review: erythroleukemia: clinical course and management, *Clin Adv Hematol Oncol* 8:288-290, 2010.

McGrattan P, Alexander HD, Humphreys MW, et al: Tetrasomy 13 as the sole cytogenetic abnormality in acute myeloid leukemia M1 without maturation, *Cancer Genet Cytogenet* 135:192-195, 2002.

Roy A, Roberts I, Norton A, Vyas P: Acute megakaryoblastic leukaemia (AMKL) and transient myeloproliferative disorder (TMD) in Down syndrome: a multi-step model of myeloid leukaemogenesis, *Br J Haematol* 147:3-12, 2009.

Santos FP, Bueso-Ramos CE, Ravandi F: Acute erythroleukemia: diagnosis and management, *Expert Rev Hematol* 3:705-718, 2010.

Scharenberg MA, Chiquet-Ehrismann R, Asparuhova MB: Megakaryoblastic leukemia protein-1 (MKL1): increasing evidence for an involvement in cancer progression and metastasis, *Int J Biochem Cell Biol* 42:1911-1914, 2010.

Sun T, Wu E: Acute monoblastic leukemia with t(8;16): a distinct clinicopathologic entity; report of a case and review of the literature, *Am J Hematol* 66:207-212, 2001.

Swerdlow SH, Campo E, Harris NL, et al: WHO classification of tumours of haematopoietic and lymphoid tissues, ed 4, Lyon, 2008, International Agency for Research on Cancer.

Thiele J, Kvasnicka HM, Schmitt-Graeff A: Acute panmyelosis with myelofibrosis, *Leuk Lymphoma* 45:681-687, 2004.

van't Veer MB: The diagnosis of acute leukemia with undifferentiated or minimally differentiated blasts, *Ann Hematol* 64:161-165, 1992.

Yin CC, Medeiros LJ, Bueso-Ramos CE: Recent advances in the diagnosis and classification of myeloid neoplasms–comments on the 2008 WHO classification, *Int J Lab Hematol* 32:461-476, 2010.

Zuo Z, Polski JM, Kasyan A, et al: Acute erythroid leukemia, *Arch Pathol Lab Med* 134:1261-1270, 2010.

Myeloid Proliferations Related to Down Syndrome

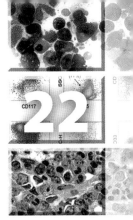

Patients with Down syndrome (trisomy of chromosome 21) have a 10- to 20-fold higher relative risk for leukemia. Approximately 10% of neonates with Down syndrome (DS) develop a transient abnormal myelopoiesis (TAM), which regresses spontaneously in the vast majority of the cases. However, about 20% of the affected neonates will eventually develop acute myelogenous leukemia (AML) before age 5. The vast majority of the AML cases in DS patients are of megakaryoblastic subtype. Mutation of the GATA-1 gene on the X chromosome, encoding for the erythroid/megakaryocytic transcription factor GATA-1, has been reported in both TAM and DS-associated megakaryoblastic leukemia.

Transient Abnormal Myelopoiesis

DS-related transient abnormal myelopoiesis (TAM) occurs in about 10% of neonates with DS. The affected neonates are usually under 1 month old (average age about 1 week) and may demonstrate asymptomatic leukocytosis to massive organomegaly and fatal liver failure. The clinicopathologic manifestations of TAM usually disappear spontaneously after several weeks (usually less than 3 months). A small proportion of patients may eventually develop AML.

MORPHOLOGY (FIGURES 22.1 AND 22.2)

- Blood smears show leukocytosis with myeloid left shift and the presence of blast cells. Blast cells are predominantly of megakaryocytic and erythroid origin. Megakaryoblasts may show cytoplasmic blebbing or coarse cytoplasmic azurophilic granules. Early erythroblasts (pronormoblasts) are often megaloblastic and show a dark blue, non-granular cytoplasm with a perinuclear pale area. Some patients may show basophilia. Leukocytosis may occasionally, exceed 100,000/µL. As usually observed in this age group, late stage nucleated and polychromatic red cells are present.
- Bone marrow shows myeloid preponderance and left shift with increased number of blasts, predominantly megakaryoblasts. Erythroid and megakaryocytic dysplasia is often present.
- Liver biopsies may show extramedullary hematopoiesis with predominance of megakaryocytes and blasts. Hepatocellular necrosis and areas of fibrosis may be present.
- Splenomegaly is often associated with necrosis, extramedullary hematopoiesis, and increased proportion of blast cells (Figure 22.2C).

IMMUNOPHENOTYPE (FIGURE 22.3)

- Immunophenotypic features of TAM by multiparametric flow cytometry (MFC) are indistinguishable from those seen in AML associated with Down syndrome.
- Blasts commonly express CD33, CD34, CD38, CD117, plus cross-lineage aberrancies of CD7 and CD56.
- In addition, there is variable expression of CD13, CD36, CD41, and CD61.
- HLA-DR is negative or partially expressed.
- Myeloperoxidase is usually negative.

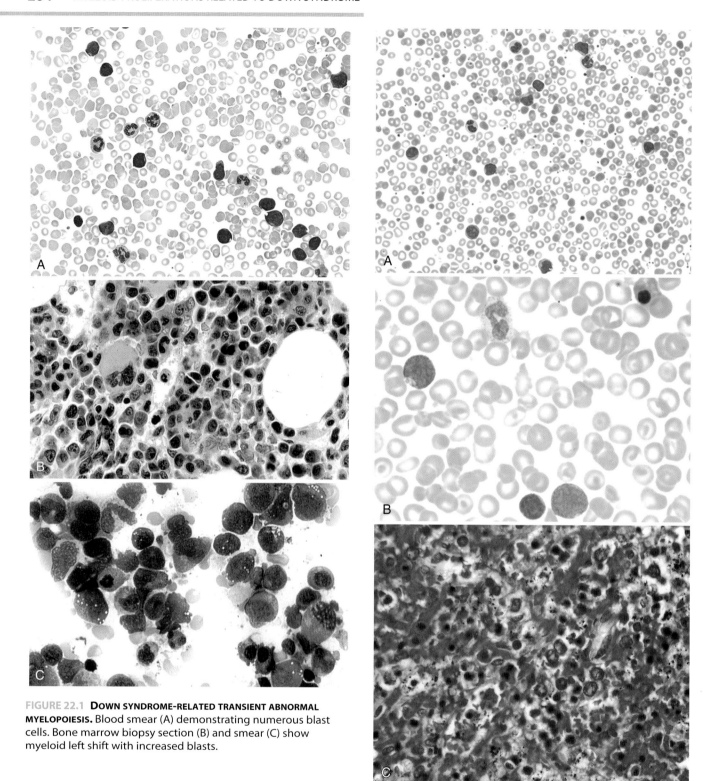

FIGURE 22.1 **DOWN SYNDROME-RELATED TRANSIENT ABNORMAL MYELOPOIESIS.** Blood smear (A) demonstrating numerous blast cells. Bone marrow biopsy section (B) and smear (C) show myeloid left shift with increased blasts.

FIGURE 22.2 **DOWN SYNDROME-RELATED TRANSIENT ABNORMAL MYELOPOIESIS.** Blood smear (A, low power; B, high power) demonstrating several blast cells. Section of spleen (C) shows small clusters and scattered blasts.

MOLECULAR AND CYTOGENETIC STUDIES

- Sequence analysis of the *GATA1* gene is performed in a small number of clinical laboratories if desired.
- The acquired mutations involved in DS-associated transient MPD should not be confused with inherited mutations in the same gene that cause thrombocytopenia (not associated with Down syndrome).
- Trisomy 21 (Figure 22.4).
- The presence of additional cytogenetic findings is consistent with an increased risk for developing subsequent AML.

DS-Associated Myeloid Leukemias

Approximately 20% of DS patients with TAM will develop AML. This occurs usually in the first 3 years of their life. Typically, evolution to AML is preceded by a prolonged myelodysplastic-like period. Approximately 10% of children with DS develop MDS (Figure 22.5), acute megakaryoblastic leukemia (Figure 22.6), or minimally differentiated AML (Figure 22.7).

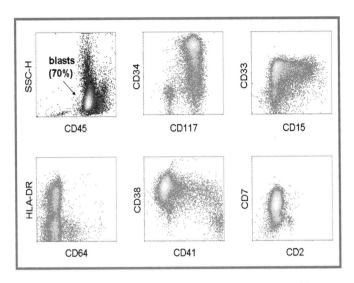

FIGURE 22.3 **FLOW CYTOMETRIC FINDINGS IN AN INFANT WITH DOWN SYNDROME-RELATED TRANSIENT ABNORMAL MYELOPOIESIS.** Open gate density plot by CD45 gating reveals 70% blasts that are CD45 dim. Density plots of the blast-enriched gate (in blue) demonstrate that the blasts express CD7, CD15 (partial), CD33, CD34 (heterogeneous, partial), CD38, CD117, and HLA-DR (partial).

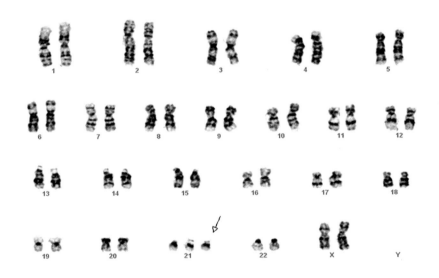

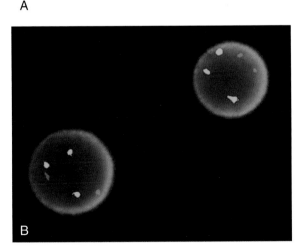

FIGURE 22.4 (A) Constitutional trisomy 21 (47,XX,+21) by karyotype in DS-TMD. (B) FISH demonstrates three copies of 21q (green signals).

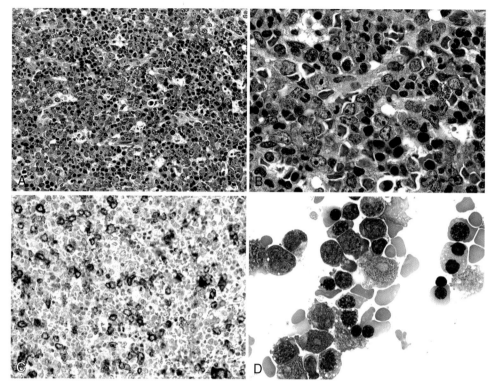

FIGURE 22.5 REFRACTORY ANEMIA WITH EXCESS BLASTS TYPE 2 IN A PATIENT WITH DOWN SYNDROME. Bone marrow biopsy section (A, low power; B, high power) is hypercellular and demonstrates increased blasts. Immnuohistochemical stain for CD34 (C) reveals numerous positive cells, and bone marrow smear (D) shows erythroid dysplasia and increased blasts.

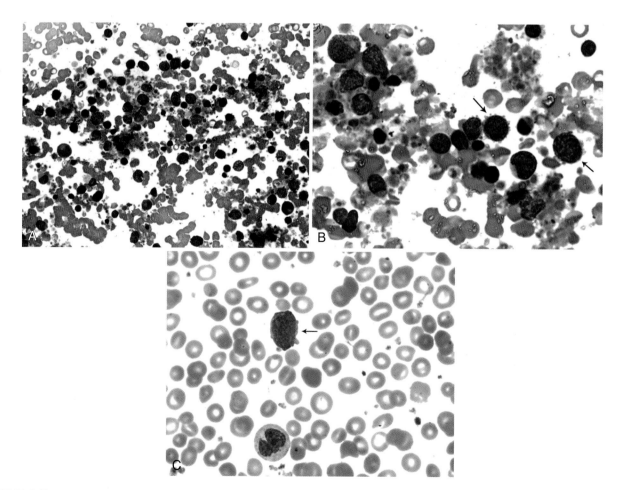

FIGURE 22.6 DOWN SYNDROME-ASSOCIATED ACUTE MEGAKARYOBLASTIC LEUKEMIA. Bone marrow smear (A, low power; B, high power) demonstrating numerous blasts, some with cytoplasmic blebs (nubs, arrows). A megakaryoblast is present in the blood smear (C, arrow).

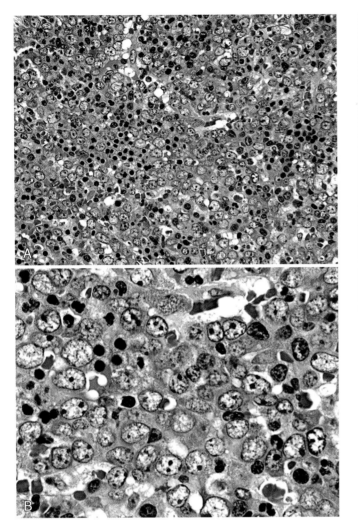

FIGURE 22.7 **DOWN SYNDROME-ASSOCIATED ACUTE MYELOID LEUKEMIA MINIMALLY DIFFERENTIATED.** Bone marrow biopsy section (A, low power; B, high power) demonstrating sheets of blast cells with fine nuclear chromatin and prominent nucleoli.

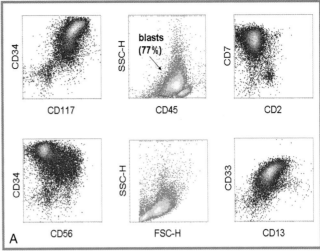
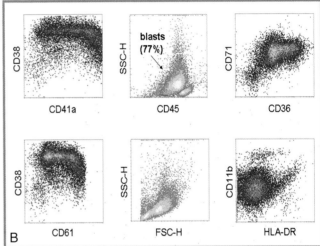

FIGURE 22.8 **FLOW CYTOMETRIC FINDINGS IN A PATIENT WITH DOWN SYNDROME-ASSOCIATED ACUTE MEGAKARYOBLASTIC LEUKEMIA.** Open gate displays (in blue) illustrate excess blasts (77% of the total) that are positive for CD45 (dim to moderate). (A) Density plots of the blast-enriched gate (in magenta) show that the blasts are abnormal with cross-lineage aberrancies of CD7 and CD56. These abnormal blasts are positive for CD13 (partial), CD33, CD34 (heterogeneous), and CD117. (B) Density plots of the blast-enriched gate (in magenta) show that the blasts are also positive for CD11b (partial, dim), CD36, CD38, CD41a (heterogeneous), CD61, and CD71. They are negative for HLA-DR.

The acute leukemia in the vast majority of cases is of the megakaryoblastic type. The DS-associated acute megakaryoblastic leukemia (AMKL) is characterized by favorable prognosis, no CNS involvement, and fewer cytogenetic abnormalities than those in non-DS children.

The event-free survival rate for DS-AMKL patients is about 80%. Almost none of the AML cases occurring in DS children older than 5 years are AMKL, and they do not show favorable prognosis. The event-free survival rate in this group of children is considerably worse and is about 15–20%.

- There is evidence of dysplastic changes in the erythroid and megakaryocytic series, such as megaloblastic changes and presence of micromegakaryocytes.
- Hepatic and splenic infiltration by blast cells is a frequent feature.

MORPHOLOGY

- Morphologic features range from MDS to a fulminant AML which, in most instances, is of the megakaryoblastic type (AMKL) (see Figures 22.5 to 22.7).
- Morphologic features of DS-associated AMKL are very similar to the TMD (see above). Blasts in the blood and/or the bone marrow account for ≥20% of the differential counts.

IMMUNOPHENOTYPE (FIGURE 22.8)

Flow cytometric features are generally similar to those described above in TAM. Blasts in DS-AMKL demonstrate:

- Expression of CD41, CD61, Factor VIII;
- Expression of CD36 and CD38;
- Highly common aberrancy of CD7 and variable CD56;

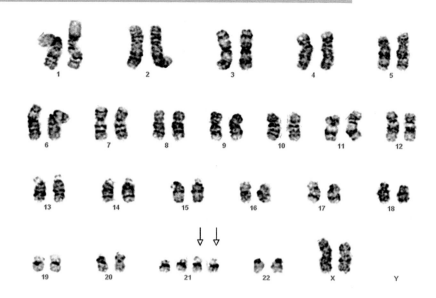

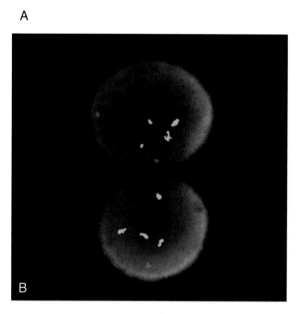

FIGURE 22.9 **DOWN SYNDROME-ASSOCIATED ACUTE MEGAKARYOBLASTIC LEUKEMIA.** (A) G-banded karyotype showing tetrasomy 21(48,XX,+21,+21). (B) FISH demonstrating four copies of 21q (green signals).

- Variable expression of and sometimes absent CD34 and/or CD117;
- Negativity for MPO and usually lack of HLA-DR;
- Negativity for or dim expression of CD45.

MOLECULAR AND CYTOGENETIC STUDIES

- Factors associated with DS leukemogenesis include somatic mutations in the chromosome X-linked hematopoietic factor, *GATA1*, which are specific and have been detected almost uniformly in all DS-AMKL and TAM cases and not detected in remission DS marrows, non-ML cases nor DS-ALL cases. The *GATA1* protein has altered transactivation activity, likely contributing to the uncontrolled proliferation of megakaryocytes.
- The acquisition of *GATA1* mutations is likely an early step in a multistep process of leukemogenesis and is thought to arise prenatally, based on several studies including their detection in fetal tissues.
- The cytogenetics of DS AML cases differs significantly from non-DS patients with AML. DS children with AML more frequently have trisomies 8, 11, and 21, dup(1p), del(6q), del(7p), dup(7q), and del(16q). A tetrasomy of chromosome 21 from a child with DS-AML is demonstrated in Figure 22.9.
- Translocations commonly seen in non-DS AML, such as t(8;21), t(15;17), inv(16), 11q23 rearrangements, are rare in patients with DS. For DS children older than 4 years of age who develop AML, the cytogenetic features, molecular biology findings, and response to therapy significantly diverge from those in younger patients, and are similar to the ones found in non-DS patients with AML

DIFFERENTIAL DIAGNOSIS

There are significant overlapping clinicopathological features between DS-associated TAM and AMKL. The major differences are that TAM is transient and is seen at an earlier age (usually in the first month of infancy), while AMKL is not transient and develops a few years later, often preceded by a prolonged MDS-like syndrome.

Additional Resources

Avet-Loiseau H, Mechinaud F, Harousseau JL: Clonal hematologic disorders in Down syndrome: A review, *J Pediatr Hematol Oncol* 17:19-24, 1995.

Lange BJ, Kobrinsky N, Barnard DR, et al: Distinctive demography, biology, and outcome of acute myeloid leukemia and myelodysplastic syndrome in children with Down syndrome: Children's Cancer Group Studies 2861 and 2891, *Blood* 91:608-615, 1998.

Crispino JD: GATA1 mutations in Down syndrome: implications for biology and diagnosis of children with transient myeloproliferative disorder and acute megakaryoblastic leukemia, *Pediatr Blood Cancer* 44:40-44, 2005.

Dixon N, Kishnani PS, Zimmerman S: Clinical manifestations of hematologic and oncologic disorders in patients with Down syndrome, *Am J Med Genet C Semin Med Genet* 142C:149-157, 2006.

Kanezaki R, Toki T, Terui K, et al: Down syndrome and GATA1 mutations in transient abnormal myeloproliferative disorder: mutation classes correlate with progression to myeloid leukemia, *Blood* 116:4631-4638, 2010.

Kruger B: Transient myeloproliferative disorder associated with trisomy 21, *Neonatal Netw* 26:7-19, 2007.

Kurahashi H, Hara J, Yumura-Yagi K, et al: Transient abnormal myelopoiesis in Down's syndrome, *Leuk Lymphoma* 8:465-475, 1992.

Malinge S, Izraeli S, Crispino JD: Insights into the manifestations, outcomes, and mechanisms of leukemogenesis in Down syndrome, *Blood* 113:2619-2628, 2009.

Lymphoblastic Neoplasms— B-Lymphoblastic Leukemia/Lymphoma

Lymphoblastic neoplasms are leukemias or lymphomas of precursor lymphoid cells—blast cells committed to lymphoid differentiation. Clinically, they are divided into two major categories: lymphoblastic lymphoma (LBL) and acute lymphoblastic leukemia (ALL). Typically, LBL represents a neoplastic process primarily involving extramedullary lymphoid tissues with or without bone marrow involvement, and ALL primarily involves bone marrow (≥20% blasts) with or without extramedullary lesions. Since LBL and ALL share considerable clinicopathologic features, and present similar biological properties, they are considered the same disease. In the WHO classification, LBL and ALL are lumped together as precursor lymphoid neoplasms and are divided into two major categories:

1. B-lymphoblastic leukemia/lymphoma
2. T-lymphoblastic leukemia/lymphoma.

B-lymphoblastic leukemia/lymphoma (B-ALL/LBL) may initially present itself as acute lymphoblastic leukemia (ALL) with bone marrow and/or blood involvement, or lymphoblastic lymphoma (LBL) with the involvement of the lymphoid and/or other extramedullary tissues. In a significant proportion of cases, however, both bone marrow and extramedullary tissues are involved. ALL and LBL are regarded as different clinical presentations of the same disease. Common clinical presentations in B-ALL patients include anemia, thrombocytopenia and/or granulocytopenia with variable leukocyte counts, frequent bone pain, and organomegaly. Patients with B-LBL without leukemia mostly are asymptomatic and demonstrate limited lymphadenopathy, particularly in the head and neck regions.

The incidence of B-ALL in the United States approaches 3 per 100,000. The peak incidence is between 2 and 5 years of age, affecting boys more than girls. Acute lymphoblastic leukemia/lymphoma is the most common form of cancer in children, comprising about 30% of all childhood malignancies. Up to 85% of childhood ALLs and 40% of childhood lymphomas are of the precursor B-cell type. B-lymphoblastic lymphoma is uncommon in adults and accounts for <1% of lymphomas. Lymphoblastic lymphoma is defined by the presence of <25% blasts in the bone marrow and evidence of a mediastinal mass or lymphadenopathy. Lymphadenopathy and cutaneous involvement are the most frequent presentations. Skin lesions may be multifocal.

The presenting clinical symptoms are often non-specific and secondary to bone marrow/lymphoid tissue infiltration and pancytopenia, such as fever, bleeding, bone pain, and lymphadenopathy.

Favorable prognostic factors include:

- Age younger than 1 or older than 10 years;
- White blood cell count in normal range or >50,000/µL;
- Hyperdiploidy: >50 chromosomes in the karyotype;
- t(12;21)(p13;q22.3) (*ETV6-RUNX1* fusion).

The overall response to therapy is significantly better in children than in adults. The current 5-year survival rate for B-ALL in children is approaching 85%, but nearing 95% in children with favorable prognostic factors.

B-ALL/LBL disorders are divided into two major categories:

1. B-lymphoblastic leukemia/lymphoma with recurrent genetic abnormalities
2. B-lymphoblastic leukemia/lymphoma, not otherwise specified.

In general, both categories share similar morphologic and cytochemical features.

Common Morphologic and Cytochemical Features

- The bone marrow biopsy and clot sections are usually hypercellular and are diffusely infiltrated by sheets of uniformly appearing blast cells (Figures 23.1 to 23.3). These cells have scanty basophilic cytoplasm with round, oval, or indented nuclei, finely dispersed nuclear chromatin, and prominent or

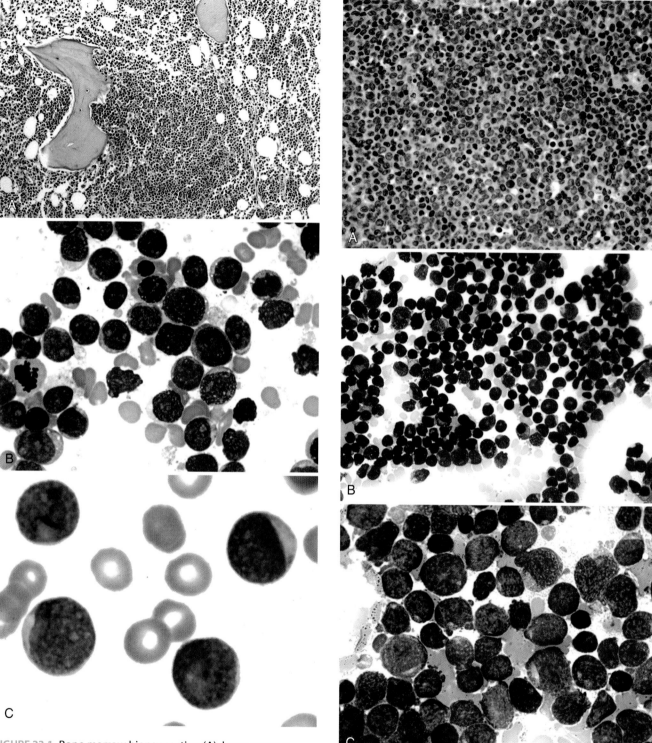

FIGURE 23.1 Bone marrow biopsy section (A), bone marrow smear (B) and blood smear (C) demonstrating lymphoblasts in a child with B lymphoblastic leukemia.

FIGURE 23.2 **B LYMPHOBLASTIC LEUKEMIA.** Bone marrow biopsy section (A) and bone marrow smear (B, low power; C, high power) demonstrating sheets of lymphoblasts.

indistinct nucleoli. Mitotic figures are variable, but often easily detectable.
- Occasionally, ALL patients may initially present with a hypoplastic marrow and pancytopenia mimicking aplastic anemia. The hypoplastic marrow contains a variable number of lymphoblasts (Figure 23.4) and may eventually become packed with lymphoblasts with an obvious acute leukemia picture.

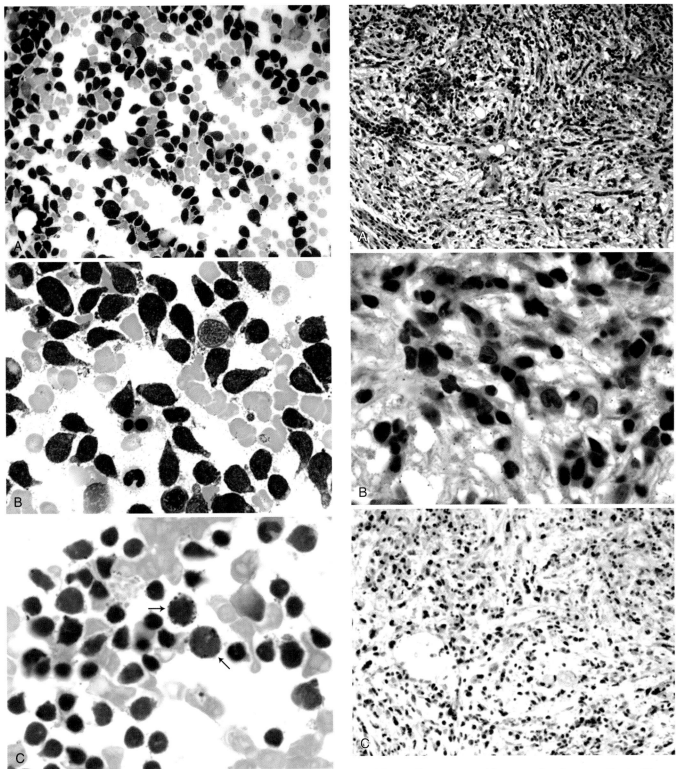

FIGURE 23.3 Bone marrow smear demonstrating the hand-mirror variant of lymphoblastic leukemia (A, low power; B, high power). PAS stain (C) shows coarse cytoplasmic PAS-positive granules.

FIGURE 23.4 Bone marrow biopsy section from a patient with hypoplastic lymphoblastic leukemia (A, low power; B, high power). Immunohistochemical stain shows numerous TdT positive cells (C).

- Bone marrow fibrosis and osteoporosis are sometimes present. Fibrosis may be mild or extensive, focal or diffuse, and may lead to unsuccessful bone marrow aspiration (dry tap). It is more frequent in B-ALL than in T-ALL.

- Bone marrow necrosis may be present in some cases. Necrosis is usually of the coagulative type with the preservation of the basic outline of the necrotic cells.
- Bone marrow smears and touch preparations show numerous blast cells which are often small with scanty non-granular blue

Table 23.1
Recurrent Genetic Abnormalities, Immunophenotype, and Prognosis In B-Lymphoblastic Leukemia/Lymphomas

Cytogenetics	Genes	Frequency	Immunophenotype	Prognosis
t(9;22)(q34;q11.2)	BCR-ABL1	25%, adults; 3–5%, children	CD10+, CD19+, CD22+, icCD79a+, TdT+, CD9+, CD34+, CD20±, and often CD13+ and/or CD33+	Poor
t(4;11)(q21;q23)	AFF1/MLL[1]	5%, children	CD19+, CD22+, icCD79a+, TdT+, CD9+, CD34+, CD15+, CD10−, CD24−	Poor
t(1;19)(q23;p13.3)	PBX1[2]/TCF3	6%, children	CD10+, CD19+, CD22+, icCD79a+, TdT+, CD9+, CD20±, CD34−	Poor[3]
t(12;21)(p13;q22)	ETV6/RUNX1	25%, children; 3–4%, adults	CD10+, CD19+, CD22+, icCD79a+, TdT+, CD34+, CD20±, CD9−	Favorable
Hyperdiploid		20–25%	CD10+, CD19+, CD22+, icCD79a+, TdT+, CD9+, CD34+, CD20±, CD45−	Favorable
Hypodiploidy		5%	CD10+, CD19+, CD22+, icCD79a+, TdT+, CD9+, CD34+, CD20±	Poor

Alternative designations:
[1] ALL or HRX
[2] PRL
[3] The unbalanced form is associated with favorable prognosis.

cytoplasm, fine chromatin, and indistinct nuclei (see Figures 23.1 and 23.2). But less frequently the blast cells are larger and more pleomorphic, show variable amounts of cytoplasm which may display vacuolization or azurophilic granules (5–10% of the cases), and one or more prominent nucleoli. Some cases may show cytoplasmic tails (pseudopods) referred to as *hand-mirror* cells (see Figure 23.3).

- In a small proportion of patients with ALL (4–7%), particularly in children, lymphoblasts contain coarse azurophilic granules.
- Lymphoblasts may also be present in the peripheral blood smears in variable numbers (see Figure 23.1C). They account for the majority of leukocytes in patients with WBC >10,000/μL. Approximately 20% of patients at the time of diagnosis present with a leukocyte count exceeding 50,000/μL.
- Anemia, granulocytopenia, and/or thrombocytopenia are common features.
- The involvement of lymph nodes and other tissues is usually diffuse with the total or partial effacement of normal architecture and morphologic features, similar to those described earlier in the bone marrow biopsy sections.
- Lymphoblasts may show coarse purple cytoplasmic granules with PAS cytochemical stains (see Figure 23.3C).

B-Lymphoblastic Leukemia/Lymphoma with Recurrent Genetic Abnormalities

This is a group of B-ALL/LBL that are characterized by recurrent genetic abnormalities and are associated with distinctive clinical and biological properties (Table 23.1). In general, this group does not present unique morphologic or cytochemical features.

B-LYMPHOBLASTIC LEUKEMIA/LYMPHOMA WITH t(9;22)(q34;q11.2)

B-ALL/LBL with t(9;22)(q34;q11.2) (*BCR-ABL1*) is observed in <5% of children and about 25% of adults. It is the most frequent genetic rearrangement in adult ALL. The prognosis is poor both in children and adults.

Immunophenotype (Figure 23.5)

- Blasts express CD10, CD19, CD34, intracellular CD22 and/or CD79a, plus TdT.
- There are often myeloid aberrancies including CD13, CD15, and CD33; but CD117 is typically absent.
- CD25 is highly associated with this entity. However, it is not routinely included in the acute leukemia panels by many flow cytometry laboratories.

Molecular and Cytogenetic Studies

- The t(9;22) can be easily identified by standard karyotyping, by *BCR-ABL1* FISH probe set commonly used in CML diagnosis (Figure 23.6), and by a specially designed dual-color FISH probe (*BCR-ABL1* Extra Signal) which identifies the *m-bcr* breakpoint-related rearrangement (Figure 23.7).
- Molecular studies of *BCR-ABL1* fusion in t(9;22) reveal two distinct subgroups giving rise to two types of fusion proteins: 185–190 kDa and 210 kDa. In some cases t(9;22) is complex and involves additional rearrangements (Figure 23.6).
- Approximately 25–30% of adult ALL patients show *BCR-ABL1* rearrangement similar to that observed in CML. The breaks usually occur within the major-bcr region (referred to as M-bcr) of the *BCR* gene and the juxtaposition to the *ABL1* gene, leading to a chimeric gene which encodes for a 210 kDa fusion protein (p210).
- In children and approximately 50–70% of adults with t(9;22), the breakpoint occurs further downstream in the *BCR* gene, referred to as the minor breakpoint cluster region (m-bcr). This fusion gene encodes smaller fusion proteins ranging from 185 to 190 kDa (p185).

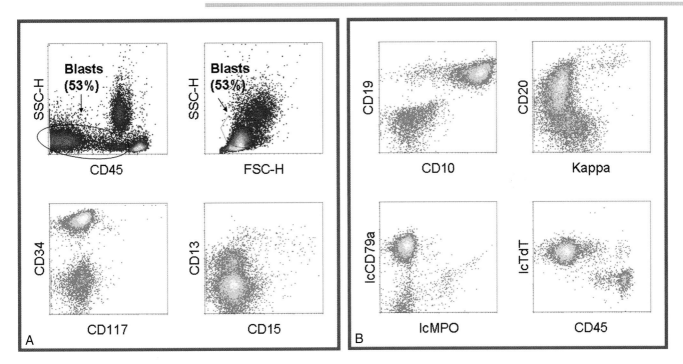

FIGURE 23.5 **FLOW CYTOMETRIC FINDINGS OF B-ALL/B-LBL WITH t(9;22).** Open gate density plots by CD45 gating and the scatter gate (in magenta) reveal excess abnormal blasts (53% of the total) (with two subsets). A major subset of the blasts is negative for CD45, while the minor one expresses dim CD45. (A) Density plots of the blast-enriched gate (in blue) show abnormal B lymphoblasts expressing CD34 with cross-lineage myeloid aberrancy of partial CD13. (B) These abnormal B lymphoblasts are also positive for CD10 (bright), CD19, CD20 (partial, dim), plus intracellular CD79a and TdT.

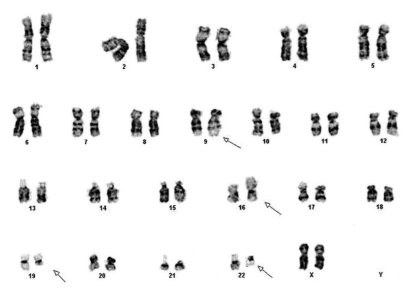

FIGURE 23.6 A complex 4-way translocation: 46,XX, t(9;22;16;19)(q34;q11.2;p13.3;q13.1).

- At the molecular level, the *BCR-ABL1* fusion event is usually detected quantitatively and extremely sensitively by real-time RT-PCR (Figure 23.8).
- The expanse of the breakpoint cluster region in ALL (as well as in CML) is too large to be covered reliably by a DNA-targeted primer set. Instead, the target chosen is the *BCR-ABL1* fusion transcript, from which long introns have been spliced out to yield a target of more manageable size. Naturally, this RNA-based test requires a reverse transcriptase (RT) step in order to generate a DNA target which can then be amplified by PCR.
- The *BCR-ABL1* quantity detected can be expressed as an absolute value or, more commonly, in relative terms, compared with the mRNA of a standard "housekeeping" gene such as *G6PDH*. Sensitivities of *BCR-ABL1* mRNA per 10,000 or 100,000 control gene mRNAs allow a rough extrapolation of the number of leukemic cells relative to normal cells in the specimen.
- Sensitivities at this level are suitable for detection of minimal residual disease in treated patients and monitoring of tumor loads over long-term therapy.

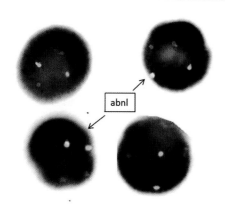

FIGURE 23.7 FISH on interphase nuclei with the *mBCR-ABL1* ES Dual Color Translocation Probe showing one green (native BCR), one large orange (native ABL1), one smaller orange (ES) and one fused orange/green (20IGIF) signal pattern in the Ph-positive abnormal cells.

B-LYMPHOBLASTIC LEUKEMIA/LYMPHOMA WITH t(v;11q23)

This is a group of B-ALL/LBL that show translocation between the *MLL* gene and other genes, such as (4;11) (q21;q23) (*AFF1-MLL*). It accounts for about 5% of acute leukemias and is the most common leukemia in infants under age 1(Figure 23.11). Lymphomatous manifestations are infrequent. The prognosis is poor. Patients have a high leukocyte count, often >200,000/μL.

Immunophenotype (Figure 23.10)

- Blasts resemble pro-B-cell immunophenotype, typically negative for CD10.
- Positive for CD19, CD34, intracellular CD22 and/or CD79a, plus TdT.
- Common myeloid aberrancies, especially CD15.

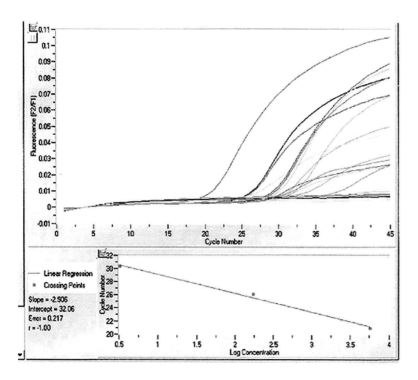

FIGURE 23.8 RT-PCR analysis of a series of patients showing presence of the *bcr-abl* fusion mRNA target, using the Roche LightCycler instrument. In general, the lower the PCR cycle number (x-axis) at which the amplification reaches its logarithmic phase, the higher the amount of starting *bcr-abl* target sequence in the specimen. Very late-rising and/or low-rising signals should be interpreted with caution.

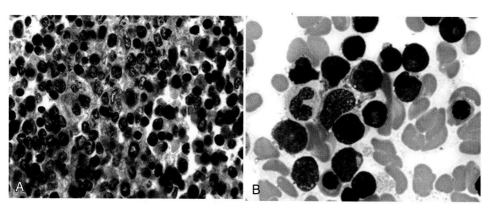

FIGURE 23.9 Bone marrow clot section (A) and smear (B) from a case of acute lymphoblastic leukemia with t(4;11). The blasts are of various sizes and show scanty cytoplasm.

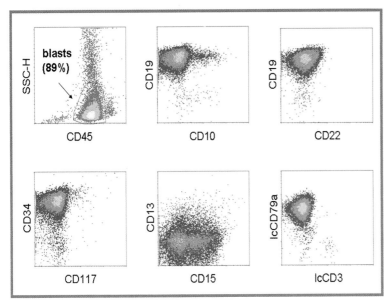

FIGURE 23.10 **FLOW CYTOMETRIC FINDINGS OF B-ALL/LBL WITH t(4;11).** Open gate density plot display by CD45 gating (in blue) reveals excess blasts (89% of the total) that are CD45 dimly positive. Density plots of the blast-enriched gate (in purple) demonstrate that the B-lymphoblasts express CD19, CD22 (partial, dim), CD34, CD38 (not shown), plus intracellular CD79a and TdT (not shown). In addition, there are myeloid aberrancies of partial CD13 and CD15. The blasts are negative for CD10, which is a characteristic feature seen in this subgroup of B-ALL/LBL.

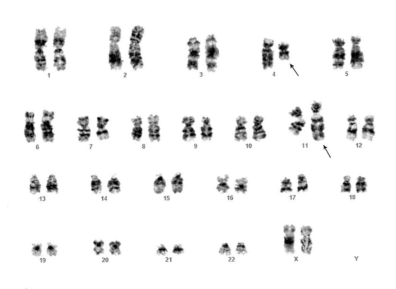

FIGURE 23.11 (A) G-banded karyotype of bone marrow of a patient with acute lymphoblastic leukemia demonstrating t(4;11)(q21;q23). (B) FISH analysis of bone marrow of a patient with acute lymphoblastic leukemia. The left image represents a normal control, and the curved arrows on the right image indicate 11q23 (MLL) rearrangement.

Molecular and Cytogenetic Studies

- Rearrangements of 11q23 are found in 70–80% of infant leukemias. In children older than 1 year the *MLL* gene rearrangements incidence is approximately 5%.
- t(4;11)(q21;q23) is the most common translocation involving 11q23 (2%), and can be detected by conventional cytogenetics, FISH, RT-PCR, or Southern blot techniques. It is associated with an unfavorable prognosis with a high relapse rate, although children 2–9 years old appear to have a much better prognosis.
- The t(4;11)(q21;q23) forms a fusion between genes *AFF1* (4q21), a transcription factor, and *MLL* (11q23) (Figure 23.11) and includes portions of the 5′ region of *MLL* and variable portions of *AFF1*. The breakpoints in *MLL* are usually in a bcr 8.3 kb region between exons 5 and 11.

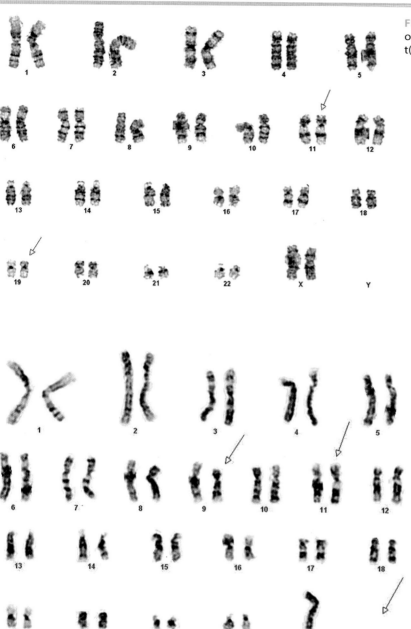

FIGURE 23.12 G-banded karyotype of bone marrow of a patient with acute lymphoblastic leukemia with t(11;19)(q23;p13.3).

FIGURE 23.13 G-banded karyotype of bone marrow of a patient with acute lymphoblastic leukemia exhibiting a loss of Y and t(9;11)(p23;q23).

- The fusion gene is transcribed into a hybrid mRNA, which can be detected by RT-PCR techniques for establishing the diagnosis or monitoring the residual disease.
- t(11;19)(q23;p13.3) is the second most common 11q23 rearrangement and also associated with unfavorable prognosis (Figure 23.12). This translocation results in a fusion between the *MLL* (11q23) and the *ENL* (eleven-nineteen-leukemia protein at 19p13) or the *MLLT1* gene, and although somewhat cryptic, can be detected by karyotype, and can be confirmed by FISH with the 11q23 "split" probe analysis.
- Other abnormalities involving 11q23 include t(9;11)(p22;q23) (*AFF1-MLL* fusion), and deletions of 11q23 which are common in Adult ALL (Figures 23.13 to 23.15).
- Because over 80 gene-partners are associated with the *MLL* gene, it is typically recommended to use an *MLL* "split" probe to screen MLL-related rearrangements. (Figure 23.14B).

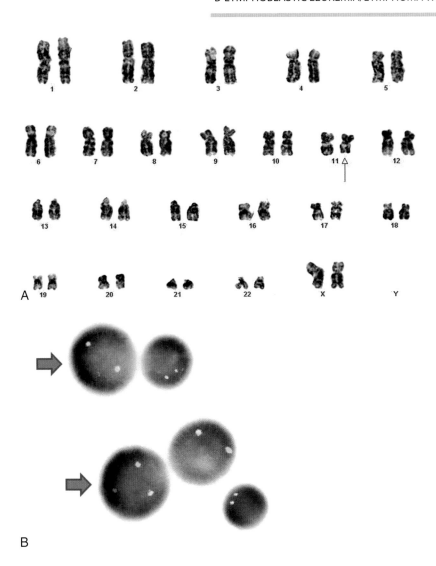

FIGURE 23.14 (A) G-banded karyotype of bone marrow of a patient with acute lymphoblastic leukemia demonstrating del(11q23). (B) FISH analysis of dual-color MLL probe showing split signals consistent with 11q23 rearrangement.

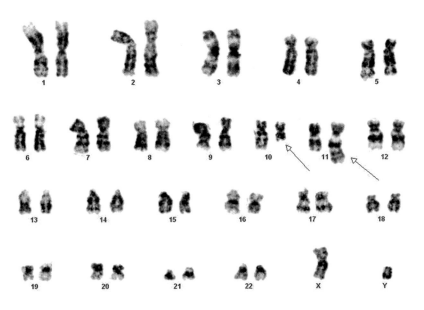

FIGURE 23.15 G-banded karyotype of bone marrow of a patient with acute lymphoblastic leukemia demonstrating t(10;11)(q22;q23).

B-LYMPHOBLASTIC LEUKEMIA/LYMPHOMA WITH t(12;21)(p13;q22)

The t(12;21) leads to the fusion of the *TEL1* (*ETV6*) gene on chromosome 12 with the *RUNX1* (*AML1*) gene on chromosome 21. This translocation is one of the most frequent genetic abnormalities in childhood ALL, accounting for about 25% of cases. It is less frequent in adults, occurring in 3–4% of patients with B-ALL/LBL. The disease has a favorable prognosis with a cure rate of >90%.

Immunophenotype (Figure 23.16)

- Blasts express CD10, CD19, CD34, intracellular CD22 and or CD79a, plus TdT.
- Myeloid aberrancies are common, especially CD13.
- Characteristically absent CD9 and CD20.

Molecular and Cytogenetic Studies

- Despite its high frequency, because of the similarity of the size and banding patterns of 12p and 21q, the t(12;21) goes undetected by routine karyotyping. Therefore, RT-PCR or FISH are the recommended techniques for the detection of this translocation (Figure 23.17)
- The dual-color dual-fusion FISH probes, in addition to finding the expected 12;21 translocation, can also detect extra *RUNX1* signals without the *ETV6-RUNX1* fusion, indicative of the existence of cells with a hyperdiploid karyotype or *ETV6* deletion (12p-) (Figure 23.18A) or *ETV6* gene amplification.
- It is important to recognize the distinction between polysomy 21 (often seen in high hyperdiploidy (Figure 23.18B) with good prognosis) and *RUNX1* amplification (Figure 23.19) (associated with poor prognosis). In the latter the gene signals are clustered, numbering greater than five or more copies.

B-LYMPHOBLASTIC LEUKEMIA/LYMPHOMA WITH t(1;19)(q23;p13.3)

The t(1;19)(q23;p13.3) leads to a chimeric gene as the result of the fusion of the *PBX1* and trancription factor 3 (*TCF3*, formerly *E2A*) genes (Figure 23.20). This translocation is primarily seen in children and accounts for about 6% of B-ALL cases. The prognosis is poor, but modern therapeutic modalities have improved the clinical outcome.

Immunophenotype

- Blasts resemble pre-B-cell immunophenotype with presence of cytoplasmic μ heavy chains.
- Blasts also express CD10, CD19, intracellular CD22 and or CD79a, plus TdT.
- There is typically strong expression of CD9.
- Expression of CD34 is variable or absent.

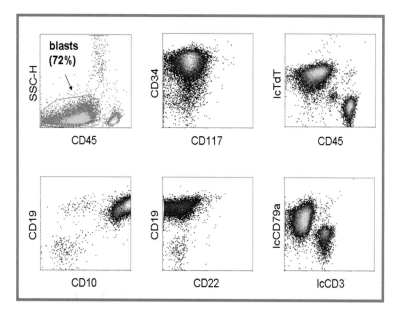

FIGURE 23.16 FLOW CYTOMETRIC FINDINGS OF B-ALL/LBL WITH t(12;21). Open gate density plot by CD45 gating (in blue) illustrates excess blasts (72% of the total) that are CD45 negative. Density plots of the blast enriched gate (in magenta) show abnormal B lymphoblasts expressing CD10 (bright), CD19, CD22 (partial), CD34, plus intracellular CD79a (dim) and TdT. Blasts were negative for CD20 (not shown). Negativity for CD20 and CD9 is typical for this subgroup.

B-LYMPHOBLASTIC LEUKEMIA/LYMPHOMA WITH RECURRENT GENETIC ABNORMALITIES

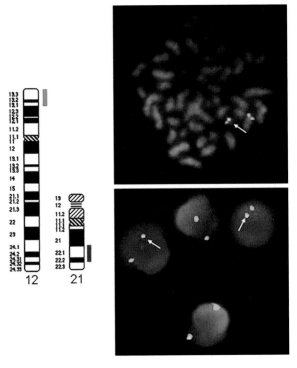

FIGURE 23.17 FISH analysis of bone marrow of a patient with acute lymphoblastic leukemia demonstrating t(12;21)(p13;q22) (arrows).

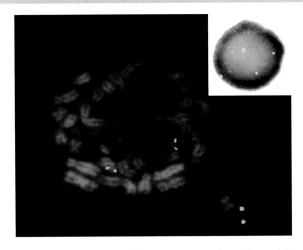

FIGURE 23.19 FISH analysis of bone marrow of a patient with acute lymphoblastic leukemia demonstrating *ETV6* (green signals) and amplification of *RUNXT1* (red signals and inset).

B-LYMPHOBLASTIC LEUKEMIA/LYMPHOMA WITH t(5;14)(q31;q32)

B-lymphoblastic leukemia/lymphoma with t(5;14)(q31;q32) (*IL3-IGH@*) is a rare disease (<1% of ALL cases) reported both in children and adults. One of the morphologic characteristic features of this disorder is the presence of a reactive eosinophilia in the bone marrow and peripheral blood (Figure 23.21).

Immunophenotype

There are no specific findings, and blasts are usually positive for CD10 and CD19.

Molecular and Cytogenetic Studies

- The t(5;14) may be the sole anomaly or accompanied by other anomalies and can be detected by karyotype studies only. This translocation with the break in the promoter region of *IL3* and in the JH region of *IGH@* results in the overexpression of IL3.

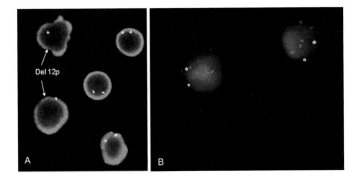

FIGURE 23.18 FISH analysis of bone marrow of a patient with acute lymphoblastic leukemia demonstrating (A) deletion of *ETV6* [del(12p13)] and (B) polysomy (five copies) of *RUNX1* (red signals).

Molecular and Cytogenetic Studies

- Two forms of t(1;19) have been reported: a reciprocal t(1;19) translocation and an unbalanced form written as der(19)t(1;19)(q23;p13) (see Figure 23.20). This translocation is one of the more common chromosomal abnormalities (5%) in B-cell precursor ALL and results in a fusion between *PBX1* (1q23) and the *TCF3* (*E2A*)(19p13) genes
- t(1;19) can be demonstrated by conventional cytogenetics, FISH, or RT-PCR.

B-LYMPHOBLASTIC LEUKEMIA/LYMPHOMA WITH HYPERDIPLOIDY

Hyperdiploidy (chromosomes >46) is reported in up to 50% of children with B-ALL, but it is far less frequent in adult patients. Hyperdiploidy is divided into two subcategories: (1) hyperdiploidy with ≤50 chromosomes (between 46 and 50 chromosomes) and (2) hyperdiploidy

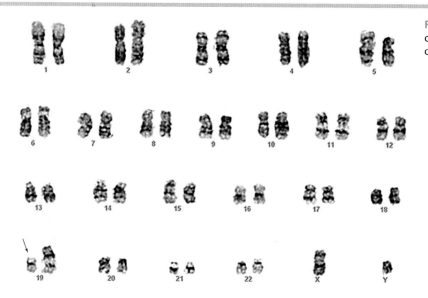

FIGURE 23.20 G-banded karyotype of bone marrow of a patient with acute lymphoblastic leukemia demonstrating der(19)(1;19)(q23;p13.3).

with >50 (usually between 51 and 65) chromosomes (Figure 23.22). B-ALL patients with hyperdioploidy have an excellent prognosis with a cure rate of >90%.

Immunophenotype (Figure 23.24)

- Blasts are positive for CD10, CD19, CD34, intracellular CD22 and or CD79a, plus TdT.
- Negative to very dimly positive for CD45.
- Myeloid aberrancies are common.

Molecular and Cytogenetic Studies

- Patients with hyperdiploidy of >50 chromosomes exhibit karyotype with numbers typically ranging from 50 to 70 chromosomes. There is no gain of specific chromosomes that is common to all hyperdiploidy, although the most common chromosomal additions include chromosomes 21 (often multiple copies), 4, 6, 10, 14, 17, 18, 20, and X. Hyperdiploidy indicates favorable prognosis, particularly in association with trisomy of chromosomes 4, 6, and 10 (see Figure 23.22B).
- Other structural aberrations, sometimes observed in hyperdiploid karyotypes (referred to as pseudo-hyperdiploid) include duplication of 1q, deletion of 6q and isochromosome 17q, and have no known prognostic impact.
- Hyperdiploidy is commonly detected by conventional karyotype studies, although these can be complemented with targeted FISH probe analyses.

B-LYMPHOBLASTIC LEUKEMIA/LYMPHOMA WITH HYPODIPLOIDY

This condition refers to B-ALL/LBL cases with chromosome numbers of <45 and DNA index of <1 (Figure 23.23). It accounts for about 5% of ALL cases. Several reports indicate poor prognosis in association with hypodiploidy in both children and adult patients with ALL.

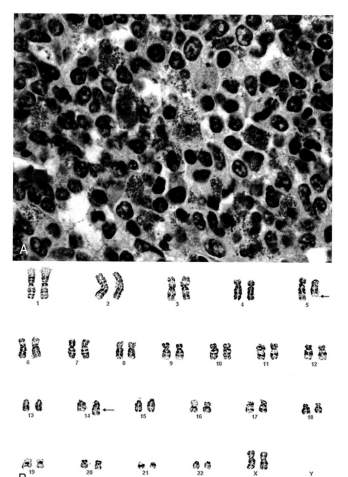

FIGURE 23.21 (A) Bone marrow biopsy section of a patient with B-acute lymphoblastic leukemia and eosinophilia. (B) G-banded karyotyping shows t(5;14)(q31;q32).

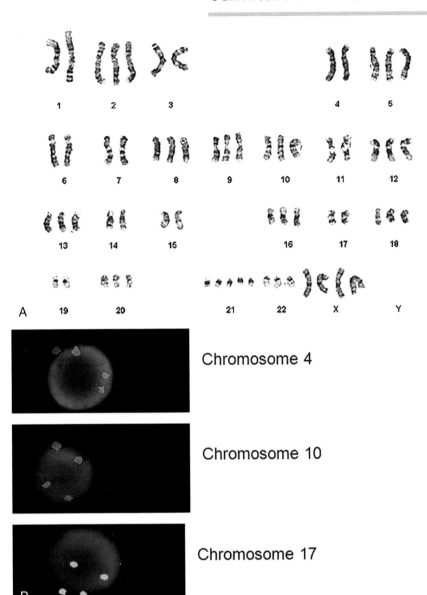

FIGURE 23.22 (A) Hyperdiploid karyotype (62,XX, +X,+X,+2,+5,+8,+9,+10,+12,+13,+16,+18,+20,+21 × 3,+22. (B) FISH analysis demonstrating gains (4 copies) of chromosomes 4, 10, and 17.

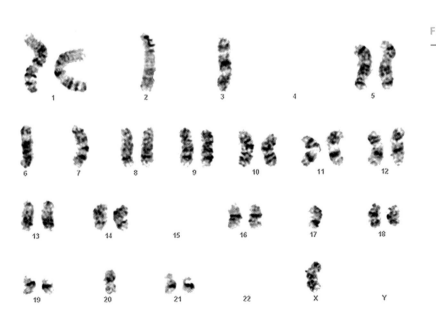

FIGURE 23.23 Karyotype 33,X,−X,−2,−3,−4, −4, −6, −7, −15, −15, −17, −20, −22,−22.

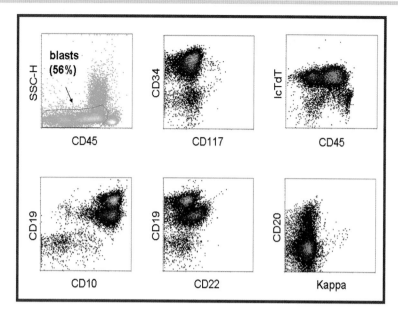

FIGURE 23.24 **FLOW CYTOMETRIC FINDINGS OF B-ALL/LBL WITH HYPERDIPLOIDY.** The blasts account for about 56% of the total analyzed events on the open gate density plot display by CD45 gating (in green), revealing two distinct subsets: one is negative for CD45, while the other is very dimly positive for CD45. Density plots of the blast-enriched gate (in red) also demonstrate two discrete subsets of B-lymphoblasts with variable levels of CD19 and CD22. The blasts are also positive for CD10, CD20 (partial, dim), CD34, as well as intracellular CD79a (not shown) and TdT. In addition, there is aberrant myeloid expression of partial CD13 (not shown).

Immunophenotype

Immunophenotypic findings are similar to the above (hyperdiploid B-LBL). There are no specific features.

Molecular and Cytogenetic Studies

- Near haploid karyotypes show a number below 30 with a typical diploid range of 23–28 chromosomes.
- The loss of specific chromosomes is not random; it appears that some chromosomes are preferentially retained (chromosomes 6, 8, 10, 14, 18, 21, X) (see Figure 23.23)
- Occasionally, these cases also exhibit a diploid chromosome number which is actually a duplication of the near-haploid clone and can be indistinguishable from a normal 46 chromosome karyotype.

B-Lymphoblastic Leukemia/ Lymphoma, Not Otherwise Specified

This term is used for B-ALL/LBL cases that are not associated with recurrent genetic abnormalities except rearrangement of the *IGH@* gene.

Immunophenotype (Figure 23.24)

- Expression of CD10, CD19, intracellular CD22 and or CD79a, plus TdT;
- Variable expression of CD34;
- Expression of CD20 in some cases, and the expression intensity is almost always partial and dim;
- Common myeloid aberrancies with CD13, CD15, and CD33;
- Negativity for surface light chains.

Molecular and Cytogenetic Studies

- B-ALL/LBL, not otherwise specified is not associated with recurrent genetic abnormalities except for *IGH@* rearrangement which has been detected in the vast majority of the cases. Because they are precursor B-cell lesions, they often will not have completed the full ontological sequence of rearrangements, from heavy to light chain. In such cases one may see rearrangement of the heavy chain genes (*IGH@*) but not of the light chains (*IGK@* or *IGL@*). Therefore, a greater proportion of these clonal cases will be detected if one uses probes or primers specific for the *IGH* region.
- The older Southern blot methods, using DNA probes typically directed at the J-region genes of the *IGH* region, have largely been replaced by PCR approaches. However, as this region is quite large, it cannot be covered comprehensively by most series of PCR primer pairs, so some rearrangements in other areas will go undetected (false negatives).
- Also, antibodies can further diversify themselves through somatic hypermutation of the variable genes, and if any of these changes occur at a primer hybridization site, they can further reduce the efficiency and sensitivity of the assay. For these reasons, many laboratories perform an initial screen by PCR, but if that is negative, reflex to the Southern blot procedure. Most PCR approaches detect 70–80% of B-cell neoplasms.
- Presence of a clonal population will produce a predominant (or at least visible) DNA pattern (peak) above the background polyclonal signal (Figure 23.25).
- A significant proportion of B-ALL/LBL, in addition to *IGH@* rearrangement, also demonstrates T-cell receptor (*TCR*) gene rearrangement.

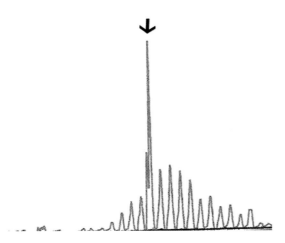

FIGURE 23.25 PCR analysis for immunoglobulin heavy chain clonality. Results are shown for the framework 1 primer set only, illustrating a clonal peak (arrow) superimposed on a polyclonal background population, a pattern often seen in leukemia specimens.

Differential Diagnosis

The differential diagnosis of B-ALL includes hematogone hyperplasia; T-ALL; various types of acute myeloid leukemia, such as minimally differentiated AML, AML without maturation, and megakaryoblastic leukemia; and metastatic small round cell tumors, such as neuroblastoma.

Hematogones account for 5–10% of the bone marrow cells in children and <5% of the bone marrow cells in adults, but they may be increased. Hematogone hyperplasia is seen in various conditions, such as iron deficiency anemia, immune-associated thrombocytopenia, and following cytotoxic chemotherapy (Figure 23.26). The distinction of hematogones from B lymphoblasts in BALL represents one of the most common challenges in the practice of hematopathology. For example, in evaluation of post-therapy effects of BALL, sometimes it can be extremely difficult to determine residual or minimal residual disease versus normal regenerative marrow, since both hematogones and residual BALL cells may display blastic morphology, and express CD34, CD10, CD19, and TdT. However, hematogones are normal bone marrow B-cell precursors and hence, by definition, consist of a heterogeneous group of B-cells that are at various stages of development and maturation. Therefore, despite the similarities, there are distinct morphologic as well as immunophenotypic features which set hematogones apart from BALL cells, the abnormal B-cell precursors.

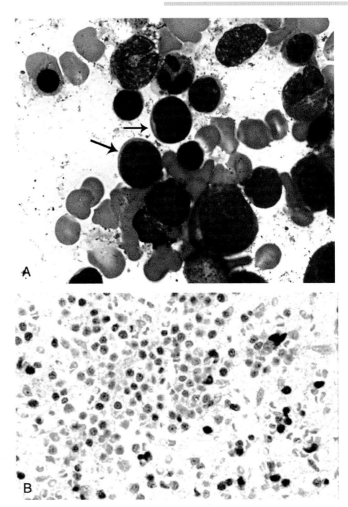

FIGURE 23.26 (A) Bone marrow smear showing scattered hematogones (arrows). (B) Immunohistochemical stain for TdT on a bone marrow biopsy section demonstrating scattered positive cells representing hematogones.

Immunophenotypic analysis by MFC is the most critical and indispensable method in distinguishing hematogones from B-lymphoblasts, because of the dynamic and unique immunophenotypic patterns of hematogones. They can be arbitrarily divided into three stages of development based upon their flow cytometric features (Figure 23.27), representing early, intermediate, and late precursors. As the B-cells mature from early to late precursors, they show increased expression of CD19 and CD45. The early precursors also express TdT and CD34, both of which are lost in the intermediate forms. At the intermediate stage, coexpression of CD10 and CD19 is most discrete and prominent, plus CD20 begins at a partial and dim level without surface light chains. The late precursors lose CD10, and acquire CD20 at moderate intensity along with expression of surface kappa or lambda light chains.

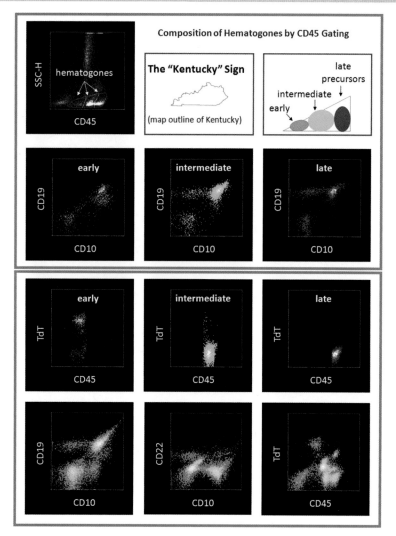

FIGURE 23.27 **CHARACTERISTIC IMMUNOPHENOTYPIC PATTERNS OF HEMATOGONE HYPERPLASIA BY FLOW CYTOMETRY.** Density plots of open gate displays (in red and blue) demonstrate the dynamic heterogeneity of hematogones. By CD45 gating (in red), hematogones are comprised of early (very low side scatter; CD45 very dim), intermediate (higher side scatter; CD45 dim to moderate), and late (CD45 moderate but slightly dimmer than mature T-cells) normal B-cell precursors. They form a unique pattern by CD45 gating, which is called by some the "Kentucky" sign, because of their resemblance to the shape of the State of Kentucky. The hematogones are displayed separately in subgroups according to their stage of maturation and development. As the normal B-cell precursors mature, the expression of CD19 is increased to normal levels. CD10 is most prominent in the intermediate precursors, and disappears gradually in the late precursors. Expression of both CD34 and TdT is similarly present only in the early precursors, which should account for less than one third of the total number of hematogones.

Morphologically, the early and some intermediate precursors of hematogones can resemble lymphoblasts, but they are always admixed with more mature lymphocytes within the same areas. In contrast to the lymphoblasts of residual BALL, the ones in hematogones are dispersed throughout the marrow and should not form large clusters (e.g., more than five blasts in one tight cluster), or aggregates, or sheets. This "dispersed" pattern of distribution by hematogones can be highlighted by immunostains for CD34 and TdT.

B-ALL is distinguished from T-ALL based on immunophenotypic and cytogenetic characteristics. Blasts in AML express myeloid-associated markers, such as CD13, CD33, and CD117, and may show positive cytochemical staining for MPO and Sudan Black B.

Metastatic round cell tumors are negative for lymphoid- and myeloid-associated CD molecules, and positive for markers that are expressed by the primary tumor.

The differential diagnosis of B-lymphoblastic lymphoma includes T-lymphoblastic lymphoma, Burkitt lymphoma, blastoid variant of mantle cell lymphoma, granulocytic sarcoma, and metastatic small round cell tumors. Burkitt and blastoid mantle cell lymphomas are TdT negative and have characteristic cytogenetic abnormalities.

Additional Resources

Hoffbrand AV, Pettit JE, Vyas P: Color Atlas of Clinical Hematology, ed 4, Philadelphia, 2010, Mosby/Elsevier.

Onciu M: Acute lymphoblastic leukemia, *Hematol Oncol Clin North Am* 23:655–674, 2009.

Cox CV, Blair A: A primitive cell origin for B-cell precursor ALL? *Stem Cell Rev* 1:189–196, 2005.

Hutter JJ: Childhood leukemia, *Pediatr Rev* 31:234–241, 2010.

Jabbour EJ, Faderl S, Kantarjian HM: Adult acute lymphoblastic leukemia, *Mayo Clin Proc* 80:1517–1527, 2005.

Cobaleda C, Sánchez-García I: B-cell acute lymphoblastic leukaemia: towards understanding its cellular origin, *Bioessays* 31:600–609, 2009.

Peters JM, Ansari MQ: Multiparameter flow cytometry in the diagnosis and management of acute leukemia, *Arch Pathol Lab Med* 135:44–54, 2011.

Digiuseppe JA: Acute lymphoblastic leukemia: diagnosis and detection of minimal residual disease following therapy, *Clin Lab Med* 27:533–549, 2007.

Haferlach T, Kern W, Schnittger S, et al: Modern diagnostics in acute leukemias, *Crit Rev Oncol Hematol* 56:223–234, 2005.

Paulsson K, Johansson B: High hyperdiploid childhood acute lymphoblastic leukemia, *Genes Chromosomes Cancer* 48:637–660, 2009.

Harrison CJ: Cytogenetics of paediatric and adolescent acute lymphoblastic leukaemia, *Br J Haematol* 144:147–156, 2009.

Teitell MA, Pandolfi PP: Molecular genetics of acute lymphoblastic leukemia, *Annu Rev Pathol* 4:175–198, 2009.

Harrison CJ, Haas O, Harbott J, et al: Detection of prognostically relevant genetic abnormalities in childhood B-cell precursor acute lymphoblastic leukaemia: recommendations from the Biology and Diagnosis Committee of the International Berlin-Frankfürt-Münster study group, *Br J Haematol* 151:132–142, 2010.

van Dongen JJ, Macintyre EA, Gabert JA, et al: Standardized RT-PCR analysis of fusion gene transcripts from chromosome aberrations in acute leukemia for detection of minimal residual disease. Report of the BIOMED-1 Concerted Action: investigation of minimal residual disease in acute leukemia, *Leukemia* 13:1901–1928, 1999.

Lymphoblastic Neoplasms— T-Lymphoblastic Leukemia/ Lymphoma

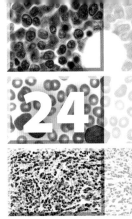

T-lymphoblastic leukemia/lymphoma (T-ALL/LBL) may initially present itself in a leukemic phase with bone marrow and/or blood involvement, or as a lymphoma with the involvement of the lymphoid and/or other extramedullary tissues. However, in a significant proportion of cases, both bone marrow and extramedullary tissues are involved.

Precursor T-cell neoplasms constitute about 15% of ALLs in children and 25% in adults. Approximately 2% of adult non-Hodgkin lymphomas are of the precursor T-cell type. Most patients are adolescent or young adults, and there is male preponderance. The vast majority of patients are at stages III or IV at the time of diagnosis. Anterior mediastinal mass and/or peripheral lymphadenopathy is detected in 50–75% of cases. Cervical, supraclavicular, and axillary lymph nodes are frequent targets. Extranodal tissues such as skin, testicle, or bone are involved less frequently. A high frequency of CNS involvement has been noted in patients with T-ALL. The overall prognosis of precursor T-cell neoplasms is worse than that of their B-cell counterparts. So far, no clear-cut correlation has been found between the prognosis and the immunophenotypic or cytogenetic results. Therapeutic modalities include consolidation with high dose chemotherapy and autologous or allogeneic stem cell transplantation.

Morphology

The morphologic features of T-ALL and T-LBL are very similar to those of B-ALL and B-LBL (see Chapter 23).

- Bone marrow biopsy and clot sections are usually hypercellular and are diffusely infiltrated by sheets of uniform appearing blast cells (Figure 24.1).
- Blast cells have scanty cytoplasm with round, oval, or indented nuclei, finely dispersed nuclear chromatin, and prominent or indistinct nucleoli. In some cases, the leukemic blast cells may appear pleomorphic with variable amounts of cytoplasm, or may show convoluted nuclei.
- Mitotic figures are more frequently observed in T-ALL than in B-ALL.
- Bone marrow fibrosis and osteoporosis may be present. Fibrosis may be mild, extensive, focal or diffuse, and may lead to unsuccessful bone marrow aspiration (dry tap). It is less frequent in T-ALL than in B-ALL, however.
- Large areas of bone marrow necrosis are infrequent. Necrosis is usually of the coagulative type with preservation of the basic outline of the necrotic cells.
- Similar to B-lymphoblastic leukemia, bone marrow smears and touch preparations show numerous blast cells that are often small with scanty non-granular blue cytoplasm, fine chromatin pattern, and indistinct nuclei. In a minority of cases, the blast cells are pleomorphic and may show azurophilic granules.
- Lymphoblasts may also be present in peripheral blood smears in variable numbers (Figure 24.1C). They account for the majority of leukocytes in patients with WBC >10,000/µL. Anemia, granulocytopenia, and/or thrombocytopenia are common features.
- Involvement of the lymph nodes and other tissues is usually diffuse with total or partial effacement of the normal architecture and morphologic features similar to those described earlier in the bone marrow biopsy sections. In some cases, the high rate of tumor cell turnover and necrosis may stimulate the macrophages. These macrophages with abundant pale, vacuolated cytoplasm, and phagocytic cell debris are dispersed throughout the lymphomatous lesion, creating a "starry sky" pattern.

Immunophenotype and Cytochemical Stains

In general, immunophenotypic features of T-ALL and T-LBL by multiparametric flow cytometry include:

- Expression of intracellular CD3 in all cases; however, surface CD3 is only observed in some cases.

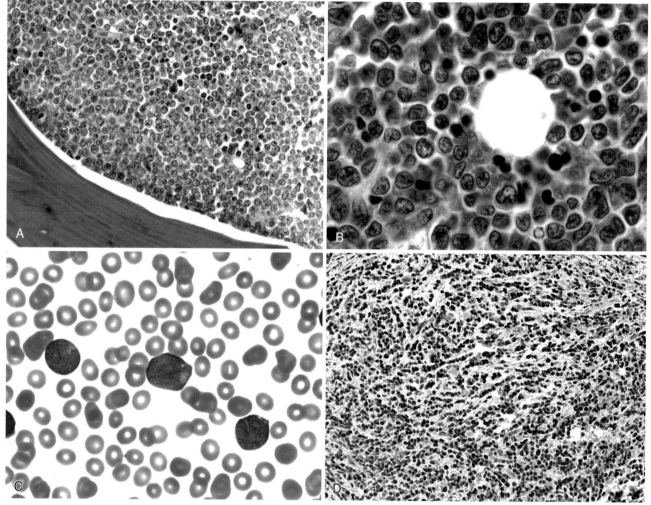

FIGURE 24.1 **T-ACUTE LYMPHOBLASTIC LEUKEMIA.** Bone marrow biopsy section showing sheets of blast cells with irregular nuclei (A, low power; B, high power). Blood smear demonstrates three blasts with scanty dark blue cytoplasm and irregular or cleaved nuclei (C). Numerous TdT-positive cells are present by immunohistochemical stain (D).

- TdT positivity in many but not all cases.
- Highly frequent aberrancies of pan-T-cell markers CD2, CD5, and CD7. These aberrancies include abnormal intensities (dim or bright), abnormal subsets, abnormal clustering profiles, and partial or complete loss of antigens.
- Variable expression of CD1a, CD4, CD8 and CD34 (See Table 24.1). Double positivity of CD4/CD8 is not uncommon, which, however, is not specific for T-ALL or T-LBL.
- Common myeloid aberrancies with expression of CD13, CD15, CD33, or CD117.
- A small proportion of cases may express other cross-lineage aberrancies like CD20 or CD79a. Expression of CD10 may be present, but is not specific for this entity.
- No lineage-specific cytochemical stains are available for T lymphoblasts. They may show focal acid phosphatase or NSE reactivity.
- Based upon their immunophenotypic resemblance to various stages of T-cell maturation, T lymphoblasts can sometimes be divided into several subcategories (Table 24.1):
  - Pro- to pre-T-cell type: positive for TdT, CD7, and cytoplasmic CD3. CD34 may be positive, as well as CD2 and or CD5. The neoplastic cells are mostly double negative for CD4 and CD8 (Figures 24.2 and 24.3).
  - Cortical T-cell type: positive for CD1a, cytoplasmic CD3, plus double positivity of CD4, and CD8. The early cortical T-ALL/LBL is the most common type.
  - Late cortical blast cells: begin to show single positivity for either CD4 or CD8 (Figure 24.2).
  - Medullary T-blasts: positive for surface CD3, are negative for CD1a, and are either CD4+ or CD8+.

Cytogenetic and Molecular Studies

- Several translocations involve common genomic sites with genes *MYC*, *TAL1*, *LMO1*, *LMO2*, and *HOX11*, located on chromosomes 8q24, 1p32, 11p13, 11p15, and 10q24, respectively.
- Precursor T-cell neoplasms show rearrangements of either or both *TRA@/TRD@* and *TCL1A* at 14q32.
- Other abnormalities (Table 24.2) include del(6q) (Figure 24.5A and B), del(9p) (Figure 24.6), and trisomy 8.

CYTOGENETIC AND MOLECULAR STUDIES   311

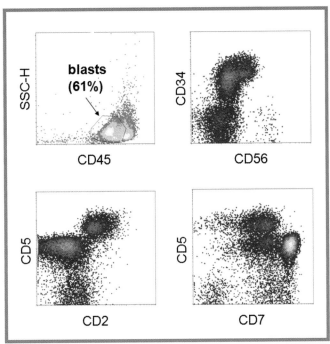

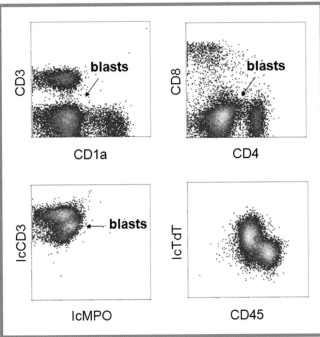

FIGURE 24.2 **FLOW CYTOMETRIC FINDINGS OF T-ALL/ LBL DOUBLE (CD4−/ CD8−) NEGATIVE.** Open gate density plot by CD45 gating (in blue) reveals 61% blasts that are CD45 dim to moderate. The blast-enriched gate is intentionally widened to include some mature T-cells as internal controls. Density plots of the blast-enriched gate (in magenta) demonstrate abnormal T lymphoblasts, which are positive for CD5 (dim), CD7 (bright; major subset), CD34, CD56 (partial), plus intracellular CD3 (dim) and TdT. The blasts are negative for CD1a, CD2, surface CD3, CD4, CD8, as well as TCR alpha/beta and gamma/delta (not shown). The phenotypic features of the abnormal T-lymphoblasts are intermediate between those of prothymocytes and subcapsular thymocytes.

Table 24.1
Molecular Abnormalities, and Immunophenotype, in Subtypes of T Lymphoblastic Leukemia/Lymphomas[1]

Type	Molecular Features	Immunophenotype
Early pro-T-cell ALL	Aberrant overexpression of *LYL1* transcription factor	CD4−, CD8−, cCD3+, CD7+, CD34+, TdT+
Early cortical T-cell ALL	Aberrant overexpression of *TLX1* (*HOX11*) transcription factor	CD4+, CD8+, cCD3+, CD7+, CD1a+, CD10+, TdT+
Late cortical T-cell ALL	Aberrant overexpression of transcription factor	CD4+, and/or CD8+, cCD3^{high}, CD7+, CD1a+, TCRα/β
Medullary T-cell ALL	Unknown	CD4+ or CD8+, sCD3+, TCRα/β, CD1a−

[1] Adapted from Jaffe ES, Harris NL, Vardiman JW, et al. Hematopathology. Saunders/Elsevier, Philadelphia, 2010.

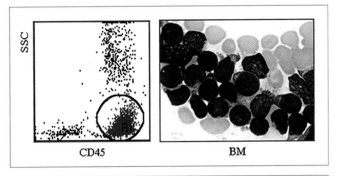

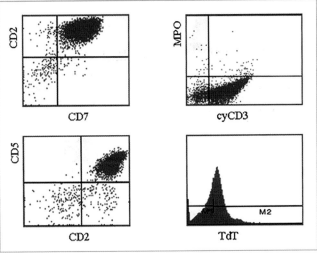

FIGURE 24.3 Flow cytometric analysis of bone marrow from a patient with early pro-T-ALL demonstrating a population of CD45+ blast cells expressing CD2, CD5, CD7, and cytoplasmic CD3 (cyCD3), and TdT. The blasts were negative for CD4, CD8, and CD1a (not shown).

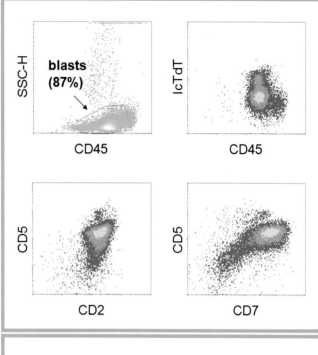

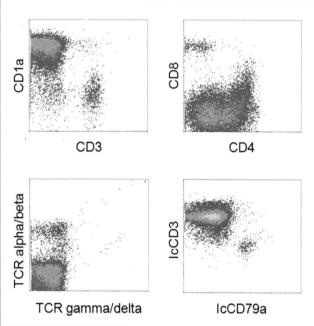

FIGURE 24.4 **FLOW CYTOMETRIC FINDINGS OF T-ALL/LBL, LATE CORTICAL.** Open gate density plot by CD45 gating (in green) reveals 87% blasts that are CD45 dim to moderate. The blast-enriched gate is intentionally expanded to include a small number of mature T-cells as internal controls. Density plots of the blast-enriched gate (in purple) show abnormal T lymphoblasts, which are positive for CD1a, CD2, CD4 (partial), CD5 (dim), CD7 (heterogeneous), CD10 (not shown), plus intracellular CD3 and TdT. The blasts are negative for CD8, surface CD3, as well as TCR alpha/beta and gamma/delta. The phenotypic features of the T lymphoblasts resemble those of late cortical thymocytes.

- Approximately 30% of cases of T-ALL/LBL show translocations involving 14q11.2 (*TRA@-TRD@*) or 7q34 (*TRB@*) (Table 24.2), such as t(11;14)(p15;q11.2), (14;21)(q11.2;q22), and t(7;14)(q34;q11.2) (Figure 24.7).

Table 24.2
Recurrent Cytogenetic Abnormalities in Precursor T-Lymphoblastic Neoplasms

Chromosomal Aberrations	Affected Genes
t(1;7)(p34;q34)	LCK/TRB@
t(1;14)(p32;q11.2)	TAL1[1]/ TRA@ or TRD@
t(7;10)(q34;q24)	TRB@/HOX11[2]
t(7;19)(q34;p13)	TRB@/LYL1
t(8;14)(q24;q11.2)	MYC/TRA@
t(11;14)(p15;q11.2)	LMO1/TRA@ or TRD@
t(11;14)(p13;q11.2)	LMO2/TRA@ or TRD@
inv(14)(q11.2q32)	TRA@ or TRD@/TCL1A
del(1p32)	TAL1
del(6q)	
del(9p)	CDKN2A/CDKN2B
+8	

Alternative designations: [1]*HOX11*, *SCL*, or *TCL5*; [2]*TAN*.

- Other aberrations such as t(10;14)(q24;q11.2) resulting in an upregulation of TLX1 (HOX11) is reported to be associated with a favorable prognosis. In contrast, high risk of early failure is common in the *TAL1*- and *LYL1*-positive groups.
- Gene expression studies indicate activation of a subset of the critical genes—*HOX11*, *TAL1*, *LYL1*, *LMO1*, and *LMO2*—in a much larger fraction of T-ALL cases than those harboring activating chromosomal translocations.
- Activating point mutations in *NOTCH1* in more than 50% of all T-ALL cases, is consistent with its being one of the central players in T-ALL pathogenesis.
- Extrachromosomal (episomal) amplification of *ABL1* has been observed in 5–6% of T-ALL and benefits from treatment with imatinib. However, this aberration is not detectable by conventional cytogenetics. The activation of tyrosine kinase by the formation of episomes is due to a cryptic fusion between *NUP214* and *ABL1* and can be identified by FISH analysis with the *ABL1* DNA probe. This fusion is associated with increased *HOX* expression and deletion of *CDKN2A* (9p−).
- Generation of T-cell receptor molecules in T-cells occurs by much the same mechanism as generation of immunoglobulin molecules in B-cells. Therefore, clonality of T-cell malignancies can be demonstrated by examining the rearrangement patterns of the T-cell receptor (*TCR*) genes, in much the same way as is done with the immunoglobulin genes in B-cell lesions (Figure 24.8). *TCR* rearrangements are somewhat more complicated, however, and there are advantages and disadvantages to the various target loci available. For example, the *TCR-β* genes produce a wider range of rearrangements and thus are more informative, but the region is so large that a high number of PCR primer sets are required to span it. The *TCR-γ* region, in contrast, is smaller and easily encompassed by a small number of primer sets, but the limited number of potential rearrangements increases the chance of the artifact known as pseudoclonality, in which amplification of a specimen in which T-cells are scant will produce an apparent clonal rearrangement pattern even though the cells are neither malignant nor clonal (Figure 24.9).

RARE VARIANTS OF LYMPHOBLASTIC LEUKEMIA/LYMPHOMA

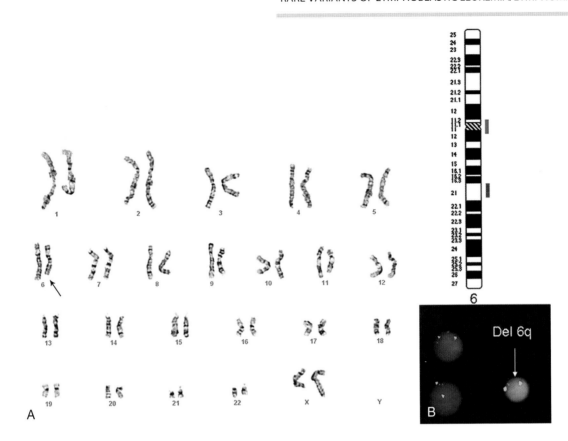

FIGURE 24.5 G-banded karyotype (A) and FISH analysis (B) demonstrating 46,XY,del(6)(q23).

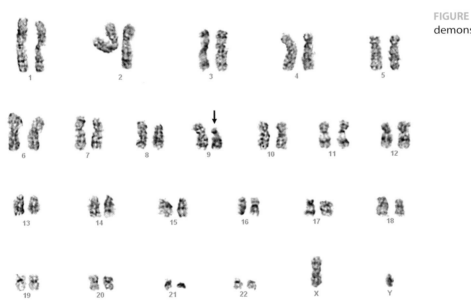

FIGURE 24.6 G-banded karyotype demonstrating 46,XY,del(9)(p21).

Rare Variants of Lymphoblastic Leukemia/Lymphoma

"CD4−/CD56+ IMMATURE NK-CELL NEOPLASM"

Rare cases of CD56+ lymphoblastic leukemia/lymphoma are reported. These cases involve both bone marrow and lymph nodes, and sometimes other organs, such as liver and spleen. In this condition, unlike the blastic plasmacytoid dendritic cell neoplasms (see Chapter 59), the skin is not involved and the neoplastic cells are negative for CD4, but express CD7. CD2, CD5, HLA-DR, and CD34 are positive is some cases. CD3 and TdT are negative, as well as myeloid and B-lymphocyte-associated markers.

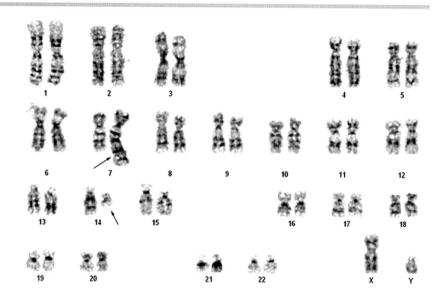

FIGURE 24.7 G-banded karyotype of bone marrow of a patient with acute T-lymphoblastic leukemia with t(7;14)(q34;q11.2).

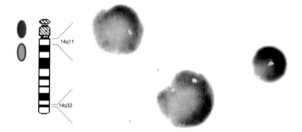

FIGURE 24.8 Dual color *TCR* alpha/delta FISH "split" probe showing rearrangement of 14q11.2 in T-ALL.

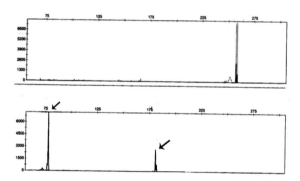

FIGURE 24.9 Pseudoclonality pattern in a TCR-γ PCR assay of a lesion with scant numbers of T-lymphocytes. One hint that the pattern represents a pseudoclonality artifact rather than true clonality is the absence of background polyclonal "noise" in addition to the strong pseudoclonal peaks.

"THYMIC LYMPHOBLASTIC T/NK-CELL LYMPHOMA"

Rare cases of mediastinal lymphoblastic lymphoma have been reported in children and young adults. The blastic tumor cells express CD56, cCD3, CD2, CD7, CD34, CD38, HLA-DR, and TdT, and are negative for surface CD3, CD5, CD16, CD57, CD30, and CD117. There is no evidence of *TCR* or *IGH* gene rearrangement.

Differential Diagnosis

The differential diagnosis of precursor T-cell neoplasms includes hematogone hyperplasia, B-ALL/LBL, a variety of acute myeloid leukemias (such as minimally differentiated AML, AML without maturation, and megakaryoblastic leukemia), and metastatic small round cell tumors (such as neuroblastoma).

Hematogones represent the normal bone marrow precursor B-cells (see Chapter 23). The earlier hematogones may express TdT, but they lack the expression of T-cell markers.

T-ALL is distinguished from B-ALL based on immunophenotypic and cytogenetic characteristics. Blasts in AML express myeloid-associated markers, such as CD13, CD33, and CD117, lack cytoplasmic CD3 expression, and may show positive cytochemical staining for MPO and Sudan Black B.

The differential diagnosis of T-lymphoblastic lymphoma includes B-lymphoblastic lymphoma, Burkitt lymphoma, blastoid variant of mantle cell lymphoma, myeloid sarcoma, and metastatic small round cell tumors. Burkitt and blastoid mantle cell lymphomas are TdT negative, express B-cell-associated CD molecules, and have characteristic cytogenetic abnormalities (see Chapter 23).

Metastatic round cell tumors are negative for lymphoid- and myeloid-associated CD molecules, and positive for markers that are expressed by the primary tumor.

Additional Resources

Burkhardt B: Paediatric lymphoblastic T-cell leukaemia and lymphoma: One or two diseases? *Br J Haematol* 149:653–668, 2010.

Chiaretti S, Foà R: T-cell acute lymphoblastic leukemia, *Haematologica* 94:160–162, 2009.

Ferrando AA: The role of NOTCH1 signaling in T-ALL, *Hematology Am Soc Hematol Educ Program* 353–361, 2009.

Gaiser T, Haedicke W, Becker MR: A rare pediatric case of a thymic cytotoxic and lymphoblastic T/NK cell lymphoma, *Int J Clin Exp Pathol* 3:437–442, 2010.

Graux C, Cools J, Michaux L, et al: Cytogenetics and molecular genetics of T-cell acute lymphoblastic leukemia: from thymocyte to lymphoblast, *Leukemia* 20:1496–1510, 2006.

Hagemeijer A, Graux C: ABL1 rearrangements in T-cell acute lymphoblastic leukemia, *Genes Chromosomes Cancer* 49:299–308, 2010.

Hoelzer D, Gökbuget N: T-cell lymphoblastic lymphoma and T-cell acute lymphoblastic leukemia: a separate entity? *Clin Lymphoma Myeloma* (Suppl 3):S214–S221, 2009.

Hoffbrand AV, Pettit JE, Vyas P: Color atlas of clinical hematology, ed 4, Philadelphia, 2010, Mosby/Elsevier.

Hutter JJ: Childhood leukemia, *Pediatr Rev* 31:234–241, 2010.

Ichinohasama R, Endoh K, Ishizawa K, et al: Thymic lymphoblastic lymphoma of committed natural killer cell precursor origin: a case report, *Cancer* 77:2592–2603, 1996.

Jabbour EJ, Faderl S, Kantarjian HM: Adult acute lymphoblastic leukemia, *Mayo Clin Proc* 80:1517–1527, 2005.

Liang X, Graham DK: Natural killer cell neoplasms, *Cancer* 112:1425–1436, 2008.

Meijerink JP: Genetic rearrangements in relation to immunophenotype and outcome in T-cell acute lymphoblastic leukaemia, *Best Pract Res Clin Haematol* 23:307–318, 2010.

Onciu M: Acute lymphoblastic leukemia, *Hematol Oncol Clin North Am* 23:655–674, 2009.

Pear WS, Aster JC: T cell acute lymphoblastic leukemia/lymphoma: a human cancer commonly associated with aberrant NOTCH1 signaling, *Curr Opin Hematol* 11:426–433, 2004.

Peters JM, Ansari MQ: Multiparameter flow cytometry in the diagnosis and management of acute leukemia, *Arch Pathol Lab Med* 135:44–54, 2011.

Pitman SD, Huang Q: Granular acute lymphoblastic leukemia: a case report and literature review, *Am J Hematol* 82:834–837, 2007.

van Dongen JJ, Langerak AW, Brüggemann M, et al: Design and standardization of PCR primers and protocols for detection of clonal immunoglobulin and T-cell receptor gene recombinations in suspect lymphoproliferations: report of the BIOMED-2 Concerted Action BMH4-CT98-3936, *Leukemia* 17:2257–2317, 2003.

Van Vlierberghe P, Pieters R, Beverloo HB, et al: Molecular-genetic insights in paediatric T-cell acute lymphoblastic leukaemia, *Br J Haematol* 143:153–168, 2008.

Acute Leukemias of Ambiguous Lineage

The addition of cytochemical and immunophenotypic techniques to the standard morphologic evaluation of leukemic blast cells has led to the growing recognition of acute leukemias with ambiguous lineage assignment. According to the WHO classification, these leukemias fall into the following categories (Figure 25.1).

1. Acute undifferentiated leukemia (AUL) which lacks sufficient evidence (such as morphologic, cytochemical, and immunophenotypic features) of lineage differentiation.
2. Mixed phenotype acute leukemia:
   a. Acute bilineage (or multilineage) leukemia which represents an acute leukemia with more than one population of blast cells. The total number of blasts from both lineages should be ≥20% of marrow or blood cells.
   b. Acute biphenotypic (multi-phenotypic) leukemia in which the population of leukemic cells coexpress more than one lineage-specific marker, such as a combination of myeloid-specific and lymphoid-specific, or B- and T-specific molecules. In rare occasions, the leukemic blasts may express a combination of myeloid-, B-cell- and T-cell-associated markers.

The following WHO criteria for lineage assignment in acute leukemia with ambiguous lineage have replaced the scoring system which was proposed by the European Group for the Immunologic Classification of Leukemia (EGIL) for the lineage assignments in acute leukemias.

- **Myeloid Lineage**: Myeloperoxidase (cytochemistry, flow cytometry, or immunohistochemistry). For *monocytic* differentiation at least two of the following markers: NSE, lysozyme, CD11c, CD14, and CD64.
- **B-Cell Lineage**: Strong CD19 expression with:
  - Strong coexpression of at least one of the following markers: CD79a, cytoplasmic CD22, and CD10, or
  - Weak CD19 expression with strong coexpression of at least two of the following markers: CD79a, cytoplasmic CD22, and CD10.
- **T-Cell Lineage**: Cytoplasmic CD3 (with an intensity approaching that of normal T cells) or surface CD3.

Acute Undifferentiated Leukemia

Acute undifferentiated leukemia (AUL) is defined as a leukemia with no morphologic, cytochemical, or specific immunophenotypic features of lymphoid or myeloid differentiation. AUL is extremely rare and probably accounts for <1% of acute leukemias.

MORPHOLOGY

- Bone marrow is often hypercellular with increased number of uniform, undifferentiated immature cells.
- Blasts consist of primitive undifferentiated cells with scant dark blue cytoplasm, no cytoplasmic granules or Auer rods, round or oval nucleus and fine nuclear chromatin (Figure 25.2). The nucleoli are often conspicuous, but in some cases are prominent. Blasts account for >20% of the bone marrow or peripheral blood differential counts.
- Blood examination may show anemia, thrombocytopenia, and/or leukopenia.

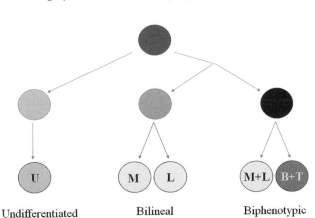

FIGURE 25.1 Scheme of clonal development of acute leukemias of undifferentiated, bilineal, and biphenotypic types.

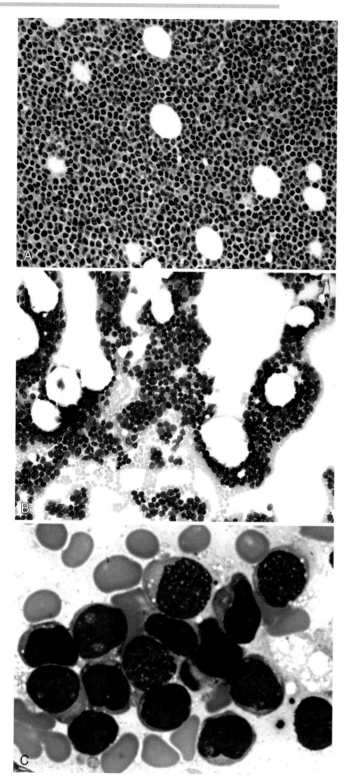

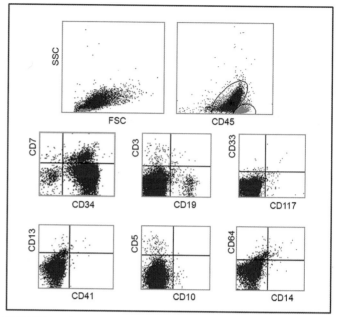

FIGURE 25.3 FLOW CYTOMETRIC FINDINGS OF UNDIFFERENTIATED ACUTE LEUKEMIA. The CD45-dim gate reveals a prominent population of blasts that are dimly positive for CD45. In addition, the blasts express CD34 and partial CD7, but are negative for CD3, CD5, CD10, CD13, CD14, CD19, CD33, CD41, CD64, and CD117. The blasts are negative for CD4 and CD56 (not shown).

FIGURE 25.2 UNDIFFERENTIATED ACUTE LEUKEMIA. Bone marrow biopsy section (A) and marrow smear (B, low power; C, high power) demonstrating undifferentiated blasts (see Figure 25.3).

IMMUNOPHENOTYPE (FIGURE 25.3)

- Blasts do not express lineage-specific antigens.
- Blasts are commonly positive for dim CD45.
- Blasts are often positive for antigens that are associated with early hematopoiesis, such as CD34, CD38, HLA-DR, and occasionally TdT.

- Variable expression of myeloid or lymphoid associated antigens can be seen, such as CD7, CD13 or CD33, and CD19, but the expression is insufficient for lineage assignment.

MOLECULAR AND CYTOGENETIC STUDIES

- By definition these leukemias are "undifferentiated" and therefore would not be expected to exhibit B-lymphoid, T-lymphoid, or myeloid markers. However, this can be a matter of semantics, dependant upon which markers one is using to arrive at such definitions. While protein markers may be absent based on immunophenotypic methods, a cell of pre-B lineage, for example, might show rearrangement of the immunoglobulin genes by sensitive PCR methods.
- Nevertheless, these leukemias are too rare to make any generalizations about particular molecular findings that might give a clue as to cell of origin.
- No recurrent or specific cytogenetic abnormalities are noted by karyotyping or FISH.

DIFFERENTIAL DIAGNOSIS

If only a limited panel of antigens is tested, morphologic and immunophenotypic features can overlap with those seen in blastic plasmacytoid dendritic cell (BPDC) neoplasm. Other differential diagnoses include ALL, AML, minimally differentiated and without maturation, pure erythroid leukemia, and megakaryoblastic leukemia.

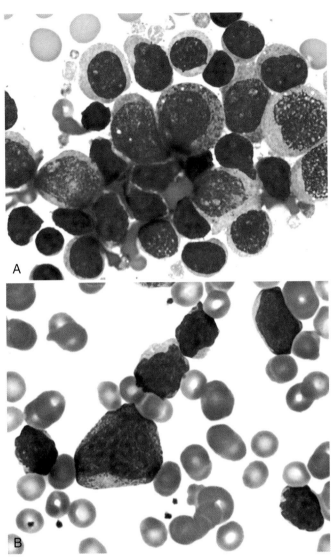

FIGURE 25.4 **ACUTE BILINEAL LEUKEMIA.** Bone marrow (A) and peripheral blood (B) smears show two distinct populations of leukemic blast cells (larger and smaller).

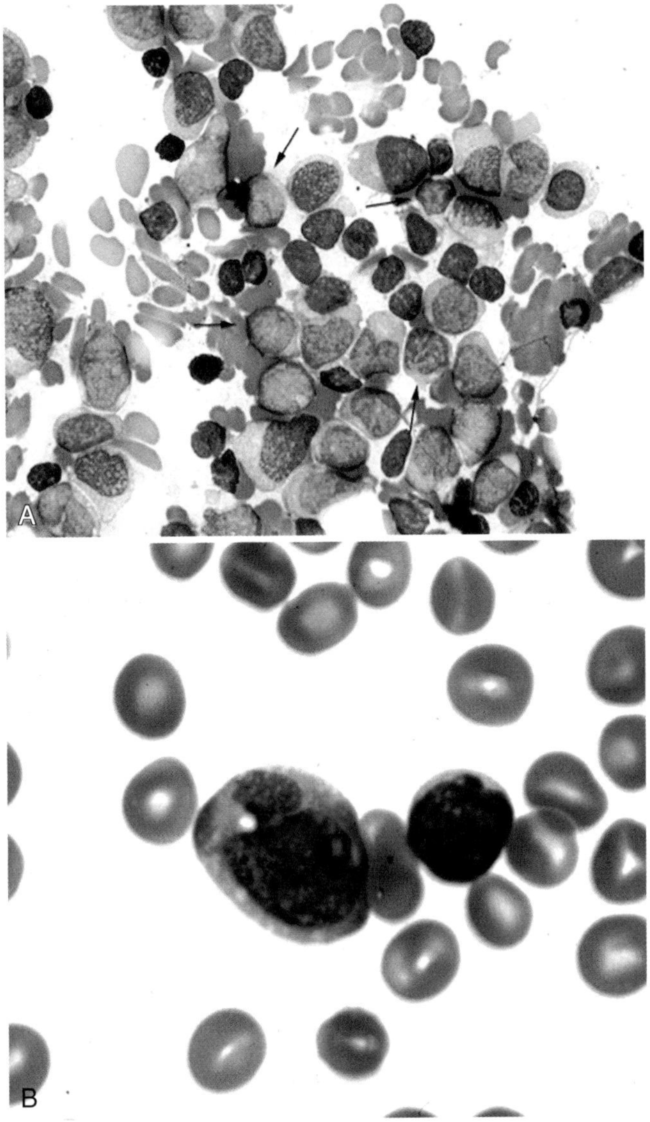

FIGURE 25.5 **ACUTE BILINEAL LEUKEMIA.** Bone marrow (A) and peripheral blood (B) smears show two distinct populations of leukemic blast cells (larger and smaller).

Mixed Phenotype Acute Leukemia

Mixed phenotype acute leukemia (MPAL) includes the following subtypes: (a) acute bilineal (or multilineage) leukemia consisting of more than one population of blast cells, and (b) acute biphenotypic (multi-phenotypic) leukemia in which the blast population coexpresses more than one lineage-specific marker, such as a combination of myeloid-specific and lymphoid-specific, or B- and T-specific molecules. In rare instances, the leukemic blasts may express a combination of myeloid-, B-cell- and T-cell-associated markers, or represent a combination of (a) and (b).

Biphenotypic acute leukemias probably represent less than 10% of all acute leukemias. The bilineal acute leukemias are less frequent. Mixed phenotype acute leukemia can occur at any age, but is more frequent in adults. They usually have an aggressive clinical course, particularly those with 11q23 abnormalities or the t(9;22) (Philadelphia chromosome). The B-cell/myeloid phenotype is the most frequent subtype.

MORPHOLOGY

- Acute bilineal leukemia may consist of two distinct morphologic populations of blast cells. In most cases these two populations represent lymphoid and myeloid lineages (Figures 25.4 and 25.5).
- Lymphoblasts are usually smaller with scant cytoplasm, no cytoplasmic granules, and less prominent nucleoli, while blasts of myeloid origin (myeloblasts, monoblasts) are larger with more abundant cytoplasm, variable amounts of cytoplasmic granules, and prominent nucleoli.
- The bilineal nature of an acute leukemia may not be distinguished on a morphologic basis, but the two populations of blast cells demonstrate distinct cytochemical and immunophenotypic properties (see below).

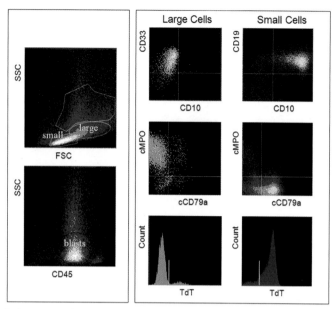

FIGURE 25.6 ACUTE BILINEAL LEUKEMIA. Flow cytometric studies reveal two distinct populations of large and small blast cells (left panel). The large blasts (right panel, in purple) express myeloid-associated antigens, while the small blasts (right panel, in red) reveal features seen in B lymphoblastic leukemia.

IMMUNOPHENOTYPE AND CYTOCHEMICAL STAINS

Bilineal Acute Leukemia (Figure 25.6)

- Two distinct populations of blasts can be detectable by either common gating strategies (e.g., CD45 gating and scatter gating) or selected back gating approaches of target populations.
- Each population expresses antigens of one specific lineage.
- Both blastic populations can also express non-lineage specific markers, such as CD34, CD38, HLA-DR, etc.
- The combination of lymphoid and myeloid phenotype bilineal acute leukemia is more common than that of B- and T-cell phenotype.
- The presence of two populations of blast cells in bilineal acute leukemia is often synchronous; both populations of blast cells are present at the same time. In rare instances, there may be two simultaneous neoplastic processes in two separate sites. For example, the bone marrow may be involved with ALL and the CNS infiltrated by AML.
- Asynchronous bilineal acute leukemias are more frequently reported and are those in which one lineage switches to another during the disease process. The predominant pattern in most studies is ALL switching to AML in relapse. The frequency of lineage switch in relapse is about 7%. Lineage switch may represent relapse of the original clone or development of a second new clone. An increased risk of development of AML has been reported in patients with ALL who receive intensive chemotherapy, particularly in patients with 11q23 abnormalities or those with t(9;22) (Philadelphia chromosome).
- There are also occasional reports of switching from bilineal to biphenotypic acute leukemias and *vice versa*.
- Cytochemical stains such as MPO, Sudan Black B, and NSE are helpful in distinguishing blasts of myeloid origin. Myeloblasts are often MPO and Sudan Black B positive and monoblasts/promonocytes usually express NSE.

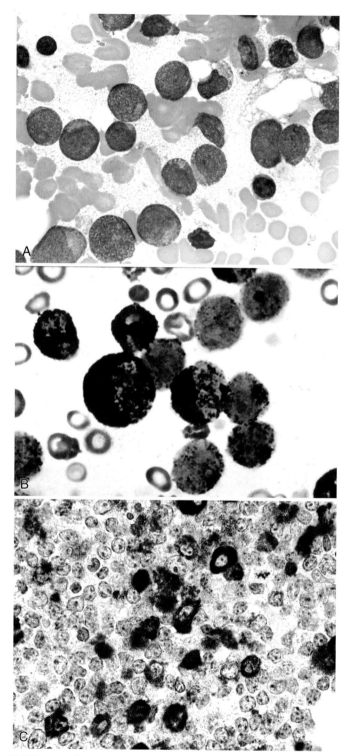

FIGURE 25.7 ACUTE BIPHENOTYPIC LEUKEMIA (MYELOID/B-LYMPHOID). Bone marrow smear shows blast cells with variable amount of non-granular cytoplasm (A). These cells are myeloperoxidase-positive (B, cytochemical stain; C, immunohistochemical stain). The blast cells show a mixed phenotype (myeloid/B-lymphoid) by flow cytometry; see Figure 25.8).

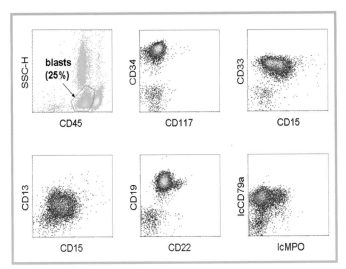

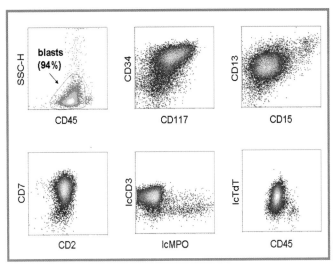

FIGURE 25.8 **FLOW CYTOMETRIC FINDINGS OF ACUTE BIPHENOTYPIC LEUKEMIA (MYELOID/B-LYMPHOID).** Open gate display by CD45 gating (in green) reveals excess blasts (25% of the total) that are dimly positive for CD45. The blast-enriched gate (in magenta) demonstrates abnormal blasts expressing both B- and myeloid-lineage-specific markers. The positive B-lineage antigens include CD19 (moderate to bright), CD22 (dim), and intracellular CD79a. The positive myeloid markers include CD11b (not shown), CD13 (dim), CD15 (partial), CD33, and intracellular myeloperoxidase (partial).

FIGURE 25.10 **FLOW CYTOMETRIC FINDINGS OF ACUTE BIPHENOTYPIC LEUKEMIA (MYELOID/T-LYMPHOID).** Open gate display by CD45 gating (in blue) shows excess abnormal blasts (94% of the total) that are CD45 dimly positive. Density plots of the blast-enriched gate (in magenta) demonstrate that the blasts express T-lineage markers CD2, CD5 (partial, dim; not shown), CD7, and intracellular CD3. In addition, the blasts are positive for myeloid antigens CD13, CD15 (partial), CD117 (heterogeneous), and intracellular myeloperoxidase (partial, heterogeneous). Positivity for myeloperoxidase is further confirmed by immunohistochemical studies. The blasts are also positive for CD34, HLA-DR (partial; not shown), and intracellular TdT.

Biphenotypic Acute Leukemia (Figures 25.7 to 25.10)

- Blasts express antigens that are specific for more than one lineage.
- Common forms include mixed B- and myeloid or T- and myeloid biphenotypes.
- Rarely, blasts may express B- and T-cell biphenotype, or mixed B-, T-, and myeloid triphenotype.

MOLECULAR AND CYTOGENETIC STUDIES

Based on molecular and cytogenetic studies, we can divide mixed phenotype acute leukemias into two major groups:

1. Mixed phenotype acute leukemia with recurrent genetic abnormalities
2. Mixed phenotype acute leukemia, otherwise not specified.

Mixed Phenotype Acute Leukemia with Recurrent Genetic Abnormalities

- These include:
  - MPAL with t(9;22)(q34;q11.2) (*BCR-ABL1*)
  - MPAL with t(v;11q23) (*MLL* rearranged)
  - MPAL B/myeloid, not otherwise specified
  - MPAL T/myeloid, not otherwise specified.
- Translocation of 11q23 and the t(9;22) are the most frequent cytogenetic abnormalities observed in biphenotypic and bilineal acute leukemias of B-precursor/myeloid type.
- More than 70 partner genes have been identified in association with 11q23 translocations, such as t(4;11), t(9;11), and t(11;19) (Figures 25.11 and 25.12).

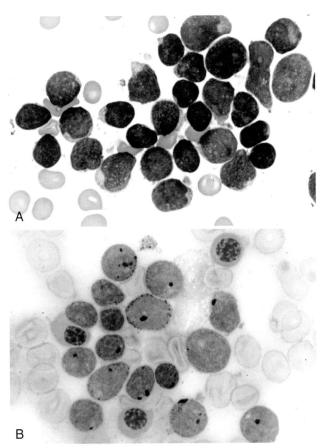

FIGURE 25.9 **ACUTE BIPHENOTYPIC LEUKEMIA (MYELOID/T-LYMPHOID).** Bone marrow smear shows blast cells with scanty cytoplasm (A). These cells show coarse PAS-positive cytoplasmic granules (B). The blast cells show a mixed phenotype (myeloid/T-lymphoid by flow cytometry; see Figure 25.10).

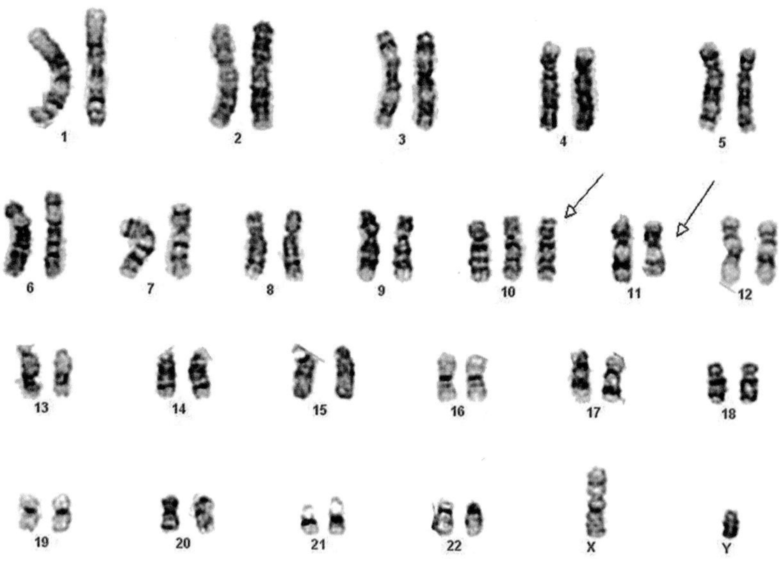

FIGURE 25.11 A karyotype with deletion of 11q23 and trisomy 10 in a biphenotypic acute leukemia.

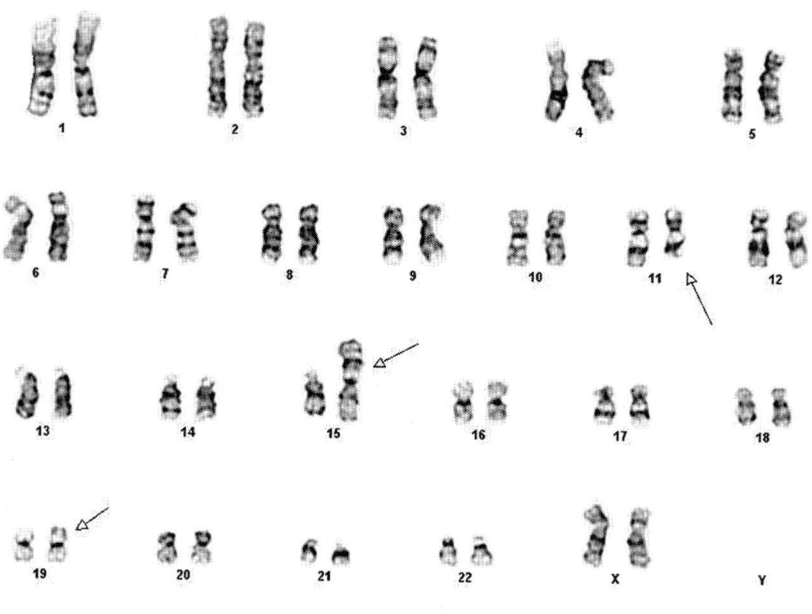

FIGURE 25.12 Karyotype with a t(11;19)(q23;p13) and dup 1q in a biphenotypic acute leukemia.

- The Philadelphia chromosome may be a part of a complex set of cytogenetic abnormalities, such as combination of t(9;22) and del(7) or t(2;9;22) (Figure 25.13).
- The T-precursor/myeloid biphenotypic or bilineal acute leukemias may be associated with t(5;18)(q31;q23) and t(3;12)(p25;q24.3).
- Also, in some cases trisomy 10 has been reported in association with acute biphenotypic leukemia (see Figure 25.11).
- Some cases of acute biphenotypic leukemia may show complex chromosomal aberrations (Figure 25.14).
- IgH and *TCRG* clonal gene rearrangements may both be seen in biphenotypic acute leukemias.
- However, unlike the markers identified by flow cytometry or immunohistochemistry, it is not possible by PCR to distinguish whether these gene rearrangements are present in two distinct sets of malignant cells, as opposed to coexisting in the same cells.

- Since a small proportion of T-cell leukemias can show secondary rearrangement of immunoglobulin genes, and vice versa for B-cell leukemias, care must be taken not to overcall a biphenotypic diagnosis based on such findings.
- Cyclin A1 and *HOXA9* gene expression have also been reported in these lesions.

Mixed Phenotype Acute Leukemia, Otherwise Not Specified

In this category the blast cells show coexpression of myeloid-specific and lymphoid-specific molecules, but lack recurrent genetic abnormalities.

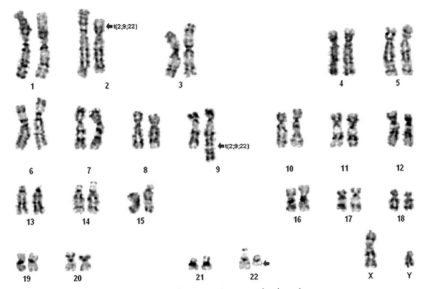

FIGURE 25.13 A karyotype with a three-way t(2;9;22) in a biphenotypic acute leukemia.

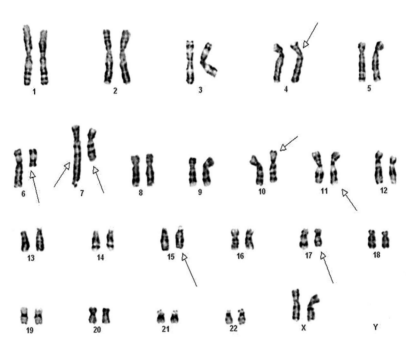

FIGURE 25.14 A complex karyotype in a biphenotypic acute leukemia: 46,XX,t(4;7)(p12;p11.2),t(6;7)(q13;q36),add(7)(p12),inv(10)(p12p15),t(11;17)(q23;q21),del(15)(q22q22),add(20)(q13.1).

Differential Diagnosis

The diagnosis of biphenotypic acute leukemia is established by immunophenotypic studies. These leukemias should be distinguished from AMLs with aberrant expression of lymphoid-associated markers and from ALLs with aberrant expression of myeloid-associated markers. Bilineal acute leukemias may show morphologic evidence of two separate leukemia populations, such as larger and smaller blasts. But diagnosis is confirmed by the evidence of two populations of blast cells distinctly expressing molecules representing more than one hematopoietic lineage.

Additional Resources

Al-Seraihy AS, Owaidah TM, Ayas M, et al: Clinical characteristics and outcome of children with biphenotypic acute leukemia, *Haematologica* 94:1682–1690, 2009.

Béné MC: Biphenotypic, bilineal, ambiguous or mixed lineage: strange leukemias! *Haematologica* 94:891–893, 2009.

Gerr H, Zimmermann M, Schrappe M, et al: Acute leukaemias of ambiguous lineage in children: characterization, prognosis and therapy recommendations, *Br J Haematol* 149:84–92, 2010.

Gluzman DF, Nadgornaya VA, Sklyarenko LM, et al: Immunocytochemical markers in acute leukaemias diagnosis, *Exp Oncol* 32:195–199, 2010.

Naghashpour M, Lancet J, Moscinski L, et al: Mixed phenotype acute leukemia with t(11;19)(q23;p13.3)/ MLL-MLLT1(ENL) B/T-lymphoid type: a first case report, *Am J Hematol* 85:451–454, 2010.

Nishiuchi T, Ohnishi H, Kamada R, et al: Acute leukemia of ambiguous lineage, biphenotype, without CD34, TdT or TCR-rearrangement, *Intern Med* 48:1437–1441, 2009.

Owaidah TM, Al Beihany A, Iqbal MA, Elkum N, Roberts GT: Cytogenetics, molecular and ultrastructural characteristics of biphenotypic acute leukemia identified by the EGIL scoring system, *Leukemia* 20:620–626, 2006.

Swerdlow SH, Campo E, Harris NL, et al: WHO classification of tumours of haematopoietic and lymphoid tissues, ed 4, Lyon, 2008, International Agency for Research on Cancer.

van den Ancker W, Terwijn M, Westers TM, et al: Acute leukemias of ambiguous lineage: diagnostic consequences of the WHO 2008 classification, *Leukemia* 24:1392–1396, 2010.

Mature B-Cell Neoplasms—Overview

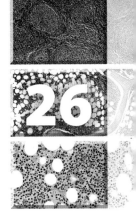

Mature B-cell neoplasms comprise a wide spectrum of lymphoid malignancies representing clonal proliferation of B-lymphocytes at various stages of maturation, from the early naïve B-cells to the end-stage mature plasma cells. These disorders may primarily involve bone marrow and peripheral blood (leukemia), lymphoid or extramedullary tissues (lymphoma), or both. They comprise >85% of all lymphoid malignancies. As shown in Box 26.1, based on the WHO classification, mature B-cell neoplasms are divided into several categories. Diffuse large B-cell and follicular lymphomas are the most frequent types of mature B-cell lymphomas (Table 26.1).

Mature B-cell neoplasms represent about 4% of all cancers. They are more common in men than in women and are more frequent in adults than in children, with a steady increase in incidence with age. In general, the lymphoid malignancies in children are more commonly extranodal than nodal, and clinically more aggressive.

The Ann Arbor staging system originally developed for the staging of Hodgkin lymphoma has been extended to non-Hodgkin lymphoid malignancies. This staging system is based on the location(s), number of involved sites, and presence or absence of systemic symptoms (Table 26.2). Since in the majority of cases of non-Hodgkin lymphoma the disease is disseminated at the time of diagnosis, the staging system is less useful in non-Hodgkin lymphoma than it is in Hodgkin lymphoma. An international prognostic index (IPI) has been proposed for patients with non-Hodgkin lymphoma. This scoring system is based on the following factors, which were found to show a reverse correlation with relapse-free survival:

- Age >60 years
- Elevated serum lactate dehydrogenase (LDH)
- Eastern Cooperative Oncology Group (ECOG) performance status 2
- Ann Arbor clinical stages III or IV
- Number of involved extranodal disease sites >1.

Box 26.1 The WHO Classification of Mature B-Cell Neoplasms

- Chronic lymphocytic leukemia/small lymphocytic lymphoma
- B-cell prolymphocytic leukemia
- Splenic marginal zone lymphoma
- Splenic lymphoma/leukemia, unclassifiable
- Hairy cell leukemia
- Lymphoplasmacytic lymphoma
- Heavy chain disease
- Plasma cell neoplasms
- Extranodal marginal zone B-cell lymphoma of musosa-associated lymphoid tissue
- Nodal marginal zone B-cell lymphoma
- Follicular lymphoma
- Primary cutaneous follicle center lymphoma
- Mantle cell lymphoma
- Diffuse large B-cell lymphoma (DLBCL), not otherwise specified
- T-cell/histiocyte-rich large B-cell lymphoma
- Primary DLBCL of the CNS
- Primary cutaneous DLBCL, leg type
- EBV positive DLBCL of elderly
- DLBCL associated with chronic inflammation
- Primary mediastinal (thymic) large B-cell lymphoma
- Intravascular large B-cell lymphoma
- ALK positive large B-cell lymphoma
- Plasmablastic lymphoma
- Large B-cell lymphoma arising in HHV8-associated multicentric Castleman disease
- Primary effusion lymphoma
- Burkitt lymphoma/leukemia
- B-cell lymphoma, unclassifiable, with features intermediate between DLBCL and Burkitt lymphoma
- B-cell lymphoma, unclassifiable, with features intermediate between DLBCL and classical Hodgkin lymphoma

Table 26.1
Frequency of Mature B-Cell Lymphomas[1]

Type of Lymphoma	Approximate Frequency
Diffuse large B-cell	31%
Follicular	22%
MALT	8%
Mature T-cell (except ALCL[2])	8%
CLL/SLL[3]	7%
Mantle cell	6%
Mediastinal large B-cell	2.5%
ALCL	2.5%
Burkitt	2.5%
Nodal marginal zone	2%
Lymphoplasmacytic	1%
Others	7.5%

[1] Adapted from Jaffe ES, Harris NL, Stein H, et al. Pathology and Genetics: Tumors of Haematopoietic and Lymphoid Tissues. IARC Press, Lyon, 2001 and The Non-Hodgkin's Lymphoma Classification Project. A clinical evaluation of the International Lymphoma Study Group classification of non-Hodgkin's lymphoma. Blood 1997; 89: 3909–3918.
[2] Anaplastic large cell lymphoma.
[3] Chronic lymphocytic leukemia/small lymphocytic lymphoma.

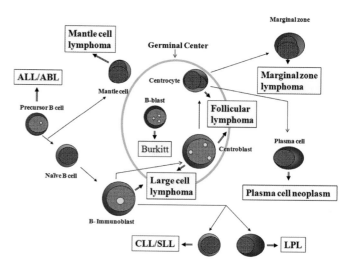

FIGURE 26.1 Scheme of B-cell differentiation and associated B-cell leukemias/lymphomas.

Table 26.2
The Ann Arbor Staging for Lymphomas

Stage	Criteria
I	Involvement of a single node region, or a single extralymphatic organ or site (Stage 1E)
II	Two or more involved lymph node regions on the same side of diaphragm, or with localized involvement of an extralymphatic organ or site (IIE)
III	Lymph node involvement on both sides of the diaphragm, or with localized involvement of an extralymphatic organ or site (IIE), or spleen (IIIS), or both (IIIES)
IV	Presence of diffuse or disseminated involvement of one or more extralymphatic organs, with or without associated lymph node involvement.

Systemic Symptoms	
A	Asymptomatic
B	Presence of fever, sweats, or weight loss >10% of body weight

Morphology

Lymphoid malignancies have extremely diverse morphologic features. This diversity, to some degree, correlates with the stage of their maturation. The precursor B-cell goes through several maturation and evolutionary steps, such as naïve lymphocyte, mantle cell, B-immunoblast, centroblast, centrocyte, marginal zone B-cell, and plasma cell, and each step depicts some characteristic morphologic features (Figure 26.1).

In general, mature B-lymphoid malignancies show two major patterns of lymph node involvement: follicular and diffuse (Figure 26.2).

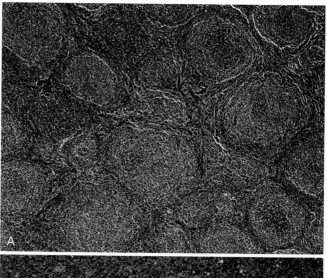

FIGURE 26.2 Lymph node sections demonstrating follicular (A) and diffuse (B) patterns of lymphomatous involvement.

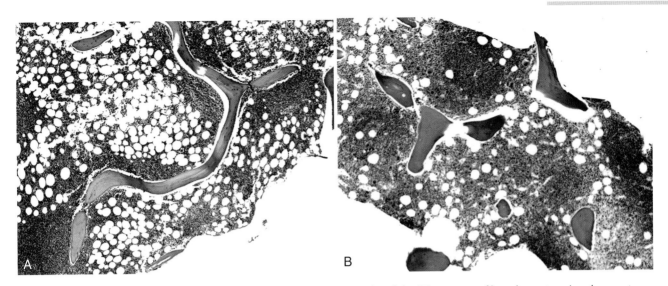

FIGURE 26.3 Bone marrow biopsy sections showing paratrabecular (A) and nodular (B) patterns of lymphomatous involvement.

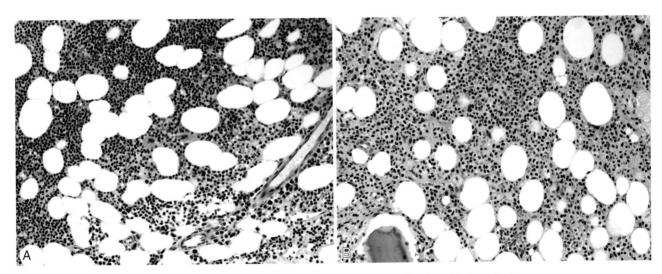

FIGURE 26.4 Bone marrow biopsy sections showing interstitial infiltrate by CLL cells (A) and hairy cells (B).

Bone marrow involvement may appear in different patterns, such as diffuse, paratrabecular, nodular, interstitial, or mixed. Diffuse involvement is defined as sheets of space-occupying lymphoid cells without formation of well-defined nodular aggregates. Paratrabecular disease refers to aggregates of lymphoma cells next to the bony trabeculae (Figure 26.3A). Nodular involvement consists of well-defined non-paratrabecular aggregates (nodules) of lymphoma cells (Figure 26.3B). Interstitial involvement is demonstrated by infiltration of lymphoma cells into the fatty tissue without obvious obliteration of the bone marrow architecture on low power examination (Figure 26.4).

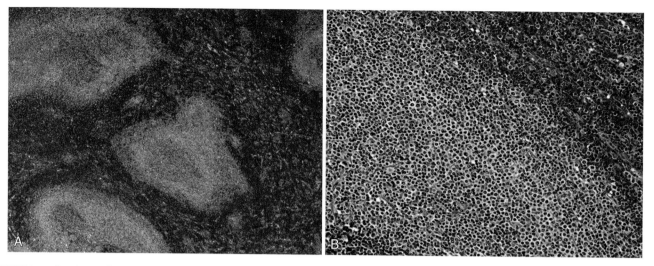

FIGURE 26.5 Sections of spleen showing expansion of the white pulp in a case of marginal zone B-cell lymphoma (A, low power; B, high power).

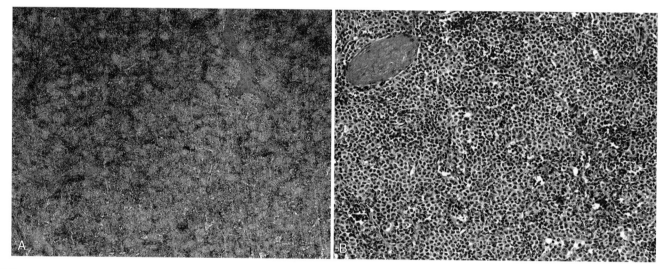

FIGURE 26.6 Sections of spleen demonstrating diffuse involvement of the red pulp in a patient with hairy cell leukemia (A, low power; B, high power).

Lymphomatous involvement of the spleen may consist of white pulp expansion, or diffuse red pulp infiltration, or both (Figures 26.5 and 26.6)

Immunophenotype

- Mature B-cell neoplasms express surface or intracellular light chain restriction. They are negative for markers seen in early B-cell precursors such as CD34 and TdT.
- Identification of monotypic B-cells through pattern recognition is the single most important immunophenotypic finding, while kappa/lambda ratio is not.
- Diagnostic pitfalls include scenarios where kappa/lambda ratio may be normal. For example, monotypic B-cells are admixed in a polytypic background (Figure 26.7), or there are dual monotypic B-cell populations (Figure 26.8).
- Cross-lineage aberrancies can be present (Figure 26.9).
- Characteristic immunophenotype of various subtypes of mature B-cell neoplasms is summarized in Table 26.3. However, immunophenotypic features alone may have only limited value in sub-classification. Multidisciplinary correlation is essential in diagnosis and prognosis.

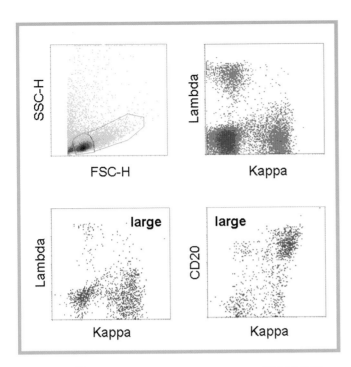

FIGURE 26.7 **MIXED MONOTYPIC AND POLYTYPIC B-CELLS DETECTED BY FLOW CYTOMETRY.** Open gate density plot (in light blue) of a lymph node sample reveals small and large lymphocytes by the scatter gate. The small lymphocytes (in green) contain polytypic B-cells. However, the large-lymphoid-enriched gate (in purple) reveals a monotypic B-cell population expressing surface kappa light chain restriction. The kappa to lambda ratio of all B-cells in this case is normal at close to 3:1.

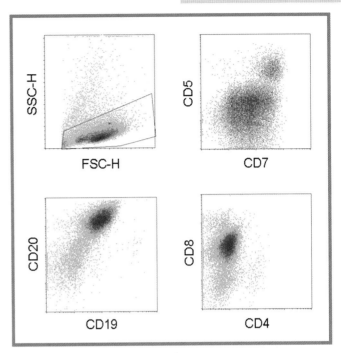

FIGURE 26.9 **FLOW CYTOMETRIC FINDINGS OF MATURE B-CELL NEOPLASM WITH CROSS-LINEAGE ABERRANCIES.** This is a case of diffuse large B-cell lymphoma. Open gate density plot display of the scatter gate reveals predominantly large lymphoids. Density plots of the lymphoid-enriched gate reveal monotypic B-cells expressing CD19, bright CD20, and surface lambda light chain restriction (not shown). The large neoplastic B-cells demonstrate cross-lineage aberrancies with T-lineage associated markers CD7 (partial, dim), and CD8 (dim).

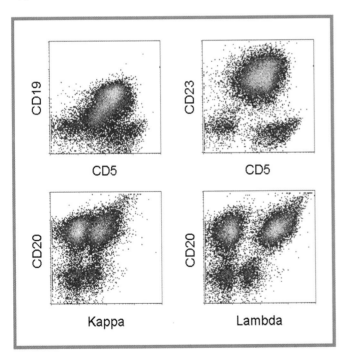

FIGURE 26.8 **DUAL MONOTYPIC B-CELL POPULATIONS DETECTED BY FLOW CYTOMETRY.** Density plots of lymphoid-enriched gate demonstrate a prominent population of abnormal B-cells expressing dim CD5, dim CD19, moderate CD20, and CD23. The abnormal B-cells are divided into two subsets by surface light chain expressions. One expresses surface kappa light chain restriction, while the other shows lambda restriction. The kappa to lambda ratio again is within normal limits, at approximately 1:1.

Table 26.3

Typical Immunophenotypic Features of Mature B-Cell Neoplasms

Type of Lymphoid Malignancy	Immunophenotype Positive	Immunophenotype Negative
CLL/SLL	CD5, CD19, CD23	CD10, FMC7, CD103, BCL-1
Mantle cell	CD5, CD19, BCL-1	CD10, CD23, BCL-6
Marginal zone	CD19, CD20, CD79a	CD5, CD10, CD23, BCL-1
Follicular	CD10, CD19, CD20, BCL-2, BCL-6	CD5, CD43, BCL-1
Lymphoplasmacytic	CD19, CD20, CD22, CD79a	CD5, CD10, CD23, BCL-1
Hairy cell	CD19, CD11c, FMC7, CD25, CD103	CD5, CD10, CD23
Prolymphocytic	CD19, CD20, CD22, CD79a, variable CD23	CD10, and usually CD5
Burkitt	CD10, CD19, CD20, CD22, BCL-6, Ki67 near 100%	CD5, CD23, CD34, TdT, BCL-2

Cytogenetics and Molecular Studies

The major common cytogenetic abnormalities observed in mature B-cell lymphoid neoplasms are shown in Table 26.4.

Mostly recurrent reciprocal translocations involve the constitutively active immunoglobulin genes, e.g., IGH@ at 14q32 (Figures 26.10 and 26.11), and oncogenes such as CCND1, BCL2, BCL6, MYC, and PAX5. These can be seen in >90% of the cases of Burkitt lymphoma, mantle cell lymphoma and follicular lymphoma. In contrast, diffuse large B cell lymphoma karyotypes exhibit heterogeneous and often complex aberrations with multiple breakpoints and translocations along with various numerical abnormalities (gains and losses). The karyotypes can be obtained from cultured tissues and exhibit both simple and complex patterns that include these translocations. Other chromosomal aberrations such as trisomies, monosomies and deletions/duplications also occur but less frequently.

Most of the chromosomal translocations are better detected by the FISH technique with specific probes that identify the fusion signal pattern and with increased sensitivity and specificity, especially in bone marrow specimens

Table 26.4
Major Genetic Features of Small Mature B-Cell Lymphoid Malignancies

Type of Lymphoid Malignancy	Chromosomal Abnormalities	Genes
CLL/SLL	del(17)(p13)	TP53
	del(11)(q22)	ATM
	Trisomy 12	
	del(13)(q12q14)	??Rb1
Mantle cell	t(11;14)(q13;q32)	CCND1-IGH@
Marginal zone		
Splenic	del(7)(q22–32)	
Extranodal	Trisomy 3	
	t(11;18)(q21;q21)	BIRC3-MALT1
Follicular	t(14;18)(q32;q21)	IGH@-BCL2
Lymphoplasmacytic	t(9;14)(p13;q32)	PAX5-IGH@
Hairy cell	Non-specific	
B-Prolymphocytic	14q32 abnormalities	IGH@
	del(11)(q22)	ATM
Burkitt	t(8;14)(q24;q32)	MYC-IGH@
	t(2;8)(p12;q24)	IGK@-MYC
	t(8;22)(q24;q11.2)	MYC-IGL@

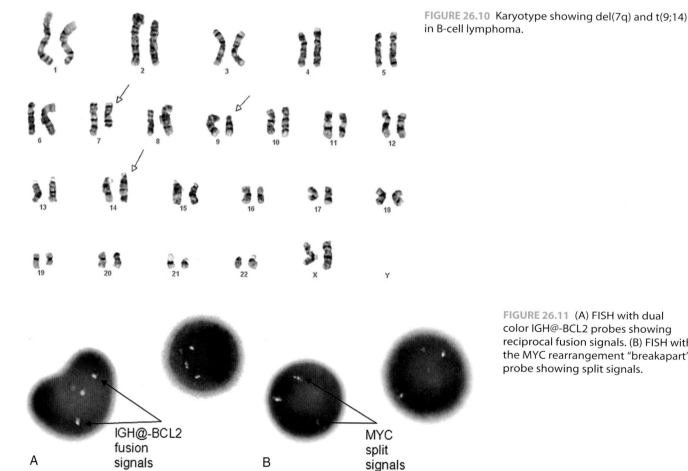

FIGURE 26.10 Karyotype showing del(7q) and t(9;14) in B-cell lymphoma.

FIGURE 26.11 (A) FISH with dual color IGH@-BCL2 probes showing reciprocal fusion signals. (B) FISH with the MYC rearrangement "breakapart" probe showing split signals.

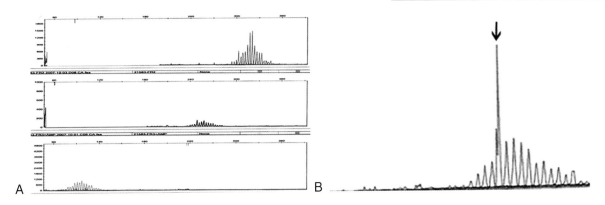

FIGURE 26.12 **EXAMPLES OF IMMUNOGLOBULIN GENE CLONALITY ANALYSIS BY PCR.** (A) Negative study showing only a polyclonal amplicon pattern with framework primers 1 (blue), 2 (black), and 3 (green). (B) Clonal peak (arrow) superimposed on a background polyclonal "smear" produced with the framework 1 primer set.

and paraffin-embedded sections. The t(14;18) translocation of *IGH@-BCL2* may be also detected by either Southern blot or PCR.

Rearrangements of the *ALK* and *NFKB2* genes are involved in anaplastic large cell lymphoma and cutaneous T-cell lymphoma, respectively, and can been seen by karyotype and FISH.

The most common tumor suppressor gene involved in lymphoid malignancies is *TP53*, the deletion of the 17p13 band which is more commonly cryptic and therefore easily detected by the FISH technique.

In addition to karyotype analysis/classical cytogenetics, and multicolor metaphase and interphase FISH, other genetics techniques such as array-CGH, SNP analysis, and gene expression profiles have also contributed to the diagnosis and classification.

The hallmark of B-cell leukemia/lymphoma diagnosis by gene rearrangement studies is the demonstration of clonal patterns of rearrangement in the DNA of the immunoglobulin loci. Most malignant lesions are assumed to be clonal, having descended from a transforming event in a single progenitor cell, and the specimen should therefore exhibit a single predominant rearrangement pattern, one present at sufficiently high proportion to stand out from the background polyclonal "smear" (Figure 26.12).

Southern blot analysis was the first method devised for detection of immunoglobulin gene rearrangements, and, in contrast to most other areas of the molecular pathology laboratory, it has yet to be entirely replaced by more modern PCR-based methods (Figure 26.13). The reason is that PCR cannot cover such a large gene family comprehensively, and will therefore miss between 10% and 50% of clonal neoplasms. Reflex-testing by Southern blot in such cases will usually be informative.

Gene rearrangement studies in B-cell lesions are performed for several purposes:

- To distinguish malignant (clonal) lesions from reactive (nonclonal, polyclonal) lesions;
- To determine likely cell of origin (B-cell versus T-cell) of the lesion;

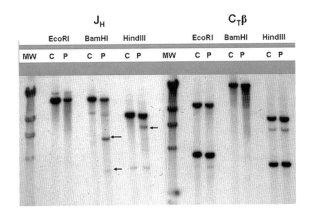

FIGURE 26.13 **SOUTHERN BLOT ANALYSIS FOR CLONALITY OF IMMUNOGLOBULIN GENE REARRANGEMENTS IN A B-CELL LYMPHOMA.** Control (human placental) DNA (C) and patient (P) DNA were digested with three different restriction endonucleases (*Eco*RI, *Bam*HI, and *Hin*dIII) and hybridized with either the JH immunoglobulin heavy chain probe (left blot) or the CTβ T-cell receptor probe (right blot) and subjected to autoradiography. Non-germline, rearranged bands are seen in the patient's DNA with two of the restriction digests (arrows) hybridized with the JH probe only; no rearranged bands are seen on the CTβ blot, consistent with clonal B-cell origin.

- To determine if a second (synchronous or metachronous) lesion represents the same or a different (new) clone;
- To detect or monitor minimal residual disease (MRD) after initial therapy.

Though one may generally assume that lymphocytes with rearranged immunoglobulin heavy chain genes are B-cells, a small percentage of T-cell malignancies may secondarily rearrange their immunoglobulin genes, and vice versa. In cases where doubt persists as to cell of origin, the analysis of J-kappa rearrangements may be informative, since these are more specific (though less sensitive) for B-cell malignancies.

Differential Diagnosis

The presence of a wide variety of mature lymphoid malignancies and numerous subcategories makes their accurate diagnosis and classification very challenging. In spite of significant clarification in diagnostic criteria and classification, advances in technology, and availability of accessory tools in numerous diagnostic centers, there is still evidence of significant discordance amongst pathologists in the diagnosis and classification of lymphoid malignancies. Some general practical points helpful in differential diagnosis of mature lymphoid neoplasms are presented as follows.

1. Morphologic features in many instances are not sufficient to separate the B-cell from the T-cell disorders. Some subcategories of small mature B-cell malignancies, such as follicular lymphoma, marginal zone B-cell lymphoma, and mantle cell lymphoma, may morphologically mimic one another. Therefore, it is highly recommended that immunophenotypic, cytogenetic, and molecular studies be part of the routine diagnostic work-up of lymphoid malignancies.
2. Node-based lymphomas should be distinguished from garden varieties of reactive lymphadenopathies, such as follicular hyperplasia, viral infections, and drug-induced lymphadenitis.
3. Cutaneous lymphoid malignancies may mimic a variety of inflammatory dermal lesions.
4. Burkitt lymphoma/leukemia should be distinguished from B- and T-cell lymphoblastic leukemias/lymphomas.

Additional Resources

Aukema SM, Siebert R, Schuuring E, et al: Double-hit B-cell lymphomas, *Blood* 117:2319–2331, 2011.

Bench AJ, Erber WN, Follows GA, et al: Molecular genetic analysis of haematological malignancies II: Mature lymphoid neoplasms, *Int J Lab Hematol* 29:229–260, 2007.

Burg G, Kempf W, Cozzio A, et al: WHO/EORTC classification of cutaneous lymphomas 2005: histological and molecular aspects, *J Cutan Pathol* 32:647–674, 2005.

Carbone A, Gloghini A, Aiello A, et al: B-cell lymphomas with features intermediate between distinct pathologic entities: From pathogenesis to pathology, *Hum Pathol* 41:621–631, 2010.

Dadi S, Le Noir S, Asnafi V, et al: Normal and pathological V(D)J recombination: Contribution to the understanding of human lymphoid malignancies, *Adv Exp Med Biol* 650:180–194, 2009.

Good DJ, Gascoyne RD: Classification of non-Hodgkin's lymphoma, *Hematol Oncol Clin North Am* 22:781–805, 2008.

Hartmann EM, Ott G, Rosenwald A: Molecular biology and genetics of lymphomas, *Hematol Oncol Clin North Am* 22:807–823, 2008.

Hsi ED, Mirza I: Update in the pathologic features of mature B-cell and T/NK-cell leukemias, *Semin Diagn Pathol* 20:180–195, 2003.

Hsi ED: The leukemias of mature lymphocytes, *Hematol Oncol Clin North Am* 23:843–871, 2009.

Jaffe ES, Harris NL, Stein H, et al: Classification of lymphoid neoplasms: the microscope as a tool for disease discovery, *Blood* 112:4384–4899, 2008.

Jaffe ES, Harris NL, Vardiman JW, et al: Hematopathology, Philadelphia, 2010, Saunders/Elsevier.

Jaffe ES: The 2008 WHO classification of lymphomas: implications for clinical practice and translational research, *Hematology Am Soc Hematol Educ Program*:523–531, 2009.

Kurtin PJ: Indolent lymphomas of mature B lymphocytes, *Hematol Oncol Clin North Am* 23:769–790, 2009.

Lenz G, Staudt LM: Aggressive lymphomas, *N Engl J Med* 362:1417–1429, 2010.

Lones MA, Raphael M, Perkins SL, et al: Mature B-cell lymphoma in children and adolescents: International group pathologist consensus correlates with histology technical quality, *J Pediatr Hematol Oncol* 28:568–574, 2006.

Philip K, Schuuring E: Molecular cytogenetics of lymphoma: where do we stand in 2010? *Histopathology* 58:128–144, 2011.

Roullet M, Bagg A: The basis and rational use of molecular genetic testing in mature B-cell lymphomas, *Adv Anat Pathol* 17:333–358, 2010.

Swerdlow SH, Campo E, Harris NL, et al: WHO classification of tumours of haematopoietic and lymphoid tissues, ed 4, Lyon, 2008, International Agency for Research on Cancer.

Walsh SH, Rosenquist R: Immunoglobulin gene analysis of mature B-cell malignancies: reconsideration of cellular origin and potential antigen involvement in pathogenesis, *Med Oncol* 22:327–341, 2005.

Chronic Lymphocytic Leukemia/Small Lymphocytic Lymphoma

Chronic lymphocytic leukemia/small lymphocytic lymphoma (CLL/SLL) is a lymphoproliferative disorder of small, mature B-lymphocytes primarily involving peripheral blood, bone marrow, and lymph nodes. The neoplastic B-cells typically coexpress CD5 and CD23 and lack expression of CD10, FMC7, and CD79b. CLL/SLL is divided into two overlapping categories:

1. CLL, primarily involving bone marrow and peripheral blood with or without lymph node involvement. The diagnosis requires an absolute B-lymphocytosis of $\geq 5000/\mu L$, sustained for at least 3 months. If absolute monoclonal B-lymphocyte count is $<5000/\mu L$ the term *monoclonal B lymphocytosis* is used. Bone marrow examination is not necessary for the diagnosis of CLL.
2. SLL, primarily involving lymph nodes with or without bone marrow involvement and no peripheral blood lymphocytosis (non-leukemic).

CLL/SLL is the most common type of leukemia in Western countries, accounting for about 40% of all leukemias in patients above 65 years of age. This disorder is extremely rare under the age of 30 years, but about 20% of patients are diagnosed under the age of 55 years. The male to female ratio is about 2:1. Although the presence of familial aggregates of CLL has been well documented, the mode of inheritance is not known. There is a sevenfold increase in risk of CLL in first-degree relatives.

Approximately 25% of patients are free of symptoms, and the CLL is an incidental finding during a routine blood examination. About 5–10% of patients show systematic symptoms, such as weight loss, fever, night sweats, and/or extreme fatigue. Physical examination may reveal lymphadenopathy, splenomegaly, and hepatomegaly in approximately 85%, 50%, and 14% of patients, respectively. Autoimmune complications, primarily hemolytic anemia and thrombocytopenia, occur in up to 25% of CLL/SLL patients.

The natural history of CLL/SLL is extremely variable, with survival times ranging from 2 to 20 years. Overall, the response rate to therapy and survival is better in women than in men. Also, patients with atypical morphologic and/or immunophenotypic features tend to have a more aggressive clinical course. In general, presence and extent of lymphadenopathy, splenomegaly, hepatomegaly, anemia, and thrombocytopenia are the major clinical parameters that correlate with prognosis (Table 27.1). Several biomarkers, such as expression of CD38 and ZAP-70, unmutated *IGVH* (V_H) genes, del(11q22.3), and del(17p13.1) are indicative of aggressive clinical course in CLL/SLL (see below).

Several treatment modalities are available for the patients with CLL/SLL. The therapeutic approaches are mainly based on the patient age and physical status, stage of the disease, and cytogenetic findings. A "watch and wait" approach may be chosen for patients in early stage of disease or low risk category, whereas patients with advanced or high risk disease usually receive treatment. Alkylating agents, purine analogs, and monoclonal antibodies (such as rituximab) are frequently utilized in the therapeutic protocols.

Table 27.1
Prognostic Factors in Chronic Lymphocytic Leukemia/Small Lymphocytic Lymphoma[1]

Factor	Low Risk	High Risk
Gender	Female	Male
Clinical stage		
Binet	A	C
Rai	0	III and IV
Lymphocyte morphology	Typical	Atypical
Bone marrow involvement	Non-diffuse	Diffuse
Elevated levels of serum $\beta 2$-macroglobulin and CD23	Not present	Present
CD38 expression	Negative	Positive
ZAP-70	Negative	Positive
IgV$_H$ gene status	Mutated	Unmutated
Cytogenetics	Normal or del(13q14)	del(17p13) or del(11q22.3)

[1] Adapted from Oscier D, Fegan C, Hillmen P, et al. Guidelines on the diagnosis and management of chronic lymphocytic leukaemia. Br J Haematol 2004; 125: 294–317.

Morphology

The blood smears show evidence of absolute lymphocytosis with >90% of lymphocytes consisting of small cells with round nucleus, coarse chromatin, indistinct or absent nucleoli, and scanty non-granular cytoplasm (Figure 27.1). A smaller proportion of lymphoid cells (<10%) consist of prolymphocytes. Prolymphocytes are larger than lymphocytes with more abundant cytoplasm and a prominent nucleolus. Smudge and basket cells are frequently present, particularly in cases with high lymphocyte count (Figure 27.1B and C). These cells are degenerated and damaged lymphocytes.

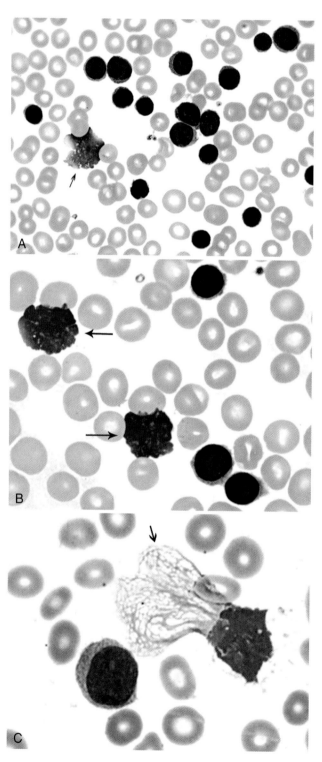

FIGURE 27.1 **CHRONIC LYMPHOCYTIC LEUKEMIA.** Blood smear (A) demonstrating lymphocytosis with the presence of smudge cells (B, arrows) and a basket cell (C).

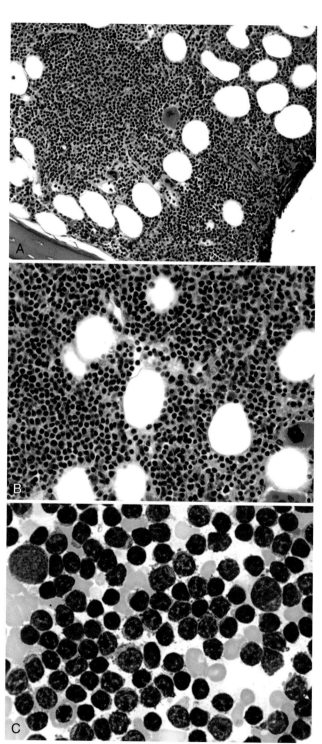

FIGURE 27.2 Bone marrow biopsy section (A and B) and bone marrow smear (C) showing involvement with chronic lymphocytic leukemia.

MORPHOLOGY

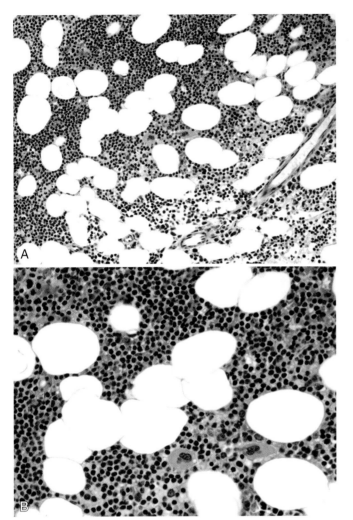

FIGURE 27.3 Bone marrow biopsy section demonstrating interstitial lymphoid infiltrate in a patient with chronic lymphocytic leukemia (A, low power; B, high power).

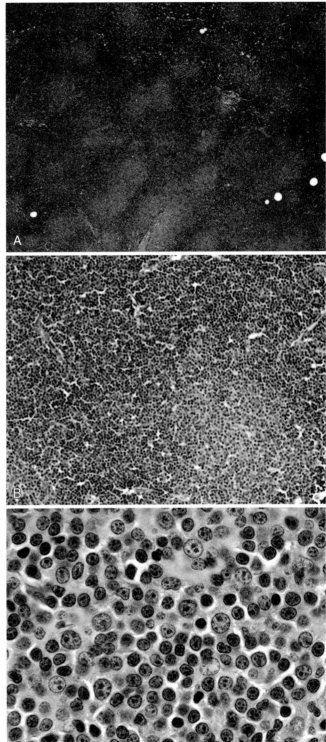

FIGURE 27.4 Lymph node section of a patient with chronic lymphocytic leukemia demonstrating proliferating centers or pseudofollicles (A and B, pale areas) consisting of a mixture of lymphocytes, prolymphocytes, and paraimmunoblasts (C).

The requirement for the diagnosis of CLL on blood examination is an absolute B-lymphocytosis (with coexpression of CD5 and CD23) of >5000/µL. About 30% of patients with CLL show a white blood cell count >100,000/µL.

Approximately 15% of CLL patients show *atypical* morphologic features, such as increased proportion of prolymphocytes (PL) (>10% but <55%), or presence of lymphocytes with cleaved nuclei, or lymphoplasmacytic morphology (see below).

The pattern of involvement in bone marrow biopsy sections is interstitial, nodular, diffuse, or a combination of these (Figures 27.2 and 27.3). The diffuse pattern is usually seen in the advanced stages of the disease. The lymphoid infiltrates consist of small, round lymphocytes with scattered prolymphocytes and larger cells called paraimmunoblasts. The lymphocyte count in bone marrow smears in CLL is usually >30% of the bone marrow cells.

The affected lymph nodes show architectural effacement with a diffuse infiltration by small lymphocytes. Characteristically, there are ill-defined paler areas with the predominance of prolymphocytes and paraimmunoblasts (Figure 27.4). These areas are called *proliferation centers* or *pseudofollicles*. Pseudofollicules are less frequently observed

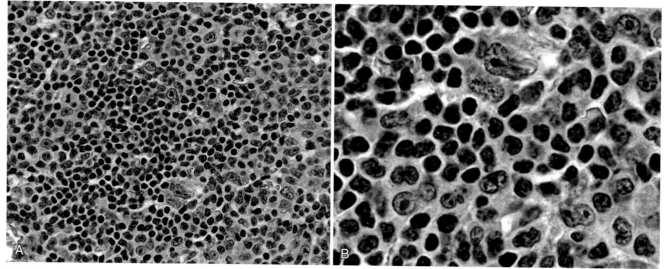

FIGURE 27.5 Reed–Sternberg-like cells in the lymph node section of a patient with small lymphocytic lymphoma (A, low power; B, high power).

in the bone marrow and spleen. In some cases, there is a predominance of atypical lymphoid cells, such as lymphocytes with irregular nuclei or lymphoplasmacytic cells. These cases may mimic mantle cell or lymphoplasmacytic lymphoma (see below). In some cases, Reed–Sternberg-like cells may be present (Figure 27.5). The lymph node involvement in some cases is mostly inter-follicular with preservation of the lymphoid follicles (inter-follicular pattern) (Figure 27.6).

The white pulp is the primary site of involvement in the spleen, but the red pulp is also frequently involved (Figure 27.7). Hepatic infiltration is usually in the portal areas (Figure 27.8).

Immunophenotype

Characteristic immunophenotypic pattern by multiparametric flow cytometry (MFC) includes (Figure 27.9):

- Dim to moderate expression of B-cell-associated antigens (commonly CD19, heterogeneously dim CD20, dim CD22, and dim or negative CD79a);
- Negativity of FMC7;
- Coexpression of dim CD5 and CD23;
- Dim surface or intracellular light chain restriction.

Immunohistochemical studies show (Figures 27.10 and 27.11):

- Coexpression of CD20, CD23, dim CD5 and BCL-2;
- Negativity of BCL-1;
- Variable negativity of markers that are dimly expressed as detected by flow cytometry.

CLINICAL CORRELATIONS

The expression of CD38 and/or ZAP-70 is associated with an aggressive clinical course.

- CD38 is an ectoenzyme 45 kD transmembrane glycoprotein which appears to contribute to proliferative potential of B-CLL cells, enhancing clinical aggressiveness of the disease.
- ZAP-70 (zeta-chain associated protein of 70 kD) is an intracellular tyrosine kinase which is involved in TCR signaling and is expressed in B-CLL cells that possess unmutated IgV_H genes. The expression of ZAP-70 is moderate in normal T- and bright in NK-cells and predominantly negative in normal B-cells. Therefore, in order to determine the expression level of ZAP70 by neoplastic B-cells, it is important to use the patient's T- and NK-cells as internal positive controls, as well as the patient's normal B-cells (CD19+/CD5−) as internal negative control. ZAP-70 is considered positive when it is expressed by at least 20% of the CLL cells (Figure 27.12).

Cytogenetic and Molecular Studies

In most CLL cases (>60%) karyotype is normal, but the abnormalities tend to increase in frequency and incidence during the course of the disease (Figure 27.13).

Chromosomal translocations are rare in CLL. Loss of the tumor suppressor genes may be the likely pathogenetic possibility.

Among patients with abnormal karyotypes, as many as 65% have one chromosome abnormality, 25% have two abnormalities, and the remainder (10%) have more complex abnormalities.

Deletion 13q14/monosomy 13 is the most common finding (36–50%), and is believed to be a primary event in

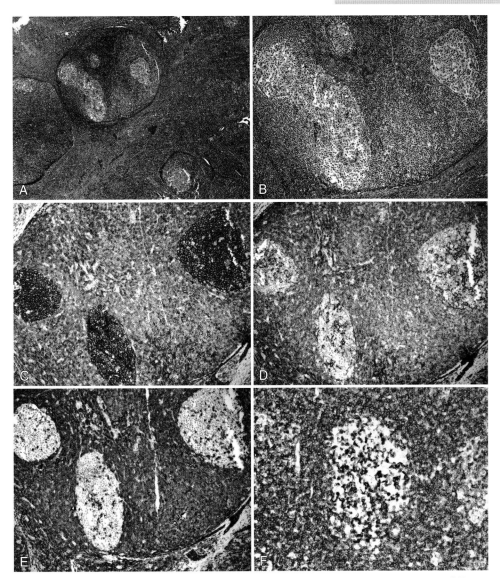

FIGURE 27.6 **SMALL LYMPHOCYTIC LYMPHOMA WITH PRIMARY INVOLVEMENT OF INTER-FOLLICULAR AREAS.** There are areas of fibrosis with several well-preserved germinal centers (A, low power; B, high power). The tumor cells in the inter-follicular areas show a dim expression of CD20, while the follicular B-cells are strongly CD20+ (C). The cells in the inter-follicular areas express CD23 (D), CD5 (E), and BCL-2 (F).

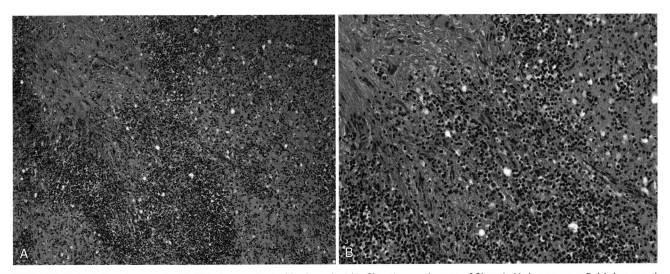

FIGURE 27.7 Splenic involvement with CLL demonstrated by lymphoid infiltration and areas of fibrosis (A, low power; B, high power).

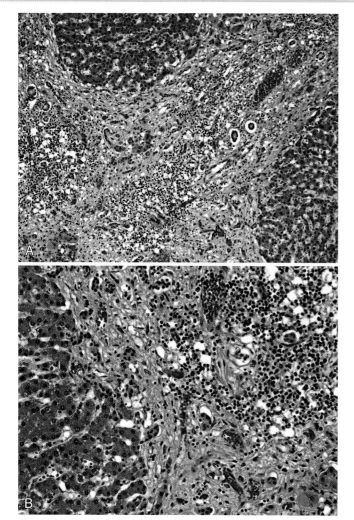

FIGURE 27.8 Hepatic involvement with CLL demonstrated by portal lymphoid infiltration and fibrosis (A, low power; B, high power).

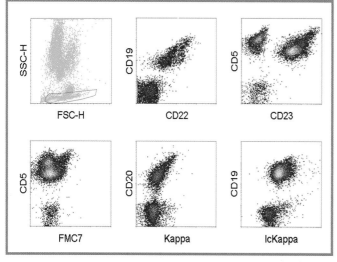

FIGURE 27.9 CHARACTERISTIC PHENOTYPE OF CLL BY FLOW CYTOMETRY. Density plots of the lymphoid-enriched gate show abnormal B-cells, which are positive for CD5 (dim), CD19, CD20 (heterogeneous, dim), CD22 (dim), CD23, and expressing no surface light chains. Cytoplasmic light chain staining reveals the monotypic nature of these abnormal B-cells with intracellular kappa light chain restriction.

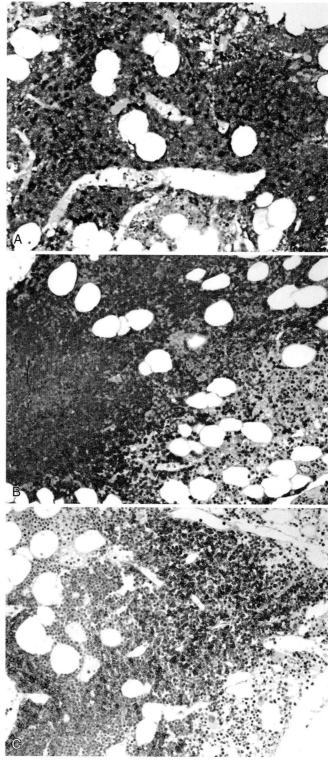

FIGURE 27.10 Immunohistochemical stains of bone marrow biopsy section from a patient with CLL. (A) Dual staining demonstrating sheets of CD20+ cells (red) and scattered CD3+ cells (brown). The tumor cells also express CD5 (B) and CD23 (C).

B-CLL, as it is present in most tumor cells and is frequently the sole abnormality (Figure 27.14).

The second most common abnormality, and the most common abnormality to be detected by conventional

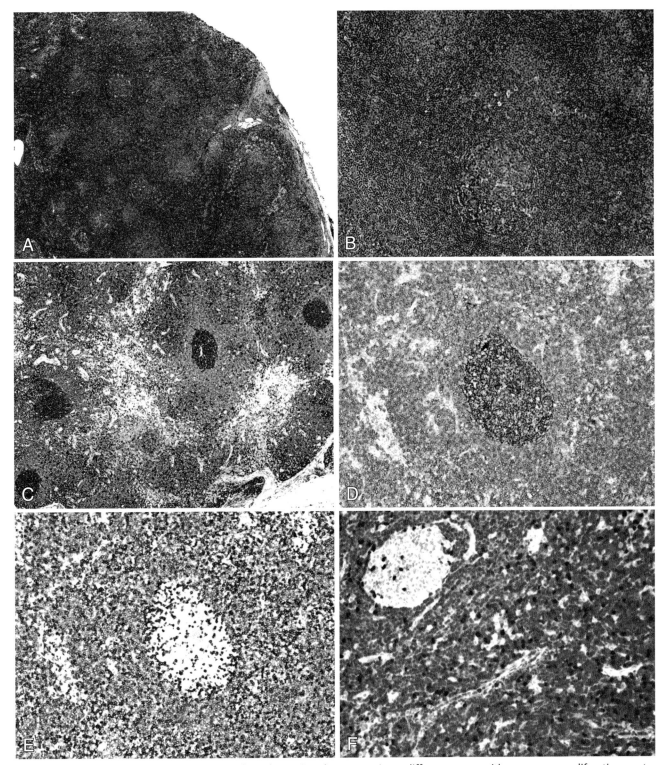

FIGURE 27.11 SMALL LYMPHOCYTIC LYMPHOMA. Lymph node section demonstrating a diffuse process with numerous proliferation centers (pseudo-follicles) and scattered well-preserved germinal centers (A, low power; B, high power). The tumor cells show a dim expression of CD20, while the follicular B-cells are strongly CD20+ (C). CD23 is expressed by the tumor cells and follicular dendritic cells (D). CD5 (E) and ZAP-70 (F) are diffusely expressed except for the follicular B-cells.

cytogenetics, is trisomy 12 (11–21%) (Figure 27.15). Trisomy 12 usually displays an excess of large lymphocytes identifying the CLL mixed-cell-type variant of the FAB classification. Trisomy 12 may be a secondary event in the course of CLL because it is typically identified in a minority of the tumor cells. Trisomy 12 is predominantly associated with unmutated *VH* genes and seems to be associated with advanced or atypical cases of CLL.

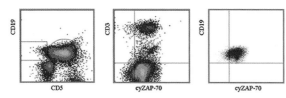

FIGURE 27.12 Flow cytometric analysis of ZAP-70. A large population of B-cells are CD5-positive, CD3-negative, and coexpress CD19 and ZAP-70.

Less frequent primary aberrations include 14q32 rearrangements (up to 21%), 11q22.3 deletion (9–15%), and a 17p13 deletion (7–12%) (Table 27.2 and Figures 27.16 and 27.17).

Other less frequent chromosome abnormalities also occur (e.g., complex karyotypes). A chromosome 6q deletion occurs in 7% of all CLL patients (as a primary event in 4%) and represents a cytogenetic and clinicobiological

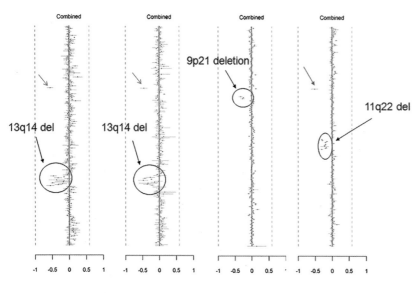

FIGURE 27.13 A set of four representative panels of Bacterial artificial chromosome (BAC)–chromosomal genomic Hybridization microarray (BAC-CGH) results obtained with DNA probes from selected genomic regions considered important in hematopoietic and lymphoid neoplasms in four cases of CLL. panels A and B show A loss of 13q14 regions (arrows) when compared with the diploid (normal) vertical line; panel C shows A loss of the 9p21 locus (arrow); and panel D represent A case with A deletion of the 11q22 locus (black arrow). three of these four cases (A, B, and D) also showed A cryptic deletion of the 8q24 (*MYC*) locus (red arrow).

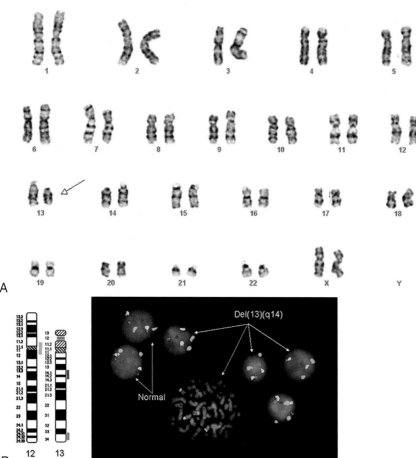

FIGURE 27.14 (A) Bone marrow karyotype of a patient with CLL demonstrating 46,XX,del(13)(q12q14). (B) FISH analysis demonstrating deletion of 13q14 (red signals), and normal signals for 13q34 (aqua), chromosome 12 (green).

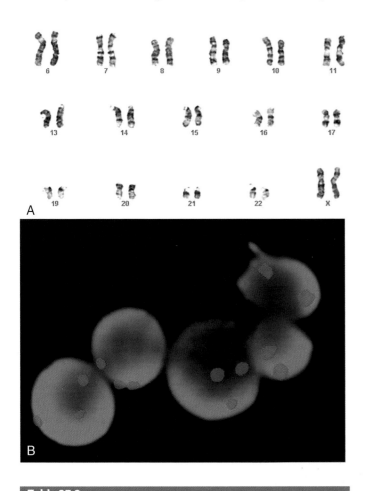

FIGURE 27.15 (A) Bone marrow karyotype and (B) FISH of a patient with CLL demonstrating trisomy 12.

Table 27.2
Correlation between Cytogenetic Abnormalities and Survival in CLL/SLL

Cytogenetics	Frequency	Survival (Years)
6q−	~4%	~3
del(17p13)	~7%	~3
del(11q22)	~10%	~6
Trisomy 12	~15%	~9
13q−	>50%	>12

entity that exhibits a distinct phenotypic and hematologic profile. Patients with del(6q) usually present with a relatively high WBC count, classical immunophenotype, and CD38 positivity, which are associated with acceleration to the more aggressive prolymphocytic leukemia (PLL). Therefore, del(6q) patients require immediate therapy to achieve remission.

Chromosomal aberrations are not always detected in CLL patients' B-cells, as B-CLL cells are unresponsive to most lymphocyte mitogens (low mitotic index) and are extremely difficult to maintain in culture.

FISH performed in conjunction with conventional cytogenetics is the method of choice. FISH techniques are more sensitive for the detection of clinically significant chromosome abnormalities than standard chromosome analysis. All molecular cytogenetic techniques (i.e., FISH, a-CGH) have increased the detection rate of CLL to 80%.

The recommended FISH panel for CLL detection consists of 11q22.3 (*ATM* gene), 13q14 (*D13S319*), the centromere of chromosome 12 (*D12Z3*), and 17p13.1 (*TP53* gene). In addition, the 6q21(*MYB*) probe and *CCND1/IGH@* t(11;14) can be added.

The mutational status of the *IGVH* genes divides CLL into two major subtypes: mutated and unmutated.

- Most CLL patients exhibit clonal rearrangement of the immunoglobulin heavy chain genes. However, such studies are usually not necessary when the diagnosis is obvious by routine pathologic examination.
- Approximately 45% of CLL patients show no evidence of *VH* gene mutation (unmutated). In general, these patients have an aggressive clinical course and advanced disease stage. This determination requires DNA sequencing and is not widely available at present. The unmutated group shows a strong association with overexpression of ZAP-70 protein, which could be detected by flow cytometry or immunohistochemistry (see Figures 27.11F and 27.12).

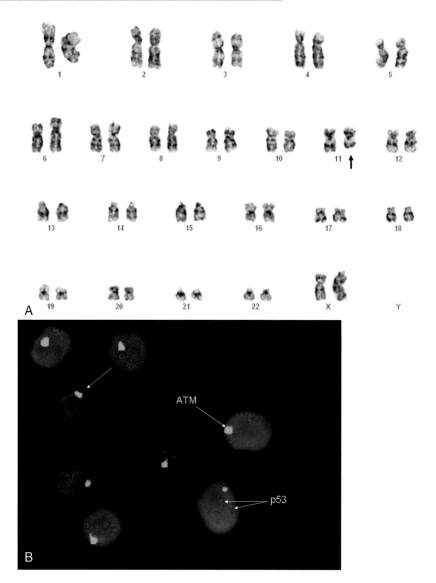

FIGURE 27.16 (A) Bone marrow karyotype of a patient with CLL showing 46,XX,del(11)(q22.1). (B) FISH analysis exhibiting deletion of 11q22 (loss of ATM-green signals).

Variants of CLL/SLL

ATYPICAL CLL

Approximately 15% of CLL patients show atypical morphologic or immunophenotypic features. The atypical morphologic features include (Figures 27.18 to 27.20):

- Presence of lymphocytes with cleaved or irregular nuclei (>10% of the lymphoid cells);
- Increased proportion of prolymphocytes (>10% but <55%). Cases with prolymphocytes >55% fall into the category of B-cell prolymphocytic leukemia.
- Plasmacytic differentiation with lymphoplasmacytic morphology.

Atypical immunophenotypic pattern by MFC (Figure 27.21) includes:

- Moderate to bright expression of B-cell-associated antigens (commonly moderate to bright CD20);
- Variable positivity of FMC7;
- Variable or lack of coexpression of CD5 and CD23;
- Bright surface light chain restriction;
- Positivity of CD10 and other aberrancies.

MU HEAVY-CHAIN DISEASE

Mu heavy-chain disease is a rare condition. Patients generally present clinicopathologic features similar to those of CLL. There is mature lymphocytosis often with infiltration of the

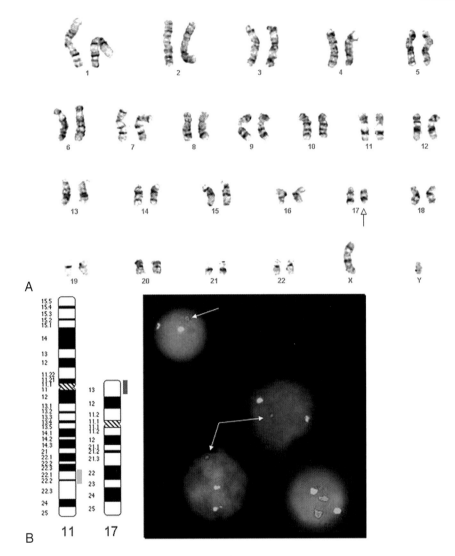

FIGURE 27.17 (A) Bone marrow karyotype and (B) FISH analysis demonstrating del(17)(p11.2).

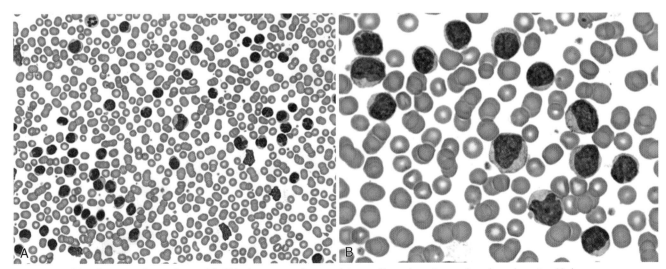

FIGURE 27.18 Blood smear of a patient with CLL demonstrating a mixture of lymphocytes and prolymphocytes (A, low power; B, high power).

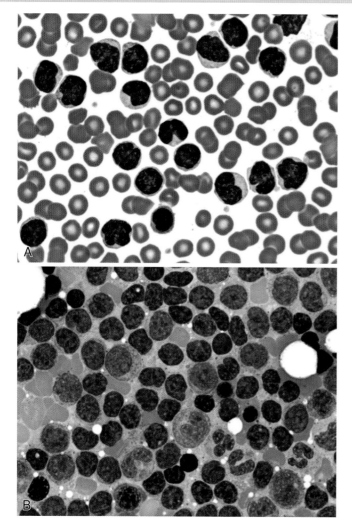

FIGURE 27.19 Blood (A) and bone marrow (B) smears of a patient with CLL demonstrating atypical lymphocytes with irregular and folded nuclei.

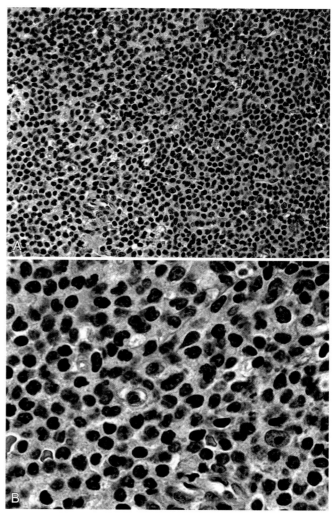

FIGURE 27.20 Lymph node biopsy section of a patient with SLL demonstrating atypical lymphoid cells with irregular and folded nuclei (A, low power; B, high power).

bone marrow, liver, and spleen, with the presence of vacuolated plasma cells. Lymphadenopathy is not a prominent clinical feature. A defective mu chain is detected in serum.

Transformation of CLL to a More Aggressive Disease (Richter Syndrome)

Development of a high grade non-Hodgkin lymphoma in patients with CLL was first described by Richter in 1928. The term Richter syndrome was later applied to the transformation of CLL to a wide variety of more aggressive lymphoid malignancies, such as large cell lymphoma (Figures 27.22 and 27.23), prolymphocytic leukemia, lymphoblastic lymphoma, Hodgkin lymphoma, and plasma cell myeloma. The incidence of Richter syndrome in CLL is

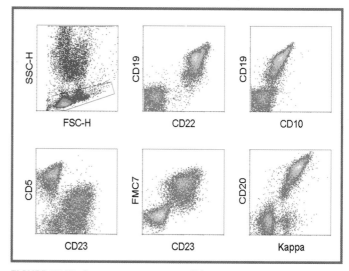

FIGURE 27.21 ATYPICAL PHENOTYPE OF CLL BY FLOW CYTOMETRY. Density plots of the lymphoid-enriched gate show monotypic B-cells that are positive for CD5 (partial, dim), CD10, CD19, CD20 (mainly moderate to bright), CD22 (moderate), CD23, FMC7, and bright surface kappa light chain restriction.

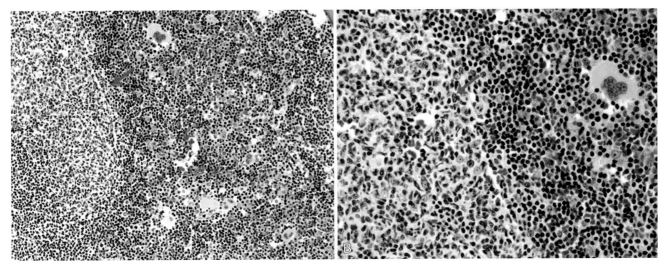

FIGURE 27.22 **RICHTER SYNDROME.** Bone marrow biopsy section from a patient with CLL demonstrating a focal area (arrow) of transformation to large cell lymphoma (A, low power; B, high power).

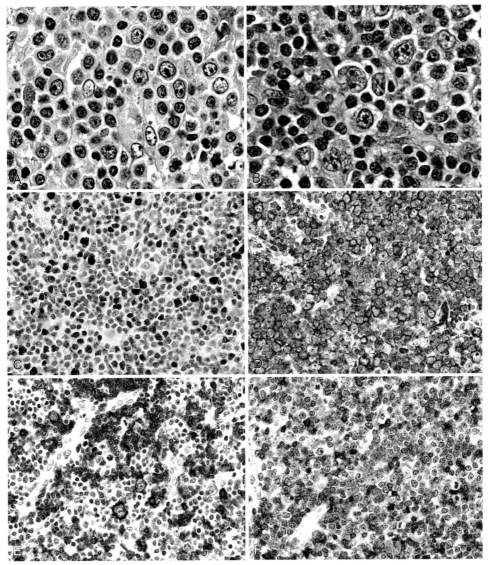

FIGURE 27.23 **RICHTER SYNDROME.** Lymph node biopsy section from a patient with CLL demonstrating a mixture of lymphocytes and large transformed cells (A and B). The large transformed cells express Ki67 (C), and, similar to the CLL cells, are positive for CD20 (D), CD23 (E) and CD5 (F).

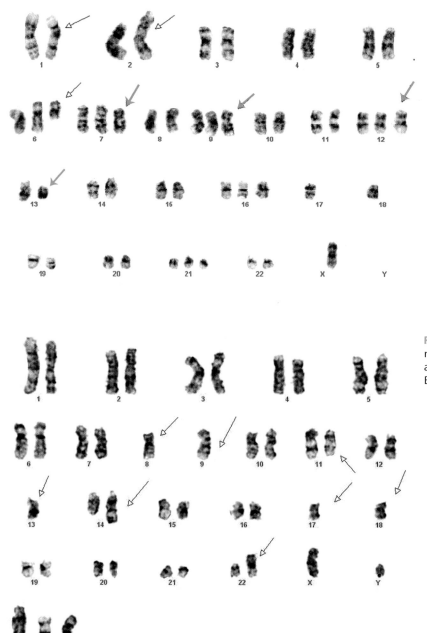

FIGURE 27.24 Karyotype of a CLL patient showing complex chromosomal aberrations including +7, +9, +12, and deletion 13q, compatible with CLL in transformation.

FIGURE 27.25 Complex karyotype with t(11;14), monosomy 13, 17 and other non-specific aberrations in a case of transformation of CLL to B-prolymphocytic leukemia.

about 5–10%, with prolymphocytic leukemia being the most frequent type of transformation. The transformed cells may arise from the original CLL clone or may represent a new neoplastic clone, and the sequence-specific quantitative PCR methods discussed earlier can be used to distinguish between these two possibilities. The exact mechanism(s) of this transformation is not well understood. Multiple genetic abnormalities such as *TP53* mutation, deletion of retinoblastoma gene (13q14), increased copy number of *MYC*, and decreased expression of *MYB* gene have been described. Trisomy 12 and 11q aberrations are more frequent in Richter syndrome than in the overall CLL population (Figures 27.24 to 27.26).

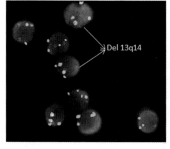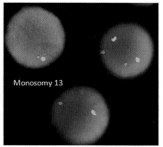

FIGURE 27.26 FISH with a CLL probe panel containing the 13q14 region probe (red), 13q34 region probe (aqua) and centromere 12 (green), exhibiting monosomy 13.

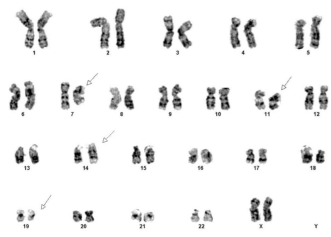

FIGURE 27.27 Karyotype of transformed CLL with 46,XX,del(7)(q32),del(11)(q21q23),t(14;19)(q32;q13)

Table 27.3
Comparison of Immunophenotypic and Cytogenetic Features in CLL/SLL, LPL, and MCL

	CLL/SLL	LPL	MCL
Immunophenotype			
CD5	+	−	+
CD10	−	−	−
CD19	+	+	+
CD20	Dim	+	+
CD22	Dim	+	+
CD23	+	± (dim)	−
CD79b	−	+	+
FMC7	−	+	+
BCL-1	−	−	+
Cytogenetics			
	del(17p13)	del(6q21–23)	t(11;14)
	del(11q22)		
	Trisomy12		
	del(13q14)		

CLL/SLL, chronic lymphocytic leukemia/small lymphocytic lymphoma; LPL, lymphoplasmacytic lymphoma; MCL, mantle cell lymphoma.

Some cases my show complex chromosomal aberrations (Figure 27.27).

In most instances, transformation is associated with the development of systemic symptoms, such as fever, weight loss, and night sweats, and/or a rapid organomegaly, such as increased lymphadenopathy, splenomegaly, and/or hepatomegaly. The site of transformation is usually lymph node or bone marrow, and occasionally extranodal/extramedullary sites, such as the skin, gastrointestinal tract, and central nervous system. Richter transformation is associated with a rapid clinical deterioration and a low response rate to therapeutic strategies.

Differential Diagnosis

The differential diagnosis comprises conditions associated with absolute peripheral blood lymphocytosis, increased proportion of CD5+ B-cells, and the presence of a monoclonal population of B-cells.

Chronic polyclonal B-cell lymphocytosis is a rare reactive lymphoproliferative disorder often observed in middle-aged women with a history of heavy smoking. The absolute blood lymphocyte count ranges from 4000 to 20,000/μL with the presence of activated and binucleated lymphocytes (see Chapter 57). The majority of the lymphocytes are B-cells. These cells, unlike CLL cells, are polyclonal and lack CD5 expression.

CD5+ B-lymphocytes comprise a subset of the B-cells in normal individuals. In most studies, they account for up to 25% of the B-cells in the blood and the bone marrow (mean of about 12%). The expression of CD5 is dim on the normal B-cells but brighter on the CLL cells. The recommended cutoff point for the detection of residual CLL in a treated patient is >25% CD5+ B-cells in the peripheral blood or in the bone marrow samples.

Monoclonal B-cell expansion in the elderly is a condition reported in about 3.5% of healthy individuals above 65 years of age with no evidence of absolute lymphocytosis or history of lymphoid malignancy. The immunophenotypic features of these monoclonal B-cells could be divided into two major groups: CLL-like and non-CLL-like. The CLL-like phenotype is characterized by CD19+, CD23+, CD5+, FMC7−, and CD10−. The CD20 expression may be dim (typical) or strong (atypical). This phenotype has been observed in up to 13.5% of healthy relatives of patients with CLL.

The non-CLL-like phenotype is characterized by CD19+, CD20+, CD23−, CD5−, FMC7+, and CD10−. This phenotype has been referred to as monoclonal B-lymphocytosis of undetermined significance (MLUS) by some investigators.

Mantle cell lymphoma (MCL) and lymphoplasmacytic lymphoma (LPL) share overlapping morphologic features with CLL. The neoplastic cells in MCL express bcl-1 protein and are usually negative for CD23. The cytogenetic hallmark for MCL is t(11;14)(q13;q32) (see Chapter 35). The immunophenotype of LPL is characterized by lack of expression of CD5, CD10, and CD23, and expression of surface and cytoplasmic IgM. A significant proportion of patients with LPL show del(6q21–23) (see Chapter 29). A comparison of immunophenotypic and cytogenetic features in CLL/SLL, LPL, and MCL is presented in Table 27.3.

Transformation to Hodgkin lymphoma (HL) should be considered when there is presence of Reed–Sternberg-like cells. As mentioned above, rare cases of CLL/SLL may show scattered Reed–Sternberg-like cells which are interspersed with sheets of small, round lymphocytes; whereas, in transformation to HL, Reed–Sternberg cells are found in a discrete region with typical HL background.

Additional Resources

Bertilaccio MT, Scielzo C, Muzio M, et al: An overview of chronic lymphocytic leukaemia biology, *Best Pract Res Clin Haematol* 23: 21-32, 2010.

Caporaso NE, Marti GE, Landgren O, et al: Monoclonal B cell lymphocytosis: clinical and population perspectives, *Cytometry B Clin Cytom* 78(Suppl 1):S115-S119, 2010.

Chanan-Khan A, Kipps T, Stilgenbauer S: Clinical roundtable monograph. New alternatives in CLL therapy: managing adverse events, *Clin Adv Hematol Oncol* 8(Suppl):1-15, 2010.

Cramer P, Hallek M: Prognostic factors in chronic lymphocytic leukemia-what do we need to know?, *Nat Rev Clin Oncol* 8:38-47, 2011.

Crowther-Swanepoel D, Houlston RS: Genetic variation and risk of chronic lymphocytic leukaemia, *Semin Cancer Biol* 20:363-369, 2010.

Damle RN, Calissano C, Chiorazzi N: Chronic lymphocytic leukaemia: a disease of activated monoclonal B cells, *Best Pract Res Clin Haematol* 23:33-45, 2010.

Goldin LR, Slager SL, Caporaso NE: Familial chronic lymphocytic leukemia, *Curr Opin Hematol* 17:350-355, 2010.

Gribben JG, O'Brien S: Update on therapy of chronic lymphocytic leukemia, *J Clin Oncol* 29:544-550, 2011.

Klein U, Dalla-Favera R: New insights into the pathogenesis of chronic lymphocytic leukemia, *Semin Cancer Biol* 20:377-383, 2010.

Lanasa MC: Novel insights into the biology of CLL, *Hematology Am Soc Hematol Educ Program*:70-76, 2010.

Matutes E, Attygalle A, Wotherspoon A, et al: Diagnostic issues in chronic lymphocytic leukaemia (CLL), *Best Pract Res Clin Haematol* 23:3-20, 2010.

Matutes E, Wotherspoon A, Catovsky D: Differential diagnosis in chronic lymphocytic leukaemia, *Best Pract Res Clin Haematol* 20: 367-384, 2007.

Molica S: A systematic review on Richter syndrome: what is the published evidence?, *Leuk Lymphoma* 51:415-421, 2010.

Montserrat E, Moreno C: Genetic lesions in chronic lymphocytic leukemia: clinical implications, *Curr Opin Oncol* 21:609-614, 2009.

Scarfò L, Dagklis A, Scielzo C, et al: CLL-like monoclonal B-cell lymphocytosis: are we all bound to have it?, *Semin Cancer Biol* 20: 384-390, 2010.

Wierda WG, Chiorazzi N, Dearden C, et al: Chronic lymphocytic leukemia: new concepts for future therapy, *Clin Lymphoma Myeloma Leuk* 10:369-378, 2010.

B-Cell Prolymphocytic Leukemia

B-cell prolymphocytic leukemia (B-PLL) is a rare lymphoproliferative disorder characterized by the clonal proliferation of prolymphocytes, primarily involving blood, bone marrow, and spleen. Prolymphocytes are medium-sized cells with variable amount of light basophilic cytoplasm, round, oval or indented nucleus, moderately condensed chromatin, and a prominent nucleolus (Figure 28.1). In prolymphocytic leukemia, prolymphocytes account for >55% of the lymphoid cells. B-cell leukemias with prolymphocytic morphology are divided into three groups:

1. *De novo* B-PLL
2. Prolymphocytic leukemia evolved from the transformation of CLL (see Chapter 27)
3. Leukemic phase of mantle cell lymphoma (MCL) with prolymphocytic morphology and evidence of t(11;14) (see Chapter 35)

B-PLL accounts for about 1% of all chronic lymphoid leukemias. It tends to affect elderly patients, usually >60 years old. The male to female ratio is >1. The characteristic features of B-PLL include a markedly elevated lymphocyte count, often >100,000/μL, massive splenomegaly, and minimal or no peripheral lymphadenopathy. Anemia and thrombocytopenia are frequent findings. Serous effusions and central nervous system involvements are infrequent. B-PLL has an aggressive clinical course with a median survival time of about 3 years. A complete response to alemtuzumab (anti-CD52 antibody) has been recently reported in a patient with B-PLL. Transformation of B-PLL to diffuse large cell lymphoma (Richter syndrome) has been observed.

FIGURE 28.1 Prolymphocytes in the blood smear of patients with B-prolymphocytic leukemia (A and B). Some prolymphocytes show cytoplasmic granules (A, green arrow) or inclusions (A, black arrow).

Morphology

- Peripheral blood, bone marrow, and spleen are the major sites of involvement.
- There is a marked peripheral blood lymphocytosis (usually >100,000/μL) with the presence of >55% prolymphocytes. Prolymphocytes are larger and contain more cytoplasm than CLL cells (Figure 28.1). In most cases, they display a round nucleus with moderately condensed chromatin and prominent nucleolus. In rare cases, however, the nucleus is irregular

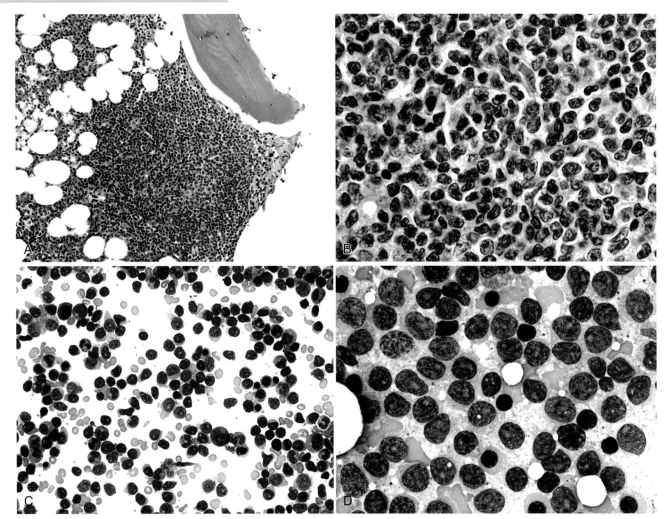

FIGURE 28.2 **PROLYMPHOCYTIC LEUKEMIA.** Bone marrow biopsy section shows a paratrabecular lymphoid aggregate consisting of medium to large lymphoid cells with irregular nuclei and a prominent nucleolus (A, low power; B, high power). Bone marrow smear shows large numbers of prolymphocytes (C, low power; D, high power).

or indented, and there may be more than one prominent nucleolus (Figure 28.2). Occasionally, a small proportion of prolymphocytes may show cytoplasmic granules or inclusions (see Figure 28.1, arrows).
- Bone marrow biopsy sections often show a diffuse infiltration by the neoplastic prolymphocytes. Bone marrow smears show clusters or sheets of prolymphocytes (see Figure 28.2).
- Splenic involvement is a frequent finding. Both white and red pulps are extensively infiltrated by prolymphocytes. The white pulp is markedly expanded and the red pulp is diffusely or patchily involved, often creating a mixture of diffuse and nodular patterns. Some of the extended white pulp nodules may show smaller lymphoid cells in the center surrounded by larger cells in the periphery.
- Lymph node involvement, although rare, is diffuse, with the effacement of nodal architecture and predominance of prolymphocytes (Figure 28.3).

Immunophenotype

Immunophenotypic features by MFC (Figure 28.4) include:

- Variable expression of B-cell associated antigens;
- Often moderate to bright CD20, and variable positivity of FMC7;
- Commonly altered or absent coexpression profile of dim CD5 and CD23 present in typical CLL;
- Commonly absent CD10;
- Usually bright surface light chain restriction;
- Common positivity of CD38.

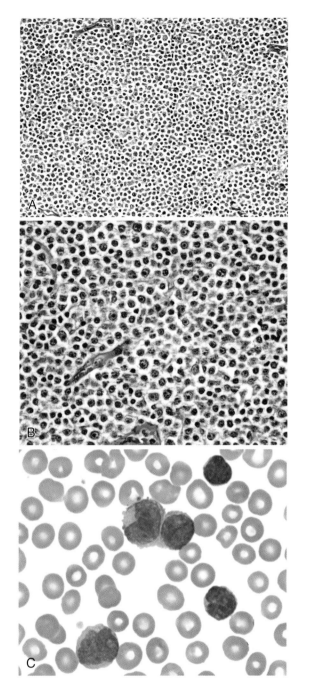

FIGURE 28.3 Lymph node biopsy section demonstrating a diffuse process with predominance of prolymphocytes (A, low power; B, high power). Prolymphocytes are present in the peripheral blood smear (C).

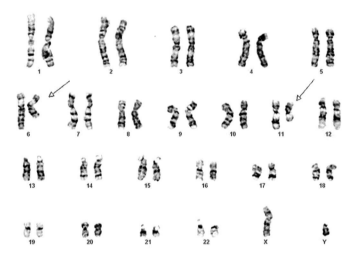

FIGURE 28.4 **FLOW CYTOMETRIC FINDINGS OF B-PROLYMPHOCYTIC LEUKEMIA.** Open gate density plot display by the scatter gate (in blue) reveals a prominent population of predominantly intermediate-sized lymphocytes. Density plots of the lymphoid-enriched gate (in purple) show monotypic B-cells expressing CD19, CD20 (moderate to bright), CD22, CD23, FMC7, and surface lambda light chain restriction. CD38 is often positive in PLL, as seen in this case. The neoplastic B-cells are negative for CD5 and CD10 (not shown).

FIGURE 28.5 Karyotype of a case of B-prolymphocytic leukemia showing deletions of 6q and 11q (arrows): 46,XY,del(6)(q21q25),del(11)(q21).

Cytogenetic and Molecular Studies

- t(11;14)(q13;q32) frequently reported in PLL in most instances, is now considered to represent the leukemic phase of a subtype of MCL (see Chapter 35).
- Trisomy 12, del(11)(q22.3), and del(13)(q14), typically common in CLL/SLL, are reported in B-PLL (Figure 28.5). However, some of these cases may represent the PLL transformation of CLL/SLL.
- Several studies report karyotypic abnormalities involving 17p13 (*TP53*) or 8q24 (*MYC*), (Figures 28.6 and 28.7, respectively).
- Loss of heterozygosity at 17p13 has been reported in 53% of B-PLL patients. Other reported cytogenetic abnormalities include del(6q), and t(8;14).
- B-PLL will usually show clonal immunoglobulin gene rearrangements just like the other B-cell malignancies.

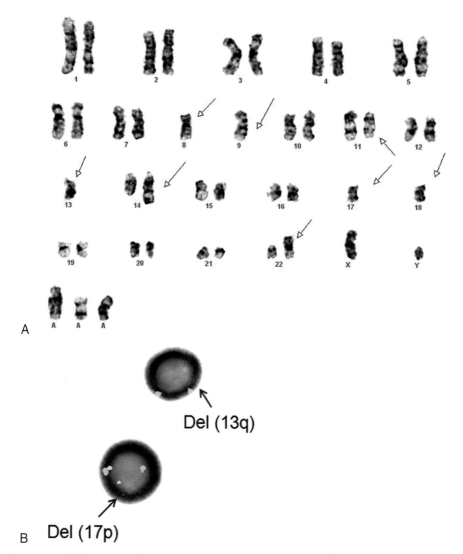

FIGURE 28.6 (A) Complex composite karyotype in a case of B-prolymphocytic leukemia showing abnormalities of chromosomes 11, 13, 14q32, and 17 among others: 43~44,XY,−8,del(11)(q22q23),add(14)(q32), −13, −17, −18, −20,add(21)(p11.2), −22,add(22)(p11.2),+1~4mar. (B) FISH analysis demonstrates deletions of 13q14 and 17p13 (*TP53* gene) loci.

Differential Diagnosis

The differential diagnosis of B-PLL is with atypical CLL (CLL/PLL), hairy cell leukemia (HCL) variant, MCL, and splenic marginal zone B-cell lymphoma. The characteristic feature of B-PLL is >55% prolymphocytes in the peripheral blood lymphocyte count, whereas in CLL the prolymphocytes account for less than 55% of the lymphoid cells. The immunophenotype of CLL/PLL in most instances is similar to that of CLL, showing coexpression of CD5 and CD23 and lack of FMC7, whereas the leukemic cells in *de novo* B-PLL lack CD5 and CD23 expression, but express FMC7. Splenomegaly, similar to B-PLL, is one of the clinical hallmarks of HCL. But lymphocyte count in HCL is low, normal, or modestly elevated. In a variant of HCL, leukemic cells morphologically mimic prolymphocytes. But the hairy cells are positive for tartrate resistant acid phosphatase (TRAP) and express CD103, whereas B-PLL cells are negative for TRAP and CD103.

As briefly discussed above, the neoplastic cells in a group of patients with splenomegaly and markedly elevated peripheral blood lymphocyte count show morphologic features identical to prolymphocytes with cytogenetic evidence of t(11;14). These disorders were originally considered as B-PLL, but now most investigators consider these as a variant of splenic mantle cell lymphoma. Splenic marginal zone B-cell lymphomas, similar to B-PLL, usually show massive splenomegaly, but the peripheral blood lymphocyte count is usually much lower and the neoplastic cells are morphologically different from prolymphocytes (see Chapter 31).

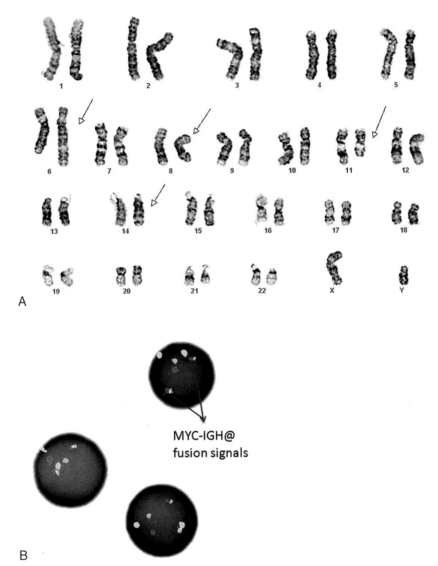

FIGURE 28.7 (A) Karyotype of a case of B-prolymphocytic leukemia showing 46,XY,+3,der(6)t(3;6)(q21;q25),t(8;14)(q24;q32),del(11)(q13q23). (B) FISH analysis exhibiting *IGH@-MYC* fusion signals (arrows) consistent with t(8;14)(q24;q32).

Additional Resources

Absi A, Hsi E, Kalaycio M: Prolymphocytic leukemia, *Curr Treat Options Oncol* 6:197–208, 2005.

Bench AJ, Erber WN, Follows GA, et al: Molecular genetic analysis of haematological malignancies II: mature lymphoid neoplasms, *Int J Lab Hematol* 29:229–260, 2007.

Del Giudice I, Osuji N, Dexter T, et al: B-cell prolymphocytic leukemia and chronic lymphocytic leukemia have distinctive gene expression signatures, *Leukemia* 23:2160–2167, 2009.

Dungarwalla M, Matutes E, Dearden CE: Prolymphocytic leukaemia of B- and T-cell subtype: a state-of-the-art paper, *Eur J Haematol* 80:469–476, 2008.

Hoffbrand AV, Pettit JE, Vyas P: *Color atlas of clinical hematology*, ed 4, Philadelphia, 2010, Mosby/Elsevier.

Krishnan B, Matutes E, Dearden C: Prolymphocytic leukemias, *Semin Oncol* 33:257–263, 2006.

Matutes E, Wotherspoon A, Catovsky D: Differential diagnosis in chronic lymphocytic leukaemia, *Best Pract Res Clin Haematol* 20:367–384, 2007.

Robak T, Robak P: Current treatment options in prolymphocytic leukemia, *Med Sci Monit* 13:RA69–80, 2007.

Ruchlemer R, Parry-Jones N, Brito-Babapulle V, et al: B-prolymphocytic leukaemia with t(11;14) revisited: a splenomegalic form of mantle cell lymphoma evolving with leukaemia, *Br J Haematol* 125:330–336, 2004.

Lymphoplasmacytic Lymphoma/ Waldenström's Macroglobulinemia

Lymphoplasmacytic lymphoma (LPL) is a clonal mature B-cell lymphoproliferative disorder consisting of small lymphocytes, plasmacytoid lymphocytes, and plasma cells. Waldenström's macroglobulinemia (WM) is defined as an LPL with any level of monoclonal IgM serum protein. The primary site of involvement is bone marrow. Lymphadenopathy and splenomegaly are reported in 5–30% of patients, but involvement of other tissues is rare.

LPL/WM is a rare disorder accounting for 1–1.5% of lymphoid malignancies. The median age is about 65 years, with <1% of patients under 40 years of age. The male to female ratio is about 1.5 Clinical symptoms include fatigue, bleeding, and hyperviscosity-related neuropathy, such as headache, vertigo, blurring or loss of vision, diplopia, or ataxia. Bone pain is rare and <5% of the patients have lytic bone lesions. Cryoglobulinemia, autoimmune hemolytic anemia and/or thrombocytopenia, and coagulopathies may occur. Amyloidosis and erythematous urticarial skin vasculitis (Schnitzler syndrome) have been reported in some LPL/WM patients.

Factors associated with high risk include age >65 years, hemoglobin <11.5 g/dL, platelet count <100,000/μL, serum α2-microglobulin level >3 mg/L, and serum IgM paraprotein >7 g/dL.

A variety of therapeutic regimens are available, including rituximab (anti-CD20) and combination chemotherapy with or without rituximab.

Morphology and Laboratory Findings (Figures 29.1 to 29.5)

- Bone marrow biopsy sections show a nodular or diffuse infiltration of the marrow by lymphocytes, plasmacytoid lymphocytes, and plasma cells in various proportions. Scattered prolymphocytes and immunoblasts are usually present.

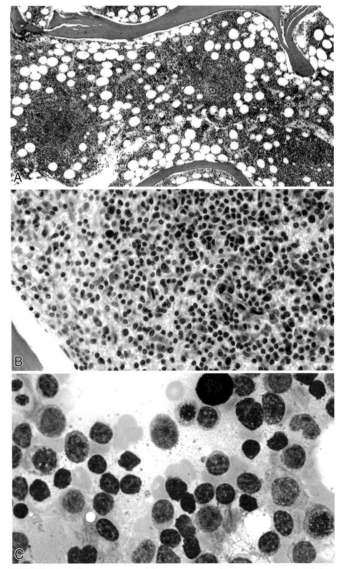

FIGURE 29.1 Bone marrow biopsy section (A, low power; B, high power) demonstrates aggregates of lymphoplasmacytic infiltrate. Bone marrow smear (C) shows a mixture of lymphocytes, plasmacytoid lymphocytes, and plasma cells.

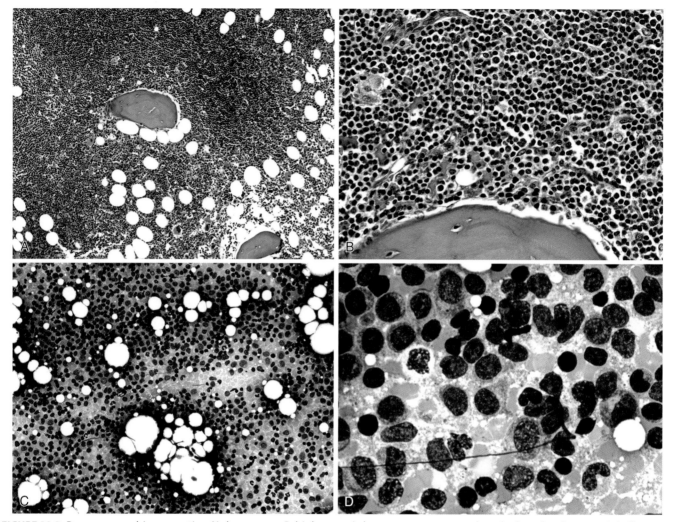

FIGURE 29.2 Bone marrow biopsy section (A, low power; B, high power) demonstrates a paratrabecular lymphoplasmacytic infiltrate. Bone marrow smear (C, low power; D, high power) shows a mixture of lymphocytes, plasma cells, and plasmacytoid lymphocytes.

- The mixed lymphoplasmacytic population is more clearly demonstrated in the bone marrow smears, which may also show increased number of mast cells. Plasma cells may show Ig-containing nuclear inclusions (Dutcher bodies) or cytoplasmic inclusions (Russell bodies) (Figure 29.6).
- Circulating neoplastic cells may be seen in peripheral blood smears, some with plasmacytoid features, but lymphocyte count is not as high as observed in CLL. The red blood cells show rouleaux formation or evidence of agglutination (cryoglobulinemia) (see Figures 29.3C and 29.4C and D).
- Moderate to severe anemia is noted in up to 80% of patients. Presence of monoclonal IgM with serum levels often >3 g/dL (75% kappa light chain restricted).
- The lymph node biopsy sections show a diffuse involvement with sheets of lymphocytes admixed with plasmacytoid lymphocytes and plasma cells. Dutcher and/or Russell bodies may be present. The sinuses are often preserved and open with scattered histiocytes engulfing immunoglobulin molecules. Transformed lymphoid cells are relatively few and are dispersed throughout the lesion, but proliferation centers, characteristic of CLL/SLL, are often lacking.
- Some cases of LPL are associated with amyloidosis (Figure 29.7).
- Splenic involvement is often diffuse, with the infiltration of both white and red pulps.
- Transformation to diffuse large B-cell lymphoma may occur in about 10–15% of cases (Figure 29.8).

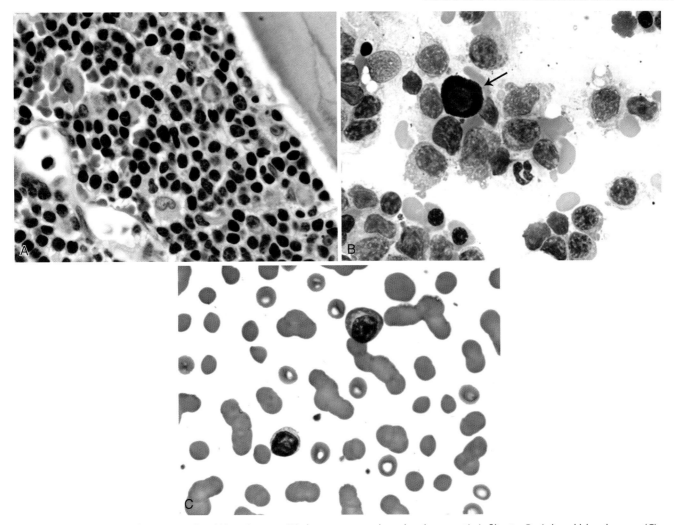

FIGURE 29.3 Bone marrow biopsy section (A) and smear (B) demonstrate a lymphoplasmacytic infiltrate. Peripheral blood smear (C) shows circulating plasma cells and red cell rouleaux formation.

Immunophenotype

Immunophenotypic features by MFC (Figure 29.9) include:

- Expression of B-cell associated antigens at various intensities;
- Usually moderate to bright expression of CD20, with associated expression of FMC7;
- Surface light chain restriction;
- Common positivity for CD38, as well as partial or dim CD25;
- Variable expression of CD23;
- Often lack of CD5 and CD10, though partial expression of either or both markers may be seen sometimes.

Immunohistochemical studies are more useful than flow cytometry in identifying the component of monotypic plasma cells, which are positive for CD138, expressing restricted cytoplasmic light and/or heavy chains (Figures 29.8 and 29.10). IgM is most common, followed by IgG. Expression of IgA is rare, and IgD is typically negative.

Cytogenetic and Molecular Studies

- Deletion of 6q21–q23 is the most common chromosomal aberration in LPL/WM, reported in 40–70% of patients (Figure 29.11).
- Some LPLs may show t(9;14)(p13;q32) (Figure 29.12), which results in the fusion of the *PAX5* and *IGH@* genes. *PAX5* encodes B-cell-specific activator protein (BSAP), which is an important regulator of B-cell proliferation and differentiation.
- Other reported chromosomal abnormalities include trisomies or structural aberrations of chromosomes 10, 11, 12, 15, 20, and 21.
- The LPL/WM cells show clonal rearrangement of immunoglobulin heavy and light chains with somatic mutation of the V-region genes.
- Documentation of somatic V-region hypermutation, a property particularly of more mature B-cells that have passed through the germinal center, requires subcloning and sequencing studies, which are not routinely available in most clinical molecular pathology laboratories.
- Mutation of the *TP53* tumor suppressor gene has also been described in some cases.

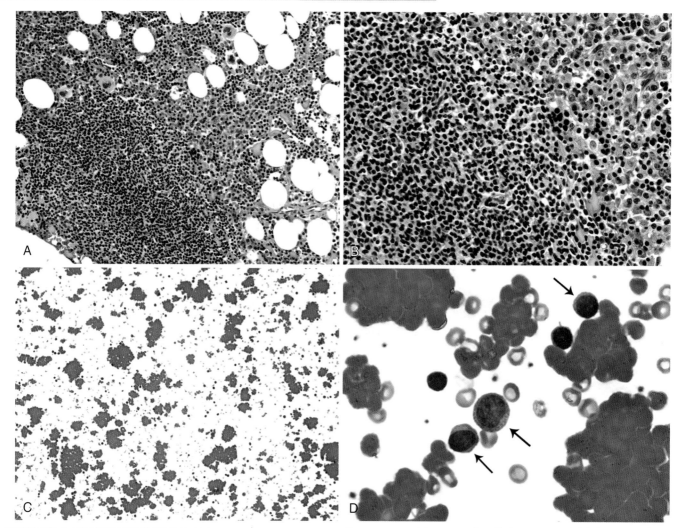

FIGURE 29.4 **LYMPHOPLASMACYTIC LYMPHOMA WITH RED CELL AGGLUTINATION.** Bone marrow biopsy section (A, low power; B, high power) shows a lymphoid aggregate primarily consisting of atypical lymphocytes with scattered plasma cells. Blood smear (C, low power; D, high power) demonstrates red cell agglutination and atypical lymphocytes (arrows).

Differential Diagnosis

The differential diagnosis of LPL includes small B-cell lymphomas, such as CLL/SLL and MCL, plasma cell myeloma, and monoclonal gammopathy of undetermined significance (MGUS). The immunophenotypic and cytogenetic features of LPL/WM, CLL, and mantle cell lymphoma (MCL) are presented in Chapter 28. Unlike CLL cells, the neoplastic cells of LPL are negative for CD5 but CD23 and may express CD138. The major distinguishing features of MCL are CD5+, bcl-1+, CD23−, and presence of t(11;14). LPL cells are CD5 negative and show no t(11;14), but frequently demonstrate del(6q21−23). Plasma cell disorders comprise predominantly of plasma cells, rarely involve lymph nodes or spleen, and usually show a non-IgM monoclonal serum paraprotein.

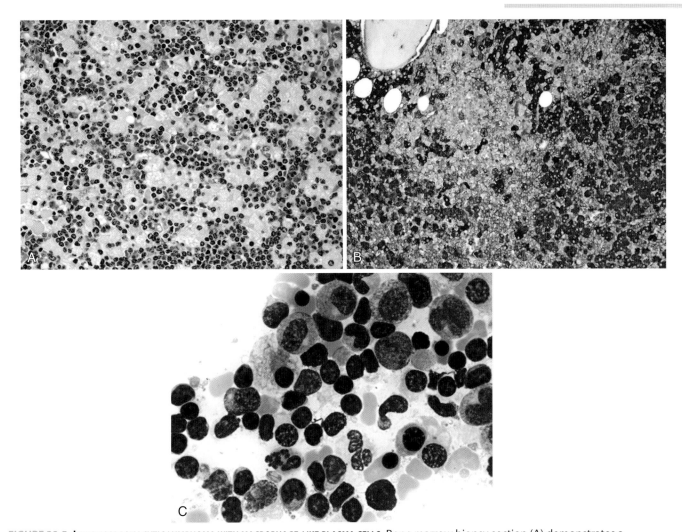

FIGURE 29.5 **LYMPHOPLASMACYTIC LYMPHOMA WITH MACROPHAGE-LIKE PLASMA CELLS.** Bone marrow biopsy section (A) demonstrates a lymphoplasmacytic infiltrate with aggregates of plasma cells with abundant pale vacuolated cytoplasm, mimicking macrophages. These plasma cells express cytoplasmic Ig kappa light chain by immunoperoxidase stain (B). Bone marrow smear (C) displays a lymphoplasmacytic infiltrate.

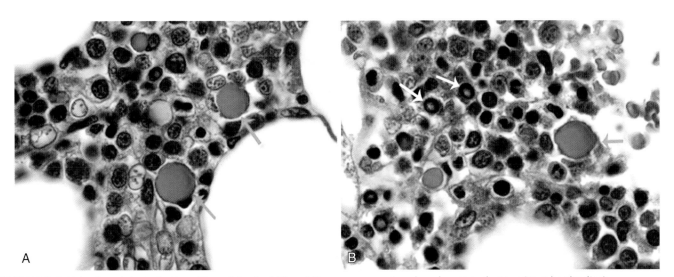

FIGURE 29.6 Cytoplasmic Ig inclusions (Russell bodies) (A and B, green arrows), and nuclear Ig inclusions (Dutcher bodies) (B, white arrows) may be present in lymphoplasmacytic lymphoma.

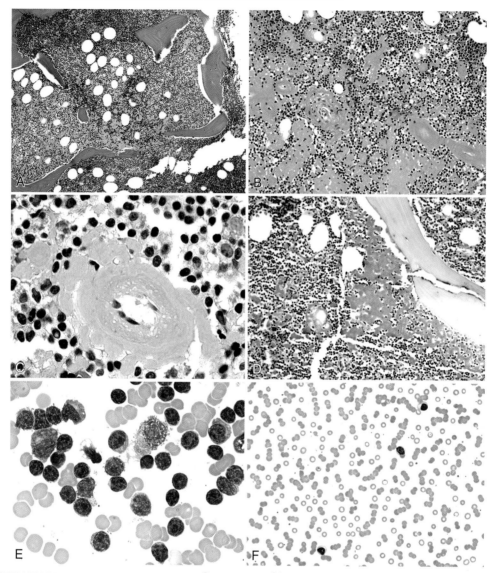

FIGURE 29.7 **LYMPHOPLASMACYTIC LYMPHOMA WITH AMYLOIDOSIS.** Bone marrow biopsy section (A, low power; B, intermediate power; C, high power) demonstrating interstitial and vascular deposits of amyloid. These deposits are positive with Congo red stain (D). Bone marrow smears show a mixture of lymphocytes, plasmacytoid lymphocytes, and plasma cells (E), and there are lymphoplasmcytic cells in the peripheral blood smear (F).

DIFFERENTIAL DIAGNOSIS 363

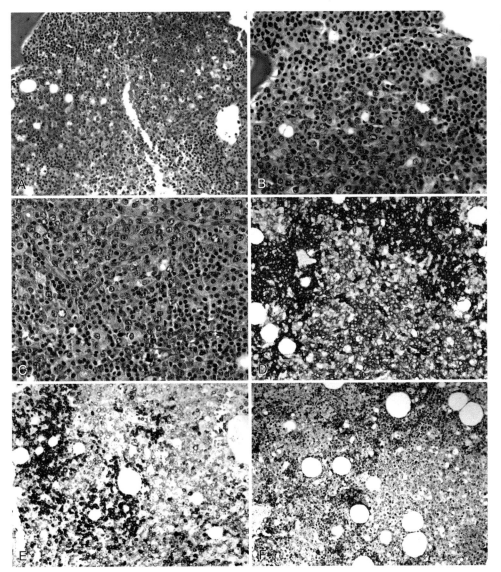

FIGURE 29.8 **Transformation of lymphoplasmacytic lymphoma to diffuse large B-cell lymphoma.** Bone marrow biopsy section (A, low power; B, intermediate power; C, high power) demonstrating aggregates of large lymphoid cells with round nuclei, open chromatin, and prominent nucleoli, surrounded by lymphoplasmacytic cells. The larger cells show weaker CD20 (D) and stronger IgM (E) expression than the smaller lymphoplasmacytic cells. The plasmacytic component is positive for CD138 (F).

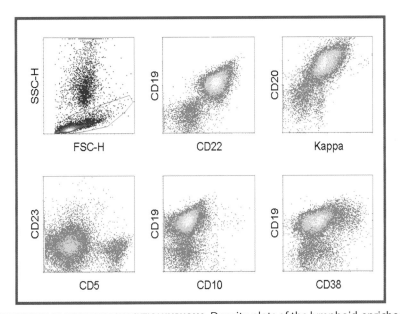

FIGURE 29.9 **Flow cytometric findings of lymphoplasmacytic lymphoma.** Density plots of the lymphoid-enriched gate (in blue) reveal a prominent population of monotypic B-cells, expressing surface kappa light chain restriction. These neoplastic B-cells are positive for CD19, CD20 (moderate), CD22, CD23 (partial), and CD38 (partial), but negative for CD5 and CD10.

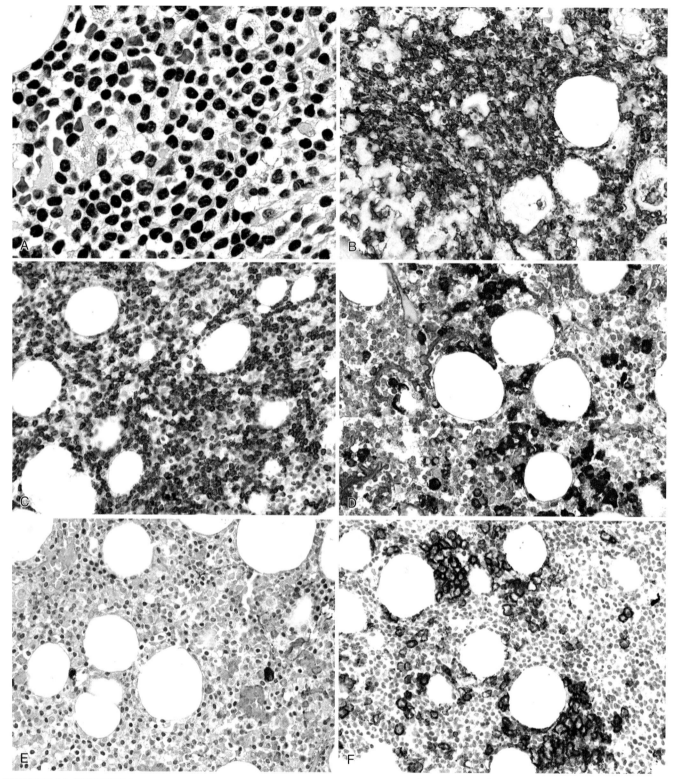

FIGURE 29.10 **IMMUNOHISTOCHEMICAL FINDINGS IN A CASE OF LYMPHOPLASMACYTIC LYMPHOMA.** Bone marrow biopsy section (A) demonstrating a lymphoplasmacytic infiltrate. These cells express CD20 (B) and BCL-2 (C), are lambda light chain restricted (D), and show rare kappa positive plasma cells (E). The plasmacytic component is positive for CD138 (F).

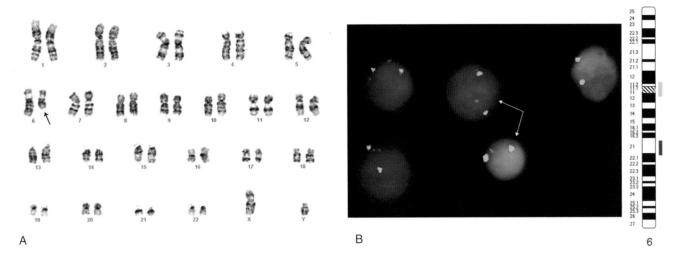

FIGURE 29.11 G-banded karyotype (A) and FISH analysis (B) demonstrating del(6)(q13q25) in a patient with lymphoplasmacytic lymphoma.

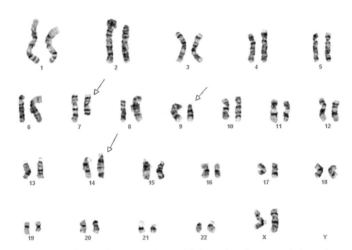

FIGURE 29.12 Karyotype showing del(7q) and t(9;14) in a patient with lymphoplasmacytic lymphoma.

Additional Resources

Ansell SM, Kyle RA, Reeder CB, et al: Diagnosis and management of Waldenström macroglobulinemia: mayo stratification of macroglobulinemia and risk-adapted therapy (mSMART) guidelines, *Mayo Clin Proc* 85:824–833, 2010.

Borgonovo G, d'Oiron R, Amato A, et al: Primary lymphoplasmacytic lymphoma of the liver associated with a serum monoclonal peak of IgG kappa, *Am J Gastroenterol* 90:137–140, 1995.

Fonseca R, Hayman S: Waldenström macroglobulinaemia, *Br J Haematol* 138:700–720, 2007.

García-Sanz R, Ocio EM: Novel treatment regimens for Waldenström's macroglobulinemia, *Expert Rev Hematol* 3:339–350, 2010.

Kristinsson SY, Koshiol J, Goldin LR, et al: Genetics- and immune-related factors in the pathogenesis of lymphoplasmacytic lymphoma/Waldenström's macroglobulinemia, *Clin Lymphoma Myeloma* 9:23–26, 2009.

Kurtin PJ: Indolent lymphomas of mature B lymphocytes, *Hematol Oncol Clin North Am* 23:769–790, 2009.

Lin P, Mansoor A, Bueso-Ramos C, et al: Diffuse large B-cell lymphoma occurring in patients with lymphoplasmacytic lymphoma/Waldenström macroglobulinemia: clinicopathologic features of 12 cases, *Am J Clin Pathol* 120:246–253, 2003.

Lin P, Medeiros LJ: Lymphoplasmacytic lymphoma/waldenstrom macroglobulinemia: an evolving concept, *Adv Anat Pathol* 12:246–255, 2005.

Lin P, Molina TJ, Cook JR, et al: Lymphoplasmacytic lymphoma and other non-marginal zone lymphomas with plasmacytic differentiation, *Am J Clin Pathol* 136:195–210, 2011.

McMaster ML, Caporaso N: Waldenström macroglobulinaemia and IgM monoclonal gammopathy of undetermined significance: emerging understanding of a potential precursor condition, *Br J Haematol* 139:663–671, 2007.

Morice WG, Chen D, Kurtin PJ, et al: Novel immunophenotypic features of marrow lymphoplasmacytic lymphoma and correlation with Waldenström's macroglobulinemia, *Mod Pathol* 22:807–816, 2009.

Neparidze N, Dhodapkar MV: Waldenstrom's macroglobulinemia: recent advances in biology and therapy, *Clin Adv Hematol Oncol* 7:677–681, 2009.

Stone MJ: Waldenström's macroglobulinemia: hyperviscosity syndrome and cryoglobulinemia, *Clin Lymphoma Myeloma* 9:97–99, 2009.

Treon SP, Hatjiharissi E, Merlini G: Waldenstrom's macroglobinemia/lymphoplasmacytic lymphoma, *Cancer Treat Res* 142:211–242, 2008.

Vijay A, Gertz MA: Waldenström macroglobulinemia, *Blood* 109:5096–5103, 2007.

Vitolo U, Ferreri AJ, Montoto S: Lymphoplasmacytic lymphoma–Waldenstrom's macroglobulinemia, *Crit Rev Oncol Hematol* 67:172–185, 2008.

Hairy Cell Leukemia

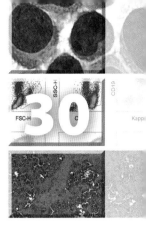

Hairy cell leukemia (HCL), or leukemic reticuloendotheliosis, is an indolent mature B-cell lymphoid leukemia characterized by the proliferation of medium-sized lymphocytes with cytoplasmic "hairy" projections, involving blood, bone marrow, spleen, and occasionally other tissues.

HCL is relatively rare and accounts for about 2% of all lymphoid leukemias. The median age is about 55 years with a marked male predominance: male to female ratio is 3–5:1. Clinical symptoms are mostly related to splenomegaly and pancytopenia and include abdominal fullness or discomfort, fatigue, weight loss, fever, bruising, and bleeding. Splenomegaly is present in about 80% of HCL cases. Hepatomegaly is observed in 20–30% of patients, but lymphadenopathy is not a major clinical feature and is found in 10–20% of patients. Table 30.1 gives a summary of the clinical and laboratory findings in HCL.

HCL is an indolent leukemia with an overall 12-year survival rate of >85%. There are reports indicating that HCL patients without splenomegaly tend to remain free from significant neutropenia, have an excellent survival rate, and are usually older than the patients with splenomegaly. Therapeutic modalities include interferon alpha, purine analogs such as pentostatin (2′-DCF) and cladribine (2-CdA), splenectomy, and rituximab (anti-CD20) therapy.

Morphology

- Hairy cells are larger than mature lymphocytes and show abundant pale blue cytoplasm, often with ill-defined border (Figure 30.1). Cells with characteristic elongated (hairy) cytoplasmic projections are frequently identified. The nuclei are round, oval, folded, indented, or dumbbell-shaped. The nuclear chromatin is condensed but finer than the CLL cells. Rare cells may show prominent nucleoli. The hairy cytoplasmic projections are easily detected in phase contrast microscopy, or in scanning and transmission electron microscopy.
- Peripheral blood examination in most patients reveals pancytopenia with the presence of various proportions of hairy cells. Only about 10% of patients show leukocytosis, and in such cases, the hairy cells account for the majority of the leukocytes. Occasionally, leukocytosis exceeds 100,000/µL. Monocytopenia is one of the characteristic features of HCL.
- Bone marrow involvement is usually interstitial or diffuse with patchy, densely cellular areas composed of neoplastic cells (Figures 30.2 to 30.4). The HCL cells are relatively uniform with abundant clear cytoplasm and round, oval, or irregular nuclei, often without prominent nucleoli or presence of mitotic figures. The presence of abundant cytoplasm creates a nuclear spacing or "fried egg" pattern. Nodular involvement is rare. In most cases, the bone marrow is hypercellular with a cellularity of >50–90%. But occasionally, the bone marrow may appear markedly hypocellular simulating aplastic anemia (Figure 30.5).
- Bone marrow biopsy sections often reveal moderate to marked increase in reticulin fibers leading to unsuccessful bone marrow aspiration (dry tap) (see Figure 30.4C). The reticulin fibers tend to surround the individual or small clusters of tumor cells.

Table 30.1

Clinical and Laboratory Findings in Hairy Cell Leukemia (HCL)

Finding	Frequency
Physical findings	
Splenomegaly	80%
Hepatomegaly	20–30%
Petechiae and ecchymosis	30%
Lymphadenopathy	10–20%
Laboratory findings	
Anemia	85%
Thrombocytopenia	80%
Neutropenia	80%
Monocytopenia	80%
Hypergammaglobulinemia	20%

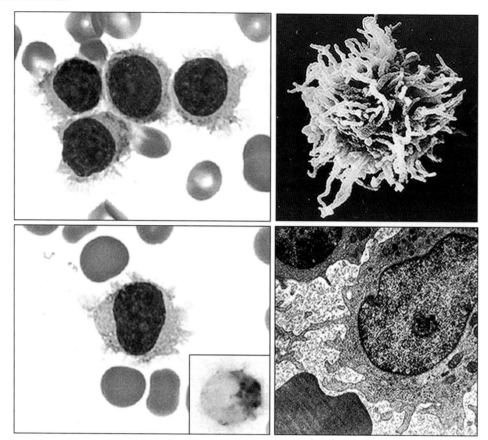

FIGURE 30.1 **HAIRY CELL LEUKEMIA.** Peripheral blood smear demonstrates neoplastic cells with cytoplasmic hairy projections (left). The inset shows a hair cell positive for tartrate resistant acid phosphatase (TRAP). Features of scanning (right top) and transmission (right bottom) electron microscopy are demonstrated.
From Naeim F. Pathology of Bone Marrow, 2nd edn. Williams & Wilkins, Baltimore, 1997 and Naeim F. Cytoskeletal redistribution of surface membrane receptors in hairy cell leukemia. Am J Clin Pathol 1980; 74: 660–663; by permission.

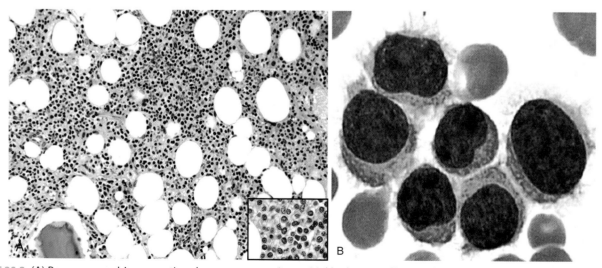

FIGURE 30.2 (A) Bone marrow biopsy section demonstrates an interstitial leukemic infiltrate. The tumor cells show nuclear spacing and a "fried egg" pattern (inset). (B) Bone marrow smear shows numerous tumor cells, some with cytoplasmic projections.

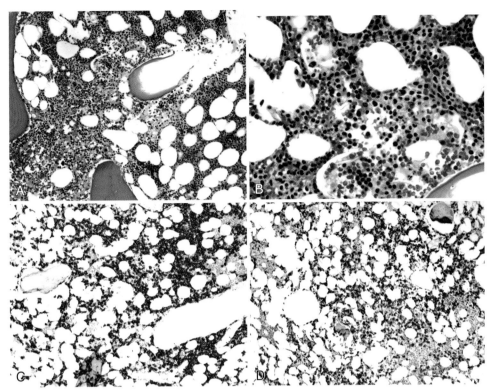

FIGURE 30.3 Bone marrow biopsy section demonstrates an interstitial leukemic infiltrate (A, low power; B, high power). These cells are positive for CD20 (C) and DBA.44 (D) by immunohistochemical stains.

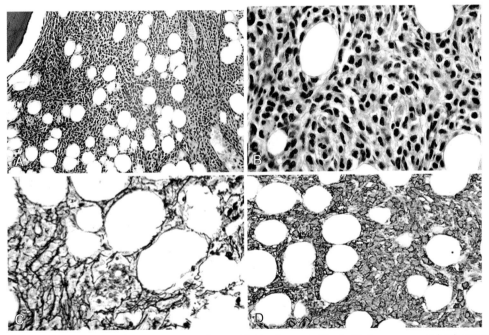

FIGURE 30.4 Bone marrow biopsy section demonstrates an interstitial leukemic infiltrate (A, low power; B, high power). The reticulin stain shows increased reticulin fibers (C), and the tumor cells express DBA.44 by immunohistochemical stain (D).

- In the vast majority of cases, diagnosis of HCL is made by examination of bone marrow biopsy sections. On rare occasions, particularly when the bone marrow involvement is focal, the initial biopsy sections may not be diagnostic and deeper cuts or additional bone marrow biopsies are required.

- Splenic involvement is one of the characteristic features of HCL, and splenomegaly is one of the most prominent clinical features. However, about 20% of patients may lack significant splenomegaly. HCL infiltrates red pulp cords and sinusoids in a diffuse pattern (Figure 30.6). Scattered red blood cell lakes

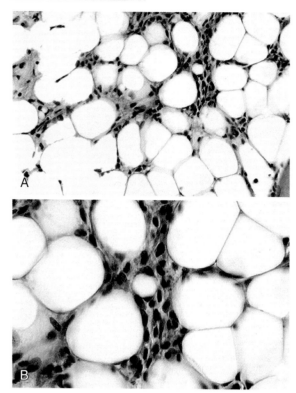

FIGURE 30.5 The hypocellular variant of hairy cell leukemia may mimic aplastic anemia in bone marrow biopsy sections. Interstitial or focal hairy cell infiltration is sometimes overlooked (A, low power; B, high power).
From Naeim F. Pathology of Bone Marrow, 2nd edn. Williams & Wilkins, Baltimore, 1997, by permission.

surrounded by hairy cells are often present (Figure 30.7). The white pulp is atrophic, and occasional small normal lymphoid aggregates may be present.

- HCL involves the liver in up to 50% and the lymph nodes in about 15% of cases. The hepatic infiltration is both portal and sinusoidal and the lymph node involvement is usually paracortical (Figure 30.8). Involvement of other tissues, such as skin and lung, is rare.

Immunophenotype and Cytochemical Stains

Immunophenotypic features by MFC (Fig. 30.9) include:

- Expression of B-cell associated antigens with bright CD20 and CD22;
- Expression of moderate to bright FMC7 and CD11c, and CD123;
- Bright surface light chain restriction;

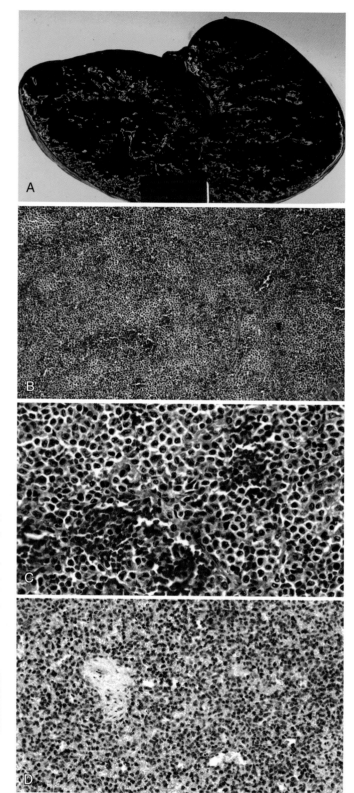

FIGURE 30.6 Splenic infiltration in hairy cell leukemia is diffuse (A) and involves the red pulp (B), with the presence of red blood cell lakes (C). The infiltrating hairy cells are positive for tartrate resistant acid phosphatase (TRAP) by immunohistochemical stain (D).

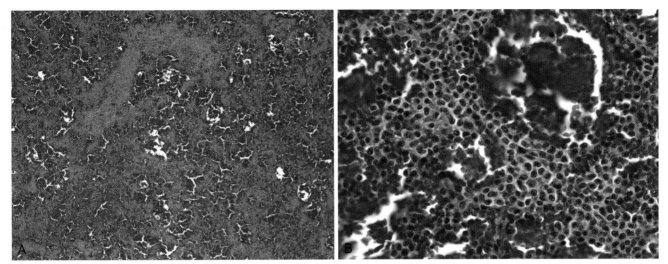

FIGURE 30.7 Splenic infiltration in hairy cell leukemia is diffuse and involves the red pulp, with the presence of red blood cell lakes (A, low power; B, high power).

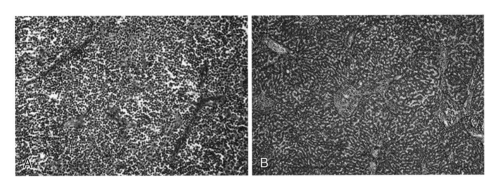

FIGURE 30.8 (A) Lymph node biopsy section showing diffuse infiltration by hairy cells. (B) Liver biopsy demonstrating infiltration of the portal areas and sinuses by the leukemic cells.

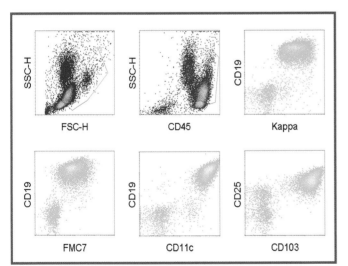

FIGURE 30.9 **FLOW CYTOMETRIC FINDINGS OF HAIRY CELL LEUKEMIA.** CD19 backgating reveals an abnormal population of B-cells on the scatter gate and by CD45 gating (in red; circled). Compared with normal lymphocytes, the abnormal B-cells demonstrate increased forward scatter, relatively high side scatter, and bright CD45. Although these CD45 and scatter characteristics resemble those of monocytes, true monocytes are often significantly reduced in HCL (monocytopenia). Density plots of the B-cell enriched gate (in green) reveal that the neoplastic B-cells are positive for CD11c (bright), CD19, CD20 (bright, not shown), CD22 (bright, not shown), FMC7 (moderate), and express surface kappa light chain restriction. In addition, there is coexpression of CD25 and CD103. The neoplastic B-cells are negative for CD5, CD10, and CD23 (not shown).

- Coexpression of CD25 and CD103;
- Common lack of CD5, CD10, and CD23, though any of these three antigens can sometimes be detected.

Immunohistochemical studies are useful in aiding the diagnosis of HCL, which is positive for Annexin A1, DBA.44 (CD72), CD123, TRAP and dim cyclin D1 (see Figures 30.3D, 30.4D, and 30.6D). Annexin A1 is most specific, and it is not seen in other B-cell lymphomas. Together with CD20, it can be used to distinguish HCL from HCL-variant and splenic marginal zone lymphoma.

Molecular and Cytogenetic Studies

- While it has generally been thought that no specific cytogenetic or molecular aberrations are associated with HCL, a recent genome-wide analysis revealed frequent mutation in the *BRAF* gene, the same V600E mutation commonly seen in colon cancer and melanoma.
- Clonal translocations involving the 14q32.3 (*IGH@*) have been described.
- Abnormalities of chromosome 5, most commonly trisomy 5 and interstitial deletions of band 5q13, have been observed in up to 40% of cases in one study.
- Deletion of the *TP53* (17p13−) occurs with a high incidence.
- As a disease of B-cells, HCL shows clonal immunoglobulin gene rearrangements.

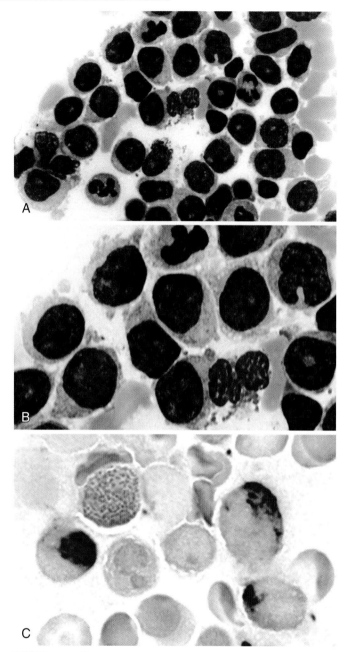

FIGURE 30.10 **HAIRY CELL LEUKEMIA, MORPHOLOGIC VARIANT.** Bone marrow smear (A, low power; B, high power) shows prolymphocyte-like hairy cells. These cells are TRAP-positive (C). From Naeim F. Pathology of Bone Marrow, 2nd edn. Williams & Wilkins, Baltimore, 1997, by permission.

- Most HCL cases show mutation of the *VH* genes and express activation-induced cytidine deaminase, a molecule essential for somatic mutation and isotype switch. Unlike most other B-cell neoplasms, HCL frequently expresses multiple Ig isotypes.

HCL Variants

A small proportion of patients with HCL (approximately 10%) may present atypical clinicopathologic features

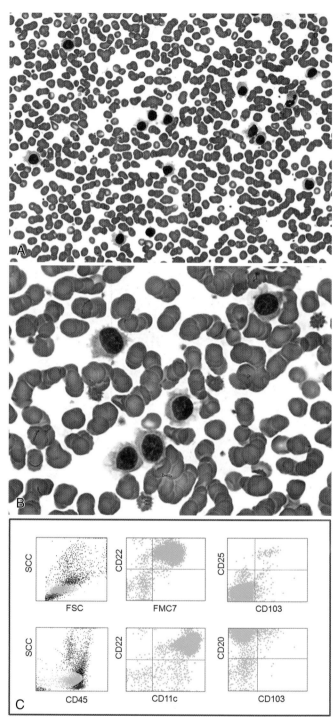

FIGURE 30.11 **HAIRY CELL LEUKEMIA, IMMUNOPHENOTYPIC VARIANT.** Blood smear demonstrating numerous medium to large lymphocytes with abundant cytoplasm and typical hairy projections (A, low power; B, high power). These cells are B-cells and express CD20 (bright), CD22, CD11c (bright), and FMC7, but are negative for CD25 and CD103 (C).

(HCL-V). These patients are usually older than typical HCL patients, show more equal distribution of male to female ratio (1.6:1), often demonstrate more massive splenomegaly and have a more aggressive clinical course. Most Japanese patients with HCL (HCl-Japanese variant) show marked lymphocytosis with leukemic cells negative for

Table 30.2
Clincopathologic Comparison of HCL, HCL-V, SMZL, and SRPL

Features	HCL	HCL-V	SMZL	SRPL
Male : female ratio	5	1.6	0.5	1.6
Media age (years)	55	71	52	77
Splenomegaly	Yes	Yes	Yes	Yes
Pattern of splenic involvement	Diffuse, red pulp	Diffuse, red pulp	Patchy, white pulp	Diffuse, red pulp
Pattern of bone marrow involvement	Mostly interstitial	Mostly interstitial	Mostly intrasinusoidal	Interstitial or intrasinusoidal
Lymphocytosis	Mild to moderate	High ($\geq$30,000/µL)	Mild to moderate	Moderate
Neutropenia	Yes	No	No	No
Monocytopenia	Yes	No	No	No
FMC7	Bright	Bright	Bright	Bright
CD22	Bright	Bright	Moderate	Bright
CD11c	Bright	Bright	±	Moderate
CD25	Bright	±	±	Faint
CD103	Bight	+/−	Negative	±
CD123	Bright	±	Negative	±
Annexin A1	Positive	Negative	Negative	Negative

HCL, hairy cell leukemia; HCL-V, hairy cell leukemia variant; SMZL, splenic marginal zone lymphoma; SRPL, splenic diffuse red pulp small B-cell lymphoma. Adapted from Traverse-Glehen A, Baseggio L, Callet-Bauchu E, et al. Hairy cell leukemia-variant and splenic red pulp lymphoma: a single entity? Brit J Haematol 2010; 150: 108–127.

CD25 expression, and weakly positive for TRAP stain. The pathologic features of HCL variants include:

- Hairy cells with prominent central nucleoli, mimicking prolymphocytes (Figure 30.10);
- Hairy cells with multilobated nuclei;
- Blastic variant with cytoplamic projection and TRAP positivity;
- Atypical immunophenotype, such as lack of coexpression of CD25 and CD103 (Figure 30.11), plus CD123−, Annexin A− and/or TRAP−;
- Lack of monocytopenia;
- High peripheral blood lymphocyte count (often $\geq$30,000/µL);
- Subtle bone marrow infiltrate with infrequent bone marrow fibrosis.

HCL-V in the current WHO classification is categorized under "splenic B-cell lymphoma/leukemia, unclassifiable".

Differential Diagnosis

The differential diagnosis of HCL includes splenic marginal zone lymphoma (SMZL), PLL, and atypical CLL/SLL (Table 30.2). HCL involves the splenic red pulp in a diffuse pattern with atrophy of the white pulp, whereas SMZL involves the white pulp and often has a nodular pattern. Bone marrow fibrosis and interstitial infiltration are common features of HCL, whereas in SMZL bone marrow fibrosis is infrequent and intrasinusoidal infiltration is the characteristic feature. Unlike SMZL cells, HCL cells are TRAP+ and express CD25, CD103, CD123, and Annexin A1. The diagnosis of HCL-V is more challenging because of the atypical morphology or lack of CD25, CD103, or TRAP positivity.

Additional Resources

Cannon T, Mobarek D, Wegge J, et al: Hairy cell leukemia: current concepts, *Cancer Invest* 26:860–865, 2008.

Cawley JC, Hawkins SF: The biology of hairy-cell leukaemia, *Curr Opin Hematol* 17:341–349, 2010.

Dunphy CH: Reaction patterns of TRAP and DBA.44 in hairy cell leukemia, hairy cell variant, and nodal and extranodal marginal zone B-cell lymphomas, *Appl Immunohistochem Mol Morphol* 16:135–139, 2008.

Fanta PT, Saven A: Hairy cell leukemia, *Cancer Treat Res* 142:193–209, 2008.

Grever MR, Lozanski G: Modern strategies for hairy cell leukemia, *J Clin Oncol* 29:583–590, 2011.

Jaffe ES, Harris NL, Vardiman JW, et al: *Hematopathology*, Philadelphia, 2010, Saunders/Elsevier.

Jöhrens K, Stein H, Anagnostopoulos I: T-bet transcription factor detection facilitates the diagnosis of minimal hairy cell leukemia infiltrates in bone marrow trephines, *Am J Surg Pathol* 31:1181–1185, 2007.

Nordgren A, Corcoran M, Sääf A, et al: Characterisation of hairy cell leukaemia by tiling resolution array-based comparative genome hybridisation: a series of 13 cases and review of the literature, *Eur J Haematol* 84:17–25, 2010.

Ravandi F: Hairy cell leukemia, *Clin Lymphoma Myeloma* 9(Suppl 3):S254–259, 2009.

Riccioni R, Galimberti S, Petrini M: Hairy cell leukemia, *Curr Treat Options Oncol* 8:129–134, 2007.

Robak T: Hairy-cell leukemia variant: recent view on diagnosis, biology and treatment, *Cancer Treat Rev* 37:3–10, 2011.

Sharpe RW, Bethel KJ: Hairy cell leukemia: diagnostic pathology, *Hematol Oncol Clin North Am* 20:1023–1049, 2006.

Swerdlow SH, Campo E, Harris NL, et al: WHO Classification of Tumours of Haematopoietic and Lymphoid Tissues, ed 4, Lyon, 2008, International Agency for Research on Cancer.

Swords R, Giles F: Hairy cell leukemia, *Med Oncol* 24:7–15, 2007.

Went PT, Zimpfer A, Pehrs A-C, et al: High specificity of combined TRAP and DBA.44 expression for hairy cell leukemia, *Am J Surg Pathol* 29:474–478, 2005.

Splenic Marginal Zone Lymphoma

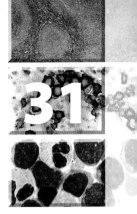

Splenic marginal zone lymphoma (SMZL), or splenic lymphoma with villous lymphocytes, is a B-cell neoplasm of small- to medium-sized lymphocytes presumably arising from the marginal zone of the splenic white pulp. Characteristic features include splenomegaly, bone marrow infiltration with sinusoidal pattern, moderate peripheral blood lymphocytosis with the presence of villous lymphocytes, and a relatively indolent course.

There are two other types of marginal zone-related lymphomas: nodal marginal zone lymphoma and extranodal marginal zone lymphoma of mucosa-associated lymphoid tissue (MALT) type. These two entities are discussed separately in Chapters 32 and 33, respectively.

SMZL accounts for about 1–2% of all non-Hodgkin lymphomas. The median age is 65 years with a male to female ratio ranges from 1:1 to 1:2. Moderate to massive splenomegaly is a common feature. Hepatomegaly is observed in some patients, but peripheral lymphadenopathy is rare. Patients usually develop mild to moderate degrees of anemia, thrombocytopenia, and neutropenia which could be attributed to bone marrow infiltration and splenic sequestration. Most patients show absolute lymphocytosis. Serum immunoglobulin studies reveal a small IgM or IgG spike, usually <3 g/dL in about half of cases.

Autoimmune conditions such as autoimmune hemolytic anemia, immune thrombocytopenia, lupus anticoagulants, rheumatoid arthritis, and biliary cirrhosis have been observed in association with SMZL.

SMZL is considered a low-risk lymphoma with an indolent clinical course. The overall survival is >70% at 10 years, with a complete remission rate of 80%. The treatment strategies include no therapy, splenic irradiation, splenectomy, chemotherapy, and treatment with anti-CD20.

Morphology

- The spleen is enlarged, with a median weight of 1750 g. The cut surface usually shows multiple gray-tan nodules of various sizes. Diffuse involvement is rare.
- In the early stages of involvement, the splenic sections show enlarged white pulps with the expansion of marginal zones merging into one another. In the center of the nodules, usually there is a remnant of a germinal center surrounded by mantle cells and expanded marginal zones (Figures 31.1 and 31.2). Mantle cells are small with scanty cytoplasm and slightly irregular nuclei. The cells surrounding mantle cells in the marginal zone are larger with more dispersed nuclear chromatin and abundant pale cytoplasm. Some of these cells resemble monocytes (monocytoid B-cells). Admixed with these cells are scattered centroblasts and immunoblasts. There is infiltration of neoplastic cells into the red pulp (Figures 31.3 and 31.4).
- In later stages, eventually, the white pulp expansion and red pulp infiltration create sheets of neoplastic cells, making the separation of white and red pulps unclear.
- The hilar splenic lymph nodes are commonly involved. The peripheral lymph nodes are affected less frequently. There is partial effacement of nodal architecture with infiltrating neoplastic nodules. Some of the nodules may contain a central reactive follicle. Sinuses are usually spared. Some cases may show complete effacement of the nodal architecture.
- Bone marrow is commonly involved. The pattern of involvement is intrasinusoidal, interstitial, nodular, paratrabecular, or a combination of these (Figures 31.5 and 31.6). The intrasinusoidal pattern is highly characteristic of SMZL.
- The bone marrow smears show the presence of atypical small- to medium-sized lymphocytes with abundant cytoplasm, round or irregular nuclei, and condensed chromatin (see Figures 31.5 and 31.6). Some of the lymphocytes may show villous cytoplasmic projections, which are often polar.
- There is usually a moderate peripheral blood lymphocytosis with various proportions of atypical lymphoid cells. These cells are morphologically similar to those described in the bone marrow smears and may or may not show polar villous

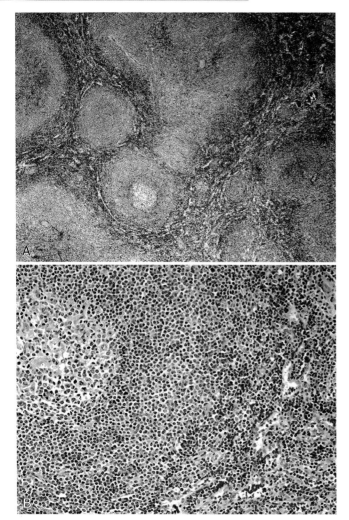

FIGURE 31.1 Splenic marginal zone B-cell lymphoma demonstrating expansion of marginal zone of the white pulps in the spleen (A, low power; B, high power).

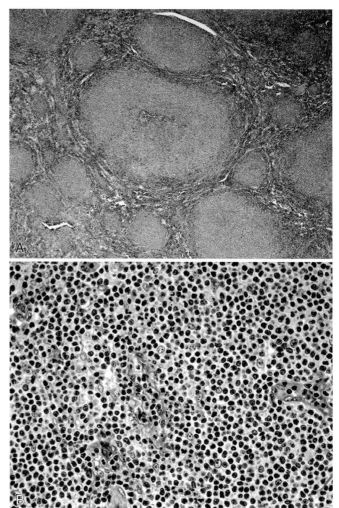

FIGURE 31.2 (A) Splenic marginal zone B-cell lymphoma showing expansion of the marginal zone into the splenic red pulp. (B) The high power view shows monocytoid lymphocytes with variable amounts of cytoplasm.

projections (see Figure 31.6C). Those without villous projections may appear plasmacytoid. Scattered larger cells with prominent nucleoli may be present.
- The liver is involved in the majority of cases. The portal tracts are the predominant sites of infiltration. Skin, pleura, and soft tissue involvements have been rarely reported.
- Transformation to large cell lymphoma occasionally occurs.

Immunophenotype

Immunophenotypic features by MFC (Figure 31.7) include:

- Expression of B-cell associated antigens at various intensities;
- Usually moderate to bright expression of CD20, and associated expression of FMC7;
- Common surface light chain restriction;
- Absent coexpression of CD25 and CD103;
- Lack of or partial expression of CD23;
- Common lack of CD5 and CD10, though partial expression of either or both markers can be seen.

By immunohistochemical studies, the neoplastic cells express surface IgM and sometimes IgD and show positivity for B-cell-associated markers, such as CD19, CD20, CD79a, FMC7, and PAX5. They are negative for BCL-1 and BCL-6, but express BCL-2. CD43 and TRAP are usually negative, while the results for DBA.44 staining are variable.

Cytogenetic and Molecular Studies

- Allelic loss of chromosome 7q (deletions or unbalanced translocation of 7q22–32) and dysregulation of the *CDK6* gene have been reported in up to 30% of SMZL. The critical region appears to be 7q32.1→q32.3 (Figure 31.8).

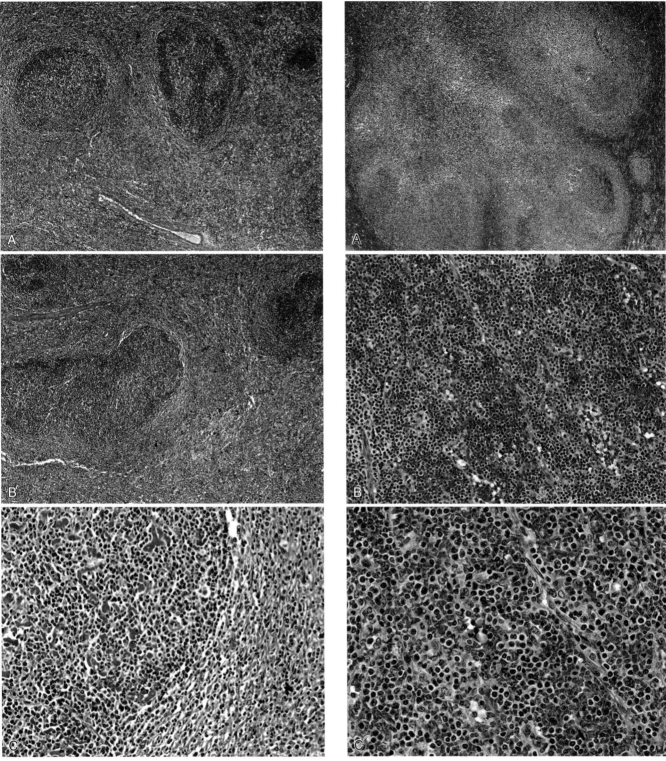

FIGURE 31.3 Splenic marginal zone B-cell lymphoma demonstrating expansion of the white pulps and infiltration into the splenic red pulp (A, low power; B, intermediate power; C, high power).

FIGURE 31.4 Splenic marginal zone B-cell lymphoma demonstrating expansion of the white pulps and infiltration into the splenic red pulp (A, low power; B, intermediate power; C, high power).

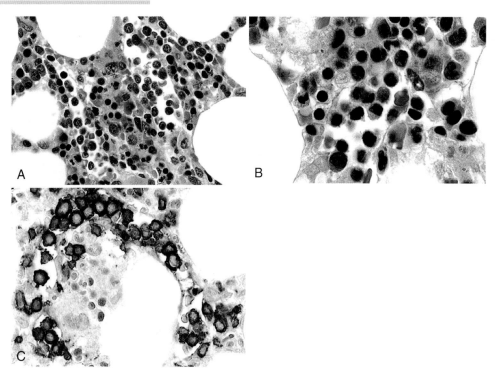

FIGURE 31.5 Bone marrow involvement in SMZL is often sinusoidal (A, low power; B, high power). Immunohistochemical stains show clusters of CD20+ cells within the sinusoids (C).
From Naeim F. Pathology of Bone Marrow, 2nd edn. Williams & Wilkins, Baltimore, 1997, by permission.

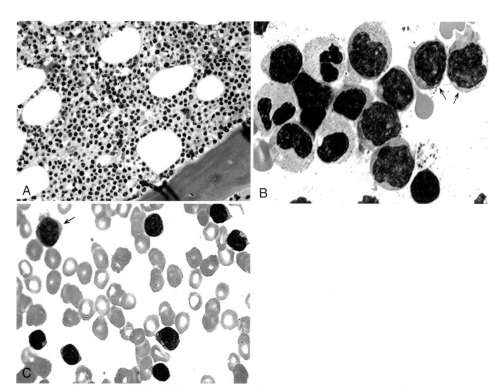

FIGURE 31.6 Bone marrow involvement in SMZL: (A) biopsy section, (B) bone marrow smear, and (C) blood smear. Some lymphocytes show polar cytoplasmic projections (arrows).

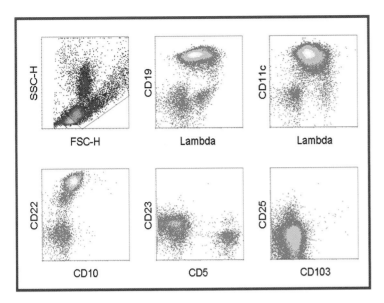

FIGURE 31.7 **FLOW CYTOMETRIC FINDINGS OF SPLENIC MARGINAL ZONE B-CELL LYMPHOMA.** Density plot of the open scatter gate (in magenta) detects an abnormal population of B-cells verified by CD19 backgating. The B-cell enriched gate (in blue) demonstrates that the neoplastic B-cells are positive for CD11c (bright), CD19, CD20 (moderate; not shown), CD22 (bright), FMC7 (not shown), and surface lambda light chain restriction. In addition, there is partial expression of CD10 and CD23. The neoplastic B-cells are negative for CD5, as well as CD25 and CD103.

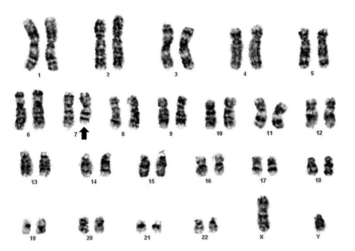

FIGURE 31.8 Karyotype showing deletion 7q in splenic marginal zone B-cell lymphoma.

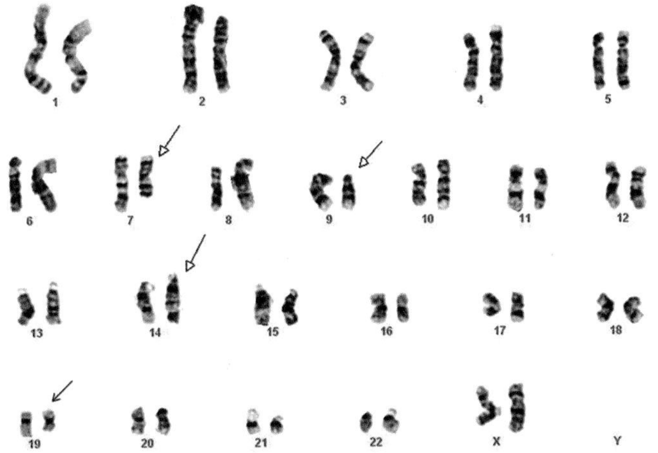

FIGURE 31.9 Karyotype with a deletion of 7q, t(9;14) and t(14;19). Deletion of 7q is mostly associated with SMZL and sometimes seen as a marker of tumor progression or transformation to a more aggressive lymphoma. t(9;14) is associated with a low-grade mature B-cell phenotype with plasmacytic differentiation.

- Abnormalities of 14q32 (*IGH@*), have also been reported in SMZL in the forms of t(6;14) and t(9;14) (Figure 31.9).
- Trisomy 3/3q is also observed in 30–50% cases of SMZL (more frequently identified by interphase FISH), while trisomy 12/12q is seen in 20–30% of cases (Figure 31.10).
- Abnormalities of the *TP53* gene are reported in about 17% of patients and is associated with aggressive disease.
- Most SMZL cases have been associated with multiple somatic IgV_H region hypermutations, but a subset without mutations has been identified that may have a different clinical course.

Other Splenic B-Cell Lymphomas

Rare cases of splenic mature B-cell lymphoma/leukemia have been reported that do not fall into the classical splenic marginal zone lymphoma, hairy cell leukemia, CLL/SLL, PLL, or LPL. In the current WHO classification, these cases have been categorized as "splenic B-cell lymphoma/leukemia, unclassifiable". The two well-defined examples of this category are hairy cell leukemia variant (HCL-V) and splenic diffuse red pulp small B-cell lymphoma. HCL-V is discussed in Chapter 30; here, we briefly discuss splenic diffuse red pulp small B-cell lymphoma.

SPLENIC DIFFUSE RED PULP SMALL B-CELL LYMPHOMA

Rare cases of splenic diffuse red pulp small B-cell lymphoma have been reported characterized by diffuse monomorphous infiltration of the red pulp by cells resembling marginal zone B-cells. There is absence of white pulp involvement, and some cases may show

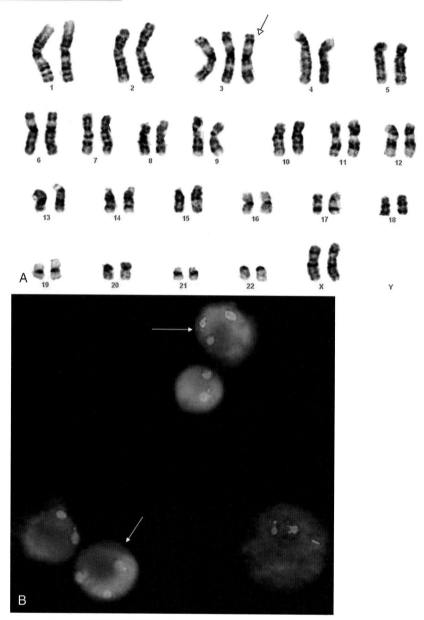

FIGURE 31.10 (A) Trisomy 3 karyotype in SMZL. (B) FISH with the centromere 3 probe showing three chromosome 3 signals in interphase nuclei.

reactive hyperplastic follicles within the white pulp. Bone marrow is often involved with sinusoidal pattern. The tumor cells express DBA-44. The peripheral blood smears may show circulating tumor cells with cytoplasmic projection (villous cells).

Differential Diagnosis

The differential diagnosis of SMZL includes HCL, LPL, MCL, FL, and CLL/SLL. A summary of morphologic, immunophenotypic, and cytogenetic characteristics of SMZL, HCL, and LPL is presented in Table 31.1. SMZL involves the splenic white pulp and often displays a nodular pattern, whereas HCL diffusely involves the red pulp with atrophy of the white pulp. Bone marrow fibrosis and interstitial infiltration are common features in HCL, whereas in SMZL marrow fibrosis is infrequent and intrasinusoidal infiltration is the characteristic feature. Unlike SMZL cells, HCL cells are TRAP-positive and express CD25 and CD103.

Overall, SMZL cells are larger and have more abundant cytoplasm than the neoplastic cells in LPL. The IgM serum levels are usually <3 g/dL in SMZL, and >3 g/dL in LPL. SMZL cells express IgM and IgD, whereas LPL cells are only IgM-positive. The primary cytogenetic abnormalities in SMZL are deletions of 7q and trisomy 3, whereas del(6q) is the most frequent cytogenetic aberration in LPL.

Table 31.1
Morphologic, immunophenotypic, and cytogenetic characteristics of LPL, SMZL, and HCL

Features	LPL	SMZL	HCL
Spleen			
Primary involved area	White pulp	White pulp	Red pulp
Pattern	Nodular	Nodular	Diffuse
Bone marrow	Nodular, diffuse, or interstitial	Mostly intrasinusoidal	Mostly interstitial
Cytology	Small lymphocytes, plasmacytoid lymphocytes, and plasma cells	Medium-sized lymphocytes, polar villous projections	Medium-sized lymphocytes, hairy projections
Immunophenotype	IgM+, CD19+, CD20+, CD22+, CD79+, FMC7+, CD23±, CD5−, CD10−, CD25−, CD103−, TRAP−	IgM+, IgD+, CD19+, CD20+, CD22+, CD79a+, CD5−, CD10−, CD23−, CD25−, CD103−, TRAP−	IgM+, CD19+, CD20+, CD22+, CD79a+, FMC7+, CD11c+, CD25+, CD103+, TRAP+, CD123+, Annexin A1+, CD5−, CD10−, CD23−
Cytogenetics	del(6q)(40–60%) t(9;14)(p13;q32) (?50%)	del(7)(q31–32) (40%) Trisomy 3/3q (30–50%)	?

HCL, hairy cell leukemia; LPL, lymphoplasmacytic lymphoma; SMZL, splenic marginal zone lymphoma.

The neoplastic cells in mantle cell lymphoma (MCL) are usually smaller, and unlike SMZL cells they express CD5 and bcl-1. MCL has a much higher frequency of peripheral lymphadenopathy than SMZL. The cytogenetic hallmark of MCL is t(11;14). The neoplastic cells in follicular lymphoma (FL) consist of a mixture of smaller centrocytes and larger centroblasts. These cells, unlike SMZL cells, express CD10 and bcl-6. The cytogenetic hallmark of FL is t(14;18).

CLL/SLL primarily consists of small mature lymphocytes with scanty cytoplasm. These cells, unlike SMZL cells, coexpress CD5 and CD23, are usually negative for FMC7, show dim expression of CD20, CD22, and CD79a, and have different cytogenetic profiles (see Table 31.1).

Additional Resources

Dogan A, Isaacson PG: Splenic marginal zone lymphoma, *Semin Diagn Pathol* 20:121-127, 2003.

Kanellis G, Mollejo M, Montes-Moreno S, et al: Splenic diffuse red pulp small B-cell lymphoma: revision of a series of cases reveals characteristic clinico-pathological features, *Haematologica* 95:1122-1129, 2010.

Kansal R, Ross CW, Singleton TP, et al: Histopathologic features of splenic small B-cell lymphomas: a study of 42 cases with a definitive diagnosis by the World Health Organization classification, *Am J Clin Pathol* 120:335-347, 2003.

Molina TJ, Lin P, Swerdlow SH, et al: Marginal zone lymphomas with plasmacytic differentiation and related disorders, *Am J Clin Pathol* 136:211-225, 2011.

Mollejo M, Algara P, Mateo MS, et al: Splenic small B-cell lymphoma with predominant red pulp involvement: a diffuse variant of splenic marginal zone lymphoma? *Histopathology* 40:22-30, 2002.

Piris MA, Mollejo M, Campo E, et al: A marginal zone pattern may be found in different varieties of non-Hodgkin's lymphoma: the morphology and immunohistology of splenic involvement by B-cell lymphomas simulating splenic marginal zone lymphoma, *Histopathology* 33:230-239, 1998.

Sagaert X, Tousseyn T: Marginal zone B-cell lymphomas, *Discov Med* 10:79-86, 2010.

Swerdlow SH, Campo E, Harris NL, et al: WHO Classification of Tumours of Haematopoietic and Lymphoid Tissues, ed 4, Lyon, 2008, International Agency for Research on Cancer.

Thieblemont C, Davi F, Noguera ME, et al: Non-MALT marginal zone lymphoma, *Curr Opin Hematol* 18:273-279, 2011.

Traverse-Glehen A, Baseggio L, Bauchu EC, et al: Splenic red pulp lymphoma with numerous basophilic villous lymphocytes: a distinct clinicopathologic and molecular entity? *Blood* 111:2253-2260, 2008.

Traverse-Glehen A, Baseggio L, Salles G, et al: Splenic marginal zone B-cell lymphoma: a distinct clinicopathological and molecular entity. Recent advances in ontogeny and classification, *Curr Opin Oncol* 23:441-448, 2011.

Zucca E, Bertoni F, Stathis A, et al: Marginal zone lymphomas, *Hematol Oncol Clin North Am* 22:883-901, 2008.

Nodal Marginal Zone Lymphoma

Nodal marginal zone lymphoma (NMZL) is a primary lymph node disease with morphologic and immunophenotypic features similar to those of splenic marginal zone lymphoma (SMZL) and extranodal marginal zone lymphoma (EMZL) of mucosa-associated lymphoid tissue (MALT). Therefore, diagnosis of NMZL requires absence of a primary extranodal or splenic involvement.

NMZL is a rare disease and accounts for <2% of all lymphoid malignancies. The median patient age is about 55, with a male to female ratio of 1:3–6. A pediatric variant exists. The prognosis is favorable, with a 5-year survival rate approaching 80%.

Morphology

See Figures 32.1 to 32.4.

- The involved lymph nodes show an interfollicular expansion by the neoplastic cells. The follicles may be absent or present. The residual follicles are reactive, regressed, or colonized by neoplastic cells. The presence of a follicular dendritic cell meshwork is suggestive of follicular colonization.
- Tumor cells are pleomorphic and usually consist of a mixture of marginal zone cells (monocytoid B-cells and cells similar to centrocytes), plasma cells and plasmacytoid cells, and scattered centroblast-like cells. Monocytoid cells or plasmacytoid cells may be the predominant cell type.
- There seems to be an association between the presence of reactive follicles and the amount of monocytoid B-cells.
- Eosinophilia may be present, which often correlates with plasmacytoid differentiation.
- Bone marrow involvement is reported in <50% of cases. It usually consists of non-trabecular lymphoid aggregates or an interstitial infiltrate. Peripheral blood involvement is rare and is demonstrated by the presence of circulating atypical small to medium lymphocytes.

Immunophenotype

Immunophenotypic features by MFC (Figure 32.5) include:

- Expression of B-cell associated antigens at various intensities;
- Common surface light chain restriction;
- Partial or dim expression of CD25 can be present;
- Absent coexpression of CD25 and CD103;
- Common lack of CD5, CD10, and CD23, though partial expression of one or more of these antigens, can be seen.

By immunohistochemical studies, BCL2 is positive, and IgD is expressed in occasional cases. Cyclin D1 and BCL6 are negative.

Molecular and Cytogenetic Studies

- As this is a B-cell neoplasm, clonal rearrangements of the immunoglobulin genes are common.
- No specific cytogenetic abnormalities have been identified. A few studies have reported trisomies of chromosome 3, 7, or 18. These trisomies are more commonly identified by FISH studies on interphase nuclei and rarely by karyotype studies.

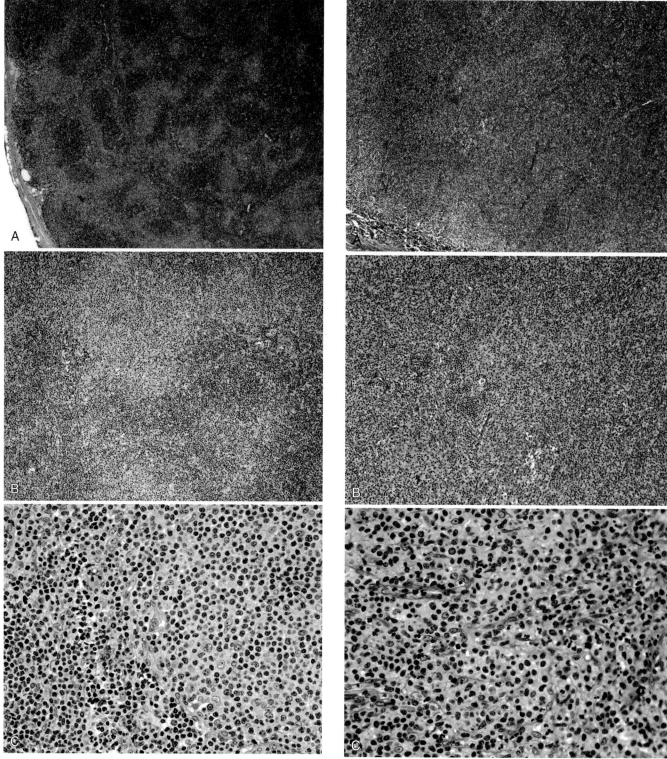

FIGURE 32.1 NODAL MARGINAL ZONE LYMPHOMA. Lymph node biopsy section demonstrating partial effacement of nodal architecture with the expansion of interfollicular areas consisting of lighter and darker areas (A, low power). There is a heterogeneous lymphoid infiltrate consisting of monocytoid B-cells, centrocyte-like cells, and centroblast-like cells (B, intermediate power; C, high power).

FIGURE 32.2 Lymph node biopsy section showing effacement of nodal architecture with the expansion of interfollicular areas consisting of lighter and darker areas (A, low power). There is a heterogeneous lymphoid infiltrate consisting predominantly of monocytoid B-cells (B, intermediate power; C, high power).

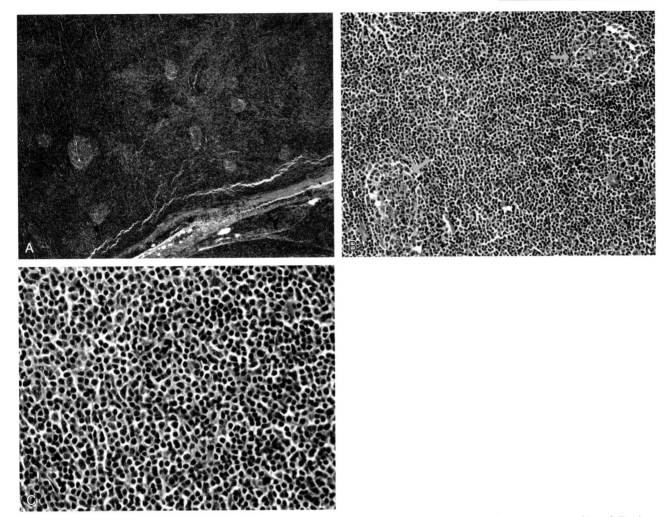

FIGURE 32.3 Lymph node biopsy section demonstrating partial effacement of nodal architecture with the expansion of interfollicular areas consisting of lighter and darker areas (A, low power). Remnants of follicular structure are present, some of which show colonization (B, intermediate power, arrows). The infiltrating cells are predominantly marginal zone monocytoid-B-cells (C, high power).

Pediatric Nodal Marginal Zone Lymphoma

Pediatric NMZL, unlike the adult type, shows a high male to female ratio (up to 20:1 under age 16). Asymptomatic localized lymphadenopathy, mostly in the head and neck region, is a common clinical presentation. It has an indolent course with a low recurrence rate after therapy.

The affected lymph nodes show marked interfollicular expansion by the marginal zone cells. Similar to adult NMZL, the neoplastic cells are pleomorphic and consist of a mixture of monocytoid cells, centrocyte-like cells, and plasma cells with scattered centroblast-like cells. Most cases show follicular expansion, some with progressive transformation of germinal centers.

Differential Diagnosis

The differential diagnosis includes reactive conditions (such as monocytoid B-cell hyperplasia), lymphoplasmacytic lymphoma (LPL), follicular lymphoma with marginal zone differentiation, and CLL/SLL with interfollicular pattern (Table 32.1). As mentioned earlier, secondary lymph node involvements in splenic and extranodal marginal zone lymphomas may mimic NMZL.

- Monocytoid B-cell hyperplasia is noted in lymphadenopathies associated with toxoplasmosis, cytomegalovirus, and HIV infections. The reactive monocytoid B-cells are found in subcapsular and medullary sinuses, and, unlike cells in NMCL, they lack BCL-2 expression.
- The distinction between NMZL and LPL is sometimes challenging (see Chapter 29). In general, in LPL, lymph node

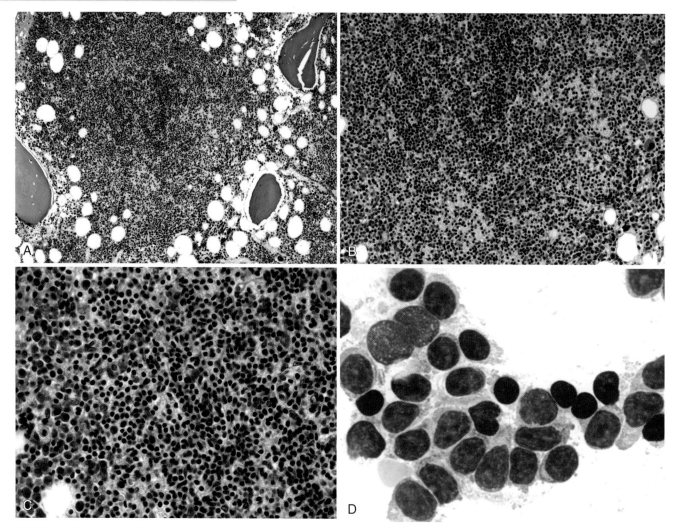

FIGURE 32.4 Bone marrow biopsy section showing a heterogenous lymphoid infiltrate consisting of small lymphocytes, monocytoid B-cells, and scattered plasma cells (A, low power; B, intermediate power; C, high power). Bone marrow smear demonstrating a cluster of atypical medium to large lymphoid cells mixed with scattered small lymphocytes (D).

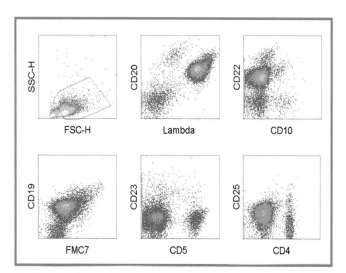

FIGURE 32.5 **FLOW CYTOMETRIC FINDINGS OF NODAL MARGINAL ZONE LYMPHOMA.** Open gate density plot (in blue) reveals a prominent population of B-cells (circled), confirmed by CD19 backgating. The neoplastic B-cells are positive for CD19 (dim), CD20 (dim), CD22 (dim), CD25 (dim), FMC7 (partial), and surface lambda light chain restriction. They are negative for CD5, CD10, and CD23.

sinuses are often preserved and the lymphoid infiltrate is more monotonous; whereas, in NMZL, sinuses are often effaced and the neoplastic infiltrate is more heterogeneous.

- In follicular lymphoma with marginal zone differentiation, the follicles are surrounded by an irregular rim of marginal zone cells. These cells, unlike follicle center cells, which express CD10 and BCL-6, are often negative or weakly positive for CD10 and BCL-6 (see Chapter 34).
- CLL/SLL with interfollicular pattern is distinguished from NMZL by more monotonous morphology, the presence of proliferating centers (pseudofollicles), and coexpression of CD5 and CD23 (see Chapter 27).

Table 32.1
Differential Diagnosis of Nodal Marginal Zone Lymphoma (NMZL)[1]

Features	NMZL	LPL	Secondary MZL	FL	CLL/SLL
Morphology					
Plasmacytoid differentiation	+	++	+	±	±
Lymph node sinuses effaced	+	−	+	+	+
Paraprotein spike	−	+	−	−	−
Dutcher bodies	±	+	±	−	−
Monocytoid cells	+	−	++	±	±
Immunophenotype					
CD43	±	+	+	−	+
CD5	−	−	−	−	+
IgD	±	−	−	±	+
CD23	−	−	−	±	+
BCL6/CD10	−	−	−	+	−
Cytogenetics	+3, +7, +18	del(6q)	t(11;18)	t(14;18)	del(13q), del(11q), del(17p), +12

CLL/SLL, chronic lymphocytic leukemia/small lymphocytic lymphoma; FL, follicular lymphoma; LPL, lymphoplasmacytic lymphoma.
[1]Adapted from Jaffe ES, Harris NL, Vardiman JW, et al. Hematopathology. Saunders/Elsevier, Philadelphia, 2010.

Additional Resources

Campo E, Miquel R, Krenacs L, et al: Primary nodal marginal zone lymphomas of splenic and MALT type, *Am J Surg Pathol* 23:59–68, 1999.

Dierlamm J, Wlodarska I, Michaux L, et al: Genetic abnormalities in marginal zone B-cell lymphoma, *Hematol Oncol* 18:1–13, 2000.

Jaffe ES, Harris NL, Vardiman JW, et al: Hematopathology, Philadelphia, 2010, Saunders/Elsevier.

McLellan M, Hasserjian R, Emerick K: Marginal zone B-cell lymphoma of the head and neck: a rare case and review, *Laryngoscope* 120(Suppl 4):S165, 2010.

Nathwani BN, Anderson JR, Armitage JO, et al: Marginal zone B-cell lymphoma: A clinical comparison of nodal and mucosa-associated lymphoid tissue types. Non-Hodgkin's Lymphoma Classification Project, *J Clin Oncol* 17:2486–2492, 1999.

Remstein ED, James CD, Kurtin PJ: Incidence and subtype specificity of API2-MALT1 fusion translocations in extranodal, nodal, and splenic marginal zone lymphomas, *Am J Pathol* 156:1183–1188, 2000.

Rizzo KA, Streubel B, Pittaluga S, et al: Marginal zone lymphomas in children and the young adult population; characterization of genetic aberrations by FISH and RT-PCR, *Mod Pathol* 23:866–873, 2010.

Salama ME, Lossos IS, Warnke RA, et al: Immunoarchitectural patterns in nodal marginal zone B-cell lymphoma: a study of 51 cases, *Am J Clin Pathol* 132:39–49, 2009.

Swerdlow SH, Campo E, Harris NL, et al: WHO Classification of Tumours of Haematopoietic and Lymphoid Tissues, ed 4, Lyon, 2008, International Agency for Research on Cancer.

Extranodal Marginal Zone Lymphoma of Mucosa-Associated Lymphoid Tissue (MALT Lymphoma)

Extranodal marginal zone B-cell lymphoma of mucosa-associated lymphoid tissue (MALT lymphoma) is a neoplastic B-cell infiltrate that involves marginal zones of reactive follicles and extends into the interfollicular areas. The neoplastic B-cells typically infiltrate epithelial tissues (e.g., mucosal epithelium and skin), forming lymphoepithelial lesions. The infiltrating cells are polymorphic, consisting of a mixture of small lymphocytes, marginal zone (centrocyte-like) cells, monocytoid B-cells, and often plasma cells. Scattered immunoblasts and centroblast-like cells are also present.

A strong association has been found between MALT lymphoma and certain autoimmune disorders and infections, such as Sjögren syndrome, Hashimoto thyroiditis, hepatitis C virus infection, *Helicobacter pylori* gastritis, and *Borrelia afzelii* infection of skin.

MALT lymphoma accounts for about 5% of all non-Hodgkin lymphomas. The most frequent site of involvement is stomach (50%), followed by lung (14%), salivary glands (14%), ocular adnexa (12%), skin (11%), thyroid (4%), and breast (4%). There is a slight female predominance. Symptoms are related to the site of involvement, such as abdominal pain (peptic ulcer disease) in the case of gastric lymphoma, or Sjögren syndrome in the case of salivary gland involvement. Systemic "B" symptoms are infrequent. Involvement of multiple mucosal sites at the time of initial diagnostic workup has been reported in up to one-third of patients.

MALT lymphomas have a high rate of complete remission, with 80% survival rate at 10 years. The therapeutic approaches include a variety of combinations of antibiotics, surgery, radiation, and chemotherapy, though universally accepted optimal therapeutic regimens based on the results of controlled studies have not yet been established.

Morphology

- In general, the involved tissues show a lymphomatous infiltrate on a background of chronic inflammation. The infiltrate starts around the reactive follicles and spreads into the surrounding areas.
- The MALT lymphoma cells are polymorphous, consisting of various proportions of small lymphocytes, marginal zone (centrocyte-like) cells, monocytoid B-cells, and plasma cells (Figures 33.1 to 33.5). Scattered and variable number of blast cells (immunoblasts or centroblast-like) are present. If the blasts form large sheets, then a diagnosis of large cell lymphoma should be made.
- Occasional follicles may show "colonization" by marginal zone or monocytoid B-cells.
- The epithelial tissue is characteristically infiltrated by the neoplastic cells, forming *lymphoepithelial lesions* (see Figure 33.1). A lymphoepithelial lesion is defined as an infiltrative aggregate of ≥3 neoplastic cells in the epithelium, with distortion or destruction of the epithelial structure.
- Gastric MALT lymphoma is strongly associated with *H. pylori* gastritis. The lymphomatous involvement may be focal or extremely florid (see Figures 33.1 and 33.2).
- In cases with salivary gland involvement a history of chronic sialadenitis (such as Sjögren's syndrome) is often present. The presence of lymphoepithelial lesions is a characteristic feature (see Figures 33.3 and 33.4).
- Lymph node or bone marrow involvement is reported in up to 25% of cases in some studies. Peripheral blood is usually not involved in initial stages.

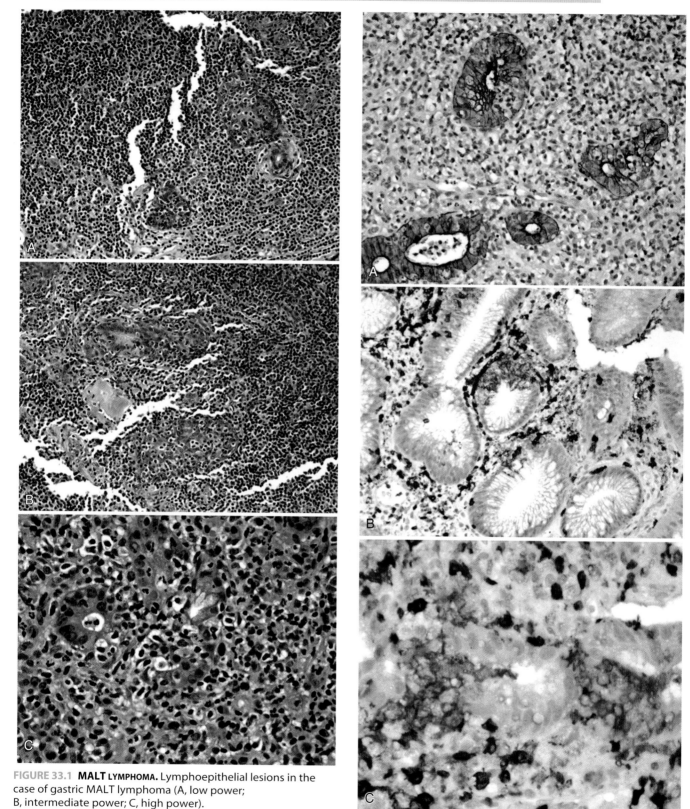

FIGURE 33.1 **MALT LYMPHOMA.** Lymphoepithelial lesions in the case of gastric MALT lymphoma (A, low power; B, intermediate power; C, high power).

FIGURE 33.2 **IMMUNOHISTOCHEMICAL STAINS IN MALT LYMPHOMA.** Cytokeratin stain highlights glandular structures infiltrated by lymphocytes (A). The infiltrating lymphocytes are predominantly CD20+ (red stain) (B, intermediate power; C, high power).

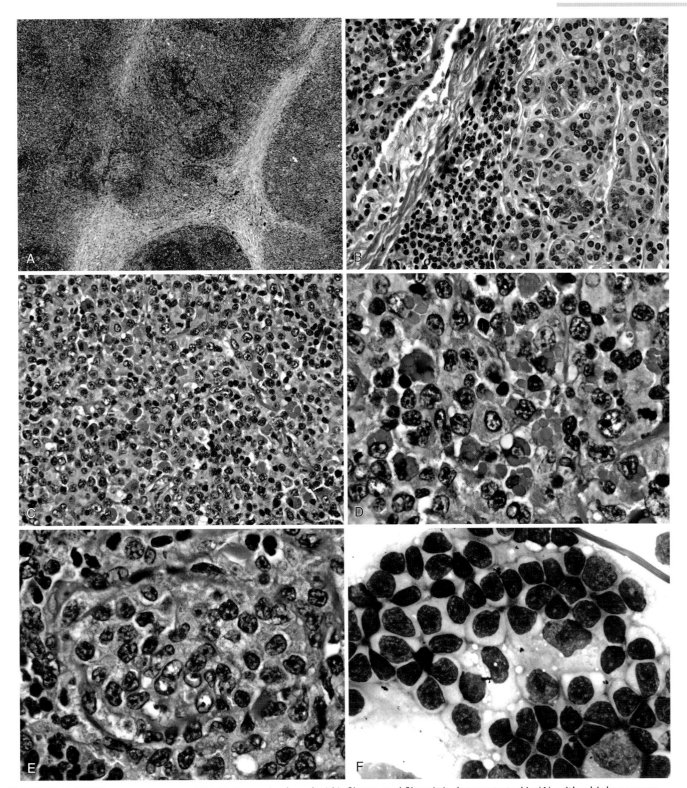

FIGURE 33.3 **MALT LYMPHOMA, SALIVARY GLAND.** A massive lymphoid infiltrate and fibrosis is demonstrated in (A), with a higher power view (B) showing entrapped salivary glands. Plasma cells account for a significant proportion of the neoplastic cells, some of which demonstrate Russell bodies (C and D, arrows). An aggregate of monocytoid B-cells is demonstrated (E). Touch preparation shows a relatively pleomorphic lymphoid population with variable amount of cytoplasm (F).

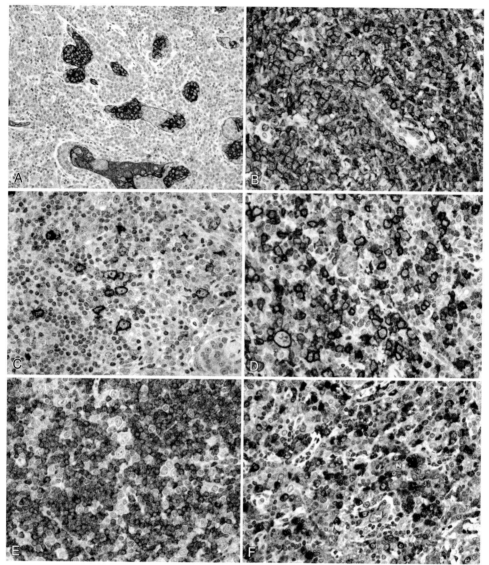

FIGURE 33.4 **IMMUNOHISTOCHEMICAL FEATURES OF MALT LYMPHOMA, SALIVARY GLAND.** Glandular structures are positive for cytokeratin (A). The neoplastic lymphoid cells are strongly CD20+ (B). Scattered larger blastic cells express CD30 (C). The plasmacytic component expresses CD138 (D), and the neoplastic cells are IgM+ (E) and express kappa light chain (F).

Immunophenotype

Flow cytometric features of MALT include (Figure 33.6):

- Expression of B-cell associated markers at various intensities;
- Surface light chain restriction;
- Monotypic B-cells are commonly admixed with reactive B-cells;
- Often lack of CD5 and CD10 expression;
- Occasional partial expression of CD23.

By immunohistochemical studies, neoplastic B-cells express CD20, BCL-2, and IgM (less frequently IgA and IgG) and often lack expression of CD5, CD10, and CD23 (see Figures 33.4 and 33.5). In some cases, monotypic plasma cells can be seen with light chain restriction. Staining for BCL1 should be negative.

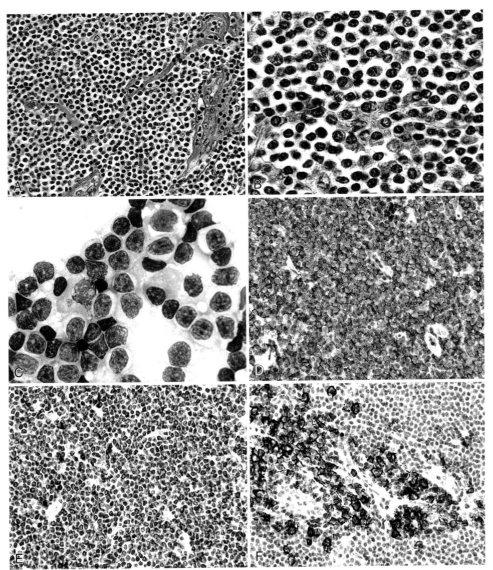

FIGURE 33.5 **ORBITAL MALT LYMPHOMA.** Large aggregates of monocytoid B-cells are demonstrated (A), admixed with scattered plasma cells (B, higher power). Touch preparation shows lymphoplasmacytic cells (C). Immunohistochemical stains demonstrate strong expression of CD20 (D) and BCL-2 (E). The plasma cell component is CD138+ (F).

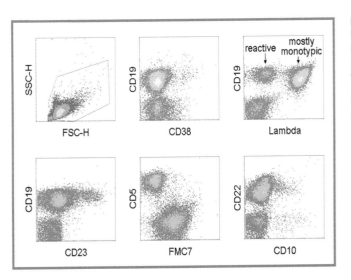

FIGURE 33.6 **FLOW CYTOMETRIC FINDINGS OF MALT LYMPHOMA.** Lymphocyte-enriched gate contains mainly monotypic B-cells (CD19 dim), which are admixed with some reactive B-cells (CD19 moderate). The neoplastic B-cells are positive for CD19 (dim), CD20 (not shown), CD22, FMC7, and surface lambda light chain restriction. They are negative for CD5, CD10, and CD23. In the background, there are variable expressions of CD10 and CD23, which are attributed to the reactive B-cells.

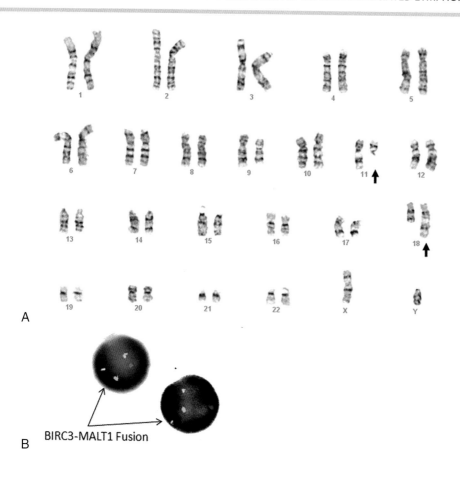

FIGURE 33.7 (A) 46,XX,t(11;18)(q21;q21) (*BIRC3-MALT1*) positive karyotype in MALT Lymphoma. (B) FISH with a dual-color (BIRC3-green; MALT1-red) probe showing two fusion signals consistent with t(11;18).

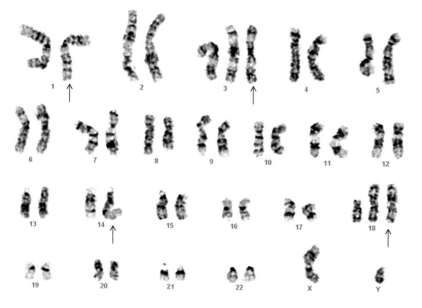

FIGURE 33.8 48,XY,t(1;14)(p22;q32), +3,−18,+iso(18)(q10)×2 in MALT lymphoma.

Cytogenetic and Molecular Studies

- The most common cytogenetic abnormality is a t(11;18)(q21;q21) (*BIRC3-MALT1*) observed in about 25–35% of cases (Figure 33.7).
- Trisomy 3 is the most common numerical cytogenetic abnormality in MALT lymphoma, reported in up to 60% of cases.
- A less frequent non-random translocation is t(1;14)(p22;q32) (*BCL10-IGH@*), seen in association with gastric and lung MALT lymphomas (Figure 33.8).
- These two translocations have not been observed in SMZL or nodal marginal zone B-cell lymphoma.
- Histological transformation of MALT lymphoma to large cell lymphoma has been associated with t(6;14)(p21;q32), indicating the alteration of cyclin D3 expression (*CCND3*) in this process.

- Important negative findings are lack of involvement of the *BCL-1* and *BCL-2* genes, and absence of t(11;14)(q13;q32) and t(14;18)(q32;q21).
- Occasional cases may show *BCL-6* rearrangements involving chromosome 3q27.
- At the molecular level, most cases will show clonal immunoglobulin rearrangements by standard assays, while *BCL-2* gene rearrangement will be negative.

Differential Diagnosis

The differential diagnosis of MALT lymphoma includes a variety of reactive lymphoproliferative disorders and small B-cell lymphomas. MALT lymphoma is distinguished from *Helicobacter pylori* gastritis, lymphoepithelial sialadenitis, and Hashimoto thyroiditis by the presence of destructive lymphoid infiltration of the epithelium and evidence of a monotypic B-cell population by immunophenotypic studies or genetic analyses. Features distinguishing MALT lymphomas from other small B-cell lymphomas are similar to those previously discussed for SMZL and NMZL (see Chapters 31 and 32).

Additional Resourses

Heeren JH, Croonen AM, Pijnenborg JM: Primary extranodal marginal zone B-cell lymphoma of the female genital tract: a case report and literature review, *Int J Gynecol Pathol* 27:243–246, 2008.

Kurtin PJ: Indolent lymphomas of mature B lymphocytes, *Hematol Oncol Clin North Am* 23:769–790, 2009.

O'Rourke JL: Gene expression profiling in Helicobacter-induced MALT lymphoma with reference to antigen drive and protective immunization, *J Gastroenterol Hepatol* 23(Suppl 2):S151–S156, 2008.

Piris MA, Arribas A, Mollejo M: Marginal zone lymphoma, *Semin Diagn Pathol* 28:135–145, 2011.

Ruskoné-Fourmestraux A, Fischbach W, Aleman BM, et al: EGILS consensus report: Gastric extranodal marginal zone B-cell lymphoma of MALT, *Gut* 60:747–758, 2011.

Sagaert X, Tousseyn T: Marginal zone B-cell lymphomas, *Discov Med* 10:79–86, 2010.

Sagaert X, Van Cutsem E, De Hertogh G, et al: Gastric MALT lymphoma: a model of chronic inflammation-induced tumor development, *Nat Rev Gastroenterol Hepatol* 7:336–346, 2010.

Serefhanoglu S, Tapan U, Ertenli I, et al: Primary thyroid marginal zone B-cell lymphoma MALT-type in a patient with rheumatoid arthritis, *Med Oncol* 27:826–832, 2010.

Stathis A, Bertoni F, Zucca E: Treatment of gastric marginal zone lymphoma of MALT type, *Expert Opin Pharmacother* 11:2141–2152, 2010.

Stefanovic A, Lossos IS: Extranodal marginal zone lymphoma of the ocular adnexa, *Blood* 114:501–510, 2009.

Voulgarelis M, Moutsopoulos HM: Mucosa-associated lymphoid tissue lymphoma in Sjögren's syndrome: risks, management, and prognosis, *Rheum Dis Clin North Am* 34:921–933, 2008.

Zucca E, Dreyling M: ESMO Guidelines Working Group Gastric marginal zone lymphoma of MALT type: ESMO clinical recommendations for diagnosis, treatment and follow-up, *Ann Oncol* 20(Suppl 4):113–114, 2009.

Zullo A, Hassan C, Andriani A, et al: Primary low-grade and high-grade gastric MALT-lymphoma presentation, *J Clin Gastroenterol* 44:340–344, 2010.

Follicular Lymphoma

Follicular lymphoma (FL) is the second most common lymphoma in North America and western Europe and is a neoplasm of follicle center B-cells consisting of a mixture of centrocytes and centroblasts. The pattern of lymph node involvement is at least partially follicular. Other terminologies used for this lesion are follicle center lymphoma and follicular center cell lymphoma.

FL accounts for 20% and 35% of the non-Hodgkin lymphomas in western Europe and the United States, respectively. It is the most common type of indolent lymphoma in Western countries. The median age at diagnosis is about 60 years with slight female predominance.

Painless peripheral lymphadenopathies in the cervical, axillary, and inguinal regions are the major clinical presentations. Mediastinal and hilar lymph nodes, bone marrow, spleen, and liver are also frequently involved. Involvement of the central nervous system is rare. Systemic "B" symptoms are present in about 20% of cases, and rare patients may present primary extranodal disease involving skin or other tissues. The clinical course is variable and primarily depends on the stage and grade of the disease (see below). The median survival is 7–10 years for patients with grade 1 or 2 and stage III to IV diseases.

The adverse prognostic factors for FL according to the international prognostic index (IPI) include:

- Age >60 years,
- Stage III or IV,
- Hemoglobin level <12 g/dL,
- Number of involved nodal areas >4, and
- Elevated serum lactate dehydrogenase (LDH).

Therapeutic modalities range from involved field radiotherapy for stages I and II to combination chemotherapy for advanced stages. Rituximab has also been added to the therapeutic regimens, particularly in relapsed or refractory lymphomas. Autologous and allogeneic stem cell transplantations have been attempted for patients with recurrent FL, though limitations such as high recurrence rate and risk of secondary MDS still exist. An anti-idiotype vaccination in FL is under evaluation.

Morphology

- The involved lymph nodes show effacement of the nodal architecture with a neoplastic lymphoid infiltrate displaying a predominantly follicular pattern (Figures 34.1 and 34.2). The

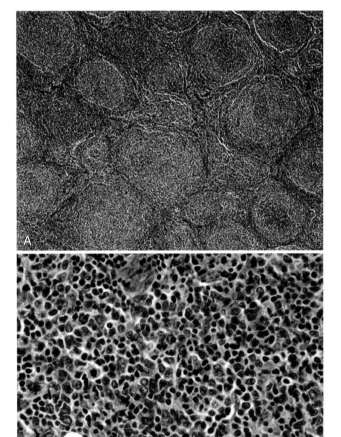

FIGURE 34.1 **FOLLICULAR LYMPHOMA, LYMPH NODE.** (A) Low power view demonstrating back-to-back follicles of various sizes and remnants of mantle zones. (B) High power view showing a mixed population of centrocytes and centroblasts.

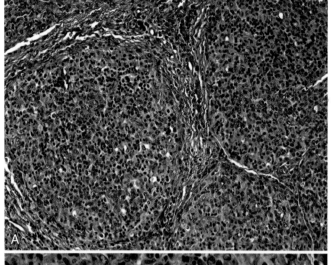
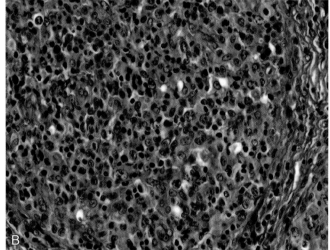

FIGURE 34.2 **FOLLICULAR LYMPHOMA, LYMPH NODE.** (A) Intermediate power view demonstrating back-to-back follicles with loss of mantle zones. (B) High power view showing a mixed population of centrocytes and centroblasts.

located nucleolus, and ill-defined pale cytoplasm. These cells express CD21 and CD23.
- Areas of diffuse involvement may be present. The neoplastic cells are usually present in the interfollicular areas and are easily identified by immunohistochemical stains, expressing CD10 and BCL-6. Discrete clusters of marginal zone monocytoid B-cells may be present in about 10% of cases.
- Primary extranodal involvements, such as gastrointestinal tract, lung, or other tissues, show similar morphologic features (Figures 34.3 to 34.5) (see below).
- Bone marrow involvement is observed in 40–45% of cases and is typically paratrabecular (Figures 34.6 and 34.7). The lymphomatous aggregates usually do not show follicular configuration and consist of a mixture of centrocytes and centroblasts. There may be discordance between morphologic findings of the involved bone marrow and the lymph node in the same patient. In such cases, the grade of bone marrow involvement is often less than that of the involved lymph node.
- Peripheral blood involvement is a frequent finding with the presence of atypical lymphoid cells with nuclear clefts or notches (see Figure 34.7B).

The pattern of involvement is divided into four categories: (a) follicular, (b) follicular and diffuse, (c) minimally follicular, and (d) diffuse (Table 34.1).

The following grading system has been recommended for FLs based on the proportion of centroblasts per high power microscopic field (hpf) (Table 34.2) (Figures 34.8 and 34.9): grade 1, 0–5 centroblasts/hpf; grade 2, 6–15 centroblasts/hpf; grade 3>15 centroblasts/hpf. Grade 3 consists of two subcategories: 3A when centrocytes are present, and 3B when solid sheets of centroblasts are present.

According to the WHO recommendations, variations in the pattern or grading observed in different areas of the involved tissue should be mentioned in the pathology report.

neoplastic follicles are densely packed against one another, often ill-defined, and show loss of polarity plus obliteration or depletion of mantle zones.
- The involved follicles show no polarity or tingible body macrophages and consist of various proportions of centrocytes and centroblasts interspersed with T-cells and follicular dendritic cells.
- Centrocytes are small- to medium-sized, show scant pale cytoplasm, irregular (angulated, twisted, convoluted) nucleus, and inconspicuous nucleoli. Centrocytes are the predominant cells in the majority of the cases.
- Centroblasts are large transformed cells with a small amount of basophilic cytoplasm, round, oval, or slightly irregular nuclei, vesicular chromatin, and one to three nucleoli located close to the nuclear membrane. In some cases, centroblasts may show significant atypical features, such as hyperchromatic or markedly irregular nuclei.
- Follicular dendritic cells (FDC) are also large and have a round nucleus and a vesicular chromatin, but, unlike centroblasts, they can be present in pairs or show double nuclei. FDC contain bland and dispersed chromatin, a small centrally

Immunophenotype

Flow cytometric findings include (Figure 34.10):

- Expression of B-cell associated antigens at variable intensities including often bright CD20;
- Surface light chain restriction;
- Coexpression of CD10 in most cases; but lack of CD10 in some cases of grade 3 FL;
- Variable expression of CD23;
- Common expression of CD38;
- Lack of CD5 expression.

By immunohistochemical studies (Figures 34.11 to 34.13):

- BCL2 characteristically stains neoplastic B-cells, though it can be negative in about half of grade 3 FL cases.
- Positivity of CD10 or positivity of BCL6 in conjunction with MUM1 negativity defines germinal center B-cell like phenotype.

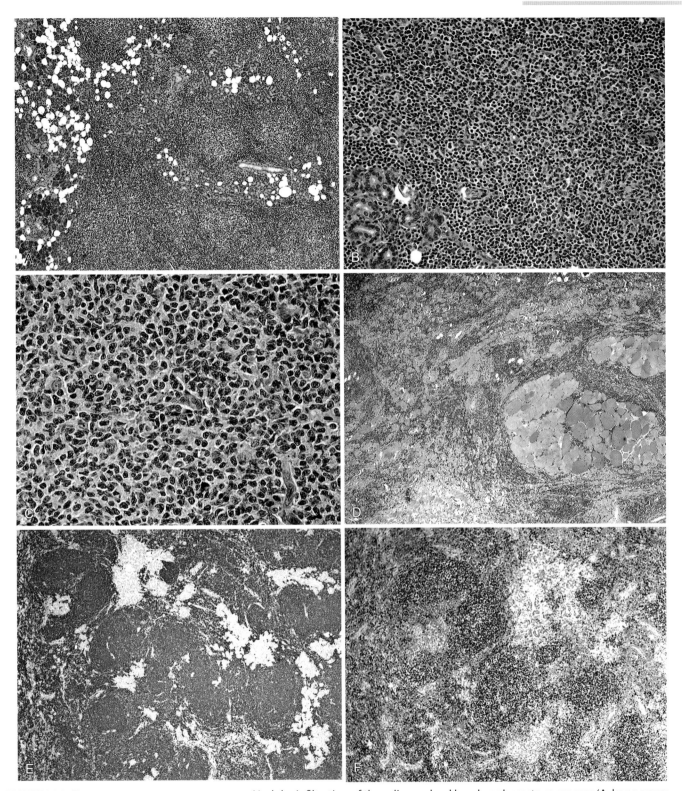

FIGURE 34.3 **FOLLICULAR LYMPHOMA, SALIVARY GLAND.** Nodular infiltration of the salivary gland by a lymphomatous process (A, low power; B, intermediate power). The neoplastic cells consist of a mixture of centrocytes and centroblasts (C, high power). There is infiltration of the adjacent skeletal muscle (D), and the lymphoid infiltrate is positive for CD20 (E) and BCL-6 (F).

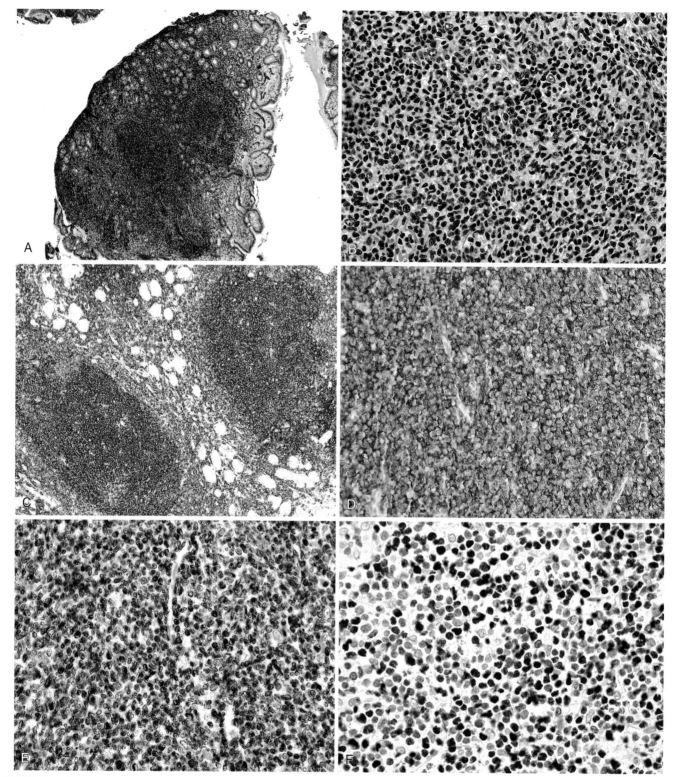

FIGURE 34.4 **FOLLICULAR LYMPHOMA, STOMACH.** Nodular infiltration of the gastric mucosa by a lymphomatous process (A, low power). The neoplastic cells consist of a mixture of centrocytes and centroblasts (B, high power). These cells express CD20 (C), CD10 (D), BCL-2 (E) and BCL-6 (F).

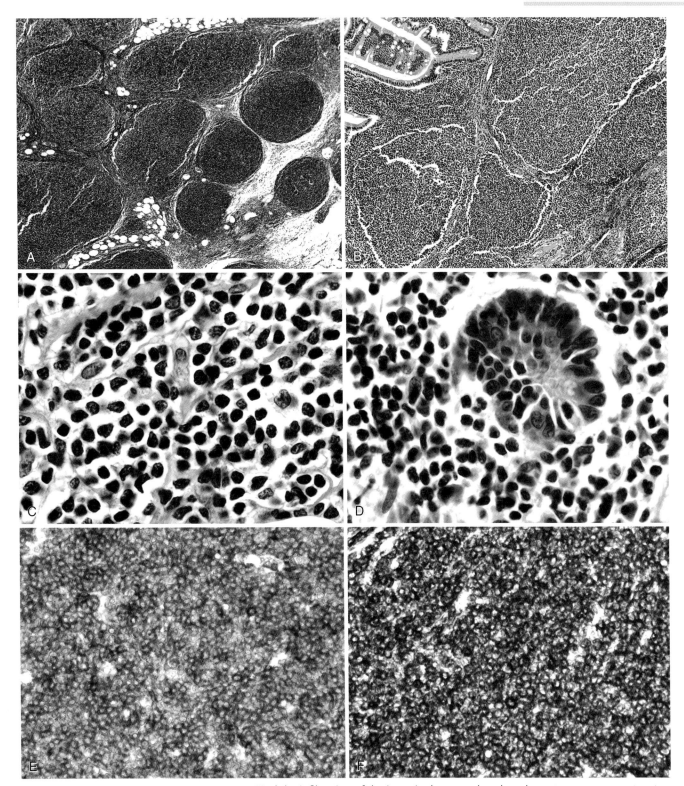

FIGURE 34.5 **FOLLICULAR LYMPHOMA, SMALL INTESTINE.** Nodular infiltration of the intestinal mucosa by a lymphomatous process (A, low power; B, intermediate power). The neoplastic cells consist primarily of centrocytes (C, high power) and infiltrate glandular structure, mimicking MALT lymphoma (D, high power). The neoplastic cells are positive for CD20 (E) and BCL-2 (F).

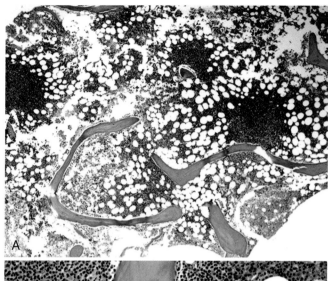
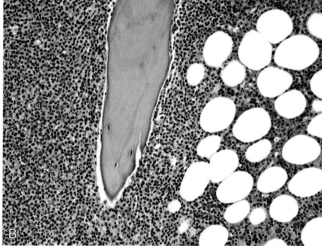
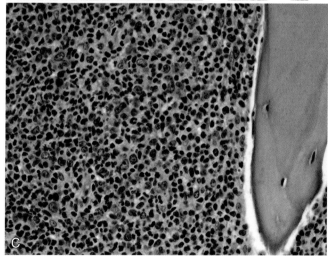

FIGURE 34.6 Bone marrow section of a patient with follicular lymphoma showing paratrabecular lymphoid infiltrates consisting of a mixture of centrocytes and centroblasts (A, low power; B, intermediate power; C, high power).

- Presence of scattered CD10+ and/or BCL-6+ cells in interfollicular areas is frequently noted (see Figure 34.13 and Table 34.3).
- BCL6 positivity is important in cases where CD10 may be negative.
- Ki67 proliferation index is usually low in grade 1 and grade 2 FL, but higher in grade 3 FL. However, occasional cases of low grade FL by morphologic standards, may have high Ki67 index. These cases are thought to behave more aggressively than their low Ki67 counterparts.
- Combined use of CD21 and CD23 can highlight the follicular dendritic meshwork.
- CD43 may be positive in some cases of high grade FL.

Cytogenetic and Molecular Studies

- The most common chromosomal translocation observed in over 80% of FL is the t(14;18)(q32;q21) (Figure 34.14). There have also been complex translocations described involving a third chromosome breakpoint in addition to 14q32 and 18q21.
- This translocation places the *BCL-2* gene on chromosome 18 next to the *IGH@* heavy chain locus on chromosome 14, resulting in the over-expression of bcl-2 which confers a survival advantage to the B-cells. In some patients, a subsequent genetic event involving a gene for cell proliferation (e.g., *RAS*) may transform indolent disease into a more aggressive lymphoma. The BCL-2 protein is not normally expressed in germinal center cells.
- Variant translocations have also been described such as t(2;18)(p12;q21) and t(18;22) (q21;q11.2), involving the *IGK@* or *IGL@* genes.
- In addition to conventional cytogenetics and FISH, t(14;18) can also be detected by molecular methods using Southern blot, qualitative PCR testing, or quantitative (real-time) PCR (Figure 34.15). Recent studies have shown prognostic correlation with the level of *BCL-2* fusion genes using quantitative PCR methods.
- Most cases demonstrate a pattern of translocation involving the major breakpoint region (MBR) of the *BCL-2* gene with a smaller proportion involving the minor cluster region (MCR). Primers specific for MBR and MCR are required for detection of each.
- About 25% of cases involve other breakpoints that will not be detected by either primer set and require additional custom primers. That is one reason why the cytogenetic test for t(14;18) will actually pick up more cases than the molecular test.
- t(14;18) is not specific for FL and has been found in 15–20% of diffuse large B-cell lymphoma (DLBCL) and MALT lymphomas. Most FL patients, in addition to t(14;18), show clonal evolution events leading to additional chromosomal aberrations, such as trisomies of 7, 8, 12, 18, or abnormalities of 3q, 6q, 13q, and 17q (Table 34.4). 3q27 aberrations involving the *BCL-6* gene is observed in about 15% of the cases.

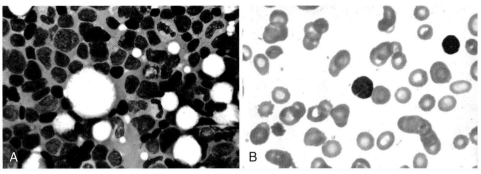

FIGURE 34.7 Bone marrow (A) and peripheral blood (B) smears of a patient with follicular lymphoma demonstrating atypical lymphocytes with irregular or cleaved nuclei.

Table 34.1
Patterns of Involvement in Follicular Lymphoma

Pattern	Follicular Proportion
Follicular	>75%
Follicular and diffuse	25–75%
Focally follicular	<25%
Diffuse	0%

Table 34.2
Grading of Follicular Lymphoma

Grading	Definition
Grade 1	0–5 centroblasts/hpf[1]
Grade 2	6–15 centroblasts/hpf
Grade 3	>15 centroblasts/hpf
3A	Centrocytes present
3B	Solid sheets of centroblasts

[1] 40× objective (18 mm of view ocular); count 10 field and divide by 10. Adapted from Swerdlow SH, Campo E, Harris NL, et al. WHO Classification of Tumours of Haematopoietic and Lymphoid Tissues, 4th edn. International Agency for Research on Cancer, Lyon, 2008.

- A more aggressive variant of FL (grade 3B) is less frequently associated with t(14;18), but often carries the t(3;14)(q27;q32) involving the *BCL-6* oncogene.
- 17p13 abnormalities involving the *TP53* gene are often associated with transformation.
- It should be kept in mind that follicular lymphomas have a high false-negative rate (approaching 50%) on immunoglobulin gene rearrangement studies performed by PCR, due to a high incidence of somatic hypermutation which interferes with hybridization of the PCR primers.
- Gene expression profiling is beginning to be applied to FL and may be useful in diagnosis in future.

Variants of Follicular Lymphoma

PEDIATRIC FOLLICULAR LYMPHOMA

Pediatric FL is characterized by a high male to female ratio, involving cervical lymph nodes (Waldeyer's ring) or testis, and by grade 3A morphology. The tumor cells are positive for B-cell-associated markers and express CD10, BCL-6, and CD43. BCL-2 is often negative, and usually there is no evidence of t(14;18). In general, the prognosis is favorable.

PRIMARY INTESTINAL FOLLICULAR LYMPHOMA

Duodenum is the most common site of extranodal FL (see Figure 34.5). It often presents as multiple small polyps, and may cause biliary obstruction. Morphologic, immunophenotypic, and genetic features are similar to those of nodal FL. IgA expression is present in a significant proportion of cases. The disease in most instances is localized and the prognosis is excellent.

INTRAFOLLICULAR NEOPLASIA (IN SITU FOLLICULAR LYMPHOMA)

Interfollicular neoplasia refers to a rare condition in which one or more follicles in an otherwise structurally well-preserved and normal-appearing lymph node show strong positivity for BCL-2 protein. This strong BCL-2 expression is sometimes associated with a monomorphous cell population in the follicle(s). The significance of this finding is not clear, but further clinical evaluation to rule out FL elsewhere in the patient is recommended.

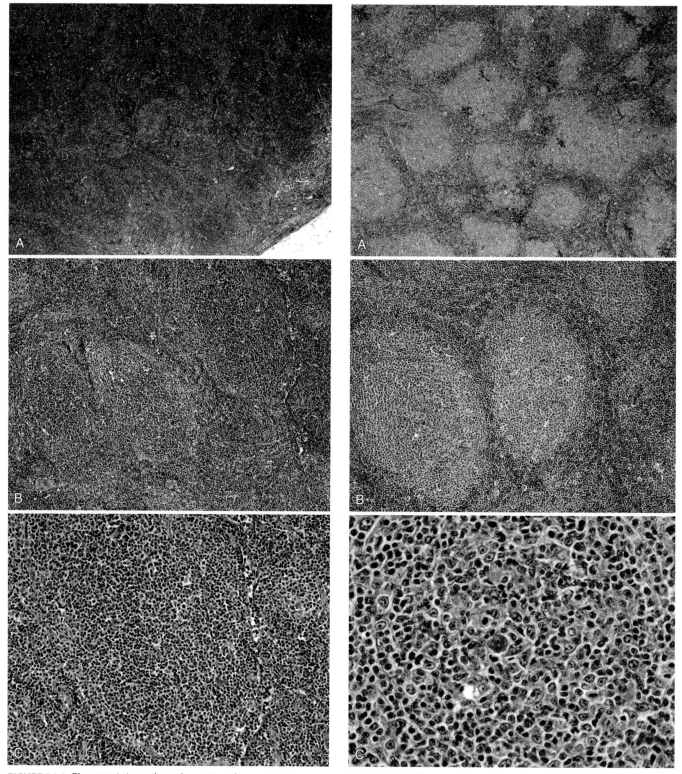

FIGURE 34.8 **FL GRADE 1.** Lymph node section demonstrating follicular lymphoma grade 1, predominantly consisting of centrocytes (A, low power; B, intermediate power; C, high power).

FIGURE 34.9 **FL GRADE 3.** Lymph node section demonstrating follicular lymphoma grade 3, predominantly consisting of centroblasts (A, low power; B, intermediate power; C, high power).

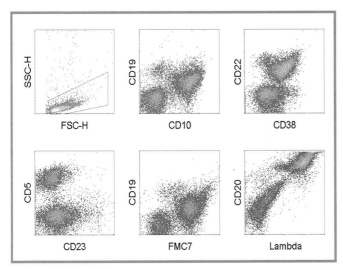

FIGURE 34.10 **FLOW CYTOMETRIC FINDINGS OF FOLLICULAR LYMPHOMA.** Lymphocyte-enriched gate demonstrates monotypic B-cells. Compared with occasional normal B-cells in the background, the neoplastic B-cells are positive for CD10, CD19 (dim), CD20 (bright), CD22 (dim), CD38, FMC7, and with surface lambda light chain restriction. They are negative for CD5 and CD23.

Differential Diagnosis

The major differential diagnosis includes reactive follicular hyperplasia, follicular colonization in marginal zone lymphomas, mantle cell lymphoma, nodular lymphocyte predominant Hodgkin lymphoma, and nodular sclerosis classical Hodgkin lymphoma.

In reactive follicular hyperplasia (RFH), reactive follicles are usually separated by interfollicular areas, are surrounded by a mantle zone, show polarity, contain tingible body macrophages, and lack bcl-2 expression. In FL, the neoplastic follicles are often back to back or merging, show minimal or no mantle zone, appear monomorphic with loss of polarity, lack tingible body macrophages, and express bcl-2 (see Table 34.3). There is evidence of interfollicular infiltration by the presence of CD10 and bcl-6 positive cells in these areas.

Follicular colonization in marginal zone lymphomas may mimic FL. However, the clinical setting and immunophenotypic features (CD10−, BCL-6−) are different from those of FL (CD10+, BCL-6+).

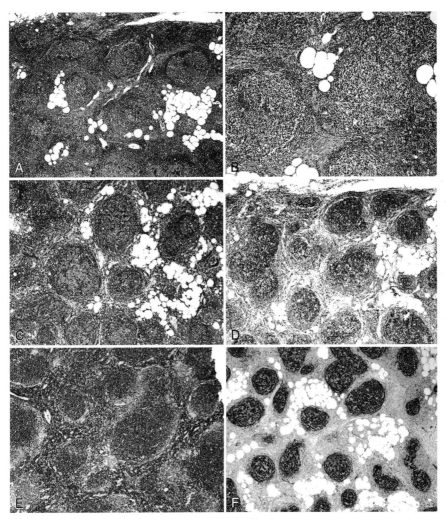

FIGURE 34.11 **FOLLICULAR LYMPHOMA MIMICKING FOLLICULAR HYPERPLASIA.** (A) Follicular structures are separated from one another with a rim of mantle cells resembling reactive follicles. (B) Higher power view demonstrates lack of tingible body macrophages and some degree of fibrosis. (C) Dual immunohistochemical staining for CD3 (brown) and CD20 (red) shows predominance of CD20+ cells within the follicles. These cells are positive for CD10 (D) and BCL-2 (E). CD21 stain shows a meshwork of interfollicular dendritic cells (F).

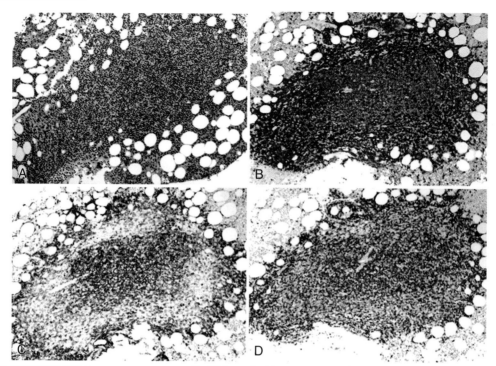

FIGURE 34.12 **BONE MARROW LYMPHOMATOUS INVOLVEMENT IN FL.** (A) Lymphoid aggregate with expansion into the surrounding fatty tissue. (B) Dual immunohistochemical staining for CD3 (brown) and CD20 (red) shows CD20+ cells in the center surrounded by CD3+ cells. The CD20+ cells also express CD10 (C) and bcl-2 (D).

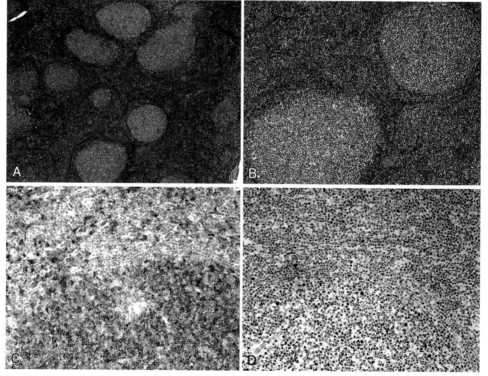

FIGURE 34.13 Lymph node section demonstrating follicular lymphoma (A, low power; B, high power). The neoplastic follicles express CD10 (C) and BCL-6 (D). There are also scattered CD10+ and BCL-6+ cells outside the follicles in the interfollicular areas.

Table 34.3
Morphologic Features Distinguishing Follicular Lymphoma (FL) from Reactive Follicular Hyperplasia (RFH)

Features	FL	RFH
Follicles		
	Back to back or merging, often with loss of mantle zone	Separated, with preservation of mantle zone
	Lack of polarity	Presence of polarity
	Lack of tingible body macrophages	Presence of tingible body macrophages
	Monotypic	Polytypic
	Commonly BCL-2 positive	Mostly BCL-2 negative
	Low or variable Ki-67 fraction	High Ki-67 fraction
Interfollicular Areas		
	Presence of CD10 positive cells	Absence of CD10 positive cells
	Presence of BCL-6 positive cells	Absence of BCL-6 positive cells

Rare cases of mantle cell lymphoma with nodular or follicular patterns may simulate FL (see the Chapter 35). Neoplastic cells in mantle cell lymphoma are small and monomorphous with lack or rare centroblasts. Unlike FL cells, they express CD5 and BCL-1 and lack CD10 expression.

Cases of FL with interfollicular sclerosis may be associated with the presence of atypical large centrocytes or centroblasts mimicking Hodgkin cells. Such cases resemble nodular sclerosis classical Hodgkin's lymphoma. However, the aytpical centrocytes/centroblasts express CD45, CD10, CD20, and BCL-6, and are negative for CD15 and CD30.

Distinction between FL and nodular lymphocyte-predominant Hodgkin lymphoma may occasionally be challenging. In general, the large cells in nodular lymphocyte-predominant Hodgkin lymphoma are negative for BCL-2 and often express EMA, and follicles are larger and variable in shape; whereas, in FL, neoplastic cells are BCL-2+, and follicles are smaller and more uniform.

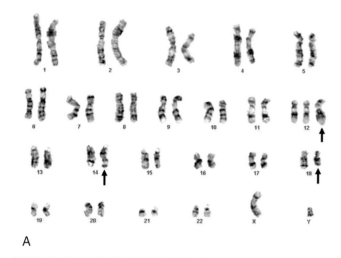

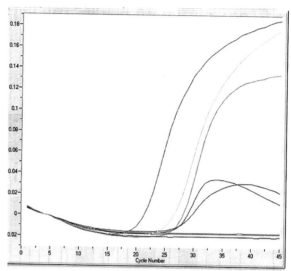

FIGURE 34.15 *BCL-2* FUSION GENE QUANTITATION BY REAL-TIME PCR. Patient 1 is positive for *BCL-2*, demonstrating amplification of the target with mid-log point at about cycle 30 (light blue curve). Patient 2 is negative for *BCL-2*, demonstrating no amplification of this target (red curve). The dark gray curves represent amplification of the control reference gene in the specimens, tPA, confirming that there are no inhibitors of PCR in the specimen.

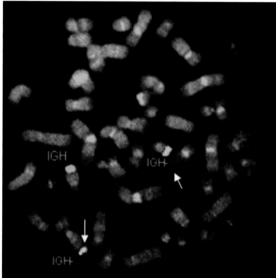

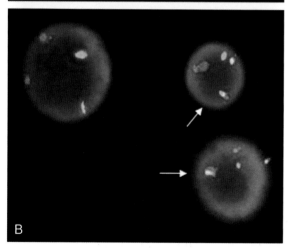

FIGURE 34.14 (A) Karyotype and (B) FISH analysis of lymphoma cells from a patient with follicular lymphoma demonstrating 47,XY, +12,t(14;18)(q32;q21) and *IGH-BCL-2* fusion.

Table 34.4	
Cytogenetic Abnormalities in Follicular Lymphoma[1]	
Abnormalities	**Frequency**
Structural	
t(14;18)(q23;q21)	78%
3q27 rearrangements	16%
17p–[2]	15%
del(6)(q23–26)[2]	13%
del(6)(q11–q15)	13%
Del(1)(q12–q21)	13%
Del(1)(p21–p22)	10%
Del(10)(q22–q24)	10%
Numerical	
+X	21%
+7	20%
+18	20%
+12 and dup 12q	10%

[1] Adapted from Tilly H, Rossi A, Stamatoullas A, et al. Prognostic value of chromosomal abnormalities in follicular lymphoma. Blood 1994; 84: 1043–1049.
[2] Associated with a worse prognosis.

Additional Resources

Agrawal R, Wang J: Pediatric follicular lymphoma: a rare clinicopathologic entity, *Arch Pathol Lab Med* 133:142–146, 2009.

Carbone A, Gloghini A, Santoro A: In situ follicular lymphoma: pathologic characteristics and diagnostic features, *Hematol Oncol* 30:1–7, 2012.

Carbone A, Santoro A: How I treat: diagnosing and managing "in situ" lymphoma, *Blood* 117:3954–3960, 2011.

Hayashi D, Lee JC, Devenney-Cakir B, et al: Follicular non-Hodgkin's lymphoma, *Clin Radiol* 65:408–420, 2010.

Jaffe ES, Harris NL, Vardiman JW, et al: Hematopathology, Philadelphia, 2010, Saunders/Elsevier.

Kenkre VP, Kahl BS: Follicular lymphoma: emerging therapeutic strategies, *Expert Rev Hematol* 3:485–495, 2010.

Luminari S, Cox MC, Montanini A, et al: Prognostic tools in follicular lymphomas, *Expert Rev Hematol* 2:549–562, 2009.

Relander T, Johnson NA, Farinha P, et al: Prognostic factors in follicular lymphoma, *J Clin Oncol* 28:2902–2913, 2010.

Solal-Céligny P, Cahu X, Cartron G: Follicular lymphoma prognostic factors in the modern era: what is clinically meaningful? *Int J Hematol* 92:246–254, 2010.

Swerdlow SH, Campo E, Harris NL, et al: WHO classification of tumours of haematopoietic and lymphoid tissues, ed 4, Lyon, 2008, International Agency for Research on Cancer.

Tan D, Horning SJ: Follicular lymphoma: clinical features and treatment, *Hematol Oncol Clin North Am* 22:863–882, 2008.

Wrench D, Montoto S, Fitzgibbon J: Molecular signatures in the diagnosis and management of follicular lymphoma, *Curr Opin Hematol* 17:333–340, 2010.

Yamamoto S, Nakase H, Yamashita K, et al: Gastrointestinal follicular lymphoma: review of the literature, *J Gastroenterol* 45:370–388, 2010.

Mantle Cell Lymphoma

Mantle cell lymphoma (MCL) is an aggressive B-cell neoplasm consisting of a monotonous small to medium-sized mature lymphocytes with scant cytoplasm and slightly irregular nuclei. MCL lacks centroblasts and immunoblasts. The characteristic cytogenetic features of MCL is t(11;14)(q13;q32) with the fusion of *BCL1 (CCND1)-IGH@* genes.

MCL accounts for 5–10% of non-Hodgkin lymphomas in the United States and Europe. The median age is around 60 years with a male to female ratio of about 3:1. Lymphadenopathy (stage III or IV), splenomegaly, hepatomegaly, and bone marrow involvement are the major clinical features and about 30% of the patients show systemic "B" symptoms. In general, MCL is considered an aggressive lymphoma with a median survival of 3–5 years. But, rare indolent types have been reported. Patients with peripheral blood involvement, high mitotic figures, blastoid variants, and complex cytogenetic abnormalities show a more aggressive clinical course. Therapeutic modalities such as combination chemotherapy and rituximab may improve the response rate, but without cure in most cases.

Morphology

- The affected lymph node shows effacement of the nodal architecture with an infiltrative process with pathologic features that may include diffuse, vaguely nodular, mantle zone, or rarely follicular patterns, or a combination of these (Figures 35.1 to 35.3). Most commonly, the involved lymph node shows transitional areas between diffuse and nodular patterns, but occasionally a follicular pattern is predominant. In the mantle zone pattern, the neoplastic cells expand the mantle zone area surrounding a germinal center. This pattern is more common in spleen. A prominent meshwork of follicular dendritic cells is usually present. Hyalinized small vascular structures are frequently present.
- The cytologic features in typical MCL cases consist of monotonous small to medium-sized lymphocytes with scant cytoplasm, slightly irregular nucleus, condensed chromatin, and inconspicuous nucleoli (see Figure 35.1C and 35.2C). Centroblasts and immunoblasts are typically absent, though centroblasts may be present in remnants of germinal centers.
- In some cases, a proportion of the neoplastic cells may show more abundant cytoplasm and appear like monocytoid B-cells. The neoplastic cells in occasional cases may mimic CLL/SLL with small lymphocytes, round nucleus, and condensed chromatin.
- A prolymphocyte-like variant with marked leukocytosis mimicking PLL has also been described. These patients usually have splenomegaly and demonstrate clinicopathological features very similar to those of PLL, except for the demonstration of t(11;14).
- The blastoid variant is an aggressive form, and the neoplastic cells resemble lymphoblasts, with dispersed chromatin, prominent nucleoli, and high mitotic figures (Figures 35.4 and 35.5) (Table 35.1).
- The pleomorphic variant consists of large cells with variable amount of cytoplasm and cleaved or oval nucleus (Figure 35.6).
- Bone marrow infiltration is reported in 50–80% of the cases. The pattern of involvement is often a combination of nodular, interstitial, and paratrabecular. Isolated paratrabecular infiltrations are rare (Figures 35.7 and 35.8).
- The neoplastic cells may also be present in the peripheral blood in various numbers creating a condition mimicking CLL.
- The spleen is affected in 30–50% of the cases, with the primary involvement of the white pulp. The white pulp nodules are expanded and often are merged, sometimes surrounding the residual germinal centers. The extent of splenic red pulp involvement is variable (Figure 35.9).
- The most common clinical presentation of gastrointestinal involvement in MCL is *lymphomatous polyposis* with the presence of multiple intestinal lymphoid polyps.

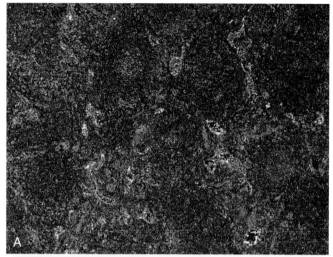
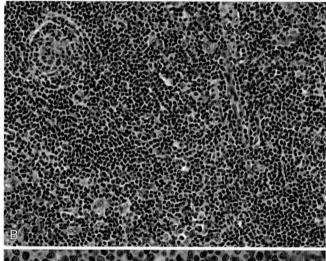
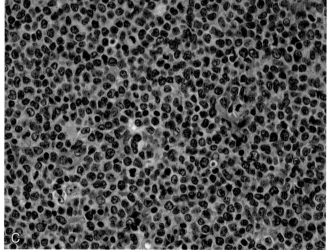

FIGURE 35.1 **MANTLE CELL LYMPHOMA.** Lymph node section demonstrates expansion of mantle zones with remnants of follicular structures (A, low power; B, intermediate power). The neoplastic lymphoid cells are monomorphic and show scattered mitotic figures (C, high power).

- In "in situ" mantle cell lymphoma, the lymphoma involvement is limited to the inner mantle zone, and lymphoma cells express cyclin D1 (BCL-1), CD5, and are weakly BCL2+ (see below).

Immunophenotype

Flow cytometric findings include (Figures 35.10 and 35.11):

- Expression of B-cell associated antigens at variable intensities;
- Surface light chain restriction;
- Coexpression of dim CD5;
- Absent to variable expression of CD23;
- Absent to occasional expression of CD10;
- In contrast to CLL, FMC7 and CD38 are commonly positive;
- In aggressive variants, CD5 can be negative and there may be expression of CD10 and/or BCL6 by immunohistochemical stains (see Figure 35.11).

By immunohistochemical studies, neoplastic B-cells express CD5, IgM, and IgD. Cyclin D1 (BCL1) can be detected in nearly all cases, and BCL2 is always positive (see Figures 35.2 to 35.9). Ki67 proliferation index is low in cases other than blastoid/pleomorphic variants.

Cytogenetic and Molecular Studies

The characteristic cytogenetic alteration in MCL is the t(11;14)(q13;q32) (Figure 35.12), seen in over 95% of cases when combined with conventional cytogenetics and FISH. This translocation is extremely rare in other lymphomas. Blastoid variants, in addition to t(11;14)(q13;q32), may show complex chromosomal aberrations (Figure 35.13).

This translocation it has been detected in occasional atypical cases of CLL associated with an aggressive clinical course, as well as some cases of PLL, but it is likely that these cases represent atypical forms of MCL. Approximately 5% of patients with plasma cell myelomas also show t(11;14) (see Chapter 41).

The main feature of MCL is overexpression of cyclin D1, which is due to the translocation of 11q13 (*BCL1* or *CCND1*) to the 14q32-heavy chain locus (*IGH@*). The over-expression induces increased cycling of the cells, but studies have shown that this is probably not the sole oncogenic feature.

The breakpoints on 11q13 (*CCND1*) occur along a wide region, with only 35% of these within a restricted region (major translocation cluster, MTC), limiting the usefulness

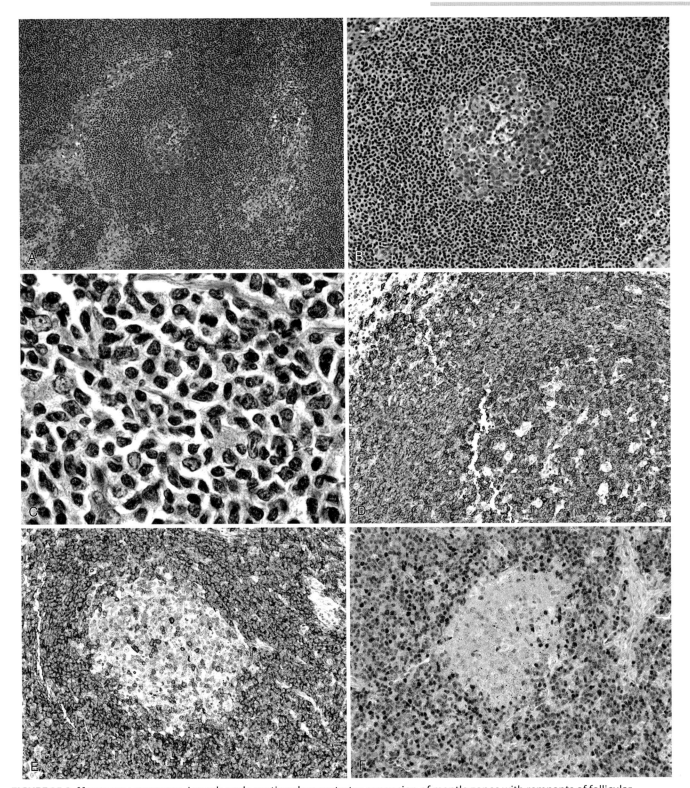

FIGURE 35.2 **MANTLE CELL LYMPHOMA.** Lymph node section demonstrates expansion of mantle zones with remnants of follicular structures (A, low power; B, intermediate power). The neoplastic lymphoid cells are small to medium size with scanty cytoplasm and irregular nuclear borders (C, high power). These cells are positive for CD20 (D), CD5 (E), and BCL-1(F) by immunohistochemistry. The germinal center is CD20+, shows scattered CD5+ cells, and is negative for BCL-1.

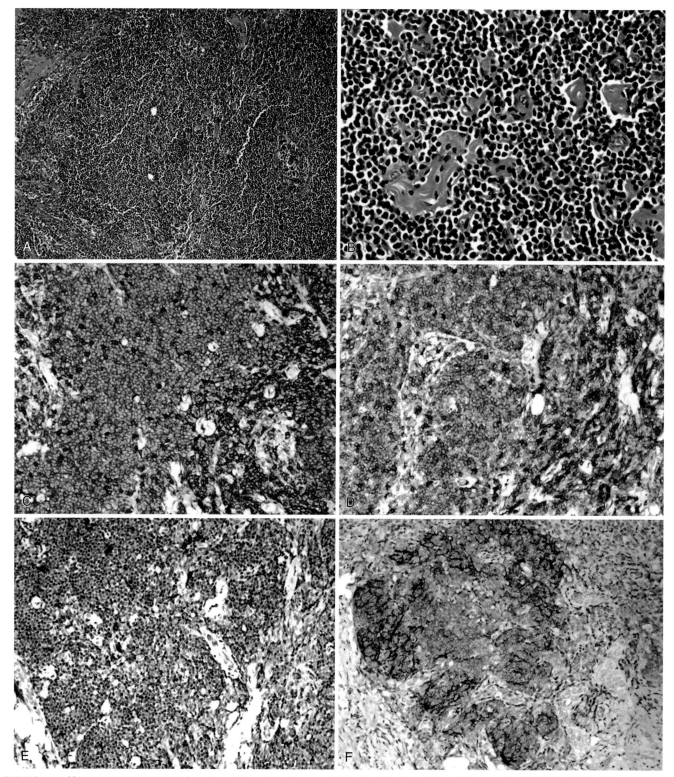

FIGURE 35.3 **MANTLE CELL LYMPHOMA.** A lymph node section demonstrates sheets of small lymphocytes in a hyalinized fibrotic stroma (A, low power; B, intermediate power). Immunohistochemical stains show that these lymphocytes are predominantly B-cells (C, CD20+ red, CD3+ brown) and express CD5 (D) and BCL-1 (E). A meshwork of follicular dendritic cells express CD21 (F).

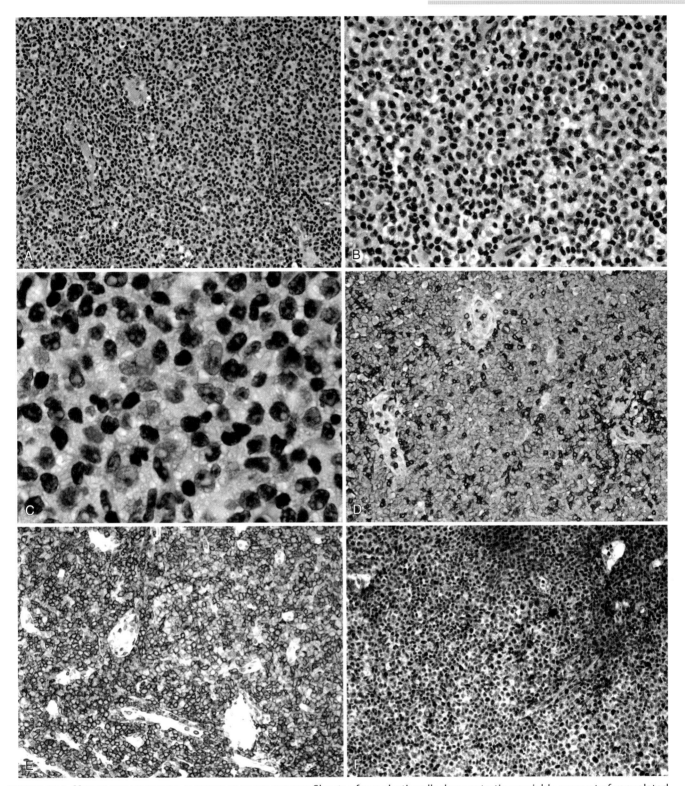

FIGURE 35.4 **MANTLE CELL LYMPHOMA, CLASSICAL BLASTOID VARIANT.** Sheets of neoplastic cells demonstrating variable amount of vacuolated cytoplasm, irregular nuclei with finely dispersed chromatin, and prominent nucleoli (A, low power; B, intermediate power; C, high power). Tumor cells express CD20 (D, CD20+ red, CD3+ brown), CD5 (E) and BCL-1 (F).

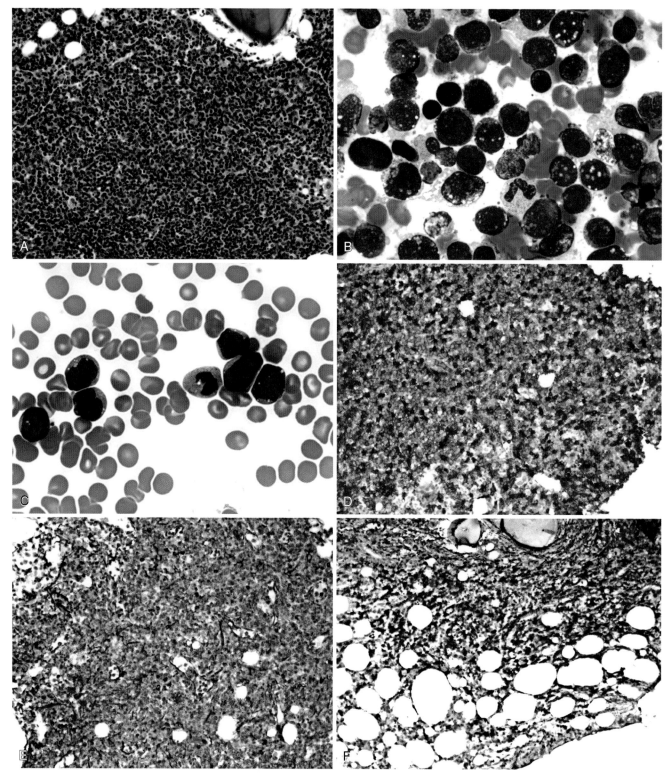

FIGURE 35.5 **MANTLE CELL LYMPHOMA, CLASSICAL BLASTOID VARIANT.** Sheets of neoplastic cells are displayed in bone marrow biopsy section (A). Bone marrow (B) and blood (C) smears show numerous blast cells with dark blue vacuolated cytoplasm, round or irregular nuclei, finely dispersed chromatin, and prominent nucleoli. These cells express CD20 (D, CD20+ red, CD3+ brown), CD5 (E) and BCL-1 (F).

Table 35.1
Cytologic Variants of Mantle Cell Lymphoma[1]

Variant	Characteristics
Typical	Small to medium size lymphocytes with scant cytoplasm, slightly irregular nucleus and condensed chromatin
Aggressive variants	
Blastoid	Cells resembling lymphoblasts with high mitotic figure (>10/10 hpf)
Pleomorphic	Pleomorphic large cells with variable amount of cytoplasm, cleaved or oval nucleus. Nucleoli may be prominent
Other variants	
CLL-like	Small lymphocytes with scant cytoplasm, round nucleus, and condensed chromatin
Marginal-zone-like	Prominent foci of medium size cells with abundant pale cytoplasm, resembling monocytoid B-cells or prolymphocytes. The pale areas may resemble proliferation centers seen in CLL/SLL

[1] Adapted from Swerdlow SH, Campo E, Harris NL, et al. WHO Classification of Tumours of Haematopoietic and Lymphoid Tissues. IARC Press, Lyon, 2010.

of a PCR-based assay. Thus t(11;14) can primarily be detected by cytogenetics studies. FISH is now considered the technique of choice for identifying the t(11;14) during the diagnostic workup.

PRAD1, located approximately 120 kb downstream of the MTC breakpoint, was first identified in studies on parathyroid adenomas with inversion in chromosome 11, and was considered the putative oncogene deregulated by t(11;14). The PRAD1 sequence was recognized as having a high degree of homology with cyclins, and the new member in that gene family was renamed CCND1 encoding for the cyclin D1 protein.

In t(11;14), the coding region of CCND1 is structurally intact, but the chromosomal rearrangement positioning the CCND1 gene adjacent to the enhancer region of the immunoglobulin heavy chain gene results in upregulation of CCND1 and in increased expression of the cyclin D1 protein.

In normal lymphoid cells, the RNA and protein levels of cyclin D1 are extremely low or absent. Although the oncogenic mechanism of cyclin D1 is not completely understood, the constant expression of cyclin D1 has an important role in the pathogenesis of MCL because cyclin D1 promotes the progression of cells through the main commitment checkpoint in G1- to S-phase of the cell cycle.

Although MYC mRNA over-expression has been found in a subset of MCLs, no structural gene alterations of MYC seem to be involved in the pathogenesis of MCL.

Some studies have shown mutations of the TP53 or its over-expression to occur in the aggressive variants of MCL, where the effect of over-expressed cyclin D1 may be enhanced by loss of cyclin-CDK inhibition due to the TP53 mutation.

Structural and numerical centrosome abnormalities have been described to take place at a much higher frequency in MCL than in other lymphoma subtypes, resulting in near-tetraploid karyotype, especially among blastoid variants.

Comparative genomic hybridization (CGH) and FISH studies have identified a characteristic profile of chromosomal changes in MCL that differ from other lymphomas. The most frequent chromosomal imbalances are gains of chromosomes 3q (49–70%), 8q (22–30%), and 12q (20–30%), and the most frequent losses of chromosomes 1p (24–33%), 6q (27–37%), 9p (16–41%), 11q (22–31%), and 13q (41–69%).

DNA amplifications and the high degree of polyploidy (tertaploidy) have been reported to be higher in the blastoid variants (35–80%) than in the common variant of MCL.

Several cell-cycle related genes are known to be deregulated when typical MCLs are compared with blastoid MCLs. For example, the CDK4 gene is upregulated in the blastoid MCL. CDK4 cooperates with cyclin D1 in the progression through the G1/S checkpoint.

Other cytogenetic abnormalities that differ when comparing typical and blastoid MCL are deletions of TP53 and p16 (9p21) in the latter. Tumor cell proliferation has been shown to be associated with decreased survival time and is predictive of blastoid transformation.

Indolent Variant of MCL

Some cases of MCL show an indolent clinical course and may not need treatment at diagnosis. The characteristic features of indolent MCL include a non-nodal leukemic disease, hypermutated IGVH genes, evidence of t(11;14), and negative SOX11 protein expression. CD5 expression is variable. Figures 35.14 to 35.16 depict a 63-year-old man with peripheral blood lymphocytosis for several years. The patient was not treated and had no lymphadenopathy or splenomegaly. Flow cytometric studies on the peripheral blood showed a monoclonal B-cell population negative for CD5, CD10, and CD23. The bone marrow biopsy sections showed an extensive lymphoid infiltrate expressing CD20, BCL-1, and BCL-2. FISH studies revealed t(11;14)(q13;q32).

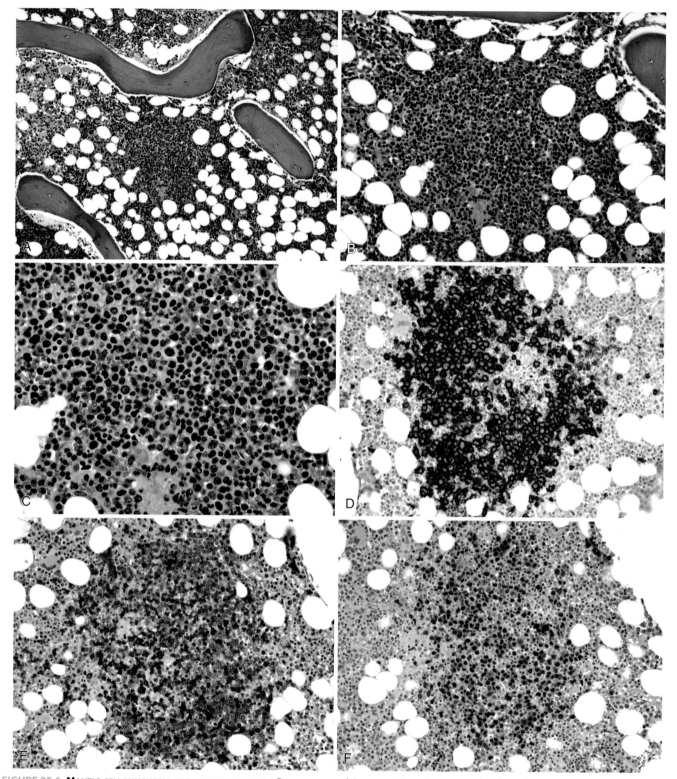

FIGURE 35.6 **MANTLE CELL LYMPHOMA, PLEOMORPHIC VARIANT.** Bone marrow biopsy section demonstrating lymphomatous involvement by an aggregate of pleomorphic atypical lymphoid cells (A, low power; B, intermediate power; C, high power). The neoplastic cells express CD20 (D), CD5 (E) and BCL-1 (F).

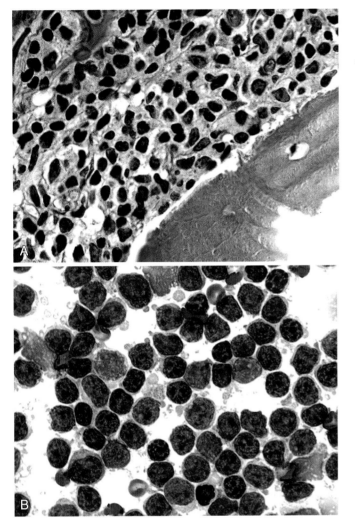

FIGURE 35.7 **BONE MARROW INVOLVEMENT WITH MANTLE CELL LYMPHOMA.** (A) Bone marrow biopsy section demonstrating an atypical paratrabecular lymphoid aggregate consisting of small to medium-sized lymphocytes with irregular nuclei. (B) Bone marrow smear showing mature lymphocytes with round to slightly irregular nuclei. Scattered prolymphocyte-like cells are present.

Differential Diagnosis

The differential diagnosis includes reactive lymphadenopathies, CLL/SLL, FL, PLL, marginal zone lymphomas, diffuse large B-cell lymphoma, and acute leukemias.

Reactive lymph nodes such as some cases of follicular hyperplasia and Castleman's disease may mimic MCL with nodular or mantle zone patterns. In these conditions, the nodal architecture is relatively preserved, and lymphocytes usually do not show irregular nuclei and do not express BCL-1. There is also lack of monoclonality or t(11;14).

There are overlapping cytologic features between CLL/SLL and MCL. Also, the vaguely nodular patterns in MCL and the pseudo-follicles in CLL/SLL may resemble one another. Unlike CLL, the neoplastic cells of MCL are monomorphic and lack prolymphocytes and paraimmunoblasts, usually lack the expression of CD23, express FMC7, CD79a, and BCL-1 nuclear protein, and demonstrate t(11;14).

The vague nodular pattern in MCL may mimic follicular pattern in FL. Unlike FL, the neoplastic cells of MCL lack centroblasts and immunoblasts, do not express CD10, are negative for t(14;18), express CD5, and demonstrate t(11;14).

MCL with monocytoid B-cell morphology may simulate marginal zone lymphomas of the spleen, lymph node, or extranodal sites. Marginal zone B-cells lack the expression of CD5, CD10, and CD23, and are negative for t(11;14). The reported cases of PLL with t(11;14) are now considered *prolymphocytoid variant* of MCL.

Diffuse large B-cell lymphoma, particularly the subtype that expresses CD5, may be confused with the pleomorphic MCL, but these cases are negative for t(11;14).

Blastoid MCL may also mimic acute lymphoid or myeloid leukemias. The neoplastic cells in blastoid MCL are negative for CD34, TdT, and CD117 and express CD5 and BCL-1 in addition to B-cell-associated markers. There is evidence of t(11;14) and/or *CCND1* rearrangement.

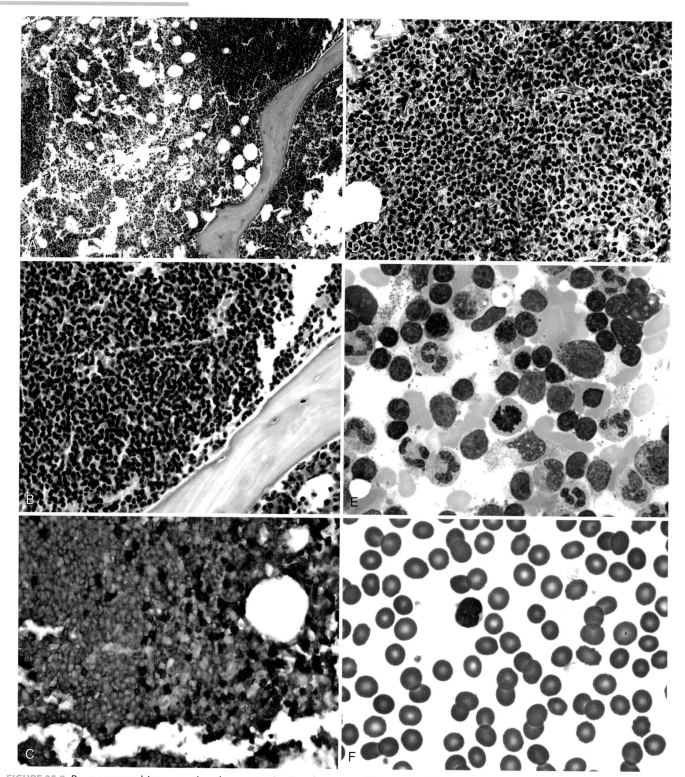

FIGURE 35.8 Bone marrow biopsy section demonstrating involvement with mantle cell lymphoma (A, low power; B, intermediate power). Immunohistochemical stains show tumor cells expressing CD20 (C, red) and BCL-1 (D). Bone marrow smear (E) shows numerous atypical lymphoid cells, and blood smear (F) demonstrates a small lymphocyte with irregular nucleus.

DIFFERENTIAL DIAGNOSIS

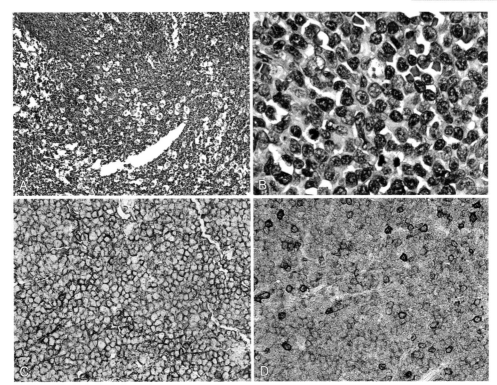

FIGURE 35.9 **SPLENIC INVOLVEMENT WITH MANTLE CELL LYMPHOMA, BLASTOID VARIANT.** The neoplastic cells infiltrate red pulp and are intermixed with tingible body macrophages (starry sky pattern) (A, low power). The tumor cells are predominantly medium to large cells with open nuclear chromatin and prominent nucleoli (B, high power). These cells express CD20 (C) and CD5 (D).

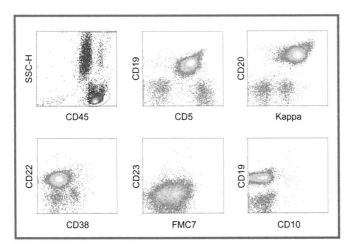

FIGURE 35.10 **FLOW CYTOMETRIC FINDINGS OF MANTLE CELL LYMPHOMA.** Lymphocyte-enriched gate (circled on CD45 open gate) shows predominantly monotypic B-cells, which are positive for CD5 (dim), CD19, CD20, CD22, CD25 (dim; not shown), FMC7 (partial), CD38 (partial), and expressing surface kappa light chain restriction. They are negative for CD10 and CD23.

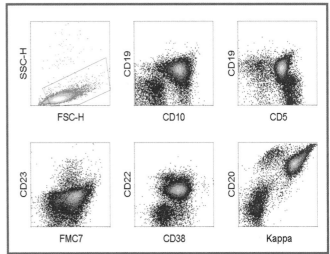

FIGURE 35.11 **FLOW CYTOMETRIC FINDINGS OF MANTLE CELL LYMPHOMA COEXPRESSING CD5 AND CD10.** Lymphocyte-enriched gate reveals mostly monotypic B-cells. The neoplastic B-cells are positive for CD5 (dim), CD10, CD19, CD20, CD22 (dim), CD38, FMC7 (partial), and expressing surface kappa light chain restriction. They are negative for CD23.

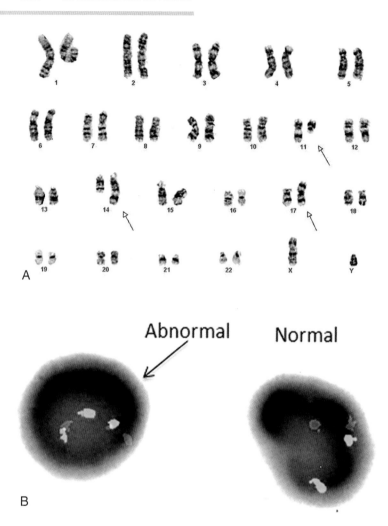

FIGURE 35.12 (A) Karyotype of mantle cell lymphoma showing the characteristic 46, XY, t(11;14)(q13;q32), and an iso(17q). (B) FISH analysis exhibits dual-fusion signals consistent with t(11;14)/IGH@-CCND fusion.

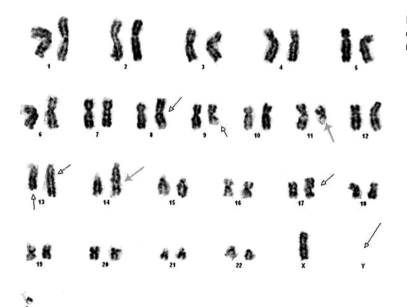

FIGURE 35.13 A karyotype of a blastoid variant of mantle cell lymphoma showing t(11;14) (green arrows) with monosomy Y, iso 8q, and deletions of 9q, 13q, and 17p13.

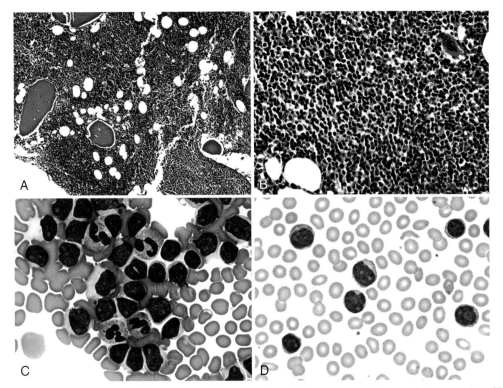

FIGURE 35.14 **INDOLENT MANTLE CELL LYMPHOMA.** Bone marrow biopsy section (A, low power; B, intermediate power) and bone marrow smear (C, high power) demonstrate bone marrow infiltration by small to medium-sized mature lymphocytes. Blood smear (D, high power) shows several atypical medium-sized lymphocytes with slightly irregular nuclei.

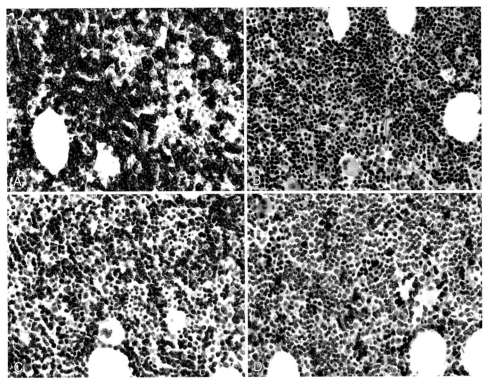

FIGURE 35.15 **INDOLENT MANTLE CELL LYMPHOMA.** Immunohistochemical stains of the bone marrow biopsy sections show the neoplastic cells expressing CD20 (A), BCL-1 (B) and BCL-2 (C). Only scattered cells are positive for CD5 (D).

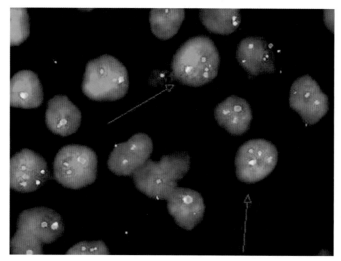

FIGURE 35.16 Mantle cell Lymphoma FFPE section showing cells (arrows) with a yellow (IGH@-CCND1) fusion signals positive for t(11;14).

Additional Resources

Carbone A, Santoro A: How I treat: diagnosing and managing "in situ" lymphoma, *Blood* 117:3954–3960, 2011.

Dreyling M, Hoster E, Bea S, et al: Update on the molecular pathogenesis and clinical treatment of Mantle Cell Lymphoma (MCL): minutes of the 9th European MCL Network conference, *Leuk Lymphoma* 51:1612–1622, 2010.

Espinet B, Solé F, Pedro C, et al: Clonal proliferation of cyclin D1-positive mantle lymphocytes in an asymptomatic patient: an early-stage event in the development or an indolent form of a mantle cell lymphoma? *Hum Pathol* 36:1232–1237, 2005.

Fernàndez V, Salamero O, Espinet B, et al: Genomic and gene expression profiling defines indolent forms of mantle cell lymphoma, *Cancer Res* 70:1408–1418, 2010.

Hartmann EM, Ott G, Rosenwald A: Molecular outcome prediction in mantle cell lymphoma, *Future Oncol* 5:63–73, 2009.

Jaffe ES, Harris NL, Vardiman JW, et al: Hematopathology, Philadelphia, 2010, Saunders/Elsevier.

Nodit L, Bahler DW, Jacobs SA, et al: Indolent mantle cell lymphoma with nodal involvement and mutated immunoglobulin heavy chain genes, *Hum Pathol* 34:1030–1034, 2003.

Ondrejka SL, Lai R, Kumar N, et al: Indolent mantle cell leukemia: clinicopathologic variant characterized by isolated lymphocytosis, interstitial bone marrow involvement, kappa light chain restriction, and good prognosis, *Haematologica* 96:1121–1127, 2011.

Pérez-Galán P, Dreyling M, Wiestner A: Mantle cell lymphoma: biology, pathogenesis, and the molecular basis of treatment in the genomic era, *Blood* 117:26–38, 2011.

Pileri SA, Falini B: Mantle cell lymphoma, *Haematologica* 94:1488–1492, 2009.

Ruskoné-Fourmestraux A, Audouin J: Primary gastrointestinal tract mantle cell lymphoma as multiple lymphomatous polyposis, *Best Pract Res Clin Gastroenterol* 24:35–42, 2010.

Williams ME, Connors JM, Dreyling MH, et al: Mantle cell lymphoma: report of the 2010 Mantle Cell Lymphoma Consortium Workshop, *Leuk Lymphoma* 52:24–33, 2011.

Williams ME, Dreyling M, Winter J, et al: Management of mantle cell lymphoma: key challenges and next steps, *Clin Lymphoma Myeloma Leuk* 10:336–346, 2010.

Xu W, Li JY: SOX11 expression in mantle cell lymphoma, *Leuk Lymphoma* 51:1962–1967, 2010.

Diffuse Large B-Cell Lymphoma

Diffuse large B-cell lymphoma (DLBCL) is the most common non-Hodgkin lymphoma in North America and western Europe, consisting of large neoplastic B-lymphocytes diffusely infiltrating the involved tissues. The neoplastic B-cells have a nuclear size more than twice the size of small lymphocyte, or ≥ a macrophage nucleus. DLBCL is vastly heterogeneous with numerous subtypes and disease entities (Box 36.1). However, these subtypes and disease entities account for a minority of the DLBCL cases. The remaining majority of the cases are referred to as DLBCL, not otherwise specified (NOS). In this chapter DLBCL, NOS is discussed. The subtypes and other disease entities are discussed in the following chapters.

DLBCL, NOS accounts for about 30% of all non-Hodgkin lymphomas. The median age is about 64 years. The male to female ratio is slightly >1. Most patients present with a rapidly enlarging mass, usually in the neck or abdomen. Extranodal involvement is observed in up to 40% of the patients, involving gastrointestinal tract, skin, CNS, bone, testis, liver, spleen, lung, and other organs. Approximately 20% of patients are at stage 1 and about 40% show disseminated disease at the time of diagnosis. Bone marrow involvement is seen in 10–20% of cases. Approximately 30% of patients show systemic "B" symptoms. A significant proportion of DLBCLs are the result of transformation of less aggressive lymphomas, such as CLL/SLL, LPL, SMZL, MALT lymphoma, and FL.

DLBCL, NOS is considered an aggressive lymphoma. The International Prognostic Index (IPI) is highly predictive of patients' clinical outcome. The clinical parameters used in the IPI include age >60, elevated serum LDH, ECOG performance status ≥2, stage III or IV, and number of involved extranodal sites >1. The 5-year survival rate is 26% and 73% for the high and low risk IPI groups, respectively. Other adverse prognostic factors include expression of CD5 and survivin protein. Survivin is the product of the *BIRC5* gene and is an inhibitor of apoptosis. A gain in different regions of chromosome 3q has been associated with shorter survival. On the contrary, expression of bcl-6

Box 36.1 WHO Classification of Diffuse Large B-cell Lymphoma[1]

Diffuse large B-cell lymphoma, not otherwise specified (DLBCL, NOS)
- Common morphologic variants
  - Centroblastic
  - Immunoblastic
  - Anaplastic
- Rare morphologic variants
- Molecular subgroups
  - Germinal center B-cell-like (GBC)
  - Activated B-cell-like (ABC)
- Immunohistochemical subgroups
  - CD5-positive DLBCL
  - Germinal center B-cell-like (GCB)
  - Non-germinal center B-cell-like (non-GCB)

Diffuse large B-cell lymphoma subtypes (see Chapter 37)
- T-cell/histiocyte-rich large B-cell lymphoma
- Primary DLBCL of the CNS
- Primary cutaneous DLBCL, leg type (see Chapter 40)
- EBV positive DLBCL of the elderly

Other lymphomas of large B-cells (see Chapter 38)
- Primary mediastinal (thymic) large B-cell lymphoma
- Intravascular large B-cell lymphoma
- DLBCL associated with chronic inflammation
- Lymphomatoid granulomatosis
- ALK-positive LBCL
- Plasmablastic lymphoma
- Large B-cell lymphoma arising in HHV8-associated multicentric Castleman disease
- Primary effusion lymphoma

Borderline cases
- B-cell lymphoma, unclassifiable, with features intermediate between DLBCL and Burkitt lymphoma (see Chapter 39)
- B-cell lymphoma, unclassifiable, with features intermediate between DLBCL and classical Hodgkin lymphoma (see Chapter 38)

[1] From Swerdlow SH, Campo E, Harris NL, et al. WHO Classification of Tumours of Haematopoietic and Lymphoid Tissues. International Agency for Research on Cancer, Lyon, 2008.

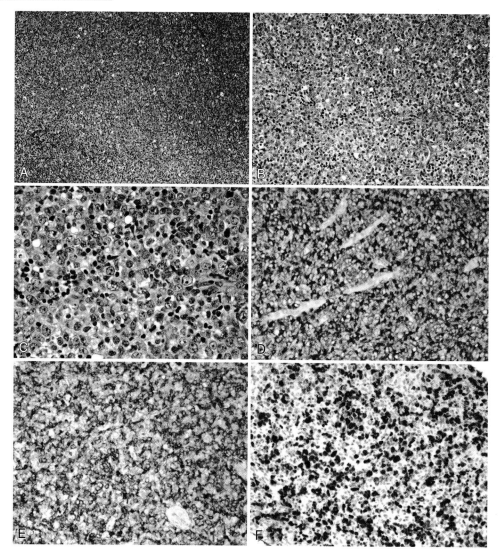

FIGURE 36.1 **DIFFUSE LARGE B-CELL LYMPHOMA.** Lymph node section (A, low power; B, intermediate power; C, high power). The neoplastic cells express CD20 (D, red), CD79a (E), and a high percentage of Ki-67 (F).

nuclear protein has been reported to correlate with longer overall survival.

Therapeutic regimens include combination chemotherapy (e.g., CHOP) with or without involved field radiation for early stages, and CHOP or other alternative combination chemotherapy regimens with or without rituximab, or autologous transplantation in advanced or recurrent disease.

Morphology and Laboratory Findings

- The involved lymph nodes show partial or total effacement of the nodal architecture, with diffuse infiltration of large atypical neoplastic cells, often with areas of necrosis (Figure 36.1). Interfollicular or sinusoidal patterns are infrequent. Occasionally, tumor cells may form cohesive aggregates mimicking metastatic carcinoma. The involved lymph nodes may show coexisting low grade lymphoma, such as CLL/SLL, FL, MZL, or nodular lymphocyte predominant Hodgkin lymphoma.
- The most common extranodal primary site is the gastrointestinal tract, but skin, CNS, bone, spleen, liver, and other organs may be involved. Extranodal involvement is interstitial with formation of a mass and destruction of the normal structures, often with mucosal ulceration.
- Bone marrow involvement can be interstitial or sinusoidal, diffuse, or nodular (Figures 36.2 and 36.3). The neoplastic cells may mimic clusters of megaloblasts or metastatic carcinoma.
- Peripheral blood smears may show the presence of neoplastic large cells (see Figures 36.2F and 36.3B). The serum LDH levels are elevated in about half of cases.

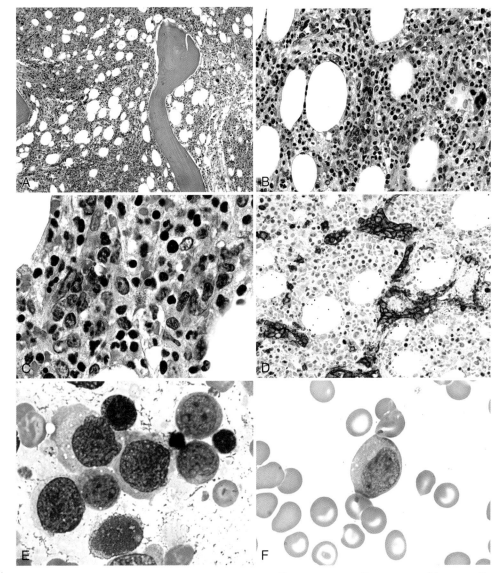

FIGURE 36.2 SINUSOIDAL PATTERN OF BONE MARROW INVOLVEMENT BY DIFFUSE LARGE B-CELL LYMPHOMA. Bone marrow biopsy section demonstrating clusters of large atypical lymphocytes in the sinusoidal spaces (A, low power; B, intermediate power; C, high power). These clusters are CD20+ (D). Bone marrow smear (E) shows large atypical cells mimicking erythroblasts. Blood smear (F) shows a large blastic lymphoid cell.

Several morphologic variants have been described including centroblastic, immunoblastic, and anaplastic types.

- **Centroblastic.** This variant is the most frequent type and consists of medium- to large-sized cells with scant amphophilic/basophilic cytoplasm, round or oval nucleus, fine nuclear chromatin, and several nucleoli bound to the nuclear membrane (Figure 36.4). Multilobated (>3 lobes) centroblasts may be present, and sometimes may create a polymorphic appearance. Some cases may show a mixture of centroblasts and immunoblasts.
- **Immunoblastic.** This variant comprises about 10% of DLBCL and represents diffuse lymphomas with >90% immunoblasts. Immunoblasts are large cells with abundant basophilic cytoplasm, round or oval nucleus, fine chromatin, and a prominent central nucleolus (Figure 36.5). Plasmacytoid features may be present. This variant is commonly seen in immunecompromised patients.
- **Anaplastic.** This variant consists of atypical, large cells with bizarre pleomorphic nuclei, some of which may resemble Reed–Sternberg cells (Figures 36.6 and 36.7). In some cases, the neoplastic large lymphoid cells are spindle-shaped, mimicking sarcoma (Figure 36.8), or appear in large clusters, resembling metastatic carcinoma. Sinusoidal pattern has been observed.

Immunophenotype

Flow cytometric findings include:

- Expression of B-cell associated antigens at variable intensities;
- Occasional absence of one or two B-cell antigens;

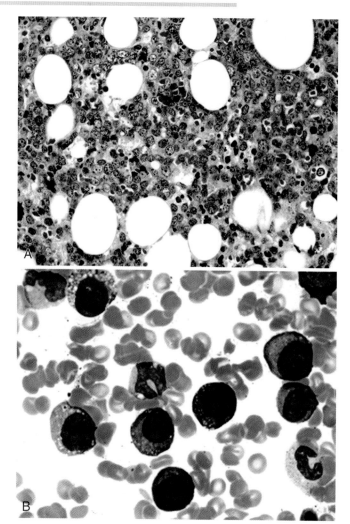

FIGURE 36.3 Bone marrow (A) and blood (B) involvement in a patient with diffuse large B-cell lymphoma.

- Surface or intracellular light chain restriction detectable in about two-thirds of cases;
- Coexpression of CD10 in most cases of germinal center B-cell-like (GCB) (Figure 36.9);
- Absent CD10 in non-germinal center B-cell-like cases (non-GCB) (Figure 36.10);
- Coexpression of CD5 in less than 10% of the *de novo* cases (Figures 36.11 and 36.12), and rarely in DLBCL arising from CLL/SLL (Figure 36.13). Patients with CD5+ DLBCL show higher incidence of CNS involvement and poor prognosis. Unlike Blastoid mantle cell lymphoma, CD5+ DLBCL is negative for BCL-1.

By immunohistochemical studies:

- The GCB subgroup is categorized by (1) expression of CD10 in >30% of the neoplastic cells, and (2) negative CD10, but positive BCL6 plus negative MUM1 (see Figures 36.1, 36.4, and 36.7).
- All other cases are regarded as the non-GCB subgroup.
- Neoplastic large B-cells are negative for BCL1, but positive for BCL2 in about half of cases.

- Expression of CD30 may be seen in cases with anaplastic features.
- Ki67 proliferation rate is usually high, and can sometimes exceed 90%.

Molecular and Cytogenetic Studies

Numerous chromosomal aberrations, point mutations, and deletions have been described in DLBCL, suggesting a heterogeneous genetic background. Because of complex karyotypes noted in DLBCL, it is often best detected by conventional cytogenetics, which, despite its low resolution, represents a genome-wide method by which many structural and numerical abnormalities can be detected simultaneously.

Clonal karyotypic abnormalities are reported in 90% of DLBCL. The most commonly involved breakpoints are 14q32 (*IGH@*), 3q27(*BCL6*), and 18q21(*BCL2*). Other frequent chromosomal abnormalities include 1p36, 1p22, 8q24, 3q21, 9p, and 6q21 chromosomal bands (Figures 36.14 to 36.17).

Translocations involving the locus for *BCL6* at 3q27 are found in ~40% of cases.

Rearrangements of *MYC* (8q24) can be found in 5–20% of these cases. However, karyotype, FISH, and array-CGH studies show DBLCL with a t(8;14) commonly exhibit a higher degree of genomic aberrations (highly complex karyotype) than Burkitt lymphomas with t(8;14), and thus the two diseases can be distinguished based on the karyotype abnormalities.

In more than half of cases, additional abnormalities, most commonly the t(14;18), are observed. These "double-hit" or "dual-hit" lymphomas often have complex cytogenetic alterations (Figures 36.18 and 36.19), including occasional breakpoints at 3q27/*BCL6* ("triple-hit" lymphomas).

However, overexpression of the bcl-2 protein is found in 25–50% of cases, suggesting other mechanisms for the bcl-2 overexpression.

Translocations affecting the immunoglobulin gene sites at 14q32 (*IGH@*), 22q11.2 (*IGL@*), and 2p12 (*IGK@*) also have been identified. These include t(14;18)(q32;q21), t(8;14)(q24;q32) or t(8;22)(q24;q11.2), t(3;14)(q27;q32), t(3;22)(q27;q11.2) (Figure 36.20), and other rearrangements involving 14q32.

Immunoglobulin heavy chain clonality can be demonstrated by PCR studies (Figure 36.21).

Genomic profiling studies show two major subgroups: (a) characteristic features of germinal center B-cells (GCB) and (b) a profile similar to the activated B-cells (ABC). Gene profiling studies of GCB, ABC, and mediastinal large B-cell lymphoma, demonstrated significant differences in the frequency of particular chromosomal aberrations. DLBCL-GCB frequently display t(14;18), deletions of 17p

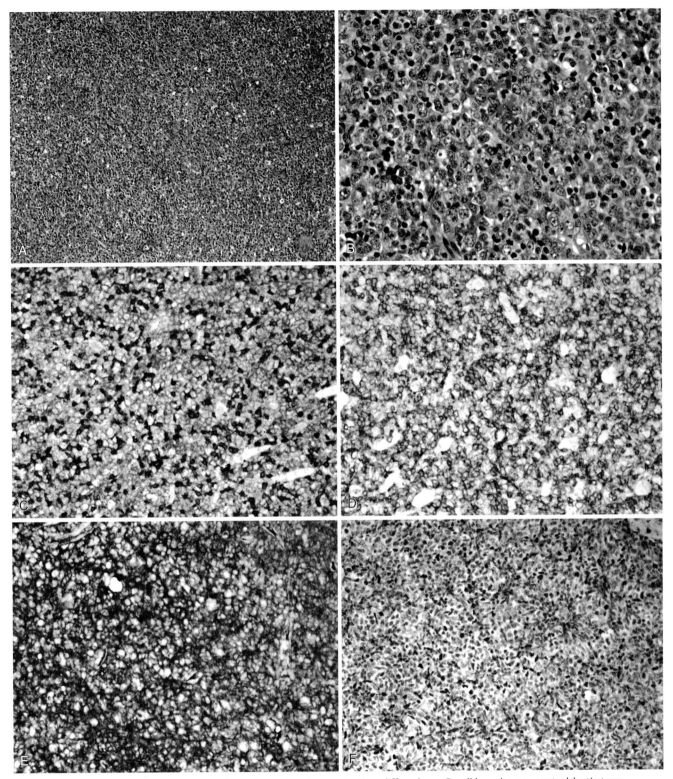

FIGURE 36.4 **DLBCL, CENTROBLASTIC TYPE.** Lymph node section demonstrating diffuse large B-cell lymphoma, centroblastic type, consisting of medium- to large-sized cells with round or oval nucleus, fine nuclear chromatin, and several nucleoli (A, low power; B, high power). The neoplastic cells express CD20 (C, red), CD79a (D), kappa (E) and BCL-6 (F).

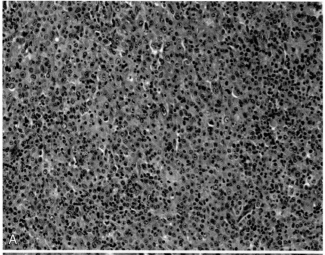

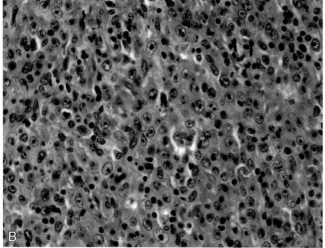

FIGURE 36.5 **DLBCL, IMMUNOBLASTIC TYPE.** Lymph node section demonstrating diffuse large cell lymphoma, immunoblastic type (A, low power; B, high power). Immunoblasts are large cells with abundant basophilic cytoplasm, round or oval nucleus, fine chromatin, and a prominent central nucleolus.

(*TP53*), and amplifications of 2p (*REL*) and gains of 12q, whereas the group of DLBCL-ABC exhibit frequent trisomy 3, gains of chromosomes 3q and 18q21–q22, and losses of chromosome 6q, 9p (*CDKN2A*).

Differential Diagnosis

The differential diagnosis includes Burkitt lymphoma, T-cell lymphomas consisting of large cells, pleomorphic variant of mantle cell lymphoma, anaplastic plasma cell tumors, myeloid sarcoma, and non-hematopoietic malignancies.

The centroblastic variant of DLBCL may show overlapping morphologic features with Burkitt lymphoma. In general, the expression of Ki67, CD10, and p53 is higher in the neoplastic cells of Burkitt lymphoma than in the DLBCL cells. Burkitt cells are BCL-2-negative (see Chapter 39).

Tumor cells in the pleomorphic variant of mantle cell lymphoma usually have scant cytoplasm, and inconspicuous nucleoli and express BCL-1 and CD5. Anaplastic large cell lymphomas are positive for CD30 and may express AKL, CD2, and/or CD5. Anaplastic plasma cells are usually CD138+ and CD20−. Blasts in myeloid sarcoma express myeloid-associated markers, such as myeloperoxidase and are negative for CD20. Reed–Sternberg and Hodgkin cells in classical Hodgkin lymphoma express CD15 and CD30 and are negative for CD45. Non-hematopoietic malignancies usually show a cohesive growth pattern and are negative for CD45 and CD20.

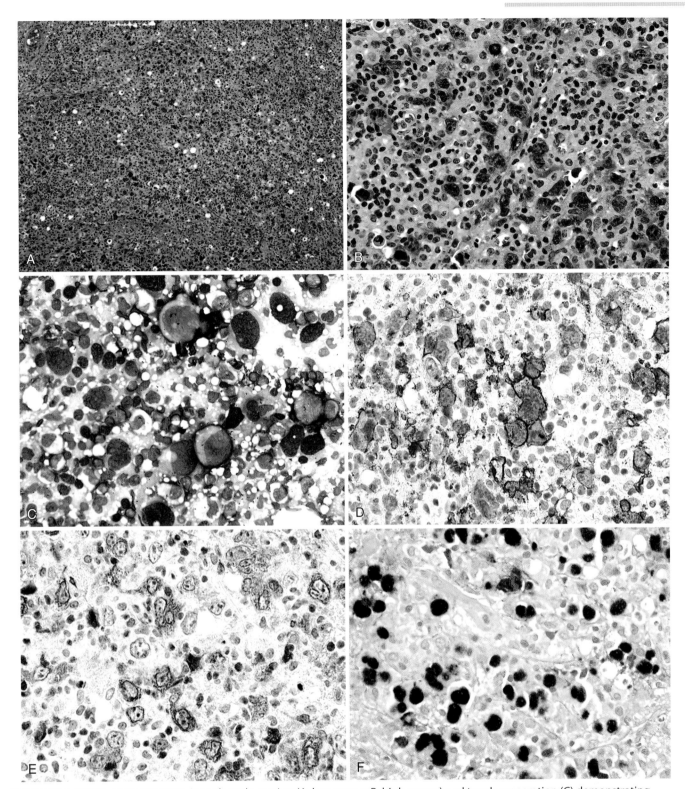

FIGURE 36.6 **DLBCL, ANAPLASTIC TYPE.** Lymph node section (A, low power; B, high power) and touch preparation (C) demonstrating diffuse large cell lymphoma, anaplastic type. This type consists of atypical, large cells with bizarre pleomorphic nuclei, some of which resemble Reed–Sternberg cells. The neoplastic large cells are positive for CD20 (D), BCL-2 (E), and EBV-EBER (F).

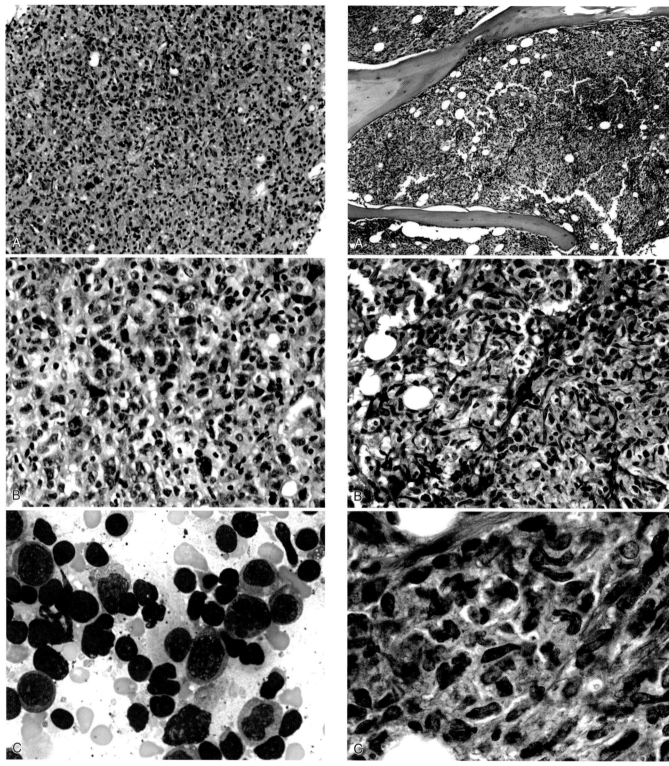

FIGURE 36.7 **DLBCL, ANAPLASTIC TYPE.** Bone marrow biopsy section demonstrating lymphomatous involvement in a patient with anaplastic large B-cell lymphoma (A, low power; B, high power). Anaplastic cells are demonstrated in bone marrow smear (C).

FIGURE 36.8 **SARCOMA-LIKE VARIANT OF DLBCL.** Bone marrow biopsy section demonstrating a sarcoma-like variant of anaplastic large B-cell lymphoma consisting of spindle-shaped, atypical large lymphocytes (A, low power; B, intermediate power; C, high power).

DIFFERENTIAL DIAGNOSIS **435**

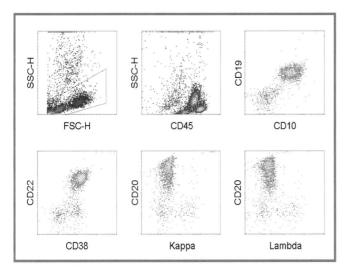

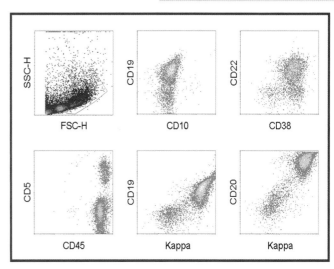

FIGURE 36.9 **FLOW CYTOMETRIC FINDINGS OF DIFFUSE LARGE B-CELL LYMPHOMA, GERMINAL CENTER B-CELL-LIKE.** On open gate density displays (in magenta), B-cells highlighted by CD19 backgating, are large cells and show dim to moderate expression of CD45 in two subsets. In addition, the large abnormal B-cells are positive for CD10, CD19, CD20, CD22, CD38, and express no surface light chains. (Immunohistochemical studies showed the large neoplastic B-cells positive for BCL-6).

FIGURE 36.10 **FLOW CYTOMETRIC FINDINGS OF DIFFUSE LARGE B-CELL LYMPHOMA, NON-GERMINAL CENTER B-CELL-LIKE.** Large-lymphocyte-enriched gate (in blue) demonstrates predominantly monotypic B-cells, which are positive for CD19, CD20 (bright), CD22, CD38 (partial), CD45 (dim), and with surface kappa light chain restriction. They are negative for CD10. (Immunohistochemical studies showed the large neoplastic B-cells expressing MUM1 and BCL2).

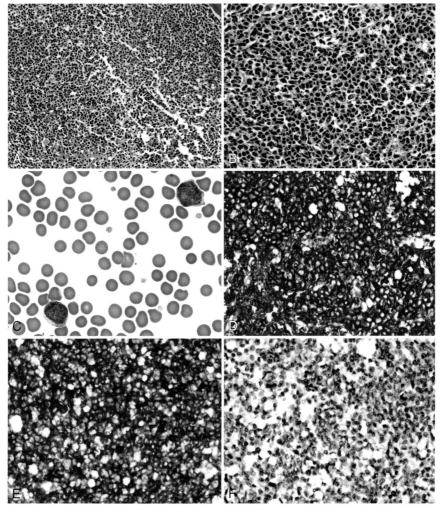

FIGURE 36.11 **CD5+ DIFFUSE LARGE B-CELL LYMPHOMA.** Bone marrow biopsy section (A, intermediate power; B, high power) showing large atypical cells with variable amount of cytoplasm and irregular nuclei. Two large atypical cells are demonstrated in the peripheral blood smear (C). The neoplastic cells express CD20 (D), and CD5 (E), but are negative for BCL-1 (F).

436 DIFFUSE LARGE B-CELL LYMPHOMA

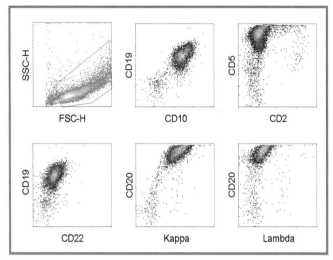

FIGURE 36.12 **FLOW CYTOMETRIC FINDINGS OF DIFFUSE LARGE B-CELL LYMPHOMA, CD5-POSITIVE.** Large-lymphocyte-enriched gate (in magenta) reveals almost exclusively monotypic B-cells. The large neoplastic B-cells are positive for CD5 (bright), CD10, CD19, CD20 (bright), CD22 (dim), and expressing surface kappa light chain restriction. (Immunohistochemical stains for BCL1 were negative).

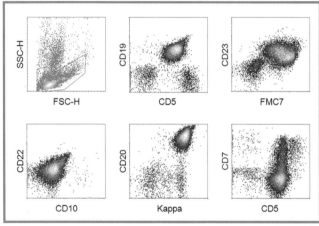

FIGURE 36.13 **FLOW CYTOMETRIC FINDINGS OF CLL TRANSFORMED TO DLBCL.** Intermediate-sized and large-lymphocyte-enriched gate (in magenta) shows predominantly monotypic B-cells, which are positive for CD5 (dim), CD19, CD20, CD22 (partial), CD23, FMC7, and expressing surface kappa light chain restriction. In addition, there is an aberrant expression of partial CD7. The neoplastic B-cells are negative for CD10.

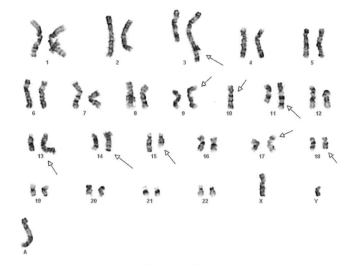

FIGURE 36.14 Karyotype of a case of DLBCL demonstrating complex chromosomal aberrations including t(14;18) and 3q27 rearrangement.

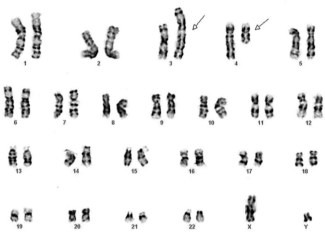

FIGURE 36.15 Karyotype exhibiting 3q27 rearrangement in a case of DLBCL.

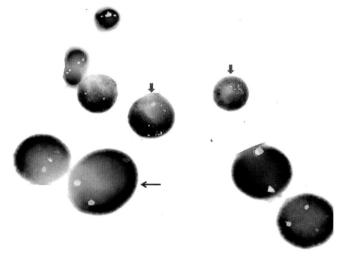

FIGURE 36.16 FISH with the dual color BCL6 probe showing "spilt" signals (red arrow) consistent with 3q27 rearrangements, and amplification (green arrows).

DIFFERENTIAL DIAGNOSIS 437

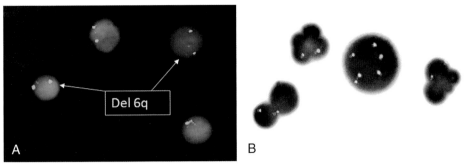

FIGURE 36.17 Losses in 6q (A, red signals) and gains in 18q (B) are common in DLBCL.

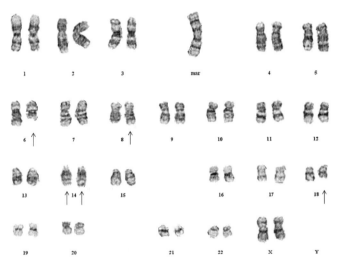

FIGURE 36.18 A "dual hit" DLBCL lymphoma with both t(8;14) and t(14;18) along with deletion 6q.

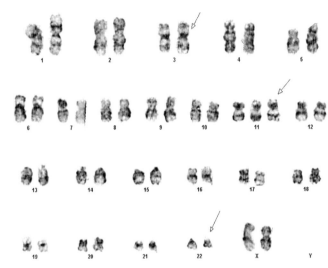

FIGURE 36.20 A DLBCL (germinal-center subgroup) karyotype: 47,XX,t(3;22)(q27;q11.2),+11.

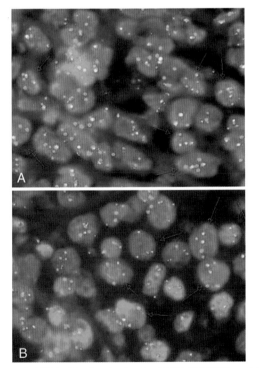

FIGURE 36.19 FISH with the dual color *IGH@-BCL2* (A) and *IGH@-MYC* (B) in a "dual hit" DLBCL positive for t(14;18) and t(8;14). Arrows point to cells with fusion signals.

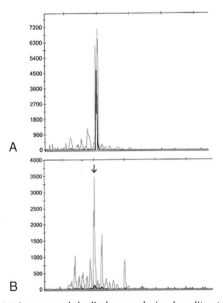

FIGURE 36.21 Immunoglobulin heavy chain clonality study in a patient with diffuse large B-cell lymphoma using PCR (framework 3 primer set results shown). (A) Clonal peak at 100 bp at the time of initial diagnosis. (B) The results at start of relapse. An amplicon peak of the same size is seen, suggesting that this is a recurrence of the initial clone, though this time there is a more prominent background polyclonal population.

Additional Resources

Bellan C, Stefano L, Giulia de F, et al: Burkitt lymphoma versus diffuse large B-cell lymphoma: a practical approach, *Hematol Oncol* 28:53–56, 2010.

Carbone A, Gloghini A, Aiello A, et al: B-cell lymphomas with features intermediate between distinct pathologic entities: from pathogenesis to pathology, *Hum Pathol* 41:621–631, 2010.

Chan WJ: Pathogenesis of diffuse large B cell lymphoma, *Int J Hematol* 92:219–230, 2010.

Ennishi D, Takeuchi K, Yokoyama M, et al: CD5 expression is potentially predictive of poor outcome among biomarkers in patients with diffuse large B-cell lymphoma receiving rituximab plus CHOP therapy, *Ann Oncol* 19:1921–1926, 2008.

Flowers CR, Sinha R, Vose JM: Improving outcomes for patients with diffuse large B-cell lymphoma, *CA Cancer J Clin* 60:393–408, 2010.

Jaffe ES: The 2008 WHO classification of lymphomas: implications for clinical practice and translational research, *Hematology Am Soc Hematol Educ Program*:523–531, 2009.

Miyazaki K, Yamaguchi M, Suzuki R, et al: CD5-positive diffuse large B-cell lymphoma: a retrospective study in 337 patients treated by chemotherapy with or without rituximab, *Ann Oncol* 22:1601–1607, 2011.

Murawski N, Zwick C, Pfreundschuh M: Unresolved issues in diffuse large B-cell lymphomas, *Expert Rev Anticancer Ther* 10:387–402, 2010.

Niitsu N: Current treatment strategy of diffuse large B-cell lymphomas, *Int J Hematol* 92:231–237, 2010.

Salaverria I, Siebert R: The gray zone between Burkitt's lymphoma and diffuse large B-cell lymphoma from a genetics perspective, *J Clin Oncol* 29:1835–1843, 2011.

Slack GW, Gascoyne RD: MYC and aggressive B-cell lymphomas, *Adv Anat Pathol* 18:219–228, 2011.

Swerdlow SH, Campo E, Harris NL, et al: WHO classification of tumours of haematopoietic and lymphoid tissues, ed 4, Lyon, 2008, International Agency for Research on Cancer.

Tzankov A, Zlobec I, Went P, et al: Prognostic immunophenotypic biomarker studies in diffuse large B cell lymphoma with special emphasis on rational determination of cut-off scores, *Leuk Lymphoma* 51:199–212, 2010.

Diffuse Large B-Cell Lymphoma Subtypes

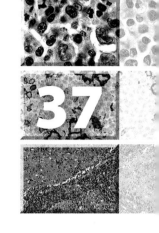

The WHO classification of DLBCL includes four subtypes:

1. T-cell/histiocyte-rich large B-cell lymphoma,
2. EBV+ DLBCL of the elderly,
3. primary DLBCL of the CNS, and
4. primary cutaneous DLBCL, leg type.

In this chapter, categories 1–3 are discussed. Primary cutaneous DLBCL, leg type is discussed in Chapter 40.

T-Cell/Histiocyte-Rich Large B-Cell Lymphoma

T-cell/histiocyte-rich large B-cell lymphoma (THRLBCL) is a variant of DLBCL characterized by scattered large B-cells interspersed in a background of small reactive T-lymphocytes and often variable numbers of histiocytes.

THRLBCL is primarily a nodal disease, reported in middle-aged men and accounting for about 10% of DLBCLs. It is an aggressive nodal lymphoma, often found in advanced stages (III or IV) at diagnosis. Splenomegaly is noted in about 25% of patients. Bone marrow involvement is frequent (30–60%).

MORPHOLOGY

- The lymph node involvement is usually diffuse, but, in rare cases, focal vague nodularity can be present. The lymphoid infiltrate is polymorphic, consisting of scattered individually dispersed large B-cells (<10% of total cells) in a background of small T-lymphocytes (Figures 37.1 and 37.2) and histiocytes. The large neoplastic B-cells do not form aggregates or sheets.
- The large B-cells show morphologic features of centroblasts or immunoblasts, or may appear anaplastic and pleomorphic, sometimes resembling Reed–Sternberg cells (Figure 37.3).
- The background small lymphocytes (T-cells) may show mild atypical features including irregular nuclear shape or slightly larger size.
- Variable numbers of histiocytes are often present; plasma cells and eosinophils are rare or absent.
- Splenic involvement consists of multiple micronodules, and the lymphoma exhibits the same cellular composition as seen in the lymph node.
- Bone marrow involvement is patchy or nodular with similar morphology as described for the involved lymph nodes.

IMMUNOPHENOTYPE

- Flow cytometry has limited value in diagnosing THRLBCL, because of the low quantity of the large neoplastic B-cells compared with the abundance of T-cells and histiocytes in the background. In addition, the cellular integrity of these large neoplastic B-cells is more likely to be compromised during the staining and washing process in setting up MFC assays. It is rare that a discrete large neoplastic B-cell population can be detected by MFC in THRLBCL.
- In contrast, immunohistochemical studies are more useful in phenotypic analysis of THRLBCL (Figures 37.1 to 37.3). The large neoplastic cells express CD45 plus pan B-cell antigens, and stain for CD20 and PAX 5 is strong. They are also positive for BCL6 and often positive for BCL2.
- Staining for CD30 may highlight scattered large cells in some cases (Figure 37.3C), but CD15 is negative. In the background, there are numerous CD3-positive T-cells and CD68-positive histiocytes.

MOLECULAR AND CYTOGENETIC STUDIES

- Because of the paucity of neoplastic cells (<5%) in a background of reactive T cells and histiocytes, it is difficult to perform routine genetic studies.
- Somatic mutations of the immunoglobulin heavy chain genes are common.
- Rearrangements of *BCL6* can be detected by FISH in about half of the cases.

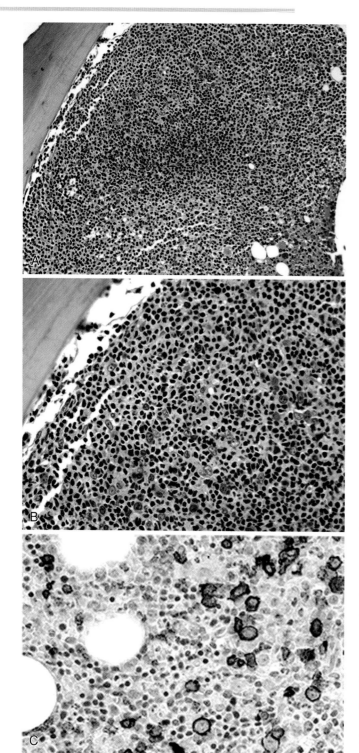

FIGURE 37.1 **T-CELL/HISTIOCYTE-RICH LARGE B-CELL LYMPHOMA.** Bone marrow biopsy section demonstrating scattered large cells mixed with a large number of small lymphocytes (A, low power; B, high power). Large cells are CD20-positive and small cells are CD20-negative (C).

- Translocations involving the 14q32 (*IHG@*) region, e.g., t(9;14)(p13;q32), are a recurrent abnormality.
- Array-CGH studies have identified frequent genomic imbalances such as duplications of Xp/Xq, 4q, and 18q, and deletions of 17p.

DIFFERENTIAL DIAGNOSIS

The differential diagnosis includes reactive lymphoid hyperplasia, classical mixed cellularity Hodgkin lymphoma, nodular lymphocyte predominant Hodgkin lymphoma (NLPHL), peripheral T-cell lymphoma, and lymphomatoid granulomatosis. A summary of distinguishing features is presented in Table 37.1.

In reactive lymphoid hyperplasia the lymph node architecture is completely or partially preserved, the large reactive lymphoid cells often appear in focal aggregates, and depict polytypic staining for Ig light chains.

The Reed–Sternberg cells and their variants in mixed cellularity Hodgkin lymphoma express CD15 and CD30 and are negative for CD45, whereas the large cells in THRLBCL express CD20 and CD45 but are negative for CD15 and mostly negative for CD30.

Unlike THRLBCL, the large B-cells in NLPHL are predominantly located within vague lymphoid nodules, and often surrounded by CD57+ T-cells. They also express EMA. The lymphoid nodules demonstrate a rich follicular dendritic meshwork as demonstrated by positive CD21 or CD23 immunohistochemical stains.

In peripheral T-cell lymphomas the large cells express pan-T-cell markers and the smaller lymphoid cells are more atypical and pleomorphic than the T-cells seen in THRLBCL. T-cell receptor (*TCR*) gene rearrangement studies are positive.

Similar to THRLBCL, large atypical B-cells in lymphomatoid granulomatosis are found in a background of reactive T cells, but they are almost always positive for EBV, and the lesion is extranodal.

EBV+ Diffuse Large B-Cell Lymphoma of the Elderly

EBV-associated DLBCL occurs in patients older than 50 years without any known history of immunodeficiency or lymphoma. Other well-defined clonal EBV-associated disorders, such as primary effusion lymphoma, lymphomatoid granulomatosis, plasmablastic lymphoma, and DLBCL associated with chronic inflammation, are excluded from this category.

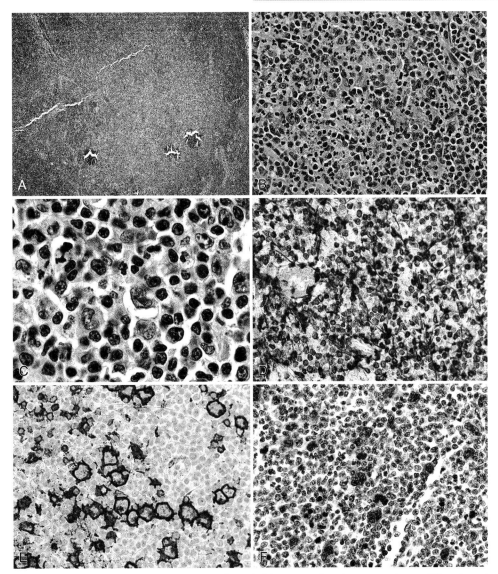

FIGURE 37.2 **T-CELL/HISTIOCYTE-RICH LARGE B-CELL LYMPHOMA.** Lymph node biopsy section demonstrating a diffuse lymphomatous process (A, low power) with scattered large cells, some mimicking Hodgkin cells (B, intermediate power; C, high power). Small lymphocytes are predominantly CD3+ (D), and large cells express CD20 (E) and Pax5 (F).

The disease shows slight male preponderance (M:F ratio 1.4) with a median age of about 70 years. Extranodal involvement is a common feature frequently affecting skin, stomach, lung, and tonsils. It is an aggressive disease, with unfavorable prognostic factors including "B" symptoms and age >70 years.

MORPHOLOGY (FIGURES 37.4 TO 37.6)

- The affected lymph nodes and/or extranodal tissues show architectural effacement by a diffuse lymphoid infiltrate consisting of either monotonous sheets of atypical large cells, or polymorphous infiltrate of mixed large atypical lymphocytes, small lymphocytes, plasma cells, and histiocytes.

- The large cells may appear as centroblasts, immunoblasts, anaplastic, or Reed–Sternberg-like. Areas of necrosis are often present.

IMMUNOPHENOTYPE (FIGURES 37.4 TO 37.6)

- Flow cytometry has limited utilities since monotypic B-cells may not be detectable, especially in the polymorphous subtype.
- By immunohistochemical studies, the neoplastic cells stain for pan B-cell markers including CD20, PAX5, and CD79a. BCL2, CD10, and BCL6 are often negative, but MUM1 positivity is common.

442 DIFFUSE LARGE B-CELL LYMPHOMA SUBTYPES

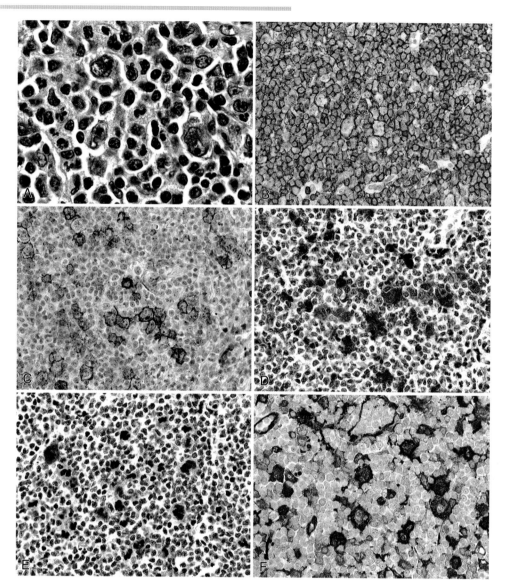

FIGURE 37.3 **T-CELL/HISTIOCYTE-RICH LARGE B-CELL LYMPHOMA; SAME CASE AS FIGURE 37.2.** Lymph node biopsy section demonstrating scattered Hodgkin/Reed–Sternberg (HRS)-like cells in a background of small lymphocytes (A). Both small and large cells express CD45 (B). Large cells, similar to HRS cells, express CD30 (C), BOB.1 (D), Oct.2 (E) and fascin (F).

Table 37.1
Differential Diagnosis of T-Cell/Histiocyte-Rich Large B-Cell Lymphoma (THRLBCL)

Pathology	Morphology	Immunophenotype	Molecular/Cytogenetics
THRLBCL	Effacement of nodal architecture; predominantly diffuse pattern, scattered large cells in a background of small T-cells and histiocytes	Large cells CD20+, CD45+; small cells CD3+, CD45+	*IGH@* gene rearranged; no evidence of *TCR* rearrangement; rarely EBV+
Reactive lymphoid hyperplasia	Preserved nodal architecture, large cells often in aggregates	Large cells are a mixture of CD3+ and CD20+ cells	No evidence of *IGH@* or *TCR* gene rearrangements
Mixed cellularity Hodgkin lymphoma	Effacement of nodal architecture; presence of RS cells and variants	RS cells and variant are CD15+, CD30+, CD45−	No evidence of *IGH@* or *TCR* gene rearrangements; frequently EBV+
Nodular lymphocyte predominant Hodgkin lymphoma	Effacement of nodal architecture with a nodular pattern and large cells within nodules	Large cells are CD20+, EMA+ and are surrounded by CD57+ T-cells; rich follicular dendritic mesh (CD21+, CD23+)	No evidence of *IGH@* or *TCR* gene rearrangements; rarely EBV+
Peripheral T-cell lymphoma	Effacement of nodal architecture; atypical small lymphocytes	Large cells are CD3+, CD20−	May show *TCR* rearrangement; usually EBV−
Lymphomatoid granulomatosis	Extranodal involvement; scattered large cells in a background of small T-cells	Large cells CD20+, CD45+; small cells CD3+, CD45+	*IGH@* gene rearranged; commonly EBV+

RS, Reed–Sternberg.

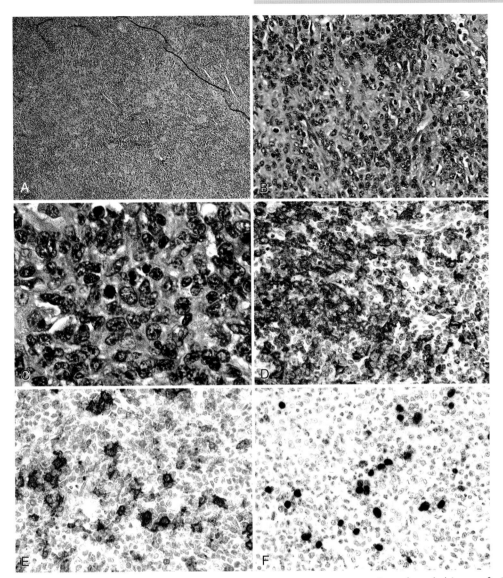

FIGURE 37.4 **EBV-POSITIVE DIFFUSE LARGE B-CELL LYMPHOMA OF THE ELDERLY, LYMPH NODE INVOLVEMENT.** Lymph node biopsy of a 79-year-old man demonstrating a diffuse lymphoma (A), primarily consisting of large atypical lymphoid cells (B, intermediate power; C, high power). These cells express CD20 (D), and some are positive for CD30 (E) and EBV-EBER (F).

- Staining for CD30 is variably positive, while CD15 is negative. In cases with plasmablastic features, cytoplasmic Ig is often detectable, though CD20 can be negative.
- EBV positivity is demonstrated using combined methods of IHC for LMP1 and in-situ hybridization for EBV-EBER.

MOLECULAR AND CYTOGENETIC STUDIES

- There are very few reports of EBV-associated DLBCL with specific genetic abnormalities.
- Translocations involving the 3q27 (*BCL6*) region can be identified by FISH studies.
- Molecular studies will exhibit clonal *IGH@* gene rearrangements.

DIFFERENTIAL DIAGNOSIS

The pleomorphic subtype with the presence of Reed–Sternberg-like cells may mimic classical Hodgkin lymphoma. Unlike classical Hodgkin lymphoma, the EBV+ DLBCL more frequently involves extranodal tissues, often shows areas of necrosis, and the atypical large cells are EBV+, CD20+, CD45+, and lack expression of CD15.

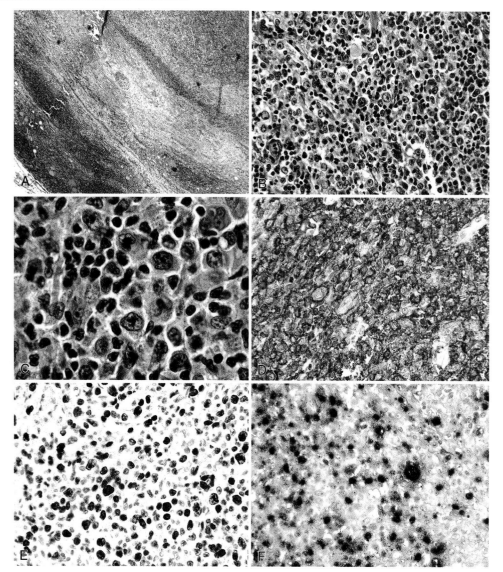

FIGURE 37.5 **EBV-POSITIVE DLBCL OF THE ELDERLY, THYROID INVOLVEMENT.** Thyroid biopsy of a 73-year-old woman demonstrate a diffuse lymphoid infiltrate consisting of mixed small and large atypical lymphocytes (A, low power; B, intermediate power; C, high power). The majority of the lymphoid cells express CD20 (D). Numerous cells are also positive for KI-67 (E) and EBV-EBER (F).

Primary DLBCL of the CNS

Primary DLBCL of CNS includes large B-cell lymphomas primarily affecting intracerebral and intraocular tissues. Intravascular large B-cell lymphomas, large B-cell lymphomas of the dura, AIDS-associated large B-cell lymphomas, and secondary CNS lymphomas are excluded from this category.

These tumors are relatively rare (about 1% of all non-Hodgkin lymphoma). The median age is around 60 years and there is a slight male preponderance. About 50–80% of patients show some neurological symptoms.

MORPHOLOGY

There is a diffuse infiltration of affected tissue often with perivascular involvement (Figure 37.7). The neoplastic cells resemble centroblasts, often mixed with reactive small lymphocytes, macrophages, astrocytes, microglial cells, and areas of necrosis.

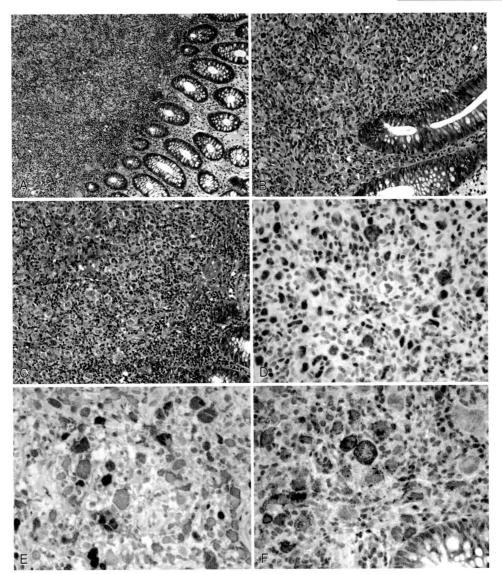

FIGURE 37.6 **EBV-POSITIVE DLBCL OF THE ELDERLY, COLON INVOLVEMENT.** Colon biopsy from a 75-year-old man demonstrating a large cell lymphoma (A, low power; B, intermediate power; C, high power). Numerous large cells are positive for Pax-5 (D), p63 (E), and EBV-LMP (F).

IMMUNOPHENOTYPE

MFC findings reveal abnormal or monotypic large B-cells, similar to those described previously in Chapter 36. By IHC studies, the large neoplastic B-cells express pan B-cell antigens, as well as commonly BCL2. CD10 is positive in a small fraction of cases, while BCL6 positivity is more common. MUM1 is positive in the vast majority of cases.

MOLECULAR AND CYTOGENETIC STUDIES

- FISH studies show *BCL6* gene rearrangements, along with gains of 18q21.
- Other aberrations include deletions of 6q, and gains of 12q, 22q.
- Negative for t(8;14), t(11;14), or t(14;18).

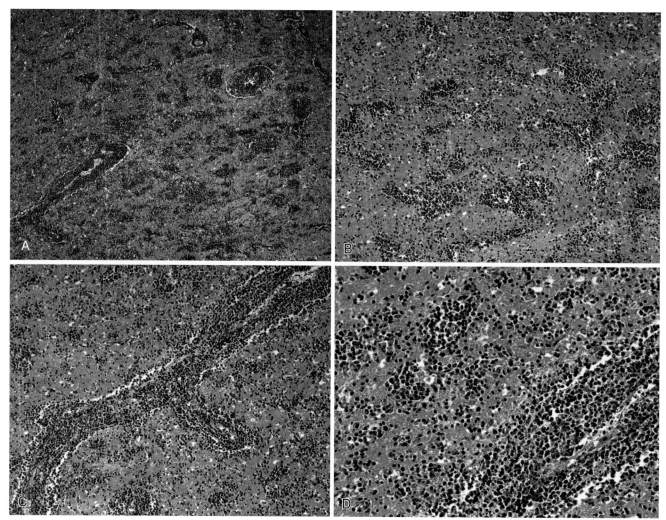

FIGURE 37.7 PRIMARY DIFFUSE LARGE B-CELL LYMPHOMA OF CNS. Brain sections demonstrate a perivascular and parenchymal infiltrate of medium to large atypical lymphoid cells (A, low power; B and C, intermediate power; D, high power).

Additional Resources

Abramson JS: T-cell/histiocyte-rich B-cell lymphoma: biology, diagnosis, and management, *Oncologist* 11:384–392, 2006.

Abrey LE: Primary central nervous system lymphoma, *Curr Opin Neurol* 22:675–680, 2009.

Aki H, Tuzuner N, Ongoren S, et al: T-cell-rich B-cell lymphoma: a clinicopathologic study of 21 cases and comparison with 43 cases of diffuse large B-cell lymphoma, *Leuk Res* 28:229–236, 2004.

Castillo JJ, Beltran BE, Miranda RN, et al: Epstein-barr virus-positive diffuse large B-cell lymphoma of the elderly: what we know so far, *Oncologist* 16:87–96, 2011.

Dojcinov SD, Venkataraman G, Pittaluga S, et al: Age-related EBV-associated lymphoproliferative disorders in the Western population: a spectrum of reactive lymphoid hyperplasia and lymphoma, *Blood* 117:4726–4735, 2011.

El Weshi A, Akhtar S, Mourad WA, et al: T-cell/histiocyte-rich B-cell lymphoma: clinical presentation, management and prognostic factors: report on 61 patients and review of literature, *Leuk Lymphoma* 48:1764–1773, 2007.

Fraga M, Sánchez-Verde L, Forteza J, et al: T-cell/histiocyte-rich large B-cell lymphoma is a disseminated aggressive neoplasm: differential diagnosis from Hodgkin's lymphoma, *Histopathology* 41:216–229, 2002.

Gerstner ER, Batchelor TT: Primary central nervous system lymphoma, *Arch Neurol* 67:291–297, 2010.

Gualco G, Weiss LM, Barber GN, et al: Diffuse large B-cell lymphoma involving the central nervous system, *Int J Surg Pathol* 19:44–50, 2011.

Hofscheier A, Ponciano A, Bonzheim I, et al: Geographic variation in the prevalence of Epstein-Barr virus-positive diffuse large B-cell lymphoma of the elderly: a comparative analysis of a Mexican and a German population, *Mod Pathol* 24:1046–1054, 2011.

Preusser M, Woehrer A, Koperek O, et al: Primary central nervous system lymphoma: a clinicopathological study of 75 cases, *Pathology* 42:547–552, 2010.

Shimoyama Y, Oyama T, Asano N, et al: Senile Epstein-Barr virus-associated B-cell lymphoproliferative disorders: a mini review, *J Clin Exp Hematopathol* 46:1–4, 2006.

Soussain C, Hoang-Xuan K: Primary central nervous system lymphoma: an update, *Curr Opin Oncol* 21:550–558, 2009.

Swerdlow SH, Campo E, Harris NL, et al: WHO classification of tumours of haematopoietic and lymphoid tissues, ed 4, Lyon, 2008, International Agency for Research on Cancer.

Wong HH, Wang J: Epstein-Barr virus positive diffuse large B-cell lymphoma of the elderly, *Leuk Lymphoma* 50:335–340, 2009.

Other Lymphomas of Large B Cells

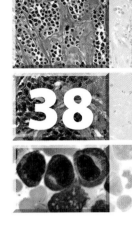

In this chapter we discuss a group of B-cell lymphomas that in the current WHO classification are classified as "other lymphomas of large B-cells". These include:

- Primary mediastinal (thymic) large B-cell lymphoma
- Intravascular large B-cell lymphoma
- DLBCL associated with chronic inflammation
- Lymphomatoid granulomatosis
- ALK-positive LBCL
- Primary effusion lymphoma
- Plasmablastic lymphoma
- Large B-cell lymphoma arising in HHV8-associated multicentric Castleman disease.

Also, HHV8- and EBV-associated germinotropic lymphoproliferative disorder and B-cell lymphoma, unclassifiable with features intermediate between DLBCL and classical Hodgkin lymphoma are briefly discussed in this chapter.

Primary Mediastinal (Thymic) Large B-Cell Lymphoma

Primary mediastinal (thymic) large B-cell lymphoma (MLBCL) is a distinct clinicopathologic subtype of DLBCL primarily involving the thymus. The major bulk of tumor at the time of diagnosis is confined to the anterior mediastinum.

Primary MLBCL constitutes about 7% of DLBCLs and 2.4% of all non-Hodgkin lymphomas. Women are more affected than men, and the median age is around 40 years. The affected patients usually show a bulky anterior mediastinal mass originating in the thymus. Superior vena cava syndrome is reported in over half of cases. Relapses are often extranodal involving liver, gastrointestinal tract, CNS, and other organs. Event-free 10-year survival is about 50%. Therapeutic regimens include CHOP or other alternative combination chemotherapy regimens followed by either involved field or modified mantle field radiation therapy.

MORPHOLOGY (FIGURES 38.1 AND 38.2)

- The involved tissue is diffusely infiltrated by large cells with variable amount of cytoplasm, which is often pale or clear. The nuclear features may resemble those of centroblasts or large centrocytes, often with the presence of large cells with multilobated nuclei.
- Immunoblast-like cells may be dominant in a minority of cases. Reed–Sternberg-like cells may be present, as well as scattered eosinophils. Mitotic figures are frequent. Rare cases may depict spindle-shaped morphology.
- Sclerosis is a common feature separating solid nests of tumor cells by thick hyalinized bands of connective tissue.

IMMUNOPHENOTYPE

- Flow cytometric results are similar to those in other DLBCLs.
- By immunohistochemical studies, the neoplastic cells express B-cell antigens, such as CD19, CD20, CD22, and CD79a. In addition, they are positive for CD45, plus commonly positive for CD30, as well as CD23 and MUM1. CD15 may be detected in rare cases. The neoplastic cells variably express BCL2 and BCL6, but expression of CD10 is less common.

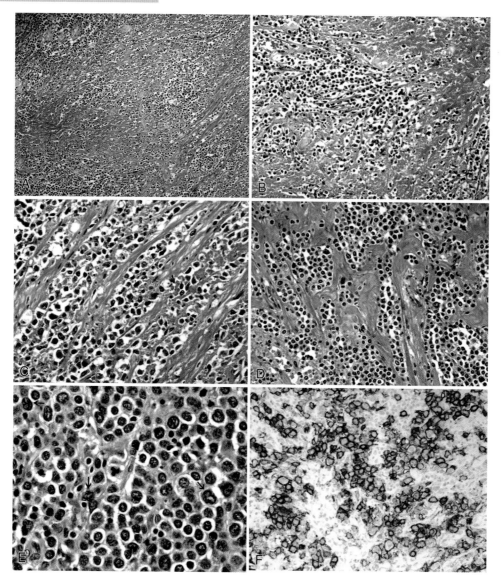

FIGURE 38.1 **PRIMARY MEDIASTINAL B-CELL LYMPHOMA.** Aggregate of neoplastic large cells with areas of necrosis are separated by thick bands of hyalinized connective tissue (A, low power; B, intermediate power; C, high power). A field devoid of necrosis is shown (D). Higher power view shows large cells with variable amount of clear cytoplasm (E). Some cells mimic lacunar cells (arrows). The neoplastic cells express CD20 (F).

CYTOGENETIC AND MOLECULAR STUDIES

- While cytogenetic testing is not as informative here, the most frequent chromosomal abnormalities are gains of chromosomes 9p21 and 2p16.
- The gains of 2p16 are due to the frequent amplification of the *REL* gene.
- Immunoglobulin gene rearrangement studies are usually positive, (Figure 38), but *BCL-2*, *BCL-6*, and *MYC* genes lack rearrangements.
- *MAL*, a gene that encodes a protein associated with lipid rafts in the T-cells and epithelial cells, is over-expressed in most cases.

DIFFERENTIAL DIAGNOSIS

The differential diagnosis of MLBCL includes nodular sclerosis Hodgkin lymphoma, thymic neoplasms, and other mediastinal masses, such as seminoma and anaplastic large cell lymphoma. The neoplastic cells of MLBCL express B-cell-associated markers and CD45, and lack expression of CD15, cytokeratin, and T-cell-associated markers (Table 38.1).

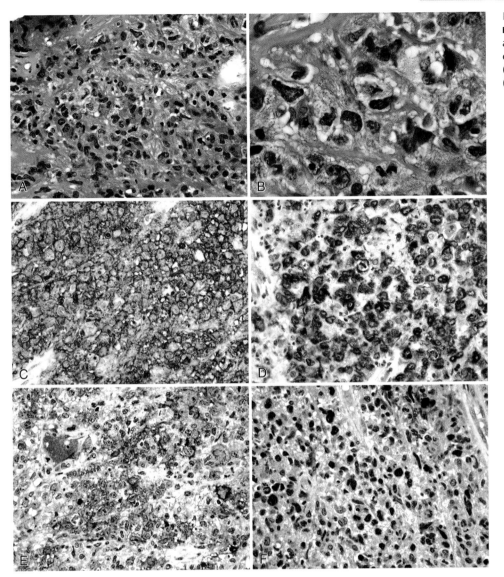

FIGURE 38.2 **PRIMARY MEDIASTINAL B-CELL LYMPHOMA.** Aggregate of atypical neoplastic large cells embedded in a fibrous stroma (A and B) express CD20 (C), BCL-2 (D), CD30 (E) and Ki-67 (F).

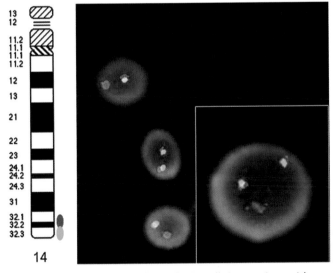

FIGURE 38.3 FISH analysis of neoplastic cells in a patient with primary mediastinal large B-cell lymphoma consistent with *IGH@* gene rearrangement.

Intravascular Large B-Cell Lymphoma

Intravascular large B-cell lymphoma (IVLBCL) is a rare and aggressive variant of extranodal large B-cell lymphoma characterized by the presence of aggregates of large to medium-size neoplastic B-cells within the lumens of small to medium-sized blood vessels. The lack of expression of adhesion molecules, such as CD29 (beta 1 integrin) and CD54 (ICAM-1), in some cases of IVLBCL suggests that the intravascular pattern is secondary to a defect in homing receptors in the tumor cells.

IVLBCL is an aggressive systemic disorder. Clinical symptoms are mostly secondary to the occlusion of the small to medium-sized vessels in various organs. Clinical manifestations may include skin lesions, neurological symptoms, nephritic syndrome, and disseminated intravascular coagulation (DIC). Hepatosplenomegaly, anemia,

Table 38.1
Differential Diagnosis of Primary Mediastinal (Thymic) Large B-Cell Lymphoma (MLBCL)

Pathology	Major Morphology	Immunophenotype	Other Remarks
MLBCL	Large cells with clear cytoplasm, sclerosis	CD45+, CD19+, CD20+, CD22+, CD10−, CD30+, CD23+	Rearranged *IGH@*
NSHL, syncytial type	Dense sheets of large cells, numerous eosinophils, sclerosis, areas of necrosis	CD45−, CD30+, Fascin+, PAX5+ CD15±, CD20±; EBV+ in 35% of the cases	
Thymic carcinoma	Cohesive growth pattern, desmoplastic stroma, may depict squamous differentiation	Cytokeratin+, CD45−, CD20−	
Seminoma	Round nuclei, glycogen+	CD45−, CD117+, placental alkaline phosphatase+	Almost exclusively in males
ALCL	Large pleomorphic anaplastic cells, hallmark cells	Pan-T±, CD30+, CD43+, ALK±, EMA±	Rearranged *TCR*

NSHL, nodular sclerosis Hodgkin lymphoma; ALCL, anaplastic large cell lymphoma.

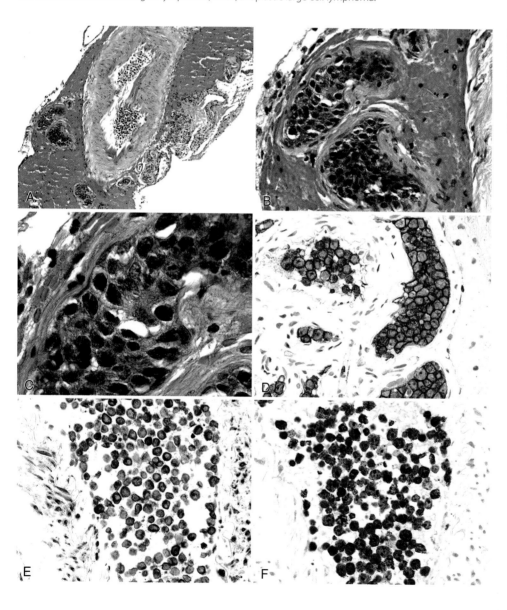

FIGURE 38.4 INTRAVASCULAR LARGE B-CELL LYMPHOMA. Aggregates of large atypical lymphoid cells are in the lumen of blood vessels (A, low power; B, intermediate power; C, high power). The neoplastic cells express CD20 (D), BCL-2 (E) and Ki-67 (F).

and thrombocytopenia are reported in over 70% of patients. A correlation has been reported between CD5+, CD10− phenotype, and poor prognosis in patients with IVLBCL.

MORPHOLOGY

- The presence of aggregates of large to medium-size neoplastic B-cells within the lumens of small to medium-sized blood vessels (Figure 38.4).

Table 38.2

Immunophenotypic Features of Intravascular Large B-cell Lymphoma[1]

Phenotype	% Positive
CD5	38
CD10	13
CD19	85
CD20	96
CD23	4
Bcl-2	91
Bcl-6	26
MUM1/IRF4	95
Bcl-1	0
κ chain	71
λ chain	18
EBER (EBV)	0

[1]Adapted from Murase T, Yamaguchi M, Suzuki R, et al. Intravascular large B-cell lymphoma (IVLBCL): a clinicopathologic study of 96 cases with special reference to the immunophenotypic heterogeneity of CD5. Blood 2007; 109: 478–485.

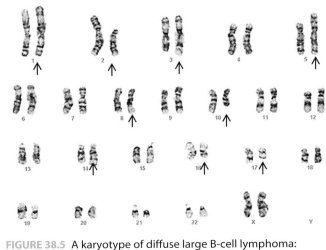

FIGURE 38.5 A karyotype of diffuse large B-cell lymphoma: 46,XX,t(2;1;8)(p13;p34.3;q24),t(3;17)(q27;q23),der(5)t(5;10)(p15.3;q22),del(8)(q13q22), add(10)(q22),t(14;16)(q32;q22).

- The neoplastic cells show a vesicular nucleus and prominent nucleoli, resembling centroblasts or immunoblasts.
- The intravascular tumor clusters may be associated with fibrin thrombi, and/or palisade along the vessel wall resembling angiosarcoma.
- The presence of neoplastic cells in peripheral blood is uncommon.

IMMUNOPHENOTYPE

The neoplastic cells are positive for B-cell-associated markers, such as Ig, CD19, CD20, CD22, and CD79a. They are commonly positive for BCL2, MUM1, and sometimes BCL6. Expression of CD5 is more common than CD10 (Table 38.2).

CYTOGENETIC AND MOLECULAR STUDIES

- No recurrent cytogenetic abnormalities have been reported.
- *BCL6* rearrangement has been reported sporadically but most show clonal IGH@ gene rearrangement (Figure 38.5).

DIFFERENTIAL DIAGNOSIS

Differential diagnosis includes acute leukemias, carcinomatosis and angiosarcoma. The neoplastic cells in ALL often express TdT and CD34, whereas cells in IVLBCL are negative for TdT and CD34. AML cells may show some cytoplasmic granularity and express myeloid-associated markers, such as MOP, CD13, CD33, and CD117. In carcinomatosis, the neoplastic cells are CD45− and cytokeratin+. The tumor cells in angiosarcoma express CD34 and are CD20-negative.

Diffuse Large B-Cell Lymphoma Associated with Chronic Inflammation

This rare subtype of DLBCL is associated with EBV and longstanding chronic inflammation. It usually involves bone marrow, pleural space or other body cavities.

DLBCL associated with chronic pleuritis (pyothorax-associated lymphoma) is often observed in tuberculous pleuritis. It is a male-predominance disorder with a male to female ratio of about 12:1. The average age is around 70 years. It is found more in Japan than in Western countries.

Other EBV-associated longstanding chronic inflammations, such as chronic osteomyelitis, chronic cutaneous ulcer, or metallic implantation, may occasionally lead to EBV-associated DLBCL.

MORPHOLOGY

The involved tissue shows diffuse and destructive infiltration of atypical large cells which are morphologically similar to conventional DLBCL cells.

IMMUNOPHENOTYPE

Immunohistochemical studies show neoplastic cells positive for CD20 and CD79a in most cases, while some cases display loss of CD20 and/or CD79a, with plasmacytic differentiation expressing MUM1 and CD138. CD30 can be positive. Occasionally, T-cell antigens can be variably positive. EBV-positivity can be demonstrated by in situ hybridization studies.

MOLECULAR AND CYTOGENETIC STUDIES

Most cases will demonstrate clonal immunoglobulin gene rearrangements, often with a background polyclonal gene rearrangement signal produced from the inflammatory infiltrate.

DIFFERENTIAL DIAGNOSIS

The differential diagnosis includes primary effusion lymphoma, anaplastic types of plasma cell neoplasms, and certain types of non-hematopoietic tumors (mesothelioma, carcinoma, sarcoma).

Primary effusion lymphoma occurs in HIV-infected patients, with large pleomorphic, plasmablastic tumor cells, suspended in pleural fluid, which express CD30, CD138, and IRF4/MUM-1, and are often negative for CD20. On the other hand, pyrothorax-associated lymphoma forms solid masses expresses CD20 and is negative for CD138.

Non-hematopoietic neoplasms are negative for CD45 and express their associated specific markers.

Lymphomatoid Granulomatosis

Lymphomatoid granulomatosis is a rare, extranodal, angiocentric lymphoproliferative disorder consisting of large, EBV-positive B-cells in a background of polymorphous reactive cells, predominantly consisting of T-lymphocytes. It is an EBV-associated lymphoproliferative

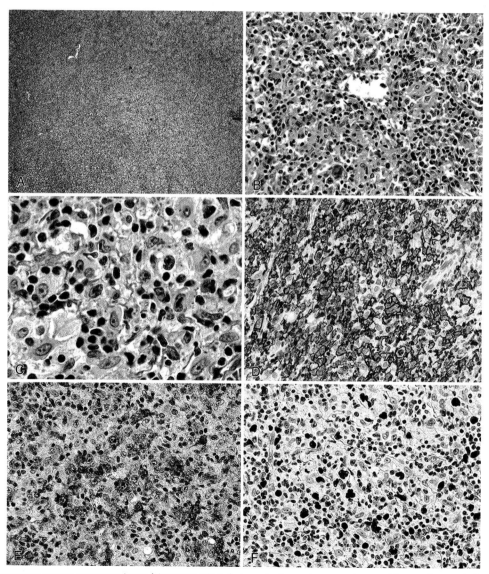

FIGURE 38.6 **LYMPHOMATOID GRANULOMATOSIS.** Lung biopsy section demonstrates a dense polymorphic, angiocentric lymphoid infiltrate (A, low power; B, intermediate power; C, high power). Numerous large cells are CD20-positive (D) and some express CD30 (E) and/or EBV-EBER (F).

disorder developed under an immunodeficiency setting, such as HIV infection, X-linked lymphoproliferative syndrome, methotrexate therapy, allogeneic organ transplantation, or Wiscott–Aldrich syndrome.

Lymphomatoid granulomatosis is usually observed in adult males. The most common sites of involvement in order of frequency are lung, skin, kidney, liver, and brain. Other sites, such as upper respiratory and gastrointestinal tracts, spleen, and lymph nodes are occasionally affected. Clinical symptoms are related to the involved organ, with respiratory symptoms being the most common presentation. The clinical course, particularly for grades II and III (see below), is often aggressive. However, some patients may show a fluctuating clinical course with occasional spontaneous remission.

MORPHOLOGY (FIGURES 38.6 AND 38.7)

Lymphomatoid granulomatosis is an angiocentric and angiodestructive polymorphous lymphoproliferative process characterized by the presence of a small number of EBV-positive large B-cells in a background of inflammatory cells. The large EBV-positive cells resemble immunoblasts, but bizarre large cells with multilobated nuclei or Reed–Sternberg-like cells may be present. The inflammatory component consists of mixed small lymphocytes, plasma cells, immunoblasts, and histiocytes, with scattered eosinophils and rare neutrophils.

This lymphoproliferative disorder characteristically infiltrates into the vascular structures and may cause vascular damage, fibrinoid necrosis, and ischemic changes in the surrounding tissues.

There are three histological grades:

- Grade I consisting of a polymorphous lymphoid infiltrate with absent or rare large, immunoblastic, EBV-positive lymphocytes (<5/hpf).
- Grade II consisting of a polymorphous lymphoid infiltrate with moderate numbers of large, immunoblastic, EBV-positive lymphocytes (5–20/hpf). Necrosis is often present.
- Grade III is considered a variant of DLBCL and consists of numerous large, immunoblastic, and EBV-positive lymphocytes (>50/hpf). In some areas, these cells may appear in clusters or small sheets. Necrosis is often extensive. Large bizarre cells with multilobed nucleus and Reed–Sternberg-like cells are often present.

IMMUNOPHENOTYPE

- The EBV-positive large cells are positive for CD20 and CD5.
- Expression of CD79a and CD30 is variable, but CD15 is negative. Background reactive lymphocytes are commonly CD4-positive T-cells.

CYTOGENETIC AND MOLECULAR STUDIES

- No recurrent chromosomal abnormalities have been reported.
- There is evidence of clonal *IGH@* gene rearrangement in most grade II and III cases. The clonal population may not be identical in different sites. Establishment of clonality in the grade I lesions may not be conclusive.
- Most cases show EBV infection demonstrated by EBER or other EBV-specific markers using either in situ hybridization or PCR.
- The EBV infection in some cases is clonal.

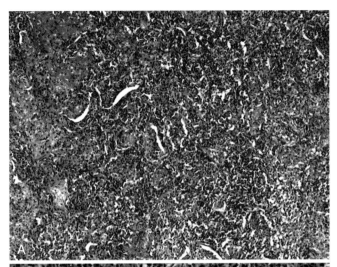
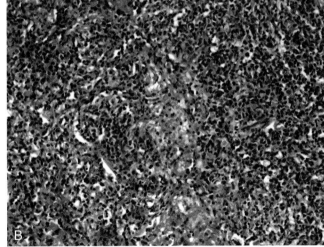
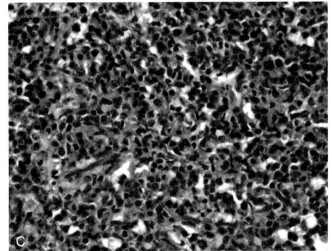

FIGURE 38.7 **LYMPHOMATOID GRANULOMATOSIS.** Lung tissue demonstrating interstitial lymphoid infiltrate and fibrosis (A, low power; B, intermediate power). The infiltrate is polymorphic consisting of small and large lymphoid cells (C, high power).

Table 38.3
Differential Diagnosis of Lymphomatoid Granulomatosis (LG)

Pathology	Major Features	Immunophenotype	Other Remarks
LG	Angiocentric, scattered or clusters of large cells in a polymorphic background of inflammatory cells, predominantly T-cells	Large cells CD20+, PAX5+, CD79a+	Large cells EBER+
Post-transplant LPD	History of immune-suppression, abundant B-cells	Predominant cells expressing pan-B-cell markers	EBV±
DLBCL with chronic inflammation	Associated with a history of chronic inflammation, usually pleural-based	Large cells CD20+, PAX5+, CD79a+	Large cells EBV+
Extranodal T/NK-cell lymphoma	Angiocentric, predominantly atypical T-cells	CD3+, CD56+	EBV+, Rearranged TCR
Classical HL	HRS cells in a polymorphic background of inflammatory cells	HRS cells CD15+, CD30+, CD45−	EBV±
Wegener granulomatosis	Patchy necrosis surrounded by palisading granulomas; inflammatory infiltrate contains neutrophils	Non-specific	

HL, Hodgkin lymphoma; HRS, Hodgkin/Reed–Sternberg; LPD, lymphoproliferative disorders.

DIFFERENTIAL DIAGNOSIS

The differential diagnosis of lymphomatoid granulomatosis includes a garden variety of reactive and neoplastic lymphproliferative disorders including post-tranplant lymphoproliferative disorders, DLBCL with chronic inflammation, extranodal T/NK-cell lymphoma, classical Hodgkin lymphoma, and Wegener granulomatosis. Table 38.3 shows the major clinicopathologic differences between these entities.

ALK-Positive Large B-Cell Lymphoma

ALK-positive LBCL is a rare, aggressive subtype of DLBCL. It tends to appear in children (30%) and young adults. The median age is about 35 years with a male to female ratio of 3:1. Patients are usually at the advanced stages (III or IV) at the time of diagnosis, and the prognosis is poor.

MORPHOLOGY

- The neoplasntic cells are large and may show centroblastic or immunoblastic morphologic features.
- Sinusoidal infiltration is a common feature, and sometimes a cohesive growth pattern may be seen, mimicking metastatic carcinoma.

IMMUNOPHENOTYPE

- The neoplastic cells express ALK with a unique cytoplasmic granular pattern.
- In addition, they are positive for CD138 and EMA, with only occasional weak staining for CD45.
- Cytoplasmic Ig and light chain restriction is common.
- The neoplastic cells are negative for B- or T-cell lineage-associated antigens, and CD30 is usually negative.

MOLECULAR AND CYTOGENETIC STUDIES

- FISH and PCR studies show the most commonly observed cytogenetic abnormality of ALK-positive DLBCL, t(2;17)(p23;q23) involving Clathrin (CLTC) on 17q23 and ALK on 2p23.
- ALK-positive DLBCL with t(2;5)(p23;q35) involving ALK on 2p23 and nucleophosmin on 5q35, as seen in the majority of T/null anaplastic large cell lymphomas, have also been rarely reported.
- These cases are negative for EBV.

DIFFERENTIAL DIAGNOSIS

Differential diagnosis of ALK+ large B-cell lymphoma includes DLBCL with sinusoidal pattern, anaplastic large cell lymphoma (see Chapter 51), and plasmablastic lymphoma (see below).

Primary Effusion Lymphoma

Primary effusion lymphoma (PEL) is a rare variant of large cell lymphoma characterized by malignant serous effusions without detectable tumor masses. Rare cases may show solid organ involvements, such as skin and heart. It has been primarily observed in AIDS patients and is associated with human herpes virus 8 (HHV8). However, rare cases of HHV-8-negative PEL have been reported. The majority of cases are also co-infected with EBV. Pleural, pericardial, and peritoneal cavities are the most frequent sites of involvement.

Most patients are men. Clinical symptoms are secondary to the effusions. Most patients do not respond to conventional chemotherapy and have a short survival time, usually <6 months.

MORPHOLOGY

- Cytocentrifuge slide preparations of the lymphomatous effusions show large neoplastic cells with immunoblastic, plasmablastic, and/or anaplastic features (Figures 38.8 and 38.9). These cells have a variable amount of basophilic cytoplasm. A paranuclear hof may be present.
- Nuclei are round or irregular and show prominent nucleoli. Binucleated or multilobated nuclei may be present, and some tumor cells may resemble Reed–Sternberg cells.

IMMUNOPHENOTYPE

Flow cytometric features (Figure 38.10) include:

- Negative to dim expression of CD45.
- Expression of CD38 and/or CD138.
- Absence of B-cell-associated antigens, including CD19, CD20, and CD79a.
- Absence of surface or intracellular light chain restriction.

By IHC studies, the neoplastic cells are also positive for EMA and CD30. They are positive for HHV8, plus EBV-EBER. In the solid form of PEL, variable expression of B-cell antigens may be seen occasionally.

CYTOGENETIC AND MOLECULAR STUDIES

- Complex karyotypes are observed. Recurrent chromosome aberrations include: partial trisomies 12, trisomies 7, and aberrations of 1q21–25.
- Although the tumor cells often lack the expression of membrane or cytoplasmic Ig, the *Ig* genes are rearranged and mutated, so molecular studies are more appropriate here than in many of the other B-cell lymphomas. However, mutations affecting PCR primer hybridization targets can cause false-negative results.
- Some cases may also show *TCR* gene rearrangement. In these situations, additional clonality testing using J-κ gene PCR may be helpful.
- HHV8 viral genomes are detected in virtually all patients, and most cases show EBV infection demonstrated by EBER using either in situ hybridization or PCR.

DIFFERENTIAL DIAGNOSIS

The differential diagnosis includes secondary involvement of body cavities with anaplastic plasma cell myeloma, subtypes of DLBCL, anaplastic large cell lymphoma, and metastatic non-hematopoietic tumors (Table 38.4).

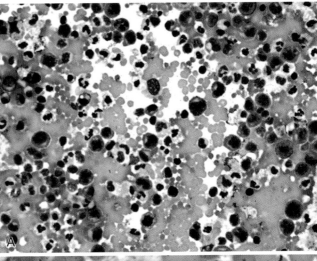
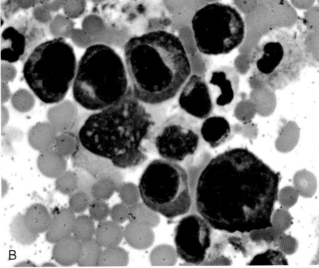
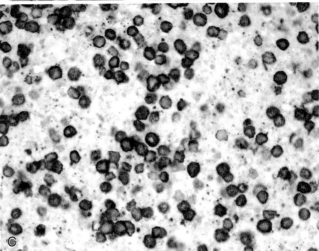

FIGURE 38.8 **PRIMARY EFFUSION LYMPHOMA.** Pericardial effusion demonstrating large plasmablastic cells with dark blue cytoplasm and irregular nuclei mixed with red cells and inflammatory cells (A, low power; B, high power). The plasmablastic cells express CD138 (C).

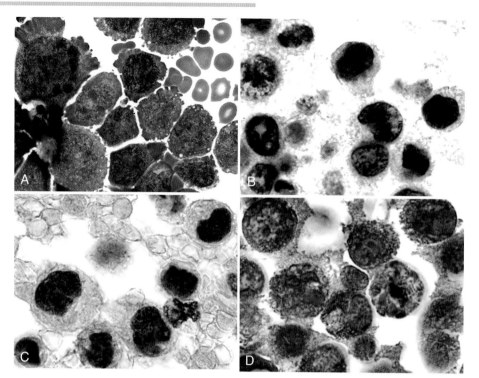

FIGURE 38.9 **PRIMARY EFFUSION LYMPHOMA.** Peritoneal fluid demonstrating large plasmablastic cells with dark blue cytoplasm and round or irregular nuclei (A). These cells are EBV-EBER+ (B), HHV8+ (C) and EMA+ (D).

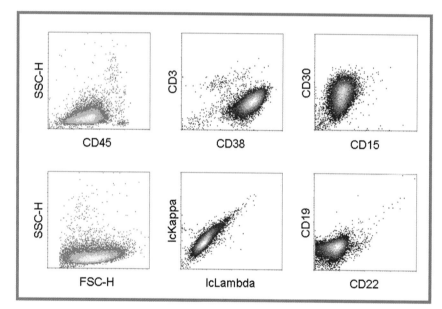

FIGURE 38.10 **FLOW CYTOMETRIC FINDINGS OF PRIMARY EFFUSION LYMPHOMA.** Open gate density displays (in blue) demonstrate a predominantly large cell population that is partially and very dimly positive for CD45. These large cells are positive for CD30 and CD38, but negative for CD3, CD15, CD19, CD22, as well as surface (not shown) and intracellular light chains.

Table 38.4
Differential Diagnosis of Primary Effusion Lymphoma (PEL)

Pathology	Major Features	Immunophenotype	Other Remarks
PEL	Anaplastic/plasmablastic morphology, rare solid organ involvement	Pan-B−, CD30+, CD138+, EBV+, HHV-8+	Rearranged *IGH@*
Anaplastic plasma cell myeloma	Primary bone marrow involvement	Pan-B−, CD30−, CD138+, cIgG+ or cIgA+, CD117+, CD56±	Rearranged *IGH@*, t(11;14), aneuploidy
LBCL arising in HHV8+ multicentric Castleman disease	Lymph node and splenic involvement	Pan-B±, CD30±, CD138−, cIgMλ, EBV+, HHV-8+	Rearranged *IGH@*
HHV8/EBV+ germinotropic lymphoproliferative disorder	Lymph node, preserved architechure, plasmablasts in germinal centers	IRF4/MUM1+, CD10−, CD20−, CD79a−, BCL-2− and BCL-6−	Favorable prognosis
Metastatic non-hematopoietic	Primary solid tissue involvement	CD45−, expression of specific non-hematopoietic marker, such as cytokeratin, S100 or others	No *IGH@* rearrangement

Plasmablastic Lymphoma

Plasmablastic lymphoma is a rare type of DLBCL with morphologic features of plasmablasts, commonly reported in HIV-infected patients. This entity does not include primary effusion lymphoma, ALK+ large B-cell lymphoma, and HHV8+ germinotropic lymphoproliferative disorder. This tumor typically involves the oral cavity of AIDS patients, though extra-oral cases have also been reported. The extra-oral sites include nasal cavity, lung, gastrointestinal tract (including the anus), skin, bone, and soft tissue. In addition to HIV-infected patients, plasmablastic lymphoma has been reported in organ transplant recipients, patients with autoimmune diseases, and individuals without immunodeficiency. The disease occurs predominantly in men and has a poor prognosis

MORPHOLOGY (FIGURES 38.11 AND 38.12)

- There is a diffuse infiltrate of large cells resembling immunoblasts or atypical immature plasma cells with variable amounts of basophilic cytoplasm, often with paranuclear hof. The nucleus is often eccentric, round, or irregular, with open nuclear chromatin and one or several prominent nucleoli.
- Starry sky pattern with tingible body macrophages, numerous mitotic figures, and the presence of variable numbers of mature lymphocytes and plasma cells are frequent features

IMMUNOPHENOTYPE (FIGURE 38.13)

- The neoplastic cells express CD38, CD138, MUM1, and are variably positive for CD79a. In addition, they are commonly positive for EMA and CD30.
- Cytoplasmic Ig and light chain restriction is often observed.
- They are negative to occasionally weak positive for CD45, CD20, and PAX5. Ki67 proliferation rate is greater than 90%.
- EBV-EBER is positive in most cases, while HHV8 is consistently negative.

CYTOGENETIC AND MOLECULAR STUDIES

- No recurrent cytogenetic abnormalities have been reported. However, plasmablastic transformation is highly associated with *MYC* translocation along with *IGH@* partners. In addition, this finding can be observed in both *de novo* plasmablastic lymphoma and in the plasmablastic transformation of a lower grade plasma cell neoplasm, including plasma cell myeloma.
- The *Ig* genes are clonally rearranged.
- HHV8 viral genomes are detected in virtually all patients.
- Most cases show molecular markers of EBV infection.

DIFFERENTIAL DIAGNOSIS

The differential diagnosis includes anaplastic plasmacytoma, LBCL arising in HHV8-associated multicentric Castleman disease (see below), primary effusion lymphoma (see above), and Burkitt lymphoma.

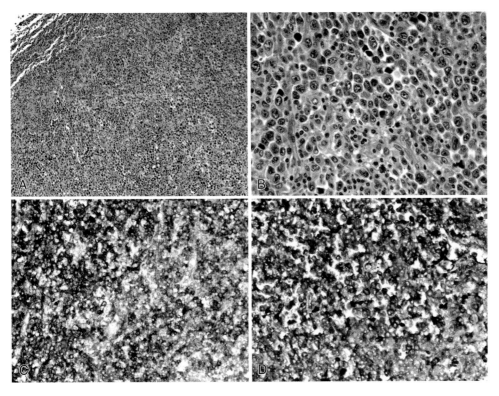

FIGURE 38.11 **Plasmablastic lymphoma in a patient with AIDS.** Aggregates of plasmablasts are present (A, low power; B, high power). These cells express CD138 (C) and Ig lambda light chain (D).

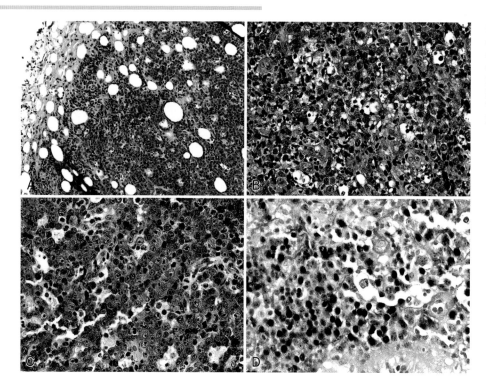

FIGURE 38.12 PLASMABLASTIC LYMPHOMA IN A PATIENT WITH AIDS. Fibroadipose tissue of breast is infiltrated by large clusters of plasmablasts and scattered tingible body macrophages (A, low power; B, intermediate power; C, high power). The plasmablasts are positive for EBV-EBER (D).

Large B-Cell Lymphoma Arising in HHV8-Associated Multicentric Castleman Disease

This rare category of LBCL arises in a background of multicentric Castleman disease and is associated with HHV8 and HIV. Patients demonstrate enlarged lymph nodes, splenomegaly, marked immunodeficiency, and often Kaposi sarcoma. It is a highly aggressive lymphoma with poor prognosis.

MORPHOLOGY

Two morphologic subtypes have been described. HHV8+ multicentric Castleman disease:

- Lymph node and splenic sections show hyalinization of the germinal centers with expansion of the mantle zone.
- Scattered or clusters of large plasmablastic cells are present within the mantle zone. These cells have abundant dark blue cytoplasm, eccentric nucleus, and one to two prominent nucleoli.

HHV8+ plasmablastic lymphoma:

- Small sheets of plasmablasts with partial or complete effacement of nodal or splenic architecture.
- In addition to the splenic involvement, neoplastic infiltrate may be present in the liver, lung, and/or GI tract.
- Occasionally tumor cells may appear in the peripheral blood (leukemic phase).

IMMUNOPHENOTYPE

- The plasmablasts demonstrate nuclear expression of LANA-1 antigen, and strong expression of cytoplasmic IgM with lambda light chain restriction.
- They are variably positive for CD20, and sometimes CD38, but negative for CD79a, CD138, and EBV-EBER.

MOLECULAR AND CYTOGENETIC STUDIES

- The karyotypes are commonly normal. Some rare non-specific aberrations such as t(1;6)(p11.2;p11.2), del(7)(q21q22), and del(8)(q12q22) have been reported.
- Most cases should demonstrate clonal immunoglobulin gene rearrangements.
- HHV8 genomic sequences are detectable by in situ hybridization or PCR.

DIFFERENTIAL DIAGNOSIS

Differential diagnosis includes all categories of LBCL which have a plasmablastic morphology, are associated with immunodeficiency, and are positive for HHV8 and/or HIV (see Table 38.4).

HHV8- and EBV- Associated Germinotropic Lymphoproliferative Disorder

This is a rare, localized, nodal disorder which has been reported in HIV seronegative patients. The overall nodal architecture is preserved and there is no feature of Castleman disease. The germinal centers are expanded and partially or totally replaced by clusters of plasmablasts. The plasmablasts are positive for IRF4/MUM1 but negative for CD10, CD20, CD79a, BCL-2, and BCL-6.

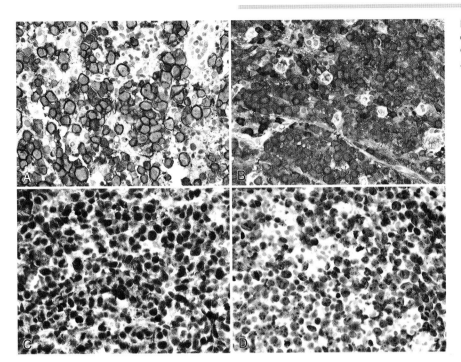

FIGURE 38.13 **PLASMABLASTIC LYMPHOMA; SAME CASE AS FIGURE 38.11.** The plasmablasts express CD138 (A), lambda light chain (B), MUM1 (C) and Ki-67 (D).

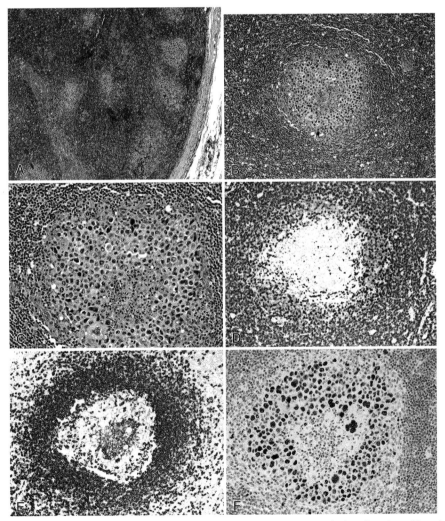

FIGURE 38.14 **GERMINOTROPIC LYMPHOPROLIFERATIVE PROCESS.** The germinal centers are expanded and replaced by plasmablasts consisting of large pleomorphic cells with round or convoluted nuclei and one or several nucleoli (A, low power; B, intermediate power; C, high power). The plasmablasts are surrounded by a mantle zone expressing CD5 (D) and CD20 (E). The plasmablasts are positive for IRF4/MUM1 (F) but negative for CD10, CD20, CD79a, BCL-2, and BCL-6 (not shown).

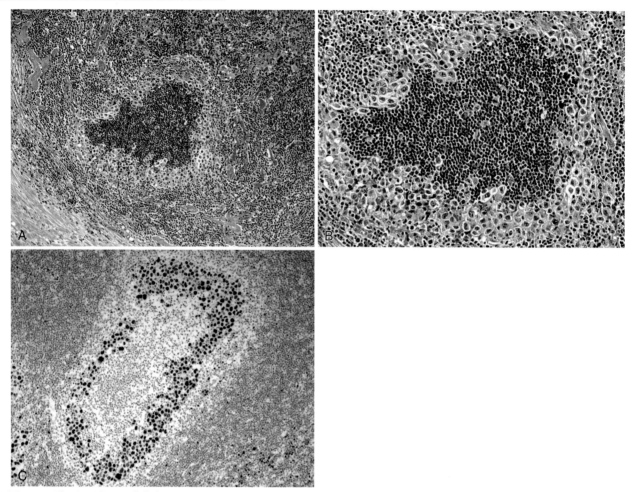

FIGURE 38.15 The "Inside-out follicle" pattern. The plasmablasts are at the periphery and surround centrally located small lymphocytes (A, intermediate power; B, high power). The plasmablasts express MUM1 (C).

They show a polyclonal or oligoclonal pattern of Ig gene rearrangements on PCR. The disorder responds well to radiation or chemotherapy.

Figures 38.14 and 38.15 present a germinotropic lymphoproliferative process in a patient with a history of nodular sclerosis Hodgkin lymphoma. However, the lesion was negative for both EBV and HHV8.

B-Cell Lymphoma, Unclassifiable, with Features Intermediate between DLBCL and Classical Hodgkin Lymphoma

This title refers to the B lineage lymphomas that show overlapping clinicopathologic features between DLBCL and classical Hodgkin lymphoma. These tumors often appear as an anterior mediastinal mass. Superior vena caval syndrome and/or respiratory distress are among clinical symptoms.

The tumor consists of sheets of large pleomorphic cells diffusely infiltrating a fibrous stroma (Figure 38.16). Some of the pleomorphic cells resemble lacunar or Hodgkin cells. Necrotic areas are frequently present, and there may be scattered eosinophils, lymphocytes, and histiocytes.

The neoplastic cells express CD20, CD79a, CD15, and CD30 and are often positive for PAX5, OCT-2, and BOB.1. ALK and CD10 are generally negative.

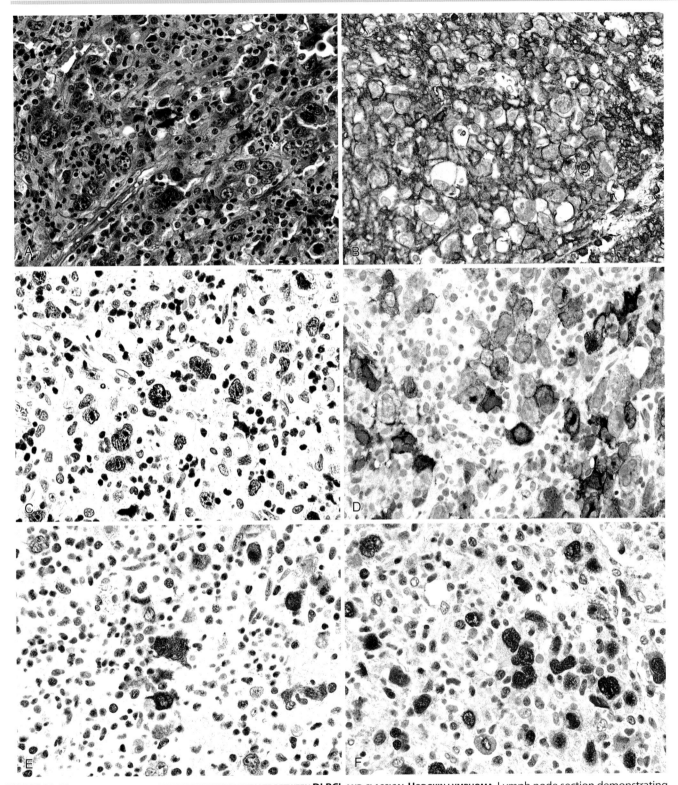

FIGURE 38.16 B-CELL LYMPHOMA WITH FEATURES INTERMEDIATE BETWEEN DLBCL AND CLASSICAL HODGKIN LYMPHOMA. Lymph node section demonstrating clusters of atypical large cells, some resembling HRS cells (A). The large cells express CD45 (B), Pax5 (C), CD30 (D), BOB.1 (E), and Oct2 (F).

Additional Resources

Ascoli V, Lo Coco F, Torelli G, et al: Human herpesvirus 8-associated primary effusion lymphoma in HIV-patients: a clinicopidemiologic variant resembling classic Kaposi's sarcoma, *Haematologica* 87:339-343, 2002.

Beltran B, Castillo J, Salas R, et al: ALK-positive diffuse large B-cell lymphoma: report of four cases and review of the literature, *J Hematol Oncol* 27:2-11, 2009.

Brimo F, Michel RP, Khetani K, et al: Primary effusion lymphoma: a series of 4 cases and review of the literature with emphasis on cytomorphologic and immunocytochemical differential diagnosis, *Cancer* 111:224-233, 2007.

Castillo J, Pantanowitz L, Dezube BJ: HIV-associated plasmablastic lymphoma: lessons learned from 112 published cases, *Am J Hematol* 83:804-809, 2008.

Castillo JJ, Reagan JL: Plasmablastic lymphoma: a systematic review, *ScientificWorldJournal* 11:687-696, 2011.

Chan JK: Anaplastic large cell lymphoma: redefining its morphologic spectrum and importance of recognition of the ALK-positive subset, *Adv Anat Pathol* 5:281-313, 1998.

Chen YB, Rahemtullah A, Hochberg E: Primary effusion lymphoma, *Oncologist* 12:569-576, 2007.

Copie-Bergman C, Niedobitek G, Mangham DC, et al: Epstein-Barr virus in B-cell lymphomas associated with chronic suppurative inflammation, *J Pathol* 183:287-292, 1997.

D'Antonio A, Addesso M, Memoli D, et al: Lymph node-based disease and HHV-8/KSHV infection in HIV seronegative patients: report of three new cases of a heterogeneous group of diseases, *Int J Hematol* 93:795-801, 2011.

Du MQ, Diss TC, Liu H, et al: KSHV- and EBV-associated germinotropic lymphoproliferative disorder, *Blood* 100:3415-3418, 2002.

Erös N, Károlyi Z, Kovács A, et al: Intravascular B-cell lymphoma, *J Am Acad Dermatol* 47(Suppl 5):S260-S262, 2002.

Ferreri AJ, Campo E, Seymour JF, et al: International Extranodal Lymphoma Study Group (IELSG). Intravascular lymphoma: clinical presentation, natural history, management and prognostic factors in a series of 38 cases, with special emphasis on the 'cutaneous variant', *Br J Haematol* 127:173-183, 2004.

Gitelson E, Al-Saleem T, Smith MR: Review: lymphomatoid granulomatosis: challenges in diagnosis and treatment, *Clin Adv Hematol Oncol* 7:68-70, 2009.

Hsi ED, Lorsbach RB, Fend F, et al: Plasmablastic lymphoma and related disorders, *Am J Clin Pathol* 136:183-194, 2011.

Hutchinson CB, Wang E: Primary mediastinal (thymic) large B-cell lymphoma: a short review with brief discussion of mediastinal gray zone lymphoma, *Arch Pathol Lab Med* 135:394-398, 2011.

Jaffe ES, Wilson WH: Lymphomatoid granulomatosis: pathogenesis, pathology and clinical implications, *Cancer Surv* 30:233-248, 1997.

Katzenstein AL, Doxtader E, Narendra S: Lymphomatoid granulomatosis: insights gained over 4 decades, *Am J Surg Pathol* 34:e35-48, 2010.

Loong F, Chan AC, Ho BC, et al: Diffuse large B-cell lymphoma associated with chronic inflammation as an incidental finding and new clinical scenarios, *Mod Pathol* 23:493-501, 2010.

Lundell RB, Weenig RH, Gibson LE: Lymphomatoid granulomatosis, *Cancer Treat Res* 142:265-272, 2008.

Nakashima MO, Roy DB, Nagamine M, et al: Intravascular large B-cell lymphoma: a mimicker of many maladies and a difficult and often delayed diagnosis, *J Clin Oncol* 29:e138-140, 2011.

Oschlies I, Burkhardt B, Salaverria I, et al: Clinical, pathological and genetic features of primary mediastinal large B-cell lymphomas and mediastinal gray zone lymphomas in children, *Haematologica* 96:262, 2011.

Pervez S, Khawaja RD: Mediastinal lymphomas: primary mediastinal (thymic) large B-cell lymphoma versus classical Hodgkin lymphoma, histopathologic dilemma solved? *Pathol Res Pract* 206:365-367, 2010.

Ponzoni M, Ferreri AJ, Campo E, et al: Definition, diagnosis, and management of intravascular large B-cell lymphoma: proposals and perspectives from an international consensus meeting, *J Clin Oncol* 25:3168-3173, 2007.

Rafaniello Raviele P, Pruneri G, Maiorano E: Plasmablastic lymphoma: a review, *Oral Dis* 15:38-45, 2009.

Steidl C, Gascoyne RD: The molecular pathogenesis of primary mediastinal large B cell lymphoma, *Blood* 118:2659-2669, 2011.

Sukpanichnant S, Sivayathorn A, Visudhiphan S, et al: Multicentric Castleman's disease, non-Hodgkin's lymphoma, and Kaposi's sarcoma: a rare simultaneous occurrence, *Asian Pac J Allergy Immunol* 20:127-133, 2002.

Van Roosbroeck K, Cools J, Dierickx D, et al: ALK-positive large B-cell lymphomas with cryptic SEC31A-ALK and NPM1-ALK fusions, *Haematologica* 95:509-513, 2010.

Zinzani PL, Piccaluga PP: Primary mediastinal DLBCL: evolving biologic understanding and therapeutic strategies, *Curr Oncol Rep* 13:407-415, 2011.

Burkitt Lymphoma

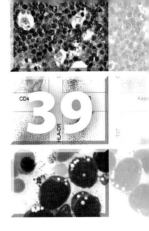

Burkitt lymphoma in the 2001 WHO classification was divided into three major categories: "Classical"; "Atypical Burkitt/Burkitt-like"; and "Burkitt lymphoma with plasmacytoid appearance". In the current WHO classification (2008), the non-classical variants are referred to as "B-Cell lymphoma, unclassifiable, with features intermediate between diffuse large B-cell lymphoma and Burkitt lymphoma" (see below).

Burkitt Lymphoma

Burkitt lymphoma/leukemia (BL) is a highly aggressive mature B-cell lymphoid malignancy consisting of endemic, sporadic, and immunodeficiency-associated variants. All variants demonstrate chromosomal rearrangements involving the *MYC* oncogene and share morphologic and immunophenotypic features, but differ in clinical and geographic presentations. The male to female ratio is about 2–4:1 in all three forms. The endemic and sporadic forms are most common in children, and affected children with the endemic form are younger than patients with the sporadic form.

- **Endemic BL** occurs in equatorial Africa. It is the most common childhood malignancy in this region with a peak incidence at 4–7 years. There is some correlation between the incidence of endemic BL and geographical distribution of endemic malaria. The disease usually presents as a mass in the jaw or facial bones and spreads to other extranodal sites, such as bone marrow, peripheral blood, meninges, testis, ovary, kidney, and breast.
- **Sporadic BL** is seen all over the world, accounting for about 1–2% of all lymphomas in the United States and western Europe. It occurs mainly in children and young adults and represents 30–50% of all childhood lymphomas. The most common clinical presentation is abdominal mass involving stomach, distal ileum, cecum, and/or mesentery. Bone marrow, peripheral blood, kidney, testis, ovary, breast, and CNS may be involved.
- **Immunodeficiency-associated BL** is primarily observed in patients with HIV infection. Post-transplant-associated BL is less frequent. This form more often involves lymph nodes but may involve bone marrow and peripheral blood presenting as acute leukemia.

A rapidly growing mass and elevated serum LDH levels are characteristic features of BL. Approximately 70% of patients are in advanced stages of the disease at diagnosis. BL is considered to be a highly aggressive tumor. Involvement of the bone marrow and CNS, tumor size >10 cm diameter, and elevated serum LDH are considered poor prognostic factors. Response to intensive combination chemotherapy with complete remission approaches 90% in early stages and 60–80% in advanced stages of the disease. Allogeneic bone marrow transplantation has been utilized in patients not responding to chemotherapy.

MORPHOLOGY

- The involved tissues are diffusely infiltrated by sheets of monotonous medium-sized neoplastic lymphoid cells.
- The typical (classical) cytologic features are uniformity, high nuclear to cytoplasmic ratio, basophilic and often vacuolated cytoplasm, round nucleus with clumped chromatin and relatively clear parachromatin, and multiple centrally located nucleoli.
- Mitotic figures are numerous as well as a high rate of apoptosis. Scattered macrophages with abundant pale cytoplasm-containing cell debris are present, creating a "starry sky" pattern (Figure 39.1).
- BL may infiltrate bone marrow and/or peripheral blood and present a leukemic picture (Figure 39.2).

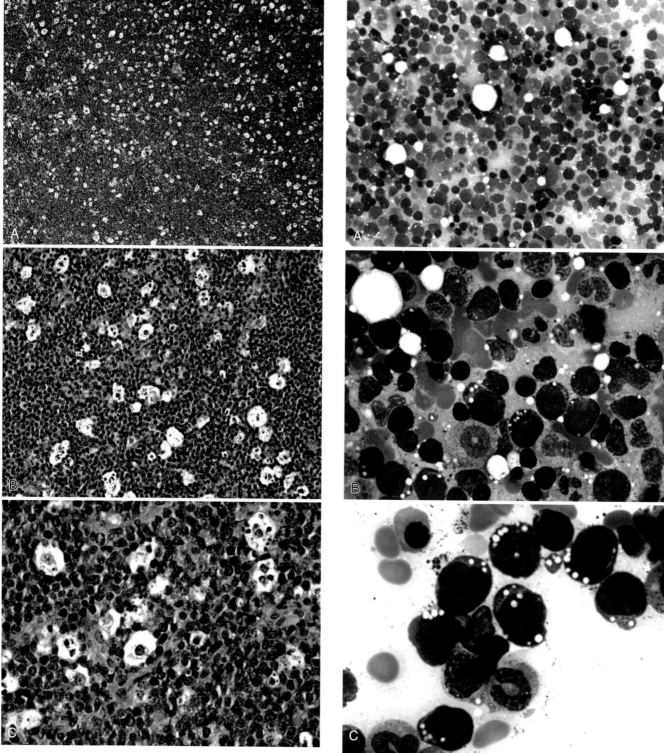

FIGURE 39.1 **BURKITT LYMPHOMA.** Lymph node section demonstrating a diffuse infiltration by monomorphic lymphoid cells with high nuclear to cytoplasmic ratio, basophilic cytoplasm, round nucleus with clumped chromatin and relatively clear parachromatin, and multiple centrally located nucleoli. Scattered macrophages with abundant pale cytoplasm-containing cell debris are present, creating a "starry sky" pattern (A, low power; B, intermediate power; C, high power).

FIGURE 39.2 **BURKITT LYMPHOMA.** Bone marrow smear showing monomorphic neoplastic lymphoid cells with high nuclear to cytoplasmic ratio, basophilic vacuolated cytoplasm, and round nuclei (A, low power; B, intermediate power; C, high power).

BURKITT LYMPHOMA

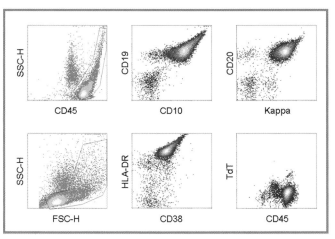

FIGURE 39.3 **FLOW CYTOMETRIC FINDINGS OF BURKITT LEUKEMIA/LYMPHOMA.** Open gate density displays (in blue) reveal that B-cells verified by CD19 backgating, are predominantly intermediate to large, and express moderate to bright CD45. The B-cell-enriched-gate (in magenta) demonstrates monotypic B-cells with germinal center B-cell like phenotype. They are positive for CD10, CD19, CD20 (moderate to bright), CD38, HLA-DR, and express bright surface kappa light chain restriction. The neoplastic B-cells were negative for TdT.

IMMUNOPHENOTYPE AND CYTOCHEMICAL STAINS

MFC findings (Figure 39.3) demonstrate germinal center B-cell like features with:

- Expression of pan B-cell markers, and often bright CD20.
- Expression of CD10 and CD38.
- Moderate to bright surface light chain restriction; and
- Lack of expression of TdT and CD34.

By immunohistochemical studies (Figure 39.4), the neoplastic cells also express BCL6 and IgM. BCL2 is usually negative, though patchy and weak staining may be seen in some cases. Ki67 proliferation rate is nearly 100%. EBV is positive in a subset of cases. Special stain Oil Red O highlights cytoplasmic lipid vacuoles (Figure 39.5).

CYTOGENETIC AND MOLECULAR STUDIES

The genetic hallmark of BL is translocation of *MYC* at chromosome 8q24 commonly with the IGH@ heavy chain region on chromosome 14 [t(8;14)(q24;q32)]

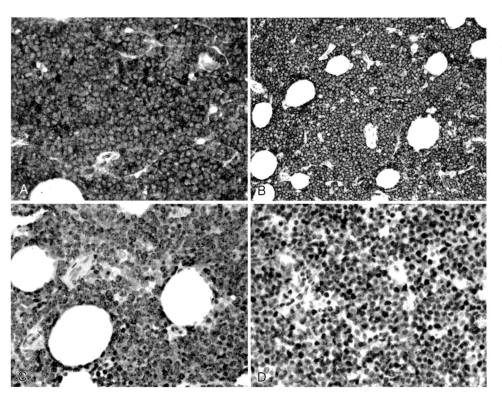

FIGURE 39.4 **BURKITT LYMPHOMA: IMMUNOHISTOCHEMICAL STAINS.** The tumor cells express CD10 (A), CD20 (B), Ki67 (C), and bcl-6 (D).

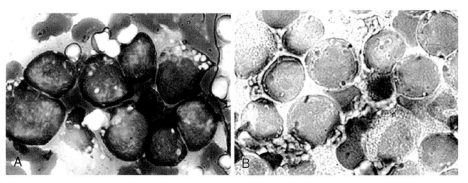

FIGURE 39.5 Touch preparation of a lymph node involved with Burkitt lymphoma. (A) Wright's stain demonstrates tumor cells with vacuolated cytoplasm. (B) Oil red O stain shows numerous orange-red dot-like deposits within the cytoplasmic vacuoles. *Naeim F. Pathology of Bone Marrow, 2nd ed. Williams & Wilkins, Baltimore, 1977, by permission.*

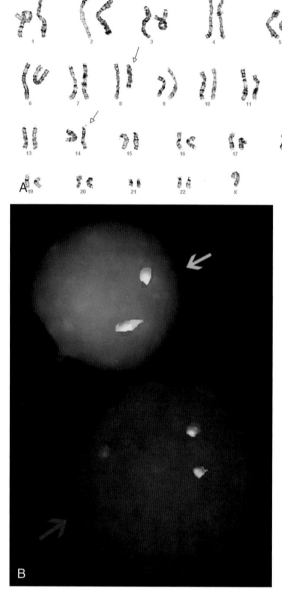

FIGURE 39.6 (A) Karyotype demonstrating 46,XY,t(8;14)(q24;q32) and (B) FISH analysis with split *MYC* signals in a patient with Burkitt lymphoma.

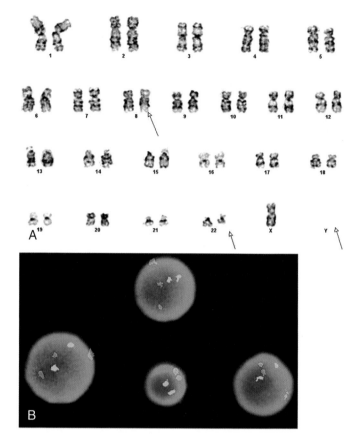

FIGURE 39.7 (A) Karyotype showing t(8;22)(q24;q11.2) and a loss of the Y-chromosome (arrows). (B) FISH with the dual-color dual fusion MYC(red) and IGH@ (green) probes showing split MYC signals but no t(8:14) fusions in a patient with Burkitt lymphoma.

(Figure 39.6), but less frequently, it may involve *IGK@* on chromosomes 2 (kappa) and 22 (*IGL@*), t(2;8)(p12;q24), and t(8;22)(q24;q11.2) (Figure 39.7). These specific rearrangements can be identified by standard karyotyping or by FISH with dual-color dual-fusion translocation probes or with generic dual-color *MYC* "split" probes (Figure 39.8). The karyotypes in adults and children with BL exhibit a core set of aberrations (a cytogenetic detection of an IGH@/MYC translocation and absence of translocations of *BCL2*, *BCL6*, and/or *CCND1*) and with minor cytogenetic differences.

In endemic cases the breakpoint on chromosome 14 involves the joining region of the *Ig* heavy chain, and the breakpoint on chromosome 8 lies outside the *MYC* gene; whereas in sporadic BL the breakpoints are in the heavy chain switch region and inside the *MYC* gene. The heterogeneity of the molecular breakpoints makes this translocation difficult to detect by standard PCR approaches, and therefore cytogenetics or FISH is preferred.

The lack of complex genetic alterations in addition to a *MYC* translocation is consistent with a "true" Burkitt lymphoma. The presence of an additional translocation as found in other types of B-NHL (such as involving *BCL2*, *BCL6*, and/or *CCND1*), or a complex karyotype with multiple gains and/or losses, represent another type of B-NHL or a secondary transformed lymphoma.

It is important to note that a *MYC* translocation is not specific for BL, but is also seen in a subset of morphological DLBCL, follicular lymphoma, or mantle cell lymphoma. However, the cytogenetic profile of the "true" BL core subset is different from that of other B-NHL cases with a *MYC* translocation. The latter usually represents transformed disease and therefore may harbor a non-*IGH@/MYC* breakpoint or additional translocations (e.g., involving *BCL2* or *CCND1*), as well as a higher number of numerical aberrations as part of a complex karyotype. In "true" BL, gains of 1q, 7, and 12 are found as common recurring events, and less frequently deletions of 6q, 17p, or 13q.

Morphologic BL without a confirmed *MYC* translocation is very rare and shows genetic differences from the core subset.

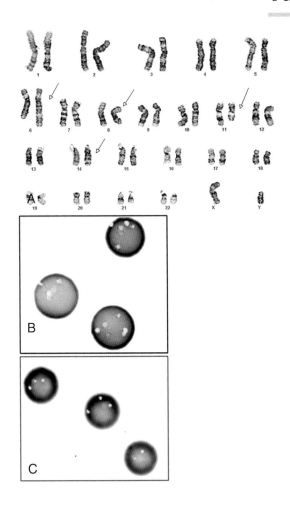

All lymphoma cases that, in addition to a *MYC* translocation, have one or more translocations involving 18q21/*BCL2*, 11q13/*CCND1* or 3q27/*BCL6* are regarded as "double-hit" (DH) lymphomas as per the WHO 2008 classification. However, double-hit BL (BL *IGH@/MYC* DH) should not be considered as "true" BL.

Cytogenetic analysis is very important in BL diagnostic work-up as it provides the information on the many relevant genetic abnormalities such as the partner of the *MYC* translocation, the presence or absence of additional translocations, and presence or absence of other structural abnormalities.

Mutation of the *BCL6* gene has been reported in 25–50% of cases. Most endemic cases and 25–40% of cases associated with AIDS contain the EBV genome.

DIFFERENTIAL DIAGNOSIS

The differential diagnosis includes B acute lymphoblastic leukemia/lymphoma and DLBCL. Precursor B-cell neoplasms are often TdT+ and CD34+ and lack the expression of membrane Ig heavy or light chains, whereas BL cells are negative for CD34 and TdT and express membrane Ig. Some cases of BL are borderline and share some of the morphologic features of DLBCL by demonstrating larger cells or an admixture of centroblast- or immunoblast-like cells. Also, a minority of DLBCL cases may demonstrate t(8;14) and *MYC* rearrangement (Table 39.1).

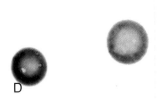

FIGURE 39.8 Complex chromosomal abnormalities with *MYC* rearrangement. (A) Karyotype demonstrating 46,XY,+3,der(6) t(3;6)(q21;q25),t(8;14)(q24;q32),del(11)(q13q23),r(16)(p13.3q24), and three color FISH analysis showing (B) MYC-IGH@ fusion signals (red/green or yellow), (C) three copies of BCL6 or trisomy 3q, and (D) deletion of 11q (API2).

B-Cell Lymphoma, Unclassifiable, with Features Intermediate Between Diffuse Large B-Cell Lymphoma and Burkitt Lymphoma

This title refers to the B-lineage lymphomas that show overlapping clinicopathologic features between DLBCL and Burkitt lymphoma. Some of these cases in the previous WHO classification were referred to as Burkitt-like

Table 39.1
Major Pathologic Features of Burkitt Lymphoma (BL), Diffuse Large B-Cell Lymphoma (DLBCL) and Intermediate BL/DLBCL[1]

Lymphoma	Morphology	Ki-67	BCL-2 expression	Molecular/Cytogenetic
BL	Monotonous round, small to medium-size cells	>90% +	Negative	*MYC* rearrangement, t(8;14), t(2;8), t(8;22)
Intermediate BL/DLBCL	Small to medium-size cells, sometimes mixed with large cells	Often >90% +, sometimes <90% +	±	Common *MYC* rearrangement plus a complex karyotype, *BCL2* or *BLC6* rearrangement
DLBCL	Large cells	Rare >90% +, common <90% +	±	Rare *MYC* rearrangement, common *BCL2* or *BCL6* rearrangement

[1] Adapted from Swerdlow SH, Campo E, Harris NL, et al. WHO Classification of Tumours of Haematopoietic and Lymphoid Tissues. International Agency for Research on Cancer, Lyon, 2008.

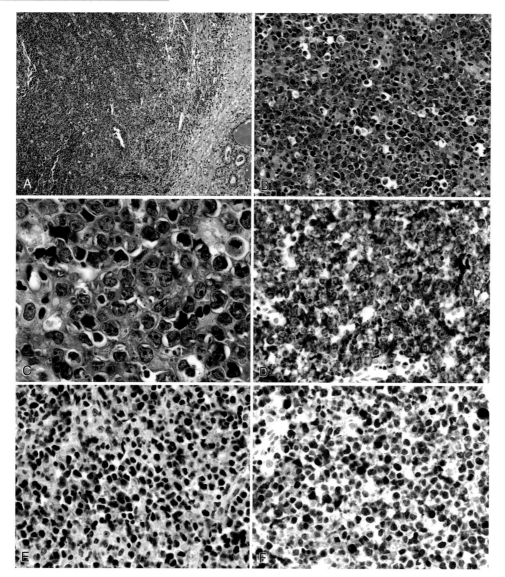

FIGURE 39.9 **B-CELL LYMPHOMA WITH FEATURES INTERMEDIATE BETWEEN DIFFUSE LARGE B-CELL LYMPHOMA AND BURKITT LYMPHOMA.** Thyroid tissue demonstrating a lymphomatous process consisting of medium to large blastic cells resembling centroblasts or immunoblasts (A, low power; B, intermediate power; C, high power). These cells strongly express CD10 (D), BCL-6 (E) and Ki67 (F). Cytogenetic studies showed complex chromosomal aberrations, including t(8;14) (see Figure 39.8).

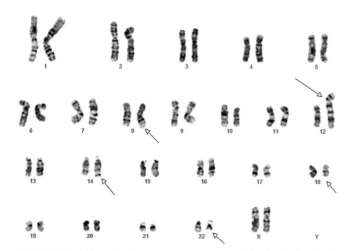

FIGURE 39.10 Complex karyotyping abnormalities including both t(8;22)(q24.1;q11.2) and t(14;18)(q32;q21) and duplication of 12q.

lymphoma. These cases commonly show starry sky pattern with the presence of tingible body macrophages, frequent mitotic figures, and numerous apoptotic cells.

These cases could morphologically be divided into two groups:

- Tumors consisting of smaller Burkitt-like cells and larger immonoblastic or centroblastic cells (Figure 39.9).
- Tumors consisting of cells similar to the classical Burkitt lymphoma cells but lacking the immunophenotypic or cytogenetic features typical of Burkitt lymphoma.

The neoplastic cells express pan-B markers and surface IG, and are positive for BCL2. Ki67 expression is high, ranging from 50% to 100% of the cells.

About 35–50% of the cases show *MYC* translocation which is either with *IG* genes or represent other translocations (Figure 39.10). *BCL2* rearrangement is noted in about 15% of the cases. *BCL6* rearrangement may occur but is less frequent.

Pathologic features of BL, DLBCL and intermediate BL/DLBCL are shown in Table 39.1.

Additional Resources

Aldoss IT, Weisenburger DD, Fu K, et al: Adult Burkitt lymphoma: advances in diagnosis and treatment, *Oncology (Williston Park)* 22:1508–1517, 2008.

Bellan C, Stefano L, Giulia de F, et al: Burkitt lymphoma versus diffuse large B-cell lymphoma: a practical approach, *Hematol Oncol* 28:53–56, 2010.

Biko DM, Anupindi SA, Hernandez A, et al: Childhood Burkitt lymphoma: abdominal and pelvic imaging findings, *AJR Am J Roentgenol* 192:1304–1315, 2009.

Brady G, Macarthur GJ, Farrell PJ: Epstein-Barr virus and Burkitt lymphoma, *Postgrad Med J* 84:372–377, 2008.

Carbone A, Gloghini A: AIDS-related lymphomas: from pathogenesis to pathology, *Br J Haematol* 130:662–670, 2005.

Forteza-Vila J, Fraga M: Burkitt lymphoma and diffuse aggressive B-cell lymphoma, *Int J Surg Pathol* 18(3 Suppl):133S–135S, 2010.

Jaffe ES: The 2008 WHO classification of lymphomas: implications for clinical practice and translational research, *Hematology Am Soc Hematol Educ Program*:523–531, 2009.

Kenkre VP, Stock W: Burkitt lymphoma/leukemia: improving prognosis, *Clin Lymphoma Myeloma* 9(Suppl 3):S231–S238, 2009.

Nagy N, Klein G, Klein E: To the genesis of Burkitt lymphoma: regulation of apoptosis by EBNA-1 and SAP may determine the fate of Ig-myc translocation carrying B lymphocytes, *Semin Cancer Biol* 19:407–410, 2009.

Orem J, Mbidde EK, Lambert B, et al: Burkitt's lymphoma in Africa, a review of the epidemiology and etiology, *Afr Health Sci* 7:166–175, 2007.

Perkins AS, Friedberg JW: Burkitt lymphoma in adults, *Hematology Am Soc Hematol Educ Program*:341–348, 2008.

Rowe M, Kelly GL, Bell AI, et al: Burkitt's lymphoma: the Rosetta Stone deciphering Epstein-Barr virus biology, *Semin Cancer Biol* 19:377–388, 2009.

Salaverria I, Siebert R: The gray zone between Burkitt's lymphoma and diffuse large B-cell lymphoma from a genetics perspective, *J Clin Oncol* 29:1835–1843, 2011.

Swerdlow SH, Campo E, Harris NL, et al: *WHO classification of tumours of haematopoietic and lymphoid tissues*, ed 4, Lyon, 2008, International Agency for Research on Cancer.

Primary Cutaneous B-Cell Lymphomas

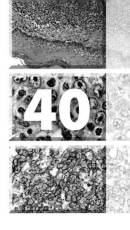

Primary cutaneous B-cell lymphomas are in general less aggressive than their counterparts involving lymphoid tissues. They represent about 25% of primary cutaneous lymphomas. The major subtypes of primary cutaneous B-cell lymphomas include follicle center lymphoma, marginal zone lymphoma, diffuse large B-cell lymphoma–leg type, and plasmacytoma.

Primary Cutaneous Follicle Center Lymphoma

Primary cutaneous follicle center lymphoma (PCFCL) is the most common B-cell lymphoma of skin. It occurs in adults, with a slight male preponderance and an average age of about 65 years. The neoplasm consists of various proportions of centrocytes and centroblasts depicting a follicular, follicular and diffuse, or diffuse pattern.

Head, neck, and upper back are the most frequent sites of involvement. The involved area usually consists of a centrally located large nodule with satellite smaller nodules. The tumor is often localized, with an estimated 5-year survival rate approaching 100%. Therapy includes excision and radiation.

MORPHOLOGY (FIGURE 40.1)

- The infiltrate is usually in mid dermis and consists of a mixture of centrocytes and centroblasts arranged in follicular, follicular and diffuse, or diffuse pattern. The number of centroblasts can be quite variable.
- Follicular structures are irregular and variable in size, often separated by fibrosis. Occasionally, neoplastic cells break out of the follicle and surround aggregates of small non-neoplastic lymphocytes (*inside-out follicles*).
- The tumor cells are admixed with variable numbers of reactive T-cells and scattered histiocytes. Plasma cells, eosinophils, and neutrophils are rare.

IMMUNOPHENOTYPE

- The neoplastic cells are positive for B-cell associated markers (CD20, PAX5, and CD79a), as well as BCL6.
- CD10 is often positive in cases with follicular growth pattern, but negative in cases with more diffuse infiltrate.
- BCL2 is negative to variably positive. However, strong coexpression of CD10 and BCL2, should raise consideration for nodal FL involving skin.
- CD5 as well as CD43 are negative, and MUM1 is negative in most cases.

MOLECULAR AND CYTOGENETIC STUDIES

- Lack of the t(14;18) that is commonly observed in nodal follicular type lymphoma.
- PCR techniques detect clonal rearrangement of immunoglobulin genes in approximately half of cases.
- Gene-expression studies of PCFCL show profiles similar to that of germinal center B-cell (GBC)-like diffuse large B-cell lymphomas.
- *FOXP1* (3p14), over-expression has been reported in PCFCL, though negative for the t(3;14)(p14;q32), but trisomy 3 is not uncommon.

DIFFERENTIAL DIAGNOSIS

The differential diagnosis for cases of PCFCL with follicular pattern includes cutaneous lymphoid hyperplasia, primary cutaneous marginal zone lymphoma (PCMZL), and secondary cutaneous follicular lymphoma. Cutaneous lymphoid hyperplasia is observed in a variety of conditions, such as, autoimmune disease, drugs, tattoos, infections, or arthropod bite. It may mimic cutaneous lymphomas, particularly

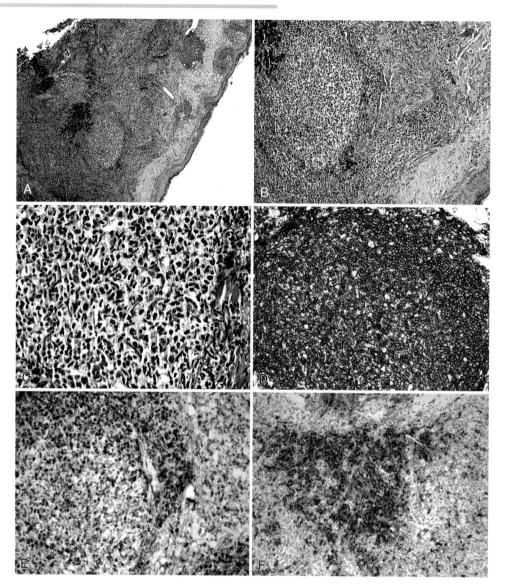

FIGURE 40.1 **Primary cutaneous follicular center lymphoma.** Skin biopsy demonstrating a dermal lymphoid infiltrate with a patchy distribution and nodular structure (A, low power; B, intermediate power; C, high power). The infiltrate consists of mixed centrocytes and centroblasts expressing CD20 (D) and BCL-2 (E). CD23 demonstrates a meshwork of follicular dendritic cells (F).

primary cutaneous marginal zone lymphoma (PCMZL). PCFCL with diffuse pattern should be distinguished from primary cutaneous large B-cell lymphoma. The reactive follicles present in PCMZL lack expression of BCL2 protein, whereas follicles in PCFCL variably express BCL2.

Primary Cutaneous Marginal Zone Lymphoma

Primary cutaneous marginal zone lymphoma (PCMZL) is the second most common B-cell lymphoma of skin, accounting for about 25% of all primary cutaneous B-cell lymphomas. The neoplastic cells, similar to extracutaneous marginal zone lymphomas, consist of various proportions of centrocye-like cells, monocytoid B cells, lymphoplasmacytoid cells, and plasma cells.

The average patient age is about 50 years, and women are more affected than men. PCMZL is more frequently seen on the trunk and upper extremities, presenting as one or several red papules, plaques, and/or nodules. Patients have no "B" symptoms, nor elevated serum LDH, or microglobulins.

PCMZL is an indolent disease with a 5-year survival approaching 100%. There is usually an excellent response to local treatment such as irradiation, interferon, and steroids. Extracutaneous relapse has been reported in up to 30% of patients.

MORPHOLOGY (FIGURE 40.2)

- The cutaneous infiltrate is located in the dermis, often extending into the subcutaneous tissue. It consists of various proportions of centrocyte-like cells, monocytoid B cells, lymphoplasmacytoid cells, and plasma cells.
- Reactive germinal centers are often present, and there may be follicular colonization (infiltration of neoplastic B-cells into the germinal center).
- Variable amounts of reactive T-cells are often admixed with the tumor cells.

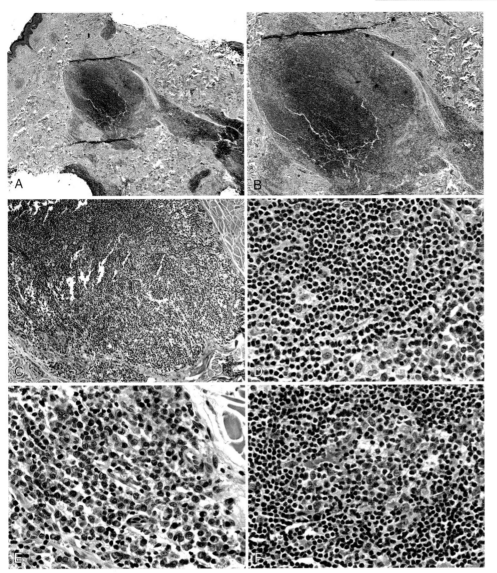

FIGURE 40.2 **PRIMARY CUTANEOUS MARGINAL ZONE LYMPHOMA.** Skin biopsy demonstrating a dermal lymphoid infiltrate extending into the deep dermis (A, low power; B, intermediate power; C, high power). The lymphoid infiltrate consists of various proportions of centrocyte-like cells, monocytoid B-cells (D), and plasma cells (E). Follicular colonization may be present (F).

IMMUNOPHENOTYPE (FIGURE 40.3)

- The neoplastic lymphocytes are positive for B-cell antigens as well as BCL2, but negative for CD5, CD10, BCL6, and CD23.
- The residual reactive germinal center B-cells are negative for BCL2.
- A monotypic component of plasma cells is detectable in most cases, where the plasma cells are positive for CD79a, CD138, MUM1, and light chain restriction.
- Ig class switching is often evident with expression of IgG, IgA, or IgE.
- When follicular colonization is present, disrupted follicular dendritic cell meshworks are highlighted by CD21 and CD23.

MOLECULAR AND CYTOGENETIC STUDIES

- The t(11;18)(q21;21) translocation, which results in the fusion of the *BIRC3* (baculoviral IAP repeat-containing 3, an apoptosis inhibitor) and *MALT1* genes, found in up to half of cases of MALT lymphomas (extranodal MZL of MALT-type) is virtually never detected in PCMZL.
- The t(14;18)(q32;q21) (*IGH@-MALT1*) fusion can be detected in both PCMZL and MALT lymphomas.
- Genetic abnormalities associated with aberrant BCL10, such as t(1;14)(p22;q32) translocation and rearrangement of BCL10 have been reported in 5% of extranodal MZL. Aberrant nuclear BCL10 has been detected in 36% of PCMZL cases.
- Other chromosomal aberrations observed at various frequencies in both PCMZL and MALT lymphomas are: t(3;14)(p14.1:q32) (*IGH@-FOXP1* fusion), trisomy 3, and trisomy 8.

DIFFERENTIAL DIAGNOSIS

The differential diagnosis of PCMZL includes cutaneous lymphoid hyperplasia and PCFCL. There are significant overlapping clinicopathologic features between cutaneous lymphoid hyperplasia and PCMZL. However, establishment of monoclonality or rearrangement of *IG* genes, presence of sheets of marginal zone cells, and aggregates of monotypic plasma cells are in support of diagnosis of PCMZL. The reactive follicles present in PCMZL lack expression of BCL2 protein, whereas follicles in PCFCL express BCL2.

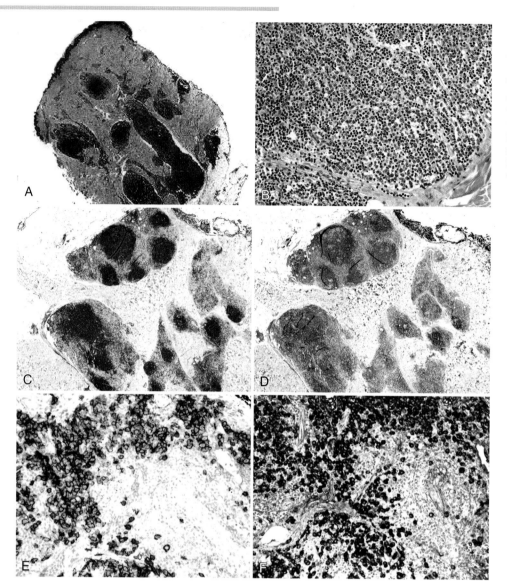

FIGURE 40.3 **PRIMARY CUTANEOUS MARGINAL ZONE LYMPHOMA, THE SAME CASE.** Skin biopsy demonstrating a lymphoid infiltrate extending into the deep dermis (A, low power), and showing various proportions of centrocyte-like cells, monocytoid B-cells and plasma cells (B, high power). The lymphoid infiltrate is strongly positive for CD20 (C) and BCL-2 (D). The plasma cells express CD138 (E) and kappa light chain (F).

Primary Cutaneous Diffuse Large B-Cell Lymphoma–Leg Type

Primary cutaneous diffuse large B-cell lymphoma (PCDLBCL)-leg type involves one or both lower legs and appears as erythematous, or bluish-red nodule(s). These lesions are often ulcerated and may eventually expand to extracutaneous sites. It has an unfavorable prognosis with a 5-year survival rate of about 50%.

MORPHOLOGY (FIGURE 40.4)

- Diffuse infiltration of large monomorphic cells in the dermis. The infiltrating cells resemble centroblasts or immunoblasts. Some cases may show compartmentalization by connective tissue.
- Epidermotropism is absent, but neoplastic cells may extend to the dermal-epidermal junction, particularly in ulcerated cases.

IMMUNOPHENOTYPE

- The immunophenotypic features are similar to those of DLBCL with non-germinal center B-cell-like. The neoplastic large cells express B-cell-associated antigens, BCL2 and MUM1. Expression of BCL6 is variable, but CD10 is negative.
- Expression of CD30 can be seen sometimes, while CD15 is negative. In cases where plasmablasts or immunoblasts are prominent, the neoplastic cells may lack CD20, and show cytoplasmic Ig.
- The Ki67 proliferation rate is usually greater than 60%. EBV may be positive in some cases.

MOLECULAR AND CYTOGENETIC STUDIES

- A great majority of these tumors show chromosomal abnormalities including gains of 18q (can cause *BCL2* over-expression) (Figure 40.5), and 7p and losses of 6q, 9p. No evidence of t(14;18).
- The genetic expression profile is similar to that of activated B-cells. Translocation of *MYC*, *BCL6*, and *IGH@* reported.

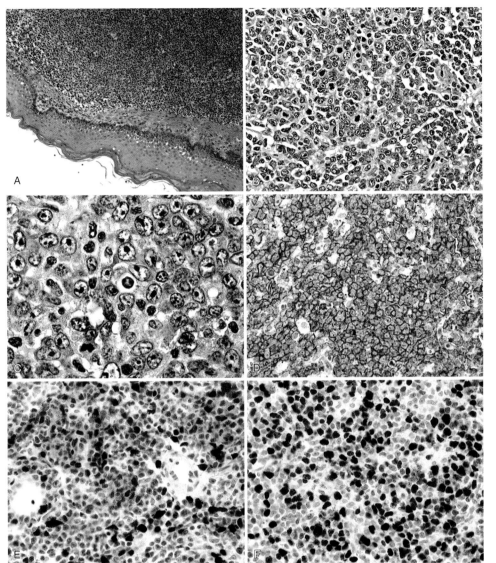

FIGURE 40.4 DIFFUSE LARGE B-CELL LYMPHOMA–LEG TYPE. Skin biopsy demonstrating a lymphoid infiltrate extending into the upper dermis with lack of epidermotropism (A, low power). The infiltrate consists of sheets of immunoblast-like cells with several mitotic figures present (B, intermediate power; C, high power). The tumor cells express CD20 (D), MUM1 (E) and a high Ki67 (F).

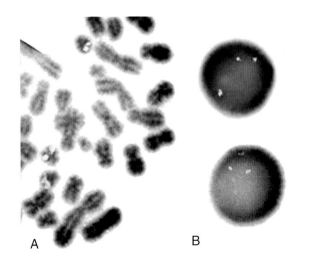

FIGURE 40.5 DIFFUSE LARGE B-CELL LYMPHOMA–LEG TYPE. Trisomy 18/18q as identified by karyotype (A) and FISH probes specific to 18q (B).

DIFFERENTIAL DIAGNOSIS

The differential diagnosis includes the diffuse form of PCFCL, and non-hematopoietic tumors of skin such as cutaneous adnexal tumors and melanomas. Diffuse PCFCL is a neoplasm consisting of centrocytes and centroblasts usually expressing BCL6, CD10, and a meshwork of CD21. These cells are negative for MUM-1/IRF4.

Plasmacytoma

Cutaneous plasmacytoma (Figure 40.6) is a rare plasma cell neoplasm that occurs without evidence of plasma cell myeloma. It may appear as solitary or multiple bluish-red

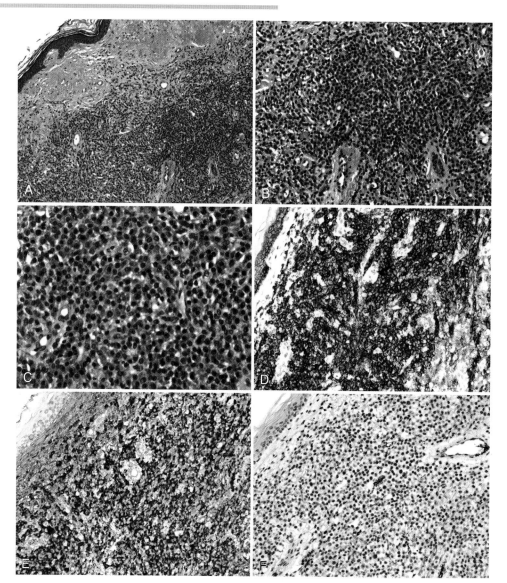

FIGURE 40.6 CUTANEOUS PLASMACYTOMA. Skin biopsy showing a dermal plasmacytic infiltrate (A, low power; B, intermediate power; C, high power). The plasma cells express CD138 (D) and lambda (E) and are negative for kappa (F).

nodules. They are effectively treated with excision and radiation therapy.

The plasmacytic infiltrate may consist of a monomorphic population of typical plasma cells or highly pleomorphic cells with anaplastic nuclei. The tumor cells show monotypic expression of Ig light chains and are often positive for CD79a, CD38, and CD138 (see Chapter 41).

Additional Resources

Breton AL, Poulalhon N, Balme B, et al: Primary cutaneous marginal zone lymphoma as a complication of radiation therapy: Case report and review, *Dermatol Online J* 16:6, 2010.

Dalle S, Thomas L, Balme B, et al: Primary cutaneous marginal zone lymphoma, *Crit Rev Oncol Hematol* 74:156–162, 2010.

Dijkman R, Tensen CP, Buettner M, et al: Primary cutaneous follicle center lymphoma and primary cutaneous large B-cell lymphoma, leg type, are both targeted by aberrant somatic hypermutation but demonstrate differential expression of AID, *Blood* 107:4926–4929, 2006.

Espinet B, García-Herrera A, Gallardo F, et al: FOXP1 molecular cytogenetics and protein expression analyses in primary cutaneous large B cell lymphoma, leg-type, *Histol Histopathol* 26:213–221, 2011.

Fitzhugh VA, Siegel D, Bhattacharyya PK: Multiple primary cutaneous plasmacytomas, *J Clin Pathol* 61:782–783, 2008.

Golling P, Cozzio A, Dummer R, et al: Primary cutaneous B-cell lymphomas – clinicopathological, prognostic and therapeutic characterisation of 54 cases according to the WHO-EORTC classification and the ISCL/EORTC TNM classification system for primary cutaneous lymphomas other than mycosis fungoides and Sézary syndrome, *Leuk Lymphoma* 49:1094–1103, 2008.

Huang CT, Yang WC, Liu YC, et al: Primary cutaneous diffuse large B-cell lymphoma, leg type, with unusual clinical presentation of bluish-reddish multicolored rainbow pattern, *J Clin Oncol* 29:e497–e498, 2011.

Kempf W, Sander CA: Classification of cutaneous lymphomas – an update, *Histopathology* 56:57–70, 2010.

Richmond HM, Lozano A, Jones D, et al: Primary cutaneous follicle center lymphoma associated with alopecia areata, *Clin Lymphoma Myeloma* 8:121–124, 2008.

Senff NJ, Noordijk EM, Kim YH, et al: European Organization for Research and Treatment of Cancer and International Society for Cutaneous Lymphoma consensus recommendations for the management of cutaneous B-cell lymphomas, *Blood* 112:1600–1609, 2008.

Tristano AG: Extramedullary orbital and cutaneous plasmacytomas, *Am J Hematol* 77:203–204, 2004.

Willemze R: Primary cutaneous B-cell lymphoma: classification and treatment, *Curr Opin Oncol* 18:425–431, 2006.

Willoughby V, Werlang-Perurena A, Kelly A, et al: Primary cutaneous plasmacytoma (posttransplant lymphoproliferative disorder, plasmacytoma-like) in a heart transplant patient, *Am J Dermatopathol* 28:442–445, 2006.

Zhao J, Han B, Shen T, et al: Primary cutaneous diffuse large B-cell lymphoma (leg type) after renal allograft: case report and review of the literature, *Int J Hematol* 89:113–117, 2009.

Plasma Cell Neoplasms

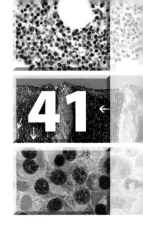

The monoclonal proliferation of plasma cells, also known as monoclonal gammopathies, plasma cell dyscrasias, or paraproteinemias, consists of several clinicopathologic entities. They are all characterized by a single class of immunoglobulin (Ig) product, referred to as "M-component" (monoclonal component), present in serum and/or urine protein electrophoresis (Figure 41.1). In addition to protein electrophoresis, other more sensitive and specific techniques, such as immunoelectrophoresis and immunofixation, have been used to detect the specific class or subtype of monoclonal Ig in these disorders (Figures 41.2 and 41.3). The following categories are discussed in this chapter (Box 41.1).

- Monoclonal gammopathy of undetermined significance (MGUS)
- Plasma cell myeloma (PCM) and variants
- Plasmacytoma and variants
- Monoclonal Ig deposit diseases
- Light and heavy chain diseases.

Waldenström's macroglobulinemia (lymphoplasmacytic lymphoma) is discussed in Chapter 29.

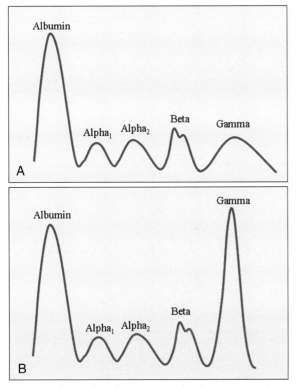

FIGURE 41.1 Schematics of serum protein electrophoresis: (A) demonstrates a normal profile, and (B) shows a spike in the gamma region.

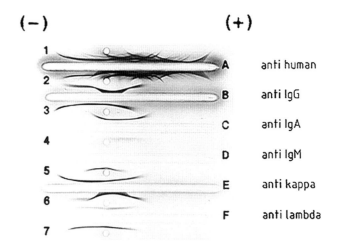

FIGURE 41.2 Serum immunoglobulin electrophoresis with a panel of IgG, IgA, IgM, kappa, and lambda antibodies demonstrating an IgG/kappa spike. Wells 1–7 contain control serum (odd number) or patient serum (even number).
From Bossuyt X, Bogaerts A, Schiettekatte G, et al. Serum protein electrophoresis and immunofixation by a semiautomated electrophoresis system. Clin Chem 1998; 44: 944–949.

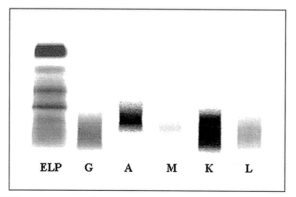

FIGURE 41.3 Immunofixation analysis of a serum demonstrating IgA-kappa monoclonal gammopathy. Courtesy of Eugene Dinovo, Ph.D., Department of Veterans Affairs, Greater Los Angeles Healthcare System.

> **Box 41.2 Clinicopathologic Features of MGUS According to the International Myeloma Working Group[1]**
>
> 1. Serum M-protein level <3g/dL
> 2. Bone marrow monoclonal plasma cells <10%
> 3. No evidence of other B-cell lymphoproliferative disorders
> 4. No evidence of organ or tissue impairment or bone lesions
>
> [1]The International Myeloma Working Group. Criteria for the classification of monoclonal gammopathies, multiple myeloma and related disorders: a report of the International Myeloma working Group. Br J Haematol 2003; 121: 749–757.

> **Box 41.1 Classification of Plasma Cell Neoplasms**
>
> 1. Monoclonal gammopathy of undetermined significance (MGUS)
> 2. Plasma cell myeloma
>   a. Asymptomatic myeloma (smoldering myeloma)
>   b. Symptomatic myeloma
>   c. Non-secretory myeloma
>   d. Plasma cell leukemia
> 3. Osteosclerotic myeloma (POEMS syndrome)
> 4. Plasmacytoma
>   a. Solitary plasmacytoma of bone
>   b. Extramedullary plasmacytoma
> 5. Immunoglobulin deposition diseases
>   a. Primary amyloidosis
>   b. Light and heavy chain deposition disease

Monoclonal Gammopathy of Undetermined Significance

The term monoclonal gammopathy of undetermined significance (MGUS) stands for a clinical condition defined by the presence of monoclonal Ig production without evidence of plasma cell myeloma, amyloidosis, Waldenström's macroglobulinemia, or other related plasma cell or lymphoproliferative disorders (see Box 41.2). Several other terms have been used for this condition, such as idiopathic, asymptomatic, non-myelomatous, cryptogenic, and benign monoclonal gammopathy. The term benign monoclonal gammopathy is inappropriate, because a significant proportion of patients with MGUS eventually develop plasma cell myeloma or other B-lymphoproliferative disorders. The cumulative probability of progression of MGUS to PCM or other lymphoproliferative disorders in one large study was 12% at 10 years, 25% at 20 years, and 30% at 25 years. MGUS and plasma cell myeloma represent different time points along the same disease spectrum, and, so far, no molecular or cytogenetic test can reliably distinguish them.

The incidence of MGUS increases by age. The prevalence of MGUS is about 3% in individuals older than 50 and 5% in those older than 70 years. It affects more African-Americans than Caucasians. MGUS is usually an incidental finding detected by elevated total protein concentration on a routine blood test, followed by demonstration of a monoclonal spike by serum protein electrophoresis. Plasma cells are usually <10% of the bone marrow cells. A higher percentage of plasma cells in bone marrow is a predictor of progression of MGUS to a more aggressive B-lymphoproliferative disorder.

MORPHOLOGY AND LABORATORY FINDINGS

- Bone marrow plasma cells in biopsy sections and smears are modestly increased and account for <10% of the total bone marrow nucleated cells (Figure 41.4).
- Evidence of serum M-component which is IgG in about 75%, IgM in 15%, and IgA in 10% of cases.
- Small quantities of Ig light chains may be detected in the urine (Bence-Jones protein) in about 25% of patients.

IMMUNOPHENOTYPE

Flow cytometric features include:

- Monotypic plasma cells coexpressing CD38 and CD138, with intracellular light chain restriction;
- Common aberrant expression of CD56;
- Negativity of CD19;
- Difficulty in detecting intracellular light chain restriction when polytypic plasma cells are present in background.

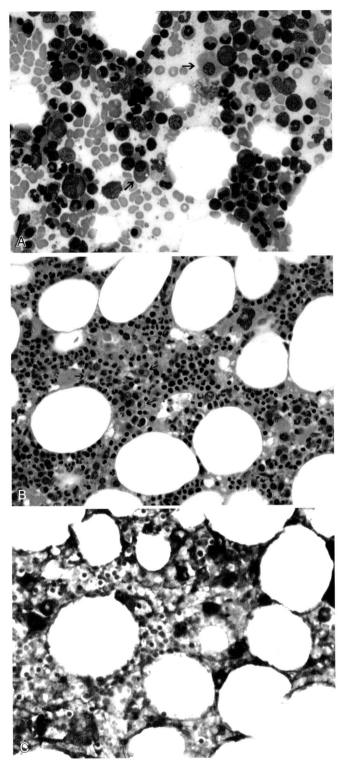

FIGURE 41.4 MONOCLONAL GAMMOPATHY OF UNDETERMINED SIGNIFICANCE. Bone marrow smear (A) and biopsy section (B) demonstrate modest plasmacytosis. Plasma cells show kappa light chain restriction (brown) by dual immunohistochemical stains (C).

Immunohistochemical studies are more useful than MFC in plasma enumeration. However, it may be difficult to demonstrate light chain restriction when the monotypic plasma cells are admixed with polytypic ones.

Plasma Cell Myeloma

Plasma cell myeloma (PCM) [multiple myeloma (MM), myelomatosis, Kahler's disease] is a multifocal bone-marrow-based plasma cell neoplasm with the production of monoclonal Ig, often associated with bone destruction and osteolytic lesions, hypercalcemia, and anemia. PCM has been divided into the following clinicopathologic entities:

- Asymptomatic myeloma (smoldering myeloma)
- Symptomatic myeloma (or symptomatic plasma cell myeloma)
- Non-secretory myeloma
- Plasma cell leukemia.

Asymptomatic (Smoldering) Myeloma

Smoldering myeloma (SM) represents the point of transition from MGUS to PCM without anemia, skeletal lesions, hypercalcemia, or renal insufficiency. The serum M-protein level is ≥3 g/dL, and clonal plasma cells are ≥10% of bone marrow cells but <30% (see Box 41.3). These patients do not need treatment but should be followed up closely, because many of them eventually become symptomatic.

Symptomatic PCM

Symptomatic PCM, or myelomatosis, is characterized by monoclonal proliferation of plasma cells in the bone marrow and an M-protein production (Figure 41.5). This neoplastic proliferation leads to bone destruction and pathological fractures, particularly in the spine and ribs.

Box 41.3 Criteria for Asymptomatic Myeloma (Smoldering Myeloma)[1]

- Serum M-protein ≥3 g/dL
- Bone marrow clonal plasma cells ≥10%
- No related tissue or organ impairment or symptoms

[1]The International Myeloma Working Group. Criteria for the classification of monoclonal gammopathies, multiple myeloma and related disorders: a report of the International Myeloma working Group. Br J Haematol 2003; 121: 749–757.

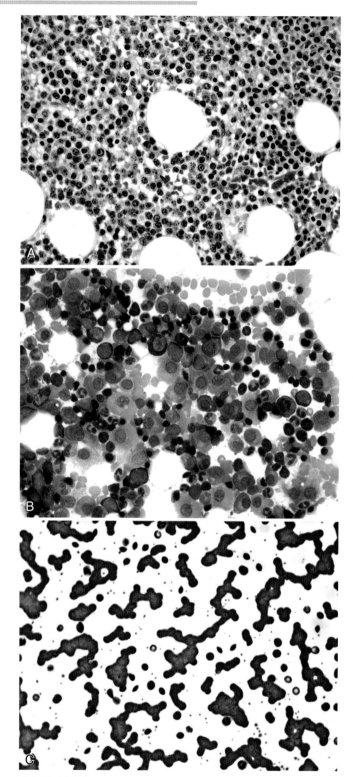

FIGURE. 41.5 **PLASMA CELL MYELOMA.** Bone marrow biopsy section (A) and bone marrow smear (B) demonstrate numerous plasma cells. Blood smear shows rouleaux formations (C).

Table 41.1
Myeloma-Related Organ or Tissue Impairment[1]

Impairment	Criteria
Elevated serum calcium levels	>0.25 mmol/L above the upper limit of normal or >2.75 mmol/L
Renal insufficiency	Creatinine >173 mmol/L
Anemia	Hemoglobin 2 g/dL below the lower limit of normal or <10 g/dL
Bone lesions	Lytic lesions or osteoporosis with compression fractures
Others: symptomatic hyperviscosity, amyloidosis, recurrent bacterial infections	

[1]The International Myeloma Working Group. Criteria for the classification of monoclonal gammopathies, multiple myeloma and related disorders: a report of the International Myeloma working Group. Br J Haematol 2003; 121: 749–757.

Box 41.4 Criteria for the Diagnosis of Plasma Cell Myeloma Set by the International Myeloma Working Group[1]

- M-protein in serum and/or urine
- Bone marrow clonal plasma cells or plasmacytoma[2]
- Related organ or tissue impairment (end organ damage, including bone lesions)[3]

[1]The International Myeloma Working Group. Criteria for the classification of monoclonal gammopathies, multiple myeloma and related disorders: a report of the International Myeloma working Group. Br J Haematol 2003; 121: 749–757.
[2]Immunophenotypic studies demonstrate monoclonal population of abnormal plasma cells.
[3]Some patients may have no symptoms but show evidence of organ or tissue impairment.

Anemia, hypercalcemia, and renal insufficiency are other common features (Table 41.1).

The criteria for the diagnosis of PCM proposed by the International Myeloma Working Group (IMWG) are shown in Box 41.4. According to the IMWG, the most critical criterion for symptomatic myeloma is the evidence of organ or tissue impairment manifested by anemia, hypercalcemia, lytic bone lesions (Figure 41.6), renal insufficiency, hyperviscosity, amyloidosis, or recurrent infections. In the diagnostic criteria proposed by the IMWG, no level of serum or urine M-protein and no minimal percentage of clonal bone marrow plasma cells are required. The adverse prognostic factors in plasma cell myeloma are shown in Box 41.5.

Non-Secretory Myeloma

Non-secretory myeloma (Figure 41.7) accounts for 1–5% of all myelomas and is characterized by the absence of detectable M-protein in the serum and urine. However, utilization of more sensitive techniques, such as serum-free light chain assay, may significantly reduce the number of these cases. The reports on non-secretory myelomas suggest a lower incidence of renal failure and hypogammaglobulinemia,

Plasma Cell Leukemia

Plasma cell leukemia (PCL) is a rare event, which is characterized by the presence of >2000/μL of circulating plasma cells and/or ≥20% of the white cell differential count (Figure 41.8).

Plasma cell leukemia is divided into two categories: primary and secondary.

- Primary plasma cell leukemia constitutes about 60% of cases and is manifested *de novo* without evidence of previous history of PCM.
- Secondary plasma cell leukemia accounts for the remaining 40% and represents leukemic transformation in patients with a history of PCM.

Patients with primary plasma cell leukemia are younger, have fewer lytic bone lesions and smaller amounts of serum M-protein, demonstrate higher incidence of hepatosplenomegaly, and show a longer survival than patients with secondary plasma cell leukemia. The immunophenotype of primary plasma cell leukemia is frequently IgD, IgE, or light chain only, but clonal gene rearrangement studies targeting the IgM region will be sufficient to establish a monoclonal plasma cell disorder if needed.

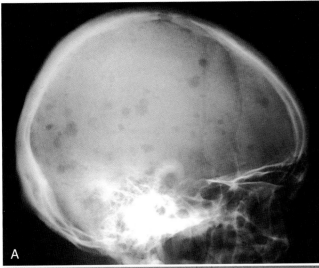
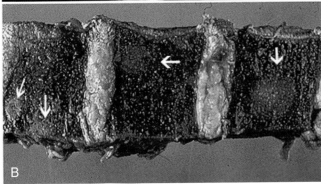

FIGURE 41.6 (A) Skull X-ray of a patient with plasma cell myeloma demonstrating punched out lytic lesions. (B) Gross specimen of vertebrae showing several foci involved with plasma cell myeloma (arrows).

MORPHOLOGY AND LABORATORY FINDINGS

- Bone marrow biopsy sections are usually hypercellular and show clusters or sheets of plasma cells (Figure 41.9).
- Occasionally the bone marrow may appear hypocellular (Figure 41.10).
- The plasma cell infiltrates are more often focal lesions/patchy than diffuse, and therefore repeated biopsies may be required to establish the diagnosis. Prominent osteoclastic activity may be evident, adjacent to the bony trabeculae.
- The cytologic features of plasma cells may vary from normal-appearing mature plasma cells to immature and anaplastic forms (Figures 41.5, 41.9, 41.11, and 41.12).
- Plasmablasts show a high nuclear to cytoplasmic ratio, deep blue cytoplasm, with or without perinuclear hof, round or irregular nucleus, fine chromatin, and one or several prominent nucleoli (Figure 41.13). Multinucleated or multilobated plasma cells may be present.
- Cells with cherry-red, round cytoplasmic (Russell bodies) or nuclear inclusions (Dutcher bodies) (Figure 41.14), as well as cytoplasmic crystals (Figure 41.15) may be present. Some plasma cells may appear like grapes and demonstrate numerous round Ig-containing cytoplasmic structures (Mott or Morula cells) (Figure 41.16).
- Occasionally, plasma cell myeloma may mimic lymphplasmacytic lymphoma ("lymphoid appearing myeloma") by morphology and expression of CD20 (Figure 41.17). But, the neoplastic cells are positive for IgG and often show BCL-1 expression and evidence of t(11;14).
- The blood smears may show Rouleaux formation of the red blood cells or the presence of plasma cells in various proportions (Figure 41.18). In plasma cell leukemia, the circulating

Box 41.5 Adverse Prognostic Factors in Plasma Cell Myeloma[1]

- Age ≥70 years
- Serum albumin <3 g/dL
- Serum creatinine ≥2 mg/dL
- Beta-2 microglobulin >4 mg/L
- Plasma cell labeling index ≥1%
- Circulating plasma cells >10 per 50,000 mononuclear cells
- Serum calcium ≥11 mg/dL
- Platelet count <15,000/μL
- Hemoglobin <10 g/dL
- Cytogenetics: t(4;14)(p16;q32.3), t(14;16)(q32.3;q23), del 17p13 (locus of *p53*), non-hyperdiploidy

[1] Adapted from Kyle RA, Gertz MA, Witzig TE, et al. Review of 1027 patients with newly diagnosed multiple myeloma. Mayo Clin Proc 2003; 78: 21–33.

lower median percentage of bone marrow plasma cells, higher incidence of neurological presentation, and longer survival than the secretory myelomas. The therapeutic approaches for non-secretory myeloma are similar to those for symptomatic PCM.

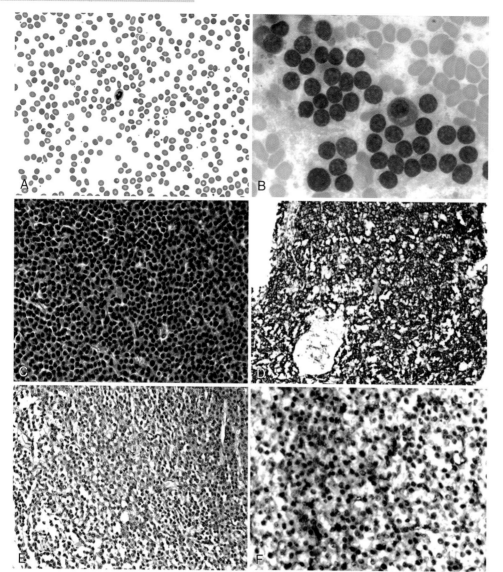

FIGURE 41.7 **NON-SECRETORY MYELOMA.** Blood smear (A) shows lack of red cell rouleaux formations. Bone marrow smear (B) and biopsy section (C) demonstrate clusters of plasma cells. These cells are CD138+ (D), but lack expression of IgG (E). In (F, dual staining), kappa light chain (red) is negative and rare cells show lambda (brown) light chain expression.

plasma cells account for ≥20% of the differential counts and >2000/μL absolute counts.
- Over 95% of patients with PCM show an M-protein in the serum or urine at the time of diagnosis. Approximately 50% of patients with PCM show IgG and 25% IgA M-proteins. About 20% demonstrate only monoclonal light chain production. IgD and IgM PCM are rare. Bence–Jones protein is detected in the urine of about 75% of patients. Approximately 20% of patients show hypercalcemia. The serum beta-2 microglobulin may be elevated.

An international staging system has been proposed based on the serum beta-2 microglobulin (β2M) and serum albumin levels:

- Stage I: Serum β2M <3.5 mg/L and serum albumin ≥3.5 g/dL
- Stage II: Neither stage I nor stage III
- Stage III: Serum β2M ≥5.5 mg/L.

Conventional radiologic studies reveal lytic bone lesions, osteoporosis, or fractures in about 80% of patients at diagnosis.

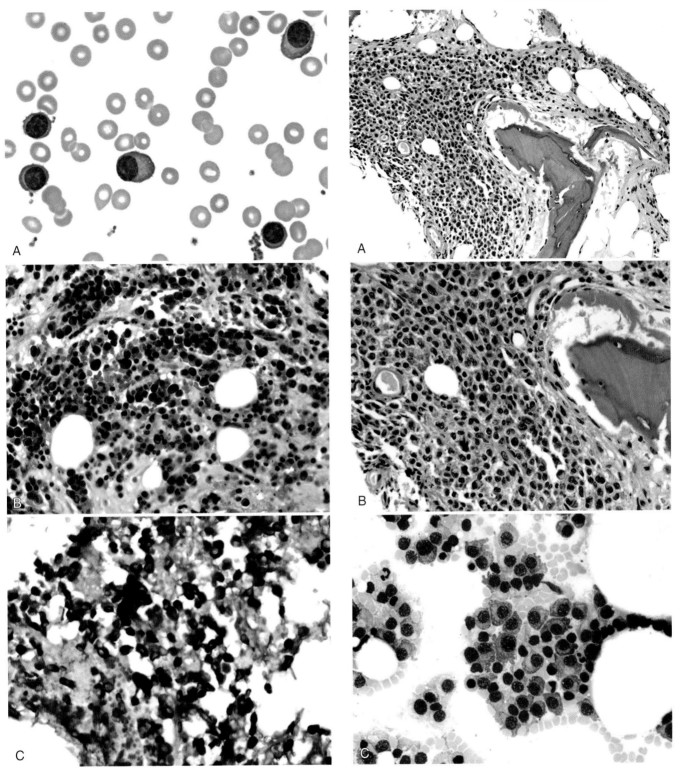

FIGURE 41.8 **PLASMA CELL LEUKEMIA.** (A) Blood smear demonstrating circulating plasma cells. (B) Bone marrow biopsy section showing plasmacytic infiltration with (C) the expression of CD138.

FIGURE 41.9 **PLASMA CELL MYELOMA.** Bone marrow biopsy section demonstrating a paratrabecular aggregate of plasma cells (A, low power; B, high power). Bone marrow smear shows clusters of plasma cells (C).

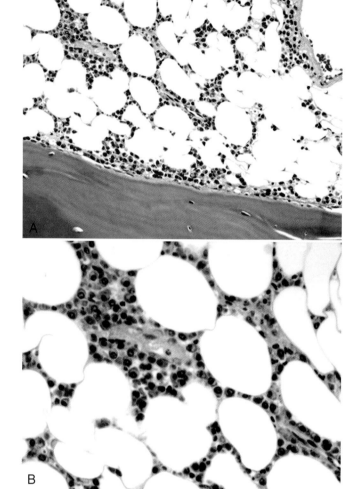

FIGURE 41.10 Hypocellular plasma cell myeloma with interstitial infiltration of plasma cells around fatty tissue (A, low power; B, high power).

IMMUNOPHENOTYPE

Compared with immunohistochemical studies, flow cytometry tends to underestimate the number of malignant plasma cells. However, it is a more sensitive measure in identifying their phenotypic aberrancies. The immunophenotypic features of plasma cell myeloma (Figure 41.19) by MFC include:

- Coexpression of CD38 (bright) and CD138;
- Intracellular light chain restriction;
- Aberrant expression of CD56 in greater than 60% of cases, which, however, is commonly absent in plasma cell leukemia (Figure 41.20);
- Variable expression of myeloid aberrancies including commonly CD117, as well as CD13, CD15, and CD33;
- Negativity to variable positivity of CD45;
- Occasional and partial positivity of CD20;
- Rare positivity of partial or dim CD19;
- Often express CD79a.

Immunohistochemical stain for CD138 is the preferred method of plasma cell enumeration. But this molecule is

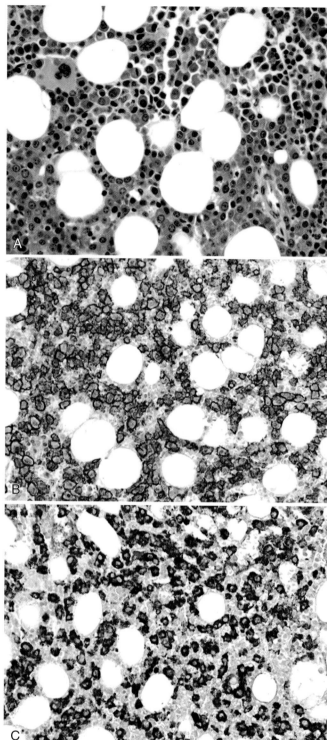

FIGURE 41.11 **Plasma cell myeloma.** Bone marrow biopsy section demonstrating a large population of plasma cells (A). Immunohistochemical stains show expression of CD138 (B) and kappa light chain (C) by the plasma cells.

unstable and is degradable rapidly. Light chain restriction can be detected in the vast majority of cases, and heavy chain restriction may be demonstrable. Expression of cyclin D1 (*BCL1*) is seen in some cases, which correlates with t(11;14) by cytogenetic/FISH studies.

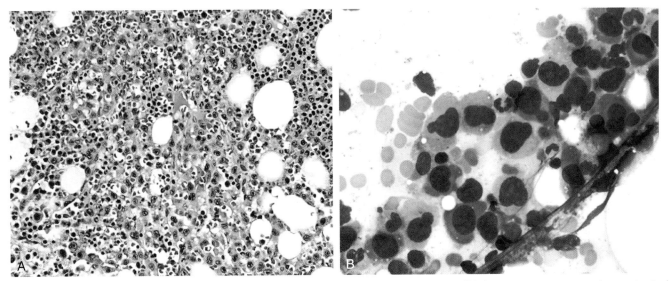

FIGURE 41.12 **PLASMA CELL MYELOMA.** Bone marrow biopsy section (A) and bone marrow smear (B) demonstrate numerous large, atypical plasma cells with abundant cytoplasm, irregular nuclear borders, and prominent nucleoli.

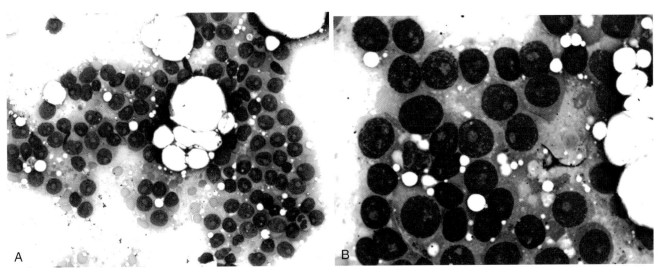

FIGURE 41.13 **PLASMABLASTIC MYELOMA.** Bone marrow smear demonstrating cells with variable amount of blue cytoplasm, round nuclei with finely dispersed chromatin, and a single prominent nucleolus (A, low power; B, high power).

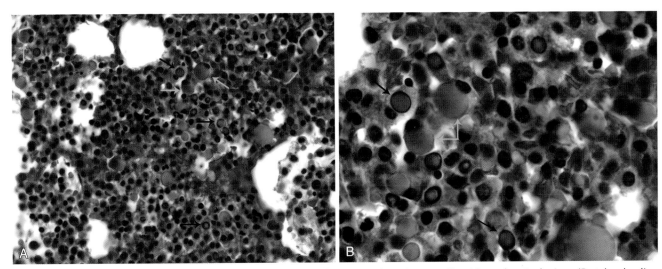

FIGURE 41.14 **PLASMA CELL MYELOMA.** Bone marrow biopsy section demonstrating plasma cells with nuclear inclusions (Dutcher bodies, black arrows) and cytoplasmic inclusions (Russell bodies, green arrows) (A, low power; B, high power).

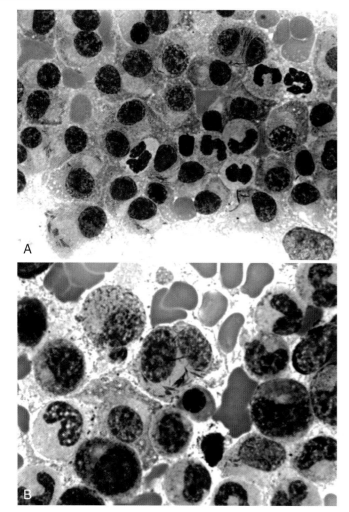

FIGURE 41.15 PLASMA CELL MYELOMA. Bone marrow smear demonstrating plasma cells with cytoplasmic needle-like immunoglobulin crystals (A, low power; B, high power).

CYTOGENETIC AND MOLECULAR STUDIES

By classical cytogenetics, only one-third of myeloma patients have a complex abnormal karyotype. The remaining two-thirds show normal karyotypes. However, the observed normal karyotypes are often derived from the other non-neoplastic hematopoietic cells, rather than from abnormal plasma cells because these cells fail to grow in culture.

Samples may fail to grow because of the low proliferative capacity of the myeloma cells.

It may not be possible to obtain abnormal metaphases, because of inadequate quality of the bone marrow aspirates received for cytogenetic studies. Aspirates frequently contain very few plasma cells compared with the corresponding smears used for morphological assessment, since the number of tumor cells in a given specimen largely depends on the level of local bone marrow infiltration and the degree of sample dilution by blood. For this reason, it is essential that the first few milliliters of the bone marrow drawn be sent for cytogenetic analysis. Also, the needle

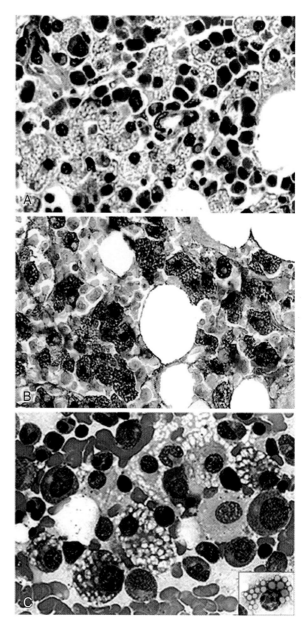

FIGURE 41.16 PLASMA CELL MYELOMA WITH CYTOPLASMIC VACUOLES. (A) Bone marrow biopsy section, (B) immunohistochemical stain for Ig kappa light chain, and (C) bone marrow smear.
From Naeim F. Pathology of Bone Marrow, 2nd edn. Williams & Wilkins, Baltimore, 1997, by permission.

should be repositioned during aspiration, rather than simply continuing to withdraw marrow from the initial puncture site, to ensure that adequate numbers of abnormal cells are submitted to the laboratory.

The biological transition of normal plasma cells to MGUS/SM to multiple myeloma consists of several genomic changes that are often overlapping and represent a series of oncogenic events. Karyotypic instability has been detected as a major pathogenetic factor. Identification of aggressive disease biology is achieved through analysis of bone marrow plasma cells, including metaphase cytogenetics and FISH studies.

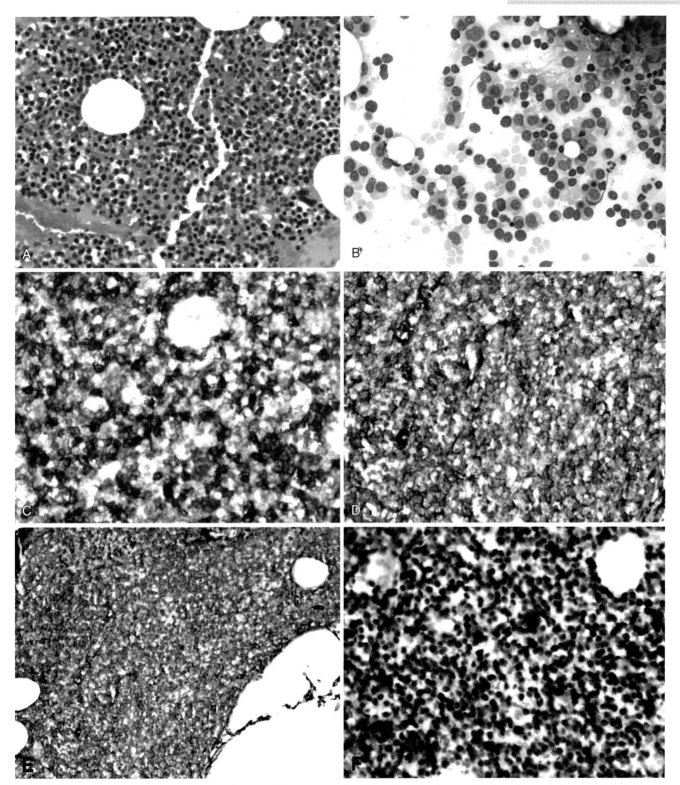

FIGURE 41.17 LYMPHOID APPEARING MYELOMA. A plasma cell myeloma mimicking lymphoplasmacytic lymphoma is demonstrated. Bone marrow clot section (A) shows a large number of lymphoid appearing cells. Bone marrow smear (B) depicts a mixture of plasma cells, lymphocytes and plasmacytoid lymphocytes. These cells express CD20 (C), CD138 (D), IgG (E) and BCL-1 (F). The patient had bone lytic lesions, and elevated IgG. Cytogenetic studies revealed t(11;14).

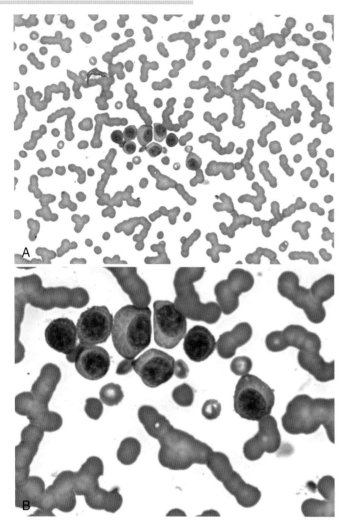

FIGURE 41.18 Red cell rouleaux formation and aggregate of plasma cells in a peripheral blood smear of a patient with plasma cell leukemia (A, low power; B, high power).

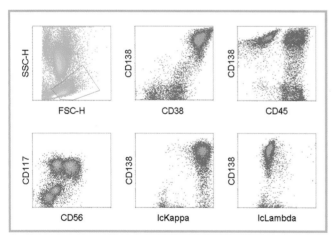

FIGURE 41.19 **FLOW CYTOMETRIC FINDINGS OF PLASMA CELL MYELOMA.** On open scatter gate (in green), a plasma-cell-enriched population is highlighted and confirmed by CD38/CD138 backgating. The plasma cells (positive for both CD38 and CD138) are abnormal, expressing aberrant CD56 (large subset) and CD117, with intracellular kappa light chain restriction. A small subset of the monotypic plasma cells is negative for CD45, while the remaining demonstrates dim to moderate positivity of CD45.

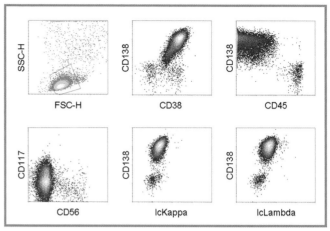

FIGURE 41.20 **FLOW CYTOMETRIC FINDINGS OF PLASMA CELL LEUKEMIA.** Plasma-cell-enriched gate of a peripheral blood sample demonstrates a prominent population of abnormal plasma cells, which are positive for CD38, CD117 (partial), CD138, and intracellular lambda light chain restriction (dim). The abnormal plasma cells are negative for CD45 and CD56.

Because of the low proliferative activity early in the disease, results from conventional cytogenetics are often limited, with most showing a normal karyotype. About 30–40% of new patients demonstrate an abnormal karyotype by conventional metaphase techniques. Early cytogenetic changes are seen among almost all patients at the level of MGUS. The presence of an abnormal karyotype generally correlates with an elevated plasma cell labeling index and tumor burden, suggestive of advanced disease status. The karyotypes, when seen, are commonly complex and unstable, presenting with both numerical and structural aberrations. It should also be noted that some of the well-established structural abnormalities, like t(4;14)(p16;q32), are cryptic and therefore go undetected by routine cytogenetics.

These limitations have been overcome by the use of FISH probes on interphase nuclei, which are now a common and important adjunct to routine karyotyping. A further improvement in the yields of interphase FISH results is now achieved by enriching the bone marrow plasma cells by auto-magnetic activated cell sorting with anti-CD138 immunobeads. The development of interphase FISH (iFISH) on sorted CD138 plasma cells, allows for the assessment of chromosomal abnormalities regardless of the number of plasma cells and the low proliferative rate (Figure 41.21). Thus with the iFISH techniques one can detect trisomies of the odd number chromosomes that are established as a surrogate to identify ploidy subtypes.

FISH with probes targeted to detect specific aberrations with established clinical significance is now routinely used in PCM (MM) studies, along with traditional metaphase chromosome studies. Both studies can also be used to monitor the therapeutic response or to help direct therapy. In addition, results obtained from both techniques appear to be independent prognosticators at diagnosis, as well as providing complementary information for generating models for risk stratification. However, cytogenetics or FISH cannot distinguish MGUS from smoldering myeloma.

The expansion of malignant monoclonal plasma cells with numerical and structural chromosome aberrations has permitted the delineation of hyperdiploid and non-hyperdiploid forms as two major pathogenetic pathways, which are approximately equally distributed. The correlations between chromosomal aberrations and risk groups are detailed in Box 41.6.

Numerical aberrations leading to a hyperdiploid or hypodiploid karyotype are present in approximately 50–60% of precursor and multiple myeloma tumors.

- The hyperdiploid group is characterized by gains of the odd number chromosomes 3, 5, 7, 9, 11, 15, 19, or 21, and associated with a better prognosis (Figures 41.22 and 41.23).
- The hypodiploid group is associated with poorer overall survival, and the abnormal clones seen in this group include hypodiploidy, pseudodiploids, and/or near-tetraploid variants, and are almost always associated with structural aberrations (Figure 41.24). The near-tetraploid karyotypes appear to be 4n duplications of cells having pseudodiploid or hypodiploid karyotypes. The hyperdiploid group shows a consistent set of trisomies and fewer structural aberrations. The most frequently lost chromosomes are 13, 14, 16, and 22.
- Both hyper- and hypodiploid clones also exhibit, in addition to the numerical gains and losses, primary *IGH@* translocations. For example, 10% of hyperdiploid PCM show *IGH@* rearrangements.
- The non-hyperdiploid group typically displays chromosomal rearrangements and deletions, and karyotypes typically have translocations involving the immunoglobulin heavy chain (*IGH@*) locus at 14q32. The hyperdiploid and the non-hyperdiploid subgroups also appear to show different molecular signatures, methylation patterns, and prognosis.

Primary translocations involving the *IGH@* locus (14q32) or one of the *IGL* loci (κ at 2p12 or λ at 22q11) have been reported to occur early in the pathogenesis of PCM

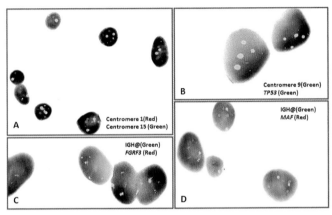

FIGURE 41.21 A panel of images showing clusters of CD138-enriched plasma cells with abnormal signal patterns: (A) gains of chromosomes 1, and 15; (B) gains of chromosome 9 and 17p(*TP53*); (C) FGFR3-IGH@ fusion consistent with t(4;14), and (D) gains of 16q (MAF).

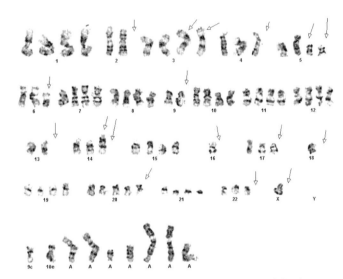

FIGURE 41.22 A complex hyperdiploid karyoytpe with both numerical and structural abnormalities and IGH@ (14q32) rearrangement in plasma cell myeloma.

Box 41.6 Risks Groups Based on Cytogenetic Abnormalities Identified by Conventional Karyotyping and FISH

High risk
- Hypodiploidy
- Del(17p)
- t(4;14), t(14;16), t(14;20), t(8;14)
- Karyotype del(13)
- 1p/1q abnormalities
- Complex abnormal karyotype

Standard risk
- Hyperdiploidy
- Trisomy 3, 5, 7, 9, 11, 15, 19
- Del(13q) (FISH)

Low risk
- t(11;14), t(6;14)
- Normal karyotype

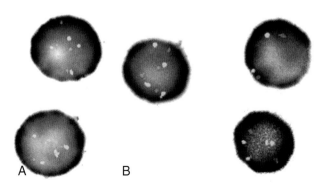

FIGURE 41.23 PLASMA CELL MYELOMA. Complex signal patterns frequently observed by FISH in plasma cell myeloma with hyperdiploidy: (A) chromosome 5, and (B) chromosome 7.

(Figure 41.25), whereas secondary translocations occur later and are involved in tumor progression. Most primary translocations are simple reciprocal translocations and juxtapose an oncogene and one of the immunoglobulin enhancers on the derivative 14 translocated chromosome, resulting in the dysregulated or increased expression of an oncogene. These translocations are mediated by errors in one of three B-cell specific DNA modification mechanisms: IGH switch recombination, errors in somatic hypermutation, and, rarely, VDJ recombination.

Large studies from several different groups show that the prevalence of IgH translocations increases with disease stage: about 50% in MGUS or SM, 55–70% for intramedullary PCM, 85% in PCL, and >90% in human myeloma cell lines (HMCL). These *IGH@* translocations can be efficiently detected by FISH analyses.

The incidence of light-chain translocations (κ, 2p12 or λ, 22q11) is far less, with about 10% being found in MGUS and 20% in advanced myeloma and HMCL. Translocations involving an Igλ locus are rare, occurring in only 1–2% of MM tumors and HMCLs.

There are seven translocations that appear to represent primary oncogenic events involving oncogenes and the *IGH@* locus and seen in approximately 40–50% of MM tumors (Tables 41.2 and 41.3). These can be divided into mainly three recurrent IgH translocation groups as follows.

1. *CYCLIN D*: 11q13 (Cyclin D1) (15%); 12p13 (Cyclin D2) (<1%); 6p21 (Cyclin D3) (2%).
2. *MMSET/FGFR3*: 4p16 (MMSET and usually FGFR3) (15%).
3. *MAF*: 16q23 (c-*MAF*) (5%); 20q12 (*MAFB*) (2%); 8q24.3 (*MAFA*) (<1%).

With the exception of *FGFR3* (especially with an activating mutation) and possibly c-*MAF*, the consequences of these translocations have not been adequately confirmed as essential for maintenance of the tumor and/or as therapeutic targets.

Virtually all PCM and MGUS tumors have cyclinD dysregulation, suggesting an early and unifying pathogenetic

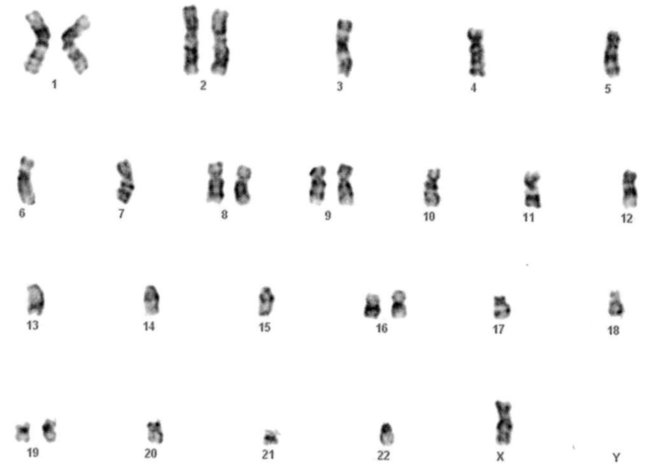

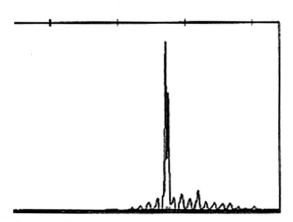

FIGURE 41.24 Bone marrow karyotype in a patient with plasma cell myeloma demonstrating hypodiploidy with 30,X,−X,−3, −4, −5,−6,−7,−10,−11,−12,−13,−14, −15, −17,del(17)(p13), −18, −20, −21, −22.

FIGURE 41.25 IgH PCR study of an extramedullary plasmacytoma showing a prominent clonal peak (framework 1 primer set) amid a weaker polyclonal background.

Table 41.2

Chromosomal Regions and Genes as Translocation Partners with 14q32 (*IGH@*) in MM/MGUS, with known frequencies

Region	Gene(s)
4p16	FGFR3, WHSC1 (15%)
4p13	ARHH
6p23–25	IRF4
6p21	CCND3 (15–20%)
8q24	MYC
11q13	CCND1
16q22–q23	MAF (5–10%)
18q21	BCL2
20q11–q13	MAFB

Table 41.3

Frequency of Recurrent Chromosomal Aberrations in Myeloma

Alterations	Percent
Recurrent and Primary Cytogenetic Abnormalities	
Hyperdiploidy	50–60
t(4:14)	15
t(11;14)	20
t(14;16)	3
t(14;20)	1
Monosomy 13 or 13q	45
del(17p)	8
Genome Instability Markers	
Gains of 1q	35
Deletion 1p	30
Gains of 5q	50
Deletion 12p	10

event. The most frequent translocation is t(11;14) (q13;q32) which is found in about 15% of both MGUS and PCM (Figure 41.26), and appears to be associated with a favorable outcome and is therefore regarded as neutral with regard to prognosis. Its prevalence is >40% in primary amyloidosis (AL), suggesting that there is a novel t(11;14) phenotype that rarely progresses beyond a minimum MGUS tumor mass and usually is not detected by conventional serum electrophoresis, so that it is not detected in the absence of pathological deposits.

Tumors with translocations affecting any of the three *MAF* genes show a distinctive gene expression profile. Many of the genes that are upregulated in these tumors are thought to be shared targets for all three *MAF* transcription factors. The t(14;16)(q32;q23) translocation involves the *IGH@* (14q32) locus and c-*MAF* (16q23) locus and is found in 6–7% of patients with MM (Figure 41.27), with a lower incidence in MGUS/SMM and is associated with poor outcome. Less frequent is the *MAFB* (20q12) translocation—the reciprocal t(14;20)(q32;q12). The prognostic outcome is assumed to be the same as for the t(14;16).

t(4;14)(p16;q32) (*MMSET/FGFR3;IGH@*) is the second most common translocation and is seen less frequently in MGUS/SM than in PCM (Figure 41.28). This translocation involves two protein-coding genes—Wolf-Hirschhorn syndrome candidate 1 gene (*WHSC1*), or multiple myeloma SET domain (*MMSET*), and fibroblast growth factor receptor 3 (*FGFR3*), an oncogenic receptor tyrosine kinase—at the 4p16 band. The *MMSET* is dysregulated and a third of the t(4;14) translocation patients do not express the *FGFR3*, because this translocation is often seen as an unbalanced rearrangement, with a loss of the derivative chromosome 14. It is unknown whether the loss of *FGFR3* expression is a primary or a secondary event, or even whether dysregulation of *FGFR3* is critical in pathogenesis. However, the dysregulation of *MMSET* in the t(4;14) PCM group suggests an important role for dysregulated *MMSET* in the initiation and maintenance of the disease.

The t(4;14)(p16;q32) is only detectable by FISH or RT-PCR methods. Cells with the t(4;14) also commonly exhibit a loss of chromosome 13 and other deletions, and this has universally been associated with poor survival.

In late disease stages, the complex karyotypes can include combinations of two translocations from the groups of multiple primary aberrations along with several secondary aberrations. In some of these cases, one of the translocations clearly is secondary, since it is found in only a subset of tumor cells, or is presumptively secondary since it is a complex translocation or insertion. These rare examples suggest that the three groups of translocations have the potential to complement one another.

Secondary translocations, sometimes involving an Ig locus, can occur at any stage of myelomagenesis. Two loci, *MYC* and *IGH@*, are more commonly involved in the development of secondary translocations. Other

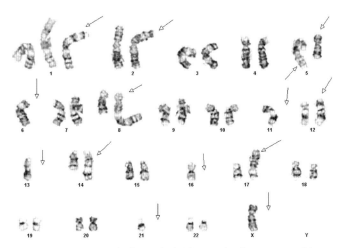

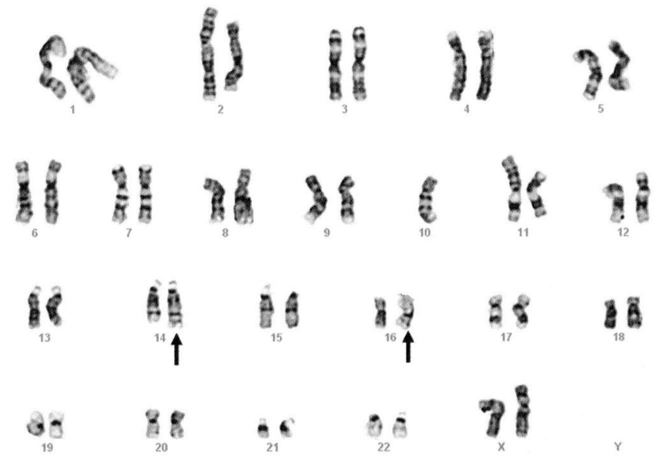

FIGURE 41.26 A pseudo-hypodiploid complex karyotype with an unbalanced t(11;14), monosomy 13, and chromosome 1 and 17p abnormality along with other aberrations.

FIGURE 41.27 Bone marrow karyotype in a patient with plasma cell myeloma demonstrating 45,XX,t(2;11)(p11.2;p11.2),−10,t(14;16)(q32;q23).

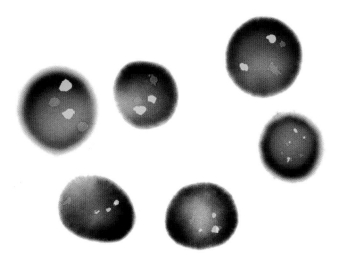

FIGURE 41.28 FISH with the dual color IGH@-FGFR3 probes shows normal, fusion, and fusion with secondary abnormalities (extra copies of 4p and 14q32).

abnormalities observed as part of the secondary aberrations include the loss or deletion of chromosome 13, deletions and/or amplifications of chromosome 1, and deletion of the *TP53* locus at chromosome 17p13.

Translocations involving the *MYC* gene (8q24) are absent or very rare in MGUS. But heterogeneous translocations involving *MYC* are seen in 15% of MM tumors, 44% of advanced tumors, and nearly 90% of Myeloma Cell Lines, and are noted commonly as part of complex rearrangements or insertions, sometimes involving three different chromosomes.

Most *MYC* gene rearrangements involve the *IGH@* locus and are thought to represent a very late progression event that occurs at a time when MM tumors are becoming less stromal-cell-dependent and/or more proliferative. *MYC* translocations are rare in primary PCL.

10–20% of the secondary translocations involving the *IGH@* locus in MGUS and PCM do not involve *MYC* or any one of the seven recurrent partners. These rearrangements are complex, unbalanced, and occur with similar frequencies in hyperdiploid and non-hyperdiploid tumors (whereas the recurrent or primary translocations occur predominantly in non-hyperdiploid tumors). Thus, most of them are likely to represent secondary Ig translocations that can occur at any stage of tumorigenesis, including MGUS.

Nearly 50% show chromosome 13 aberrations; most (85%) exhibit monosomy 13, with 15% showing a deletion of the 13q14 (Figures 41.29 and 41.30).

While earlier studies associated monosomy and/or del(13q) with a poor prognosis, there is recent evidence to suggest that the prognostic relevance is more associated with other genetic aberrations. Monosomy 13 is more frequently seen in association with t(4;14)(p16;q32), suggesting that chromosome 13 deletions may play an important role in the clonal expansion of myeloma tumors.

Deletion 17p13, resulting in the heterozygous loss or inactivation of the tumor-suppressor *TP53* gene, is an indicator of very poor prognosis. It is more commonly identified by interphase FISH studies (~10%) and is considered to be a late event. Deletion 17p patients (see Figure 41.29) have more aggressive disease, a higher prevalence of extramedullary disease, and overall shorter survival. It is rarely seen as a sole abnormality but frequently seen in subclones along with other chromosomal aberrations.

Cytogenetic abnormalities t(4;14), t(14;16), t(14;20), 17p deletion, and high (>3%) proliferation index identify a subgroup of about 25% of patients with aggressive (high risk) disease who are unlikely to do well with conventional therapy, and in whom alternative (experimental) approaches can therefore be justified.

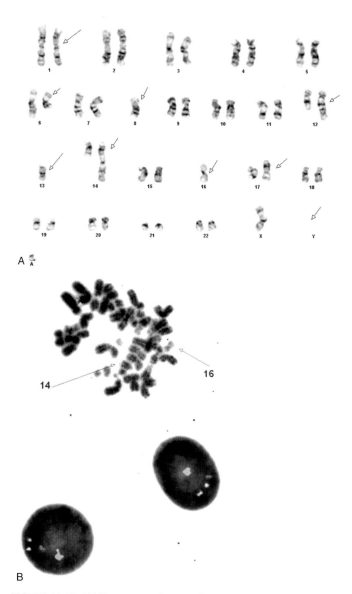

FIGURE 41.29 Bone marrow karyotype in a patient with plasma cell myeloma demonstrating 46,XX,del(13)(q13q22).

FIGURE 41.30 (A) Karyotype abnormalities include monosomy 13, rearrangement of 14q32, and deletions of 6q and 17p. (B) Concurrent FISH studies confirm −13, and 17p−, but also identify a complex three-way t(14;16)(q32;q23).

DNA-based techniques, such as array comparative genomic hybridization (AACGH) have identified the role of the NFκB pathway in myeloma.

ACGH studies also identify chromosome 1 aberrations as the most common structural aberrations in PCM and mostly involve interstitial or terminal deletions of 1p and whole-arm gains or amplification of 1q, particularly the 1q21 locus. Deletions of 1p are associated with a poor prognosis.

The 1q arm, particularly the 1q12→q23 region, contains a large number of possible candidate genes that show amplification and/or deregulated expression important in myeloma. The whole-arm segments of 1q translocate randomly to non-homologous chromosomes—"jumping translocations" of 1q. Such instability likely offers proliferative advantage during tumor progression, and thus amplification of 1q is associated with poor prognosis.

It is important to note that post-therapy PCM may also exhibit chromosome anomalies that are typical of MDS such as +8, 5q–, 7q–, 11q– and 20q–.

Gene-expression profiling (GEP) of bone marrow plasma cells is a powerful alternative way to identify cytogenetic abnormalities and provides insights into other biological processes such as cell proliferation. GEP analysis can accurately classify patients into high risk and low risk groups with favorable or unfavorable prognosis and identifies patients with high risk PCM more efficiently than does conventional testing.

Osteosclerotic Myeloma

Osteosclerotic myeloma, or POEMS syndrome, is characterized by a combination of peripheral neuropathy (P), organomegaly (O), endocrinopathy (E), monoclonal plasma cell disorder (M), and skin changes (S). Other frequent features include sclerotic bone lesions, Castleman disease, papilledema, serous effusions, and thrombocytosis.

The bone marrow biopsy sections show sclerotic bone and a monoclonal population of plasma cells. Plasmacytosis is usually modest (median 5%), but the bone marrow is often hypercellular with myeloid preponderance, increased megakaryocytes, and thick bone trabeculae. The M-protein is typically small and sometimes undetectable by routine serum protein electrophoresis.

Plasmacytoma

Plasmacytoma is a solitary neoplasm of plasma cells involving bone or extramedullary sites. Plasmacytoma demonstrates identical morphologic and immunophenotypic features to those of PCM (Box 41.7).

Box 41.7 Characteristics of Plasmacytoma of Bone and Extramedullary Sites[1]

- No M-protein in serum or urine[2]
- Solitary bone or extramedullary tumor of monoclonal plasma cells
- Bone marrow not consistent with plasma cell myeloma
- Normal skeletal survey
- No related organ or tissue impairment

[1] The International Myeloma Working Group. Criteria for the classification of monoclonal gammopathies, multiple myeloma and related disorders: a report of the International Myeloma working Group. Br J Haematol 2003; 121: 749–757.
[2] A small amount of M-component may be present in some cases.

SOLITARY PLASMACYTOMA OF BONE

Solitary plasmacytoma of bone is a rare condition and accounts for about 3–5% of the plasmacytic neoplasms. The median age is around 55 years and the male to female ratio is about 2:1. It often involves the axial skeleton, particularly thoracic vertebrae and ribs. Involvement of distal bones, particularly below the knees or elbows, is extremely rare. Bone pain at the site of the lesion is one of the most common presenting symptoms. Infiltration of the tumor cells into the surrounding soft tissue may result in a palpable mass. Radiologic studies show no evidence of additional lesions. Morphologic and immunophenotypic features are identical to those of PCM, but no serum or urine M-protein is detected, and there is no evidence of anemia, hypercalcemia, renal insufficiency, or other organ or tissue impairment. Multiple solitary plasmacytomas without evidence of PCM occur in up to 5% of cases. Approximately half of patients with solitary plasmacytoma of bone eventually develop PCM. Plasmacytomas of >5 cm diameter have a greater chance of conversion to PCM. Radiotherapy is the treatment of choice.

EXTRAMEDULLARY PLASMACYTOMA

Extramedullary plasmacytoma is a monoclonal plasma cell neoplasm that arises outside the bone marrow. The most frequent site of involvement is the upper respiratory tract, including the nasal cavity and sinuses, nasopharynx, and larynx, but any organ or tissue may be involved, such as gastrointestinal tract and urinary tracts, thyroid, male and female reproductive systems, parotid gland, lymph nodes, testicles, and central nervous system (Figures 41.31 and 41.32). The diagnosis is made based on the monoclonality of the plasma cell tumor and lack of evidence for PCM and serum or urine M-protein. IgA is the most frequent immunophenotype, but again it is IgM that is typically assessed at the molecular level for monoclonality. Approximately 15% of the patients may eventually develop symptomatic PCM. Surgery and/or radiation are the treatments of choice.

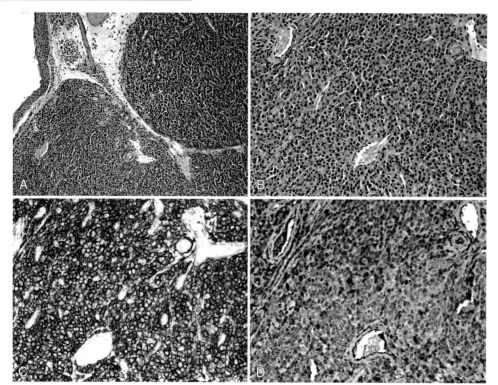

FIGURE 41.31 **Solitary plasmacytoma of conjuctiva.** Biopsy section demonstrating a large aggregate of plasma cells (A, low power; B, intermediate power). Immunohistochemical stains for CD138 (C) and kappa light chain (D) demonstrate a monoclonal population of plasma cells. *Courtesy of G. Pezeshkpour, M.D., Department of Pathology, VA Greater Los Angeles Healthcare System.*

Deposition Diseases

Monoclonal immunoglobulin deposition diseases are monoclonal gammopathies characterized by the deposition of Ig-derived proteins in the organs and tissues, causing impairment of their function. These disorders are divided into two major groups: (1) disorders with the deposition of fibrillary proteins (primary amyloidosis), and (2) disorders with the deposition of an amorphous, non-fibrillary protein, known as monoclonal light and heavy chain deposition diseases.

PRIMARY AMYLOIDOSIS

"Amyloidosis" is a general term referring to a heterogeneous group of disorders characterized by the extracellular deposition of fibrillary proteins with antiparallel beta-pleated sheet configuration on X-ray diffraction. These fibrillary structures are identified on biopsy sections by intense yellow-green fluorescence by thioflavine T and by binding to Congo Red stain, leading to apple green birefringence under polarized light. The major categories of amyloidosis include (1) primary amyloidosis, (2) secondary amyloidosis, (3) dialysis-related amyloidosis, (4) heritable amyloidosis, and (5) senile amyloidosis (Table 41.4).

Primary amyloidosis (AL) refers to a specific type of amyloidosis in which the fibrillary protein is derived from monoclonal Ig light chains. Primary amyloidosis is considered a variant of monoclonal plasma cell proliferative disorder, and in about 10% of cases it is associated with symptomatic PCM.

In approximately 75% of cases, the fibrillary protein is derived from the variable region of lambda light chain. The kappa light chain is involved in the remaining 25%. There appears to be a correlation between the site of involvement and the involved variable region of the light chain. For example, the amyloid deposit in patients with dominant renal involvement is often derived from V lambda IV, whereas in patients with dominant cardiac involvement the amyloid deposit is derived from V lambda II or III.

Morphology and Laboratory Findings

- The diagnosis of amyloidosis is based on biopsies obtained from the affected organs. Liver and kidney biopsies are positive in over 90% of cases, followed by abdominal fat pad aspirate and biopsies of rectum, bone marrow (Figure 41.33), and skin.
- A serum or urinary monoclonal protein can be detected in over 85% of patients, using immunofixation techniques. Also, a protein known as serum amyloid P component (SAP) is detected by scintigraphy in patients with primary amyloidosis.

Molecular and Cytogenetic Studies

Monosomy of chromosome 18 is the most common abnormality in primary amyloidosis followed by t(11;14)(q13;q32) and del(13q14). The most frequent clinical symptoms in primary amyloidosis are (1) nephrotic

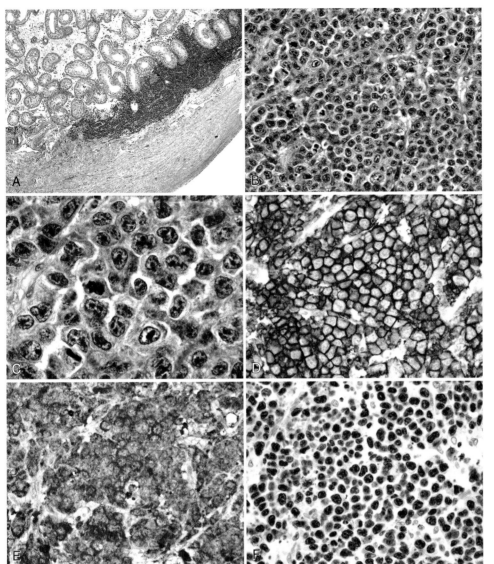

FIGURE 41.32 **PLASMABLASTIC PLASMACYTOMA OF TESTIS.** Biopsy section (A, low power; B, intermediate power; and C, high power). The neoplastic cells express CD138 (D), lambda light chain (E, red) and KI67 (F).

Table 41.4		
Major Categories of Amyloidosis[1]		
Amyloid Protein	**Precursor**	**Clinical Status**
AL	Ig light chain	Primary amyloidosis, local or systemic, associated with monoclonal plasma cell disorders
AA	Apolipoprotein AA	Secondary amyloidosis, systemic, associated with chronic infections
Aβ2M	Beta-2 microglobulin	Hemodyalysis, systemic
AApoAI, AApoAII, AGel, ALys, ACys, others	Apolipoprotein AI and AII, gelsolin lysozyme, crystatin C, and others	Familial, systemic
Ab, APro, ATTR, AMed	Ab protein precursor, prolactin, transthretin, lactadherin	Senile, local or systemic

[1] Adapted from Buxbaum JN. The systemic amyloidoses. Curr Opin Rheumatol 2004; 16: 67–75.

syndrome with or without renal insufficiency, (2) cardiomyopathy, (3) peripheral neuropathy, (4) hepatomegaly, and (5) macroglossia. Elevated serum β2 microglobulin and bone marrow plasma cells ~10%, dominant cardiac involvement, and circulating plasma cells ~1% correlate with poor prognosis.

The actuarial survival for 810 patients studied at the Mayo Clinic was 51% at 1 year, 16% at 5 years, and 4.7%

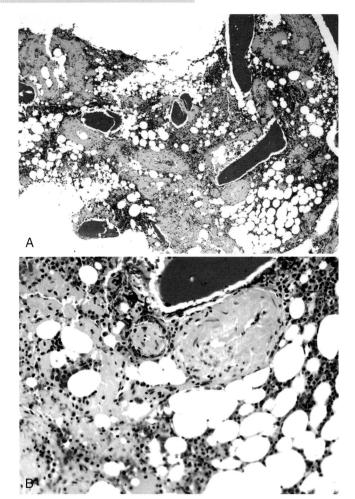

FIGURE 41.33 **PRIMARY AMYLOIDOSIS.** Bone marrow biopsy section (A, low power; B, high power).

Table 41.5
Comparison between Primary Amyloidosis and Light Chain Deposition Disease

Type	Ig	Deposition	Congo Red
Primary amyloidosis (AL)	Light chain, often λ	Fibrillary	Positive
Light chain deposition (LCDD)	Light chain, often κ	Non-fibrillary	Negative

at 10 years. Progression to PCM is rare, and in one large study it was reported in only 0.4% of the patients between 10 and 81 months. Therapeutic approaches include chemotherapy, such as melphalan with or without prednisone or dexamethasone, and stem cell transplantation.

The most common form of heritable amyloidosis is familial Mediterranean fever, an autosomal recessive autoinflammatory disorder characterized by periodic fevers, abdominal pain (peritonitis), pleuritis, arthritis, pericarditis, and skin rash. It is most prevalent in individuals of Armenian descent, in whom the carrier frequency approaches 1 in 7, but is also found in Arabs, Jews, Persians, Italians, and Greeks. In addition to the periodic attacks of pain and fever, the life-threatening complication of the disease is amyloid deposition, particularly in the kidneys. However, this form of amyloidosis is to be distinguished from the others discussed here since it is neither related to plasma cell proliferation nor is it composed predominantly of Ig protein components.

Differential diagnosis is made by the clinical history and genetic testing. The causative gene, *MVFV*, will often demonstrate mutations in the homozygous or compound heterozygous state. Since over 60 different mutations have been reported, practical testing is usually limited to a subset of the more common ones found in Mediterranean populations. Technical approaches include DNA sequencing and allele-specific DNA probe hybridization.

MONOCLONAL LIGHT AND HEAVY CHAIN DISEASES

Light Chain and Heavy Chain Deposition Disease

The light chain deposition disease (LCDD) and heavy chain deposition disease (HCDD) are clinical variants of monoclonal plasma cell disorders characterized by the deposition of abnormal light chain, heavy chain, or both in the tissues or organs. The deposits, unlike primary amyloidosis, are not fibrillary, do not bind Congo Red, and do not contain SAP. LCDD is more common than primary amyloidosis and often consists of kappa light chain (Table 41.5). The primary defect in LCDD appears to be mutations in the Ig light chain variable region, with predominant involvement of V kIV of kappa light chain. Deletion of the *CH1* constant domain and point mutation of variable regions of the heavy chain are the primary events in HCDD. These events lead to premature secretion of heavy chain binding protein and increased tendency for tissue deposition. HCDD of IgG1 and IgG3 isotypes is associated with reduced complement activities. Many organs may be involved, including kidneys, liver, heart, nerves, and blood vessels. Approximately 85% of cases show a detectable serum M-component.

Heavy Chain Diseases

The heavy chain disease is a monoclonal lymphoplasmacytic disorder characterized by the production of incomplete Ig molecules because of the lack of gamma chain binding sites for light chains. There are three major categories of heavy chain diseases: α, γ, and μ.

The α **heavy chain disease** (αHCD, Mediterranean lymphoma) is considered a variant of MALT-type lymphoma. It occurs in older children and young adults and is associated with gastrointestinal symptoms, such as malabsorption, intestinal obstruction, and diarrhea. αHCD is the most frequent heavy chain disease, with most reported cases being from the Middle East, North and South Africa, and the Far East. The pathologic features are more or less similar to those described in MALT lymphoma, depicted by a mucosal infiltrate of centrocyte-like lymphocytes and plasma cells. An abnormal heavy chain protein is detected in the serum of 20–90% of patients. Antibacterial therapy may completely resolve the disease in early stages. Some cases may eventually transform to large B-cell lymphoma. Despite the similar ethnic appellation, this disorder has nothing to do with familial Mediterranean fever, which is not a B-cell disorder, with molecular studies demonstrating mutation of the *MEFV* gene (Figure 41.34).

The γ heavy chain disease (γHCD, Franklin disease) is a rare condition presenting with lymphadenopathy, splenomegaly, and hepatomegaly, with lymphoplasmacytic infiltration similar to lymphoplasmacytic lymphoma. Immunofixation studies may show the presence of serum IgG without light chain. Some patients may demonstrate autoimmune disorders or chronic lymphocytic leukemia.

The μ heavy chain disease is a rare lymphoproliferative disorder with clonal IgM molecules with defective variable region. Clinically, it resembles chronic lymphocytic leukemia and is often associated with hepatosplenomegaly. The bone marrow is infiltrated by mature small lymphocytes admixed with vacuolated plasma cells. Lymphadenopathy is unusual.

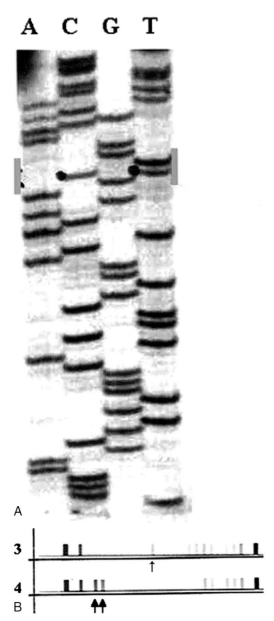

FIGURE 41.34 (A) DNA sequencing gel showing heterozygosity for the V726A mutation in the 2 *MEFV* gene of a patient with familial Mediterranean fever (the second mutation was M694V). The red dots mark the nucleotide position showing both T (normal) and C (mutant). (B) Reverse hybridization strips using allele-specific oligonucleotide probes directed against the most prevalent mutations in the *MEFV* gene associated with familial Mediterranean fever. Patient #3 is homozygous for mutation K695R (single arrow), and patient #4 is compound heterozygous for mutations P369S and E148Q (double arrows).

Additional Resources

Albarracin F, Fonseca R: Plasma cell leukemia, *Blood Rev* 25:107–112, 2011.

Anderson KC: New insights into therapeutic targets in myeloma, *Hematology Am Soc Hematol Educ Program*:184–190, 2011.

Buxbaum J: Mechanisms of disease: monoclonal immunoglobulin deposition. Amyloidosis, light chain deposition disease, and light and heavy chain deposition disease, *Hematol Oncol Clin North Am* 6:323–346, 1992.

Dimopoulos M, Kyle R, Fermand JP, et al: Consensus recommendations for standard investigative workup: report of the International Myeloma Workshop Consensus Panel 3, *Blood* 117:4701–4705, 2011.

Ge F, Bi LJ, Tao SC, et al. Proteomic analysis of multiple myeloma: current status and future perspectives, *Proteomics Clin Appl* 5:30–37, 2011.

Giralt S: Stem cell transplantation for multiple myeloma: current and future status, *Hematology Am Soc Hematol Educ Program*:191–196, 2011.

Hamidah NH, Azma RZ, Ezalia E, et al: Non-secretory multiple myeloma with diagnostic challenges, *Clin Ter* 161:445–448, 2010.

Hillengass J, Moehler T, Hundemer M: Monoclonal gammopathy and smoldering multiple myeloma: diagnosis, staging, prognosis, management, *Recent Results Cancer Res* 183:113–131, 2011.

Klein B, Seckinger A, Moehler T, et al: Molecular pathogenesis of multiple myeloma: chromosomal aberrations, changes in gene expression, cytokine networks, and the bone marrow microenvironment, *Recent Results Cancer Res* 183:39–86, 2011.

Korde N, Kristinsson SY, Landgren O: Monoclonal gammopathy of undetermined significance (MGUS) and smoldering multiple myeloma (SMM): novel biological insights and development of early treatment strategies, *Blood* 117:5573–5581, 2011.

Landgren O, Kyle RA, Rajkumar SV: From myeloma precursor disease to multiple myeloma: new diagnostic concepts and opportunities for early intervention, *Clin Cancer Res* 17:1243–1252, 2011.

Lorsbach RB, Hsi ED, Dogan A, et al: Plasma cell myeloma and related neoplasms, *Am J Clin Pathol* 136:168–182, 2011.

Ma ES, Shek TW, Ma SY: Non-secretory plasma cell myeloma of the true non-producer type, *Br J Haematol* 138:561, 2007.

Moehler T, Goldschmidt H: Therapy of relapsed and refractory multiple myeloma, *Recent Results Cancer Res* 183:239–271, 2011.

Palumbo A, Anderson K: Multiple myeloma, *N Engl J Med* 364:1046–1060, 2011.

Roussel M, Facon T, Moreau P, et al: Firstline treatment and maintenance in newly diagnosed multiple myeloma patients, *Recent Results Cancer Res* 183:189–206, 2011.

Sawyer JR: The prognostic significance of cytogenetics and molecular profiling in multiple myeloma, *Cancer Genet* 204:3–12, 2011.

Mature T- and NK-Cell Neoplasms—Overview

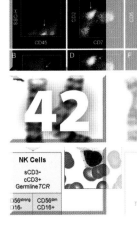

T-Cells

T-cells are derived from pluripotent hematopoietic stem cells in the bone marrow. The precursor T-cells (CD34+, TdT+, CD1a−/+, IcCD3+, CD5+, and CD7+) leave the bone marrow and enter the thymus for further maturation and functional development. This maturation process starts from the subcapsular zone of the thymus and extends to the thymic cortex and then thymus medulla (Figure 42.1). In the subcapsular zone, the precursor T-cells (prothymocytes) proliferate and rearrange their TCRα chains. They express TdT, CD1a, IcCD3, CD5, and CD7, but are negative for CD4 and CD8 (double negative). When the T-cells enter the cortex they develop expression of CD4 and CD8 (double positive) and rearrange their TCRβ chains. Therefore, the structure of surface TCR/CD3 complex is completed in the cortex. Approximately 65% of the thymic cortical T-cells retain their expression of TdT. Lymphocytes with higher receptor (TCR/CD3) affinity for self major histocompatibility complex (MHC) molecules are selected to proceed, and those with low affinity are eliminated by apoptosis. The T-cells that are selected on class I MCH become CD4−CD8+ (cytotoxic), and those which are selected by class II MHC become CD4 + CD8− (helper). The thymic medulla contains a mixture of helper (CD4+) and cytotoxic (CD8+) T-cells. Approximately 60–70% of the T-cells are CD4+ and 30–40% CD8+. Medullary T-cells do not express TdT.

The CD4+ T-cells differentiate into Th1, Th2, and Th17. The Th1 cells produce IL-2, INF-γ, and lymphotoxin. Th2 cells produce IL-4, IL-5, IL-9, IL-13, and GM-CSF, and Th17 cells produce IL-6 and IL-17.

Approximately 90–95% of the circulating T-cells carry TCRαβ receptors. The remainder 5–10% develop a TCR receptor with γ and δ chains. A small proportion of γδ T-cells is generated in the thymus, but a larger proportion is made in the GI tract. The proportion of γδ T-cells in the GI tract and skin is about 25–30%. The γδ T-cells are negative for CD4 and CD8 (double negative).

Most T-cell neoplasms are of helper (CD4+) phenotype. Abnormal expression of T-cell-associated markers is a frequent finding in T-cell neoplasms (Figure 42.2). For example, loss of CD7 is a common feature in mycosis fungoides/Sézary syndrome and adult T-cell leukemia/lymphoma, CD26 expression is lost in mycosis fungoides/Sézary syndrome, and coexpression of CD4 and CD8 is noted in a proportion of T-prolymphocytic leukemias, and peripheral T-cell lymphoma, NOS.

Clonality in T-cell neoplasms is established by demonstration of *TCR* rearrangement and cytogenetic abnormalities (Figure 42.3). The most frequent cytogenetic and abnormalities include loss or addition of 4q, 5q, 6q, 7q, 8q, 9q, 10q, 12q, 13q, 17q, and X, and trisomy 3 and 5. Isochromosome 7q is associated with hepatosplenic T-cell lymphoma (Figure 42.4), t(14;14)(q11;q32) (*TCL1A-TCL1B*) is reported in T-prolymphocytic leukemia (Figure 42.5),

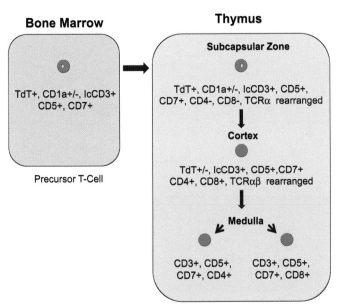

FIGURE 42.1 Scheme of T-cell differentiation.

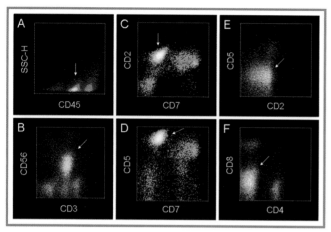

FIGURE 42.2 ABERRANCIES OF MATURE T-CELL NEOPLASMS DETECTED BY FLOW CYTOMETRY. Identification of distinct T-cell aberrancies by MFC is important in evaluating mature T-cell neoplasms. In general, various aberrancies can be detected, and here are some examples of abnormal mature T-cells. (arrows) (A) Dim expression of CD45; (B) Coexpression of dim CD3 and CD56; (C) Bright CD2 and complete loss of CD7; (D) Bright CD5 and complete loss of CD7; (E) Partial dim expression of CD2 and loss of CD5; (F) Partial dim expression of CD8.

and a garden variety of translocation of the *ALK* gene on chromosome 2p23 is observed with other genes in anaplastic large cell lymphoma (Figure 42.6). Clonal integration of HTLV-1 is detected in adult T-cell leukemia/lymphoma. In T-cell malignancies, often the karyotype can be very complex involving multiple genomic regions (Figure 42.7).

NK-cells

Natural killer (NK) cells are a subtype of T-cells and play a critical role in host defense against invading infectious pathogens and malignant transformation. NK-cells originate in the bone marrow from hematopoietic progenitor cells and migrate to the peripheral blood, spleen, lymph nodes, and other tissues. They are characterized by expressing CD56 with or without CD16. They do not develop a complete TCR/CD3 complex receptor, and thus do not express surface CD3. But activated NK-cells express the cytoplasmic ε and ζ chains of CD3.

Two types of NK-cells are recognized: (1) $CD56^{bright}CD16^-$ and (2) $CD56^{dim}CD16^+$. The $CD56^{bright}CD16^-$ subset is primarily concentrated in lymphoid tissue, secretes cytokines to help coordinate adaptive immunity, and is the major subtype recruited to sites of inflammation and malignancies. In general, neoplastic proliferation of this subtype leads to an aggressive clinical course. The $CD56^{dim}CD16^+$ subset constitutes about 90% of NK-cells, primarily circulates in peripheral blood, and demonstrates potent cytotoxicity. Colonal proliferation of this subtype is usually associated with a chronic indolent clinical course.

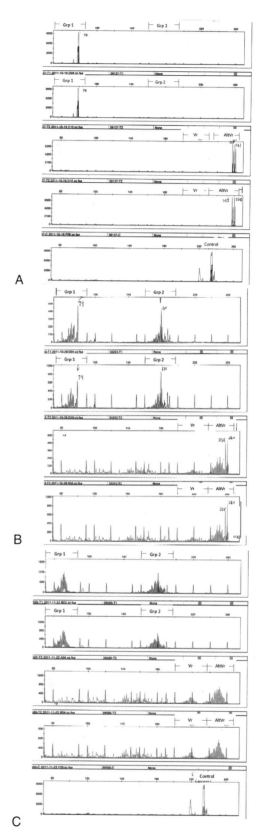

FIGURE 42.3 Time-course of T-cell receptor-γ clonality studies by PCR in a patient recently diagnosed and then treated for T-cell leukemia. (A) PCR profile at time of initial diagnosis, showing only clonal peaks. (B) Same study on a specimen from 2 weeks later, still showing the clonal signals but now with a background of recovering polyclonal (non-malignant) cells. (C) Results 1 month later, now showing only a polyclonal pattern.

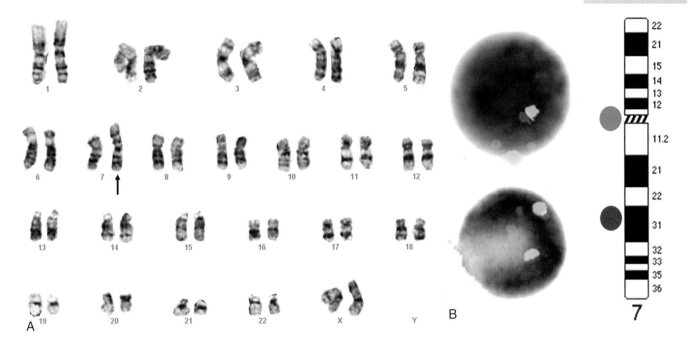

FIGURE 42.4 (A) Karyotype and (B) FISH study of tumor cells in a patient with hepatosplenic lymphoma demonstrating 46,XX,iso(7)(q10).

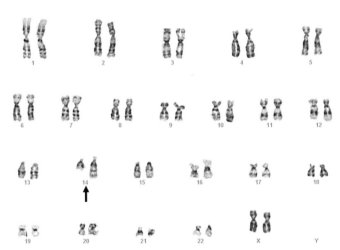

FIGURE 42.5 Karyotype of tumor cells in a patient with T-prolymphocytic leukemia, demonstrating t(14;14)(q11;32).

There is evidence suggesting that CD56^{bright} NK-cells may differentiate into CD56^{dim} NK-cells under certain conditions. NK-cells account for 5–25% of peripheral blood lymphocytes.

Morphologically, NK-cells are considered a variant of large granular lymphocytes—medium to large cells with abundant cytoplasm and azurophilic cytoplasmic granules. Large granular lymphocytes (LGL) are of two major types: (1) NK-type and (2) T-type (Figure 42.8). The neoplasms of NK-cells do not express surface CD3, are often positive for CD56, and show germline *TCR*, whereas, tumors of T-LGL express surface CD3, are often CD56−, but express CD16 and CD57 (Figure 42.9), and show *TCR* gene rearrangement.

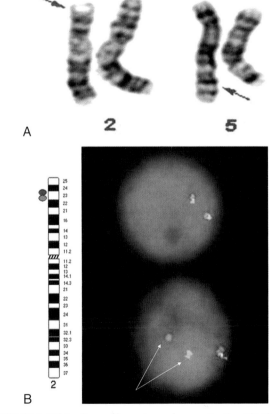

FIGURE 42.6 (A) Section of karyotype demonstrating t(2;5) (p32;q35) in a patient with anaplastic large cell lymphoma. (B) FISH with ALK (2p23) dual-color, breakapart rearrangement probe shows split signals.

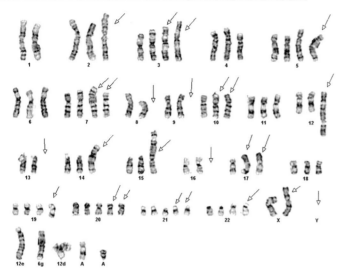

FIGURE 42.7 Complex karyotype abnormalities observed in a patient with peripheral T-cell lymphoma.

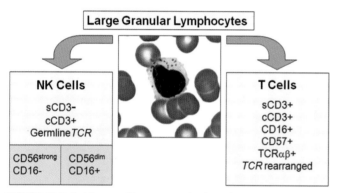

FIGURE 42.8 Subtypes of large granular lymphocytes.

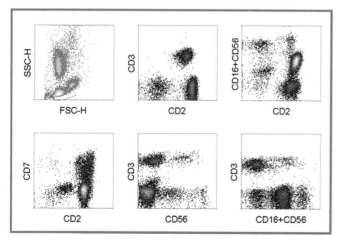

FIGURE 42.9 **MATURE NK-CELL NEOPLASM DETECTED BY FLOW CYTOMETRY.** Open scatter gate (in blue) highlights a neoplastic NK-cell enriched population, which is defined by CD2 backgating. The neoplastic NK-cells are positive for CD2 (bright), CD16 (dim), and CD45 (not shown), but negative for CD3, CD7, and CD56. In addition (not shown), they are negative for CD4, CD5, and CD8.

NK-cell neoplasms show germline *TCR*, but may demonstrate clonal cytogenetic abnormalities, such as deletion of 6q, 7q, 7p, or 11q. Clonal integration of EBV is frequent.

Classification

Mature T- and NK-cell neoplasms represent a wide spectrum of lymphoid malignancies developed from clonal proliferation of mature T- and NK-cells. These disorders may involve bone marrow and peripheral blood (leukemia), lymphoid or extramedullary tissues (lymphoma), or both. Mature T- and NK-cell neoplasms comprise 15% of all lymphoid tumors. They are divided into three major clinicopathological groups:

- **Leukemic or disseminated**, such as T-cell prolymphocytic leukemia, T-cell large granular lymphocyte leukemia, chronic lymphoproliferative disorders of NK-cells, aggressive NK-cell leukemia, and adult T-cell lymphoma/leukemia;
- **Nodal**, such as peripheral T-cell lymphoma, NOS, angioimmunoblastic T-cell lymphoma, and anaplastic large cell lymphoma;
- **Extranodal**, such as hepatosplenic T-cell lymphoma, extranodal NK/T-cell lymphoma of nasal type, enteropathy-type T-cell lymphoma, mycosis fungoides/Sézary syndrome, primary cutaneous anaplastic large cell lymphoma, subcutaneous panniculitis-like T-cell lymphoma, and other primary cutaneous lymphomas.

The current WHO classification of mature T- and NK-cell neoplasms is presented in Table 42.1. The neoplsms in order of frequency are: peripheral T-cell lymphoma, NOS (about 30%); angioimmunoblastic T-cell lymphoma (18.5%); ALK+ and ALK– anaplastic large cell lymphoma (12%); extranodal NK/T-cell lymphoma of nasal type (about 10.5%); adult T-cell leukemia/lymphoma (about 9.5%); and enteropathy-type T-cell lymphoma (about 5%).

T- and NK-cell neoplasms show significant geographical variations in their incidence. For example, adult T-cell leukemia/lymphoma is more frequent in Japan and the Caribbean basin, or extranodal NK/T-cell lymphoma, nasal type is much more frequent in Hong Kong than Europe and North America. Enteropathy-associated T-cell lymphoma most commonly is seen in people of Welsh or Irish descent.

Diagnosis of T- and NK-cell neoplasms, similar to the other hematopoietic/lymphoid tumors, requires a multidisciplinary approach based on a combination of morphologic evaluation, immunophenotyping, molecular analysis, and cytogenetic studies.

Table 42.1
Classification of T- and NK-cell Neoplasms

WHO Classification	Predominant Immunophenotype	Molecular/Cytogenetics
T-cell large granular lymphocytic leukemia	CD3+, CD8+, TCRαβ++, CD16+, CD57+	Rearranged *TCR*
Chronic lymphoproliferative disorders of NK-cells	sCD3−, cCD3+, CD8±, CD16+, CD56±, CD57±	None
Aggressive NK-cell leukemia	CD2+, sCD3−, cCD3+, CD56+, CD16±	del(6)(q21q25)
Extranodal NK/T-cell lymphoma, nasal type	CD2+, sCD3−, cCD3+, CD56+, CP+[1], EBV+	del(6)(q21q25), i(6)(p10)
T-cell prolymphocytic leukemia	CD3+, CD4+ or CD4+/CD8+, CD7+, TCRαβ++, TCL1+	Rearranged *TCR*, t(14;14)(q11;q32), del(11q)
Adult T-cell leukemia/lymphoma	CD2+, CD3+, CD4+, CD5+, CD7−, CD25+, FOXP3+	Rearranged *TCR*, HTLV-1
Hepatosplenic T-cell lymphoma	CD3+, TCRγδ+, CD4−, CD5−, CD8±, CD56±, TIA-1±	Rearranged *TCR*, i(7)(q10)
Enteropathy-associated T-cell lymphoma	CD3+, CD7+, CD8−, CP+, CD4−, CD5−, CD103	Rearranged *TCR*, 1q+, 5q+
Mycosis fungoides	CD2+, CD3+, CD4+, CD5+, CD7−	Rearranged *TCR*, *STAT3* activation
Sézary syndrome	CD2+, CD3+, CD4+, CD5+, CD7−, CD26−	Rearranged *TCR*, deletion of 1p, 6q, 10q, 17p, 19
Subcutaneous panniculitis-like T-cell lymphoma	TCRαβ+, CD8+, CP+, CD56−	Rearranged *TCR*
Primary cutaneous CD30+ T-cell lymphoid disorders	CD3+, CD4+, CD30+, CD8−, CP+	Rearranged *TCR*
Primary cutaneous γδ T-cell lymphoma	TCRγδ+, βF1−, CD3+, CD56+, CP+, CD4−, CD5−, CD8−	Rearranged *TCR*
Angioimmunoblastic T-cell lymphoma	CD3+, CD4+, CD7±, CD8−, CD10+, PD1+, CXCL13+	Rearranged *TCR*, 3+, 5+, and X+
Anaplastic large cell lymphoma, ALK-positive	CD30+, ALK+, EMA+, CD3±, CD2±, CD4±, CD5±, CP+	t(2;5)(p23;q35); *ALK-NPM*, other variants
Anaplastic large cell lymphoma, ALK-negative	CD30+, ALK−, CD43+, CD3±, CD2±, CD4±, CD5±, CP+	Rearranged *TCR*, **lack** of t(2;5)(p23;q35)
Peripheral T-cell lymphoma, NOS	CD3+, CD4+, CD5±, CD7±, CD8−, TCRαβ++	Rearranged *TCR*, chromosomal gains and losses
EBV+ T-cell lymphoproliferative disorder of childhood	EBV+, CD2+, CD3+, CD56−, TIA-1+	Rearranged *TCR*

[1] Cytoplasmic proteins.

Additional Resources

Alam R, Gorska M: 3: Lymphocytes, *J Allergy Clin Immunol* 111(2 Suppl):S476–S485, 2003.

Balato A, Unutmaz D, Gaspari AA: Natural killer T cells: an unconventional T-cell subset with diverse effector and regulatory functions, *J Invest Dermatol* 129:1628–1642, 2009.

Chaplin DD: Overview of the immune response, *J Allergy Clin Immunol* 125(2 Suppl):S3–S23, 2010.

de Leval L, Gaulard P: Pathology and biology of peripheral T-cell lymphomas, *Histopathology* 58:49–68, 2011.

Dunleavy K, Piekarz RL, Zain J, et al: New strategies in peripheral T-cell lymphoma: understanding tumor biology and developing novel therapies, *Clin Cancer Res* 16:5608–5617, 2010.

Jaffe ES, Harris NL, Vardiman JW, et al: *Hematopathology*, Philadelphia, 2010, Saunders/Elsevier.

Liang X, Graham DK: Natural killer cell neoplasms, *Cancer* 112:1425–1436, 2008.

Rolink AG, Massa S, Balciunaite G, et al: Early lymphocyte development in bone marrow and thymus, *Swiss Med Wkly* 136:679–683, 2006.

Swerdlow SH, Campo E, Harris NL, et al: WHO classification of tumours of haematopoietic and lymphoid tissues, ed. 4, Lyon, 2008, International Agency for Research on Cancer.

Large Granular Lymphocytic Neoplasms and Related Disorders

The large granular lymphocytes (LGLs) account for 5–25% of the peripheral blood lymphocytes and are characterized by abundant cytoplasm with azurophilic granules (Figure 43.1). The azurophilic granules contain cytolytic components, such as perforin and granzymes. Perforin is a cytolytic protein that induces apoptosis by creating pores in the plasma membrane of the target cells. Granzymes are proteases that induce apoptosis in virus-infected cells.

The LGLs are divided into two major categories: T-cell and NK-cell. The T-LGL cells typically express CD3, CD8, CD16, CD57, and show *TCR* gene rearrangements, whereas the NK-cells express CD56, are negative for surface CD3, may express CD8 and CD16, and do not show *TCR* gene rearrangement (see Figure 42.8 in Chapter 42). The neoplastic LGL disorders are characterized by persistent (>6 months) large granular lymphocytosis and/or evidence of infiltration of various organs, such as upper respiratory tract, bone marrow, spleen, and liver.

There are two types of NK-cells:

- The $CD56^{bright}CD16^-$ subset is primarily concentrated in lymphoid tissue, secretes cytokines to help coordinate adaptive immunity, and is the major subtype recruited to sites of inflammation and malignancies. In general, neoplastic proliferation of this subtype leads to an aggressive clinical course.
- The $CD56^{dim}CD16^+$ subset comprises about 90% of the NK cells, primarily circulates in peripheral blood and demonstrates potent cytotoxicity. Colonal proliferation of this subtype is usually associated with an indolent clinical course (see Figure 42.8 in Chapter 42).

Reactive (non-clonal) large granular lymphocytosis is a relatively frequent phenomenon and has been observed in a variety of conditions, such as viral infections, collagen vascular disorders, myelodysplastic syndromes (MDS), non-Hodgkin lymphomas, hemophagocytic syndrome, and in patients with solid tumors. However, in some cases of MDS an abnormal population of CD8+ T-LGL cells is detected by flow cytometry, which is clonal by PCR, and may play a critical role in the development of MDS.

According to the WHO, the large granular lymphocytic neoplasms and related disorders are classified as follows:

1. T-cell large granular lymphocytic (T-LGL) leukemia
2. Chronic lymphoproliferative disorders of NK-cells (Provisional Entity)
3. Aggressive NK-cell leukemia
4. Extranodal NK/T-cell lymphoma, nasal type.

T-Cell Large Granular Lymphocytic Leukemia

T-cell large granular lymphocytic (T-LGL) leukemia is a chronic lymphoproliferative disorder characterized by persistent T-cell large granular lymphocytosis (usually >2000/μL), anemia and or neutropenia. There is a strong association with rheumatoid arthritis.

T-LGL leukemia accounts for 2–3% of mature T-cell leukemias and represents the vast majority of leukemias with large granular lymphocytes. The median age is around 60 years with only 10% of patients younger than 40 years. The clinical symptoms are primarily related to the patient's neutropenia and anemia, such as recurrent infections, fever, night sweats, and fatigue. Approximately 30% of patients are asymptomatic at the time of diagnosis. T-LGL leukemia has been frequently observed in association with connective tissue diseases, primarily rheumatoid arthritis.

Rheumatoid factor, antinuclear antibodies, and circulating immune complexes are detected in 40–60% of patients. Rheumatoid arthritis has been reported in about 25% of patients with T-LGL leukemia. Many of these patients present the triad combinations of neutropenia, rheumatoid arthritis, and splenomegaly (Felty's syndrome). Other associated disorders include B-cell lymphoid malignancies, non-Hodgkin lymphoma, thymoma, monoclonal gammopathies, and MDS.

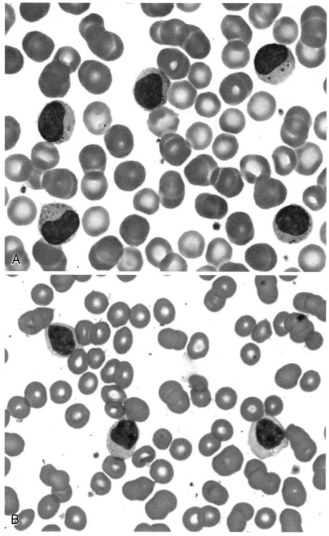

FIGURE 43.1 Large granular lymphocytes. Cytoplasmic granules in LGL cells are usually clearly visible (A) but are sometimes difficult to detect (B).

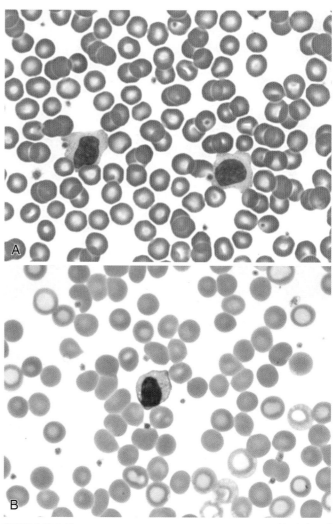

FIGURE 43.2 T-CELL LARGE GRANULAR LYMPHOCYTIC LEUKEMIA. Peripheral blood smear shows large lymphocytes with abundant cytoplasm and variable numbers of azurophilic cytoplasmic granules (A and B).

T-LGL leukemia is considered a chronic and indolent disorder with reported median survival of >10 years. However, patients with severe neutropenia, "B" symptoms, and CD56+ tumor cells have a less favorable prognosis and require treatment. Therapeutic approaches include low-dose chemotherapy with methotrexate, cyclophosphamide, or cyclosporin as single agents or in combination with prednisone.

MORPHOLOGY

- T-LGL leukemia typically shows a persistent (>6 months) absolute large granular lymphocytosis of >2000/µL (Figure 43.2).
- The total lymphocyte count in most patients is modestly elevated (5000–10,000/µL), but in about one-fourth of the cases it is within normal limits. In a minority of patients (about 5%), the absolute LGL count is <1000/µL, or the large lymphocytes lack cytoplasmic azurophilic granules, despite their CD3, CD16 and CD57 coexpression.
- Granulocytopenia is observed in over 80% of patients, with approximately half of the patients demonstrating <500/µL absolute neutrophil counts. Neutropenia is attributed to different possible mechanisms, such as induction of apoptosis in neutrophils by Fas ligand secreted by the leukemic LGL cells, bone marrow infiltration, splenomegaly, or an autoimmune process.
- Anemia is observed in about half of patients, which may be severe and transfusion dependent (about 20%). The possible mechanisms of anemia include an autoimmune process, splenomegaly, bone marrow infiltration, or pure red cell aplasia. T-LGL leukemia is reported as the most common underlying cause of the pure red cell aplasia. The inhibition of erythroid colony-forming units (CFU-E) and burst-forming units (BFU-E) by the T-LGL leukemic cells has been observed in patients with pure red cell aplasia.
- Moderate thrombocytopenia may be present, due to an autoimmune process, bone marrow infiltration, or secondary to splenomegaly.
- The bone marrow is involved in about 90% of cases. The pattern of leukemic infiltration is usually interstitial and/or sinusoidal (Figure 43.3).

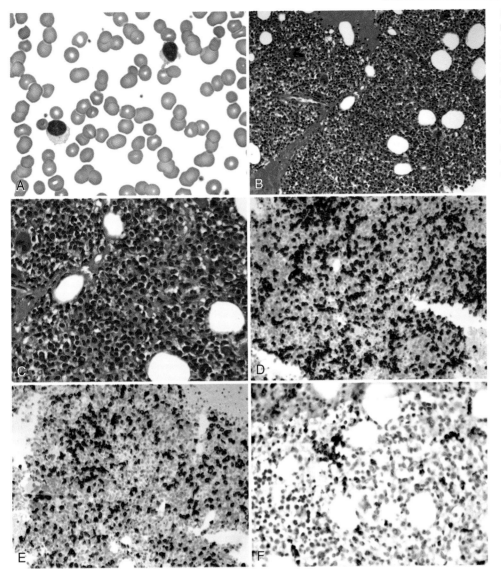

FIGURE 43.3 **T-CELL LARGE GRANULAR LYMPHOCYTIC (T-LGL) LEUKEMIA.** Peripheral blood smear shows large granular lymphocytes (A). Bone marrow biopsy section demonstrates a hypercellular marrow with the presence of scattered lymphocytes and plasma cells (B, low power; C, high power). Immunohistochemical stains reveal small aggregates and scattered cells expressing CD3 (D), CD8 (E) and granzyme B (F).

- Involvement of the spleen is a common feature, with infiltration of the red pulp, often associated with white pulp hyperplasia secondary to the presence of an autoimmune condition (Felty's syndrome).
- The liver infiltration may involve portal and sinusoidal areas. Involvement of the lymph nodes and other organs is unusual.

IMMUNOPHENOTYPE

Multiparametric flow cytometry (MFC) features of T-LGL leukemia include:

- Cytotoxic mature T-cell phenotype in the vast majority of cases, with expression of CD8, surface CD3, and T-cell receptor alpha/beta (Figure 43.4A);
- Pan T-cell aberrancies with variably reduced expression of CD5 and or CD7;
- Expression of CD16 and CD57 in the vast majority of cases;
- CD56 positivity in some cases;
- Rare variants of CD4+ mature T-cell phenotype (Figure 43.4B) or gamma/delta T-cells (Figure 43.4C).

Immunohistochemical studies demonstrate the following.

- Most cases show a mature post-thymic phenotype and typically express CD3, CD8, TCR-αβ, CD16, CD57, CD122 (IL-2 receptor-β).
- The tumor cells are positive for cytotoxic granule markers TIA-1 and granzyme B.
- A minor subtype is characterized by the expression of CD4, CD26, CD56, or TCR-γδ.
- The CD56+ cases are often associated with a more aggressive clinical course.
- Several monoclonal antibodies are raised against the TCR variable domain and are available for immunophenotypic analysis of TCR. Of these, monoclonal antibodies against the Vβ 13.1 region have been reported to be highly associated with T-LGL leukemia.

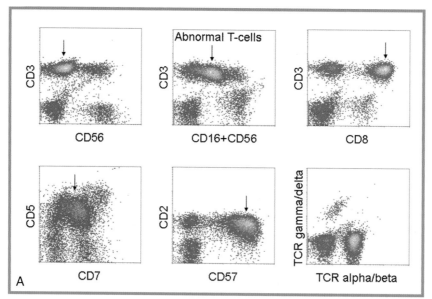

FIGURE 43.4A **FLOW CYTOMETRIC FINDINGS OF T-LGL LEUKEMIA, CD8 POSITIVE.** Lymphocyte-enriched gate of peripheral blood reveals mostly abnormal T-cells (arrows). Compared with the background normal T-cells, the neoplastic T-cells display aberrant phenotype with positivity of CD2 (dim), CD3, CD5 (dim), CD7 (partial and dim), CD8, CD16 (dim), CD56 (small subset), CD57, and expression of TCR alpha/beta.

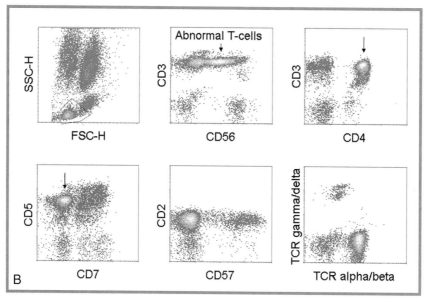

FIGURE 43.4B **FLOW CYTOMETRIC FINDINGS OF T-LGL LEUKEMIA, CD4 POSITIVE.** Open scatter gate (in purple) of peripheral blood highlights T-cell enriched population (circled), which is verified by CD3 backgating. Most of the T-cells are abnormal (arrows), expressing CD2, CD3, CD4, CD5 (dim), CD56 (dim), TCR alpha/beta, and show complete loss of CD7. In addition to negativity of CD57, the abnormal T-cells are negative for CD8, CD16, and CD25 (not shown).

MOLECULAR AND CYTOGENETIC STUDIES

- The T-LGL leukemia cells show TCR-β and/or TCR-γ chain rearrangement by Southern blot or PCR. TCR-β probes are used for Southern blot analysis (Figure 43.5), whereas the TCR-γ region is most often used as the primary PCR target (Figures 43.6 and 43.7) because it is much smaller and thus easier to amplify with just a few primer sets.

- Most T-LGL leukemia cases will manifest a normal karyotype. Fewer than 10% of cases will exhibit karyotypic abnormalities, which include trisomies of chromosomes 3, 8, and 14, deletions of chromosomes 6 and 5q, and 14q (Figure 43.8).

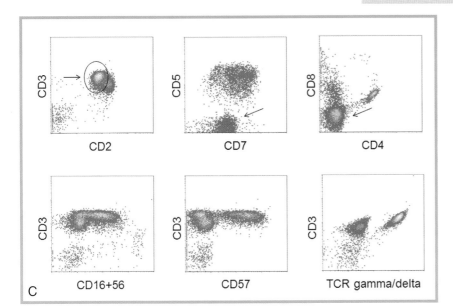

FIGURE 43.4C **FLOW CYTOMETRIC FINDINGS OF T-LGL LEUKEMIA WITH EXPRESSION OF GAMMA/DELTA.** Most of the T-cells are abnormal (circle and arrows), expressing CD2 (dim), CD3 (bright), CD7 (dim), CD16 + 56 (partial), CD57 (major subset), and TCR gamma/delta. They are negative for CD4, CD5, and CD8.

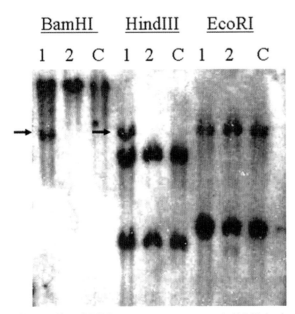

FIGURE 43.5 Clonal *TCRB* gene rearrangement in T-LGL leukemia detected by Southern blotting using the T C β probe. Non-germline, rearranged hybridization bands are seen with two of the three restriction enzymes used (BamHI, HindIII, arrows), which is sufficient to diagnose T-cell clonality.

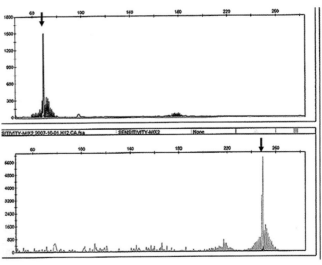

FIGURE 43.6 Clonal *TCRG* gene rearrangement in T-LGL leukemia. Discrete clonal PCR peaks are demonstrated above the polyclonal background smear in two of the four *TCRG* gene regions targeted (Group 1 and AltVγ, arrows).

Chronic Lymphoproliferative Disorder of NK Cells

Chronic lymphoproliferative disorder is characterized by a persistent NK-cell lymphocytosis (>2000/μL, >6 months) with an indolent clinical course. In the vast majority of patients, there are no clinical symptoms and NK-lymphocytosis is an incidental finding. However, anemia, neutropenia, peripheral neuropathy, nephrotic

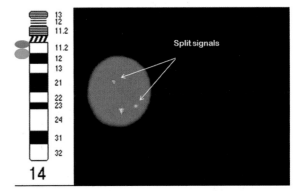

FIGURE 43.7 T-cell receptor gene rearrangement confirmed by a dual-color TCR-α/δ FISH probe at 14q11.2 in an interphase nucleus.

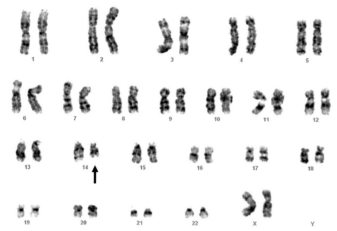

FIGURE 43.8 Karyotype of tumor cells in a patient with T-cell large granular leukemia showing an interstitial deletion of 14q.

syndrome, and/or splenomegaly have been observed in rare instances. The neoplastic nature of this disorder is uncertain.

Asymptomatic patients are followed with no specific medication, and the minority of symptomatic patients receive immunosuppressive therapy. In rare instances, transformation to an aggressive phase occurs, but such cases lack EBV association.

MORPHOLOGY

Persistent large granular lymphocytosis (>2000/μL). The large granular lymphocytes show abundant cytoplasm with azurophilic granules (Figure 43.9).

IMMUNOPHENOTYPE

MFC studies (Figure 43.10) demonstrate:

- Absent surface CD3 expression;
- Expression of CD16 and common expression of dim CD56;
- Aberrant expression profiles of CD2, CD7, and CD57, compared with normal NK-cells.

In addition to the above features, immunohistochemical studies are positive for cytotoxic granule markers TIA1 and granzyme B.

MOLECULAR AND CYTOGENETIC STUDIES

- Vast majority of the cases show normal karyotype.
- Gene rearrangement studies for *TCR* and *IGH@* are negative.
- EBV is negative.
- X-chromosome inactivation studies may suggest clonality in female patients.

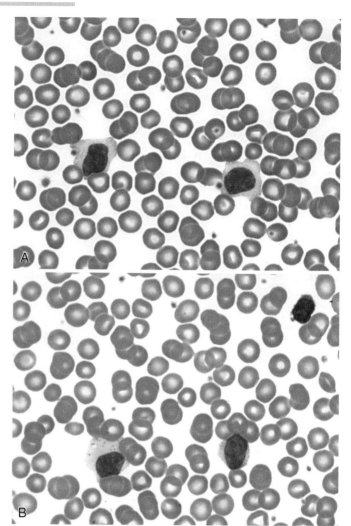

FIGURE 43.9 **Chronic lymphoproliferative disorder.** (A, B) Peripheral blood smear showing large granular lymphocytes.

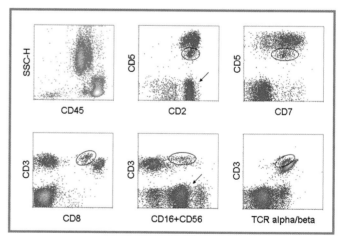

FIGURE 43.10 **Flow cytometric findings of chronic lymphoproliferative disorder of NK-cells with concurrent abnormal T-cell subset.** Lymphocyte-enriched gate (in magenta) of peripheral blood shows mostly neoplastic NK-cells (arrows), positive for CD2, CD16+56 (dim), and CD45; but negative for CD7, CD8, and surface CD3. In addition to the neoplastic NK-cells, a small subset (circles) of the T-cells is abnormal, expressing CD2, CD3 (bright), CD5 (dim), CD7 (dim), CD8 (dim), CD16/CD56 (dim), and TCR alpha/beta (dim). Molecular studies have revealed clonal T-cell gene rearrangement several times over the years, which led to the misinterpretation of T-LGL leukemia in the past, overlooking the neoplastic NK-cells.

Aggressive NK-Cell Leukemia

Aggressive NK-cell leukemia is characterized by systemic proliferation of LGLs of NK type (NK-LGL), strong Epstein–Barr virus (EBV) association, and an aggressive clinical course.

It affects younger individuals with a median age of about 40 years. The disease presents with fever, night sweats, weight loss, anemia or pancytopenia, and often massive hepatosplenomegaly. Multiorgan failure is the major cause of death, with a survival of <1 year in most instances. Chemotherapy has not been effective.

MORPHOLOGY

- The peripheral blood shows an absolute lymphocytosis (usually >10,000/μL) with increased proportion of LGLs. These cells have abundant light blue cytoplasm containing azurophilic granules (Figure 43.11). The nuclei are round, oval, or irregular and may appear pleomorphic or hyperchromatic. The nuclear chromatin is condensed and nucleoli are indistinct. The amount and size of the azurophilic granules are variable. According to some observers, these cells are slightly larger than the normal LGLs seen in the peripheral blood.
- Anemia is common and usually severe, and thrombocytopenia is frequent. In contrast to T-LGL leukemia, severe neutropenia is less common.
- Bone marrow is almost always infiltrated by the neoplastic cells. The involvement is diffuse, focal, interstitial, or sinusoidal (Figure 43.12). The infiltrating lymphoid cells may be mixed with normal hematopoietic cells and are sometimes difficult to detect. The neoplastic cells may appear pleomorphic with large immature forms (Figure 43.13). Scattered reactive histiocytes may be present, some showing hemophagocytosis.
- The extramedullary infiltrations may mimic *extranodal NK/T-cell lymphoma of the nasal type* (see below) and often show vascular involvement (angiocentric pattern) and areas of necrosis.
- Most patients show splenic and hepatic involvement, often with massive hepatosplenomegaly. The leukemic infiltration in the spleen involves the red pulp.
- Other frequently involved organs include the liver, gastrointestinal tract, and lymph nodes. The cerebrospinal and peritoneal fluids may be involved.

IMMUNOPHENOTYPE

MFC findings (Figure 43.14) are:

- Negative for surface CD3;
- Positive for CD2 and CD56;
- Commonly positive for CD16 and occasionally positive for CD11b, but often negative for CD16 and CD57.

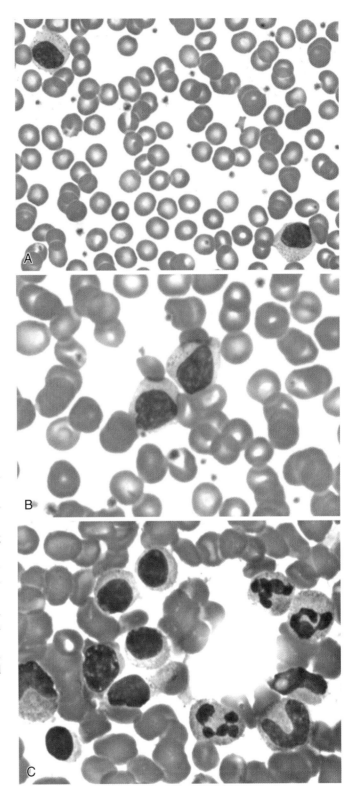

FIGURE 43.11 **AGGRESSIVE NK-CELL LEUKEMIA.** Peripheral blood (A, B) and bone marrow (C) smears demonstrating large granular lymphocytes.

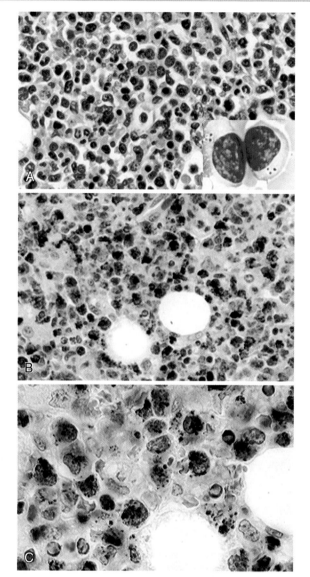

FIGURE 43.12 BONE MARROW INVOLVEMENT IN A PATIENT WITH AGGRESSIVE NK-CELL LEUKEMIA. An interstitial atypical lymphoid infiltrate is noted (A). The tumor cells are highlighted by a TIA-1 histochemical stain (B and C). The inset of (A) shows two LGL cells by touch preparation.
Adapted from Naeim F. Pathology of Bone Marrow, 2nd edn. Williams & Wilkins, Baltimore, 1997, by permission.

- Evidence of EBV infection in clonal episomal form has been observed in the majority of cases, and EBNA-1 and EBER-1 can be detected by in situ hybridization techniques.
- The non-random cytogenetic abnormalities noted include: gains of chromosomes 1p, 6p, 8, 11q, 12q, 17q, 19p, 20q, and Xp, losses of 6q (mostly q21→q25 by FISH), 7p, 11q, 13q, and 17p (Figure 43.16).
- A possible involvement of 8p22–p23, due to translocations involving 8p23, has been reported in NK-cell neoplasms, with the partner chromosomes being 8q13, 17q24, and 1q10.

Extranodal NK/T-Cell Lymphoma, Nasal Type

Extranodal NK/T-cell lymphoma, nasal type is characterized by strong EBV association and angiocentric infiltration, leading to vascular destruction and necrosis. The nasopharynx is the most common site of involvement, but other extranodal organs such as skin, gastrointestinal tract, testis, and soft tissues may be affected.

Extranodal NK/T-cell lymphomas are more prevalent in Asia and Central and South America, and predominantly affect males in the fifth decade of their age. They are clinically divided into two categories: nasal and non-nasal.

- The nasal NK/T-cell lymphomas occur in the nose and the upper respiratory and oral cavities, including nasopharynx, paranasal sinuses, tonsils, and larynx. Symptoms are local and may include nasal obstruction, bleeding, or destruction and perforation of the hard palate. These tumors are locally invasive but infrequently show distant metastasis. Less than 10% of patients demonstrate bone marrow involvement.
- The non-nasal NK/T-cell lymphomas are often multifocal, and dissemination occurs in the early stage of the disease. The primary sites of involvement include the skin, digestive system, spleen, and testis. Nodal involvement is rare.

In general, the non-nasal tumors are more aggressive than the nasal types and respond poorly to therapy. The therapeutic approaches include radiation and combination chemotherapy. The reported complete remission rate for early stages of the disease ranges from 60–80% for radiotherapy and 40–60% for chemotherapy, but about 50% of the patients may show relapse within the first year.

MORPHOLOGY

- The lymphomatous infiltrate is polymorphic with a diffuse or patchy involvement and an angiocentric and angiodestructive growth in over 85% of cases (Figures 43.17 to 43.19). The ulceration of the overlying epithelium is common and may be associated with atypical hyperplasia of the adjacent epithelium, mimicking squamous cell carcinoma.
- The neoplasm consists of a mixture of small to large atypical lymphoid cells with variable amount of cytoplasm and irregular nuclei. The predominant infiltrating lymphoid cells may

In addition to the expression of the above phenotypic features, immunohistochemical studies show positivity of cytotoxic granule markers TIA1 and granzyme B (Figure 43.15). EBV-EBER is positive by in-situ hybridization studies.

MOLECULAR AND CYTOGENETIC STUDIES

- As NK-cells develop from progenitor lymphocytes and do not express either immunoglobulins or T-cell receptors, this lesion does not demonstrate clonal T- or B-cell gene rearrangements.

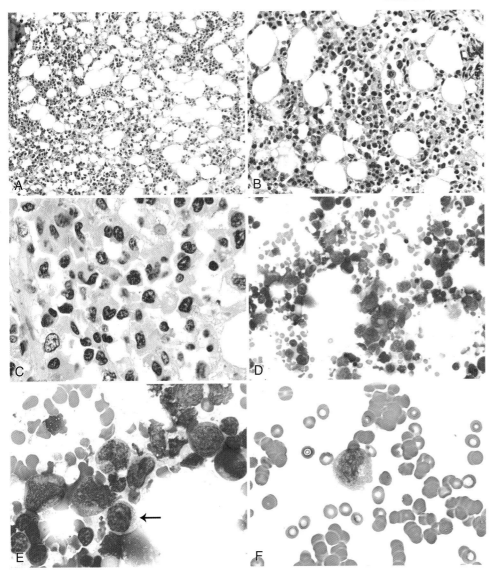

FIGURE 43.13 **AGGRESSIVE NK-CELL LEUKEMIA.** Bone marrow biopsy section showing an interstitial and sinusoidal infiltration by large, pleomorphic cells (A, low power; B, intermediate power; C, high power). Bone marrow (D, intermediate power; E, high power) and blood (F, high power) smears demonstrate large, pleomorphic cells with prominent nucleoli and cytoplasmic granules.

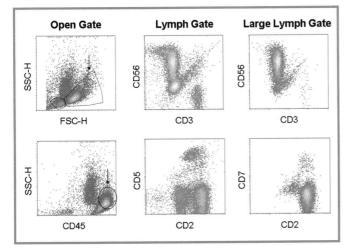

FIGURE 43.14 **FLOW CYTOMETRIC FINDINGS OF AGGRESSIVE NK-CELL LEUKEMIA.** Open gate density displays (in red) of bone marrow aspirate demonstrate a prominent population of neoplastic NK-cells (arrows) that are intermediate-sized to large with increased side scatter and expression of bright CD45. These characteristics are verified by CD56 backgating. The neoplastic NK-cells are positive for CD2 (bright) and CD56 (heterogeneous), but negative for surface CD3 as well as CD5 and CD7.

be large or small. The larger cells may show prominent nucleoli. The lymphomatous infiltrate is often heavily admixed with inflammatory cells, such as lymphocytes, plasma cells, eosinophils, and histiocytes.

- Focal or confluent coagulative necrosis is common, often with the presence of apoptotic bodies. Some of the tumor cells, particularly in cytologic preparations (such as touch preparation), may show cytoplasmic azurophilic granules.

IMMUNOPHENOTYPE

MFC findings (Figure 43.20) are:

- Negative for surface CD3;
- Positive for CD2 and CD56 in most cases, with occasional positivity for CD7;
- Negative for CD16 and CD57.

In addition to the expression of above-mentioned markers, immunohistochemical studies are positive for cytotoxic granule markers TIA1, granzyme B, and perforin.

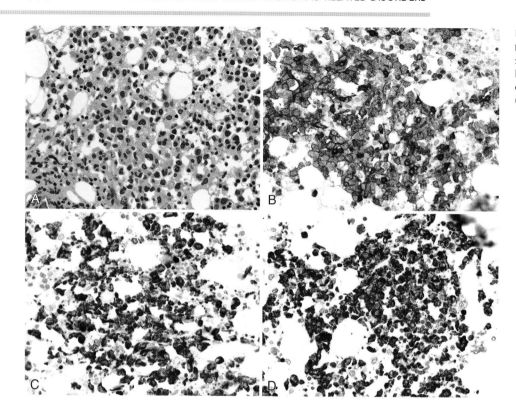

FIGURE 43.15 AGGRESSIVE NK-CELL LEUKEMIA. Bone marrow biopsy section showing infiltration by large, pleomorphic cells (A) and expressing CD2 (B), granzyme B (C), and TIA-1 (D).

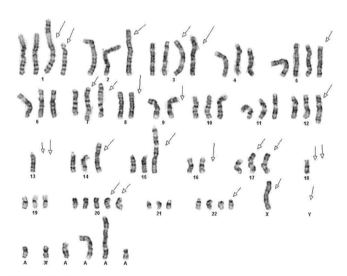

FIGURE 43.16 A karyotype of an aggressive NK-cell leukemia exhibiting gains and losses of several chromosomes, in addition to the many structural aberrations that are commonly observed (arrows).

EBV is positive in all cases, and EBV-EBER is the stain of choice.

MOLECULAR AND CYTOGENETIC STUDIES

- The neoplastic cells do not demonstrate *TCR* gene rearrangements.
- Evidence of EBV infection in clonal episomal form has been observed in the vast majority of cases, and EBNA-1 and EBER-1 can be detected by in situ hybridization techniques. Although the in situ hybridization technique will demonstrate the presence of EBV RNA in essentially all of the tumor cells, proof of clonality, in the absence of clonal *TCR* gene rearrangements, requires examination of Southern blot patterns (fingerprints) of the EBV genome. PCR approaches should be used with caution, given the high frequency of latent EBV infection in the general population.
- Array-CGH studies show that recurrent regions characteristic of the extranodal NK/T lymphoma, nasal-type group, were gains of 2q, and loss of 1p, 2p, 4q, 5p, 5q, 6q, and 11q (Figure 43.21).

EBV-Associated T/NK-Cell Lymphoproliferative Disorder of Children and Young Adults

EBV-associated T/NK-cell lymphoproliferative disorder (EBV-T/NK LPD) of children and young adults is a rare disorder associated with high morbidity and mortality, which is more prevalent in East Asians and Native Americans from Central and South America. The disease has a wide spectrum of clinical presentation from indolent to aggressive course, and has been subdivided by the chronic active EBV infection (CAEBV) Study Group (2008) into the following categories

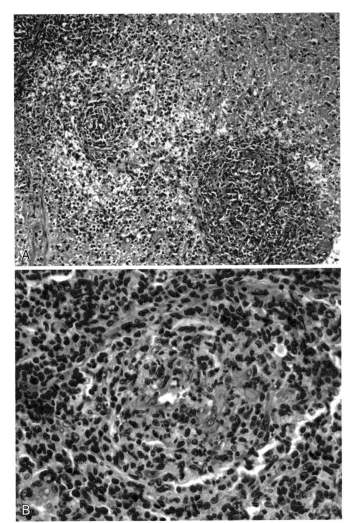

FIGURE 43.17 EXTRANODAL NK/T-CELL LYMPHOMA, NASAL TYPE. This is characterized by angiocentric and angiodestructive infiltration (A, low power; B, high power). The vascular infiltration involves all layers of the vascular wall.

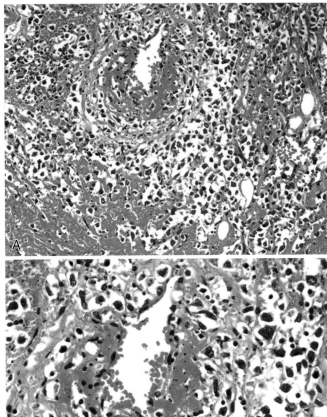

FIGURE 43.18 EXTRANODAL NK/T-CELL LYMPHOMA, NASAL TYPE. This is characterized by an angiodestructive process leading to obstruction and necrosis (A, low power; B, high power).

- Category A1: polymorphic LPD without clonal proliferation of EBV-infected cells;
- Category A2: polymorphic LPD with clonality;
- Category A3: monomorphic LPD (T-cell or NK-cell lymphoma/leukemia) with clonality;
- Category B: monomorphic LPD (T-cell lymphoma) with clonality and fulminant course.

Categories A1, A2, and A3 appear to represent a continuous spectrum and are considered equivalent to CAEBV. Category B is the exact equivalent of infantile fulminant EBV-associated T-LPD, or, according to the WHO classification (2008), systemic EBV-positive T-cell lymphoproliferative disease of childhood.

Systemic EBV-positive T-cell lymphoproliferative disease of childhood is a rare, rapidly progressive disease with sepsis and multisystem failure leading to death. It has clinicopathologic overlapping features with aggressive NK-cell leukemia.

The neoplastic T-cells in the involved organs, such as liver, spleen, or lymph nodes, show variable morphologic features from case to case, ranging from small round mature lymphocytes to pleomorphic medium to large cells (Figure 43.22). Hemophagocytosis is often a prominent feature. The tumor cells are CD2+, CD3+, CD8+, TIA1+, EBV-EBER+ and CD56−. Rare cases may show coexpression of CD4 and CD8. The *TCR* gene is rearranged, and the neoplastic cells carry type A EBV.

Differential Diagnosis

The clinicopathologic features of T-LGL leukemia, chronic lymphoproliferative disorder of NK cells, aggressive NK-cell leukemia, and extranodal NK/T-cell lymphoma are

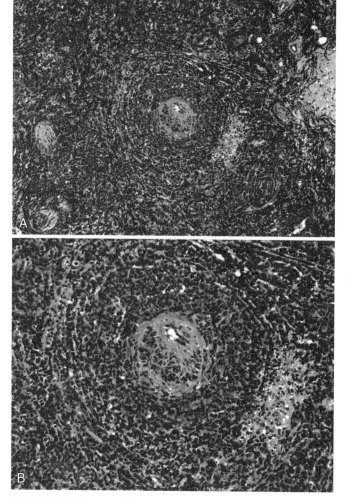

FIGURE 43.19 Extranodal NK/T-cell lymphoma, nasal type is characterized by angiocentric and angiodestructive infiltration (A, low power; B, intermediate power; C, high power).

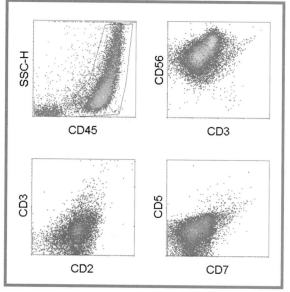

FIGURE 43.20 FLOW CYTOMETRIC FINDINGS OF EXTRANODAL NK/T-CELL LYMPHOMA, NASAL TYPE. Open gate display (in blue) of nasal tissue by CD45 gating demonstrates a prominent cell population that is positive for CD45 (moderate to bright) with highly variable side scatter. These cells are neoplastic NK-cells, expressing CD2 and CD56 but negative for surface CD3 plus CD5 and CD7.

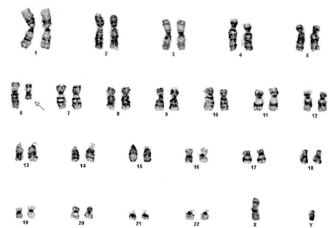

FIGURE 43.21 Karyotype of tumor cells in a patient with extranodal NK/T-cell lymphoma, nasal type demonstrating 46,XY,del(6)(q21).

summarized in Table 43.1. T-LGL leukemia, chronic lymphproliferative disorder of NK cells, and aggressive NK-cell leukemia should be distinguished from secondary, reactive large granular lymphocytosis and NK cell lymphocytosis. The reactive T-LGL expansions have been reported in autoimmune connective tissue disorders, inflammatory skin disorders, lymphomas, hemophagocytic syndrome, and a variety of viral infections, such as EBV, HIV, and CMV. Reactive large granular lymphocytosis is polyclonal and is not associated with chromosomal aberrations.

Increased number of circulating NK-cells has been observed in patients with viral infection, solid tumors, MDS, and atomic bomb survivors.

The differential diagnosis for extranodal NK/T-cell lymphomas of nasal type includes CD56+ T-cell lymphomas, such as hepatosplenic T-cell lymphoma and other peripheral T-cell lymphomas. The CD56+ T-cell lymphomas are negative for EBV, express surface CD3 and TCR, and show evidence of *TCR* gene rearrangement.

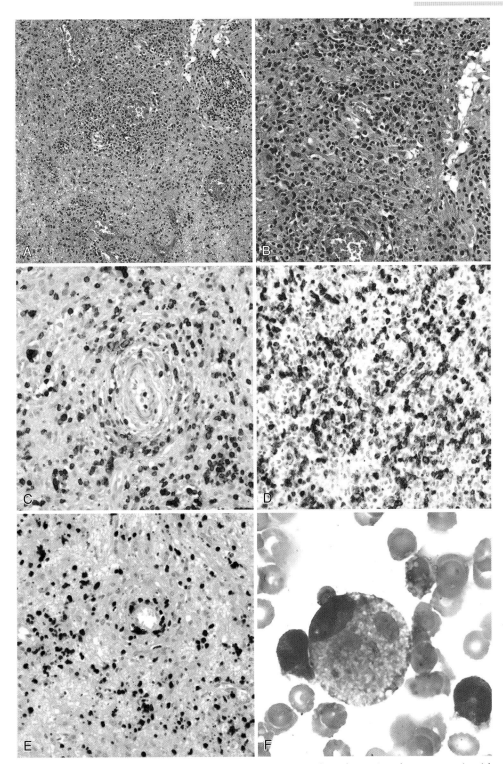

FIGURE 43.22 SYSTEMIC EBV+ T-CELL LYMPHOPROLIFERATIVE DISEASE OF CHILDHOOD. Lymph node section shows necrosis with a perivascular cellular infiltrate (A). Atypical lymphoid cells are present, and there is abundant apoptotic debris (B). Atypical lymphocytes are CD3+ (C), CD8+ (D), and EBER+ (E). Hemophagocytic histiocytes are present in the bone marrow aspirate (F).
Adapted from Jaffe ES, Harris NL, Vardiman JW, et al. Hematopathology. Saunders/Elsevier, Philadelphia, 2010, by permission.

Table 43.1
Clinicopathologic Features of T-LGL Leukemia, CLNK, ANKCL, and ENK/T

Features	T-LGL Leukemia	CLNK	ANKCL	ENK/T
Median age (years)	60	60	40	50
Male:Female	1	1	1	>1
Association	Rheumatoid arthritis	None	EBV	EBV
Immunophenotype	sCD3+, CD8+, TCR+, CD16+, CD57+	cCD3ε+, CD16+, TCR−, CD56$^{dim/-}$	cCD3ε+, CD56^{strong}, TCR−, CD16±	cCD3ε+, CD56^{strong}, TCR−, CD16−
Prognosis	Indolent	Indolent	Aggressive	Aggressive

ANKCL, aggressive NK-cell leukemia; CLNK, chronic lymphoproliferative disorder of NK cells; ENK/T, extranodal NK/T-cell lymphoma, nasal type; T-LGL, T-cell large granular lymphocytic.

NK-cell enteropathy is an entity reported in rare cases with a benign NK-cell lymphoproliferative disorder mimicking intestinal lymphoma. The patients show vague gastrointestinal symptoms with lesions involving stomach, duodenum, small intestine, and colon. Biopsies reveal a mucosal infiltrate of atypical cells with an NK-cell phenotype (CD56+/TIA-1+/Granzyme B+/cCD3+), which do not invade the glandular epithelium. T-cell receptor-γ gene rearrangement and EBV-encoded RNA in situ hybridization are negative. The patients do not develop progressive disease and do not need aggressive chemotherapy.

Additional Resources

Al-Hakeem DA, Fedele S, Carlos R, et al: Extranodal NK/T-cell lymphoma, nasal type, *Oral Oncol* 43:4-14, 2007.

Chen YH, Chadburn A, Evens AM, et al: Clinical, morphologic, immunophenotypic, and molecular cytogenetic assessment of CD4-/CD8-γδ T-cell large granular lymphocytic leukemia, *Am J Clin Pathol* 136:289-299, 2011.

Dearden C: Large granular lymphocytic leukaemia pathogenesis and management, *Br J Haematol* 152:273-283, 2011.

Fortune AF, Kelly K, Sargent J, et al: Large granular lymphocyte leukemia: natural history and response to treatment, *Leuk Lymphoma* 51:839-845, 2010.

Gualco G, Domeny-Duarte P, Chioato L, et al: Clinicopathologic and molecular features of 122 Brazilian cases of nodal and extranodal NK/T-cell lymphoma, nasal type, with EBV subtyping analysis, *Am J Surg Pathol* 35:1195-1203, 2011.

Hervier B, Rimbert M, Maisonneuve H, et al: Large granular lymphocyte leukemia with pure red cell aplasia associated with autoimmune polyendocrinopathy-candidiasis-ectodermal dystrophy: an unfortuitous association? *Int J Immunopathol Pharmacol* 23:947-949, 2010.

Jaffe ES, Harris NL, Vardiman JW, et al: Hematopathology, Philadelphia, 2010, Saunders/Elsevier.

Lamy T, Loughran TP Jr: Clinical features of large granular lymphocyte leukemia, *Semin Hematol* 40:185-195, 2003.

Lamy T, Loughran TP Jr: How I treat LGL leukemia, *Blood* 117:2764-2774, 2011.

Liu X, Loughran TP Jr: The spectrum of large granular lymphocyte leukemia and Felty's syndrome, *Curr Opin Hematol* 18:254-259, 2011.

Mansoor A, Pittaluga S, Beck PL, et al: NK-cell enteropathy: a benign NK-cell lymphoproliferative disease mimicking intestinal lymphoma: clinicopathologic features and follow-up in a unique case series, *Blood* 117:1447-1452, 2011.

Ohgami RS, Ohgami JK, Pereira IT, et al: Refining the diagnosis of T-cell large granular lymphocytic leukemia by combining distinct patterns of antigen expression with T-cell clonality studies, *Leukemia* 25:1439-1443, 2011.

Ohshima K, Kimura H, Yoshino T, et al: CAEBV Study Group. Proposed categorization of pathological states of EBV-associated T/natural killer-cell lymphoproliferative disorder (LPD) in children and young adults: overlap with chronic active EBV infection and infantile fulminant EBV T-LPD, *Pathol Int* 58:209-217, 2008.

Rezk SA, Huang Q: Extranodal NK/T-cell lymphoma, nasal type extensively involving the bone marrow, *Int J Clin Exp Pathol* 4:713-717, 2011.

Rodríguez-Pinilla SM, Barrionuevo C, García J, et al: Epstein-Barr virus-positive systemic NK/T-cell lymphomas in children: report of six cases, *Histopathology* 59:1183-1193, 2011.

Sevilla DW, El-Mallawany NK, Emmons FN, et al: Spectrum of childhood Epstein-Barr virus-associated T-cell proliferations and bone marrow findings, *Pediatr Dev Pathol* 14:28-37, 2011.

Spears MD, Harrington AM, Kroft SH, et al: Immunophenotypic stability of T-cell large granular lymphocytic leukaemia by flow cytometry, *Br J Haematol* 151:97-99, 2010.

Yok-Lam K: The diagnosis and management of extranodal NK/T-cell lymphoma, nasal-type and aggressive NK-cell leukemia, *J Clin Exp Hematop* 51:21-28, 2011.

T-Cell Prolymphocytic Leukemia

T-cell prolymphocytic leukemia (T-PLL) is a sporadic and aggressive lymphoproliferative disorder of post-thymic T-cells characterized by a high peripheral blood lymphocyte count and infiltration of the bone marrow, spleen, liver, lymph nodes, and skin.

T-PLL is a rare T-cell lymphoproliferative disorder often presented with hepatosplenomegaly and generalized lymphadenopathy. Most patients are older than 50 years. The peripheral lymphocyte count is markedly elevated (usually >100,000/μL), commonly associated with anemia and thrombocytopenia. Skin infiltration and serous effusions may be observed. In general, T-PLL has an aggressive clinical course with a median survival of <1 year, though occasional cases have shown spontaneous remission. An indolent form of T-PLL has also been reported with t(3;2)(q21;q11.2) and elevated serum β2-microglobulin. Combination chemotherapy, and more recently treatment with monoclonal anti-CD52 antibodies (CAMPATH-1H, alemtuzumab), and allogeneic stem cell transplantation have been used with some responses.

Morphology

T-PLL refers to a group of mature T-cell leukemias with a diverse morphology but similar clinical outcome. The observed morphologic variations include:

- Cells with typical prolymphocytic features; medium-sized lymphocytes with variable amount of non-granular basophilic cytoplasm, round, oval, or irregular nucleus, coarse chromatin, and a single prominent nucleolus (Figure 44.1). T-prolymphocytes often show cytoplasmic blebs (Figure 44.2). The prolymphocytic morphology accounts for about 70% of cases.
- Small lymphocytes often with irregular nuclei and indistinct nucleolus. This morphologic subtype was previously referred to as T-CLL, but now it is included in T-PLL because of similar biological behavior. It constitutes about 25% of cases.

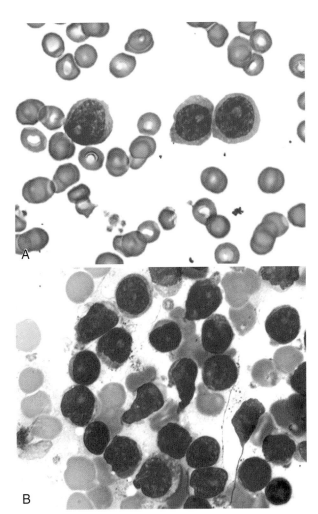

FIGURE 44.1 **T-PROLYMPHOCYTIC LEUKEMIA.** Blood (A) and bone marrow (B) smears demonstrating numerous prolymphocytes.

- Approximately 5% of cases may show lymphoid cells with cerebriform (Sézary-like) nuclei (Figure 44.3).
- There is marked peripheral blood lymphocytosis, usually >100,000/μL, and often with anemia and thrombocytopenia.

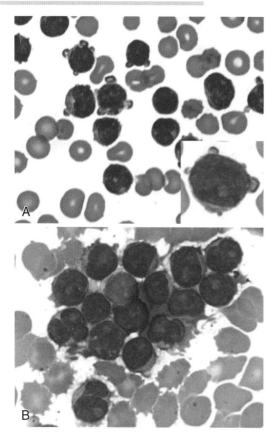

FIGURE 44.2 **T-PROLYMPHOCYTIC LEUKEMIA.** Blood (A) and bone marrow (B) smears demonstrating numerous prolymphocytes with cytoplasmic blebs.

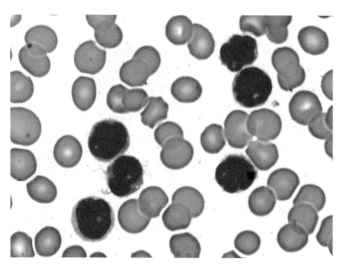

FIGURE 44.3 Peripheral blood smear of a patient with T-prolymphocytic leukemia demonstrating prolymphocytes mimicking Sézary cells.

- The bone marrow is commonly infiltrated in a diffuse or nodular pattern. Splenic infiltration consists of involvement of both white and red pulps (Figure 44.4). The skin is affected in about 20% of cases, with dense infiltration of the dermis without epidermal infiltration. The involved lymph nodes are diffusely infiltrated, primarily in the paracortical areas. The remnants of follicular structures may be present.

Immunophenotype

MFC findings demonstrate:

- Peripheral T-cell phenotype negative for CD34, CD1a or TdT;
- Variably aberrant expression profile of CD2, CD3, CD5, and CD7;
- Expression of CD4 in most cases (Figure 44.5);
- Double CD4/CD8 positivity in about 25% of cases (Figure 44.6), a relatively unique feature observed in T-PLL;
- Expression of CD8 in occasional cases (Figure 44.7).

In addition to the above-mentioned markers, immunohistochemical studies show overexpression of TCL1.

Cytogenetic and Molecular Studies

Analogous to the immunoglobulin (IGH@) receptor loci that are frequently affected by translocations in B-cell lymphomas, the T-cell receptor (*TCR*) gene loci are targeted by chromosomal breakpoints involving various translocation partners.

Classical cytogenetic studies reveal complex karyotypes and some recurrent chromosomal abnormalities, the most frequent being t(14;14)(q11.2;q32) (Figure 44.8), inv(14)(q11.2q32),t(X;14)(q28;q11.2),i(8)(q10) and t(8;8)(p12;q11.2).

T-cell receptor (*TCR*)-α/δ genes located in 14q11.2 are involved in most of the chromosomal abnormalities in cases of T-PLL. The TCL-1 oncoprotein is expressed in approximately 70% of T-PLL patients. T-PLL patients show the translocation t(X;14)(q28;q11.2) (Figure 44.9), which results in rearrangement of the *MTCP-1* gene (a member of the *TCL-1* gene family) located at Xq28.

The i(8)(q10) chromosome is another common finding in T-PLL. The genes identified in the 8q rearrangement breakpoints include *PLEKHA2*, *NBS1*, *NOV* and *MYST3* (Figure 44.10).

The *TCL1* and *MTCP1* loci rearrange with the *TCR*-α/δ chain locus or, less commonly, with the *TCR*-β chain locus on chromosome 7 [t(7;14)(q35;q32.1)], leading to their activation.

Mutations in the ataxia telangiectasia mutated (*ATM*) gene, located in the 11q22-23 chromosomal region have been associated with inactivation or significantly reduced expression of the ATM protein, which is believed to function as a tumor suppressor (Figure 44.10). Aberrations of the *ATM* gene, are known to also play a role in malignant transformation of T-cells in patients with T-PLL.

In the commonly deleted region on chromosome 11q, recurrent microdeletions targeting the microRNA 34b/c and the transcription factors ETS1 and FLI1 have been observed.

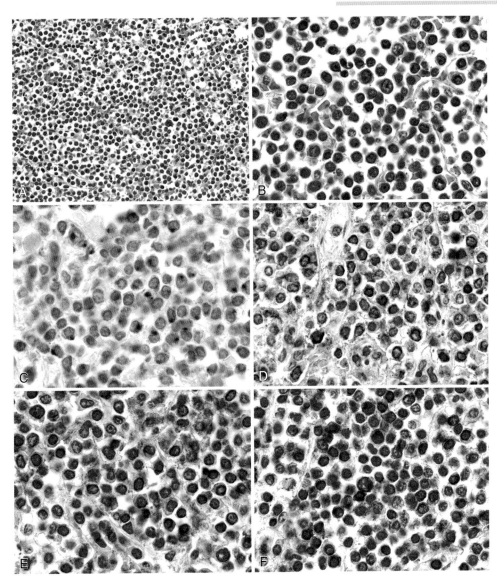

FIGURE 44.4 **SPLENIC INVOLVEMENT IN A PATIENT WITH T-PROLYMPHOCYTIC LEUKEMIA.** There is a diffuse infiltration of the red pulp by the tumor cells (A, low power; B, high power). The tumor cells express CD2 (C), CD3 (D), CD4 (E), and CD7 (F).

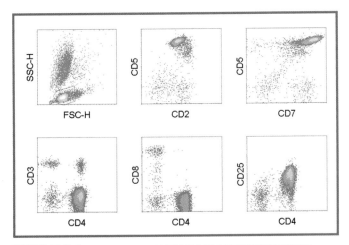

FIGURE 44.5 **FLOW CYTOMETRIC FINDINGS OF T-CELL PROLYMPHOCYTIC LEUKEMIA, CD4 POSITIVE.** Lymphocyte-enriched gate of peripheral blood reveals predominantly abnormal T-cells. Compared with normal T-cells in the background, the neoplastic T-cells display an aberrant phenotype with positivity of CD2 (dim), CD4, CD5 (homogeneously bright), CD7 (bright), and CD25. They are negative for surface CD3, CD8, as well as TdT and TCR alpha/beta or gamma/delta (not shown).

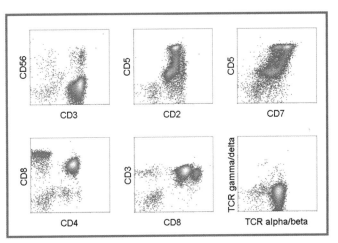

FIGURE 44.6 **FLOW CYTOMETRIC FINDINGS OF T-CELL PROLYMPHOCYTIC LEUKEMIA, CD4/CD8 DOUBLE POSITIVE.** Lymphocyte-enriched gate of peripheral blood demonstrates a distinct and prominent population of double positive T-cells expressing surface CD3, CD4, dim CD8, and TCR alpha/beta. These abnormal T-cells display dim to moderate expression of CD2, plus heterogeneous loss of CD5 and CD7.

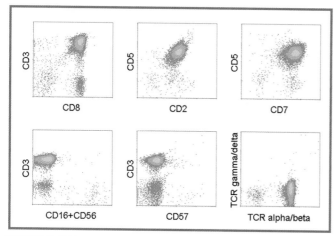

FIGURE 44.7 **FLOW CYTOMETRIC FINDINGS OF T-CELL PROLYMPHOCYTIC LEUKEMIA, CD8 POSITIVE.** Lymphocyte enriched gate of peripheral blood reveals near exclusively abnormal T-cells, which are positive for CD2 (dim), CD3 (major subset), CD5, CD7 (bright), CD8, and TCR alpha/beta. The neoplastic T-cells are negative for CD16/CD56 and CD57, as well as CD4 (not shown). In addition, a small subset of these neoplastic T-cells shows loss of CD3.

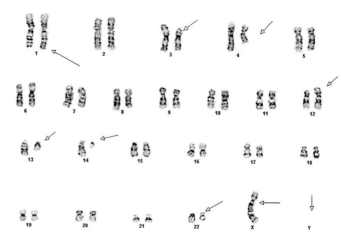

FIGURE 44.9 Karyotype of tumor cells in a patient with T-prolymphocytic leukemia showing complex abnormalities including deletion 13q, and t(X;14)(q28;q11.2).

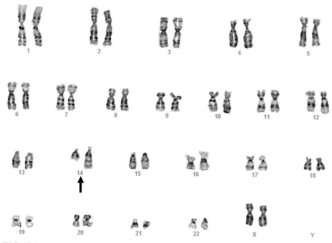

FIGURE 44.8 Karyotype of tumor cells in a patient with T-prolymphocytic leukemia with 46,XX,t(14;14)(q11.2;q32).

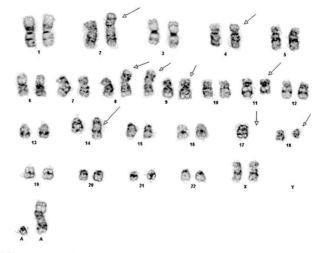

FIGURE 44.10 Karyotype of tumor cells in a patient with T-prolymphocytic leukemia with complex abnormalities including del(11q), and isochromosome (8q).

Other chromosomal abnormalities less frequently described are deletions or translocations of 6q, 12p, 13q, 17p and monosomy 22 (Figures 44.11 and 44.12).

At the molecular level, the rearrangements of *TCR* and the involvement of *TCL1*, *MTCP1-B1*, and *ATM* genes are common findings. *ATM* mutations have been detected in over half of T-PLL cases, suggesting that *ATM* acts as a tumor suppressor gene. Unfortunately, the *ATM* gene is extremely large, and detection of mutations, which requires extensive gene sequencing, is not routinely available for this purpose.

TCR gene rearrangements are detected in the same general manner as for immunoglobulin gene rearrangements in B-cell malignancies. They are often of more crucial importance to the case, however, since one does not have the advantage of surface immunoglobulin immunophenotyping (light-chain restriction) as ancillary evidence of clonality.

The Southern blot method will pick up a greater proportion of clonal rearrangements because it is capable of surveying a larger span of the target gene region. Most laboratories use a probe directed to the constant region of the beta-chain genes ($T_C\beta$).

Unlike the J_H region targeted in B-cell lesions, the TCR-β region is very large and complex, and to cover it adequately (i.e., with a pick-up rate approaching that of Southern blot) requires use of a large number of primers.

A common compromise is to target the *TCR*-γ locus, which has far fewer V and J genes, and thus a less complex array of rearrangements needs to be detected. However, the relatively more limited number of *TCR*-γ rearrangements can produce a type of false-positive result known as *pseudoclonality*: if the total number of T-cells in the submitted

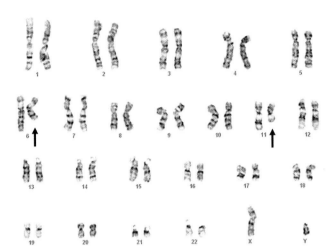

FIGURE 44.11 Karyotype of tumor cells in a patient with T-prolymphocytic leukemia demonstrating 46,XY,del(6)(q13q23),del(11)(q21).

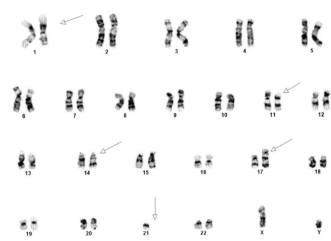

FIGURE 44.12 Karyotype of tumor cells in a patient with T-prolymphocytic leukemia demonstrating 45,XY,add(1)(q21),del(11)(p11.2p13),inv(14)(q11.2q32),add(17)(q25),−21.

specimen is scant, preferential amplification of a small number of them that happen to have the same rearrangement pattern can give the appearance of a clone when there really is none. Pseudoclonal PCR signals can sometimes be distinguished from true clonal signals by their lower intensity, lack of reproducibility, and paucity or absence of background polyclonal signal.

Differential Diagnosis

The differential diagnosis includes all leukemic lymphoproliferative disorders that have prolymphocytic morphology or cerebriform nuclei. B-prolymphocytic leukemia and prolymphocytic variant of mantle cell lymphoma are of B-cell lineage with their own characteristic immunophenotypic features. The neoplastic cells of Sézary syndrome (SS) often lack CD7 expression, whereas T-PLL cells are typically CD7 positive. The tumor cells of adult T-cell leukemia/lymphoma (ATL) may mimic the cerebriform variant of T-PLL, but, unlike T-PLL cells, they are positive for human T-lymphotropic virus type I (HTLV-I).

Additional Resources

Dearden CE, Khot A, Else M, et al: Alemtuzumab therapy in T-cell prolymphocytic leukemia: comparing efficacy in a series treated intravenously and a study piloting the subcutaneous route, *Blood* 118:5799–5802, 2011.

Dearden CE: T-cell prolymphocytic leukemia, *Clin Lymphoma Myeloma*(Suppl 3):S239–S243, 2009.

Dearden CE: T-cell prolymphocytic leukemia, *Med Oncol* 23:17–22, 2006.

Dungarwalla M, Matutes E, Dearden CE: Prolymphocytic leukaemia of B- and T-cell subtype: a state-of-the-art paper, *Eur J Haematol* 80:469–476, 2008.

Khot A, Dearden C: T-cell prolymphocytic leukemia, *Expert Rev Anticancer Ther* 9:365–371, 2009.

Krishnan B, Matutes E, Dearden C: Prolymphocytic leukemias, *Semin Oncol* 33:257–263, 2006.

Matutes E, Brito-Babapulle V, Swansbury J, et al: Clinical and laboratory features of 78 cases of T-prolymphocytic leukemia, *Blood* 78:3269–3274, 1991.

Ravandi F, O'Brien S, Jones D, et al: T-cell prolymphocytic leukemia: a single-institution experience, *Clin Lymphoma Myeloma* 6:234–239, 2005.

Thorat KB, Gujral S, Kumar A, et al: Small cell variant of T-cell prolymphocytic leukemia exhibiting suppressor phenotype, *Leuk Lymphoma* 47:1711–1713, 2006.

Tse E, So CC, Cheung WW, et al: T-cell prolymphocytic leukaemia: spontaneous immunophenotypical switch from CD4 to CD8 expression, *Ann Hematol* 90:479–481, 2011.

Adult T-Cell Leukemia/Lymphoma

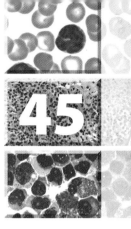

Adult T-cell leukemia/lymphoma (ATL) is an HTLV-I-associated peripheral T-cell lymphoid malignancy often presenting as an acute leukemic onset and aggressive clinical course.

The geographic areas with highest prevalence of HTLV-I include Japan, Africa, Caribbean islands, South America, and the southern part of the United States. Patients are adults with a median age of about 50 years with male to female ratio of about 3:2. Clinical symptoms may include hypercalcemia, lytic bone lesions, cutaneous lesions simulating mycosis fungoides (MF), lymphadenopathy, pulmonary lesions, and hepatosplenomegaly. Hypercalcemia is observed in over 70% of the cases during the clinical course, which appears to result from osteoclastic proliferation and increased bone resorption. ATL cells have been shown to express receptor activator of nuclear factor-kappaB (RANK) ligand which plays a role in differentiation of hematopoietic precursors to osteoclasts. There are four types of clinical presentation:

1. Acute onset, which is the most common type and occurs in approximately 60% of cases. It has an aggressive clinical course with 4-year survival rate of 5–12%.
2. The lymphomatous type represents about 20% of cases and is characterized by prominent lymphadenopathy and no blood involvement but also aggressive clinical course.
3. The chronic type, constituting about 15% of the cases, with skin lesions and absolute lymphocytosis but no hypercalcemia.
4. The smoldering type, representing 5% of cases, with normal blood lymphocyte counts and <5% circulating neoplastic cells and frequent skin or pulmonary lesions. There is no hypercalcemia.

Approximately one quarter of cases of chronic or smoldering types eventually progress to an acute phase. This transition is often associated with specific changes on gene expression profiling.

The clinical outcome in most cases is very poor, with a median survival of <1 year despite advances in chemotherapy. Combination chemotherapies such as cyclophosphamide, adriamycin, vincristine, and prednisone (CHOP), nucleoside analogs, topoisomerase inhibitors, interferon, anti-CD25, and zudovudine are among the therapeutic possibilities.

Morphology

- The neoplastic cells in the peripheral blood and bone marrow smears are pleomorphic, ranging from medium-sized to large, with variable amount of amphophilic or basophilic non-granular cytoplasm, and hyperlobated (clover leaf) or convoluted nuclei. The nuclear chromatin is condensed and nucleoli may be present (Figure 45.1).
- Multinucleated anaplastic giant cells with convoluted or cerebriform nuclei and cells resembling Reed–Sternberg cells may be seen in tissue infiltrations.
- A small proportion of blast-like cells with dispersed chromatin and prominent nucleoli are usually present.
- Bone marrow involvement is usually patchy (Figures 45.2 and 45.3) and is often associated with osteoclastic activities, leading to hypercalcemia.
- The involved lymph nodes show diffuse infiltration with effacement of nodal architecture and proliferation of endothelial venules (Figure 45.4).
- Skin infiltration usually involves the upper dermis, with frequent epidermal involvement and formation of tumor cell aggregates resembling Pautrier microabscesses (Figure 45.5).
- Other sites of involvement include lung, pleura (Figure 45.6), liver, spleen, gastrointestinal tract, and CNS.

Immunophenotype

MFC features (Figures 45.7 and 45.8) include:

- Typically helper T-cell phenotype expressing CD4;
- Moderate to bright expression of CD25;

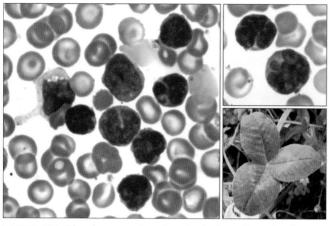

FIGURE 45.1 Blood smears showing atypical medium-sized to large cells with amphophilic or basophilic non-granular cytoplasm and hyperlobated (clover leaf) nuclei.

- Variably aberrant expression of CD2, CD3, and CD5;
- Absent CD7 in most cases.

In addition to the above-mentioned markers, immunohistochemical studies are positive for FOXP3 (see Figure 45.3F). CD30 can be expressed in large transformed cells, but ALK and cytotoxic granule markers are negative.

Molecular and Cytogenetic Studies

HTLV-I has been implicated in the development of ATL. A subpopulation of patients infected by HTLV-I (6% of males, 2% of females) eventually develop ATL after a long latent period.

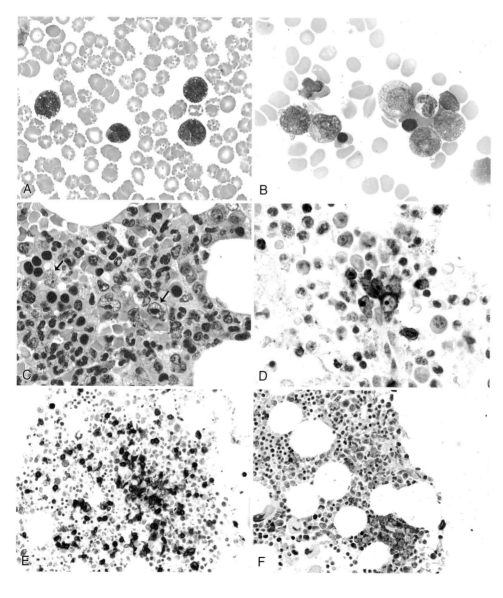

FIGURE 45.2 **ADULT T-CELL LEUKEMIA, ACUTE ONSET.** Blood (A) and bone marrow (B) smears demonstrating large, atypical lymphoid cells with lobulated nuclei. Bone marrow biopsy section (C) shows scattered interstitial infiltration of large, atypical lymphoid cells with irregular nuclear border and prominent nucleolus (arrows). Immunohistochemical stains highlight clusters of the large atypical lymphocytes expressing CD2 (D), CD3 (E) and CD4 (F).

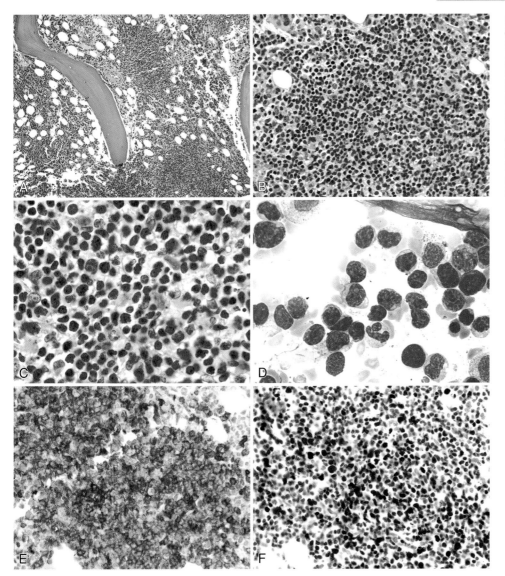

FIGURE 45.3 **ADULT T-CELL LEUKEMIA, CHRONIC VARIANT.** Bone marrow biopsy section demonstrating patchy infiltration of bone marrow by atypical lymphoid cells in large aggregates (A, low power; B, intermediate power; C, high power). Medium-sized atypical lymphoid cells with indented or convoluted nuclei are present in the bone marrow smear (D). Immunohistochemical stains are strongly positive for CD5 (E) and FOXP3 (F).

The transmission of HTLV-I from infected cells to noninfected cells is via cell–cell interaction which is apparently facilitated by ICAM-1 (CD56). The infected cells enter the human body through three major routes: (1) sexual transmission, (2) breast feeding, and (3) parenteral transmission. The HTLV-I *TAX* gene plays an important role in leukemogenesis of the infected cells. The Tax protein (p40) induces proliferation and inhibits apoptosis of the HTLV-I infected cells. However, ATL cells do not always need *TAX* expression; *TAX* transcription has been detected in only 34% of ATL cases by RT-PCR. Therefore, multistep genetic and epigenetic changes are implicated in ATL leukemogenesis. For example, mutation of *p53*, deletion of *p16*, and upregulation of *TSLC1* genes are reported in ATL.

The aberrant methylation of certain genes such as *MELIS* and *EGR3* provides examples of epigenetic changes in ATL.

Recent reports suggest that the HTLV-I *HBZ* gene may play an important role in the regulation of viral replication and proliferation of infected T-cells.

The neoplastic cells demonstrate *TCR* gene rearrangements with clonally integrated HTLV-I.

The detection and clonal pattern analysis of HTLV-I are quite specialized and are available only in selected reference laboratories. The deletion of the *p16* (multiple tumor suppressor 1) gene and the mutation of the *p53* gene have been reported. HTLV-1 gag/pol mRNA has been detected by real-time PCR (Figure 45.9).

There is no distinct karyotypic or molecular genetic abnormality in ATL. Cytogenetic analyses often show complex karyotypes with various aberrations, particularly in the leukemic forms. The karyotypes exhibit a number of breaks and multiple numerical rearrangements, aneuploidy, hypotetraploidy, derivative chromosomes, balanced and unbalanced, translocations, and deletions.

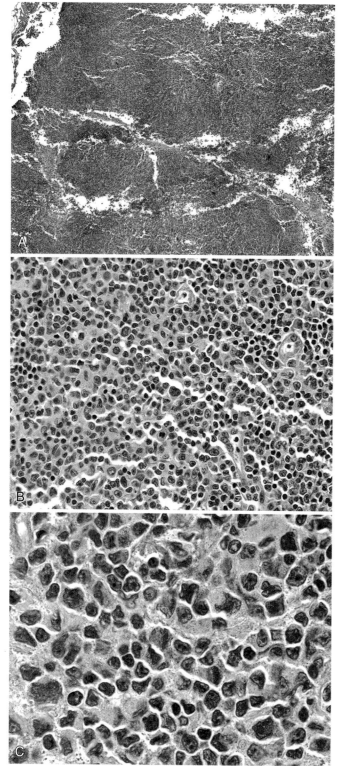

FIGURE 45.4 **ADULT T-CELL LEUKEMIA/LYMPHOMA.** Lymph node biopsy section demonstrating diffuse infiltration by medium-sized to large atypical lymphocytes with highly irregular or lobulated nuclei (A, low power; B, intermediate power; C, high power).

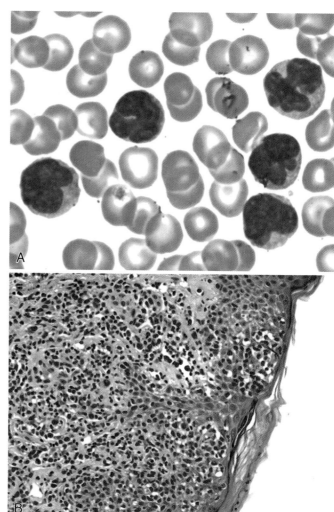

FIGURE 45.5 **ADULT T-CELL LEUKEMIA/LYMPHOMA.** Blood smear (A) demonstrates atypical lymphocytes with lobulated nuclei. Skin biopsy section (B) shows a dense lymphoid infiltrate involving the upper dermis with prominent epidermotropism.

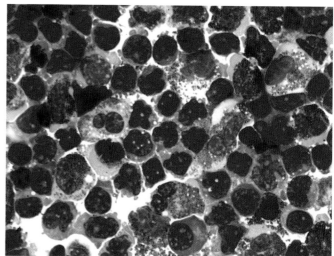

FIGURE 45.6 Pleural effusion in a patient with adult T-cell leukemia/lymphoma demonstrating numerous medium-sized to large atypical lymphocytes with highly irregular or lobulated nuclei. Several eosinophils are also present.

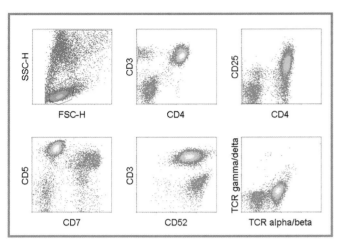

FIGURE 45.7 **FLOW CYTOMETRIC FINDINGS OF ADULT T-CELL LEUKEMIA/LYMPHOMA, ACUTE ONSET.** Lymphocyte-enriched gate of pleural effusion shows a prominent population of abnormal T-cells in a background of occasional reactive T-cells. The neoplastic T-cells are positive for CD3, CD4, CD5 (bright), CD25 (bright), CD52, and TCR alpha/beta, but display complete loss of CD7.

- chromosome 14 anomalies at q11.2 and at 7q35, the site of the *TCRB* gene region, seem to be specific to ATL.
- Recurrent abnormalities include trisomies of chromosomes 3, 8, 9, 21, and X, monosomies of chromosomes 4, 10, and 22, and abnormalities of chromosomes 6 and 14q (at 14q11.2 and 14q32 breakpoints) (Figures 45.10 and 45.11).
- Of these abnormalities, involvement of 7q, 13q, 14q, and i(18) have been indicated to be specific for T-cell leukemias/lymphomas including ATL (Figure 45.12).
- Gains of 14q32 are common and may be a recurrent specific abnormality in ATL.
- Aggressive forms display more genomic aberrations than chronic forms.
- The number of chromosomal imbalances correlates with clinical outcome. Different genomic profiles are observed when analyzing material from different sites or material taken at different time points in the same patient, suggestive of clonal evolution.

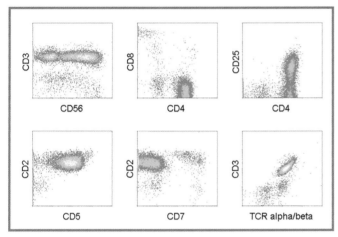

FIGURE 45.8 **FLOW CYTOMETRIC FINDINGS OF ADULT T-CELL LEUKEMIA/LYMPHOMA, CHRONIC VARIANT.** Lymphocyte-enriched gate of bone marrow reveals predominantly abnormal T-cells, which are positive for CD2, CD3, CD4, CD5 (dim), CD25, CD56 (major subset), and TCR alpha/beta. The neoplastic T-cells demonstrate complete loss of CD7.

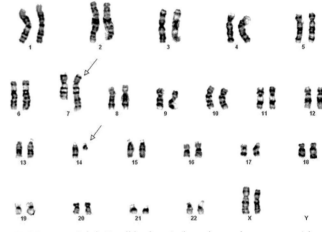

FIGURE 45.10 Adult T-cell leukemia/lymphoma karyotype with trisomy X and inv(14).

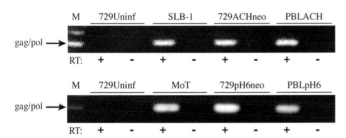

FIGURE 45.9 Detection of HTLV-1 gag/pol mRNA by real-time RT-PCR. SLB-1, PBL-ACH, and 729ACHneo represent HTLV-1 cell lines. Cell line 729 is a negative control.
Adapted from Li M, Green PL. Detection and quantitation of HTLV-1 and HTLV-2 mRNA species by real-time RT-PCR. J Virol Methods. 142:159-68, 2007, by permission.

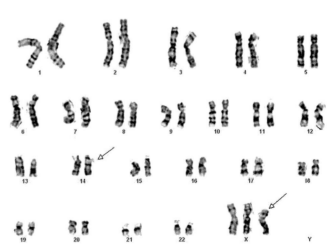

FIGURE 45.11 Karyotype of a patient with Adult T-cell leukemia/lymphoma demonstrating trisomy X and inv(14).

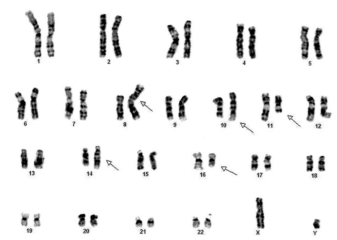

FIGURE 45.12 Karyotype of a patient with Adult T-cell leukemia/lymphoma showing 46,XY,i(8)(q10),add(10)(q24),del(11)(q21),inv(14)(q11.2q32),del(16)(q22).

Differential Diagnosis

The neoplastic cells in peripheral blood and bone marrow smears may mimic Sézary cells and may share the same immunophenotypic features (expressing CD4 and pan-T-cell markers except CD7). In general, nuclear convolution in Sézary cells is more delicate and finer than in the neoplastic ATL cells. In all ATL cases, the neoplastic cells are positive for HTLV-I, whereas most patients with SS are HTLV-I negative. Also, the median age for ATL is lower than that of MF/SS. Hypercalcemia and lytic bone lesions are frequent findings in ATL but absent in MF/SS.

The tissue infiltrates in ATL may contain anaplastic large cells and/or Reed–Sternberg-like cells mimicking anaplastic large cell lymphoma (ALCL) or Hodgkin lymphoma.

Additional Resources

Aifantis I, Raetz E, Buonamici S: Molecular pathogenesis of T-cell leukaemia and lymphoma, *Nat Rev Immunol* 8:380–390, 2008.

Dasanu CA: Newer developments in adult T-cell leukemia/lymphoma therapeutics, *Expert Opin Pharmacother* 12:1709–1717, 2011.

Ishida T, Ueda R: Antibody therapy for adult T-cell leukemia-lymphoma, *Int J Hematol* 94:443–452, 2011.

Mahieux R, Gessain A: Adult T-cell leukemia/lymphoma and HTLV-1, *Curr Hematol Malig Rep* 2:257–264, 2007.

Matutes E: Adult T-cell leukaemia/lymphoma, *J Clin Pathol* 60:1373–1377, 2007.

Meijerink JP: Genetic rearrangements in relation to immunophenotype and outcome in T-cell acute lymphoblastic leukaemia, *Best Pract Res Clin Haematol* 23:307–318, 2010.

Miyagi T, Nagasaki A, Taira T, et al: Extranodal adult T-cell leukemia/lymphoma of the head and neck: a clinicopathological study of nine cases and a review of the literature, *Leuk Lymphoma* 50:187–195, 2009.

Takahashi T, Tsukuda H, Itoh H, et al: Primary and isolated adult T-cell leukemia/lymphoma of the bone marrow, *Intern Med* 50:2393–2396, 2011.

Yasunaga J, Matsuoka M: Human T-cell leukemia virus type I induces adult T-cell leukemia: from clinical aspects to molecular mechanisms, *Cancer Control* 14:133–140, 2007.

Yoshida M: Molecular approach to human leukemia: isolation and characterization of the first human retrovirus HTLV-1 and its impact on tumorigenesis in adult T-cell leukemia, *Proc Jpn Acad Ser B Phys Biol Sci* 86:117–130, 2010.

Hepatosplenic T-Cell Lymphoma

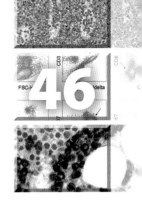

Hepatosplenic lymphoma is a rare extranodal peripheral T-cell lymphoma characterized by sinusoidal infiltration of liver, spleen, and bone marrow. In most reported cases, the neoplastic cells represent the TCR-γδ T-cell subtype, though rare cases express TCR-αβ.

Hepatosplenic T-cell lymphoma is a rare disease, accounting for about 5% of all peripheral T-cell lymphomas. The median age is about 35 years, with a male to female ratio of around 3:1. Splenomegaly and thrombocytopenia are common, followed by hepatomegaly, anemia, and leukopenia. Lymphadenopathy and extranodal involvements other than those mentioned above are rare. There appears to be an association between incidence of hepatosplenic T-cell lymphoma and the use of infliximab, adalimumab, and AZA/6-MP in patients with inflammatory bowel disease.

Hepatosplenic T-cell lymphoma has an aggressive clinical course with a median survival of about 16 months. Combination chemotherapy is the common therapeutic approach, often with unsatisfactory results. Rare patients respond to interferon-alpha.

Morphology

- The neoplastic lymphoid cells are usually monomorphic, medium sized, with moderate amount of cytoplasm, round to slightly irregular nuclei, condensed chromatin, and inconspicuous nucleoli. Occasional cases may show a highly pleomorphic cell population.
- Liver infiltration is sinusoidal with various degrees of portal tract involvement (Figure 46.1).
- The pattern of infiltration in the spleen is diffuse with the involvement of the red pulp (Figures 46.2 and 46.3).
- Lymph node involvement is rare.
- Bone marrow involvement is reported in 75–100% of cases. The lymphoid infiltration is often subtle and sinusoidal (Figure 46.4). Occasionally, the number of histiocytes is increased, with features of hemophagocytosis.

- Scattered abnormal cells may be seen in the peripheral blood (Figure 46.4A). A fulminant leukemic picture can occur. Leukemic cells may be large with blastic features.

There are other variants of γδ T-cell lymphomas which mostly involve extranodal tissues, such as skin, subcutaneous tissue, or intestine, which are not considered as hepatosplenic lymphoma.

Immunophenotype

MFC features (Figure 46.5) include:

- Expression of CD3 and TCR gamma/delta;
- Rare variant of TCR alpha/beta type (Figure 46.6);
- Double negativity of CD4/CD8, but some cases with partial dim expression of CD8;
- Variable aberrancies of CD2 and CD7, and common loss of CD5;
- Expression of CD56 in some cases.

In addition to the above-mentioned markers, immunohistochemical studies are positive for TIA1 (see Figure 46.4F), but granzyme B and perforin are often negative.

Molecular and Cytogenetic Studies

- The neoplastic cells typically show *TCR* gene rearrangement by Southern blotting or PCR techniques (Figure 46.7).
- In situ hybridization studies for EBV are negative.
- The most frequent cytogenetic abnormalities observed are extra copies of 7q, mostly in the form of isochromosome 7q, and trisomy 8. Isochromosome 7q is currently viewed as a pathognomonic genetic alteration in hepatosplenic T-cell lymphoma and can therefore serve as a diagnostic tool for this entity (Figure 46.8). Also, del(13)(q12q14) has been reported in patients with hepatosplenic lymphoma (Figure 46.9).

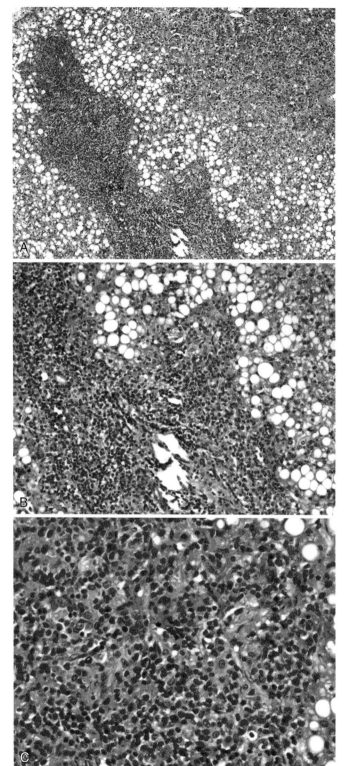

FIGURE 46.1 **HEPATIC INVOLVEMENT WITH HEPATOSPLENIC T-CELL LYMPHOMA.** Patchy lymphoid infiltration is evident in a liver with fatty degeneration (A, low power; B, intermediate power; C, high power).

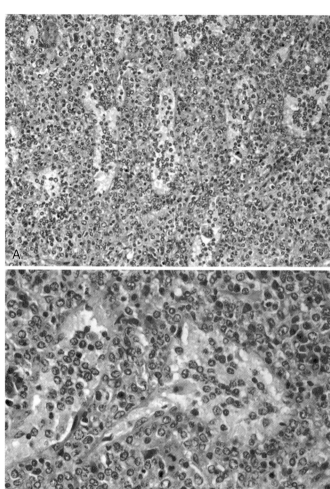

FIGURE 46.2 **HEPATOSPLENIC T-CELL LYMPHOMA WITH THE INVOLVEMENT OF SPLENIC RED PULP.** (A, low power; B, high power).

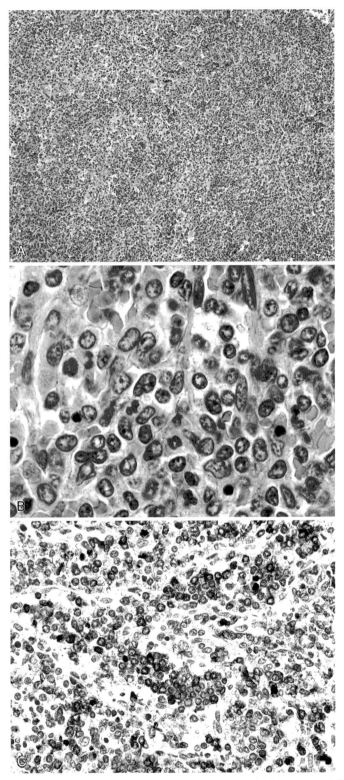

FIGURE 46.3 HEPATOSPLENIC T-CELL LYMPHOMA WITH DIFFUSE INVOLVEMENT OF SPLENIC RED PULP. (A, low power; B, high power). The neoplastic cells present in the sinusoidal spaces express CD7 (C).

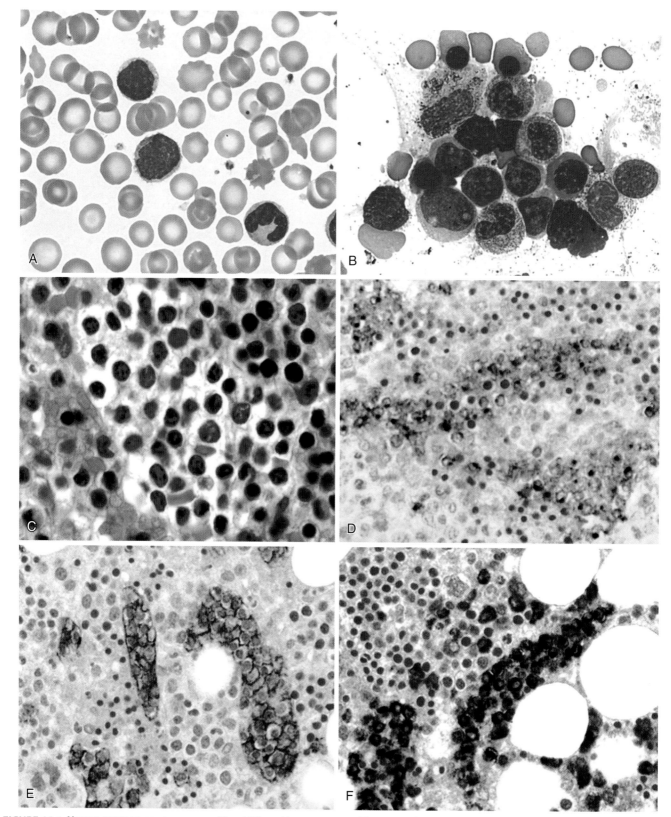

FIGURE 46.4 **HEPATOSPLENIC T-CELL LYMPHOMA.** Blood (A) and bone marrow (B) smears show atypical lymphocytes. Bone marrow involvement is often sinusoidal (C) demonstrated by accumulation of CD3+ (D), CD56+ (E), and TIA1+ (F) cells.

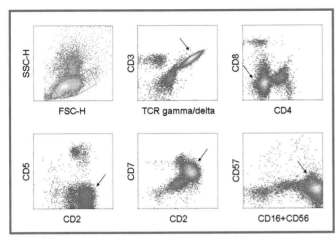

FIGURE 46.5 **FLOW CYTOMETRIC FINDINGS OF HEPATOSPLENIC T-CELL LYMPHOMA, GAMMA/DELTA TYPE.** Open gate scatter display (in blue) reveals a prominent population of intermediate to large T-cells, which is verified by CD3 backgating. The lymphocyte-enriched-gate (in magenta) contains mostly abnormal T-cells (arrow) that are positive for CD2 (bright), CD3, CD7 (dim), CD16/CD56 (bright), and TCR gamma/delta. The neoplastic T-cells are double negative for CD4 and CD8, with complete loss of CD5.

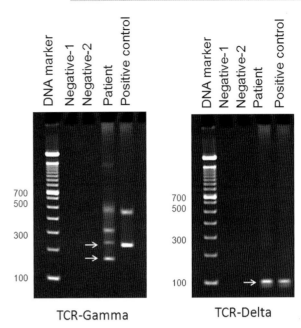

FIGURE 46.7 Southern blot analysis demonstrating *TCRG* and *TCRD* gene rearrangements in a patient with hepatosplenic T-cell lymphoma.
Courtesy of Ryan Phan, Ph.D., Department of Pathology, VA Greater Los Angeles Healthcare System.

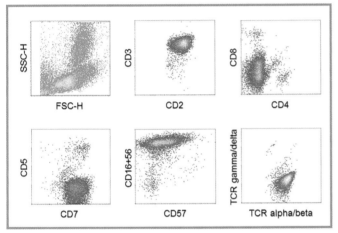

FIGURE 46.6 **FLOW CYTOMETRIC FINDINGS OF HEPATOSPLENIC T-CELL LYMPHOMA, ALPHA/BETA TYPE.** Open gate scatter display (in blue) reveals a prominent population of T-cells that are intermediate-sized to large, which is verified by CD3 backgating. The T-cells display abnormal phenotype expressing CD2, CD3, CD7, CD8 (partial, dim), CD16 + 56, CD57 (partial), and TCR alpha/beta. There is lack of expression of CD4, CD5, and TCR gamma/delta.

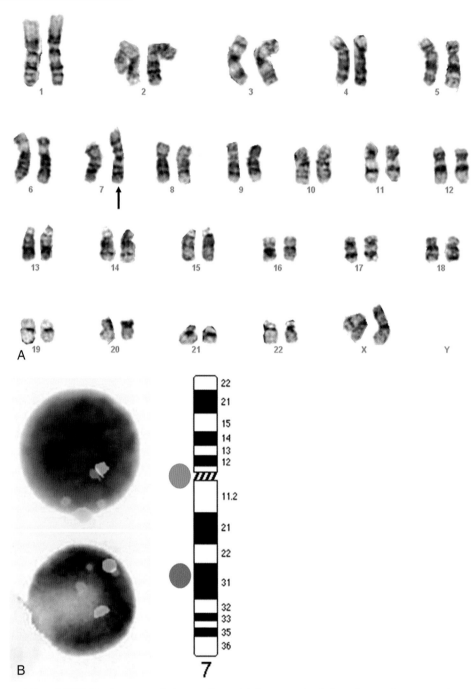

FIGURE 46.8 Karyotype (A) and FISH (B) of tumor cells in a patient with hepatosplenic lymphoma demonstrating 46,XX,iso(7)(q10).

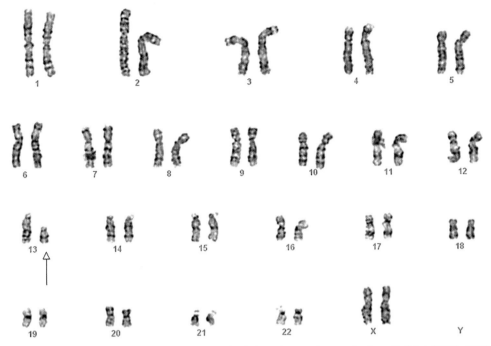

FIGURE 46.9 Karyotype of tumor cells in a patient with hepatosplenic lymphoma demonstrating 46,XX,del(13)(q12q22).

Differential Diagnosis

The differential diagnosis includes neoplasms of large granular lymphocytes, such as aggressive NK-cell and T-LGL leukemias. Aggressive NK-cell leukemia and hepatosplenic lymphoma share an overlapping morphologic pattern (sinusoidal involvement) and immunophenotypic features (CD56 and TIA1 expression). However, neoplastic cells of hepatosplenic lymphoma are surface CD3+, CD4−, CD8−, CD5− and granzyme B−, and in many cases demonstrate isochromosome 7q.

T-LGL leukemia cells are CD3+ and may occasionally express $TCR\text{-}\gamma\delta$, but unlike lymphoid cells of hepatosplenic lymphoma are CD5+, CD57+, and CD56−.

Additional Resources

Beigel F, Jürgens M, Tillack C, et al: Hepatosplenic T-cell lymphoma in a patient with Crohn's disease, *Nat Rev Gastroenterol Hepatol* 6:433–436, 2009.

Belhadj K, Reyes F, Farcet JP, et al: Hepatosplenic gammadelta T-cell lymphoma is a rare clinicopathologic entity with poor outcome: report on a series of 21 patients, *Blood* 102:4261–4269, 2003.

Gaulard P, Belhadj K, Reyes F: Gammadelta T-cell lymphomas, *Semin Hematol* 40:233–243, 2003.

Humphreys MR, Cino M, Quirt I, et al: Long-term survival in two patients with hepatosplenic T cell lymphoma treated with interferon-alpha, *Leuk Lymphoma* 49:1420–1423, 2008.

Kotlyar DS, Osterman MT, Diamond RH, et al: A systematic review of factors that contribute to hepatosplenic T-cell lymphoma in patients with inflammatory bowel disease, *Clin Gastroenterol Hepatol* 9:36–41, 2011.

Lu CL, Tang Y, Yang QP, et al: Hepatosplenic T-cell lymphoma: clinicopathologic, immunophenotypic, and molecular characterization of 17 Chinese cases, *Hum Pathol* 42:1965–1978, 2011.

Mandava S, Sonar R, Ahmad F, et al: Cytogenetic and molecular characterization of a hepatosplenic T-cell lymphoma: report of a novel chromosomal aberration, *Cancer Genet* 204:103–107, 2011.

Minauchi K, Nishio M, Itoh T, et al: Hepatosplenic alpha/beta T cell lymphoma presenting with cold agglutinin disease, *Ann Hematol* 86:155–157, 2007.

Taguchi A, Miyazaki M, Sakuragi S, et al: Gamma/delta T cell lymphoma, *Intern Med* 43:120–125, 2004.

Tripodo C, Iannitto E, Florena AM, et al: Gamma-delta T-cell lymphomas, *Nat Rev Clin Oncol* 6:707–717, 2009.

Weidmann E: Hepatosplenic T cell lymphoma: a review on 45 cases since the first report describing the disease as a distinct lymphoma entity in 1990, *Leukemia* 14:991–997, 2000.

Enteropathy-Associated T-Cell Lymphoma

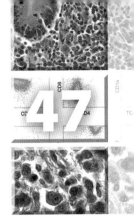

Enteropathy-associated T-cell lymphoma (EATL) is a rare intraepithelial T-cell intestinal lymphoma consisting of a polymorphic lymphoid infiltrate of medium to large atypical lymphocytes.

Most patients are over 60 years of age, with a male to female ratio of about 1. Patients have a history of celiac disease and present with symptoms of malabsorption, abdominal pain, and occasionally evidence of intestinal perforation. The clinical outcome is usually poor with frequent recurrences.

Morphology

- The most frequent site of involvement is the jejunum, but other parts of small intestine or gastrointestinal tract, such as stomach and colon, may be involved. The tumor is often multifocal and appears as ulcerating nodules. The mesenteric lymph nodes are usually affected.
- The neoplasm consists of a pleomorphic lymphoid infiltrate. There is a predominance of atypical medium sized to large lymphoid cells with variable amount of pale cytoplasm, round or irregular vesicular nuclei, and prominent nucleoli (Figures 47.1 and 47.2). Anaplastic large cells or multinucleated giant cells may be present, mimicking anaplastic large cell lymphoma. The infiltrate commonly is mixed with inflammatory cells, such as histiocytes and eosinophils. Areas of necrosis are often present. Granulomas may be present.
- The histology of the non-involved mucosa, remote from the neoplasm, commonly shows evidence of villous atrophy and crypt hyperplasia, features associated with celiac disease.
- The lymph node involvement may be intrasinusoidal or paracortical, or both. Areas of necrosis are often present.
- A monomorphic variant (type II EATL) has been described in which the tumor cells are slightly larger than normal lymphocytes and broadly infiltrate in the submucosa and muscularis propria (Figure 47.3).

Immunophenotype

MFC findings (Figure 47.4) include:

- Expression of CD3;
- Variably aberrant profile of CD2 and CD7;
- Absent CD4 and CD5;
- Expression of CD8 and CD56 uncommon in EATL;
- Expression of CD8, CD56, and TCR beta in monomorphic form (type II EATL) tumors.

By immunohistochemical studies (Figure 47.5), the neoplastic cells:

- Are positive for cytotoxic-associated proteins, such as TIA1 and granzyme B;
- Express CD3, CD7, TCR-γδ, and CD103;
- Are negative for CD4 and CD5.
- The monomorphic (type II EATL) tumors may express CD8 and/or CD56.
- In most cases, some proportion of tumor cells are also positive for CD30.

Molecular and Cytogenetic Studies

- The *TCR* genes, most commonly γ and δ, are clonally rearranged.
- Loss of heterozygosity (LOH) at chromosome 9p21 with a deletion of the *CDKN2a* gene (Cyclin-dependent kinase inhibitor 2A or *p16*) is seen in almost a fifth of the cases.
- Gains of chromosome 9q33–q34, 7q31, 5q33–34, and 1q have been identified by CGH studies. Deletions of 6p24, 7p21, 17p, and 17q are also reported.

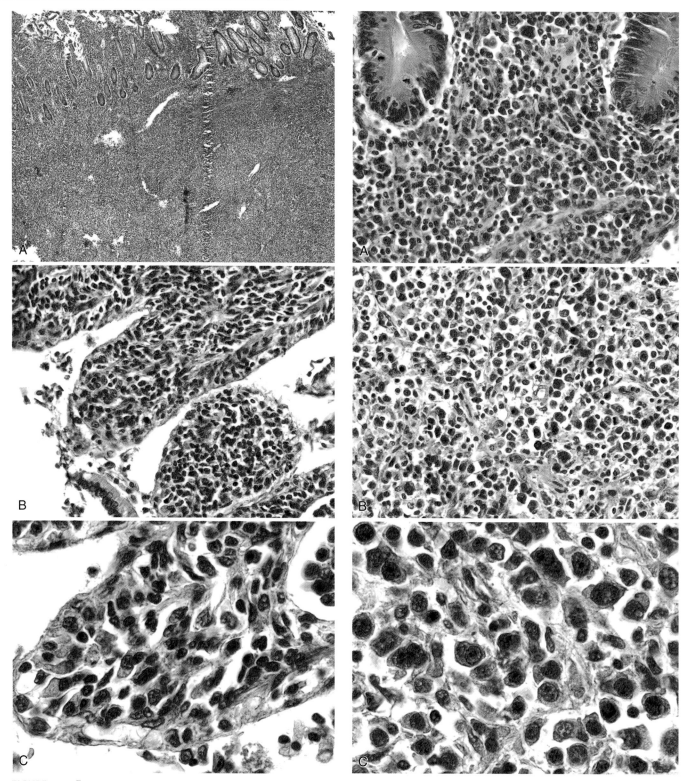

FIGURE 47.1 **ENTEROPATHY-ASSOCIATED T-CELL LYMPHOMA OF SMALL INTESTINE.** Diffuse infiltration of mucosa and submucosa by atypical, pleomorphic lymphocytes. Numerous anaplastic large cells are present (A, low power; B, intermediate power; C, high power).

FIGURE 47.2 **ENTEROPATHY-ASSOCIATED T-CELL LYMPHOMA OF SMALL INTESTINE.** Diffuse infiltration of mucosa and submucosa with large lymphoid cells. The neoplastic cells are blastoid with large nucleus, open nuclear chromatin, and one prominent nucleolus (A, low power; B, intermediate power; C, high power; D, high power, oil emersion).

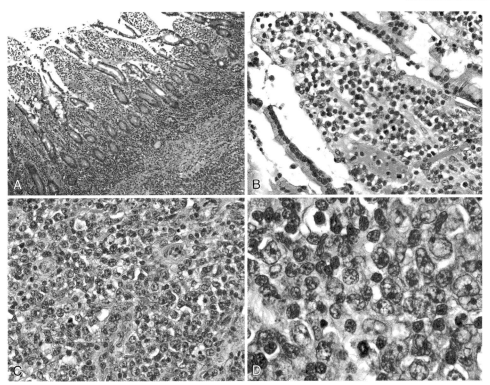

FIGURE 47.3 **TYPE II EATL.** Enteropathy-associated T-cell lymphoma of small intestine demonstrating a diffuse infiltration of atypical, monomorphic lymphocytes in the mucosa and submucosa (A, low power; B, intermediate power; C, high power).

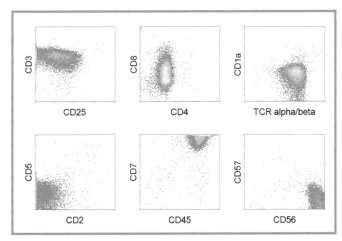

FIGURE 47.4 **FLOW CYTOMETRIC FINDINGS OF ENTEROPATHY-ASSOCIATED T-CELL LYMPHOMA.** Lymphocyte-enriched gate of an abdominal wall mass reveals near exclusively abnormal T-cells, which are positive for CD3, CD7 (bright), CD8 (partial, dim), CD25 (partial), CD45, CD56, and TCR alpha/beta. The neoplastic T-cells demonstrate complete loss of CD2 and CD5. They are also negative for CD1a, CD4, and CD57.

Differential Diagnosis

The differential diagnosis includes inflammatory bowel disorders and other types of lymphomas, particularly diffuse large B-cell lymphoma and anaplastic large cell lymphoma. Inflammatory bowel diseases primarily consist of polymorphic inflammatory cells, without infiltrating clusters or sheets of atypical neoplastic lymphoid cells. They do not show TCR rearrangement and lack chromosomal aberrations.

In EATL, the intestinal mucosa remote from the involved area commonly shows evidence of villous atrophy and crypt hyperplasia, features associated with celiac disease. Also the tumor cells commonly express CD103 and TCR-γδ, whereas, anaplastic large cell lymphoma lacks these expressions and diffuse large B-cell lymphoma expresses pan B-cell-associated markers.

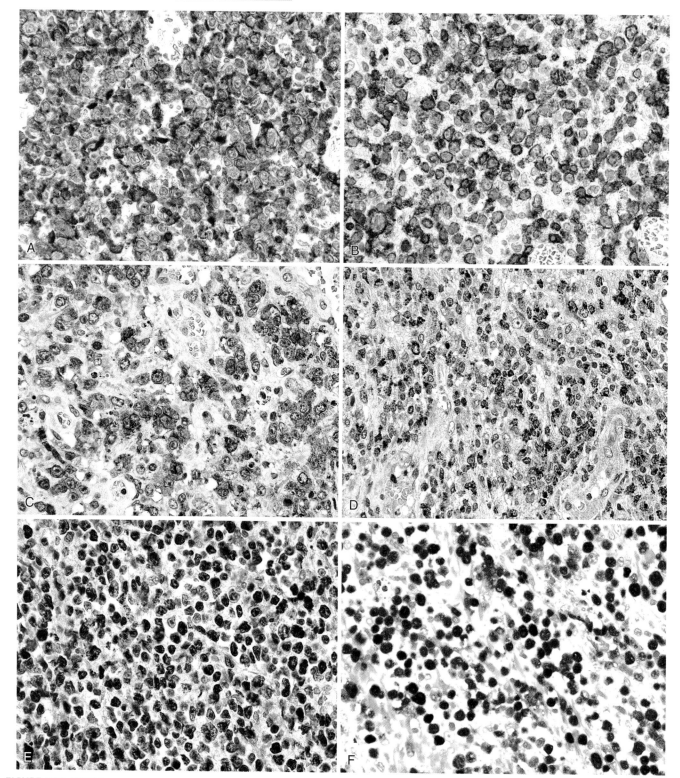

FIGURE 47.5 **IMMUNOHISTOCHEMICAL STUDIES IN A PATIENT WITH ENTEROPATHY-ASSOCIATED T-CELL LYMPHOMA.** Sections of affected small intestine demonstrate expression of CD2 (A), CD3 (B), CD30 (C), TIA-1 (D), Ki67 (E) and EBV-EBER (F).

Additional Resources

Chan JK, Chan AC, Cheuk W, et al: Type II enteropathy-associated T-cell lymphoma: a distinct aggressive lymphoma with frequent γδ T-cell receptor expression, *Am J Surg Pathol* 35:1557–1569, 2011.

Chandesris MO, Malamut G, Verkarre V, et al: Enteropathy-associated T-cell lymphoma: a review on clinical presentation, diagnosis, therapeutic strategies and perspectives, *Gastroenterol Clin Biol* 34:590–605, 2010.

Delabie J, Holte H, Vose JM, et al: Enteropathy-associated T-cell lymphoma: clinical and histological findings from the international peripheral T-cell lymphoma project, *Blood* 118:148–155, 2011.

Ferreri AJ, Zinzani PL, Govi S, Pileri SA: Enteropathy-associated T-cell lymphoma, *Crit Rev Oncol Hematol* 79:84–90, 2011.

Jaffe ES, Harris NL, Vardiman JW, et al: *Hematopathology*, Philadelphia, 2010, Saunders/Elsevier.

Sieniawski MK, Lennard AL: Enteropathy-associated T-cell lymphoma: epidemiology, clinical features, and current treatment strategies, *Curr Hematol Malig Rep* 6:231–240, 2011.

van de Water JM, Cillessen SA, Visser OJ, et al: Enteropathy associated T-cell lymphoma and its precursor lesions, *Best Pract Res Clin Gastroenterol* 24:43–56, 2010.

Zettl A, deLeeuw R, Haralambieva E, et al: Enteropathy-type T-cell lymphoma, *Am J Clin Pathol* 127:701–706, 2007.

Mycosis Fungoides and Sézary Syndrome

Mycosis fungoides (MF) is a cutaneous peripheral T-cell lymphoma (CTCL) characterized by an indolent course and skin manifestations ranging from patches to plaques and tumor formation. Sézary syndrome (SS) is a closely related neoplasm which is associated with cutaneous involvement, erythroderma, and circulating neoplastic cells in the peripheral blood.

Mycosis Fungoides

Mycosis fungoides is an epidermotropic T-cell lymphoma with the infiltration of upper dermis resulting in patches, plaques, and nodules. The vast majority of the cases are of helper T-cell phenotype.

MF is the most common primary cutaneous T-cell lymphoma accounting for about 45% of lymphomas present in the skin. The peak age is 55–60 years, with a male to female ratio of about 2:1. MF usually presents as indolent cutaneous erythematous scaly patches or plaques, often with pruritus, mimicking common skin disorders, such as eczema or psoriasis. These lesions may wax and wane for many years and may eventually progress to cutaneous tumor formation, erythroderma, and the infiltration of the neoplastic cells into the circulation (SS).

Extracutaneous involvement is relatively uncommon in the early stages of the disease but becomes more frequent in the advanced stages, comprising 8% in the plaque stage compared with 30–40% in the erythrodermatous stage.

The regional lymph nodes are the most frequent sites of involvement followed by lungs, spleen, and liver. The clinical staging is based on the extent of cutaneous lesions (T) and the involvement of the lymph nodes (L), viscera (M), and blood (B). The overall 5-year survival for MF has been reported as 87%, compared with 33% for SS. Therefore, transformation from MF to SS is indicative of poor prognosis.

Transformation to CD30+ large cell lymphoma is relatively frequent. In two large studies of patients with MF, the cumulative probability of transformation to large cell lymphoma was 39% in 12 years, with a median time interval of 1–6.5 years after diagnosis of MF. Transformation to CD30+ large cell lymphoma is associated with an aggressive clinical course. In one study, the median survival was 3 years in transformed MF patients compared with 14 years in untransformed patients.

Treatment includes a broad spectrum of options. The topical therapeutic measures, such as nitrogen mustard, carmustine, electron beam therapy, and phototherapy, are used for the early stages of the disease. Systemic chemotherapy, alone or in combination with topical therapy, is used in advanced stages of the disease. Purine and pyrimidine analogs are the primary chemotherapy agents.

MORPHOLOGY

Skin biopsies in early stages of patch formation demonstrate a perivascular infiltrate of lymphohistiocytic cells with scattered atypical lymphocytes in the upper dermis. Epidermotropism characterized by the presence of single or linear arrangement of haloed lymphoid cells in the basal layer of the epidermis may or may not be present (Figure 48.1).

The atypical cells are usually medium to large size with variable amount of cytoplasm and convoluted (cerebriform) nuclei.

In the stage of plaque formation, the rete ridges are elongated, epidermotropism is more pronounced, and aggregates

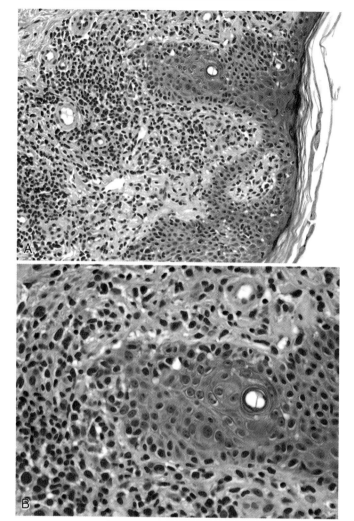

FIGURE 48.1 **MYCOSIS FUNGOIDES, EARLY STAGE.** Skin biopsy section demonstrating an upper dermis lymphohistiocytic infiltration with evidence of epidermotropism (A, low power; B, intermediate power).

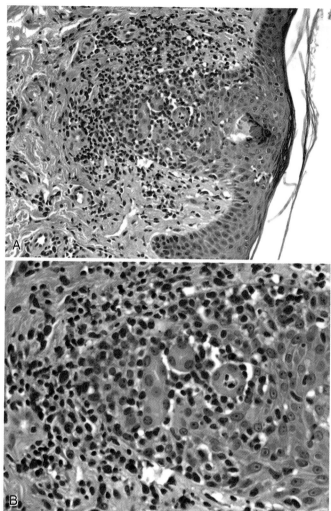

FIGURE 48.2 **MYCOSIS FUNGOIDES.** Skin biopsy section demonstrating epidermotropism and the presence of lymphocytes within the epidermis (A, low power; B, high power).

of atypical cells referred to as "Pautrier microabscesses" may be present in the epidermis (Figures 48.2 and 48.3). Pautrier microabscesses are characteristic of MF, but are observed in only about one-third of the patients.

In more advance stages, dermal infiltration expands, with tumor formation and an increased proportion of larger atypical lymphoid cells (Figure 48.4). The epidermotropism may disappear.

Lymphadenopathy is frequent in advanced cases (30–40%) and is divided into three categories as follows.

- Category 1: Dermatopathic lymphadenitis with no atypical, cerebriform lymphocytes (no involvement).
- Category 2: Dermatopathic lymphadenitis with early involvement and scattered atypical, cerebriform lymphocytes.
- Category 3: Partial or complete effacement of nodal architecture with diffuse infiltration of atypical, cerebriform lymphocytes.

Transformation to large CD30+ T-cell lymphoma is characterized by formation of microscopic nodules of large transformed cells or >25% large cells in the neoplastic infiltrate. Elevation of serum lactate dehydrogenase (LDH) and β2-microglobulin are predictive of transformation.

IMMUNOPHENOTYPE

Immunophenotypic features (Figures 48.5 and 48.6) include:

- Mostly phenotype of mature T-helper cells with expression of CD4 and TCR-αβ;
- Expression of pan-T-cell antigens CD2, CD3, and CD5, often with variably aberrant patterns;
- Common loss of CD7, especially within the epidermotropic lymphocytes;
- Common expression of CD25 (IL-2 receptor-α) with variable patterns;

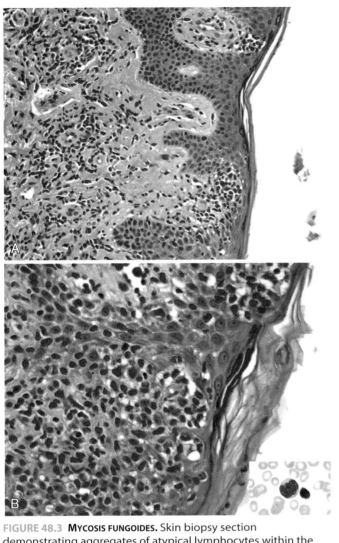

FIGURE 48.3 **MYCOSIS FUNGOIDES.** Skin biopsy section demonstrating aggregates of atypical lymphocytes within the epidermis (Pautrier microabscesses) (A, low power; B, high power). The inset shows a large, atypical lymphocyte with convoluted nuclei (Sézary cell) in the peripheral blood smear.

- Occasional CD8 phenotype, which is more common in pediatric cases;
- Lack of expression of CD26;
- Expression of CD158.
- Cytotoxic granule associated proteins may be detected in advanced stages of MF.

CYTOGENETIC AND MOLECULAR STUDIES

- Patients with cutaneous T-cell lymphoma commonly show a wide variety of clonal or non-clonal numerical and structural chromosomal aberrations in their blood or skin cultures.
- The neoplastic blood cells have a low mitotic index, and stimulation with phytohemagglutinin and interleukin (IL-2, IL-7) is used.
- Although standard karyotyping studies of MF are difficult, the probability of detecting an abnormal clone correlates with the clinical stage, with 50% exhibiting abnormal chromosomes.

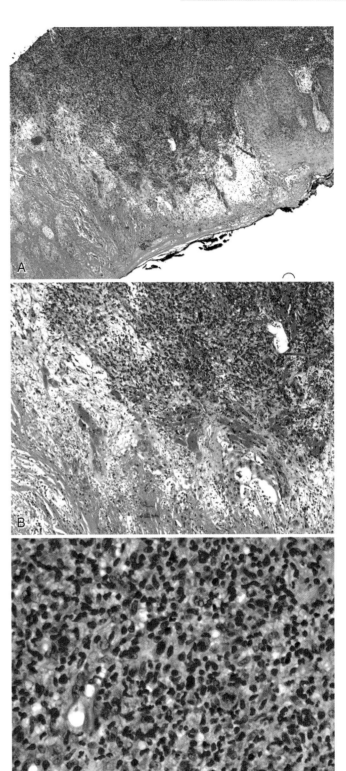

FIGURE 48.4 **MYCOSIS FUNGOIDES, ADVANCED STAGE.** Skin biopsy section demonstrating areas of fibrinoid necrosis and massive lymphoid infiltration of dermis (A, low power; B, intermediate power). The infiltrate is pleomorphic, consisting of atypical small to large lymphocytes showing irregular nuclear borders (C, high power).

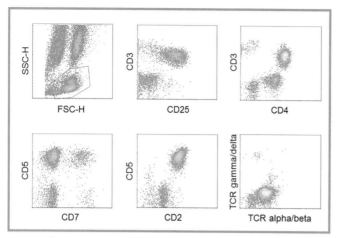

FIGURE 48.5 FLOW CYTOMETRIC FINDINGS OF MYCOSIS FUNGOIDES. Open scatter gate (in purple) demonstrates small to intermediate-sized lymphocytes (circled). The lymphocyte-enriched-gate (in blue) reveals mostly abnormal T-cells, which express CD2, CD3 (dim), CD4, CD5, CD25, and TCR alpha/beta (dim). These neoplastic T-cells demonstrate complete loss of CD7 expression.

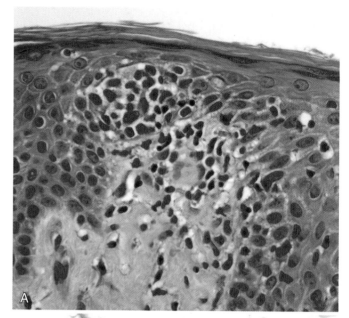
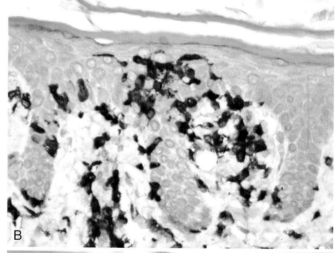
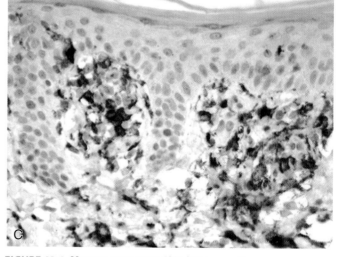

FIGURE 48.6 MYCOSIS FUNGOIDES. Skin biopsy section demonstrating Pautrier microabscesses (A) and expressing CD3 (B) and CD4 (C).

- No recurrent or specific abnormality has been found in MF, leading to a hypothesis of genetic instability.
- The most common structural abnormalities involve chromosomes 1p36, deletion 2q, 6, 7, 4, deletion 9p, 9q34, 10q, 12, 14, 15, 17q24, 19p, and most common numerical abnormalities involve chromosomes 11, 21, 22, 8, 9, 15, 16, 17, and 19 in descending order of frequency.
- DNA content analysis of skin tumor cells shows evidence of aneuploidy in up to one-third of patients (Figure 48.7). DNA ploidy ranges from hyperdiploid to hypotetraploid confirming that chromosomal imbalances are usually associated with hypotetraploidy
- The MF cells show *TCR* gene rearrangement.
- Additional *ERBB2* (*Her2/neu*) gene copies and inactivation of *CDKN2A/p16* have been reported in MF cells.

VARIANTS OF MF

Over fifteen variants of MF have been described. Here, we briefly describe three clinically significant types, *folliculotropic MF, pagetoid reticulosis,* and *granulomatous slack skin.* Compared with classical MF, folliculotropic MF has a more aggressive clinical course and unfavorable prognosis, while the latter two have excellent prognosis.

Folliculotropic MF is a rare condition characterized by the infiltration of atypical MF cells in the hair follicles. The primary sites of involvement are head and neck, and the lesions may be associated with alopecia.

Pagetoid reticulosis is characterized by interepidermal growth and expansion of the MF cells. Pagetoid reticulosis is usually localized and has an excellent prognosis. The epidermal infiltrate is patchy, diffuse, or band-like and is often associated with reactive inflammatory cells such as lymphocytes and eosinophils. The immunophenotype is CD4+ and often CD30+.

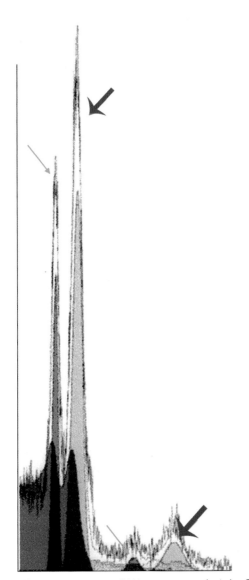

FIGURE 48.7 **HYPERDIPLOIDY IN MF.** DNA content analysis by flow cytometry demonstrates a hyperdiploid population (green arrow=diploid, red arrow=hyperdiploid).

Granulomatous slack skin is an extremely rare type involving major skin folds, such as axilla and groin. It consists of granulomatous formation with clonal infiltration of CD4+ T-cells.

Sézary Syndrome

Sézary syndrome (SS) is characterized by cutaneous T-cell lymphoma, erythroderma, presence of atypical cerebriform lymphocytes (Sézary cells) in the peripheral blood, and lymphadenopathy. In general, Sézary syndrome is considered the disseminated (leukemic) phase of MF. Criteria for the diagnosis of SS are:

- An absolute Sézary cell count of ≥1000/µL;
- An expanded CD4:CD8 ratio of >10 or evidence of a T-cell clone;
- Immunophenotypic abnormalities, such as loss of one or more T-cell-associated antigens.

Clinical features include erythroderma, palmar and plantar hyperkeratosis, and generalized lymphadenopathy. Pruritis and alopecia may be present. SS is an aggressive disease with a 5-year survival rate of 10–20%.

MORPHOLOGY

- The Sézary cells have a variable amount of non-granular cytoplasm and show the characteristic delicately convoluted, cerebriform nucleus with condensed chromatin and inconspicuous nucleoli (Figures 48.8B inset and 48.8A). These cells may vary in size, with the smaller forms referred to as Lutzner cells.
- The skin biopsies may resemble MF or may be non-diagnostic. Epidermotropism may be absent.
- Lymph nodes show effacement of nodal architecture and heavy infiltration of relatively monotonous Sézary cells.
- Bone marrow involvement, if present, is often sparse and interstitial (Figure 48.8).

IMMUNOPHENOTYPE

Immunophenotypic features are similar to those of MF and include (Figure 48.9):

- Expression of CD2, CD3, CD5, and TCR-αβ with variably aberrant patterns;
- CD4 in most cases;
- Characteristic absence of CD7 and CD26;
- Common expression of CD25;
- Uncommon phenotypes of CD8, double positive, or double negative.

CYTOGENETIC AND MOLECULAR STUDIES

- Chromosome aberrations are detectable by conventional cytogenetic analysis in up to 60% of cases with PB involvement (Figure 48.10)
- The most frequent genetic lesions include monosomy 10, losses of 10q and 17p, gains of 8q24 and 17q, and diverse structural alterations involving these regions.
- Expression patterns in regions of genomic imbalance show that a large number of genes are deregulated, and may play a causative role.
- Overall, chromosomal instability is characteristic of this lymphoma and related to a poor prognosis, but no specific abnormalities that may be directly involved in development of the disease have yet been found.

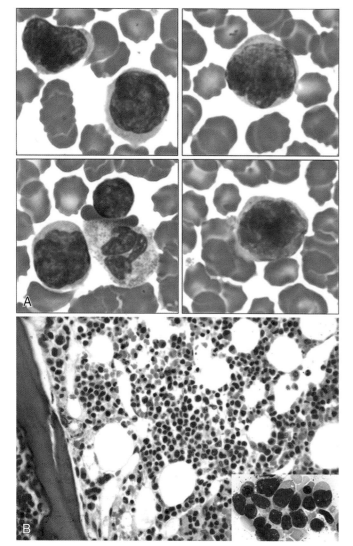

FIGURE 48.8 SÉZARY SYNDROME. Several Sézary cells with convoluted nuclei are demonstrated in the blood smear (A). Bone marrow biopsy section (B) and smear (B, inset) show an aggregate of atypical lymphocytes with irregular nuclei. *Adapted from Naeim F. Pathology of Bone Marrow, 2nd edn. Williams & Wilkins, Baltimore, 1997, by permission.*

- CGH studies show chromosome imbalances in greater than 50%: losses at 1p, 17p, 10q, and 19; gains at 4q, 18, and 17q. 1p33–36 and 10q26 may represent regions of minimal recurrent deletion.
- The MF cells show clonal *TCR* gene rearrangement.
- Since the molecular target may be a relatively small number of T-cells in a skin biopsy, care must be taken to assure optimal sampling of the lesion. Some laboratories have resorted to laser-capture microdissection to address this problem.
- Adding to the technical challenge is the fact that most of these specimens are received in the molecular pathology laboratory already formalin-fixed and paraffin-embedded. In that case there is no recourse to Southern blotting if the PCR study turns out negative.

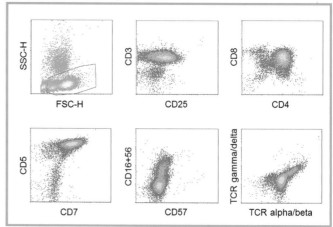

FIGURE 48.9 FLOW CYTOMETRIC FINDINGS OF SÉZARY SYNDROME. Open scatter gate density display (in green) of peripheral blood highlights a prominent population of intermediate-sized to large lymphoid cells. These lymphoid cells are predominantly abnormal T-cells (in purple) displaying double positivity of CD4 and CD8. In addition, the neoplastic T-cells are positive for CD3, CD5 (homogeneous clustering profile), CD7, CD25 (partial), CD16+56 (partial), and TCR alpha/beta.

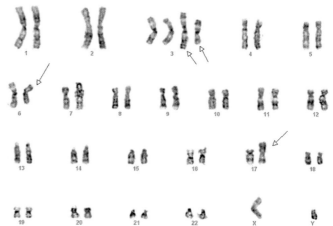

FIGURE 48.10 Karyotype of a patient with Sezary Syndrome showing deletion 6q and 17p: 48,XY,+der(3)t(1;3)(q21;q12),+add(3)(q12),del(6)(q13q21),der(17)t(3;17)(q21;p13).

Differential Diagnosis

The differential diagnosis of MF includes a garden variety of benign reactive skin disorders, such as psoriasis, eczema, parapsoriasis, drug reactions, contact dermatitis, and photodermatitis.

These distinctions can be quite challenging and often settled only by the results of *TCR* gene rearrangement studies. As most referred skin biopsies will be paraffin-embedded, PCR analysis is the primary approach for these lesions, since formalin-fixed tissue does not yield DNA of

Table 48.1

Clinicopathologic Features of T-prolymphocytic Leukemia (T-PLL), Adult T-cell Leukemia/Lymphoma (ATL), and Mycosis Fungoides (MF)

Features	T-PLL	ATL	MF
Median age (years)	>50	50	55–60
Male:Female	?	1.5	2
Association	ATM[1]	HTLV-1	Unknown
Skin involvement	15%	15%	100%
Immunophenotype	CD3+, CD4+, CD7+, some CD4+/CD8+	CD3+, CD4+, CD7−	CD3+, CD4+, CD7−, CD26−
Overall prognosis	Aggressive	Aggressive	Indolent

[1] Ataxia telangiectasia gene.

high enough quality for Southern blot analysis to be reliable. However, it is in this setting that the potential for false-positive results due to spurious amplification of a small number of T-lymphocytes in the specimen (pseudoclonality) comes to the fore. In our laboratory, we include a disclaimer to this effect when the skin biopsy contains only scattered or scant T-lymphocytes and/or when we see an isolated clonal spike in the PCR profile in the absence of any polyclonal background signal; this is a hint that the signal may be artifactual.

MF should be distinguished from other primary cutaneous lymphomas. MF and SS share overlapping morphologic and immunophenotypic features with ATL (see Chapter 45). ATL occurs in a younger age group and is associated with HTLV-I (see Table 48.1) The neoplastic cells in a minority of T-PLL cases may be Sézary cell-like (see Chapter 44). The T-PLL cells are usually CD7+.

Additional Resources

Döbbeling U: The molecular pathogenesis of mycosis fungoides and Sézary syndrome, *G Ital Dermatol Venereol* 143:385–394, 2008.

Fraser-Andrews EA, Mitchell T, Ferreira S, et al: Molecular staging of lymph nodes from 60 patients with mycosis fungoides and Sézary syndrome: correlation with histopathology and outcome suggests prognostic relevance in mycosis fungoides, *Br J Dermatol* 155: 756–762, 2006.

Galper SL, Smith BD, Wilson LD: Diagnosis and management of mycosis fungoides, *Oncology (Williston Park)* 24:491–501, 2010.

Gerami P, Rosen S, Kuzel T, et al: Folliculotropic mycosis fungoides: an aggressive variant of cutaneous T-cell lymphoma, *Arch Dermatol* 144:738–746, 2008.

Hristov AC, Vonderheid EC, Borowitz MJ: Simplified flow cytometric assessment in mycosis fungoides and sézary syndrome, *Am J Clin Pathol* 136:944–953, 2011.

Humme D, Lukowsky A, Sterry W: Diagnostic tools in mycosis fungoides, *G Ital Dermatol Venereol* 145:375–384, 2010.

Hwang ST, Janik JE, Jaffe ES, et al: Mycosis fungoides and Sézary syndrome, *Lancet* 371:945–957, 2008.

Kempf W, Ostheeren-Michaelis S, Paulli M, et al: Granulomatous mycosis fungoides and granulomatous slack skin: a multicenter study of the Cutaneous Lymphoma Histopathology Task Force Group of the European Organization For Research and Treatment of Cancer (EORTC), *Arch Dermatol* 144:1609–1617, 2008.

Kim EJ, Lin J, Junkins-Hopkins JM, et al: Mycosis fungoides and sézary syndrome: an update, *Curr Oncol Rep* 8:376–386, 2006.

Lansigan F, Choi J, Foss FM: Cutaneous T-cell lymphoma, *Hematol Oncol Clin North Am* 22:979–996, 2008.

Lenane P, Powell FC, O'Keane C, et al: Mycosis fungoides—a review of the management of 28 patients and of the recent literature, *Int J Dermatol* 46:19–26, 2007.

Meyerson HJ: Flow cytometry for the diagnosis of mycosis fungoides, *G Ital Dermatol Venereol* 143:21–41, 2008.

Möbs M, Knott M, Fritzen B, et al: Diagnostic tools in Sézary syndrome, *G Ital Dermatol Venereol* 145:385–391, 2010.

Olsen E, Vonderheid E, Pimpinelli N, et al: Revisions to the staging and classification of mycosis fungoides and Sézary syndrome: a proposal of the International Society for Cutaneous Lymphomas (ISCL) and the cutaneous lymphoma task force of the European Organization of Research and Treatment of Cancer (EORTC), *Blood* 110:1713–1722, 2007.

Olsen EA, Rook AH, Zic J, et al: Sézary syndrome: immunopathogenesis, literature review of therapeutic options, and recommendations for therapy by the United States Cutaneous Lymphoma Consortium (USCLC), *J Am Acad Dermatol* 64:352–404, 2011.

Prince HM, Whittaker S, Hoppe RT: How I treat mycosis fungoides and Sézary syndrome, *Blood* 114:4337–4353, 2009.

Reddy K, Bhawan J: Histologic mimickers of mycosis fungoides: a review, *J Cutan Pathol* 34:519–525, 2007.

Vaughan J, Harrington AM, Hari PN, et al: Immunophenotypic stability of sézary cells by flow cytometry: usefulness of flow cytometry in assessing response to and guiding alemtuzumab therapy, *Am J Clin Pathol* 137:403–411, 2012.

Zinzani PL, Ferreri AJ, Cerroni L: Mycosis fungoides, *Crit Rev Oncol Hematol* 65:172–182, 2008.

Other Primary Cutaneous T-Cell Lymphoproliferative Disorders

Primary Cutaneous CD30-Positive T-Cell Lymphoproliferative Disorders

Primary cutaneous CD30-positve T-cell lymphoproliferative disorders are divided into two major subtypes: (1) primary cutaneous anaplastic large cell lymphoma, and (2) lymphomatoid papulosis.

PRIMARY CUTANEOUS ANAPLASTIC LARGE CELL LYMPHOMA

Primary cutaneous anaplastic large cell lymphoma (C-ALCL) is a pleomorphic T-cell lymphoma in which >75% of the tumor cells express CD30. This entity should be distinguished from transformation of mycosis fungoides to CD30+ large cell lymphoma, and secondary cutaneous involvement in systemic ALCL.

Most patients show solitary skin lesions presenting as nodules or papules which may be ulcerated. Multifocal skin lesions are observed in about 20% of cases, and extranodal involvement may occur in about 10% of cases. Spontaneous regression has been observed.

C-ALCL usually occurs in elderly patients with a median age of about 60 years. The male to female ratio is about 1.5–2:1. The prognosis for localized lesions is very good with a 5-year survival rate of over 90%, compared with 50% for generalized cutaneous ALCL. The conventional treatment for localized lesions is excision with or without radiation. Combination chemotherapy is recommended for disseminated skin disease.

Morphology (Figure 49.1)

- There is dense dermal infiltration with clusters or sheets of anaplastic large cells, surrounded by lymphocytes. Numerous Reed–Sternberg-like cells and/or multinucleated giant cells may be present.
- Epidermal involvement is rare and may be associated with ulceration.
- Extension into subcutaneous tissue may be present.
- A rare "pyogenic" variant has been described with numerous neutrophils.

Immunophenotype

Immunophenotypic features include (Figure 49.2):

- Common expression of CD4 T-helper cell phenotype, and positivity for CD30;
- Variable loss of pan-T-cell antigens;
- Common expression of cytotoxic granule associated proteins TIA1, granzyme B, and perforin;
- Absent ALK, EMA, and CD15.

Molecular and Cytogenetic Studies

Most cases (>90%) show *TCR* gene rearrangement.

LYMPHOMATOID PAPULOSIS

Lymphomatoid papulosis (LyP) is a chronic papular or nodular skin lesion consisting of large anaplastic lymphoid cells embedded in an inflammatory background, and often a central necrosis. The disease is recurrent and shows spontaneous regression with the disappearance of the individual lesions in 3–12 weeks. The male to female ratio is about 2.5:1. Most patients under age 19 are males and most over age 19 are females and have a high incidence of thyroiditis. Approximately 20% of patients may develop or show coexisting T-cell cutaneous lymphoma or Hodgkin lymphoma. LyP has an excellent prognosis.

Morphology (Figure 49.3)

- In early stages, there is perivascular and patchy dermal infiltration of large atypical lymphoid cells mixed with variable numbers of inflammatory cells. The atypical cells may be multinucleated or resemble Reed–Sternberg cells.
- In advanced stages, there are sheets or large clusters of monotonous large atypical cells admixed with a small number of inflammatory cells.

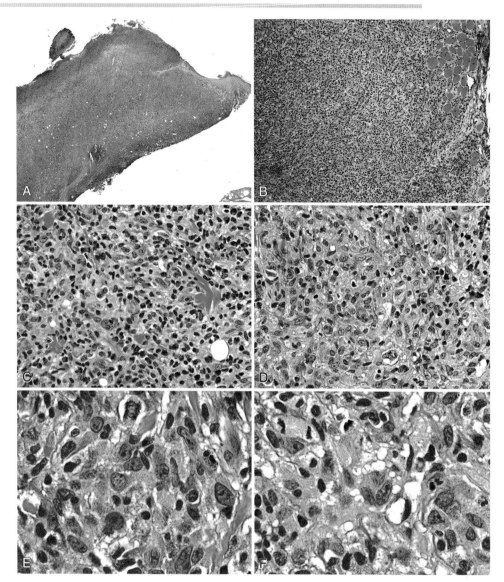

FIGURE 49.1 **Primary cutaneous anaplastic large cell lymphoma.** Skin ulceration with extensive infiltration of dermis and subcutaneous tissue by pleomorphic large cells (A to F from low to high power).

There are three morphologic types:

- Type A: Clusters of CD30 positive large cells are found in a background of inflammatory cells.
- Type B: Small atypical lymphoid cells with convoluted nuclei infiltrate upper dermis and epidermis (dermatotropic), resembling mycosis fungoides. This type is uncommon.
- Type C: Sheets or large clusters of large atypical CD30-positive cells with scattered inflammatory cells.

Immunophenotype

Immunophenotypic features include:

- Type A and type C lesions: Large atypical cells have phenotypic features similar to those seen in C-ALCL (see above);
- Type B: CD4 T-cell phenotype but lack of CD30;

Molecular and Cytogenetic Studies

- *TCR* gene rearrangement has been reported in about 60% of cases.
- The profile of the clonal rearrangement by PCR can be used to distinguish primary LP lesions from metastases of other T-cell lymphomas.

Subcutaneous Panniculitis-Like T-Cell Lymphoma

Subcutaneous panniculitis-like T-cell lymphoma is a rare lymphoma which involves subcutaneous fat without dermal or epidermal infiltration, leading to erythematous or violaceous nodules, plaques, or both.

The clinical manifestation of subcutaneous panniculitis-like T-cell lymphoma is variable, ranging from indolent course to a rapidly fatal hemophagocytic process. The systemic hemophagocytosis is characterized by fever,

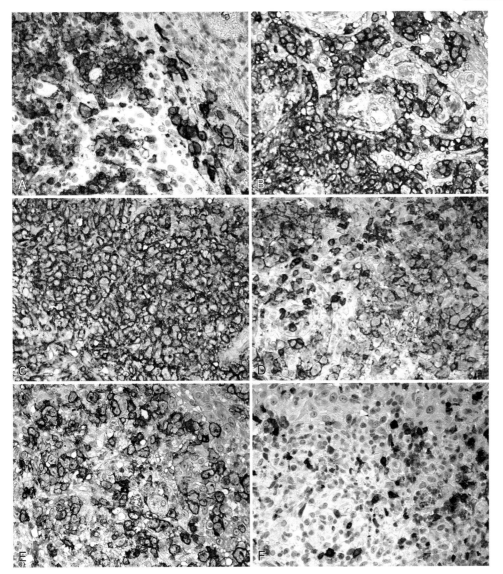

FIGURE 49.2 **PRIMARY CUTANEOUS ANAPLASTIC LARGE CELL LYMPHOMA, IMMUNOPHENOTYPE.** The neoplastic cells express CD2 (A), CD3 (B), CD4 (C), CD5 (D), and CD30 (F) by immunohistochemical stains. They are negative for CD7 (F).

hepatosplenomegaly, lung infiltration, liver dysfunction, coagulation abnormalities, and pancytopenia. This hemophagocytic syndrome may develop before or during the manifestation of T-cell lymphoma.

Local radiation therapy and/or systemic chemotherapy are used. The 5-year survival rate has been reported as 80% in a recent study.

MORPHOLOGY (FIGURES 49.4 AND 49.5)

- The subcutaneous infiltrate consists of a mixture of small, medium to large atypical cells with areas of necrosis. The neoplastic T-cells have a tendency to rim around adipocytes.
- The infiltrate often contains reactive histiocytes which may show hemophagocytosis.

IMMUNOPHENOTYPE

Immunophenotypic features include:

- Mature cytotoxic T-cell phenotype with expression of CD3, CD8, and TCR-αβ;
- Expression of cytotoxic markers TIA1, granzyme B, and perforin;
- Negativity for CD56.

MOLECULAR AND CYTOGENETIC STUDIES

The tumor cells in most patients show clonal *TCR* gene rearrangement.

Cutaneous γδ T-Cell Lymphoma

Cutaneous γδ T-cell lymphoma was previously considered a subtype of subcutaneous panniculitis-like T-cell lymphoma, accounting for 25% of cases. Morphologic features

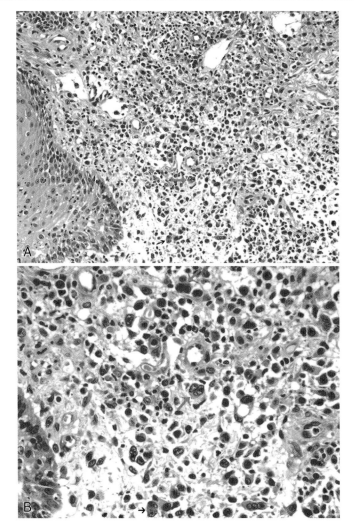

FIGURE 49.3 **LYMPHOMATOID PAPULOSIS.** Skin biopsy demonstrates a polymorphous infiltrate with atypical large lymphocytes and Reed–Sternberg-like cells (arrows) (A, low power; B, high power).

are similar to those of subcutaneous panniculitis-like, except that dermal and epidermal involvement may be present. Cutaneous γδ T-cell lymphoma is a more aggressive disease than subcutaneous panniculitis-like T-cell lymphoma. The disease has a poor prognosis with a median survival of less than 2 years.

The neoplastic cells are of TCR-γδ type, positive for CD2, CD3, and CD56, with strong expression of cytotoxic markers. They are mostly negative for both CD4 and CD8, though some cases may be positive for CD8. The tumor cells are also negative for CD5 with variable loss of CD7.

Southern blot analysis using the TCR-β probe is negative in these cases, but PCR analysis for *TCR-γ* will usually be informative. However, negativity for TCR-β clonality alone cannot always be used to infer γδ origin.

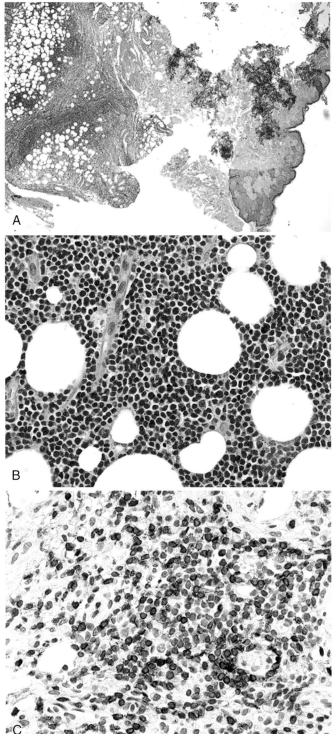

FIGURE 49.4 **SUBCUTANEOUS PANNICULITIS-LIKE T-CELL LYMPHOMA.** Skin biopsy (A, low power; B, high power). The infiltrating lymphocytes are CD3+ (C).

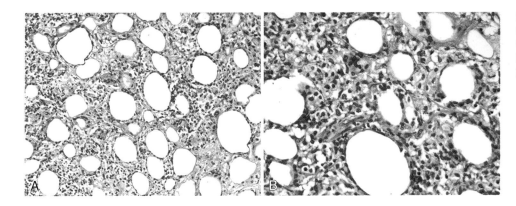

FIGURE 49.5 **SUBCUTANEOUS PANNICULITIS-LIKE T-CELL LYMPHOMA.** Subcutaneous fatty tissue shows infiltrate (A, low power; B, high power).

Primary Cutaneous Aggressive Epidermotropic CD8+ Cytotoxic T-Cell Lymphoma

Primary cutaneous aggressive epidermotropic CD8+ cytotoxic T-cell lymphoma is characterized by localized or disseminated eruptive skin lesions (papules, nodules, tumors) with epidermal infiltration of CD8+ cytotoxic T-cells and an aggressive clinical course. Tumor cells vary from small to large, show pleomorphic or blastic nuclei, and express CD3, CD8, βF1, plus cytotoxic proteins. The neoplastic cells are negative for CD4, with variable loss of CD2, CD5, and CD7. There is evidence of clonal *TCR* gene rearrangement.

Primary Cutaneous CD4+ Small/Medium T-Cell Lymphoma

This cutaneous lymphoma consists of a pleomorphic small to medium sized lymphocytes, often appearing as a solitary lesion on the face, neck, or upper trunk (Figure 49.6). Unlike MF, there is absence of patches. There is a dense nodular infiltration of the dermis with a tendency to extend into the subcutis. Epiderrmotropism may be focally present.

The neoplastic cells express CD3 and CD4, with variable loss of pan-T-cell markers. They are negative for CD8, CD30, plus cytotoxic proteins. The *TCR* genes are clonally rearranged.

Epstein–Barr Virus-Associated Hydroa Vacciniforme-Like Cutaneous Lymphoma of Childhood

Hydroa-like cutaneous T-cell lymphoma (hydroa-like CTCL) is a rare EBV-associated T-cell lymphoma of childhood presenting as cutaneous rash characterized by edema, blisters, ulcers, crusts, and scars, resembling hydroa vacciniforme. It is seen mainly on the face and sometimes on the extremities.

The lesion consists of T-cell infiltration of the skin and subcutis with variable exocytosis and angiocentricity (Figure 49.7). There is often ulceration of the overlying epidermis. The neoplastic cells are small to medium size, often with no significant atypia or increased mitosis. They are either cytotoxic T-cells or NK-cells and express EBV-EBER. Cases with cytotoxic T-cell phenotype show *TCR* gene rearrangement.

Differential Diagnosis

The main distinguishing morphologic, immunophenotypic, and molecular/cytogenetic features of different types of primary cutaneous T-cell lymphomas are summarized in Table 49.1. The differential diagnosis of primary cutaneous T-cell lymphomas also includes a garden variety of benign reactive skin disorders and secondary cutaneous lymphomas.

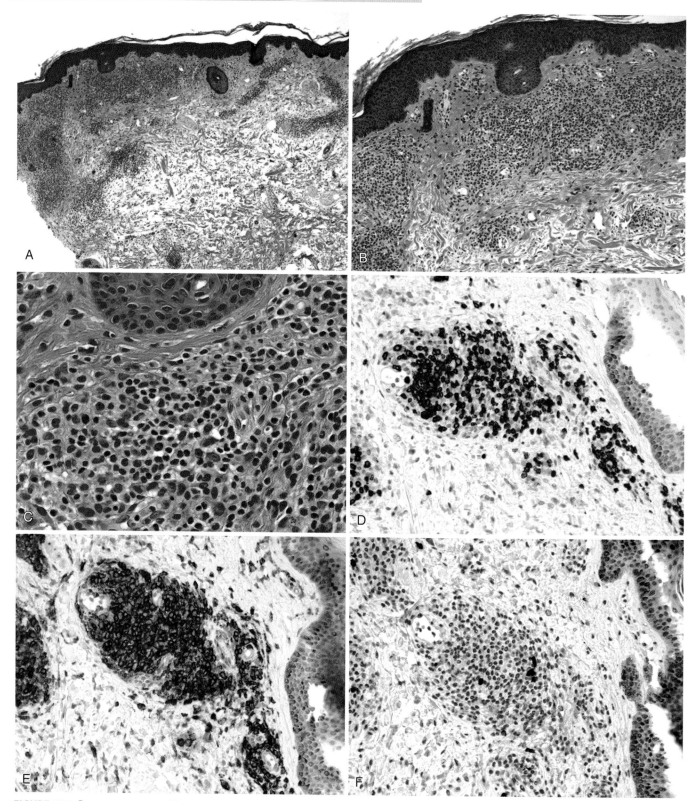

FIGURE 49.6 PRIMARY CUTANEOUS CD4+ T-CELL LYMPHOMA. Skin biopsy section demonstrating a patchy dermal lymphoid infiltration consisting of small to medium sized lymphocytes (A, low power; B, intermediate power; C, high power). The lymphocytes are positive for CD3 (D), and CD4 (E), and negative for CD8 (F).

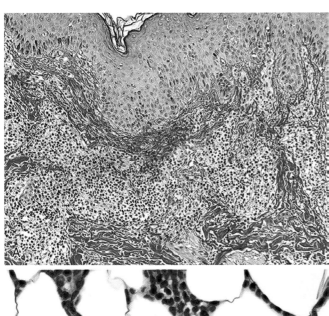
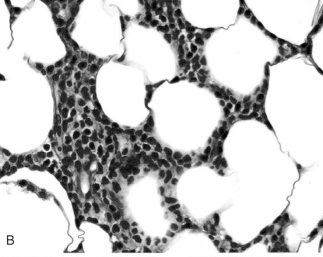
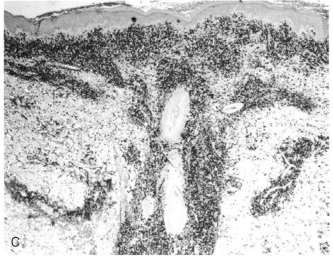

FIGURE 49.7 HYDROA VACCINIFORME-LIKE T-CELL LYMPHOMA. Small to medium sized lymphoid cells infiltrate the dermis (A). Infiltrate extends into subcutaneous tissue (B). Nearly all lymphoid cells are EBV-EBER+ (C).
From Jaffe ES, Harris NL, Vardiman JW, et al. Hematopathology. Saunders/Elsevier, Philadelphia, 2010, by permission.

Table 49.1
Differential Diagnosis of Primary Cutaneous T-Cell Lymphomas

Type	Morphology	Immunophenotype	Molecular/Cytogenetics
Mycosis fungoides	Patches, plaques, tumors, epidermotropism, Pautrier microabscesses, cells with cerebriform nuclei	CD3+, CD4+, βF1+, CD7−, CD8−	TCR rearranged
C-ALCL[1]	Often solitary tumor with ulceration, pleomorphic infiltrate of large atypical cells, including multinucleated forms	CD3+, CD4+, CD30+, CD8−, EMA−, ALK−	TCR rearranged
SPTCL[2]	Subcutaneous nodules, tumor cells rim individual fat cells, prominent apoptosis	CD3+, CD8+, βF1+, CD8−	TCR rearranged
γδ T-cell lymphoma	Subcutaneous, dermal, epidermal, tumor cells rim individual fat cells, prominent apoptosis	CD3+, CD56+, TCR-γ-1+, CD4−, CD8−	TCR-γ rearranged
CD8+ aggressive epidermotropic	Pagetoid epidermotropism, adenexal destruction	CD3+, CD8+, βF1+, CD4−, CD5−, CD56−	TCR rearranged
CD4+ small/medium T-cell lymphoma	Solitary or localized lesions, monotonous dermal infiltrates of small to medium sized lymphocytes	CD3+, CD4+, βF1+, CD8−	TCR rearranged

[1]C-ALCL, cutaneous anaplastic large cell lymphoma.
[2]SPTCL, subcutaneous panniculitic T-cell lymphoma.
Adapted from Jaffe ES, Harris NL, Vardiman JW, et al. Hematopathology. Saunders/Elsevier, Philadelphia, 2010.

Additional Resources

Akilov OE, Pillai RK, Grandinetti LM, et al: Clonal T-cell receptor β-chain gene rearrangements in differential diagnosis of lymphomatoid papulosis from skin metastasis of nodal anaplastic large-cell lymphoma, *Arch Dermatol* 147:943–947, 2011.

Barrionuevo C, Anderson VM, Zevallos-Giampietri E, et al: Hydroa-like cutaneous T-cell lymphoma: a clinicopathologic and molecular genetic study of 16 pediatric cases from Peru, *Appl Immunohistochem Mol Morphol* 10:7–14, 2002.

Belloni-Fortina A, Montesco MC, Piaserico S, et al: Primary cutaneous CD30+ anaplastic large cell lymphoma in a heart transplant patient: case report and literature review, *Acta Derm Venereol* 89:74–77, 2009.

Beltraminelli H, Leinweber B, Kerl H, et al: Primary cutaneous CD4+ small-/medium-sized pleomorphic T-cell lymphoma: a cutaneous nodular proliferation of pleomorphic T lymphocytes of undetermined significance? A study of 136 cases, *Am J Dermatopathol* 31:317–322, 2009.

Cetinözman F, Jansen PM, Willemze R: Expression of Programmed Death-1 in primary cutaneous CD4-positive small/medium-sized pleomorphic T-cell lymphoma, cutaneous pseudo-T-cell lymphoma, and other types of cutaneous T-cell lymphoma, *Am J Surg Pathol* 36:109–116, 2012.

Diamantidis MD, Myrou AD: Perils and pitfalls regarding differential diagnosis and treatment of primary cutaneous anaplastic large-cell lymphoma, *ScientificWorldJournal* 11:1048–1055, 2011.

Garcia-Herrera A, Song JY, Chuang SS, et al: Nonhepatosplenic γδ T-cell lymphomas represent a spectrum of aggressive cytotoxic T-cell lymphomas with a mainly extranodal presentation, *Am J Surg Pathol* 35:1214–1225, 2011.

Go RS, Wester SM: Immunophenotypic and molecular features, clinical outcomes, treatments, and prognostic factors associated with subcutaneous panniculitis-like T-cell lymphoma: a systematic analysis of 156 patients reported in the literature, *Cancer* 101:1404–1413, 2004.

Gormley RH, Hess SD, Anand D, et al: Primary cutaneous aggressive epidermotropic CD8+ T-cell lymphoma, *J Am Acad Dermatol* 62:300–307, 2010.

Guitart J, Querfeld C: Cutaneous CD30 lymphoproliferative disorders and similar conditions: a clinical and pathologic prospective on a complex issue, *Semin Diagn Pathol* 26:131–140, 2009.

Jaffe ES, Harris NL, Vardiman JW, et al: *Hematopathology*, Philadelphia, 2010, Saunders/Elsevier.

Jang MS, Baek JW, Kang DY, et al: Subcutaneous panniculitis-like T-cell lymphoma: successful treatment with systemic steroid alone, *J Dermatol* 39:96–99, 2012.

Kadin ME: Pathobiology of CD30+ cutaneous T-cell lymphomas, *J Cutan Pathol* 33(Suppl 1):10–17, 2006.

Kagaya M, Kondo S, Kamada A, et al: Localized lymphomatoid papulosis, *Dermatology* 204:72–74, 2002.

Morimura S, Sugaya M, Tamaki Z, et al: Lymphomatoid papulosis showing γδ T-cell phenotype, *Acta Derm Venereol* 91:712–713, 2011.

Ohmatsu H, Sugaya M, Fujita H, et al: Primary cutaneous CD8+ aggressive epidermotropic cytotoxic T-cell lymphoma in a human T-cell leukaemia virus type-1 carrier, *Acta Derm Venereol* 90:324–325, 2010.

Parveen Z, Thompson K: Subcutaneous panniculitis-like T-cell lymphoma: redefinition of diagnostic criteria in the recent World Health Organization–European Organization for Research and Treatment of Cancer classification for cutaneous lymphomas, *Arch Pathol Lab Med* 133:303–308, 2009.

Plaza JA, Ortega P, Lynott J, et al: CD8-positive primary cutaneous anaplastic large T-cell lymphoma (PCALCL): case report and review of this unusual variant of PCALC, *Am J Dermatopathol* 32:489–491, 2010.

Querfeld C, Khan I, Mahon B, et al: Primary cutaneous and systemic anaplastic large cell lymphoma: clinicopathologic aspects and therapeutic options, *Oncology (Williston Park)* 24:574–587, 2010.

Rosen ST, Querfeld C: Primary cutaneous T-cell lymphomas, *Hematology Am Soc Hematol Educ Program*:323–330, 513, 2006.

Shunmugam M, Chan E, O'Brart D, et al: Cutaneous γδ T-cell lymphoma with bilateral ocular and adnexal involvement, *Arch Ophthalmol* 129:1379–1381, 2011.

Sim JH, Kim YC: CD8+ lymphomatoid papulosis, *Ann Dermatol* 23:104–107, 2011.

Tripodo C, Iannitto E, Florena AM, et al: Gamma-delta T-cell lymphomas, *Nat Rev Clin Oncol* 6:707–717, 2009.

Wang Y, Li T, Tu P, et al: Primary cutaneous aggressive epidermotropic CD8+ cytotoxic T-cell lymphoma clinically simulating pyoderma gangrenosum, *Clin Exp Dermatol* 34:e261–e262, 2009.

Xu Z, Lian S: Epstein-Barr virus-associated hydroa vacciniforme-like cutaneous lymphoma in seven Chinese children, *Pediatr Dermatol* 27:463–469, 2010.

Yamane N, Kato N, Nishimura M, et al: Primary cutaneous CD30+ anaplastic large-cell lymphoma with generalized skin involvement and involvement of one peripheral lymph node, successfully treated with low-dose oral etoposide, *Clin Exp Dermatol* 34:e56–e59, 2009.

Yip L, Darling S, Orchard D: Lymphomatoid papulosis in children: experience of five cases and the treatment efficacy of methotrexate, *Australas J Dermatol* 52:279–283, 2011.

Angioimmunoblastic T-Cell Lymphoma

Angioimmunoblastic T-cell lymphoma (AITL) is a peripheral T-cell lymphoma characterized by generalized lymphadenopathy, hepatosplenomegaly, anemia, hypergammaglobulinemia, and a polymorphic infiltrate involving germinal center T-helper (GC-Th) cells and follicular dendritic cells.

AITL accounts for about 2% of all non-Hodgkin lymphomas. The peak incidence is between the sixth and seventh decades with no significant sex predilection and wide geographical distribution. The clinical presentation often mimics an infectious process characterized by "B" symptoms and generalized lymphadenopathy. Hepatosplenomegaly has been reported in 50–70% of patients. Around 50% of patients complain of pruritus and/or show skin rashes (Table 50.1).

Table 50.1
Clinical Features and Laboratory Findings in Angioimmunoblastic T-Cell Lymphoma (AITL)[1]

	Frequency (%)
Symptoms and signs	
B symptoms	68–85
Generalized lymphadenopathy	94–97
Splenomegaly	70–73
Hepatomegaly	52–72
Skin rash	48–58
Effusions	23–37
Laboratory findings	
Anemia	40–57
Hypergammaglobulinemia	50–83
Autoantibodies	66–77
Elevated LDH	70–74
Bone marrow involvement	61
Clonal cytogenetic aberrations, such as +3, +5, and +X	70
TCR rearrangement	~100

[1] Adapted from Dogan A, Attygalle AD, Kyriakou C. Angioimmunoblastic T-cell lymphoma. Br J Haematol 2003; 121: 681–691.

A garden variety of autoimmune disorders have been observed in association with AITL, such as autoimmune hemolytic anemia, polyarthritis, rheumatoid arthritis, autoimmune thyroiditis, and vasculitis. Single agent chemotherapy (steroids or methotrexate) and various combinations of chemotherapeutic regimens have been tried, with an overall discouraging outcome and a 5-year survival rate of 30–35%.

Morphology and Laboratory Findings

The involved lymph nodes display partial or total effacement of nodal architecture by a polymorphic infiltrate, predominantly involving paracortical areas. The lymph node sinuses are usually well preserved. The following morphologic variations appear to correspond to the various stages of the disease (Figures 50.1 to 50.4).

- In the early stages of the disease, there is preservation of nodal architecture, follicular hyperplasia, poorly developed mantle zones, and expanded paracortex with a polymorphic infiltrate consisting of lymphocytes, eosinophils, plasma cells, macrophages, transformed large lymphoid blasts, occasional Reed–Sternberg-like cells, and vascular proliferation with abundant endothelial venules.
- In more advanced stages, the normal architecture is almost completely lost, except for occasional depleted follicles showing concentrically arranged follicular dendritic cells. In some cases proliferation of the follicular dendritic cells may extend beyond the follicles. A polymorphic infiltrate with numerous transformed blast cells and vascular proliferation are present.
- In the late stages of the disease (the most commonly observed morphology), there is complete effacement of nodal architecture, prominent proliferation of follicular dendritic cells, extensive vascular proliferation, and, in many cases, perivascular collections of atypical medium-sized to large lymphoid cells with clear or pale cytoplasm.

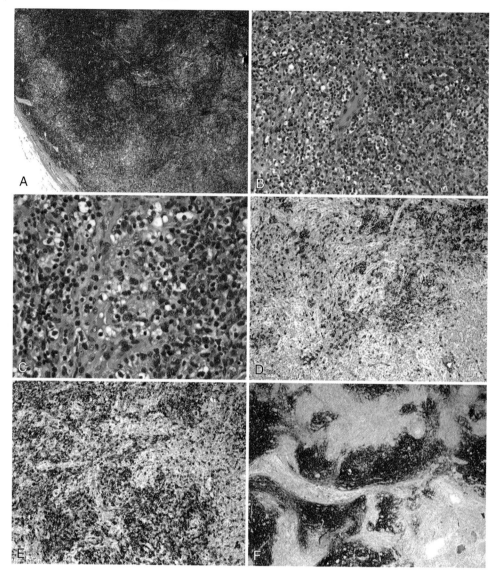

FIGURE 50.1 **ANGIOIMMUNOBLASTIC T-CELL LYMPHOMA.** Lymph node section showing effacement of nodal architecture with the presence of large cells with clear cytoplasm, primarily adjacent to vascular structures (A, low power; B, intermediate power; C, high power). The large cells with clear cytoplasm are follicular center CD10+ T-cells (D). Large aggregates of CD4+ T-cells (E) and CD21+ follicular dendritic cells (F) are demonstrated.

- Bone marrow, spleen, liver, skin, and lung are the most frequent extranodal sites of involvement (Figures 50.5 and 50.6). The involvement of the extranodal sites is often non-specific and consists of a polymorphic infiltrate mimicking an inflammatory process.
- Laboratory findings include anemia or pancytopenia, hypergammaglobulinemia, circulating autoantibodies, and elevated serum LDH (see Table 50.1).

Immunophenotype

MFC features include (Figure 50.7):

- Characteristic expression of normal TFH (follicular helper T-cell) phenotype with positivity of CD10;
- Expression of CD4;
- Expression of pan-T-cell markers with variable aberrancies.

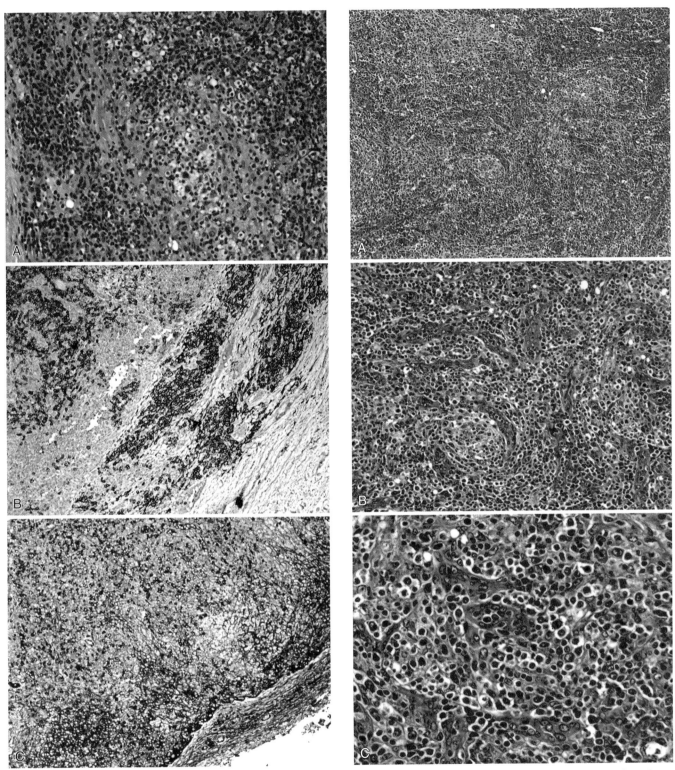

FIGURE 50.2 **ANGIOIMMUNOBLASTIC T-CELL LYMPHOMA.** Lymph node section demonstrating aggregates of plasma cells (arrow) and clusters of large cells with clear cytoplasm (A). Plasma cell aggregates express CD138 (B), and consist of a mixed population of kappa+ (brown) and lambda+ (red) cells (C).

FIGURE 50.3 **ANGIOIMMUNOBLASTIC T-CELL LYMPHOMA.** Lymph node section demonstrating hypervascularity and presence of medium to large cells with clear cytoplasm (A, low power; B, intermediate power; C, high power).

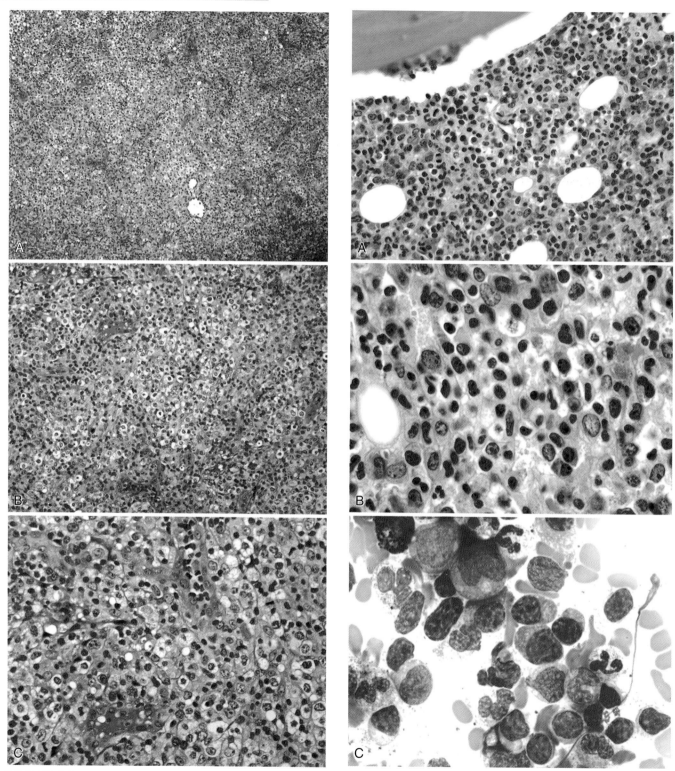

FIGURE 50.4 **ANGIOIMMUNOBLASTIC T-CELL LYMPHOMA.** Lymph node section demonstrating predominance of large cells with clear cytoplasm (A, low power; B, intermediate power; C, high power).

FIGURE 50.5 **ANGIOIMMUNOBLASTIC T-CELL LYMPHOMA.** The detection of bone marrow involvement is often challenging. Ill-defined polymorphic aggregates of lymphocytes, plasma cells, eosinophils and histiocytes may be detected in biopsy sections (A, intermediate power; B, high power) or bone marrow smears (C).

MOLECULAR AND CYTOGENETIC STUDIES   573

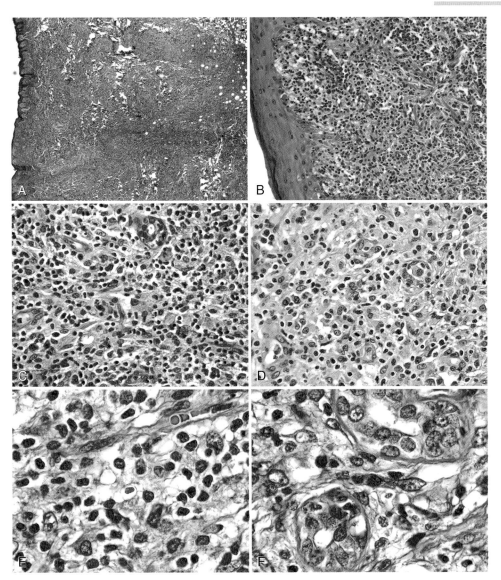

FIGURE 50.6 **ANGIOIMMUNOBLASTIC T-CELL LYMPHOMA OF THE SKIN.** Extensive infiltration of the dermis by a pleomorphic and vascular tumor (A, low power; B, intermediate power; C, high power). Numerous large cells with abundant clear cytoplasm are present (D and E), and there are high endothelial venules (F).

By immunohistochemical studies, the neoplastic cells are positive for additional TFH markers, such as BCL6, PD1, and CXCL13. Staining for CD21, CD23, and CD35 highlights expanded follicular dendritic meshworks. EBV staining by in situ hybridization studies is commonly positive.

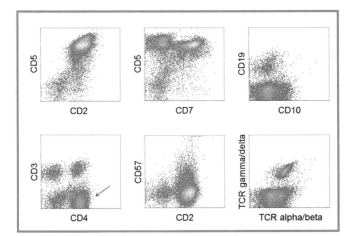

FIGURE 50.7 **FLOW CYTOMETRIC FINDINGS OF ANGIOIMMUNOBLASTIC T-CELL LYMPHOMA.** Lymphoid-enriched gate of axillary lymph node reveals mostly abnormal T-cells expressing CD4 (arrow) but negative for surface CD3, in a background of occasional reactive T-cells. The neoplastic T-cells are positive for CD2, CD5 (moderate to bright), CD10 (partial), CD57 (partial), and TCR alpha/beta (partial). They display dim CD7 expression.

Molecular and Cytogenetic Studies

- *TCR* rearrangement studies reveal clonal rearrangement in over 75% of cases (Figure 50.8).
- The affected lymph nodes in approximately 10% of cases may show an expanded monoclonal B-cell population, often in association with increased number of EBV-infected large B-cells. This phenomenon is considered secondary to EBV stimulation and not reflective of the primary malignant process.

- Approximately 70% of AITL patients show clonal chromosomal aberrations. The most frequent recurrent abnormalities include trisomy 3, trisomy 5, and an additional X chromosome (see Table 50.1). Other frequently observed recurrent cytogenetic abnormalities are gains of 11q13, 19, 22q, and rearrangements of 1p, 3p, 14q.

Differential Diagnosis

The differential diagnosis includes various viral infections and collagen vascular disorders. Bone marrow biopsy and fine needle aspiration or needle core biopsy of the enlarged lymph node usually do not yield a definitive diagnosis. Diagnosis is achieved by the morphologic examination of the entire lymph node.

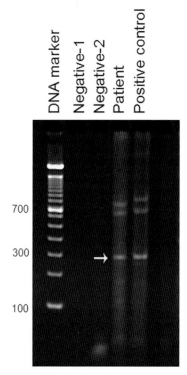

FIGURE 50.8 Southern blot analysis demonstrating *TCRB* gene rearrangement in a patient with angioimmunoblastic T-cell lymphoma.
Courtesy of Ryan Phan, Ph.D., Department of Pathology, VA Greater Los Angeles Heathcare System.

Additional Resources

Alizadeh AA, Advani RH: Evaluation and management of angioimmunoblastic T-cell lymphoma: a review of current approaches and future strategies, *Clin Adv Hematol Oncol* 6:899–909, 2008.

de Leval L, Gisselbrecht C, Gaulard P: Advances in the understanding and management of angioimmunoblastic T-cell lymphoma, *Br J Haematol* 148:673–689, 2010.

Dogan A, Attygalle AD, Kyriakou C: Angioimmunoblastic T-cell lymphoma, *Br J Haematol* 121:681–691, 2003.

Dunleavy K, Wilson WH, Jaffe ES: Angioimmunoblastic T cell lymphoma: pathobiological insights and clinical implications, *Curr Opin Hematol* 14:348–353, 2007.

Gaulard P, de Leval L: Follicular helper T cells: implications in neoplastic hematopathology, *Semin Diagn Pathol* 28:202–213, 2011.

Good DJ, Gascoyne RD: Atypical lymphoid hyperplasia mimicking lymphoma, *Hematol Oncol Clin North Am* 23:729–745, 2009.

Grogg KL, Morice WG, Macon WR: Spectrum of bone marrow findings in patients with angioimmunoblastic T-cell lymphoma, *Br J Haematol* 137:416–422, 2007.

Iannitto E, Ferreri AJ, Minardi V, et al: Angioimmunoblastic T-cell lymphoma, *Crit Rev Oncol Hematol* 68:264–271, 2008.

Jaffe ES, Harris NL, Vardiman JW, et al: *Hematopathology*, Philadelphia, 2010, Saunders/Elsevier.

Swerdlow SH, Campo E, Harris NL, et al: *WHO classification of tumours of haematopoietic and lymphoid tissues*, ed 4, Lyon, 2008, International Agency for Research on Cancer.

Zaki MA, Wada N, Kohara M, et al: Presence of B-cell clones in T-cell lymphoma, *Eur J Haematol* 86:412–419, 2011.

Anaplastic Large Cell Lymphomas

Anaplastic large cell lymphomas (ALCL) are a group of T-cell malignancies consisting of large anaplastic, CD30+ cells with pleomorphic and often horseshoe-shaped nuclei and abundant cytoplasm. The tumor cells have a tendency to grow cohesively in the lymph node sinuses, mimicking metastatic tumors. These neoplasms are divided into two major groups:

1. Anaplastic large cell lymphoma, ALK-positive.
2. Anaplastic large cell lymphoma, ALK-negative.

Anaplastic Large Cell Lymphoma, ALK-Positive

Anaplastic large cell lymphoma, ALK-positive is associated with translocation of the *ALK* gene and expression of ALK protein. This translocation in over 80% of cases is t(2;5)(p23;q35), creating a hybrid gene as the result of fusion of the *NPM1* (nucleophosmin) gene on chromosome 5 with the *ALK* gene on chromosome 2. The NPM1-ALK protein activates the antiapoptotic PI3K–Akt pathway and a number of signal transducers and activators of transcription proteins which are all important in cellular transformation. The oncogenic properties of *NMP1-ALK* have been supported by *in vivo* studies in experimental animals. The remaining <20% of ALK+ ALCL cases demonstrate translocations that result in the fusion of the *ALK* gene with other genes (see below).

ALCL, ALK+ constitutes about 10–20% of lymphomas in children and approximately 3% of all non-Hodgkin lymphomas in adults. The vast majority of the ALK+ ALCLs occur in patients under 30 years of age. The male to female ratio is about 6:1. Approximately 70% of patients present with constitutional symptoms (mostly high fever and weight loss) and are in stage III/IV. Extranodal involvement includes skin, bone, lung, liver, and soft tissues. Approximately 10% of patients may show bone marrow involvement. Leukemic presentation is uncommon.

ALK+ ALCL has a significantly better prognosis than ALK-negative ALCL, with a 5-year overall survival rate ranging from 70% to 90% compared with 15% to 37% in ALK-negative patients. Factors associated with poor prognosis include involvement of mediastinum, spleen, lung, or liver, CD56 expression, and small cell variant.

Multi-agent chemotherapy and autologous or allogeneic bone marrow transplantation are among the routine therapeutic approaches. Anti-CD30 therapy and vaccination against ALK protein are under investigation.

MORPHOLOGY

The affected lymph nodes are partially or totally effaced due to the infiltration of the tumor cells that often have a tendency to grow cohesively in the sinuses, resembling metastatic tumors (Figure 51.1). The tumor cells are usually admixed with inflammatory cells which predominantly consist of histiocytes and plasma cells. Sclerotic thick capsule and well-formed fibrous bands are infrequent.

Three major morphologic variants have been described: (1) common type, (2) small cell variant, and (3) lymphohistiocytic variant.

- The *common variant* (70% of cases) is characterized by the predominance of large pleomorphic cells with abundant clear to light blue cytoplasm, often eccentric, horse-shoe, kidney-shaped, or multilobulated nuclei, and multiple small nucleoli. The large cells which have a horse-shoe or kidney-shaped nucleus next to the Golgi area are called the "hallmark cells," because they are detected in almost all morphologic types. The neoplastic cells may demonstrate cytoplasmic vacuoles in touch preparations. The multilobulated cells may show pseudonuclear inclusions (doughnut cells) due to invagination of the nuclear membrane (Figures 51.2 to 51.7).
- The *small cell variant* (5–10% of cases) consists of a mixture of large, medium-sized, and small pleomorphic cells. Small and

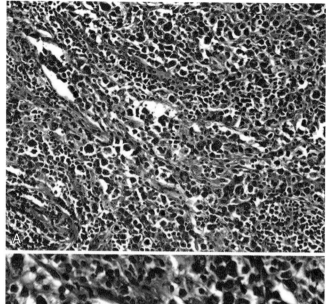
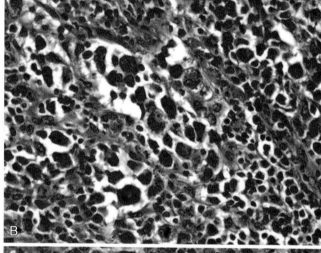
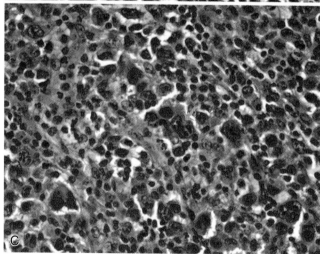

FIGURE 51.1 ANAPLASTIC LARGE CELL LYMPHOMA. Lymph node section demonstrating sinusoidal involvement. Horse-shoe (hallmark) cells (arrow) and Hodgkin-like cells are present (A, low power; B, intermediate power; C, high power).

medium-sized cells are predominant, depicting a clear cytoplasm and an irregular nucleus (fried egg cells). The large neoplastic cells tend to cluster around small vessels.
- The *lymphohistiocytic variant* (5–10% of cases) consists of small and large neoplastic cells including "hallmark" cells as well as large numbers of reactive histiocytes. Hemophagocytic macrophages may be present. The abundance of histiocytes may mask the neoplastic cell population in the H&E stains, but immunohistochemical stains for CD30 and ALK help to identify the tumor cells.

Other morphologic variants, such as Hodgkin-like pattern with the presence of Reed–Sternberg-like multinucleated cells, sarcomatoid form with large, bizarre spindle-shaped tumor cells, and giant cell-rich type with numerous multinucleated giant cells, have been reported.

An involved lymph node may show more than one morphologic pattern of involvement (composite pattern).

IMMUNOPHENOTYPE

MFC characterization may be difficult due to fragility of the large neoplastic cells. When neoplastic cells can remain intact after processing and staining, variable T-cell aberrancies are commonly detected (Figure 51.8). By immunohistochemical studies, the large neoplastic cells are (see Figures 51.3, 51.5 and 51.6):

- Positive for CD30 with membrane and Golgi staining;
- Positive for ALK with variable staining patterns (cytoplasmic, nuclear, or both);
- Variably positive for EMA in most cases;
- Negative for CD3 in more than 75% of cases;
- Commonly positive for CD2, CD4, and CD5; CD8 phenotype is rare;
- Variably positive for CD45 and strongly positive for CD25;
- Variably positive for TIA1 and granzyme B;
- Positive for CD43 and clusterin in most cases;
- Negative for EBV.

MOLECULAR AND CYTOGENETIC STUDIES

- Most cases of ALCL demonstrate clonal *TCR* rearrangements and the *NPM1-ALK* fusion gene.
- EBV sequences are not detected.
- The *NPM1-ALK* fusion is associated with a reciprocal balanced t(2;5)(p32;q35) (Figure 51.9).
- This translocation is the most common *ALK*-related chromosomal aberration, accounting for about 75% of all ALK-positive cases of ALCL. It is most readily and sensitively detected by FISH, but PCR-based molecular testing, similar to that used to detect the *BCR-ABL1* fusion in CML, can be done.
- The remaining 25% show a variety of *ALK*-related rearrangements including t(1;2)(q25;p23) [*TPM3-ALK*], t(2;3)(p23;q21) [*TFG-ALK*] (Figure 51.10), t(2;17)(p23;q23) [*CLTC-ALK*], t(X;2)(q11–12;p23) [*MSN-ALK*], and inv(2)(p23q35) (Table 51.1).
- The t(2;5) is commonly observed as part of a complex karyotype, with various structural and/or numerical anomalies (+7, +9, and +X).
- The expression of ALK in hematologic neoplasms is largely limited to ALCL tumors of T- or null-cell immunophenotype.
- Because multiple chromosomal regions are involved in the translocations with the 2p23 (*ALK* locus), the rearrangements can be easily identified by a 2p23-specific dual-color

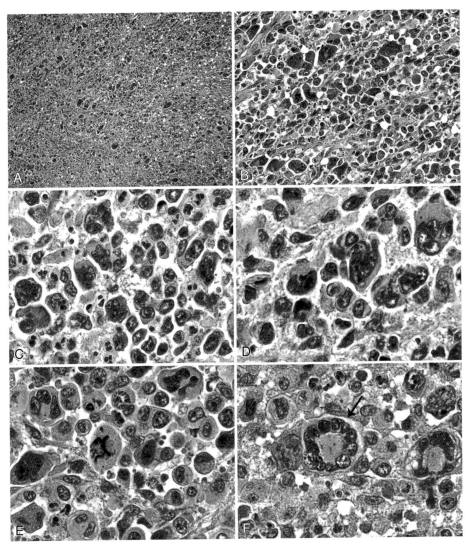

FIGURE 51.2 **ANAPLASTIC LARGE CELL LYMPHOMA.** Lymph node section demonstration a diffuse infiltration with highly anaplastic large cells in a background of lymphocytes and histiocytes. Hodgkin-like cells are present (B–E). A hallmark cell (green arrow) and a wreath cell (black arrow) are demonstrated (F).

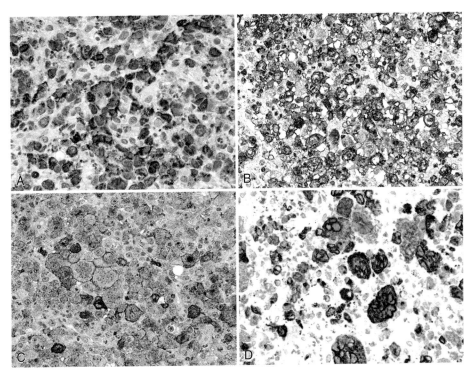

FIGURE 51.3 Immunohistochemical stains show anaplastic large cells expressing ALK (A), CD30 (B), EMA (C), and granzyme B (D).

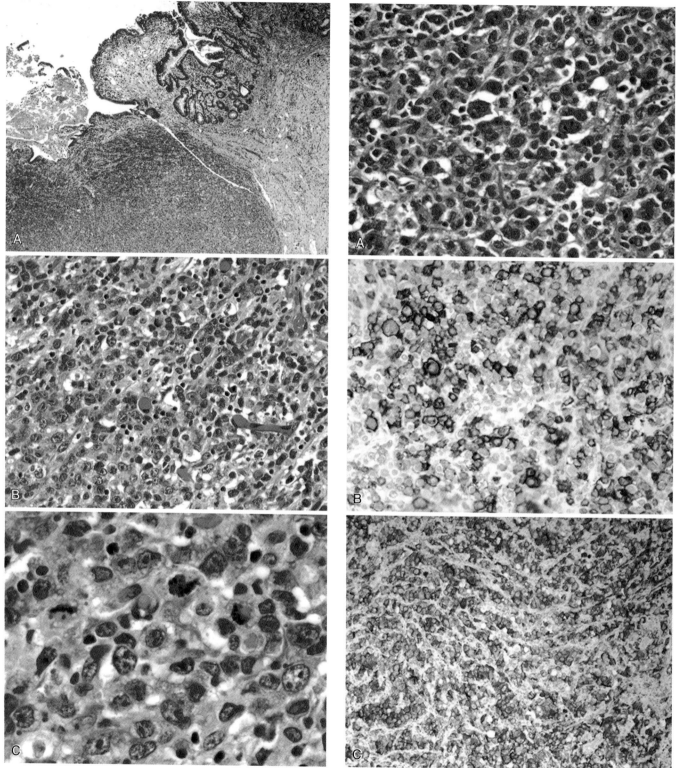

FIGURE 51.4 Gastric involvement with anaplastic large cell lymphoma (A, low power; B, intermediate power; C, high power).

FIGURE 51.5 **ANAPLASTIC LARGE CELL LYMPHOMA.** H&E (A) and immunohistochemical stains demonstrating expression of CD30 (B), and ALK (C). Arrow shows a Reed–Sternberg-like binucleated cell.

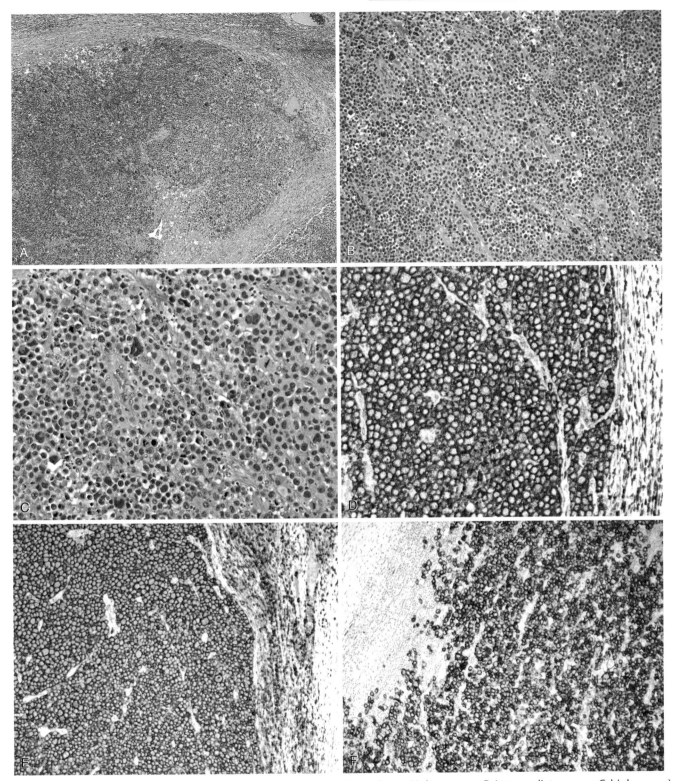

FIGURE 51.6 Lymph node section demonstrating anaplastic large cell lymphoma (A, low power; B, intermediate power; C, high power). The neoplastic cells express CD2 (D), CD4 (E), and CD30 (F).

"breakapart" FISH probe (Figure 51.11). The ALK over-expression is detected by immunohistochemical stains with anti-ALK antibodies. ALK-positive staining in the classic t(2;5) is detected in nucleolus, nucleus, and cytoplasm, whereas other translocations lead to only cytoplasmic staining.

- Recent studies of gene expression profiling of systemic ALCL have revealed differences between the ALK-positive and the ALK-negative subgroups. The ALK-positive tumors showed over-expression of *BCL6, PTPN12, CEBPB,* and *SERPINA1* genes, whereas *CCR7, CNTFR, IL22,* and *IL21* genes were over-expressed in the ALK-negative group.

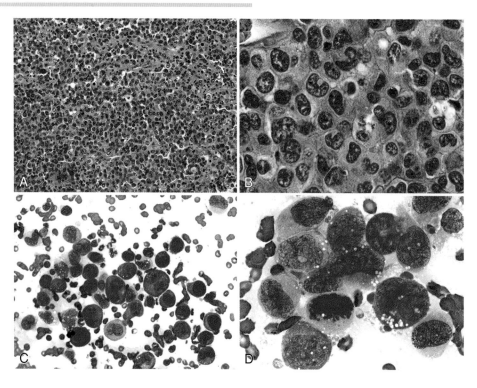

FIGURE 51.7 Lymph node sections (A and B) and touch preparations (C and D) demonstrating anaplastic large cell lymphoma.

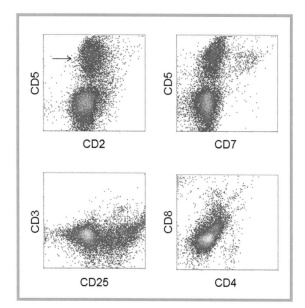

FIGURE 51.8 **Flow cytometric findings of anaplastic large cell lymphoma.** Lymphoid-enriched gate of a neck mass shows a discrete population of abnormal T-cells, which are positive for CD2 and CD5 (arrow), as well as CD25. They are negative for CD3, CD4, and CD8, revealing complete loss of CD7 expression.

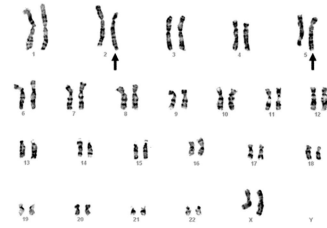

FIGURE 51.9 Karyotype of an anaplastic large cell lymphoma with a reciprocal t(2;5)(p23;q35)(arrows) resulting in NPM1-ALK fusion.

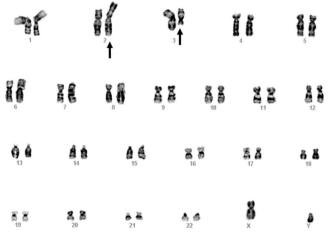

FIGURE 51.10 Karyotype of tumor cells of a patient with anaplastic large cell lymphoma demonstrates 46,XY,t(2;3)(p23;q21).

Table 51.1
Chromosomal Aberrations in ALK-Positive Anaplastic Large Cell Lymphoma[1]

Aberrations	Involved Genes[2]	ALK Protein Expression	Frequency (%)
t(2;5)(p23;q35)	ALK-NPM	Nuclear, diffuse cytoplasmic	84
t(1;2)(q21;p23)	TPM3-ALK	Diffuse cytoplasmic	13
t(2;3)(p23;q21)	ALK-TFG	Diffuse cytoplasmic	1
t(2;17)(p23;q23)	ALK-CLTC	Granular cytoplasmic	<1
t(2;17)(p23;q25)	ALK-ALO17	Diffuse cytoplasmic	<1
t(X;2)(q11-12;p23)	MSN-ALK	Membrane staining	<1
t(2;19)(p23;p13.1)	ALK-TMP4	Diffuse cytoplasmic	<1
t(2;22)(p23;q11.2)	ALK-MYH9	Diffuse cytoplasmic	<1
inv(2)(p23;q35)	ALK/ATIC	Diffuse cytoplasmic	1

[1]Adapted from Kutok JL, Aster JC. Molecular biology of anaplastic lymphoma kinase-positive anaplastic large-cell lymphoma. J Clin Oncol 2002; 20: 3691–3702, and Jaffe ES, Harris NL, Vardiman JW, et al. Hematopathology. Saunders/Elsevier, Philadelphia, 2010.

[2]ALK, anaplastic lymphoma kinase; ALO17, ALK oligomerization partner on chromosome 17; ATIC, 5-aminoimidazole-4-carboxamide ribonucleotide formyltransferase /IMP cyclohydrolase; CLTC, clathrin heavy chain; MSN, moesin; MYH9, myosin heavy chain 9; NPM, nucleophosmin; TFG, TRCK fusion gene; TPM, tropomyosin.

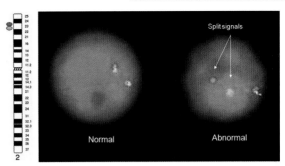

FIGURE 51.11 **ANAPLSTIC LARGE CELL LYMPHOMA.** FISH with ALK(2p23) dual-color, breakapart rearrangement probe shows split signals.

- Unlike ALK+ ALCL, the small cell variant has not been reported.

IMMUNOPHENOTYPE

- Staining for CD30 is strong and uniform.
- Most cases express one or more T-cell-associated markers, such as CD2, CD3, and CD4; CD8 expression is less frequent.
- A significant proportion of cases may be positive for EMA, TIA1 and granzyme B.
- EBV and ALK are is negative.

GENETICS AND MOLECULAR FINDINGS

- TCR genes are clonally rearranged in most cases.
- Complex chromosomal abnormalities have been found to be associated with poor prognosis.

Anaplastic Large Cell Lymphoma, ALK-Negative

ALCL, ALK-negative consists of large anaplastic, pleomorphic CD30+ cells, often with the presence of hallmark cells. The tumor cells have a tendency to grow cohesively in the lymph node sinuses, mimicking metastatic tumors. These tumors are morphologically similar to the ALCL, ALK-positive variant, except that they do not express ALK protein and are not associated with translocation of the ALK gene.

The peak incidence of ALK− ALCL is in individuals 40–65 years old with a male to female ratio of about 1. Extranodal involvement is less frequent than in ALCL, ALK+.

ALK-negative ALCL is more aggressive that ALK+ ALCL, with a 5-year overall survival rate ranging from 15% to 37% compared with 70% to 90% in ALK+ patients.

MORPHOLOGY

- Morphologic features are similar to those of ALCL, ALK+ with pleomorphic, anaplastic tumor cells, presence of the hallmark cells, and sinusoidal growth pattern.

Borderline Cases

There are reports of cases with morphologic features of ALCL, but coexpression of CD15 and CD30, resembling classical Hodgkin lymphoma (Figure 51.12). These cases are either ALK-positive or ALK-negative. There is a debate whether CD15+ CD30+ T-cell lymphomas should be considered a subtype of ALCL or peripheral T-cell lymphoma, not otherwise specified. Occasional cases of ALCL may express PAX-5 (Figure 51.13).

Differential Diagnosis

All large cell CD30+ neoplasms should be included in the differential diagnosis (Table 51.2). ALCL may share some morphologic features with classical Hodgkin lymphoma. Sclerotic thick capsule and well-formed fibrous bands which are frequently seen in Hodgkin lymphoma are infrequent in ALCL. The Reed–Sternberg cells are CD45−, CD15+, CD30+, and ALK−, and may express CD20 and/or show EBV positivity, whereas the neoplastic cells of ALK+ ALCL demonstrate ALK gene translocation, are

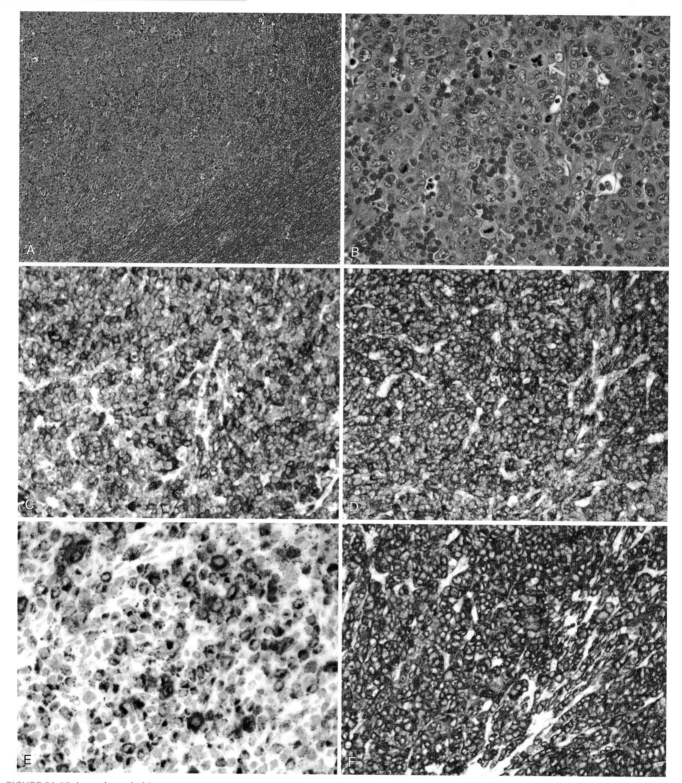

FIGURE 51.12 Lymph node biopsy section demonstrating a borderline neoplasm sharing features with anaplastic large cell lymphoma and classical non-Hodgkin lymphoma (A, low power; B, high power). There are multilobated nuclei (green arrows) and an atypical mitotic figure (yellow arrow). The neoplastic large cells express CD45 (C), CD4 (D), CD15 (E) and CD30 (F).

CD45+, CD15−, EMA+, and express one or more pan-T-cell markers. The diagnosis of ALCL, ALK-negative is more challenging because of the lack of specific immunophenotypic, molecular, or cytogenetic markers.

Sinusoidal involvements in the lymph nodes and bone marrow of patients with ALCL may mimic metastatic carcinoma. Metastatic carcinomas are cytokeratin+ and CD45−, whereas ALCLs are cytokeratin− and CD45+.

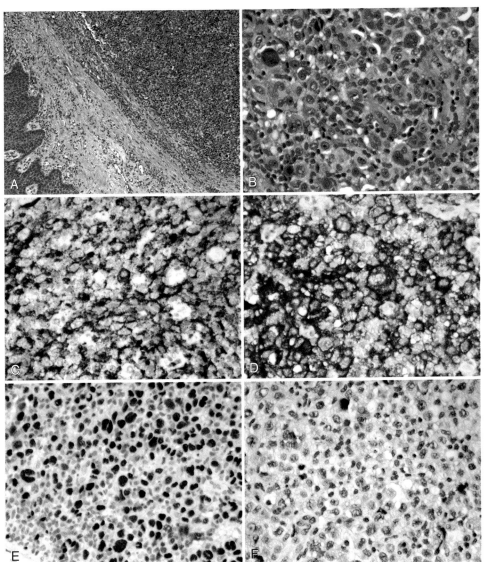

FIGURE 51.13 A case of anaplastic large cell lymphoma of tongue expressing CD30 and PAX5. H&E (A, low power; B, high power) and immunohistochemical stains showing expression of CD2 (C), CD30 (D), PAX5 (E), and lack of expression of ALK (F).

Table 51.2			
Differential Diagnosis of Anaplastic Large Cell Lymphoma (ALCL)[1]			
Entity [2]	Morphology	Immunophenotype	Molecular/Cytogenetics
ALCL, ALK+	Hallmark cells, sinusoidal growth	CD30+, ALK+, EMA+, CD43+, Pan-T±, TIA1±, granzyme B±	*TCR* rearranged, rearrangement of *ALK* gene
ALCL, ALK−	Hallmark cells, sinusoidal growth	CD30+, ALK−, EMA±, CD43+, Pan-T±, TIA1±, granzyme B±	*TCR* rearranged, no *ALK* translocation
DLBCL, ALK+	Hallmark cells, sinusoidal growth	CD30−, CD138+, ALK+, EMA+, cIgA+, CD20−, CD79a−, CD3−	*IGH* rearranged, t(2;17)(p23;q23)
DLBCL, anaplastic	Anaplastic large cells	CD20+, CD79a+, CD30±, EMA±, ALK−	*IGH* rearranged, no *ALK* translocation
PTCL, NOS, predominantly large cells	Pleomorphic cells	Pan-T+, CD30±, EMA±, ALK−, BCL-2+, TIA-1±	*TCR* rearranged, no *ALK* translocation
Hodgkin lymphoma	Reed–Sternberg cells	CD15+, CD30+, CD45−, ALK−, CD20±, EBV±	Non-specific
Metastatic carcinoma	Atypical cohesive cells, often with sinusoidal growth	Cytokeratine+, CD45−, CD30±	No *TCR* or *IGH* rearrangement

[1]Adapted from Jaffe ES, Harris NL, Vardiman JW, et al. Hematopathology. Saunders/Elsevier, Philadelphia, 2010.
[2]DLBCL, diffuse large B-cell lymphoma; PTCL, NOS, peripheral T-cell lymphoma, not otherwise specified.

Additional Resources

Amin HM, Lai R: Pathobiology of ALK+ anaplastic large-cell lymphoma, *Blood* 10:2259-2267, 2007.

Barry TS, Jaffe ES, Sorbara L, et al: Peripheral T-cell lymphomas expressing CD30 and CD15, *Am J Surg Pathol* 27:1513-1522, 2003.

Cheng M, Ott GR: Anaplastic lymphoma kinase as a therapeutic target in anaplastic large cell lymphoma, non-small cell lung cancer and neuroblastoma, *Anticancer Agents Med Chem* 10:236-249, 2010.

Feldman AL, Law ME, Inwards DJ, et al: PAX5-positive T-cell anaplastic large cell lymphomas associated with extra copies of the PAX5 gene locus, *Mod Pathol* 23:593-602, 2010.

Gorczyca W, Tsang P, Liu Z, et al: CD30-positive T-cell lymphomas co-expressing CD15: an immunohistochemical analysis, *Int J Oncol* 22:319-324, 2003.

Gustafson S, Medeiros LJ, Kalhor N, et al: Anaplastic large cell lymphoma: another entity in the differential diagnosis of small round blue cell tumors, *Ann Diagn Pathol* 13:413-427, 2009.

Inghirami G, Pileri SA: European T-Cell Lymphoma Study Group. Anaplastic large-cell lymphoma, *Semin Diagn Pathol* 28:190-201, 2011.

Jaffe ES, Harris NL, Vardiman JW, et al: *Hematopathology*, Philadelphia, 2010, Saunders/Elsevier.

Jewell M, Spear SL, Largent J, et al: Anaplastic large T-cell lymphoma and breast implants: a review of the literature, *Plast Reconstr Surg* 128:651-661, 2011.

Kutok JL, Aster JC: Molecular biology of anaplastic lymphoma kinase-positive anaplastic large-cell lymphoma, *J Clin Oncol* 20:3691-3702, 2002.

Medeiros LJ, Elenitoba-Johnson KS: Anaplastic large cell lymphoma, *Am J Clin Pathol* 127:707-722, 2007.

Nguyen JT, Condron MR, Nguyen ND, et al: Anaplastic large cell lymphoma in leukemic phase: extraordinarily high white blood cell count, *Pathol Int* 59:345-353, 2009.

Summers TA, Moncur JT: The small cell variant of anaplastic large cell lymphoma, *Arch Pathol Lab Med* 134:1706-1710, 2010.

Swerdlow SH, Campo E, Harris NL, et al: *WHO classification of tumours of haematopoietic and lymphoid tissues*, ed. 4, Lyon, 2008, International Agency for Research on Cancer.

ten Berge RL, Oudejans JJ, Ossenkoppele GJ, et al: ALK-negative systemic anaplastic large cell lymphoma: differential diagnostic and prognostic aspects—a review, *J Pathol* 200:4-15, 2003.

Younes A: CD30-targeted antibody therapy, *Curr Opin Oncol* 23:587-593, 2011.

Peripheral T-Cell Lymphoma, Not Otherwise Specified

Peripheral T-cell lymphoma, not otherwise specified (PTCL, NOS) includes all the T-cell lymphomas that are not included in the well-defined clinicopathologic entities previously described in this book. PTCL, NOS is predominantly nodal and constitutes the most common T-cell lymphoma in Western countries.

PTCL, NOS is the most frequent peripheral T-cell lymphoma, constituting approximately 30% of all peripheral T-cell lymphomas. It occurs in middle-aged to elderly patients, with a median age of 54 years and a male to female ratio of about 1–2:1. Most patients are in advanced stages (stage III/IV) at diagnosis. Extranodal disease, B symptoms, and elevated LDH are frequent findings and are associated with unfavorable prognosis. Other risk factors include age >60 years and a decline in performance status. The most frequent involved extranodal sites are bone marrow, spleen, liver, Waldeyer ring, and skin (Table 52.1).

Four prognostic groups have been identified:

- Group 1 with no risk factors and 5-year survival rate of about 60%;
- Group 2 with one risk factor and 5-year survival rate of about 50%;
- Group 3 with two risk factors and 5-year survival rate of 35%;
- Group 4 with three or four risk factors and 5-year survival rate of about 20%.

Molecular and immunophenotypic studies have demonstrated that *p53* mutation and over-expression of p53 protein correlate with treatment failure and unfavorable prognosis. In general, PTCL, NOS is an aggressive disease with a poor response to therapy and frequent relapses.

Morphology

The normal nodal architecture is often effaced with a diffuse infiltration of neoplastic lymphoid cells. A garden variety of morphologic features have been described, but in most cases the predominant cells are medium-sized to large with irregular nuclei. The nuclei are hyperchromatic or vesicular with prominent nucleoli. Mitotic figures are frequent. Large cells with clear cytoplasm and Reed–Sternberg-like cells may be present. Vascular proliferation is frequently noted.

The infiltrating neoplastic cells are often mixed with inflammatory cells, such as small lymphocytes, eosinophils, and histiocytes. Histiocytes may appear in aggregates or may show hemophagocytosis.

Three major morphologic variants have been described in the WHO classification: (1) lymphoepithelioid, (2) follicular, and (3) T-zone.

- The **lymphoepithelioid variant** (Lennert lymphoma) is characterized by the presence of numerous small aggregates of epithelioid histiocytes mixed with a lymphocytic infiltrate predominantly consisting of small lymphocytes with slightly irregular nuclei (Figure 52.1). Scattered larger lymphocytes with clear cytoplasm may be present. The pattern of lymphoid infiltration is diffuse but less frequently may be interfollicular. Vascular proliferation and presence of inflammatory cells are common features.
- The **follicular variant** demonstrates aggregates of atypical neoplastic T-cells forming interfollicular aggregates. This pattern simulates follicular lymphoma or nodular lymphocyte predominant Hodgkin lymphoma. Most cases contain medium to large cells with abundant pale cytoplasm.
- The **T-zone variant** is characterized by the expansion of interfollicular spaces as the result of infiltration of predominantly small- to medium-sized lymphocytes (Figure 52.2). Lymphoid follicles are usually well preserved or even hyperplastic. Clusters of lymphoid cells with clear cytoplasm are often present and scattered Reed–Sternberg-like cells may be present. There is vascular proliferation with predominance of high endothelial venules. Inflammatory cells, such as eosinophils, plasma cells, and histiocytes, are commonly present.

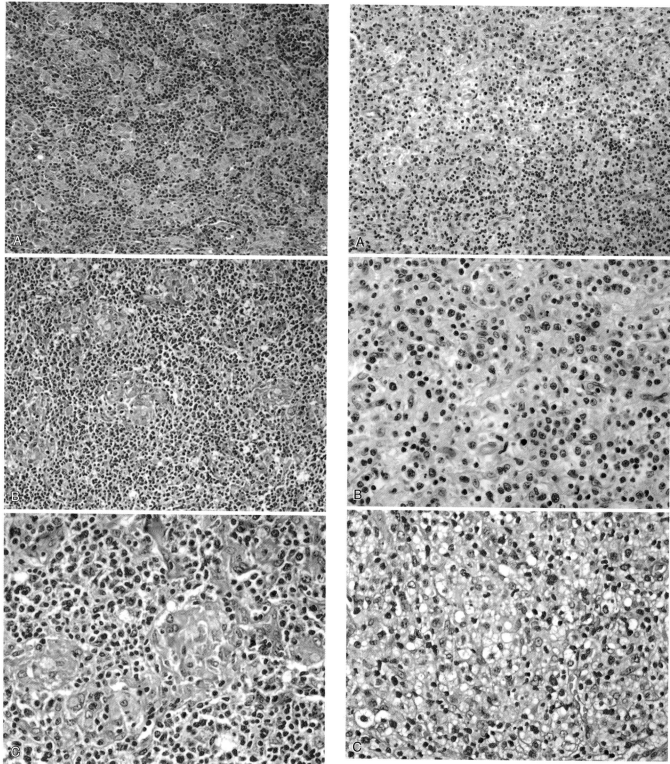

FIGURE 52.1 **PERIPHERAL T-CELL LYMPHOMA, LYMPHOEPITHELIOID CELL VARIANT (LENNERT LYMPHOMA).** Epithelioid histiocytes are mixed with infiltrating lymphocytes in small (A) or large (B and C) clusters.

FIGURE 52.2 **PERIPHERAL T-CELL LYMPHOMA, T-ZONE VARIANT.** This demonstrates a mixture of small to large lymphocytes (A and B) and the presence of lymphoid cells with clear cytoplasm (C).

Bone marrow involvement shows focal or diffuse infiltration of bone marrow with atypical, often pleomorphic cells. Focal infiltrates are usually non-paratrabecular. The infiltrates are highly vascular, are admixed with inflammatory cells and may show increased reticulin fibers.

Splenic infiltrates are often patchy and may primarily involve the periarterial T-cell zones or extend to the red pulp.

Cutaneous involvement is frequent and may appear as diffuse or nodular patterns with or without epidermotropism.

Immunophenotype

Immunophenotypic features include (Figure 52.3):

- Pan-T-cell aberrancies with variable loss of CD5 and/or CD7;
- Expression of TCR-β;
- Common CD4 phenotype; but CD8, double positive, or double negative in some cases;
- Expression of CD56 and/or cytotoxic markers in some cases;
- Positivity of CD30 in some cases, but rare expression of CD15;
- Occasional aberrant expression of B-cell-associated markers CD20 or CD79a;
- Lack of follicular T-helper (FTH)-associated markers.

Molecular and Cytogenetic Studies

- Approximately 90% of patients with PTCL, NOS show clonally rearranged TCR genes, and 70–90% demonstrate cytogenetic aberrations.
- A novel t(5;9)(q33;q22), that creates a fusion gene between ITK and SYK genes, has been reported in 17% of PTCL, NOS.
- Complex karyotypes consistent with clonal evolution are a frequent finding. Breaks involving the TCR loci are common. The chromosomes most frequently altered in structural aberrations are 1, 6, 2, 4, 11, 14, and 17. Additionally, trisomies of 3 or 5 and an extra X-chromosome are also common.
- CGH studies have found recurrent losses on chromosomes 5q, 6q, 9p, 10q, 12q, and 13q (Figure 52.4). Recurrent gains were found on chromosome 7q.

High-level amplifications of 12p13 have been observed in a few PTCL-u cases with cytotoxic phenotype. These results suggest that certain genetic alterations may indeed exist in PTCL, NOS, but definitive clinicopathologic subgroups have not yet been identified and, thus far, a variety of molecular and cytogenetic findings do not allow for a consistent model for pathogenesis to be constructed.

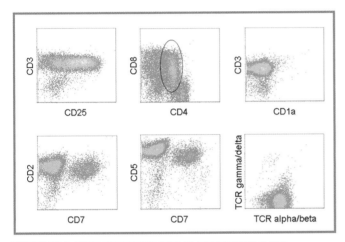

FIGURE 52.3 **FLOW CYTOMETRIC FINDINGS OF PERIPHERAL T-CELL LYMPHOMA.** Lymphoid-enriched gate of a neck lymph node reveals predominantly abnormal T-cells, which are double positive (circled) for CD4 (dim) and CD8 (heterogeneous). In addition, the neoplastic T-cells are positive for CD2, CD3, CD5 (bright), CD25 (heterogeneous), and TCR alpha/beta. They demonstrate complete loss of CD7 expression, and are negative for CD1a.

Differential Diagnosis

The differential diagnosis includes a variety of reactive lymphadenopathies, ALCL, AITL, and Hodgkin lymphoma (Table 52.1). Presence of clusters of epithelioid histiocytes in the lymphoepithelioid cell (Lennert) variant may

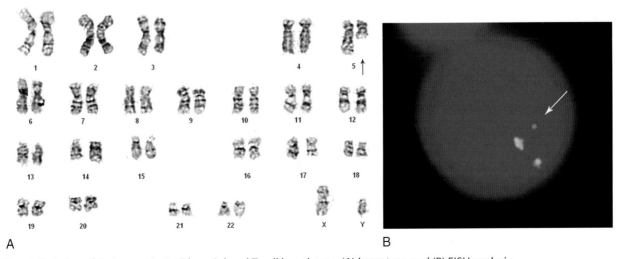

FIGURE 52.4 Deletion of 5q in a patient with peripheral T-cell lymphoma: (A) karyotype and (B) FISH analysis.

Table 52.1

Differential Diagnosis of Peripheral T-Cell Lymphoma, Not Otherwise Specified (PTCL, NOS)

Entity	T-Cells	Large HRS-Like Cells	Epitheliod Histiocytes	Molecular/ Cytogenetics
PTCL, NOS	Atypical, pleomorphic	May be present, express pan-T markers	Often present, prominent in Lennert type	Rearranged *TCR*
Classical HL	Small, no atypia	CD15+, CD30+, CD45−, EBV±	Variable	*IGH* rearrangement not routinely detected
NLPHL	Small, no atypia, interfollicular CD57+	CD15−, CD30−, CD45+, CD20+, EMA+	Variable	*IGH* rearrangement not routinely detected
THR-LBCL	Small, no atypia	CD15−, CD30−, CD45+, CD20+	Abundant	Rearranged *IGH*
AITL	Atypical, some with clear cytoplasm and CD10+	Mostly reactive B-cells, CD30+, EBV+	Variable	Rearranged *TCR*
ALCL	Atypical, pleomorphic	T-cells, CD30+, ALK±	Variable	Rearranged *ALK, TCR*
Reactive nodes	Small and large	Mostly reactive B-cells, CD30+, EBV±	Variable	No *TCR* or *IGH* rearrangement

AITL, angioimmunoblastic T-cell lymphoma; ALCL, anaplastic large cell lymphoma; HL, Hodgkin lymphoma; HRS, Hodgkin and Reed–Sternberg; NLPHL, nodular lymphocyte predominant HL; THR-LBCL, T-cell/histiocyte rich large B-cell lymphoma.

simulate toxoplasmosis, sarcoidosis, or other types of granulomatous lymphadenitis. The follicular variant may mimic follicular lymphoma or nodular lymphocyte predominant Hodgkin lymphoma. The interfollicular expansion and the presence of inflammatory cells in the T-zone variant may mimic T-zone hyperplasia. Cases with the presence of anaplastic large cells or the presence of Reed–Sternberg-like cells may resemble T-cell/histiocyte-rich large B-cell lymphoma, ALCL. Vascular proliferation and polymorphous infiltrate are among the overlapping features between PTCL, NOS and AITL.

Additional Resources

de Leval L, Gaulard P: Pathology and biology of peripheral T-cell lymphomas, *Histopathology* 58:49–68, 2011.

Foss FM, Zinzani PL, Vose JM, et al: Peripheral T-cell lymphoma, *Blood* 117:6756–6767, 2011.

Howman RA, Prince HM: New drug therapies in peripheral T-cell lymphoma, *Expert Rev Anticancer Ther* 11:457–472, 2011.

Jaffe ES, Harris NL, Vardiman JW, et al: *Hematopathology*, Philadelphia, 2010, Saunders/Elsevier.

Piccaluga PP, Agostinelli C, Gazzola A, et al: Prognostic markers in peripheral T-cell lymphoma, *Curr Hematol Malig Rep* 5:222–228, 2010.

Piccaluga PP, Agostinelli C, Tripodo C, et al: European T-cell lymphoma study group. peripheral T-cell lymphoma classification: the matter of cellular derivation, *Expert Rev Hematol* 4:415–425, 2011.

Roncolato F, Gazzola A, Zinzani PL, et al: Targeted molecular therapy in peripheral T-cell lymphomas, *Expert Rev Hematol* 4:551–562, 2011.

Savage KJ, Ferreri AJ, Zinzani PL, et al: Peripheral T-cell lymphoma–not otherwise specified, *Crit Rev Oncol Hematol* 79:321–329, 2011.

Savage KJ: Update: peripheral T-cell lymphomas, *Curr Hematol Malig Rep* 6:222–230, 2011.

Swerdlow SH, Campo E, Harris NL, et al: *WHO classification of tumours of haematopoietic and lymphoid tissues*, ed. 4, Lyon, 2008, International Agency for Research on Cancer.

Nodular Lymphocyte Predominant Hodgkin Lymphoma

Overview of Hodgkin Lymphoma

Hodgkin lymphoma (HL) is a clonal B-cell neoplasm as demonstrated by the detection of clonal immunoglobulin (Ig) V-gene rearrangements in isolated tumor cells using microdissection- and single-cell polymerase chain reaction (PCR) techniques. It is recognized by the current WHO classification as a malignant lymphoma with unique clinicopathologic features. In general, HL affects more often young adults, with a median age of 38 years at diagnosis. It primarily involves lymph nodes commonly found in the supradiaphragmatic areas, and the neoplastic tissue comprises a minor population of large pleomorphic neoplastic cells—Hodgkin and Reed–Sternberg (HRS) cells, plus their variants—that are admixed with an inflammatory background (Figures 53.1 and 53.2).

The incidence of HL is about 3 per 100,000 in western Europe and the United States and is consistently lower than that of non-Hodgkin lymphoma (NHL). It accounts for about 10–15% of all lymphomas in Europe and the United States. The median age at diagnosis is 38 years, and the male to female ratio is slightly >1. Patients with a history of infectious mononucleosis (IM), AIDS, or autoimmune diseases have higher incidence of the disease. Most HL patients present with asymptomatic lymph node enlargement in the supradiaphragmatic regions. Systemic symptoms with fever, night sweats, and weight loss are reported in about 30% of cases. Approximately 50% of patients have advanced stage disease at diagnosis. Splenic involvement is seen in about 20%, whereas bone marrow is involved in about 5% of HL cases.

Staging of HL is based on the Ann Arbor system with the addition of definition of bulky disease which is currently favored by the maximum diameter of the largest single tumor mass. A limited-stage HL is usually defined by non-bulky stage IA or IIA disease, and a cure rate of >90% can be expected regardless of prognostic factor model. Seven independent prognostic factors have been identified, which are used to assess risks of primary treatment failure and possible intensified treatment in advanced stage HL patients. They include sex, age, stage, hemoglobin, WBC, lymphocyte count, and serum albumin.

Treatment of HL is one of the great success stories in modern medicine, and about 80–90% of patients in all stages achieve long term survival. The current understanding of the appropriate therapy is established on the basis of recent results of many large randomized clinical trials with focus on maximizing effectiveness while minimizing toxicity. For limited stage HL, treatment involves brief, combined modality chemotherapy only augmented with involved field irradiation if an early complete response is not achieved. Patients with advanced stage HL require an extended course of chemotherapy without radiation therapy. High dose chemotherapy and irradiation plus autologous hematopoietic stem cell transplantation can be an effective treatment for patients with relapsed or refractory HL. In addition to chemotherapy and radiotherapy, some studies suggest that immunotherapy including monoclonal antibodies plays a positive role in the treatment of HL. Results of targeted immunotherapy using anti-CD30 chimeric receptors bound to EBV-specific cytotoxic T-cells may help in achieving complete remission in certain cases.

Based upon its clinical behaviors, as well as morphologic, immunophenotypic, and genotypic profiles, HL is divided into two entities of nodular lymphocyte predominant HL (NLPHL) and classical HL (CHL) with the following subtypes according to the current WHO classification:

- Nodular lymphocyte predominant Hodgkin lymphoma (NLPHL)
- Classical Hodgkin lymphoma
    - Nodular sclerosis classical Hodgkin lymphoma (NSHL)
    - Mixed cellularity classical Hodgkin lymphoma (MCHL)
    - Lymphocyte-rich classical Hodgkin lymphoma (LRCHL)
    - Lymphocyte-depleted classical Hodgkin lymphoma (LDHL).

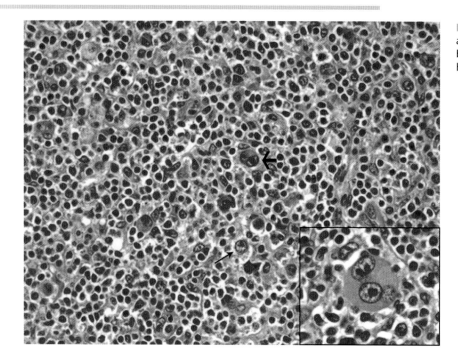

FIGURE 53.1 Reed–Sternberg cells (thick arrow and inset) and Hodgkin cells (thin arrow) in a background of lymphocytes, plasma cells, and histiocytes.

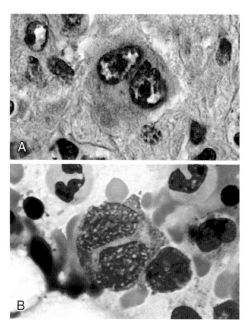

FIGURE 53.2 Classic Reed–Sternberg cells are large binucleated cells with prominent round nucleoli and perinucleolar halos displaying an "owl-eye" appearance: (A) bone marrow biopsy section and (B) bone marrow smear.

Nodular Lymphocyte Predominant Hodgkin Lymphoma

Nodular lymphocyte predominant Hodgkin lymphoma (NLPHL) is a distinct but rare subtype of HL. It constitutes 5–7% of all HL cases in the United States and Europe, with an estimated 500 new cases each year in the United States. NLPHL is a monoclonal B-cell lymphoma characterized by nodular pattern, presence of sparse large pleomorphic lymphoid (LP) cells (or popcorn cells, HRS cell variants) admixed with abundant B-lymphocytes that reside in an expanded meshwork of follicular dendritic cells. It most commonly presents as limited nodal disease involving peripheral lymph nodes above or below the diaphragm.

Patients with NLPHL are typically asymptomatic and present with localized peripheral lymphadenopathy frequently involving the cervical and axillary regions. Less than 20% of cases may demonstrate stage III or IV disease. Bulky disease and mediastinal involvement are rare. Compared with classical Hodgkin lymphoma (see Chapter 54), NLPHL is a more indolent disease with more frequent relapse, but good response to therapy, and overall more favorable prognosis. It is male predominant with a male to female ratio of 3–4:1 and has two peaks in age distribution: one that of children and the other that of 30- to 40-year-old adults.

The treatment of NLPHL with standard HL protocols leads to complete remission in >95% of patients. Prognosis is worse in advanced stage patients. A very small proportion (2–7%) of NLPHL transforms to DLBCL, which appears to have a more indolent course and more favorable prognosis than *de novo* DLBCL.

MORPHOLOGY

- Nodal architecture is partially to completely effaced by the neoplastic infiltrate, which, by definition, demonstrates at least partially nodular growth pattern (Figure 53.3). Residual

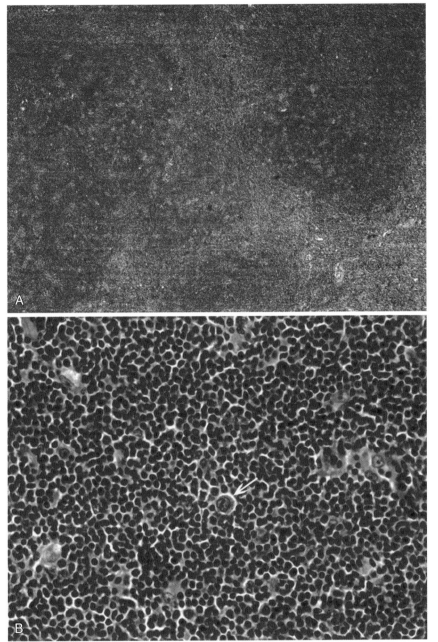

FIGURE 53.3 **NODULAR LYMPHOCYTE PREDOMINANT HODGKIN LYMPHOMA.** (A) Low power view of a lymph node section demonstrating a nodular pattern. (B) High power view showing a lymphocyte predominant (LP) or "popcorn" cell (arrow) in a background of small lymphocytes.

follicles, follicular hyperplasia, and progressive transformation of germinal centers can be present simultaneously within the lymph node that is involved.
- The neoplastic nodules are usually large and poorly defined and comprise a minor population of scattered popcorn or lymphocyte predominant cells (LP cells, formerly called L & H cells), which are admixed with occasional single or clusters of histiocytes plus abundant small lymphocytes that are typically B-cells (Figure 53.4).
- The LP cells are large and pleomorphic and often contain a single nucleus with folding and multilobation, resembling popcorn, and hence the name "popcorn" cells. These cells have abundant basophilic cytoplasm, a vesicular nuclear chromatin, and usually multiple small nucleoli.
- A follicular dendritic cell meshwork can be highlighted within the neoplastic nodules using special stains.
- When diffuse areas are present, they predominantly comprise T-cells as well as histiocytes with occasional LP cells. Neutrophils, eosinophils, and plasma cells are not commonly seen. According to the current criteria, the detection of one nodule showing the typical features of NLPHL in an otherwise diffuse growth pattern is sufficient to make the diagnosis of NLPHL.

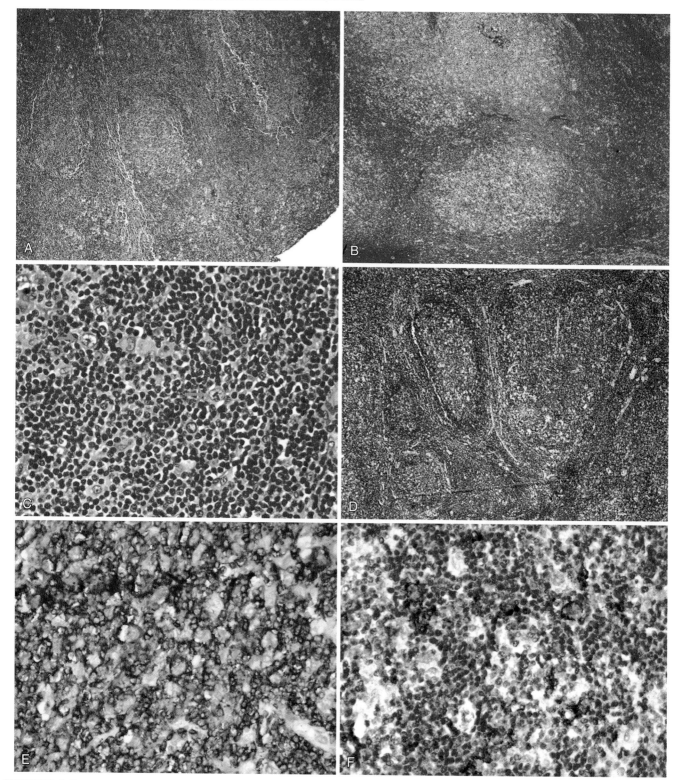

FIGURE 53.4 **NODULAR LYMPHOCYTE PREDOMINANT HODGKIN LYMPHOMA.** Lymph node section demonstrates a nodular pattern (A, low power; B, intermediate power). High power view shows two LP cells (arrows) in a background of small lymphocytes (C). Immunohistochemical stains show a predominance of CD20+ (red) cells inside the nodules, surrounded by CD3+ (brown) cells (D, low power; E, high power). The LP cells express EMA (F).

IMMUNOPHENOTYPE

The immunophenotypic features of LP cells include:

- Positivity of CD45 and B-cell associated antigens including CD20, CD22, PAX5, and CD79a;
- Coexpression of B-cell transcription factors OCT-2 and BOB.1;
- Expression of EMA in about 50% of cases (see Figure 53.4);
- Other positive staining for immunoglobulin light and heavy chains, CD75, BCL6, plus Ki67;
- Nearly always negative staining for CD15 and CD30, with only weak CD30 positive LP cells in rare cases;
- Negative for EBV-EBER.

The background shows (Figure 53.5):

- CD3+ and/or CD57+ T-cells ringing LP cells;
- Follicular dendritic meshworks that can be highlighted by CD21 and/or CD23 staining;
- Large lymphoid nodules with predominance of small B-cells plus T-cells of germinal center phenotype;
- Cells negative for CD8 and cytotoxic granule markers;
- Occasional double positive T-cells (CD4+/CD8+).

CYTOGENETIC AND MOLECULAR STUDIES

- There are only few reported cytogenetic and molecular mutations in NLPHL, including common rearrangements of the *BCL-6* gene predominantly with the *IGH@*, along with monosomy 13.
- Because of the scattered nature of the neoplastic cells admixed with non-neoplastic cells in HL, standard clinical molecular diagnostic assays (gene rearrangement clonality studies, etc.) are not valuable or, if performed, are usually negative because of the dilution effect of the background cell population. However, they are sometimes ordered to assist in differential diagnosis between HL and NHL.
- As noted at the start of this chapter, single-cell and microdissection PCR techniques have been used to prove the B-cell origin of neoplastic cells (Figure 53.6), but this is a research modality and of largely academic interest.
- Rearrangements of *BCL-6* are usually detected by immunohistochemical techniques rather than molecular tests, whereas the other chromosomal changes are best detected by cytogenetic techniques.

DIFFERENTIAL DIAGNOSIS

The differential diagnosis of NLPHL includes T-cell/histiocyte-rich large B-cell lymphoma (T/HRLBCL), lymphocyte-rich classical Hodgkin lymphoma, and follicular lymphoma (Table 53.1).

T-cell/histiocyte-rich large B-cell lymphoma (T/HRLBCL). Although NLPHL and T/HRLBCL are two distinct lymphomas requiring different clinical management,

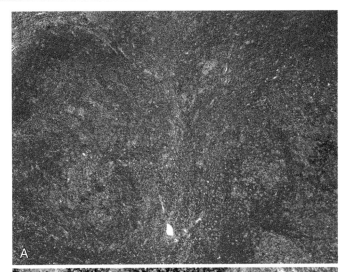
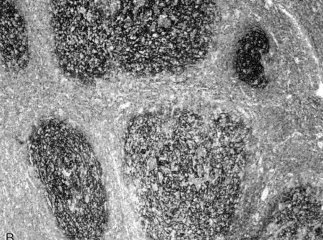
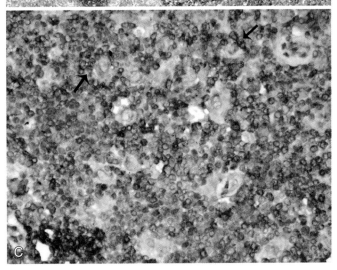

FIGURE 53.5 **NODULAR LYMPHOCYTE PREDOMINANT HODGKIN LYMPHOMA.** Lymph node section (A) demonstrating CD21+ meshwork of follicular dendritic cells (B), and CD57+ T-cells surrounding the LP cells (C).

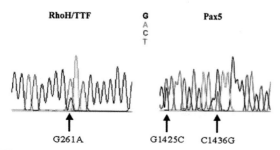

FIGURE 53.6 MICRODISSECTION AND MUTATIONAL ANALYSIS OF A REPRESENTATIVE PATIENT WITH NLPHL. Electropherogram showing mutations of the *RhoH/TTF* and *PAX5* genes in microdissected LP cells.
From Liso A, Capello D, Marafioti T, et al. Aberrant somatic hypermutation in tumor cells of nodular-lymphocyte-predominant and classic Hodgkin lymphoma. Blood 2006; 108: 1013–1020, by permission.

Table 53.1

Differential Diagnosis of Nodular Lymphocyte Predominant Hodgkin Lymphoma (NLPHL)

	Tumor Cells		
Entity	Morphology	Immunophenotype	Molecular/Cytogenetic
NLPHL	At least partially nodular, LP-type HRS cells	CD45+, CD20+, EMA±, Oct-2+, BOB.1+, BCL6+, CD15−, CD30−	No *TCR* or *IGH* rearrangement by routine techniques
T/HRLBCL	Diffuse pattern, scattered large cells	CD45+, CD20+, OCT-2+, BOB.1+, EMA±, CD15−, CD30−	*IGH* rearranged,
LRCHL	Diffuse or nodular growth	CD15+, CD30+, CD45−, EMA−, CD20±, EBV±	No *TCR* or *IGH* rearrangement by routine techniques
FL	Nodular growth, mixture of centrocytes and centroblasts	CD20+, CD10+, BCL2+, CD45+, CD15−, CD30−	*IGH* rearranged, t(14;18)(q32;q21)

FL, follicular lymphoma; LRCHL, lymphocyte-rich classical Hodgkin lymphoma; T/HRLBCL, T-cell/histiocyte-rich large B-cell lymphoma.

they share overlapping morphologic and immunophenotypic features. Some studies suggest a biologic continuum between NLPHL and T/HRLBCL, which implies that they may represent different spectrums of the same disease. Both lymphomas may have concurrent nodular and diffuse growth patterns, though, by definition, at least partial nodular pattern has to be present for NLPHL. The immunophenotypic profile of LP cells cannot be clearly distinguished from that of the neoplastic cells in T/HRLBCL. However, immunophenotypic studies of the background composition are helpful, with common findings of abundant small B-cells and a prominent follicular dendritic cell meshwork along with CD3+ CD4+ CD57+ T-cells in NLPHL, while lack of small B-cells but abundance of CD8+ T-cells and histiocytes in T/HRLBCL.

Lymphocyte-rich classical Hodgkin lymphoma (LRCHL). NLPHL and LRCHL cannot be distinguished morphologically or clinically. LRCHL frequently presents with a nodular growth pattern and a background of abundant small lymphocytes. The neoplastic cells in LRCHL sometimes display cytologic features of LP cells seen in NLPHL. However, the immunophenotypic profile of the neoplastic cells in LRCHL differs from that of the LP cells. The neoplastic cells in LRCHL show immunophenotypic features of HRS cells in other subtypes of CHL, including expression of CD30, CD15, Fascin, and EBV-LMP positivity in about 50% of the cases. Unlike NLPHL, T-cell rosettes around the neoplastic cells generally do not express CD57 in LRCHL.

Follicular Lymphoma. Follicular lymphomas consist of a mixture of centrocytes and centroblasts in various proportions. The neoplastic cells are monoclonal and express CD10 and CD20. The characteristic "popcorn" cells are absent.

Additional Resources

Eberle FC, Mani H, Jaffe ES: Histopathology of Hodgkin's lymphoma, *Cancer J* 15:129–137, 2009.

Fanale MA, Younes A: Nodular lymphocyte predominant Hodgkin's lymphoma, *Cancer Treat Res* 142:367–381, 2008.

Jaffe ES, Harris NL, Vardiman JW, et al: *Hematopathology*, Philadelphia, 2010, Saunders/Elsevier.

Lee AI, LaCasce AS: Nodular lymphocyte predominant Hodgkin lymphoma, *Oncologist* 14:739–751, 2009.

Liso A, Capello D, Marafioti T, et al: Aberrant somatic hypermutation in tumor cells of nodular-lymphocyte-predominant and classic Hodgkin lymphoma, *Blood* 108:1013–1020, 2006.

Schmitz R, Stanelle J, Hansmann ML, et al: Pathogenesis of classical and lymphocyte-predominant Hodgkin lymphoma, *Annu Rev Pathol* 4:151–174, 2009.

Swerdlow SH, Campo E, Harris NL, et al: *WHO classification of tumours of haematopoietic and lymphoid tissues*, ed. 4, Lyon, 2008, International Agency for Research on Cancer.

Tsai HK, Mauch PM: Nodular lymphocyte-predominant hodgkin lymphoma, *Semin Radiat Oncol* 17:184–189, 2007.

Classical Hodgkin Lymphoma

Classical Hodgkin lymphoma (CHL) constitutes about 95% of all HL cases in the United States and Europe. CHL is a clonal B-cell lymphoma characterized by the proliferation of a minor population of mononuclear Hodgkin and multinucleated Reed–Sternberg (HRS) cells admixed with an abundance of reactive infiltrate including various inflammatory cells. The diagnostic (classic) Reed–Sternberg cell is a large binucleated cell with prominent round nucleoli and perinucleolar halos displaying an "owl-eye" appearance (Figures 54.1 and 54.2). Reed–Sternberg cell variants and mononuclear Hodgkin cells may display particular morphologic features in certain subclasses, such as "lacunar" cells in nodular sclerosis HL. The HRS cells typically express CD15 and CD30 and are negative for CD45.

CHL has a bimodal age distribution in industrialized countries. The first peak is in early adulthood (15–35 years), and the second peak is in patients >55 years old. The male to female ratio is about 1.5:1. CHL is rare in children. CHL most commonly presents as lymphadenopathy involving the cervical, mediastinal, axillary, and para-aortic regions. Primary extranodal involvement is rare.

There is an increased risk of developing CHL in siblings. One study has reported a high risk of CHL development in monozygotic twins.

CHL is divided into four major subclasses:

- Nodular sclerosis CHL
- Mixed cellularity CHL
- Lymphocyte-rich CHL
- Lymphocyte-depleted CHL.

Nodular Sclerosis Classical Hodgkin Lymphoma

Nodular sclerosis CHL (NSCHL) is the most frequent subtype of CHL. It accounts for 50% to 80% of all CHL cases in Western countries, with a peak incidence ranging from 15 to 35 years. The male to female ratio is about 1:1, and the most frequent site of lymph node involvement is mediastinum (80%). Bone marrow involvement is noted in about 5% of cases. The prognosis is better than for other types of CHL.

MORPHOLOGY

- Lymph node sections show irregular nodules of different sizes separated by collagen bands (Figure 54.3). These bands often depict birefringence under polarized light. The extent of fibrosis varies from case to case, from a minimal amount to extensive obliterating sclerosis. The cases with minimal amount of fibrosis are referred to as **"cellular phase"** of NSCHL.
- The nodules show variable numbers of "lacunar cells" in a background of inflammatory cells consisting of lymphocytes, plasma cells, neutrophils, and eosinophils. Lacunar cells are a variant of HRS cells characterized by abundant pale cytoplasm, lobulated nuclei, and small nucleoli. In formalin-fixed tissues the cytoplasm of these cells shrinks and the cells appear to be placed in lacunae (Figure 54.4).

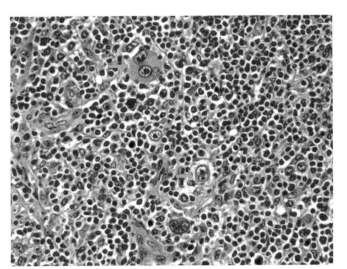

FIGURE 54.1 Reed–Sternberg cells (thick arrows) and Hodgkin cells (thin arrows) in a background of lymphocytes, plasma cells, and histiocytes.

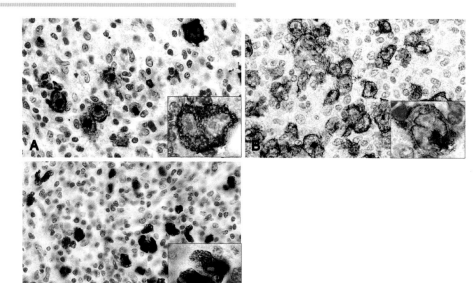

FIGURE 54.2 REED–STERNBERG AND HODGKIN CELLS IN CLASSICAL HODGKIN LYMPHOMA (CHL). The cells express: (A) CD15 and (B) CD30. They may also express EBV-encoded RNA (C). *From Naeim F. Pathology of Bone Marrow, 2nd edn. Williams & Wilkins, Baltimore, 1997, by permission.*

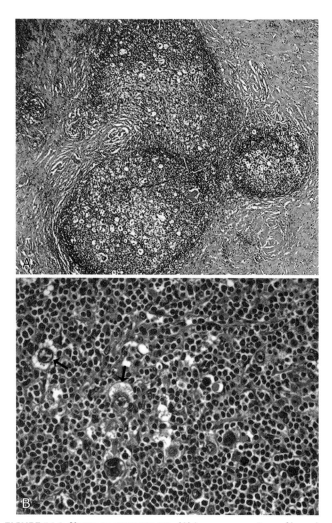

FIGURE 54.3 NODULAR SCLEROSIS CHL. (A) Low power view of lymph node section demonstrating a nodular pattern with numerous lacunar cells in the neoplastic nodules surrounded by thick collagen bands. (B) High power view showing lacunar cells (black arrows) and Reed–Sternberg cells (green arrows) mixed with lymphocytes and other inflammatory cells.

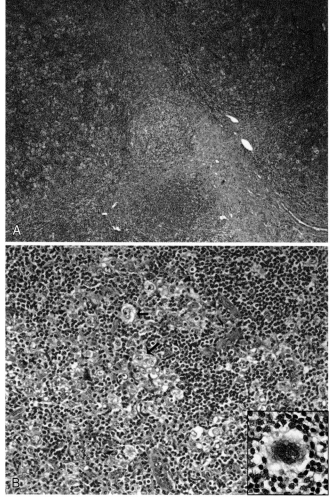

FIGURE 54.4 NODULAR SCLEROSIS CHL. Nodules are separated by thick collagen bands and contain numerous lacunar cells (A, low power; B, high power). Inset demonstrates a lacunar cell.

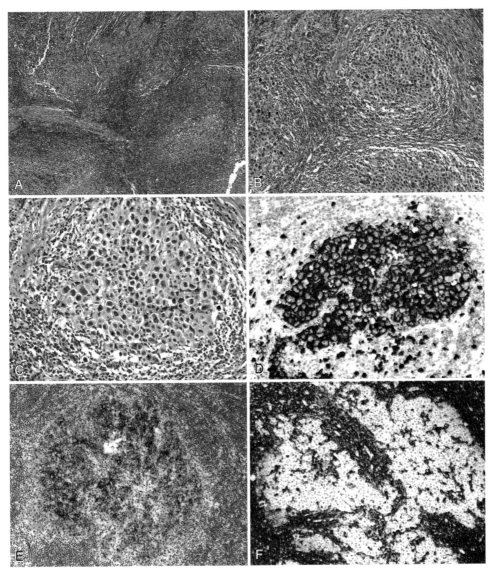

FIGURE 54.5 **NODULAR SCLEROSIS CHL, SYNCYTIAL TYPE.** Nodules are separated by thick collagen bands and primarily consist of large aggregates of lacunar cells (A, low power; B, intermediate power; C, high power). The neoplastic cells express CD15 (D) and CD30 (E), but are negative for CD45 (F).

- In a minority of the cases, the lacunar cells appear in large aggregates. These cases are referred to as "**syncytial variant**" (Figures 54.5 and 54.6).
- Areas of necrosis and histiocytic reactions may be present, simulating necrotizing granuloma.

IMMUNOPHENOTYPE

Similar to NSCLH, the HRS cells and their variants are (see Figures 54.5 and 54.6):

- Negative for CD45;
- Positive for CD30 in nearly all cases, with membrane and Golgi staining pattern;
- At least focally positive for CD15 in about 80% of cases;
- Positive for fascin and PAX5 in the vast majority of cases (>90%);
- Positive for either OCT-2 or BOB.1 (not both) in about 90% of cases;
- Positive for CD20 in 30–40% of cases; staining is usually weaker than that of the background reactive B-cells;
- Positive for IRF4/MUM1 in the vast majority of cases;
- Negative for EMA and ALK.
- Staining for EBV-EBER is positive in some cases of nodular sclerosis CHL, but less frequent than in mixed cellularity CHL.

MOLECULAR AND CYTOGENETIC STUDIES

Conventional cytogenetics and FISH mostly show numerical abnormalities i.e. aneuploid or hyperdiploid karyotype; however, no recurrent aberrations are observed in any subtype of CHL.

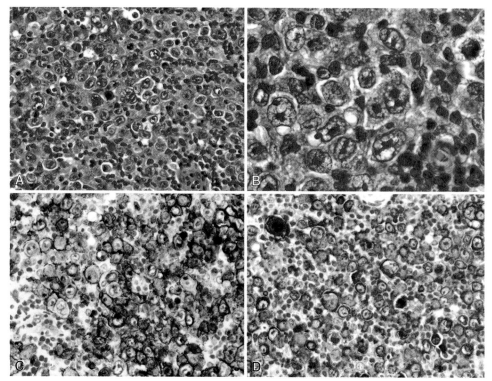

FIGURE 54.6 **NODULAR SCLEROSIS CHL, SYNCYTIAL TYPE.** Aggregate of large atypical cells with prominent nucleoli (A and B) expressing CD30 (C) and fascin (D).

- The *IGH* breakpoints have been involved in some HRS cells in CHL. The frequent partner regions are 2p16 (*REL*), 3q27 (*BCL6*), and 8q24 (*MYC*), 16p13 (*C2TA*), 17q12, and 19q13 (*BCL3*).
- CGH studies show frequent copy number alterations such as gains of 2p, 9p, 16p, 17q, 19q, 20q, and losses of 6q, 11q, and 13q. However, cytogenetic studies often yield a normal karyotype because of the rare and scattered neoplastic cells.
- Combination of FISH and immunophenotyping have shown HRS to have complex karyotypes with multiple complete and segmental gains and losses (Figure 54.7).

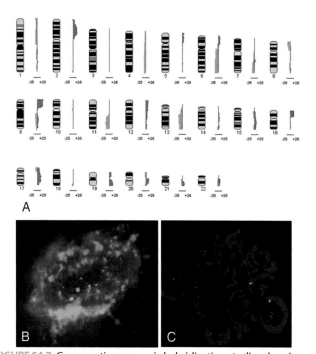

FIGURE 54.7 Comparative genomic hybridization studies showing recurrent copy number changes (imbalances) found in HRS cells of 53 classic Hodgkin lymphoma samples(A).The composite frequency plot summarizes the relative frequencies of chromosomal gains as red bars to the right and losses as green bars to the left aligned to each of the 22 autosomes. FISH interphase (B) and metaphase (C) showing increased number of ABCC1 loci signals (red) on 16p.
Adapted from Steidl C, Telenius A, Shah SP, et al. Genome-wide copy number analysis of Hodgkin Reed-Sternberg cells identifies recurrent imbalances with correlations to treatment outcome. Blood 2010; 116: 418–427, by permission.

Mixed Cellularity Classical Hodgkin Lymphoma

Mixed cellularity CHL (MCCHL) is the second most frequent subtype of CHL, accounting for about 25–30% of CHL. It is more commonly seen in developing countries and has a frequent association with EBV infection and HIV/AIDS. There is no bimodal age distribution, the median age is about 40 years, and the male to female ratio is about 1:1.

Mediastinal involvement is infrequent. The spleen, bone marrow, and liver are involved in about 30%, 10%, and 3% of cases, respectively.

MIXED CELLULARITY CLASSICAL HODGKIN LYMPHOMA

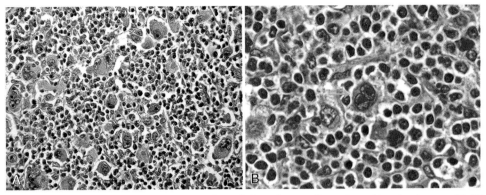

FIGURE 54.8 **MIXED CELLULARITY CHL.** (A) Numerous Hodgkin and Reed–Sternberg cells are found in a background infiltrate predominantly consisting of lymphocytes. (B) A classical Reed–Sternberg cell (arrow).

MORPHOLOGY

- The lymph node architecture is typically effaced, though some cases may show interfollicular involvement. Scattered classical HRS cells are present in an inflammatory background (Figures 54.8 to 54.11).
- The inflammatory component consists of various proportions of lymphocytes, histiocytes, plasma cells, eosinophils, and neutrophils. Histiocytes can form granuloma-like aggregates
- The pattern of involvement is diffuse, but sometimes vaguely nodular. There are no thickened fibrous capsules or broad collagen bands dividing the involved lymph node.

IMMUNOPHENOTYPE

The HRS cells and their variants are (see Figures 54.10 and 54.11):

- Negative for CD45;
- Positive for CD30 in nearly all cases, with membrane and Golgi staining pattern;
- At least focally positive for CD15 in about 80% of cases;
- Positive for fascin and PAX5 in the vast majority of cases (>90%);
- Positive for either OCT-2 or BOB.1 (not both) in about 90% of cases;
- Positive for CD20 in 30–40% of cases; staining is usually weaker than that of the background reactive B-cells;
- Positive for IRF4/MUM1 in the vast majority of cases;
- Negative for EMA and ALK.
- Staining for EBV-EBER is most frequently positive in this subtype of CHL, in about 75% of the cases.

MOLECULAR AND CYTOGENETIC STUDIES

No recurrent cytogenetic aberrations have been reported in LRCHL.

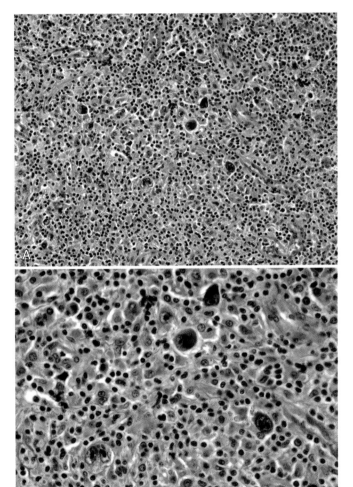

FIGURE 54.9 **MIXED CELLULARITY CHL.** Numerous Hodgkin and Reed–Sternberg cells are found in a background consisting of lymphocytes, eosinophils, and histiocytes (A and B).

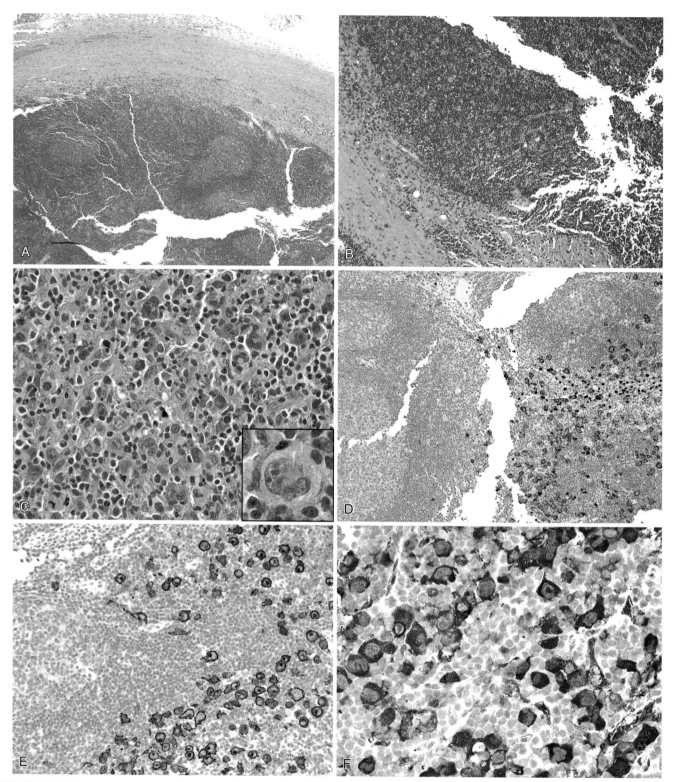

FIGURE 54.10 **MIXED CELLULARITY CHL.** Lymph node section shows a thick fibrotic capsule and interfollicular involvement with numerous Hodgkin and Reed–Sternberg cells (A, low power; B, intermediate power; C, high power). These cells express CD15 (D), CD30 (E), and fascin (F).

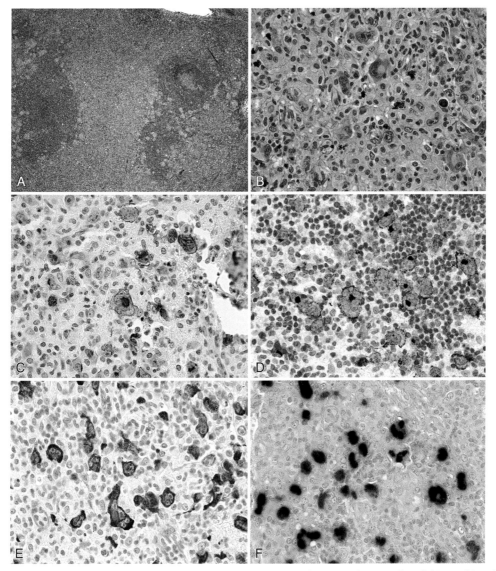

FIGURE 54.11 **MIXED CELLULARITY CHL.** Lymph node section shows an histiocyte-rich area with several Hodgkin and Reed–Sternberg cells (A, low power; B, high power) expressing CD15 (C), CD30 (D), and fascin (E). Scattered cells are positive for EBV-EBER (F).

Lymphocyte-Rich Classical Hodgkin Lymphoma

Lymphocyte-rich classical Hodgkin lymphoma (LRCHL) is a rare subtype of CHL which has clinical and morphologic features overlapping with those of NLPHL. Occasionally, the neoplastic cells can resemble LP cells seen in NLPHL, but their immunophenotypic features are different. There is often nodular, but sometimes diffuse growth pattern of lymph node involvement with scattered HRS cells admixed with small lymphocytes. Eosinophils and neutrophils are not present.

The median age ranges from 30 to 50 years with a male to female ratio of about 2.5:1. Most patients show localized disease with no "B" symptoms. Mediastinal involvement is rare. The prognosis is excellent.

MORPHOLOGY (FIGURE 54.12)

- The most frequent morphologic pattern of growth is nodular with total or partial effacement of nodal structures. The nodules consist of small lymphocytes with scattered HRS cells and lack of eosinophils or neutrophils. Nodular structures may contain small, regressed germinal centers. The HRS cells are usually outside the germinal centers.
- Diffuse LRCHL consists of a diffuse background of small lymphocytes and scattered HRS cells. Lymphocytes in some cases are mixed with histiocytes.

IMMUNOPHENOTYPE

- Although there are clinical and morphologic overlaps between NLPHL and LRCHL, their immunophenotypic features are different.
- The HRS cells and their variants in LRCHL have similar immunophenotype to those in the other subtypes of CHL

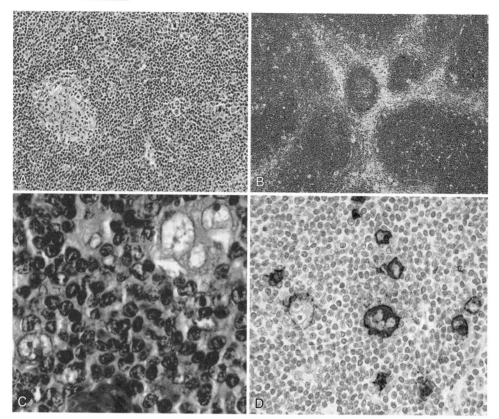

FIGURE 54.12 **LYMPHOCYTE-RICH CHL, NODULAR VARIANT.** (A) The lymph node is dominated by small lymphocytes arranged in a nodular pattern, frequently with atrophic germinal centers. The tumor cells are found in the expanded mantle zones of these B-cell nodules. (B) Immunohistochemistry for CD20 highlights the nodular pattern with a predominance of B cells. (C) A classic Reed–Sternberg (RS) cell is indicated by the arrow. (D) Immunohistochemistry for CD15 reveals strong positivity of the neoplastic cells, including a classic binucleate RS cell.
Adapted from Jaffe ES, Harris NL, Vardiman JW, et al. Hematopathology. Saunders/Elsevier, Philadelphia, 2010, by permission.

- The small lymphocytes in the nodules represent the expansion of the mantle zone area and express IgM, IgD, CD20, and CD5 (see Figure 45.12D).
- The regressed or ill-defined follicles are highlighted by a mesh of CD21+ follicular dendritic cells.
- In the diffuse subtype, small lymphocytes are of T-cell type and express CD3.

MOLECULAR AND CYTOGENETIC STUDIES

Conventional cytogenetics and FISH show aneuploidy and hyperdiploidy; however, no recurrent aberrations are observed in any subtype of CHL (see above).

Lymphocyte-Depleted Classical Hodgkin Lymphoma

Lymphocyte-depleted classical Hodgkin lymphoma (LDCHL) is the least common subtype of CHL and is more frequently seen in developing countries and in persons with HIV/AIDS. The median age ranges from 30 to 40 years with a male to female ratio of about 2.5:1. LDCHL shows a diffuse growth pattern with abundant HRS cells.

LDCHL is an aggressive tumor and usually presents in advanced stage and with "B" symptoms. Retroperitoneal lymph nodes are the most frequent site of involvement often with the involvement of the bone marrow and other intra-abdominal organs.

MORPHOLOGY

- The pattern of involvement is diffuse with preponderance of HRS cells. HRS cells may sometimes appear in aggregates and display sarcomatoid appearance (Figure 54.13).
- The background contains relatively few small lymphocytes.
- In some cases, diffuse fibrosis may be present (Figure 54.14).

IMMUNOPHENOTYPE

The HRS cells and their variants are:

- Negative for CD45;
- Positive for CD30 in nearly all cases, with membrane and Golgi staining pattern;
- At least focally positive for CD15 in about 80% of cases;

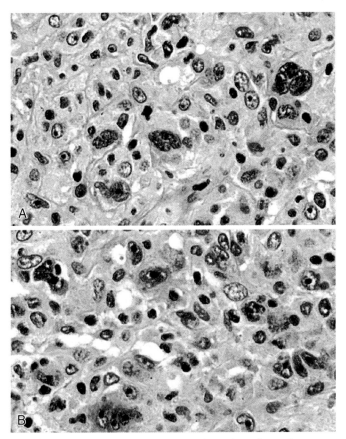

FIGURE 54.13 **LYMPHOCYTE-DEPLETED CHL.** Lymph node sections demonstrating numerous HRS cells mixed with scattered lymphocytes.
From Naeim F. Pathology of Bone Marrow, 2nd edn. Williams & Wilkins, Baltimore, 1997, by permission.

- Positive for fascin and PAX5 in the vast majority of cases (>90%);
- Positive for either OCT-2 or BOB.1 (not both) in about 90% of cases;
- Positive for CD20 in 30–40% of cases; staining is usually weaker than that of the background reactive B-cells;
- Positive for IRF4/MUM1 in the vast majority of cases;
- Negative for EMA and ALK.

Staining for EBV is positive in most HIV+ cases.

MOLECULAR AND CYTOGENETIC STUDIES

No recurrent cytogenetic aberrations have been reported in any subtype of CHL.

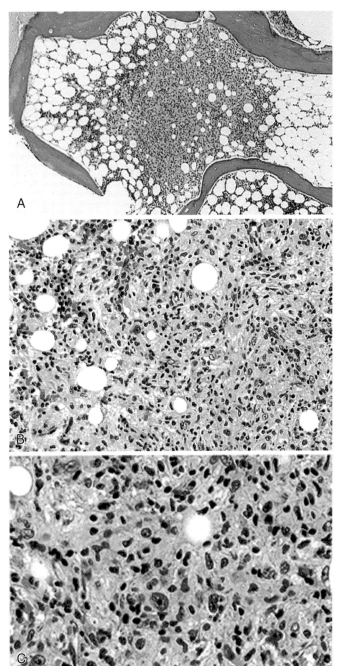

FIGURE 54.14 **LYMPHOCYTE-DEPLETED CHL.** Bone marrow section demonstrating focal involvement with scattered HRS cells in a background of fibrosis and inflammatory cells (A, low power; B, intermediate power; C, high power).
From Naeim F. Pathology of Bone Marrow, 2nd edn. Williams & Wilkins, Baltimore, 1997, by permission.

Differential Diagnosis

The differential diagnosis includes nodular lymphocyte predominant Hodgkin lymphoma (NLPHL), various subtypes of non-Hodgkin lymphoma, reactive lymphadenopathies, and non-hematopoietic malignancies. There are significant overlapping clinicopathologic features between NLPHL and LRCHL. The HRS cells in LRCHL express CD15, CD30, are CD45-negative, and may show EBV infection, whereas the neoplastic cells in NLPHL are negative for CD15 and CD30, and express pan-B-cell markers, EMA and CD45 (Table 54.1). LRCHL should be distinguished from T-cell/histiocyte-rich large B-cell lymphoma (T/HLBCL). T/HLBCL is diffuse, and the neoplastic cells are embedded in a mixture of T-cells and histiocytes and express pan-B-cell markers and CD45, whereas LRCHL is nodular, the HRS

Table 54.1
Differential Diagnosis of Classical Hodgkin Lymphoma (CHL)[1]

Entity	Architecture	Neoplastic Cells	Phenotype	Genotype
CHL				
NS	Nodular	HRS cells, lacunar cells	CD15+, CD30+, CD20±, Fascin+, PAX5+, CD45−, OCT-2+ or BOB.1+, EMA−, ALK1−	Mostly polyclonal, minority B-cell clone
MC	Diffuse	HRS cells		
LR	Often nodular, with expanded mantle zones in the nodules	HRS cells		
LD	Diffuse, sometimes with fibrosis	Abundant HRS cells	EBV±	
Other lymphomas				
NLPHL	Nodular, small B-cells and CD4+/CD57+ T-cells in the nodules	LP (L & H) cells	CD20+, CD79a+, CD45+, BOB.1+ and OCT-2+, EBV−, CD15−, CD30−	Polyclonal
T/HRLBCL	Diffuse with abundant T-cells and/or histiocytes	Scattered large atypical lymphocytes, RS-like cells	CD20+, CD79a+, EMA+, CD45+, CD15−, CD30−, EBV−	Rearranged *IGH*
ALCL	Diffuse or sinus pattern	Anaplastic large cells, hallmark cells, RS-like cells	CD30+, CD15±, CD45+, CD2+, CD4+, EMA+, ALK1±, LMP-1−, CD20−, PAX5−	Rearranged *TCR/ALK*

ALCL, anaplastic large cell lymphoma; LD, lymphocyte-depleted; LR, lymphocyte-rich; MC, mixed cellularity; NLPHL, nodular lymphocyte predominant Hodgkin lymphoma; NS, nodular sclerosis; T/HRLBCL, T-cell/histiocyte-rich large B-cell lymphoma.
[1]Adapted from Jaffe ES, Harris ML, Vardiman JW, et al. Hematopathology. Saunders/Elsevier, Philadelphia, 2010.

cells are surrounded by mantle cells and express CD15 and CD30 and are CD45-negative (see Table 54.1).

Primary mediastinal large B-cell lymphoma (PMLBCL) may show sclerosis and contain cells resembling lacunar cells or RS cells. The neoplastic cells in PMLBCL may express CD30, and CD15, and may mimic the syncytial variant of NSCHL.

Neoplastic cells in anaplastic large cell lymphoma (ALCL) express CD30 and morphologically may mimic HRS cells, but their cohesive growth pattern, hallmark cells, expression of ALK1 and CD45, and TCR rearrangement distinguish them from CHL (see Table 54.1)

Reactive lymphadenopathies, such as infectious mononucleosis, may demonstrate scattered RS-like cells. These cells express CD20, may express CD30 and LMP-1, but are usually negative for CD15. Metastatic carcinoma or melanoma may mimic the syncytial variant of NSCHL. Tumor cells of metastatic carcinoma are cytokeratin+ and CD45−, and melanoma cells are S-100+ and CD45−.

Additional Resources

Advani R: Optimal therapy of advanced hodgkin lymphoma, *Hematology Am Soc Hematol Educ Program*:310–316, 2011.

Blum KA: Upcoming diagnostic and therapeutic developments in classical Hodgkin's lymphoma, *Hematology Am Soc Hematol Educ Program*:93–100, 2010.

Eberle FC, Mani H, Jaffe ES: Histopathology of Hodgkin's lymphoma, *Cancer J* 15:129–137, 2009.

Farrell K, Jarrett RF: The molecular pathogenesis of Hodgkin lymphoma, *Histopathology* 58:15–25, 2011.

Fraga M, Forteza J: Diagnosis of Hodgkin's disease: an update on histopathological and immunophenotypical features, *Histol Histopathol* 22:923–935, 2007.

Jaffe ES, Harris NL, Vardiman JW, et al: *Hematopathology*, Philadelphia, 2010, Saunders/Elsevier.

Küppers R: The biology of Hodgkin's lymphoma, *Nat Rev Cancer* 9:15–27, 2009.

Mani H, Jaffe ES: Hodgkin lymphoma: an update on its biology with new insights into classification, *Clin Lymphoma Myeloma* 9:206–216, 2009.

Steidl C, Connors JM, Gascoyne RD: Molecular pathogenesis of Hodgkin's lymphoma: increasing evidence of the importance of the microenvironment, *J Clin Oncol* 29:1812–1826, 2011.

Steidl C, Telenius A, Shah SP, et al: Genome-wide copy number analysis of Hodgkin Reed-Sternberg cells identifies recurrent imbalances with correlations to treatment outcome, *Blood* 116:418–427, 2010.

Immunodeficiency Disorders

Primary Immunodeficiency Syndromes

The primary immunodeficiencies are rare congenital disorders with defective function of the immune system leading to increased susceptibility to infection, autoimmunity, and development of malignant neoplasms. The primary immunodeficiencies include different subtypes representing T-cell (cellular) or B-cell (humoral) defects or combined deficiencies (Table 55.1).

The T-cell defects are characterized by opportunistic viral and pneumocystis infections, whereas B-cell immunodeficiencies usually lead to bacterial infections. In this section, severe combined immunodeficiency (SCID), Wiskott–Aldrich syndrome (WAS), DiGeorge syndrome, agammaglobulinemia, ataxia telangiectasia, and X-linked lymphoproliferative syndrome are briefly discussed as examples of primary immunodeficiencies.

SEVERE COMBINED IMMUNODEFICIENCY SYNDROMES

Severe combined immunodeficiency syndromes (SCID) consist of a heterogeneous group of disorders primarily arising from molecular T-cell defects leading to developmental and functional disturbances of T-cells and sometimes natural killer (NK) cells or B-cells (Table 55.2).

There are two different modes of inheritance: X-linked and autosomal recessive. The X-linked form is associated with the mutation of the gamma chain of the IL-2 receptor gene (*IL2RG*) and is characterized by reduced numbers of circulating T- and NK-cells and normal numbers of B-cells. However, B-cells do not mature to plasma cells, leading to a virtually non-existent serum immunoglobulin in these patients. Other variants of SCID are listed in Table 55.2.

Of the many possible genes involved in SCID and related disorders, targeted DNA sequencing to detect mutations is available for selected ones, such as *IL2RG*, *ADA*, *RAG1*, and *RAG2*. However, the advent of next-generation sequencing and whole-exon analysis now allows for genome-wide testing for all of the immune deficiency genes, both known and unknown.

Newborn screening for SCID is starting to be implemented, using a test for T-cell receptor excision circles (TRECs) in DNA extracted from routine newborn screening blood spots.

Adenosine deaminase deficiency (ADD) is a subtype of SCID and accounts for about 50% of the autosomal recessive forms. ADD is characterized by a marked decline in the absolute number of both T- and B-cells in the peripheral blood and reduced number of hematogones and plasma cells in the bone marrow.

The typical symptoms of SCID are recurrent viral and bacterial infections, chronic diarrhea, and failure to thrive in newborns. These findings are associated with lack of thymic shadow on chest X-ray studies. In some cases, clinical symptoms may be delayed by several months due to the protective effects of the maternally derived antibodies.

About 1–5% of SCID patients develop lymphoid malignancies, such as primarily non-Hodgkin luymphomas. Rare cases are EBV-associated

WISKOTT–ALDRICH SYNDROME

Wiskott–Aldrich syndrome (WAS) is a rare congenital disorder characterized by the triad of immunodeficiency, eczema, and abnormal platelets (thrombocytopenia, small platelets, and platelet dysfunction) (see Chapter 58).

Mutation of the *WASP* gene located on the chromosome Xp11.22–23 is the primary molecular defect. Numerous mutations have been reported, mostly found in exons 1 and 2. These mutations lead to the absence or aberrant expression of the WAS protein in lymphocytes and megakaryocytes. Lack of WAS protein expression may lead to decreased platelet size and defective T-cell and platelet function.

Wiskott–Aldrich syndrome (WAS) is an X-linked recessive disorder primarily involving males with an average age

Table 55.1
Primary Immunodeficiency Syndromes Associated with Defective Lymphocytes[1]

T-Cell Defects		
SCID		
Reticular dysgenesis	AR	?
Alymphocytosis	AR	RAG-1; RAG-2
Deficit of T and NK cells	X-linked; AR	γ-chain-IL-2R; Jak3
PNP deficiency	AR	PNP
Omenn syndrome	AR	5′-nucleotidase?
T-cell activation defects (CID)		
HLA deficiency class-I	AR	TAP2
HLA deficiency class-II	AR	CIITA/RFX5
CD3 deficiency	AR	γε CD3
Zap-70 deficiency	AR	Zap-70
Calcium influx deficiency	AR	?
Defect in IL-2 synthesis	AR	?
Defects in DNA repair		
Ataxia telangiectasia	AR	ATM
Bloom syndrome	AR	BLM
Nijmegen syndrome	AR	NBS
Xeroderma pigmentosum	AR	7 complementation groups (XPA to G)
Others		
Wiskott–Aldrich syndrome	X-linked; AR	WASP; ?
DiGeorge syndrome	Sporadic	del(22q11), TBX1
Hyper-IgM syndrome	X-linked; AR	CD40 L; ?
B-Cell Defects		
Lymphoproliferative syndrome	X-linked; AD	?; fas
Bruton agammaglobulinemia	X-linked	btk
Common variable immunodeficiency	Sporadic	?
IgA deficiency	AD; AR; sporadic	?
IgG subclass deficiency	AR	?
Hyper IgE syndrome	AR	?

[1] Adapted from Ten RM. Primary immunodeficiencies. Mayo Clin Proc 1998; 73: 865–872.
AD, autosomal dominant; AR, autosomal recessive; btk, Bruton tyrosine kinase; CID, combined immunodeficiency; PNP, purine nucleoside phosphorylase; SCID, severe combined immunodeficiency; WASP, Wiskott–Aldrich syndrome protein.

Table 55.2
Gene Defects and Blood Lymphocyte Alterations in Certain Subtypes of Severe Combined Immunodeficiency (SCID)[1]

Subtype	Gene	CD4	CD8	B	NK
X-Linked, common γ-chain	IL2RG	↓	↓	N	↓
Janus kinase 3	JAK3	↓	↓	N	↓
IL-2 receptor-α (CD25)	IL2RA	↓	↓	N	↓
IL-7 receptor-α (CD127)	IL7RA	↓	↓	N	N
CD3 complex	CD3D	↓	↓	N	N
Recombinase activating genes	RAG1, RAG2	↓	↓	↓	N
Adenosine deaminase	ADA	↓	↓	↓	±
MHC class II	MHCIID	↓	N	N	N
Zeta-associated protein	ZAP70	N	↓	N	N
Protein tyrosine phosphatase	PTPRC	↓	↓	N	(CD45)

[1] Adapted from Bonilla F. Combined immune deficiencies. *UpToDate* 2007.

of about 21 months at diagnosis. An autosomal dominant mode of inheritance has also been reported in some families, and complete sequencing of the WAS gene is available in a number of specialized laboratories to detect the mutations.

Clinical manifestations include bleeding secondary to thrombocytopenia and abnormal platelet function, recurrent infections, autoimmune manifestations, and eczema. The platelet counts are often <50,000/μL, and platelet size is usually half the normal. There is an inverse correlation between the severity of the clinical manifestations and the detectable levels of the WAS protein in the peripheral blood.

Approximately, 10–20% of WAS patients develop lymphoid malignancies, 75% being large B-cell lymphomas. Hodgkin lymphoma accounts for less than 5% of the tumors.

The differential diagnosis includes immune-associated thrombocytopenia, other primary immunodeficiencies, myelodysplastic syndromes, and hematopoietic malignancies.

DIGEORGE SYNDROME

DiGeorge syndrome (DS) is a rare congenital immunodeficiency disorder characterized by the deletion of 22q11.2, developmental abnormalities of the third and fourth pharyngeal pouches, including thymic hypoplasia and hypoparathyroidism, and midline cardiac defects. The suggested criteria for definitive diagnosis of DS consist of reduced numbers of total T-cells plus the presence of at least two of the following three findings:

1. Deletion of chromosome 22q11.2 (Figures 55.1 and 55.2)
2. Hypocalcemia
3. Congenital cardiac defects.

Up to 90% of patients with DS show microdeletion of 22q11.2, and about 75% of the patients demonstrate congenital heart disease including tetralogy of Fallot, ventricular septal defect, or interrupted aortic arch. As testing for the 22q11.2 deletion by fluorescence in situ hybridization (FISH) has become more widespread (see Figure 55.2), the range of cardiac defects associated with the syndrome has expanded. This region of the chromosome contains about 30 genes, including the TBX1 gene. In a small number of affected individuals without a chromosome 22 deletion, mutations in the TBX1 gene are thought to be responsible for the characteristic signs and symptoms of the syndrome.

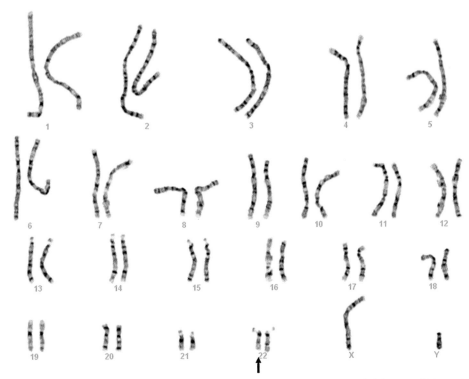

FIGURE 55.1 The characteristic karyotypic abnormality in DiGeorge syndrome is an interstitial deletion of 22q11.2 region (arrow shows the deleted 22q).

Some centers have begun checking for the deletion in almost every patient with even an isolated congenital heart defect, and the diagnostic yield has been significant. Discovery of the deletion can then raise alertness to the possibility of other immunologic and endocrine complications (which are not present in all cases).

The deletion can also be detected in the context of a whole-genome scan for copy number variants (CNVs) using array comparative genomic hybridization (aCGH). The parents should also be tested to distinguish *de novo* from inherited deletions, and to distinguish benign CNVs from pathologic deletions and insertions.

Despite much work aimed at identifying the causative gene(s) in the deleted critical region, targeted molecular testing for mutations is not yet routine.

AGAMMAGLOBULINEMIA

Agammaglobulinemia is one of the primary humoral immunodeficiencies and consists of two congenital types: X-linked and autosomal recessive.

The *X-linked* variant (Bruton agammaglobulinemia) affects boys, with clinical manifestations between 6 and 18 months of age. This disorder is caused by mutation in a tyrosine kinase gene called *BTK* (Bruton tyrosine kinase) mapped at Xq21.

Complete sequencing of the *BTK* gene to detect these variants is available in several reference laboratories. The *BTK* product is a signal transduction molecule expressed in B-cells and other hematopoietic cells.

Affected patients show profound hypogammaglobulinemia, reduced number of peripheral blood B-cells (<1% CD19+ or CD20+ cells), and recurrent bacterial and viral infections. An atypical form of X-linked agammaglobulinemia has been reported with less severe lymphopenia and hypogammaglobulinemia.

The clinical presentations of the *autosomal recessive* variant are similar to those of the X-linked variant except for the lack of *BTK* mutation and for manifestation of the disease in both genders. The autosomal recessive type is caused by mutations in several genes that regulate B-cell development, such as *IGHM, CD179B, CD79A, BLNK*, and *LRRCB* genes.

ATAXIA TELANGIECTASIA

Ataxia telangiectasia (AT) is an autosomal recessive disease characterized by immunodeficiency, ataxia, oculocutaneous telangiectasia, defective DNA repair, and increased susceptibility to ionizing radiation. The severity of immunodeficiency is variable and may affect both B and T cells. All AT patients show mutation of the *ATM* gene located on chromosome 11q22–23.

Patients demonstrate thymic atrophy and hypoplasia of the follicles in the lymph nodes. There is peripheral blood lymphopenia predominantly involving the T-cells. There is progressive loss of cerebellum Purkinje cells.

Approximately 10% of homozygote AT patients eventually develop leukemia, lymphoma, or non-lymphoid malignancies. Development of T-cell leukemia/lymphoma, particularly T-prolymphocytic leukemia, is frequent. Hodgkin lymphoma accounts for about 10% of AT-associated malignancies.

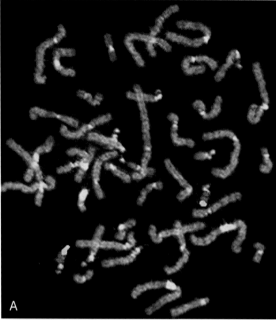
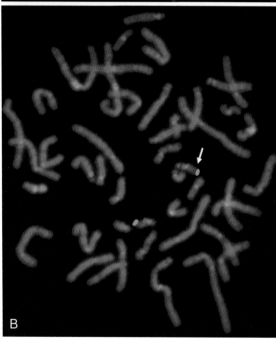

FIGURE 55.2 FISH analysis demonstrating a normal metaphase (A) and an abnormal metaphase (B) with the 22q11.2 deletion (arrow) in a patient with DiGeorge syndrome.

AT is caused by mutations in the *ATM* gene. The gene is quite large, so testing for mutations by DNA sequencing is not routine for primary diagnosis, which is largely based on clinical features and functional DNA repair assays.

Karyotyping of peripheral lymphocytes from AT homozygotes shows non-random chromosomal rearrangements, preferentially involving chromosomal breakpoints at 14q11.2, 14q32, 7q35, 7p14, 2p12, and 22q11.2, and correlate with the regions of the T-cell and B-cell receptor gene complexes.

> **Box 55.1 Diagnostic Criteria for X-Linked Lymphoproliferative syndrome (XLP)[1]**
>
> **Definitive**
> Male patient with lymphoma, immunodeficiency, aplastic anemia, lymphohistiocytic disorder or fatal EBV infection, and mutation in the *SAP* gene.
>
> **Probable**
> Male patient with lymphoma, immunodeficiency, aplastic anemia, and lymphohistiocytic disorder or fatal EBV infection. Patient has maternal cousins, uncles or nephews with a history of similar disorder.
>
> **Possible**
> Male patient with lymphoma, aplastic anemia, or lymphohistiocytic disorder, resulting in death, following EBV infection.
>
> [1] Seemayer TA, Gross TG, Egeler RM, et al. X-linked lymphoproliferative disease: twenty-five years after the discovery. Pediatr Res 1995; 38: 471–478, and Gaspar HB, Sharifi R, Gilmour KC, et al. X-linked lymphoproliferative disease: clinical, diagnostic and molecular perspective. Br J Haematol 2002; 119: 585–595.

Telomere shortening and fusions, with normal telomerase activity, have been observed in peripheral blood lymphocytes of AT patients, especially in pre-leukemic T-cell clones.

X-LINKED LYMPHOPROLIFERATIVE SYNDROME

X-linked lymphoproliferative syndrome (XLP) is a rare inherited immunodeficiency characterized by lymphocytosis, dysgammaglobulinemia, fatal infectious mononucleosis (IM), or lymphoma usually developing in response to infection with EBV.

XLP is a rare X-linked inherited disorder affecting boys, with age of onset ranging from 2 to 19 years. The most common clinical presentation is a fatal IM following EBV infection which has a very high mortality rate and a survival rate of <5%. However, in approximately one third of affected patients, EBV infection is not fatal. These patients eventually develop dysgammaglobulinemia and/or lymphoma. The definitive diagnostic criteria include a male patient with lymphoma, immunodeficiency, aplastic anemia, lymphohistiocytic disorder or fatal EBV infection, and mutation in the *SAP* gene (Box 55.1).

The treatment of choice is allogeneic hematopoietic stem cell transplantation. Antiviral agents, such as aciclovir or foscarnet, high dose immunoglobulin, immunosuppressive drugs, interferons α and γ, and HLH 94 have been tried with debatable outcomes. The HLH 94 (an anti-histiocytic regimen) has been shown to induce long term remissions.

Morphology and Laboratory Findings

- The most common pathologic finding is a massive, systemic lymphohistiocytosis associated with a clinical picture of fatal IM. The proliferating cells are EBV-infected B-cells and cytotoxic T-cells along with histiocytes.

- The lymphohistiocytic proliferation involves lymph nodes as well as other tissues such as bone marrow, brain, heart, and kidney, and is accompanied by hemophagocytosis and dysregulated cytokine release resulting in extensive tissue damage, such as hepatic necrosis and profound bone marrow hypoplasia.
- Abnormal production of serum immunoglobulin is a common finding and is often associated with a defective cellular immune function, presenting a picture of common variable immunodeficiency. The degree of hypogammaglobulinemia ranges from moderately decreased levels of IgG to severe panhypogammaglobulinemia.
- Approximately 35% of the affected children develop lymphoma, usually of B-cell type. Lymphoma is often extranodal and involves ileocecum, central nervous system, liver, and kidney. Burkitt lymphoma is the most frequent subtype (53%) followed by immunoblastic lymphoma (12%) and follicular lymphoma (12%). HL is rare.

Molecular and Cytogenetic Studies

- Mutations in two X-linked genes, *SAP* and *XIAP*, have been reported in XLP patients. *SAP* (*SH2D1A*), the most commonly affected gene, is located in the Xq25 region and encodes a signaling lymphocyte activation molecule (SLAM)-associated protein.
- SLAM has a number of functions including regulation of T-cell cytotoxicity, T-cell/B-cell co-stimulation, and induction of interferon-γ in the Th1 cells. *SAP* mutations probably result in the defective T- and NK-cell responses and dysregulated cytokine release.
- Mutation in the *XIAP* (or *BIRC4*) gene has been recently reported in some patients with XLP who showed no evidence of *SAP* mutation. *XIAP* encodes the X-linked inhibitor of apoptosis.
- *XIAP*-deficient patients with XLP have low numbers of NK-cells, suggesting that *XIAP* is required for the survival and/or differentiation of NK-cells.

Human Immunodeficiency Virus/ Acquired Immunodeficiency Syndrome

The human immunodeficiency virus (HIV)/acquired immunodeficiency syndrome (AIDS) has spread throughout the world. AIDS is associated with lymphocytopenia and defective cell-mediated immunity leading to recurrent infections, Kaposi sarcoma, lymphoma, and cervical cancer. All HIV-positive patients with a CD4 cell count of <200/µL are considered to have AIDS regardless of the presence or absence of clinical symptoms.

AIDS is an HIV-induced illness with a broad spectrum of clinical manifestations. There are two strains of HIV: HIV-1 and HIV-2. HIV-1 was discovered first and is widespread in the United States, Europe, and other parts of the world, whereas HIV-2 is virtually homologous to the simian immunodeficiency virus (SIV) and is primarily detected in West Africa.

HIV infection is usually through three major routes: sexual intercourse (70–80%), exposure to contaminated blood (5–10%), and prenatal transmission (5–10%). Viral penetration of mucosal epithelium is followed by infection of CD4+ T-cells, dendritic cells, and macrophages, with subsequent spread to the lymph nodes and blood (viremia). The entry to the CD4+ cells is mediated by the chemokine receptor CCR5. Patients homozygous for a 32-base pair deletion in CCR5 are resistant to HIV infection, whereas heterozygous-infected individuals may have a less aggressive clinical course. The heterozygote frequency in the general population is about 10%.

HIV infection is divided into several stages advancing from viral transmission to primary (acute) HIV infection, seroconversion, clinical latent period, early symptomatic stage, and finally to AIDS. The primary HIV infection is characterized by acute non-specific flu-like symptoms, such as fever, headache, sore throat, myalgia/arthralgia, diarrhea, nausea/vomiting, and lymphadenopathy. Clinical latent period refers to the 6-month asymptomatic period (except lymphadenopathy) after seroconversion. In this period, HIV is trapped by the follicular dendritic cells in the lymphoid tissues. The early symptomatic stage (Class B), previously called "AIDS-related complex," is associated with a number of clinical symptoms and opportunistic infections, such as fever, chronic diarrhea, oral leukoplakia, peripheral neuropathy, thrombocytopenia, herpes zoster infection, vaginal candidiasis, and cervical dysplasia.

The most frequent clinical presentation associated with AIDS is *pneumocystis carinii* pneumonia followed by esophageal candidiasis, wasting, and Kaposi sarcoma. The spectrum and natural course of these complications has changed dramatically with the advent of anti-retroviral therapy.

Numerous drugs have been approved by the Food and Drug Administration (FDA) for the treatment of HIV infection, including nucleoside analogs, protease inhibitors, and fusion inhibitors. Nucleoside analogs are nucleoside reverse transcriptase inhibitors, such as azidothymidine (AZT), zalcitabine (ddC), and stavudine (d4T). Examples of protease inhibitors are ritonavir (Norvir), indinavir (Crixivan), and saquinivir (Invirase). Fuzeon (Enfuvirtide) is the first approved fusion inhibitor drug.

MORPHOLOGY AND LABORATORY FINDINGS

- The peripheral blood reveals lymphopenia (often <500/µL) with a reversed CD4:CD8 ratio (Figure 55.3). Scattered reactive lymphocytes may be present. Pancytopenia is observed in over 50% of cases. Anemia is common and is usually normocytic normochromic, with decreased reticulocyte count and elevated serum iron and ferritin levels. Mild to moderate neutropenia is a common feature and approximately 30% of

patients demonstrate monocytopenia. Thrombocytopenia is frequent and in some cases is autoimmune associated (ITP-like syndrome).
- HIV antibodies are raised against various viral components, such as envelope glycoproteins GP-120 and GP-41 and core protein p24. These antibodies are detected by a variety of techniques such as enzyme-linked immunosorbent assay (ELISA), Western blot, radioimmunoassay, and immunofluorescence. Many patients show positive HIV serology (seroconversion) within 4–10 weeks after exposure to the virus, and >95% are seroconvert within 6 months.
- Bone marrow specimens from AIDS patients reveal a normocellular to hypercellular marrow, often with myeloid preponderance and left shift, and mild to severe dysplastic changes in one or more hematopoietic lines. An increased frequency of naked megakaryocyte nuclei has been observed in bone marrow samples (biopsy sections and marrow smears) of patients with AIDS. A polyclonal lymphoplasmacytosis is common, sometimes with reactive lymphocytes and immature plasma cells (Figure 55.4). CD4+ cells are reduced, particularly those of the CD45RA-negative phenotype, whereas the proportion of CD8+ cells is increased, especially the HLA-DR+/CD45RA+ subtype. Reticulin fibrosis and gelatinous transformation of bone marrow may be present.
- The affected lymph nodes show a variety of morphologic changes. Florid follicular hyperplasia is the most frequent pattern, particularly at the early stages of the disease (Figure 55.5). Numerous, large, irregular follicles with expanded germinal centers (GC), numerous mitotic figures, and tingible body macrophages are present. Mantle zones are often ill-defined or effaced. The interfollicular areas show prominent vascularity and a mixed cellular component consisting of plasma cells, immunoblasts, histiocytes, lymphocytes, and monocytoid B-cells. The CD4:CD8 ratio is reversed. Sinus histiocytosis is a frequent feature and may be associated with hemophagocytosis. In the later stages of the disease, there is evidence of lymphoid depletion by reduced number of the lymphoid follicles, decreased number of lymphocytes in the interfollicular areas, and the presence of amorphous eosinophilic deposits and/or fibrosis. The affected lymph nodes may show signs of opportunistic infection, Kaposi sarcoma, or lymphoma.

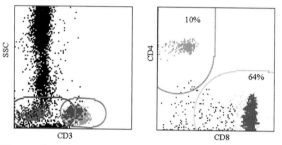

FIGURE 55.3 Flow cytometry of peripheral blood demonstrating a population of CD3+ lymphocytes (absolute count = 980/μL) with reversed CD4:CD8 ratio (CD4 orange and CD8 green).

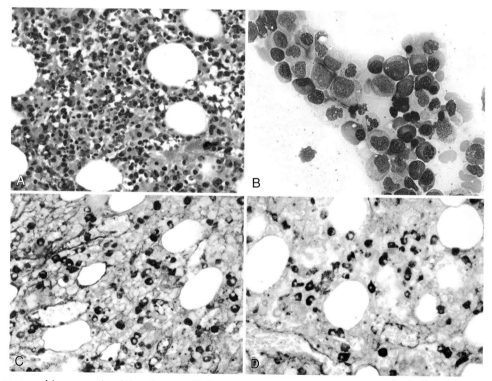

FIGURE 55.4 Bone marrow biopsy section (A) and smear (B) from a patient with AIDS showing a polyclonal plasmacytosis. Immunohistochemical stains for kappa light chain (C) and lambda light chain (D) demonstrate a polyclonal plasma cell population.

MOLECULAR AND CYTOGENETIC STUDIES

HIV RNA levels are detected in early stages of viremia by a sensitive reverse transcriptase-polymerase chain reaction (RT-PCR) assay.

The quantitative RT-PCR assay, now performed at high throughput using FDA-approved reagents, is especially useful for monitoring residual viral load in patients on anti-retroviral therapy.

HIV-ASSOCIATED LYMPHOMAS

Advances in anti-retroviral HIV treatments have significantly reduced the incidence of lymphomas in AIDS patients. In a population study the risk of development of a lymphoid malignancy in AIDS patients was reported as 113-fold, 7.6-fold, and 4.5-fold for non-Hodgkin lymphoma, Hodgkin lymphoma, and plasma cell myeloma, respectively. Active anti-retrovrial therapies have also changed the incidence of lymphoma subtypes, with an increased proportion of diffuse large B-cell lymphoma (DLBCL) and reduced incidence of Burkitt lymphoma (BL).

Most HIV-associated lymphomas are aggressive, are of B-cell type, and frequently involve extranodal tissues. The CNS, GI tract, bone marrow, and liver are the most frequent extranodal sites. Unusual sites of involvement include body cavities, bile duct, oral cavities, appendix, kidney, and lung. EBV or HHV8 infection is a frequent finding.

According to the WHO classification, HIV-associated lymphomas are divided into three major categories:

- Lymphomas that are also seen in immunocompetent patients, such as DLBCL (see Chapter 36), BL (see Chapter 39), extra-nodal marginal zone lymphoma (see Chapter 31), plasma cell myeloma (see Chapter 41) and Hodgkin lymphoma (see Chapters 53 and 54).
- Lymphomas occurring more specifically in HIV-positive patients, such as primary effusion lymphoma, plasmablastic lymphoma, and lymphoma arising in HHV8-associated multi-centric Calstleman disorders (see Chapter 38).
- Lymphomas arising in HIV patients as well as other immuno-deficient states, such as post-transplant-associated lymphoproliferative disease (see Chapter 56).

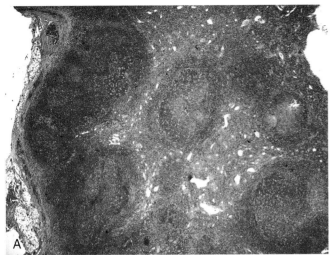
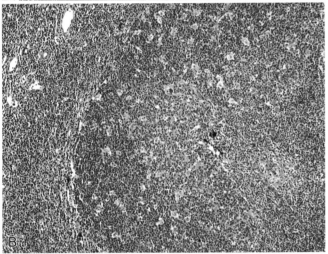
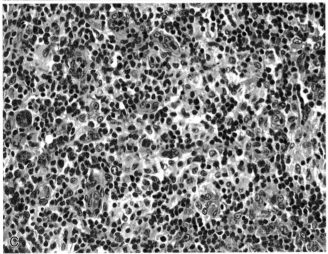

FIGURE 55.5 Lymph node biopsy section from a patient with AIDS demonstrating overall preservation of nodal architecture and expansion of follicular structures (A). Some of the follicles lack the mantle zone area (bare follicles) (B). Some of the macrophages contain melanin pigments (dermatophatic) (C).

Idiopathic CD4+ T-Lymphocytopenia

Idiopathic CD4+ T-lymphocytopenia is a condition characterized by acquired immunodeficiency, depletion of CD4+ lymphocytes, and opportunistic infections *without* evidence of HIV infection. These patients have an overall better prognosis than patients with AIDS, and a proportion of them may show a spontaneous regression. The peripheral blood absolute CD4+ count is usually <300/μL.

Additional Resources

Aloj G, Giardino G, Valentino L, et al: Severe combined immunodeficiencies: new and old scenarios, *Int Rev Immunol* 31:43–65, 2012.

Bassiri H, Janice Yeo WC, Rothman J, et al: X-linked lymphoproliferative disease (XLP): a model of impaired anti-viral, anti-tumor and humoral immune responses, *Immunol Res* 42:145–159, 2008.

Blundell MP, Worth A, Bouma G, et al: The Wiskott-Aldrich syndrome: The actin cytoskeleton and immune cell function, *Dis Markers* 29:157–175, 2010.

Borte S, Wang N, Oskarsdóttir S, et al: Newborn screening for primary immunodeficiencies: beyond SCID and XLA, *Ann N Y Acad Sci* 1246:118–130, 2011.

Bouma G, Burns SO, Thrasher AJ: Wiskott-Aldrich Syndrome: Immunodeficiency resulting from defective cell migration and impaired immunostimulatory activation, *Immunobiology* 214:778–790, 2009.

Brower V: AIDS-related cancers increase in Africa, *J Natl Cancer Inst* 103:918–919, 2011.

Cleland SY, Siegel RM: Wiskott-Aldrich Syndrome at the nexus of autoimmune and primary immunodeficiency diseases, *FEBS Lett* 585:3710–3714, 2011.

Dezube BJ: Acquired immunodeficiency syndrome-related Kaposi's sarcoma: clinical features, staging, and treatment, *Semin Oncol* 27:424–430, 2000.

Gatti R: Ataxia-Telangiectasia. In Pagon RA, Bird TD, Dolan CR, Stephens K, editors: *Gene reviews*, Seattle (WA), 1993–1999, University of Washington, Seattle. [updated 2010]

Kurisu S, Takenawa T: The WASP and WAVE family proteins, *Genome Biol* 10:226, 2009.

Lederman MM, Margolis L: The lymph node in HIV pathogenesis, *Semin Immunol* 20:187–195, 2008.

Luo L, Li T: Idiopathic CD4 lymphocytopenia and opportunistic infection—an update, *FEMS Immunol Med Microbiol* 54:283–289, 2008.

McCusker C, Warrington R: Primary immunodeficiency, *Allergy Asthma Clin Immunol* 7(Suppl 1):S11, 2011.

McDonald-McGinn DM, Sullivan KE: Chromosome 22q11.2 deletion syndrome (DiGeorge syndrome/velocardiofacial syndrome), *Medicine (Baltimore)* 90:1–18, 2011.

Mesti T, Setina TJ, Vovk M, et al: Eleven years of experience with AIDS-related lymphomas at the Institute of Oncology Ljubljana, *Med Oncol* 29:1217–1222, 2012.

Niehues T, Perez-Becker R, Schuetz C: More than just SCID—the phenotypic range of combined immunodeficiencies associated with mutations in the recombinase activating genes (RAG) 1 and 2, *Clin Immunol* 135:183–192, 2010.

Pai SY, Notarangelo LD: Hematopoietic cell transplantation for Wiskott-Aldrich syndrome: advances in biology and future directions for treatment, *Immunol Allergy Clin North Am* 30:179–194, 2010.

Perlman SL, Boder Deceased E, Sedgewick RP, et al: Ataxia-telangiectasia, *Handb Clin Neurol* 103:307–332, 2012.

Puck JM: The case for newborn screening for severe combined immunodeficiency and related disorders, *Ann N Y Acad Sci* 1246:108–117, 2011.

Rezaei N, Mahmoudi E, Aghamohammadi A, et al: X-linked lymphoproliferative syndrome: a genetic condition typified by the triad of infection, immunodeficiency and lymphoma, *Br J Haematol* 152:13–30, 2011.

Scambler PJ: 22q11 deletion syndrome: a role for TBX1 in pharyngeal and cardiovascular development, *Pediatr Cardiol* 31:378–390, 2010.

Sissolak G, Sissolak D, Jacobs P: Human immunodeficiency and Hodgkin lymphoma, *Transfus Apher Sci* 42:131–139, 2010.

Tripathi AK, Misra R, Kalra P, et al: Bone marrow abnormalities in HIV disease, *J Assoc Physicians India* 53:705–710, 2005.

Utsuki S, Oka H, Abe K, et al: Primary central nervous system lymphoma in acquired immune deficiency syndrome mimicking toxoplasmosis, *Brain Tumor Pathol* 28:83–87, 2011.

Iatrogenic Immunodeficiency-Associated Lymphoproliferative Disorders

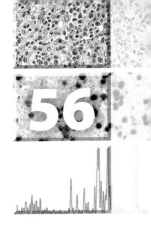

In this chapter, post-transplant lymphoproliferative disorders and lymphoproliferative disorders associated with immunosuppressive medications in autoimmune disorders are discussed.

Post-Transplant Lymphoproliferative Disorders

Post-transplant lymphoproliferative disorders (PTLD) are benign or malignant lymphoid disorders which develop after solid organ or bone marrow allogeneic transplantation. They represent a complex group with a wide spectrum of clinicopathologic features ranging from lymphoid hyperplasia to full blown lymphoma. The World Health Organization (WHO) classification defines four major categories (Table 56.1):

- Early lesions
- Polymorphic PTLD
- Monomorphic PTLD
- Classical Hodgkin lymphoma type PTLD.

The major risk factors for the development of PTLD are immunosuppression and EBV infection. EBV infection is documented in 50–80% of patients with PTLD. The more severe the immunosuppression the greater is the risk for PTLD. The incidence of PTLD is significantly higher in EBV-seronegative patients than in EBV-seropositive ones, because of the higher risk of EBV infection in recipients who have no pre-transplant immunity to EBV.

Other risk factors include age of recipient under 25 years, fewer HLA matches, and history of pre-transplant malignancy. The risk of PTLD is highest in the first year of the post-transplant period.

The incidence of PTLD in solid organ transplants is highest in intestinal and multiorgan transplants, ranging from 11% to 33%, followed by 2–9% for lung, 2–6%

Table 56.1

WHO Categories of Post-transplant Lymphoproliferative Disorders (PTLD)[1]

Type	Major Characteristic Features
Early lesions	
Reactive plasmacytic hyperplasia	Preserved nodal architecture, increased plasma cells and rare immunoblasts
IM-like lesions[2]	Preserved nodal architecture, paracortical expansion with numerous immunoblasts
Polymorphic PTLD	Destructive infiltrates consisting of a mixture of small and large lymphoid cells and plasma cells. Scattered atypical cells, areas of necrosis, and frequent mitotic figures may be present. Rearrangement of Ig or presence of EBV genome. Lack of mutations of *MYC*, *RAS*, and *TP53* genes
Monomorphic PTLD	Destructive infiltrates of monomophic atypical lymphoid cells consistent with lymphoma. Most frequent B-cell types are diffuse large B-cell, Burkitt lymphoma, and plasma cell myeloma. Rearrangement of Ig or presence of EBV genome. Mutations of *MYC*, *RAS*, and *TP53* genes may be present. Most frequent T-cell type is peripheral T-cell lymphoma, not otherwise specified. *TCR* gene rearrangement and up to 25% clonal EBV genome
Classical Hodgkin lymphoma type PTLD	Presence of HRS cells and morphology consistent with CHL; EBV+.

[1] Adapted from Jaffe ES, Harris NL, Stein H, et al. Pathology and Genetics: Tumors of Haematopoietic and Lymphoid Tissues. IARC Press, Lyon, 2001, and Swerdlow SH, Campo E, Harris NL, et al. WHO Classification of Tumours of Haematopoietic and Lymphoid Tissues. International Agency for Research on Cancer, Lyon, 2008.
[2] IM, infectious mononucleosis.

for heart, 1–3% for kidney, and 1–2% for liver transplantations. The incidence of PTLD in bone marrow allograft recipients is about 1%, except for those who receive HLA-mismatched or T-cell-depleted bone marrow or are treated by immunosuppressive drugs for graft versus host disease. In this group of patients, the risk of PTLD is up to 20%.

Patients typically demonstrate local or generalized lymphadenopathy, sometimes with graft dysfunction or other organ failures due to extranodal lymphoid infiltrate. EBV-negative and late-occurring cases have a higher tendency to be monomorphic. The therapeutic approaches include reduction in immunosuppression, antiviral therapy, chemotherapy and/or radiation therapy (for the treatment of monotypic PTLD and HL). A significant proportion of early PTLD lesions and polymorphic forms may regress by the reduction of immunosuppression. The overall prognosis is poor, particularly in monomorphic PTLD.

MORPHOLOGY

There are four morphologic subtypes of PTLD: (1) early lesions, (2) polymorphic PTLD, (3) monomorphic PTLD, and (4) classical Hodgkin lymphoma type PTLD.

Early Lesions

- Early lesions consist of plasmacytic hyperplasia and infectious mononucleosis-like disorders (Figure 56.1). These lesions mainly occur in oropharynx and lymph nodes.
- They are characterized by preservation of the nodal structures, open sinuses, and residual reactive follicles with diffuse interfollicular proliferation of plasma cells, transformed cells, and immunoblasts mixed with T-cells. Immunoblasts are commonly positive for EBV.

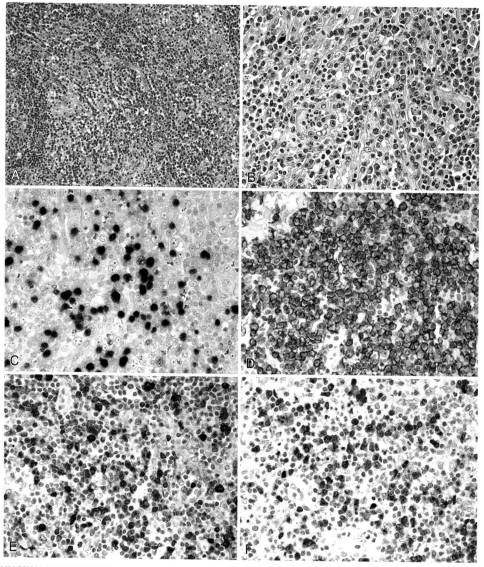

FIGURE 56.1 **AN EARLY LESION OF POST-TRANSPLANT LYMPHOPROLIFERATIVE DISORDER.** H&E stain showing plasmacytic hyperplasia. (A, low power; B, high power). Numerous cells are positive for EBV-EBER (C). Plasma cells express CD138 (D) and are polytypic, showing a mixture of kappa+ (E) and lambda + (F) plasma cells.

Polymorphic PTLD

- Polymorphic PTLD lesions represent a diffuse infiltrative process leading to the effacement of nodal architecture and/or destructive extranodal tissues.
- The infiltrate is polymorphic and consists of a mixture of lymphocytes, plasma cells, transformed lymphocytes, and immunoblasts (Figure 56.2). Scattered atypical immunoblasts may resemble RS cells. Necrotic areas may be present, often associated with histiocytes and neutrophils. Some cases may show frequent mitosis.

Monomorphic PTLD

- Monomorphic PTLD demonstrates significant architectural alteration, monomorphic features, and cellular atypia consistent with the diagnosis of lymphoma (Figure 56.3).
- In some cases, the affected organs may show both polymorphic and monomorphic infiltrates in the same tissue section.
- Monomorphic PTLD should be classified according to the WHO guidelines for the classification of lymphomas.
- B-cell NHLs are much more common than T-cell NHLs. The most frequent subtype is DLBCL (Figure 56.4) followed by Burkitt lymphoma.
- T/NK-cell types are less frequent (4% to 14%) and represent the entire spectrum of T- and NK-cell tumors. Peripheral T-cell lymphoma, NOS is the most frequent type followed by hepatosplenic T-cell lymphoma.
- In some cases, the affected organs may show both polymorphic and monomorphic infiltrates in the same tissue section, and at the molecular level they often evolve from polyclonal through oligoclonal (Figure 56.5) and finally to monoclonal when a particular clone achieves predominance.

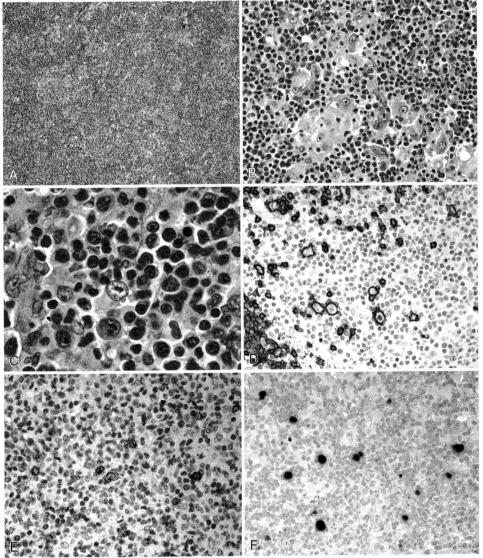

FIGURE 56.2 **POST-TRANSPLANT LYMPHOPROLIFERATIVE DISORDER, POLYMORPHIC TYPE.** Lymph node biopsy sections demonstrating effacement of nodal architecture by a diffuse polymorphic infiltrate consisting of mixed lymphocytes, plasma cells, histiocytes, transformed lymphocytes, and immunoblasts (A, low power; B, intermediate power; C, high power). The large transformed lymphocytes express CD20 (D), CD30 (E), and EBV-EBER (F).

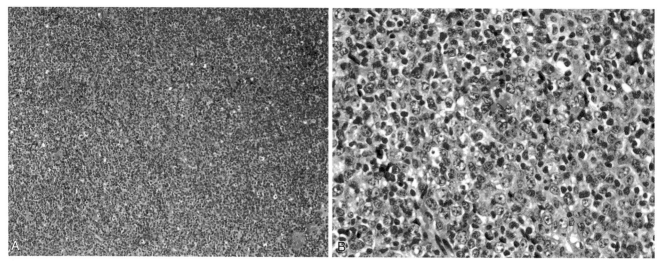

FIGURE 56.3 **POST-TRANSPLANT LYMPHOPROLIFERATIVE DISORDER, MONOMORPHIC TYPE.** Lymph node section demonstrating total effacement of nodal architecture by a diffuse infiltrate consisting of monomorphic large atypical lymphocytes (diffuse large B-cell lymphoma) (A, low power; B, high power).

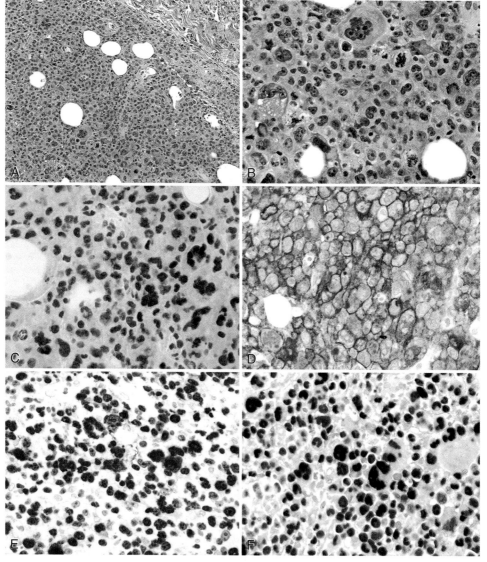

FIGURE 56.4 **ANAPLASTIC DIFFUSE LARGE B-CELL LYMPHOMA IN A POST-TRANSPLANT PATIENT.** (A, low power; B, high power). The neoplastic cells express PAX5 (C), CD30 (D), and Ki67 (E), and are positive for EBV-EBER (F).

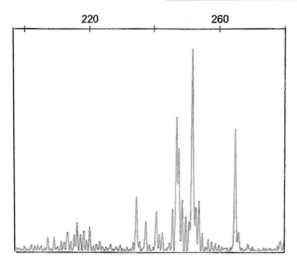

FIGURE 56.5 **AN EXAMPLE OF OLIGOCLONAL POST-TRANSPLANT LYMPHOPROLIFERATIVE DISORDER BY PCR ANALYSIS.** The number of clonal peaks (IgM heavy chain) is too great to be considered monoclonal, but they are discreet enough to represent several clonal subsets (oligoclonal proliferation).

Classical Hodgkin Lymphoma Type PTLD

- CHL type PTLD is uncommon and is characterized by the presence of Hodgkin and Reed–Sternberg (HRS) cells. The HRS cells typically express CD15 and CD30, may be EBV+, and lack CD45 expression.
- The common morphologic type is mixed cellularity CHL.

MOLECULAR AND CYTOGENETIC STUDIES

Given the complexity of PTLD, its transient and evolving nature, and our incomplete understanding of the underlying biology, caution must be observed to avoid over-interpretation of clonal gene rearrangement studies. Serial testing is often necessary to clarify the diagnosis.

- Cytogenetic abnormalities are identified mostly in monomorphic B- and T-PTLDs
- Clonal abnormalities commonly seen are trisomies of chromosomes 9 or 11, rearrangements of the 3q27 locus involving the *BCL6* gene, 8q24 involving *MYC*, and the 14q32 region mapping to the *IGH@* locus.
- *MYC* rearrangement and T-cell-associated chromosomal abnormalities correlated with poor outcome and short survival.
- Monomorphic PTLD often shows clonal *IGH@* gene rearrangement and/or EBV genome, frequently of type A, and mutations of oncogenes and tumor suppressor genes, such as *MYC*, *RAS*, and *TP53* are also identified.

DIFFERENTIAL DIAGNOSIS

The differential diagnosis includes infectious and other reactive lymphoplasmacytic hyperplasias that may be seen in post-transplant patients. Morphologic features supporting a lymphomatous process, IGH or TCR gene rearrangements, and extensive EBV infection are in favor of PTLD. Distinction between different categories of PTLD, particularly between polymorphic and monomorphic variants, is often difficult. The major morphologic, immunophenotypic, and molecular genetic differences between the PTLD subtypes are presented in Table 56.1.

Other Iatrogenic Immunodeficiency-Associated Lymphoproliferative Disorders

These include lymphoproliferative disorders (LPD) that are associated with immunosuppressive drugs in non-transplant patients, such as patients with autoimmune diseases (rheumatoid arthritis, dermatomyositis, Crohn's disease, psoriasis). The immunosuppressive drugs include methotrexate and antagonists of TNFα (such as infliximab, adalimumab, and etanercept). Methotrexate is the best known and the first drug reported in association with LPD. Rheumatoid arthritis patients who develop LPD typically have a longstanding disease and have been treated with methotrexate for an average of 3 years. The therapy duration for development of LPD is significantly shorter in patients treated with antagonists of TNFα (averaging 6–8 weeks).

The morphologic spectrum of LPD associated with immunosuppressive drugs is similar to those of PLTD (see above), including polymorphic, monomorphic (NHL), and CHL types. In patients treated with methotrexate, DLBCL, CHL, and polymorphic lymphoplasmacytic infiltrate are reported in 35–60%, 12–25%, and 15% of LPD cases, respectively. There is a strong association between Crohn's disease treated with infliximab and hepatosplenic T-cell lymphoma, particularly in young patients.

A significant proportion of patients with methotrexate-associated LPD show spontaneous partial or complete regression after withdrawal of drug therapy. The majority of LPD regressions have occurred in EBV+ patients.

Additional Resources

Albrecht J, Fine LA, Piette W: Drug-associated lymphoma and pseudolymphoma: recognition and management, *Dermatol Clin* 25:233–244, 2007.

Blaes AH, Morrison VA: Post-transplant lymphoproliferative disorders following solid-organ transplantation, *Expert Rev Hematol* 3:35–44, 2010.

Evens AM, Roy R, Sterrenberg D, et al: Post-transplantation lymphoproliferative disorders: diagnosis, prognosis, and current approaches to therapy, *Curr Oncol Rep* 12:383–394, 2010.

Jaffe ES, Harris NL, Vardiman JW, et al: Hematopathology, Philadelphia, 2010, Saunders/Elsevier.

Jagadeesh D, Woda BA, Draper J, et al: Post transplant lymphoproliferative disorders: Risk, classification, and therapeutic recommendations, *Curr Treat Options Oncol* 13:122–136, 2012.

Kfoury HK, Alghonaim M, Al Suwaida AK, et al: Nasopharyngeal T-cell monomorphic posttransplant lymphoproliferative disorders and combined IgA nephropathy and membranous glomerulonephritis in a patient with renal transplantation: a case report with literature review, *Transplant Proc* 42:4653–4657, 2010.

Smith EP: Hematologic disorders after solid organ transplantation, *Hematology Am Soc Hematol Educ Program*:281–286, 2010.

Swerdlow SH, Campo E, Harris NL, et al: WHO classification of tumours of haematopoietic and lymphoid tissues, ed 4, Lyon, 2008, International Agency for Research on Cancer.

Yagi T, Ishikawa J, Aono N, et al: Epstein-Barr virus-associated posttransplant lymphoproliferative disorders after allogeneic peripheral blood stem cell transplantation for Hodgkin-like adult T-cell leukemia/lymphoma, *Int J Hematol* 95:214–216, 2011.

Lymphocytopenia and Lymphocytosis

Lymphocytopenia

Lymphocytopenia, or lymphopenia (absolute total blood lymphocyte count <1500/µL), is one of the hallmarks of the primary and acquired immunodeficiency syndromes (AIDS) (see Chapter 56). It also occurs in a wide variety of conditions, such as aplastic anemia, tuberculosis, zinc deficiency, systemic lupus erythematosus, sarcoidosis, Hodgkin lymphoma (HL), toxic shock, and renal failure (Box 57.1). Administration of glucocorticoids, antithyomcyte globulin, and anti-CD20 antibody (rituximab), cancer chemotherapy, and radiotherapy, and thoracic duct drainage are frequently associated with lymphocytopenia.

Lymphocytosis

Lymphocytosis refers to increased number of peripheral blood lymphocytes, >4500/µL, in individuals older than 12 years of age. The absolute lymphocyte count in children <12 years may be as high as 8000/µL in normal conditions. Lymphocytosis is a common finding in most viral infections, certain bacterial infections, X-linked lymphoproliferative disease (see Chapter 56), post-splenectomy, thyrotoxicosis, certain lymphoid malignancies, and a number of other disorders (Box 57.2). Reactive lymphocytosis, particularly in viral infections, is associated with the presence of large, activated, or atypical lymphocytes. Atypical lymphocytosis is one of the characteristic features of infectious mononucleosis (IM) but has also been observed in other viral infections, such as cytomegalovirus (CMV), varicella-zoster, rubella, and hepatitis.

INFECTIOUS MONONUCLEOSIS

Infectious mononucleosis (IM) is the clinical manifestation of Epstein–Barr virus (EBV) infection and is characterized by fever, oropharyngitis, lymphadenopathy, and lymphocytosis with the presence of atypical lymphocytes in the peripheral blood.

Epstein–Barr virus (EBV) primarily spreads through saliva (kissing), by infecting the epithelial cells of the oropharynx. The virus is replicated in the epithelial cells and released in the lymphoid-enriched surrounding environment, infecting B-cells through the EBV receptor CD21 (CR2, C3d receptor). The entry of EBV into the B-lymphocytes causes polyclonal B-cell proliferation. The EBV-transformed B-cells are able to induce a massive T-cell proliferation, primarily CD8+ and CD45RO+ cytotoxic T-cells.

EBV infects over 90% of the human population worldwide, but the vast majority of primary EBV infections are not clinically detected. Symptoms often begin with malaise, headache, and fever followed by lymphadenopathy and pharyngitis. Lymphadenopathy usually involves posterior cervical chains but could become systematic. It peaks in the first week and then gradually disappears within 2–3 weeks. Other clinical findings include splenomegaly, neurologic symptoms such as facial palsies or meningoencephalitis, hepatitis, acute renal failure, and hemophagocytic lymphohistiocytosis. IM is usually a self-limited disease and clinical symptoms disappear within 3 to 4 weeks. However, EBV infection in patients with X-linked lymphoproliferative syndrome may be fatal or lead to non-Hodgkin lymphoma (see Chapter 56). Also, rare cases of fatal T-cell lymphoproliferative disorders have been reported in association with EBV infection.

Supportive therapy is the recommended approach in treating patients with IM. Administration of antiviral drugs, such as aciclovir, helps to protect people from EBV infection but has no effect on curing the infection.

Morphology and Laboratory Findings

- Lymphocytosis (>4500/µL) with the presence of more than 10% atypical lymphocytes is the characteristic morphologic

Box 57.1 Conditions Associated with Lymphocytopenia

- Immunodeficiency syndromes
  - Primary
  - Secondary
- Autoimmune diseases, such as
  - Systemic lupus erythematosus
  - Rheumatoid arthritis
  - Insulin-dependent diabetes mellitus
  - Crohn's disease
  - Sjögren's disease
- Cytotoxic drugs, such as
  - Epirubicin
  - Methotrexate
  - Paclitaxel
  - Purine nucleoside analogs
- Corticosteroids
- Antibodies, such as
  - CAMPATH-1H
  - Rituximab
  - Antithymocyte globulin
- Immunoregulatory molecules, such as
  - Thymodulin
  - Thymostimulin
  - Interferons
- Other drugs, such as
  - Calcitonin
  - Carbamazepine
  - Cimetidine
  - Opioids
- Bone marrow hypoplasia, such as
  - Aplastic anemia
  - Hematopoietic malignancies
  - Cancer chemotherapy and radiation therapy
- Other conditions
  - Exposure to silica
  - Zinc deficiency
  - Toxic shock
  - Renal failure
  - Thoracic duct drainage
  - Advanced malignancies
  - Tuberculosis

Box 57.2 Conditions Associated with Lymphocytosis[1]

- Lymphoid malignancies
- Virus-associated lymphocytosis
  - EBV (infectious mononucleosis)
  - CMV
  - HIV-1
  - Hepatitis
  - Influenza
  - Measles
  - Mumps
  - Rubella
  - Varicella
- Lymphocytosis associated with other infectious agents
  - *Babesia microti*
  - *Bartonella henselae* (cat scratch fever)
  - *Bordetella pertussis* (Whooping cough)
  - *Brucella*
  - *Mycobacterium tuberculosis*
  - *Toxoplasma gondii*
  - *Treponema pallidum* (syphilis)
- X-linked lymphoproliferative disease
- Chronic polyclonal B-cell lymphocytosis
- Polyclonal immunoblastic proliferation
- Idiopathic lymphocytosis
- Lymphocytosis associated with hypersensitivity reactions
  - Acute serum sickness
  - Drug-induced (e.g., ceftriaxone or carbamazepine)
- Stress-induced lymphocytosis
  - Cardiac emergencies
  - Sickle cell anemia
  - Status epilepticus
  - Trauma
- Other conditions associated with lymphocytosis
  - Autoimmune disorders
  - Cigarette smoking
  - Post-splenectomy
  - Thyrotoxicosis

[1] Adapted from Hoffbrand A.V., Pettit, and Vyas P.: Color atlas of clinical hematology, 4th ed. Mosby/Elsevier, 2010.

feature of EBV infection. Atypical lymphocytes are large, pleomorphic cells with abundant shady gray-blue cytoplasm with or without vacuoles (Figures 57.1 to 57.3). They may show scalloping of the cytoplasmic membrane around red blood cells. The nucleus is round, oval, or irregular; the chromatin is clumped; and the nucleoli are often small or inconspicuous.
- Anemia, granulocytopenia, and thrombocytopenia may occur, and leukocyte alkaline phosphatase (LAP) activity tends to be low.
- Bone marrow examination reveals lymphocytosis with the presence of atypical lymphocytes. Small granulomas may be present, and there may be evidence of hemophagocytosis.
- The affected lymph nodes show expansion of paracortical regions with a population of polymorphous lymphoid cells ranging from small lymphocytes to large immunoblasts, mixed with tingible body macrophages and plasma cells (Figure 57.4). Reed–Sternberg-like cells may be present. Other morphologic findings include small areas of necrosis, reactive follicles, dilated sinuses containing polymorphous lymphoid cells, and the presence of a polymorphous lymphoid infiltrate in the perinodal tissue.
- The heterophil antibody test is positive, indicating the presence of cross-reacting antibodies to antigens from phylogenetically unrelated species, such as sheep (Paul–Bunnell test), equine (Monospot test), ox, and goat red blood cells.

Immunophenotype

- The atypical lymphocytes represent activated lymphocytes that are predominantly CD8+, CD45RO+, and HLA-DR+.
- The EBV-transformed immunoblasts are also present in the paracortical regions.

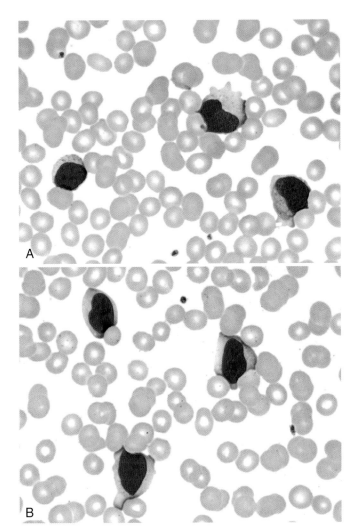

FIGURE 57.1 Blood smear from a patient with infectious mononucleosis demonstrates large, pleomorphic atypical (activated) lymphocytes with abundant cytoplasm, round, oval, or irregular nucleus, and dense chromatin. The cytoplasm demonstrates some degree of basophilia and scalloping of the cytoplasmic membrane around erythrocytes. Some cells show cytoplasmic granules.

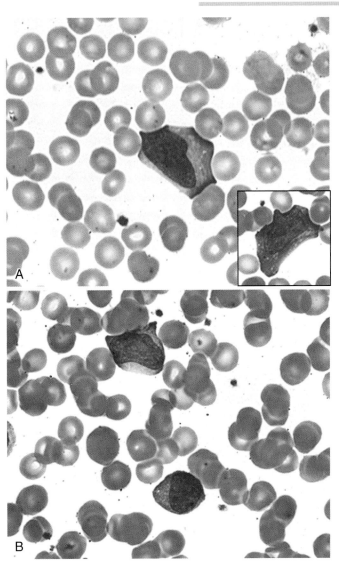

FIGURE 57.2 Blood smear from a patient with infectious mononucleosis demonstrates large, pleomorphic atypical (activated) lymphocytes with abundant basophilic cytoplasm, round, oval, or irregular nucleus, and dense chromatin. One cell (A) shows scalloping of the cytoplasmic membrane around erythrocytes.

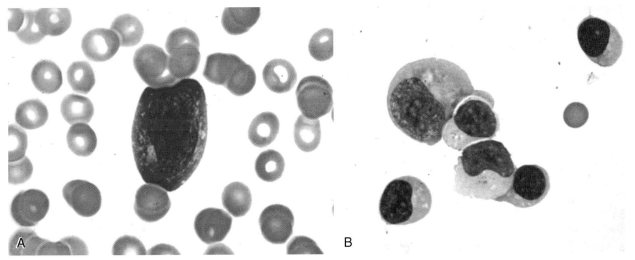

FIGURE 57.3 Blood smear (A) and pleural effusion (B) from a patient with infectious mononucleosis showing immunoblasts and atypical lymphocytes.

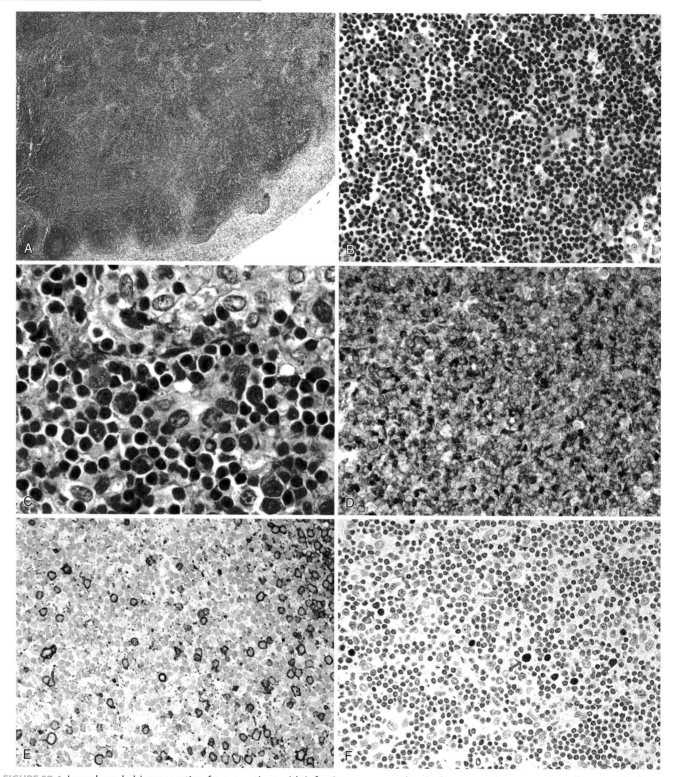

FIGURE 57.4 Lymph node biopsy section from a patient with infectious mononucleosis demonstrating expansion of paracortical regions with a population of polymorphous lymphoid cells ranging from small lymphocytes to large immunoblasts (A, low power; B, intermediate power; C, high power). These cells are predominantly CD3+ (D), mixed with scattered CD20+ (E) and EBV-EBER+ (F, arrows) cells.

- The Reed–Sternberg-like cells may express CD30 and are also positive for CD45, but negative for CD15 and EMA.
- Specific antibodies against EBV antigens such as viral capsid antigen (VCA), EBV nuclear antigen (EBNA), and early antigen (EA) are detected (Figure 57.4F).

Molecular and Cytogenetic Studies

EBV-encoded RNA (EBER) can be detected in the transformed cells by molecular techniques, such as PCR assays and in situ hybridization in the atypical lymphocytes.

Differential Diagnosis

Lymphocytosis with the presence of atypical lymphocytes is found in a variety of conditions. Approximately 10% of patients with the clinical symptoms of IM (atypical lymphocytes, fever, pharyngitis, and lymphadenopathy) are EBV-negative, and the condition is caused by other infectious agents, such as toxoplasmosis, CMV, human herpes virus 6 (HHV-6), and hepatitis B. Some drugs such as phenytoin, carbamazepine, isoniazid, and minocycline may also cause atypical lymphocytosis. The diagnosis of EBV infection is confirmed by heterophil and/or specific EBV antibody tests or the identification of EBV by molecular studies, such as PCR assays. However, one must exercise caution in interpretation of PCR results, given the high proportion of healthy EBV carriers in the population.

STRESS-INDUCED LYMPHOCYTOSIS

A transient atypical absolute lymphocytosis with lymphocyte counts of up to 13,000/µL has been observed in adult patients with cardiac emergencies, trauma, status epilepticus, or sickle cell anemia crisis. The absolute lymphocytosis in these cases is usually the result of the increased numbers of B-, T-, and NK-cells.

PERSISTENT POLYCLONAL B-CELL LYMPHOCYTOSIS

Persistent (chronic) polyclonal B-cell lymphocytosis (PPBL) is a rare condition that has been reported in young to middle-aged women. An association with heavy smoking and HLA-DR 7 has been reported, suggesting that both environmental and genetic factors are involved. Reports of familial occurrences further support underlying genetic defects in this disorder. These individuals have absolute lymphocytosis ranging from 5000 to 15,000/µL, with the presence of binucleated and/or atypical lymphocytes (Figure 57.5). There is a polyclonal increase in serum IgM levels with no lymphadenopathy or splenomegaly. The polyclonal B-cells express pan-B-cell markers such as CD19, CD20, and CD22 and show lack of or dim expression of CD5, CD10, and CD23. They may also express FMC7, CD11c, and CD25. In spite of its polyclonal nature and benign clinical behavior, PPBL, in some cases, has been associated with multiple bcl-2/Ig gene rearrangements and chromosomal abnormalities such as +i(3q), del (6q), and +8, respectively.

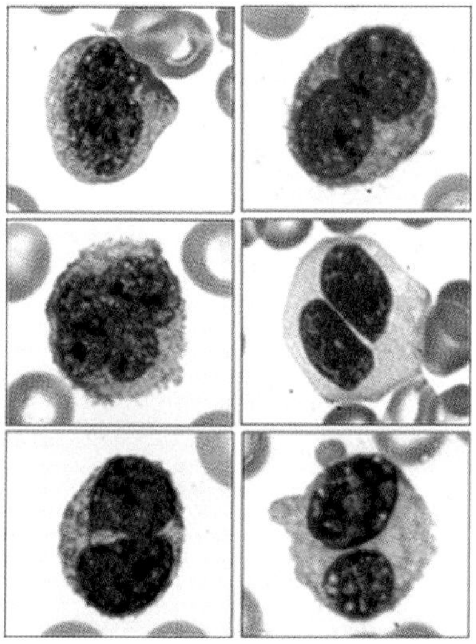

BONE MARROW BENIGN LYMPHOID AGGREGATES

Benign lymphoid aggregates (lymphoid nodules, lymphoid follicles) are relatively frequent in bone marrow sections. They appear to be more frequent in older individuals and in women. The presence of lymphoid aggregates in younger individuals usually indicates an underlying cause, such as autoimmune disorder, drug reaction, or viral infection (Box 57.3). Benign lymphoid aggregates have also been reported in association with aplastic anemia, myeloproliferative disorders, myelodysplastic syndromes, mastocytosis, and HL and non-HL.

Lymphoid aggregates consist of small, well-defined clusters of mature lymphocytes that are sometimes mixed with scattered plasma cells, eosinophils, mast cells, or histiocytes (Figure 57.6). Lymphoid aggregates are usually interstitial, surrounded by fat or hematopoietic cells. They are usually distant from bone trabeculae.

FIGURE 57.5 Binucleated lymphocytes and lymphocytes with lobated nuclei are frequently seen in patients with persistent (chronic) polyclonal B-cell lymphocytosis.
From Naeim F. Atlas of Bone Marrow and Blood Pathology. WB Saunders, Philadelphia, 2001, by permission.

Box 57.3 Conditions Associated with Benign Lymphoid Aggregates in the Bone Marrow

- Autoimmune disorders
  - Rheumatoid arthritis
  - Systemic lupus erythematosus
  - Autoimmune hemolytic anemia
  - Idiopathic thrombocytopenia
  - Hashimoto thyroiditis
- Myelodysplastic syndromes
- Myeloproliferative disorders
- Mastocytosis
- Aplastic anemia
- Lymphoid malignancies
- Viral infections
- Drugs
- Unknown

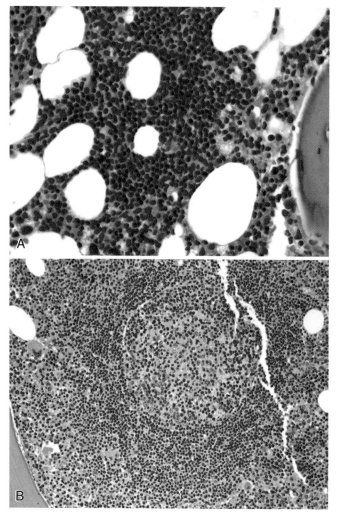

FIGURE 57.6 Benign lymphoid aggregates in bone marrow. Biopsy sections demonstrating lymphoid aggregates (A and B), one with a germinal center (B).

Table 57.1 Features of Benign and Malignant Lymphoid Aggregates in Bone Marrow Sections	
Benign Lymphoid Aggregates	**Malignant Lymphoid Aggregates**
Often well-defined and circumscribed	Usually irregular and infiltrating into the adjacent marrow
Usually interstitial	Frequently paratrabecular
Infrequent cellular atypia	Common cellular atypia
Germinal centers may be present	Germinal centers are not present
Polymorphous lesions lack RS cells and RS variants	Presence of RS cells and variants in Hodgkin lymphoma
Lack of significant fibrosis	May be associated with significant fibrosis
B-cells in aggregates are usually negative for bcl-2, CD5, CD10, and CD23	Malignant B-cells often express bcl-2 and may also express CD5, CD10, or CD23
No evidence of monoclonality by immunophenotypic, molecular, and/or cytogenetic studies	Non-Hodgkin lymphomas often show evidence of monoclonality

Approximately 5% of lymphoid aggregates may show germinal centers. The presence of germinal centers may indicate reaction to a marked or prolonged immunologic stimulation. The term *lymphoid nodular hyperplasia* is used when four or more lymphoid aggregates are seen in a low power microscopic field, or if an aggregate exceeds 0.6 mm in its greatest dimension. *Reactive polymorphous lymphohistiocytic lesion* refers to aggregates consisting of a mixture of lymphocytes, histiocytes, and other inflammatory cells. These lesions may be large, ill-defined, and paratrabecular.

Differentiation of benign lymphoid aggregates from lymphomatous involvement in the bone marrow is at times problematic (Table 57.1). Benign lymphoid aggregates are often well defined, lack an infiltrative pattern, are not paratrabecular, and are primarily composed of small, mature lymphocytes with round nuclei and condensed chromatin. They consist of a mixture of B- and T-cells with no evidence of monoclonality based on immunophenotypic, molecular, and/or cytogenetic studies. The B-cell component of the lymphoid aggregates is negative for bcl-2 and usually lacks the expression of CD5, CD10, and CD23.

Additional Resources

Casassus P, Lortholary P, Komarover H, et al: Cigarette smoking-related persistent polyclonal B Lymphocytosis: a premalignant state, *Arch Pathol Lab Med* 111:1081, 1987.

Gandhi AM, Ben-Ezra JM: Do Bcl-2 and survivin help distinguish benign from malignant B-cell lymphoid aggregates in bone marrow biopsies? *J Clin Lab Anal* 18:285–288, 2004.

Groom DA, Kunkel LA, Brynes RK, et al: Transient stress lymphocytosis during crisis of sickle cell anemia and emergency trauma and medical conditions. An immunophenotyping study, *Arch Pathol Lab Med* 114:570–576, 1990.

Gulley ML, Tang W: Laboratory assays for Epstein-Barr virus-related disease, *J Mol Diagn* 10:279–292, 2008.

Klein G, Klein E, Kashuba E: Interaction of Epstein-Barr virus (EBV) with human B-lymphocytes, *Biochem Biophys Res Commun* 396:67–73, 2010.

Luzuriaga K, Sullivan JL: Infectious mononucleosis, *N Engl J Med* 362:1993–2000, 2010.

Opeskin K, Burke M, Firkin F: Bone marrow lymphoid aggregates simulating histological features of non-Hodgkin lymphoma in Felty syndrome, *Pathology* 37:82–84, 2005.

Reimer P, Weissinger F, Tony HP, et al: Persistent polyclonal B-cell lymphocytosis—an important differential diagnosis of B-cell chronic lymphocytic leukemia, *Ann Hematol* 79:327–331, 2000.

Thiele J, Zirbes TK, Kvasnicka HM, et al: Focal lymphoid aggregates (nodules) in bone marrow biopsies: differentiation between benign hyperplasia and malignant lymphoma—a practical guideline, *J Clin Pathol* 52:294–300, 1999.

Troussard X, Cornet E, Lesesve JF, et al: Polyclonal B-cell lymphocytosis with binucleated lymphocytes (PPBL), *Onco Targets Ther* 1:59–66, 2008.

Troussard X, Valensi F, Debert C, et al: Persistent polyclonal lymphocytosis with binucleated B lymphocytes: a genetic predisposition, *Br J Haematol* 88:275–280, 1994.

Vouloumanou EK, Rafailidis PI, Falagas ME: Current diagnosis and management of infectious mononucleosis, *Curr Opin Hematol* 19:14–20, 2012.

ABNORMALITIES IN PLATELET ARACHIDONIC ACID PATHWAYS

Abnormalities of arachidonic acid pathways are extremely rare and are of two major types: (1) defect in the release of arachidonic acid from phospholipids and (2) deficiencies of cyclooxygenase or thromboxane synthetase. Affected patients are usually adults and often demonstrate mild to moderate hemorrhages. Severe bleeding is rare.

Formation of thromboxane A_2 is one of the major responses of platelets during activation. Thromboxane A_2 is necessary for platelet secretion during the stimulation of platelets with ADP, epinephrine, and low concentration of collagen and thrombin. Thromboxane A_2 is also a potent vasoconstrictor.

Acquired Platelet Disorders

DRUG-INDUCED DISORDERS

Many commonly used drugs are known to affect platelet function. Although some of these agents were developed mainly for treating patients at risk for thromboembolism because of their ability to inhibit specifically one (or more) of the several distinct molecular events required for normal platelet function, and therefore to impair primary hemostasis, most were developed for clinical indications unrelated to their hemostatic effects and were found subsequently to non-specifically inhibit platelet function. Because the mechanism(s) by which the agents in the former group impair platelet function have been studied extensively and are well established, they will be discussed only briefly here. However, because much of our knowledge in this area on drugs from the latter, much larger group comes from *in vitro* studies performed on platelets exposed to one pharmacologic agent at a time, the overall impact on the hemostasis system *in vivo* has not been established for most of them. Furthermore, the clinical relevance of this knowledge is not clear since most patients are administered more than one drug simultaneously. Inhibitors of platelet cyclooxygenase-1 (COX-1), including aspirin and other non-steroidal anti-inflammatory drugs (NSAIDs), are among the most commonly used medications.

Aspirin

Aspirin irreversibly inactivates cyclooxygenase and thereby inhibits production of thromboxane A_2 from arachidonic acid and impairs platelet secretion. The end result is defective platelet aggregation and prolonged bleeding time. Prolongation of the bleeding time may last up to 4 days after administration of aspirin is stopped. Ethanol ingestion may enhance prolongation of the bleeding time in patients who take aspirin.

Beta-Lactam Antibiotics

Beta-lactam antibiotics, such as penicillin and cephalosporin derivatives, may prolong bleeding time and induce abnormal platelet aggregation. These antibiotics seem to interfere with the function of platelet membrane integrins, such as GPIIb-IIIa and GPIa-IIa. The effect is dose and duration dependent.

Others

Excessive garlic ingestion may induce platelet dysfunction and inhibits cyclooxygenase activity. Long-term dietary supplementation with marine oils reduces the platelet content of arachidonic acid and may cause abnormal platelet aggregation and slight prolongation of the bleeding time. Dextran may slightly prolong the bleeding time without increasing operative or post-operative bleeding. Therefore, it has been used for the prevention of post-surgical thromboembolic complications.

PLATELET DYSFUNCTION ASSOCIATED WITH PATHOLOGIC CONDITIONS

Cardiopulmonary Bypass

Prolonged bleeding time, abnormal platelet aggregation, and thrombocytopenia are some of the common features of cardiopulmonary bypass. During bypass surgery, platelets adhere to fibrinogen absorbed by the bypass circuit. Bypass procedures also enhance thrombin and ADP generation and complement activation. Mechanical trauma from the bypass pump may also degranulate platelets.

Chronic Renal Failure

Uremia may lead to platelet dysfunction and abnormal aggregation. Bleeding time is often prolonged, and there may be bleeding manifestations such as purpura, epistaxis, menorrhagia, gastrointestinal bleeding, and hematuria.

Hematologic Disorders

Abnormal platelet function and morphology may occur in association with myelodysplastic syndromes, myeloproliferative disorders, and acute myeloid leukemia. Abnormal platelet functions include decreased platelet aggregation and secretion in response to ADP, epinephrine, and collagen, and reduced platelet procoagulant activity. Morphologic changes include abnormal shapes, giant forms, and hypogranularity. A case of hairy cell leukemia with abnormal platelet morphology and severe platelet dysfunction has been reported.

Additional Resources

Bain BJ, Bhavnani M: Gray platelet syndrome, *Am J Hematol* 86:1027, 2011.

Ballmaier M, Germeshausen M: Congenital amegakaryocytic thrombocytopenia: clinical presentation, diagnosis, and treatment, *Semin Thromb Hemost* 37:673–681, 2011.

Blanchette V, Bolton-Maggs P: Childhood immune thrombocytopenic purpura: diagnosis and management, *Hematol Oncol Clin North Am* 24:249–273, 2010.

Blundell MP, Worth A, Bouma G, et al: The Wiskott-Aldrich syndrome: the actin cytoskeleton and immune cell function, *Dis Markers* 29:157–175, 2010.

Cleland SY, Siegel RM: Wiskott-Aldrich Syndrome at the nexus of autoimmune and primary immunodeficiency diseases, *FEBS Lett* 585:3710–3714, 2011.

Di Paola J, Johnson J: Thrombocytopenias due to gray platelet syndrome or THC2 mutations, *Semin Thromb Hemost* 37:690–697, 2011.

Franchini M, Favaloro EJ, Lippi G: Glanzmann thrombasthenia: an update, *Clin Chim Acta* 411:1–6, 2010.

Geddis AE: Congenital amegakaryocytic thrombocytopenia, *Pediatr Blood Cancer* 57:199–203, 2011.

George JN, Aster RH: Drug-induced thrombocytopenia: pathogenesis, evaluation, and management, *Hematology Am Soc Hematol Educ Program*:153–158, 2009.

Gunay-Aygun M, Huizing M, Gahl WA: Molecular defects that affect platelet dense granules, *Semin Thromb Hemost* 30:537–547, 2004.

Hodgson K, Ferrer G, Pereira A, et al: Autoimmune cytopenia in chronic lymphocytic leukaemia: diagnosis and treatment, *Br J Haematol* 154:14–22, 2011.

Islam MS, Alamelu J: Morphological and electron microscopic characteristics of grey platelet syndrome, *Br J Haematol* 152:1–5, 2011.

Jalas C, Anderson SL, Laufer T, et al: A founder mutation in the MPL gene causes congenital amegakaryocytic thrombocytopenia (CAMT) in the Ashkenazi Jewish population, *Blood Cells Mol Dis* 47:79–83, 2011.

Lanza F: Bernard-Soulier syndrome (hemorrhagiparous thrombocytic dystrophy), *Orphanet J Rare Dis* 1:46–48, 2006.

McCrae K: Immune thrombocytopenia: no longer 'idiopathic', *Cleve Clin J Med* 78:358–373, 2011.

Nurden AT, Fiore M, Nurden P, et al: Glanzmann thrombasthenia: a review of ITGA2B and ITGB3 defects with emphasis on variants, phenotypic variability, and mouse models, *Blood* 118:5996–6005, 2011.

Nurden P, Nurden AT: Congenital disorders associated with platelet dysfunctions, *Thromb Haemost* 99:253–263, 2008.

Pham A, Wang J: Bernard-Soulier syndrome: an inherited platelet disorder, *Arch Pathol Lab Med* 131:1834–1836, 2007.

Tsai HM: Pathophysiology of thrombotic thrombocytopenic purpura, *Int J Hematol* 91:1–19, 2010.

White JG: Platelet granule disorders, *Crit Rev Oncol Hematol* 4:337–377, 1986.

Zipfel PF, Heinen S, Skerka C: Thrombotic microangiopathies: new insights and new challenges, *Curr Opin Nephrol Hypertens* 19:372–378, 2010.

Post-Therapy Changes

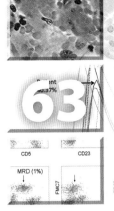

The complex and multidisciplinary therapeutic approaches that are currently used in a significant proportion of hematologic disorders are often associated with post-therapy pathologic changes. In this chapter, we briefly discuss and demonstrate some of the most frequent post-therapy changes in these disorders.

Changes Associated with Hematopoietic Stem Cell Transplantation

Hematopoietic stem cell transplantation (HSCT) is based on the principle of replacing the hematopoietic stem cells and restoring hematopoiesis in patients who receive myeloablative high-dose chemotherapy and/or radiation therapy for the treatment of neoplastic disorders or in patients who have defective hematopoiesis (Box 63.1). The source of the stem cells for transplantation is the patient's own cells (autologous transplant), or cells from another person who is an identical twin (syngeneic transplant) or who is HLA compatible with the patient (allogeneic transplant). The stem cells are harvested form the donor's bone marrow, peripheral blood, or umbilical cord blood. Few patients have an identical twin donor, and fewer than 30% of the patients have an HLA-compatible sibling. Therefore, the vast majority of stem cell transplants are allogeneic with unrelated donors being the source of stem cells. The major complications of stem cell transplantation (particularly allogeneic) include infection, veno-occlusive liver disease, graft-versus-host disease (GVHD), recurrent malignancy, post-transplant lymphoproliferative disorder, and graft rejection/failure (Box 63.2).

Box 63.1 Applications of Hematopoietic Stem Cell Transplantation

Allogeneic/Syngeneic Transplants
- Aplastic anemia
- Leukemias/lymphomas
- Plasma cell myeloma
- Paroxysmal nocturnal hemoglobinuria
- Myelodysplastic syndromes
- Autoimmune disorders
- Congenital disorders
  - Primary immunodeficiencies
  - Metabolic disorders
  - Hemoglobinopathies
  - Fanconi's anemia
  - Hereditary lymphohistiocytic hemophagocytosis

Autologous Transplants
- Leukemias/lymphomas
- Plasma cell myeloma
- Solid neoplasms

Box 63.2 Complications of Hematopoietic Stem Cell Transplantation

- Graft rejection or failure
- Graft-versus-host disease
- Veno-occlusive liver disease
- Post-transplant immunodeficiency
- Opportunistic infections
- Interstitial pneumonitis
- Infertility
- Recurrent malignancy
- Post-transplant lymphoproliferative disorder
- Secondary malignancies

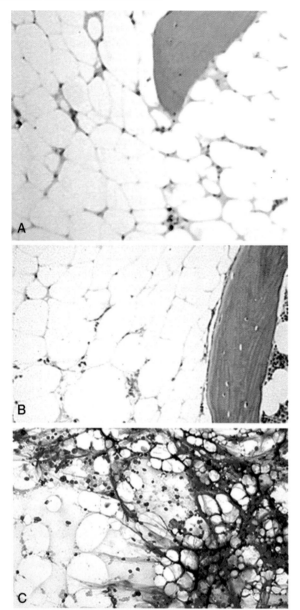

FIGURE 63.1 **POST-TRANSPLANT BONE MARROW, 1–3 WEEKS.** Biopsy sections (A and B) are hypocellular and show foci of hematopoietic cells. Bone marrow smears (C) are usually hypocellular and display scattered and small clusters of hematopoietic cells.
From Naeim, F. Atlas of Bone Marrow and Blood Pathology. W. B. Saunders, Philadelphia, 2001, by permission.

POST-TRANSPLANT BONE MARROW CHANGES

During the first week after stem cell transplantation the bone marrow appears markedly hypocellular. There may be extensive marrow damage evidenced by fat necrosis, edema and an increased number of foamy histiocytes. Occasional hematopoietic precursors and scattered lymphocytes and plasma cells are present (Figure 63.1).

Bone marrow samples obtained between 1 and 3 weeks after transplantation reveal regenerating hematopoietic cells with scattered well-defined clusters of erythroid cells within the fatty tissue, usually away from bone trabeculae. In contrast, granulocytic precursors tend to spread around

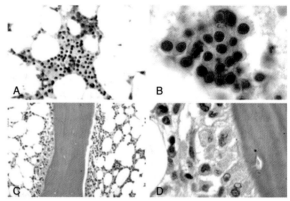

FIGURE 63.2 **POST-TRANSPLANT BONE MARROW, 1–3 WEEKS.** Biopsy sections are hypocellular and show foci of hematopoietic cells. Erythroid clusters (A and B) are often far from bone trabeculae, while myeloid clusters are usually close to the bone trabeculae (C and D).
From Naeim, F. Atlas of Bone Marrow and Blood Pathology. W. B. Saunders, Philadelphia, 2001, by permission.

fatty tissue and are often concentrated in paratrabecular areas (Figure 63.2). Megakaryocytes are rare. Dysplastic changes and left shift are often present in both lineages and evidence of hematogone regeneration. Sometimes, the presence of immature myeloid cells or hematogones may mimic residual/relapse disease in transplanted patients with the diagnosis of AML or ALL, respectively.

Bone marrow samples obtained 4–8 weeks after transplantation show increasing cellularity with large clusters of hematopoietic cells composed of mixed-lineage cellular elements (Figures 63.3 and 63.4). Dysplastic changes and myeloid/erythroid left shift are minimal. Along with the improved bone marrow cellularity, peripheral blood displays a progressive increase in hemoglobin levels and in white cells and platelet counts. Platelets are the last cellular elements to come back to normal. Normocellularity is usually achieved 8–12 weeks following transplantation.

Engraftment can be established even earlier by molecular studies, using DNA fingerprinting methods to distinguish donor from recipient cells in either bone marrow or peripheral blood specimens. In earlier days this was done by Southern blot, using restriction-fragment-length polymorphism (RFLP) markers. Now it is typically done by PCR analysis of short tandem repeat (STR) polymorphisms, which provide a wider source of targets and more rapid turnaround time (Figures 63.5 and 63.6). The sensitivity is down to 1–5% donor or recipient cells in mixed-cell (chimeric) specimens. With bone marrow specimens, it is not advisable to pursue any lower sensitivity because there are usually non-neoplastic stromal elements present which would give the appearance of residual recipient DNA.

In situations where the donor and recipient are of opposite gender, a FISH study with chromosome XX/XY-specific probes can also establish engraftment success. A quantitative analysis of 200–500 nuclei with XX/XY mismatch signal patterns can provide an estimate of engraftment (Figure 63.7). The ability of FISH to screen large numbers of cells with fluorophores for sex chromosomes

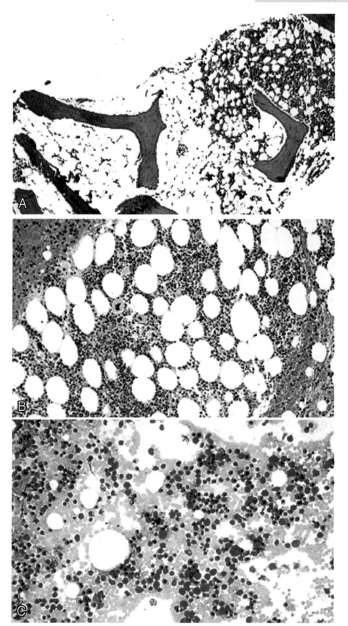

FIGURE 63.3 POST-TRANSPLANT BONE MARROW, 4–8 WEEKS. Biopsy sections show increased cellularity, which at early stages appears patchy (A). The bone marrow cells are distributed more evenly at later stages (B, clot section; C, smear).
From Naeim, F. Atlas of Bone Marrow and Blood Pathology.
W. B. Saunders, Philadelphia, 2001, by permission.

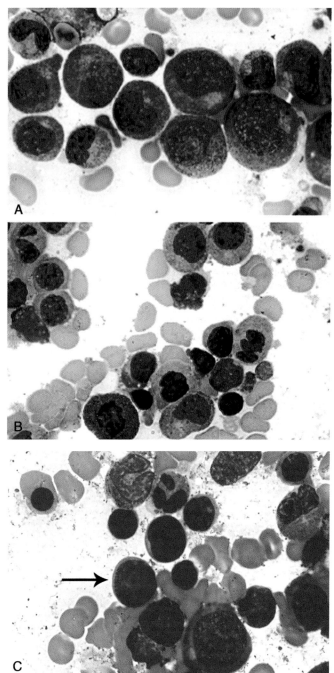

FIGURE 63.4 POST-TRANSPLANT BONE MARROW, 4–8 WEEKS. Smears show myeloid and erythroid left shift with dysplastic changes (A and B). There is often evidence of increased hematogones (C).
From Naeim, F. Atlas of Bone Marrow and Blood Pathology.
W. B. Saunders, Philadelphia, 2001, by permission.

permits reliable detection of residual recipient hematopoiesis with sensitivity of <1%.

Karyotype analysis of HSCs post-transplant can be performed after a standard 48 hour culture. Interestingly, while the majority of the cells are chromosomally normal, multiple unrelated clonal abnormalities in host bone marrow cells after allogeneic stem cell transplantation can be seen (Figure 63.8). These cells could be transient with no apparent effect on normal hematopoiesis. In addition, the breakpoints involved in these aberrations do not appear to be cancer-related and the non-recurring balanced rearrangements are also not commonly seen in therapy-related leukemia. It is possible that these chromosomal anomalies are non-constitutional but most likely represent stable stem cell damage resulting from cytotoxic therapy (Figure 63.9).

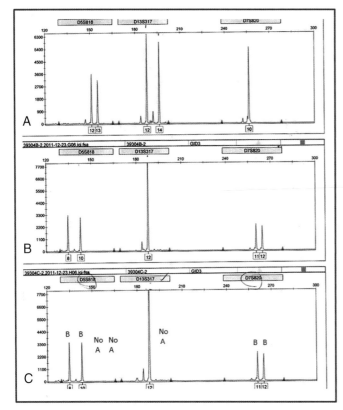

FIGURE 63.5 BONE MARROW ENGRAFTMENT ANALYSIS USING A SERIES OF THREE SHORT-TANDEM REPEAT POLYMORPHISMS. (A) Recipient DNA before transplant; (B) donor DNA; (C) recipient DNA from blood after transplant. The post-transplant specimen completely matches the donor DNA genotype, indicating successful engraftment and no residual recipient cells down to the level of sensitivity of the assay.

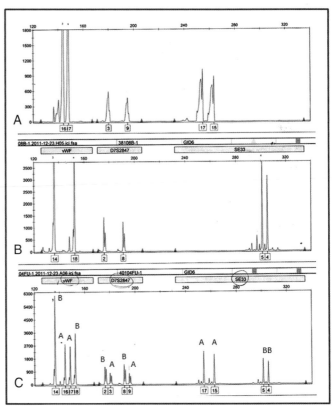

FIGURE 63.6 BONE MARROW ENGRAFTMENT ANALYSIS USING A SERIES OF THREE SHORT-TANDEM REPEAT POLYMORPHISMS. (A) Recipient DNA before transplant; (B) donor DNA; (C) recipient DNA from blood after transplant. In this case the post-transplant specimen exhibits genotypic elements of both the donor and recipient samples, indicating partial engraftment with residual recipient cells still present (mixed chimerism). The proportions are about 55% donor DNA, 45% recipient.

Graft failure (5–20% in patients transplanted for aplastic anemia) is a complication that may follow bone marrow engraftment. The bone marrow cellularity declines and there may be some non-specific changes, such as fat necrosis, increased proportion of lymphocytes and plasma cells, and presence of foamy histiocytes.

Donor cell-derived leukemia/MDS is a rare event that has been reported in allogeneic bone marrow transplant patients as a secondary neoplasm. AML and MDS are the most frequent types followed by ALL, T-LGL, and CLL. The most commonly reported cytogenetic abnormalities are monosomy 7, del(7q), rearrangements of 11q23 involving the MLL gene region and AML1 locus at 21q22. The 11q23 aberrations are more common in HSCT patients who have been exposed to topoisomerase II inhibitors or alkylating agents. Donor cell leukemia can be confirmed by karyotype studies. Figure 63.10 demonstrates a case of a 32-year-old female diagnosed with paroxysomal nocturnal hemoglobinuria and with a normal bone marrow karyotype at diagnosis. Long after transplantation, her karyotype showed trisomy 11 and t(11;21)(q23;q22) in all her donor

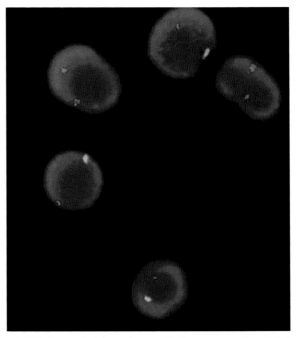

FIGURE 63.7 A panel of interphase cells from a sex-mismatch bone marrow transplant patient showing a mixture of XX (red) and XY (red/green) signal pattern, consistent with partial engraftment.

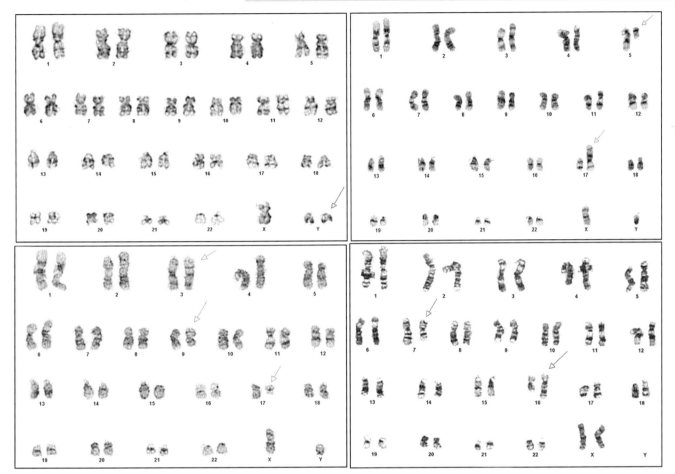

FIGURE 63.8 A panel of unrelated (non-clonal) abnormal metaphases observed in the cultured stem cells of a single individual after allogeneic transplantation.

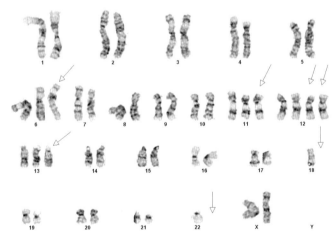

FIGURE 63.9 A non-clonal abnormal metaphase cell from a cultured HSC post-transplant, exhibiting hyperdiploid karyotype with gains and losses of chromosomes (arrows), reflecting genomic instability.

brother's cells. In another case (Figure 63.11) a 55-year old female was diagnosed with AML characterized by t(8;21)(q22;q22). Two years after transplant, she relapsed and developed pancytopenia with a t(11;19)(q23;p13) and trisomy 21 in all male donor cells. Both were DCL associated with sex-mismatched bone-marrow transplants with chromosomal aberrations involving the MLL (11q23) gene locus.

The mechanisms of the leukemic transformation of previously healthy donor HSCs is not well known. Oncogenesis of donor cell leukemia is probably a multifactorial process. It is also possible that an undetected malignant clone was present within the donor at the time of donation, or that healthy donor HSCs might have a genetic premalignant potential.

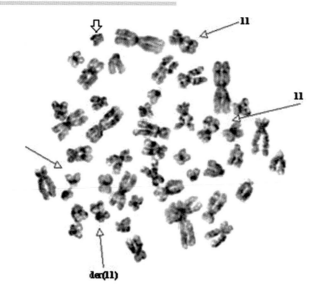

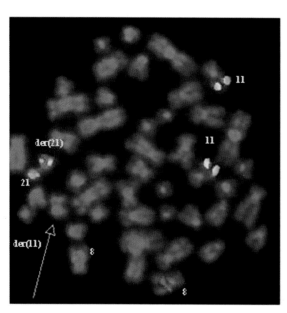

FIGURE 63.10 A metaphase cell (top) from a female recipient showing trisomy 11 and an unbalanced t(11;21)(q23;q22) in the donor cells (long arrow, Y-chromosome), consistent with donor cell leukemia. A subsequent FISH analysis on the same cell (below) with the MLL dual-color "spilt" probe confirms the trisomy 11 and the unbalanced t(11;21).

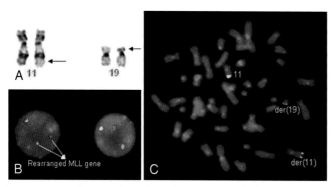

FIGURE 63.11 A partial karyotype (A) showing a balanced t(11;19) from a female recipient and confirmed by dual color MLL gene FISH probe on an interphase (B) and the male (donor) metaphase cell, consistent with donor cell leukemia (C).

Post Chemotherapy and Irradiation Changes

The morphologic changes of the bone marrow after chemotherapy and/or irradiation are the result of rapidly progressive cellular death and a transient ineffective hematopoiesis.

These changes include marked hypocellularity, fibrinoid necrosis, edema, dilated sinuses, multilobulated adipocytes, new bone formation, mild to moderate increase in reticulin fibers, and increased number of macrophages, frequently with phagocytic particles (Figures 63.12 and 63.13).

Morphologic evidence of post-therapy bone marrow regeneration usually appears 1–2 weeks after therapy. Usually, erythroid and myeloid precursors appear sooner than megakaryocytes. Myeloid precursors are usually adjacent to bone, whereas erythroid clusters are far from bone trabeculae and are surrounded by fatty tissue. Rapid bone marrow regeneration is often associated with left-shifted hematopoiesis and increased hematogones.

Radiation and/or chemotherapy effects on lymph nodes include rapidly progressive tumor necrosis with marked depletion of the lymphocytic population and

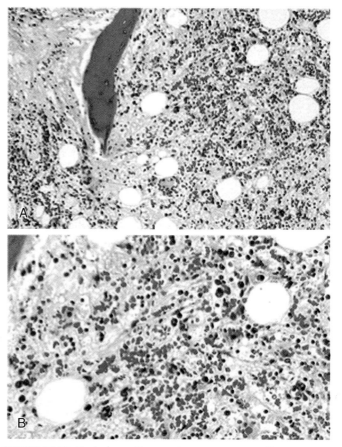

FIGURE 63.12 Post-chemotherapy bone marrow sections are hypocellular with areas of necrosis, scattered hematopoietic precursors, stromal tissue, and minimal amount of fat (A, low power; B, high power).

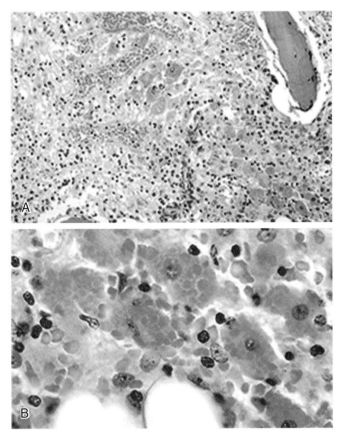

FIGURE 63.13 Bone marrow biopsy sections after chemotherapy or irradiation are hypocellular and may demonstrate necrosis, edema, and increased number of histiocytes (A). Higher power view (B) shows hemophagocytic histiocytes.

edema (Figure 63.14). Plasma cells, histiocytes/macrophages, stromal cells and vascular structures remain intact. Scattered endothelial and stromal cells may show atypical features, particularly in post-radiation. Post-therapy lymph node regeneration, based on experimental animal studies, begins first with the collection of cortical lymphocytes followed by the development of germinal centers.

Cytokine-Associated Bone Marrow Changes

The production of hematopoietic cells is regulated by a group of hematopoietic cytokines. These cytokines initiate various cellular responses, such as proliferation, differentiation, maturation, survival, and functional activities (see Table 1.2 in Chapter 1). These regulatory activities can be multilineage or lineage-specific. Some of these cytokines, such as erythropoietin, granulocyte colony-stimulating factor (G-CSF), and granulocyte–macrophage colony-stimulating factor (GM-CSF), have been used in routine clinical use to stimulate hematopoietic cell production.

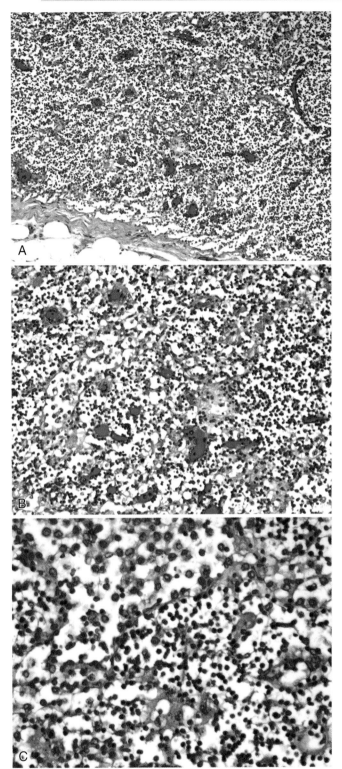

FIGURE 63.14 LYMPH NODE BIOPSY SECTION OF A PATIENT RECEIVING CHEMOTHERAPY FOR ADENOCARCINOMA OF COLON. There is marked lymphoid depletion with lack of follicular structures, increased proportion of histiocytes, and no evidence of metastasis (A, low power; B, intermediate power; C, high power).

Erythropoietin (EPO) is a renal cytokine which inhibits the apoptosis of erythroid progenitor cells, allowing their proliferation and maturation. EPO production is stimulated by hypoxia or decline in levels of hemoglobin. EPO

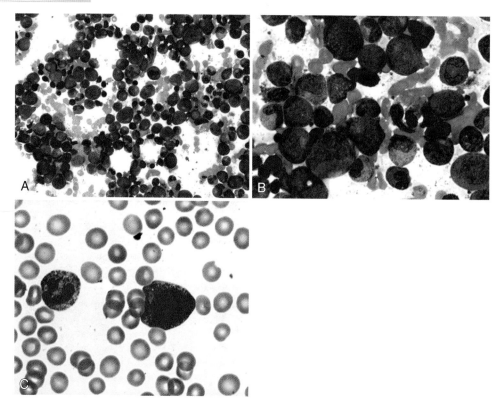

FIGURE 63.15 **BONE MARROW EFFECTS OF G-CSF THERAPY.** Marrow smears show myeloid left shift with dysplastic and hypergranular myeloid precursors (A, low power; B, high power). A hypergranular promyelocyte and a neutrophil with heavy toxic granulation are shown in blood smear (C).

has been widely utilized in anemia of chronic diseases and MDS. EPO therapy is associated with erythroid hyperplasia in the bone marrow, and increased reticulocyte count and hemoglobin/hematocrit levels in the peripheral blood.

Granulocyte-colony stimulating factor (G-CSF) is produced by bone marrow stromal cells, macrophages, endothelial cells, fibroblasts, and astrocytes and is expressed on hematopoietic progenitor cells, endothelial cells, neurons, and glial cells. G-CSF regulates proliferation, differentiation, and survival of hematopoietic progenitor cells. It expands circulating pools of neutrophils, mobilizes hematopoietic stem cells in the peripheral blood, and enhances neutrophil phagocytic function. It has been used to facilitate recovery after bone marrow transplantation and cancer chemotherapy, to increase peripheral blood progenitor cells for harvesting, and to treat severe congenital neutropenia.

Granulocyte–macrophage colony stimulating factor (GM-CSF) shares significant overlapping functional features with G-CSF. In addition, it accelerates production of monocytes and eosinophils, and increases the release of pro-inflammatory cytokines. GM-CSF is produced by macrophages, endothelial cells, fibroblasts, bone marrow stromal cells, lymphocytes, mast cells, and eosinophils.

Bone marrow changes following G-CSF and GM-CSF therapy usually consist of increased cellularity, myeloid preponderance and left shift, and often various degrees of dysplastic changes and cytoplasmic hypergranularity (Figures 63.15 and 63.16). GM-CSF therapy may lead to marked eosinophilia.

Changes Associated with Application of Monoclonal Antibodies

During the past decade and so, the Food and Drug Administration (FDA) has approved several monoclonal antibodies for the treatment of various cancers. Clinical trials of monoclonal antibody therapy have been carried on almost every type of cancer, particularly in hematologic malignancies (Table 63.1).

Two types of monoclonal antibodies are used in cancer treatments: (1) naked monoclonal antibodies and (2) conjugated monoclonal antibodies.

Naked monoclonal antibodies are those without any drug or radioactive material attached to them. They are the most commonly used antibodies at this time. Some naked monoclonal antibodies attach to malignant cells and act as a marker for the body's immune system to destroy them, such as anti-CD20 monoclonal antibodies rituximab (Rituxan) or ofatumumab (Arzerra) and anti-CD52 monoclonal antibody alemtuzumab (Campath) (see Table 63.1). Others attack the specific cellular components that are functionally important for the growth or survival of the cancer cells, such as bevacizumab (Avastin) which targets VEGF protein.

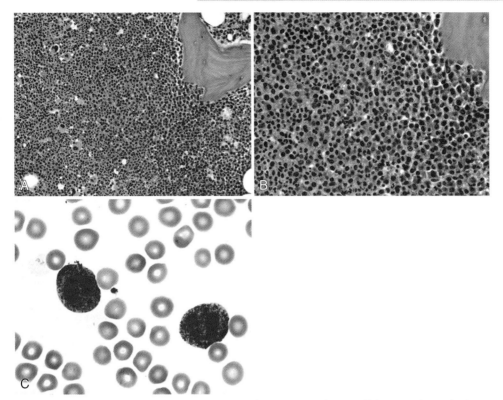

FIGURE 63.16 **EFFECTS OF G-CSF THERAPY.** Bone marrow biopsy section demonstrates hypercellularity with myeloid preponderance and left shift (A, low power; B, high power). Neutrophils with heavy toxic granulation are shown in blood smear (C).

Table 63.1

Examples of Monoclonal Antibodies Used to Treat Hematologic Malignancies

Antibody Name	Trade Name	Antigen	Used to treat
Rituximab	Rituxan	CD20	Non-Hodgkin lymphoma, B-cell type
Ofatumumab	Arzerra	CD20	Chronic lymphocytic leukemia
Alemtuzumab	Campath	CD52	B- and T-cell malignancies
Ibritumomab tiuxetan (radiolabelled)	Zevalin	CD20	Non-Hodgkin lymphoma, B-cell type
Tositumomab (radiolabelled)	Bexxar	CD20	Non-Hodgkin lymphoma, B-cell type
Gemtuzumab ozogamicin (radiolabelled)	Mylotarg	CD33	Acute myeloid leukemia

Conjugated monoclonal antibodies are those joined to a radioactive particle or a toxin to destroy the cancer cells. Examples of radiolabelled antibodies are ibritumomab tiuxetan (Zevalin) and tositumomab (Bexxar) used for the treatment of some cases of non-Hodgkin lymphoma. Gemtuzumab ozogamicin (Mylotarg) comprises an immunotoxin, *calicheamicin*, attached to an anti-CD33 antibody, and was used for treatment of AML, but further studies did not support its effectiveness.

In general, monoclonal antibody therapy, alone or in conjunction with chemotherapy, reduces the tumor mass and may help to achieve molecular remission. Rituximab has become the treatment of choice in a significant proportion of B-cell lymphoid malignancies, in combination with CHOP (cyclophosphamide, doxorubicin, vincristine, and prednisone) therapy. However, approximately 20% of patients with B-cell lymphoma treated with rituximab demonstrate a CD20-negative disease in relapse. It is important to repeat immunophenotypic studies in the biopsy specimens, such as bone marrow, obtained for the establishment of relapse diagnosis.

- If the lymphoid infiltrate is CD20-negative but expresses CD3 it signifies a non-neoplastic T-cell response.
- If the lymphoid infiltrate is CD20+ it signifies relapse.
- If the lymphoid infiltrate is CD20-negative but expresses PAX5 and/or CD79a it indicates a CD20-negative relapse.

Changes Associated with Imatinib Mesylate (Gleevec) Therapy

Imatinib mesylate (Gleevec) binds to a cleft between the N-terminal adenosine triphosphate binding domain and the C-terminal activation loop that forms the catalytic site

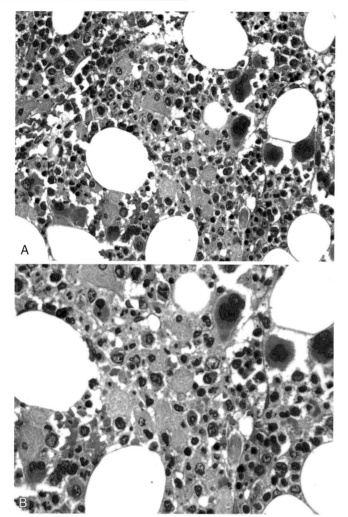

FIGURE 63.17 **BONE MARROW EFFECTS OF GLEEVEC THERAPY.** Biopsy section shows clusters of pseudo-Gaucher cells (A, low power; B, high power).

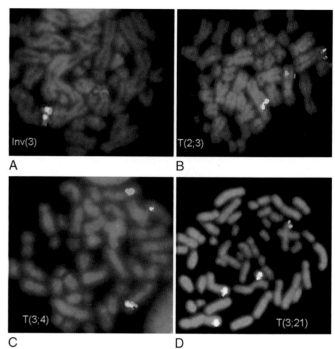

FIGURE 63.18 FISH analysis on metaphases of TKI Resistant CML BC Patients with EVI1-dual-color "breakapart" 3q26.2 probe. The red and green probes lie on either side of EVI1, so when it is rearranged the signals will appear split red or green. (A) inv(3)(p21;q26.2) showing a deletion of the 5′ region of the EVI1 locus; (B) t(2;3)(p23;q26.2) showing the break occurring in the 5′ region of the EVI1 locus; (C) t(3;4)(q26.2;q31) split signals of the 3q26.2 dual-color probe showing rearrangement at the 5′ EVI1 locus to 4q31 band; and (D) t(3;21)(q25.2;q22.1) triple-color EVI1 probe showing a translocation of the 5′ EVI1 gene region translocated to 21q (green).

of the Abl tyrosine kinase, locking the protein into the inactive conformation. It is the treatment of choice for chronic myelogenous leukemia (CML), but it has also been used in treating hypereosinophilic syndromes associated with *PDGFRA*, *PDGFRB*, or *FGFR1* rearrangements and kit-positive gastrointestinal stromal tumors (GIST). In CML the patients' complete response to imatinib therapy is about 98%, with 89% 5-year survival rate and about 17% relapse rate.

Evaluation of post-imatinib therapy bone marrow samples shows progressive changes toward normal morphology. However, the post-therapy bone marrow samples may show certain morphologic features including:

- Frequent presence of non-diagnostic lymphoid aggregates, sometimes paratrabecular, consisting of a mixture of B- and T-lymphocytes;
- Frequent presence of histiocytic aggregates (pseudo-Gaucher cells) (Figure 63.17);
- Bone marrow hypocellularity, particularly in cases with a long history of treatment. The degree of hypocellularity in some instances is so severe that the bone marrow biopsy sections resemble aplastic anemia.

MOLECULAR AND CYTOGENETIC CHANGES

The standard molecular method for monitoring Gleevec treatment is the quantitative *BCR-ABL* fusion gene assay, performed using real-time PCR. This method is sensitive down to 1 CML cell in 100,000 normal cells (or even lower).

The amount of *BCR-ABL* mRNA is given as a ratio to the mRNA of an internal control gene, or as an absolute amount based on international standards.

If the *BCR-ABL* concentration begins to trend upward in a patient on imatinib, it usually indicates the beginning of relapse due to selection of an imatinib-resistant clone of CML cells. The mechanism of resistance is the occurrence of point mutations in the region of the fusion gene that codes for the imatinib binding site in the fusion protein. Methods are available to detect these mutations, of which about 20 are recurring in numerous patients (Figure 63.18). Some mutations are resistant to second- and third-generation tyrosine kinase inhibitor drugs, while others are sensitive. Thus, the test can be used to guide medical management.

Limited reports exist regarding Philadelphia chromosome-negative, chromosomally abnormal cells arising during imatinib therapy. Based on the function and target of imatinib, it is expected that Ph-positive cells will decrease in numbers as the drug is being administered. As these cells are eliminated, the Ph-negative leukemic progenitors expand selectively, a sign of genomic instability, and by clonal evolution present as cytogenetically unrelated clones.

Although approximately 3% of responders to imatinib develop new clonal chromosomal abnormalities in Ph-negative cells, there are reports of Ph-negative acute leukemia in Ph-positive CML patients treated with imatinib, despite the presence of a cytogenetic response and without secondary chromosomal abnormalities. The appearance of additional chromosomal abnormalities in Ph-negative cells, a phenomenon affecting 5% to 10% of patients may signify risk for developing treatment-associated myelodysplastic syndrome, but longer follow-up is needed. These abnormalities include monosomy 7 which rapidly transforms to AML.

The reported frequency of additional chromosomal abnormalities is around 5% in CML chronic phase, and this increases to 50–80% in advanced phases. While the onset of new clonal chromosome abnormalities in Ph-negative cells during treatment has been described, their origin and clinical significance remain to be clarified. Patients who do not respond to imatinib therapy will progress to the blastic phase (BP), usually through the accelerated phase.

Minimal Residual Disease

Recent advances in technology have improved the specificity and sensitivity of detection of malignant hematopoietic cells in patients who were considered to be in remission by clinical and/or morphologic standards. These advanced techniques are able to detect malignant cells at 10^{-4} to 10^{-5} sensitivity at subclinical levels, a condition referred to as "minimal residual disease".

The following methods are used for the detection of minimal residual disease (MRD).

CYTOGENETICS AND MOLECULAR STUDIES

Various literature data show that interphase FISH is more sensitive than conventional cytogenetics and may potentially be useful for monitoring patients who have achieved complete cytogenetic response by conventional cytogenetic analysis. However, because established response categories are based on conventional cytogenetics and because FISH does not detect other clonal chromosomal abnormalities, conventional testing remains the recommended approach for establishing complete cytogenetic response.

Immunoglobulin (Ig) and T-cell receptor (TCR) gene rearrangements generate patient-specific DNA length and sequences which represent ideal molecular markers for detection in remission. However, this technology is susceptible to false-negative results due to clonal evolution during natural history of the disease, leading to relapse with a clone different from the original one in some cases. The PCR quantitative methods have sensitivities ranging from 1×10^{-2} to 1×10^{-4}.

Real-time quantitative polymerase chain reaction (RQ-PCR) measures specific sequences of DNA, and reverse transcription PCR (RT-PCR) measures specific sequences of mRNA. Since viable tumor cells are the source of mRNA, the target RNA sequences may correlate with the rate of cell proliferation. A sensitivity of 1×10^{-4} to 1×10^{-5} is achievable by these techniques. Molecular remission is defined as the failure of the detection of tumor cells in three sequential samples, one month apart, by the most sensitive molecular method available.

- **Chronic myelogenous leukemia.** Quantitative PCR (Q-PCR) has largely replaced cytogenetic or FISH studies in closely monitoring CML patients' response to treatment and detection of emerging tyrosine kinase inhibitor-resistant clones (Figure 63.19). Patients who achieve less than 1-log reduction 3 months post-therapy have only a 13% chance of achieving major molecular response (MMR) compared with >70% patients who have a greater depletion at 3 months. MMR is defined as 3-log tumor reduction.
- **Acute myeloid leukemia.** Q-PCR for the detection of MRD in AML is largely limited to the fusion genes secondary to chromosomal aberrations, such as *PML-RARA* in t(15;17)(q22;q21), *RUNX1-RUNX1T1* in t(8;21)(q22;q22), and *CBFB-MYH11* in inv (16)(p13.1q22) or t(16;16)(p13.1;q22). These studies may help to predict the risk of relapse in treated leukemic patients. For example, it has been shown that in patients with acute promyelocytic leukemia (PML) (Figure 63.20), those with MRD $\geq 1\times10^{-3}$ have 10-fold higher chance of relapse in 5 years than those with MRD $<1\times10^{-3}$.
- **Acute lymphoid leukemia.** Early clearance or low levels of MRD are favorable prognostic indicators in childhood ALL, whereas high levels of MRD at the end of induction therapy are often associated with a high-risk of relapse. Two main categories of targets are used for Q-PCR:
  - Gene fusion events, such as *BCR-ABL1* [t(9;22)(q34;q11.2)], *TCF3-PBX1* [t(1;19)(q23;p13.3)], and *ETV6-RUNX1* [t(12;21)(p13;q22)].
  - The clonal rearrangement of immunoglobulin (Ig) and T-cell receptor (TCR) genes.
- **Chronic lymphocytic leukemia.** Molecular detection of MRD in CLL is based on the identification of DNA sequences that are unique to CLL cells in a patient, such as rearranged immunoglobulin gene components including variable (V), diversity (D), and joining (J) regions. RQ-PCR techniques can generate very sensitive results by using specific primers for each CLL clone. Fortunately, somatic hypermutations are low (2%) in CLL, and therefore do not play a significant role in changing the primer binding sites which would cause PCR failure. It should be noted that these sequence-specific techniques are highly specialized and not available in most centers.

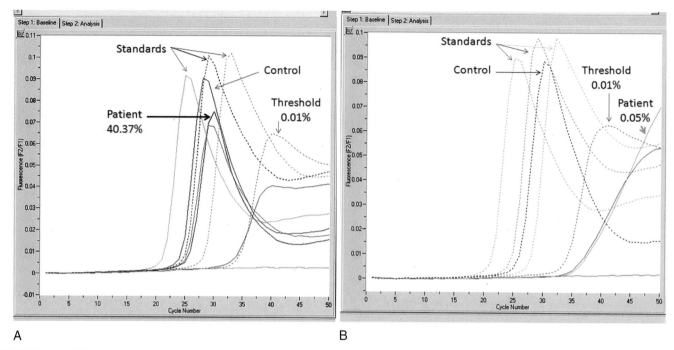

FIGURE 63.19 **RT-PCR STUDIES ON A PATIENT WITH CHRONIC MYELOGENOUS LEUKEMIA BEFORE (A) AND AFTER (B) THERAPY.** The level of *BCR-ABL* fusion is dramatically reduced from 40.37% (A) to 0.05% after therapy (B), but it is still above the threshold level of 0.01%, indicating minimal residual disease.
Courtesy of Jimin Xu, Ph.D., Department of Pathology and Laboratory Medicine, VA Greater Los Angeles Healthcare System.

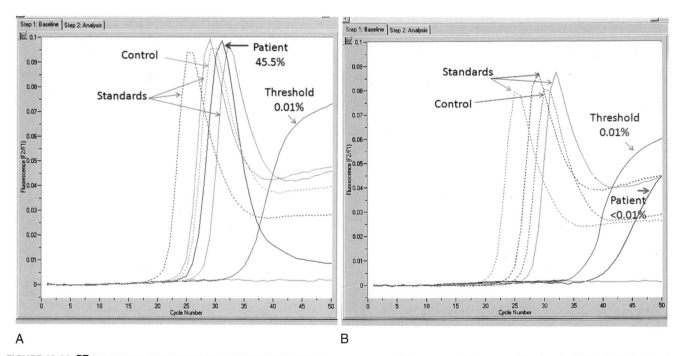

FIGURE 63.20 **RT-PCR STUDIES ON A PATIENT WITH ACUTE PROMYELOCYTIC LEUKEMIA BEFORE (A) AND AFTER (B) THERAPY.** The level of *PML-RARA* fusion is dramatically reduced from 45.5% (A) to <0.01% after therapy (B), below the threshold level.
Courtesy of Jimin Xu, Ph.D., Department of Pathology and Laboratory Medicine, VA Greater Los Angeles Healthcare System.

MULTIPARAMETRIC FLOW CYTOMETRY (MFC)

Recent technical advancement has enabled multiparametric assays, and 6–8 color MFC is now commonly applied in clinical practice. Multiparametric flow cytometry (MFC) is now considered, along with PCR analysis, the most important test for MRD detection. The molecular basis for identifying MRD by MFC is expression of abnormal or leukemia-associated immunophenotypes (LAIPs) by leukemic cells. It is also known as phenotypic aberrancies, which have been described previously in Chapter 17 (AML overview).

Using the approach of pattern recognition, a sensitivity level of 10^{-4} to 10^{-5} can be achieved by multiparametric assays. Results in the literature have demonstrated prognostic significance of MRD detected by MFC, and suggested that the level of MRD post-induction or post-consolidation may predict relapse-free survival and overall survival.

No universal panel or consensus approach is available in achieving the desired sensitivity level. However, the following are considerations that may be helpful in setting up the MRD assays and interpreting results by MFC.

- Design proper panels to maximize the frequency of common phenotypic aberrancies using the least number of tests.
- Set up at least 1 million cells per tube, and collect minimally 50–100 events per target population (leukemic cell population) utilizing live gate of target populations.
- Reference the original diagnostic histograms whenever possible for visual comparison of LAIPs.
- When phenotype switch occurs, look for other phenotypic aberrancies that may help distinguish abnormal from normal blasts.
- Report level of MRD as well as real-time sensitivity.

Figures 63.21 to 63.23 illustrate examples of MRD detection by MFC.

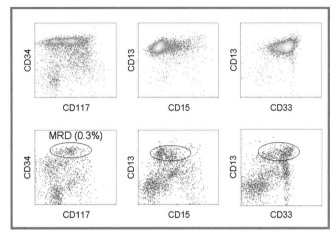

FIGURE 63.22 MRD OF ACUTE MYELOID LEUKEMIA DETECTED BY MFC. Compared with the diagnostic histograms of the pre-therapy marrow (upper panel, in blue), a small population of abnormal myeloblasts (0.3% of the total) is identified in the post-therapy sample (lower panel, in magenta), revealing similar phenotypic patterns as those seen previously with expression of CD13 (tight cluster), CD15 (partial), CD33 (heterogeneous), CD34 (bright and tight cluster), and CD117.

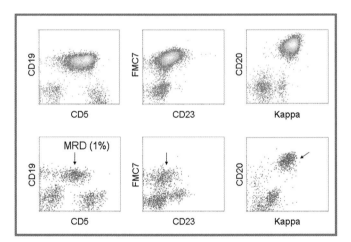

FIGURE 63.21 MINIMAL RESIDUAL DISEASE (MRD) OF MANTLE CELL LYMPHOMA DETECTED BY MFC. Compared with the original diagnostic histograms (upper panel, in blue) of mantle cell lymphoma, a small population of neoplastic B-cells (1% of the total) is detected in the post-therapy sample (lower panel, in magenta), displaying identical patterns with expression of CD5 (dim), CD19, CD20 (bright), CD23 (partial), FMC7, and surface kappa light chain restriction.

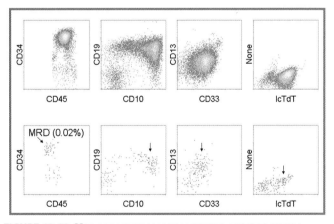

FIGURE 63.23 MRD OF B LYMPHOBLASTIC LEUKEMIA DETECTED BY MFC. Compared with the diagnostic histograms of the pre-therapy marrow (upper panel, in blue), a minute population of abnormal B lymphoblasts (0.02% of the total) is detected in the post-therapy sample (lower panel, in magenta), demonstrating identical phenotypic patterns to those observed previously with expression of CD10, CD19, CD34, intracellular TdT, plus aberrant CD13 (partial) and CD33 (partial).

Additional Resources

Al-Mawali A, Gillis D, Lewis I: The role of multiparameter flow cytometry for detection of minimal residual disease in acute myeloid leukemia, *Am J Clin Pathol* 131:16–26, 2009.

Bruggemann M, Raff T, Flohr T, et al: Clinical significance of minimal residual disease quantification in adult patients with standard-risk acute lymphoblastic leukemia, *Blood* 107:1116–1123, 2006.

Buccisano F, Maurillo L, Del Principe MI, et al: Prognostic and therapeutic implications of minimal residual disease detection in acute myeloid leukemia, *Blood* 119:332–341, 2012.

Campana D: Minimal residual disease in acute lymphoblastic leukemia, *Educ Prog Am Soc Hematol* 7–12, 2010.

Campana D: Role of minimal residual disease monitoring in adult and pediatric acute lymphoblastic leukemia, *Hematol Oncol Clin North Am* 23:1083–1098, 2009.

Campana D: Minimal residual disease in acute lymphoblastic leukemia, *Semin Hematol* 46:100–106, 2009.

Congdon CC: The destructive effect of radiation on lymphatic tissue, *Cancer Res* 26:1211–1220, 1966.

Guerrasio A, Pilatrino C, De Micheli D, et al: Assessment of minimal residual disease (MRD) in CBFbeta/MYH11-positive acute myeloid leukemias by qualitative and quantitative RT-PCR amplification of fusion transcripts, *Leukemia* 16:1176–1178, 2002.

Liu YJ, Grimwade G: Minimal residual disease evaluation in acute myeloid leukaemia, *Lancet* 360:160–162, 2002.

Shook D, Coustan-Smith E, Ribeiro RC, et al: Minimal residual disease quantitation in acute myeloid leukwmia, *Clin Lymphoma Myeloma* 9(Suppl 3):S281–S285, 2009.

Tobal K, Newton J, Macheta M, et al: Molecular quantitation of minimal residual disease in acute myeloid leukemia with t(8;21) can identify patients in durable remission and predict clinical relapse, *Blood* 95:815–819, 2000.

Uhrmacher S, Erdfelder F, Kreuzer KA: Flow cytometry and polymerase chain reaction-based analyses of minimal residual disease in chronic lymphocytic leukemia, *Adv Hematol* 1–11, 2010.

van der Velden V, Cazzaniga G, Schrauder A, et al: Analysis of minimal residual disease by Ig/TCR gene rearrangements: guidelines for interpretation of real-time quantitative PCR data, *Leukemia* 21:604–611, 2007.

Wang E, Hutchinson CB, Huang Q, et al: Donor cell-derived leukemias/myelodysplastic neoplasms in allogeneic hematopoietic stem cell transplant recipients: a clinicopathologic study of 10 cases and a comprehensive review of the literature, *Am J Clin Pathol* 135:525–540, 2011.

Wiseman DH: Donor cell leukemia: a review, *Biol Blood Marrow Transplant* 17:771–789, 2011.

Index

Note: Page numbers followed by "*f*", "*t*", and "*b*" refer to figures, tables, and boxes respectively.

A

ABL, *see* Acute basophilic leukemia
Abnormal localization of immature precursors (ALIP) 113*f*
Acanthocyte 677*f*
Acanthocytosis 696
ACD, *see* Anemia of chronic disease
aCML, *see* Atypical chronic myeloid leukemia
Acquired aplastic anemia
    clinical features 103–105
    molecular/cytogenetic studies 104–105, 104*f*
    morphology 103–104, 103*f*
Acquired immunodeficiency syndrome, *see* Human immunodeficiency virus
Acute basophilic leukemia (ABL)
    differential diagnosis 277
    immunophenotype 277
    molecular/cytogenetic studies 277
    morphology 276–277, 276*f*
    overview 276–277
Acute bilineal leukemia, *see* Mixed phenotype acute leukemia
Acute erythroid leukemia (AML-M6)
    classification 269–273
    differential diagnosis 273–274
    immunophenotype 271, 273*f*
    molecular/cytogenetic studies 273–276
    morphology
        erythroleukemia 270, 271*f*
        pure erythroid leukemia 271, 272*f*
Acute megakaryoblastic leukemia (AMKL)
    differential diagnosis 276
    immunophenotype 274–276, 275*f*
    molecular/cytogenetic studies 276
    morphology 273–274, 274*f*, 275*f*
    t(1;22)(p13;q13) 239–240, 239*f*, 240*f*
Acute monoblastic leukemia (AML-5a)
    classification 266–269
    differential diagnosis 269
    immunophenotype 267–268, 270*f*
    molecular/cytogenetic studies 268–269
    morphology 266–267, 267*f*
Acute monocytic leukemia (AML-5b)
    classification 266–269
    differential diagnosis 269
    immunophenotype 267–268
    molecular/cytogenetic studies 268–269
    morphology 266–267, 268*f*
Acute myeloid leukemia (AML)
    CEBPA mutation 242–243
    classification 219, 219*b*
    cytochemical stains
        myeloperoxidase 223, 223*f*
        naphthol AS-D acetate esterase 223
        naphthol AS-D chloroacetate 222*f*, 224
        α-napthyl butyrate esterase 223, 224*f*
        periodic acid–Schiff reaction 223, 223*f*
        Sudan Black B 223, 223*f*
    cytogenetics overview 225–226
    differential diagnosis 242*t*, 243
    Down syndrome
        differential diagnosis 288
        forms of leukemia 285–289
        immunophenotype 287–288, 287*f*
        molecular/cytogenetic studies 288, 288*f*
        morphology 286*f*, 287, 287*f*
    flow cytometry overview 223*f*, 224–225
    mastocytosis 198*f*
    minimal residual disease detection 725
    molecular studies 225
    morphology 220–222, 221*f*, 222*f*
    myelodysplasia-related changes
        differential diagnosis 249
        immunophenotype 246
        molecular/cytogenetic studies 246–249, 248*f*
        morphology 245–246, 245*f*, 247*f*, 248*f*
    myelodysplastic syndrome progression 130*t*
    not otherwise specified, *see also* Acute basophilic leukemia; Acute erythroid leukemia; Acute megakaryoblastic leukemia; Acute monoblastic leukemia; Acute monocytic leukemia; Acute myelomonocytic leukemia; Acute panmyelosis with myelofibrosis
    AML with maturation
        differential diagnosis 264–266
        immunophenotype 262–263, 264*f*
        molecular/cytogenetic studies 263, 264*f*
        morphology 262, 263*f*
    AML without maturation
        differential diagnosis 262
        immunophenotype 261
        molecular/cytogenetic studies 261–262, 262*f*
        morphology 261, 262*f*
    classification 260*t*
    hypoplastic acute myeloid leukemia 271, 279*f*
    minimally differentiated disease
        differential diagnosis 260
        immunophenotype 259–260
        molecular/cytogenetic studies 260, 261*f*
        morphology 259, 261*f*
    myeloid sarcoma 279, 280*f*
    NPM1 mutation 242
    t(3;3)(q21;q26.2)
        immunophenotype 239
        morphology 239
    t(6;9)(p23;q34)
        immunophenotype 238
        molecular/cytogenetic studies 238
        morphology 238
    t(8;21)(q22;q22)
        immunophenotype 228, 228*f*
        molecular/cytogenetic studies 228, 229*f*
        morphology 227–228, 228*f*
    t(8;16)(p11.2;p13.3) 240, 241*f*
    t(9;11)(p22;q23)
        immunophenotype 237–238
        molecular/cytogenetic studies 238, 238*f*
        morphology 237, 237*f*
    t(9;22)(q34;q11.2) 240
    t(16;16)(p13;q22)
        immunophenotype 235
        molecular/cytogenetic studies 235
        morphology 233–235
    therapy-related neoplasms
        chemotherapy or radiation therapy
            immunophenotype 253
            molecular/cytogenetic studies 253, 253*f*
            morphology 251–253, 252*f*
        clinicopathologic features 251*t*
        differential diagnosis 254
        topoisomerase inhibitor therapy
            immunophenotype 253
            molecular/cytogenetic studies 253–254, 256*f*
            morphology 253, 254*f*
Acute myelomonocytic leukemia (AML-M4)
    differential diagnosis 277
    immunophenotype 277
    molecular/cytogenetic studies 277
    morphology 277, 278*f*
    overview 263
Acute panmyelosis with myelofibrosis (APMF)
    differential diagnosis 277
    immunophenotype 277
    molecular/cytogenetic studies 277
    morphology 276–277, 276*f*
    overview 276–277
    overview 277
Acute promyelocytic leukemia (APL)
    overview 228–233
    t(5;17)(q35;q21.1);(NPM1;RARA) 233
    t(11;17)(q23;q21.1);(PLZF;RARA) 233
    t(11;17)(q13;q21.1);(NuMA;RARA) 233
    t(15;17)(q22;q21.1)

729

Acute promyelocytic leukemia (APL) (*Continued*)
  immunophenotype 231–233, 233f
  molecular/cytogenetic studies 233, 234f
  morphology
    hypergranular form 230f, 231
    hypogranular form 231, 232f
  t(15;17)(q22;q21.1);(PML;RARA) 233
Acute undifferentiated leukemia (AUL)
  classification 317f
  differential diagnosis 318
  immunophenotype 318, 318f
  molecular/cytogenetic studies 318
  morphology 317–318, 318f
Adipocyte, bone marrow cell characteristics 14, 14f
Adult T-cell leukemia/lymphoma (ATL)
  clinical features 531
  differential diagnosis 536
  immunophenotype 531–532, 535f
  molecular/cytogenetic studies 532–536, 535f, 536f
  morphology 531, 532f, 533f, 534f
Agammaglobulinemia 615
Aggressive natural killer cell leukemia
  clinical features 515–516
  differential diagnosis 519–522, 522t
  immunophenotype 515–516, 517f
  molecular/cytogenetic studies 516, 518f
  morphology 515, 515f, 516f, 517f
AIHA, *see* Autoimmune hemolytic anemia
AITL, *see* Angioimmunoblastic T-cell lymphoma
AITP, *see* Autoimmune thrombocytopenic purpura
ALCL, ALK+, *see* Anaplastic large cell lymphoma, anaplastic lymphoma kinase-positive
Alder-Reilly anomaly, granulocyte 663–664, 664f
ALIP, *see* Abnormal localization of immature precursors
ALK, *see* Anaplastic lymphoma kinase
Alkaline phosphatase 670f
Amegakaryocytosis
  clinical features 102
  molecular/cytogenetic studies 103
  morphology 102–103, 102f
AMKL, *see* Acute megakaryoblastic leukemia
AML, *see* Acute myeloid leukemia
AML-5a, *see* Acute monoblastic leukemia
AML-5b, *see* Acute monocytic leukemia
AML-M4, *see* Acute myelomonocytic leukemia
AML-M6, *see* Acute erythroid leukemia
Amyloidosis
  bone marrow 70, 71f
  classification 499t
  differential diagnosis 500t
  molecular/cytogenetic studies 498–500
  morphology and laboratory findings 498, 499f
  overview 498–500
Anaplastic large cell lymphoma, anaplastic lymphoma kinase-negative
  borderline cases 483
  clinical features 583
  differential diagnosis 583–584, 585t
  immunophenotype 583
  molecular/cytogenetic studies 583

morphology 583, 584f, 585f
Anaplastic large cell lymphoma, anaplastic lymphoma kinase-positive (ALCL, ALK+)
  borderline cases 483
  clinical features 577–583
  differential diagnosis 583–584, 585t
  immunophenotype 578, 579f, 580f, 582f
  inv(2)(p23q35) 578
  molecular/cytogenetic studies 578–583, 582f, 583t
  morphology 577–578, 578f, 579f, 582f
  t(1;2)(q25;p23) [TPM3-ALK] 578
  t(2;3) (p23;q21) [TFG-ALK] 578
  t(2;17)(p23;q23) [CLTC-ALK] 578
  t(X;2)(q11-12;p23) [MSN-ALK] 578
Anaplastic lymphoma kinase (ALK)-positive large B-cell lymphoma
  clinical features 456
  differential diagnosis 456
  immunophenotype 456
  molecular/cytogenetic studies 456
  morphology 456
Anemia of chronic disease (ACD) 703–704
Anemia, *see specific diseases*
Aneuploidy, cytogenetics 52, 52f, 53f
Angioimmunoblastic T-cell lymphoma (AITL)
  clinical features 569, 569t
  differential diagnosis 574
  immunophenotype 570–573, 573f
  molecular/cytogenetic studies 561, 574f
  morphology 569–570, 570f, 571f, 572f
Annexin-A1
  B-cell marker 34
  hairy cell leukemia 369f
APL, *see* Acute promyelocytic leukemia
APMF, *see* Acute panmyelosis with myelofibrosis
Arachidonic acid, platelet metabolism 713
Aspirin, platelet dysfunction 713
AT, *see* Ataxia telangiectasia
Ataxia telangiectasia (AT) 615–616
ATL, *see* Adult T-cell leukemia/lymphoma
Atypical chronic myeloid leukemia (aCML)
  diagnostic criteria 207b
  differential diagnosis 211, 211t
  molecular/cytogenetic studies 207
  morphology 207, 208f
  overview 207–208
Auer rods, granulocyte 664, 664f
AUL, *see* Acute undifferentiated leukemia
Autoimmune hemolytic anemia (AIHA)
  classification 699b
  clinical features 698–701, 699f
  cold-reacting antibodies 700, 700f
  differential diagnosis 700
  warm-reacting antibodies 699–700
Autoimmune thrombocytopenic purpura (AITP)
  clinical features 706–708
  differential diagnosis 708
  morphology 708

B

B-ALL/LBL, *see* B-lymphoblastic leukemia/lymphoma
Banding techniques 49
Basket cell 336f

Basophil, bone marrow cell characteristics 7
Basophilia 672
B-cell, markers
  CDs 25–33
  miscellaneous markers 33–34
B-cell lymphoma, unclassifiable
  features intermediate between diffuse large B-cell lymphoma and Burkitt lymphoma 469–470, 470f
  features intermediate between diffuse large B-cell lymphoma and classical Hodgkin lymphoma 462
B-cell prolymphocytic leukemia (B-PLL)
  classification 351
  differential diagnosis 354
  immunophenotype 352–353, 353f
  molecular/cytogenetic studies 353–354, 353f, 354f, 355f
  morphology 351–352, 351f, 352f
BCL-1 413f, 414f, 415f, 416f, 418f, 435f
BCL-2 364f, 408f, 451f, 452f, 474f
BCL-6 470f
Bernard–Soulier syndrome (BSS)
  clinical features 710–711
  differential diagnosis 710–711
  molecular studies 710
  morphology 710, 711f
Beta-lactam antibiotics, platelet dysfunction 713
Biphenotypic acute leukemia, *see* Mixed phenotype acute leukemia
BL, *see* Burkitt lymphoma
Blastic plasmacytoid dendritic cell (BPDC) neoplasm
  clinical features 653–655, 654f
  differential diagnosis 654–655, 658t
  immunophenotype 654, 655f, 656f, 657t
  molecular/cytogenetic studies 654, 658f
  morphology 653–654, 654f
Blood smear
  leukocyte morphology 15, 15f
  platelet morphology 15–16, 16f
  red blood cell morphology 14–15
  white blood cell counts 16t
B-lymphoblastic leukemia/lymphoma (B-ALL/LBL)
  clinical features 291
  differential diagnosis 305–306, 306f
  hyperdiploidy-associated disease
    immunophenotype 302, 305f
    molecular/cytogenetic studies 302, 303f
    overview 301–302
  hypodiploidy-associated disease
    immunophenotype 304
    molecular/cytogenetic studies 304, 304f
    overview 302–304
  morphology overview 291–294, 292f, 293f
  not otherwise specified
    immunophenotype 304, 305f
    molecular/cytogenetic studies 304–305, 305f
  recurrent genetic abnormalities
    immunophenotype 286f, 294
    molecular/cytogenetic studies 294–296, 295f, 296f
    overview 294–304, 294t
  t(1;19)(q23;p13.3)
    immunophenotype 300–301
    molecular/cytogenetic studies 301, 301f

t(5;14)(q31;q32)
    immunophenotype 301
    molecular/cytogenetic studies 301, 302f
t(5;14)(q31;q32)
    immunophenotype 301
    molecular/cytogenetic studies 301, 302f
t(12;21)(p13;q22)
    immunophenotype 300, 301f
    molecular/cytogenetic studies 300
t(12;21)(p13;q22)
    immunophenotype 300, 301f
    molecular/cytogenetic studies 300
t(v;11q23)
    immunophenotype 296–297, 297f
    molecular/cytogenetic studies 297–300, 297f, 298f, 299f
    morphology 296–300, 296f
BOB.1 442f, 463f
Bone, biopsy and repair 76, 77f
Bone marrow
    accessory cells 1f
    amyloidosis 70, 71f
    aplasia, see also specific diseases
        classification 105t
        differential diagnosis 108, 108t
    benign lymphoid aggregates 631–632, 631b, 632f, 632t
    cells, see also specific cells
        adipocyte 14, 14f
        basophil 7
        counts 6t
        dendritic cell 8, 9f
        endothelial cell 14
        eosinophil 7
        hematogone 11, 11f, 12f
        lymphocyte 10
        macrophage 8, 9f
        mast cell 8, 8f
        megakaryoblast 10, 10f
        metamyelocyte 7
        monocyte 8, 8f
        myeloblast 7
        myelocyte 6f, 7
        neutrophilic bands 7
        osteoblast 11–12, 13f
        osteoclast 12–14, 14f
        plasma cell 11, 12f, 13f
        prolymphocyte 11, 13f
        promyelocyte 6f, 7
        rubriblast 9, 9f, 10f
        segmented cells 7
    chemotherapy effects 720–721, 720f, 721f
    cytokine therapy effects 721–722, 722f
    examination
        biopsy sections 4
        clot sections 4
        glass slide preparations 4f, 5f
        smears 4–6, 6f
        touch preparation 6
    extracellular matrix components 2t
    fibrosis 76–77, 77b, 77f, 78f
    gelatinous transformation 69, 69f
    granuloma 70–74, 71f, 72f, 73b, 73f, 74f
    metastasis 74–76, 75f, 76f
    microvascular circulation 1–2, 4f
    monoclonal antibody therapy effects 722–725, 723t, 724f
    necrosis 70, 70f

previous biopsy site 76, 77f
radiation therapy effects 720–721, 721f
regulatory cytokines 1f, 3t
stromal changes 76–79
transplantation, see Hematopoietic stem cell transplantation
vascular changes 77, 78f
BPDC neoplasm, see Blastic plasmacytoid dendritic cell neoplasm
B-PLL, see B-cell prolymphocytic leukemia
Break-apart probe 59
BSS, see Bernard–Soulier syndrome
Burkitt lymphoma (BL), see also B-cell lymphoma, unclassifiable
    clinical features 465–469
    differential diagnosis 469, 469t
    immunophenotype 467, 467f
    molecular/cytogenetic studies 467–469, 468f, 469f
    morphology 465–467, 466f
Burr cell, see Echinocyte

C

Cabot ring 676, 678f
C-ALCL, see Primary cutaneous anaplastic large cell lymphoma
Cardiopulmonary bypass, platelet dysfunction 713
Castleman's disease, see also Large B-cell lymphoma arising in human herpesvirus-8-associated multicentric Castleman disease
    hyaline vascular type 83–86, 84f
    multicentric 85f
    plasma cell type 86, 86f
Cat-scratch disease, granulomatous lymphadenitis 92, 92f, 93f
CD1
    overview 34
    T-lymphoblastic leukemia/lymphoma 311f
CD1a, enteropathy-associated T-cell lymphoma 549f
CD2
    acute megakaryoblastic leukemia 275f
    adult T-cell leukemia/lymphoma 532f
    aggressive natural killer cell leukemia 517f, 518f
    anaplastic large cell lymphoma 582f, 585f
    angioimmunoblastic T-cell lymphoma 573f
    bilineal acute leukemia 321f
    chronic lymphoproliferative disorder of natural killer cells 514f
    chronic myelomonocytic leukemia 205f
    diffuse large B-cell lymphoma 436f
    enteropathy-associated T-cell lymphoma 543f, 549f
    extranodal natural killer cell/T-cell lymphoma, nasal type 520f
    hepatosplenic T-cell lymphoma 543f
    mastocytosis 195b
    mature natural killer cell neoplasm 506f
    overview 34
    peripheral T-cell lymphoma, not otherwise specified 589f
    primary cutaneous anaplastic large cell lymphoma 563f
    small lymphocytic lymphoma 340f

T-cell large granular lymphocytic leukemia 513f
T-cell polymphocytic leukemia 527f
T-lymphoblastic leukemia/lymphoma 311f, 312f
CD3
    acute undifferentiated leukemia 318f
    adult T-cell leukemia/lymphoma 532f, 535f
    aggressive natural killer cell leukemia 517f
    anaplastic large cell lymphoma 582f
    blastic plasmacytoid dendritic cell neoplasm 657f
    chronic lymphoproliferative disorder of natural killer cells 514f
    enteropathy-associated T-cell lymphoma 543f, 549f
    extranodal natural killer cell/T-cell lymphoma, nasal type 520f
    follicular lymphoma 405f, 406f
    hepatosplenic T-cell lymphoma 542f, 543f
    infectious mononucleosis 88f, 630f
    mantle cell lymphoma 414f, 415f, 416f
    mature natural killer cell neoplasm 506f
    mycosis fungoides 556f
    nodular lymphocyte predominant Hodgkin lymphoma 596f
    overview 34, 34f
    peripheral T-cell lymphoma, not otherwise specified 589f
    primary cutaneous anaplastic large cell lymphoma 563f
    primary cutaneous CD4+ small/medium T-cell lymphoma 566f
    primary effusion lymphoma 458f
    Sézary syndrome 558f
    subcutaneous panniculitis-like T-cell lymphoma 564f
    T-cell large granular lymphocytic leukemia 511f, 512f, 513f
    T-cell polymphocytic leukemia 527f
CD4
    adult T-cell leukemia/lymphoma 532f, 535f
    blastic plasmacytoid dendritic cell neoplasm 655f, 656f, 657f
    enteropathy-associated T-cell lymphoma 549f
    hepatosplenic T-cell lymphoma 543f
    mature B-cell neoplasms 329f
    mycosis fungoides 556f–553f
    overview 35
    peripheral T-cell lymphoma, not otherwise specified 589f
    primary cutaneous anaplastic large cell lymphoma 563f
    primary cutaneous CD4+ small/medium T-cell lymphoma 566f
    Sézary syndrome 558f
    T-cell/histiocyte-rich large B-cell lymphoma 441f
    T-cell large granular lymphocytic leukemia 512f
    T-cell polymphocytic leukemia 527f
CD5
    acute undifferentiated leukemia 318f
    adult T-cell leukemia/lymphoma 535f
    aggressive natural killer cell leukemia 517f
    anaplastic large cell lymphoma 582f
    angioimmunoblastic T-cell lymphoma 573f
    atypical CLL 346f

CD5 (*Continued*)
  B-prolymphocytic leukemia 353*f*
  diffuse large B-cell lymphoma 436*f*
  enteropathy-associated T-cell lymphoma 549*f*
  extranodal natural killer cell/T-cell
    lymphoma, nasal type 520*f*
  follicular lymphoma 405*f*
  germinotropic lymphoproliferative disorder
    461*f*
  hepatosplenic T-cell lymphoma 543*f*
  lymphoplasmacytic lymphoma 363*f*
  MALT lymphoma 393*f*
  mantle cell lymphoma 413*f*, 416*f*, 418*f*, 421*f*,
    423*f*
  mature B-cell neoplasms 329*f*
  minimal residual disease 727*f*
  mycosis fungoides 556*f*
  myelodysplastic syndrome 119*f*
  overview 26*t*–32*t*, 33, 35
  primary cutaneous anaplastic large cell
    lymphoma 563*f*
  Richter syndrome 347*f*
  Sézary syndrome 558*f*
  small lymphocytic lymphoma 339*f*
  small lymphocytic lymphoma 340*f*
  T-cell large granular lymphocytic leukemia
    512*f*, 513*f*
  T-cell polymphocytic leukemia 527*f*
  T-lymphoblastic leukemia/lymphoma 311*f*,
    312*f*
CD7
  acute erythroleukemia 273*f*
  acute megakaryoblastic leukemia 275*f*
  acute undifferentiated leukemia 318*f*
  adult T-cell leukemia/lymphoma 535*f*
  aggressive natural killer cell leukemia 517*f*
  anaplastic large cell lymphoma 582*f*
  angioimmunoblastic T-cell lymphoma 573*f*
  bilineal acute leukemia 321*f*
  chronic lymphoproliferative disorder of
    natural killer cells 514*f*
  chronic myelogenous leukemia 160*f*
  chronic myelomonocytic leukemia 205*f*
  Down syndrome-associated acute
    megakaryoblastic leukemia 287*f*
  enteropathy-associated T-cell lymphoma 549*f*
  extranodal natural killer cell/T-cell
    lymphoma, nasal type 520*f*
  hepatosplenic T-cell lymphoma 542*f*, 543*f*
  mature B-cell neoplasms 329*f*
  mature natural killer cell neoplasm 506*f*
  mycosis fungoides 556*f*
  myelodysplastic syndrome 112*f*, 119*f*, 139*f*
  overview 35
  peripheral T-cell lymphoma, not otherwise
    specified 589*f*
  Sézary syndrome 558*f*
  T-cell large granular lymphocytic leukemia
    512*f*, 513*f*
  T-cell polymphocytic leukemia 527*f*
  T-lymphoblastic leukemia/lymphoma 311*f*,
    312*f*
CD8
  adult T-cell leukemia/lymphoma 535*f*
  chronic lymphoproliferative disorder of
    natural killer cells 514*f*
  chronic lymphoproliferative disorder of
    natural killer cells 514*f*
  enteropathy-associated T-cell lymphoma 549*f*
  hepatosplenic T-cell lymphoma 543*f*
  mature B-cell neoplasms 329*f*
  myelodysplastic syndrome 112*f*
  overview 35
  T-cell large granular lymphocytic leukemia
    511*f*, 512*f*, 513*f*
  T-cell polymphocytic leukemia 527*f*
  T-lymphoblastic leukemia/lymphoma 311*f*,
    312*f*
CD10
  acute undifferentiated leukemia 318*f*
  angioimmunoblastic T-cell lymphoma 573*f*
  atypical CLL 346*f*
  bilineal acute leukemia 320*f*
  B-lymphoblastic leukemia/lymphoma with
    BCR-ABL1 fusion 295*f*
  B-lymphoblastic leukemia/lymphoma with
    hyperplody 305*f*
  B-lymphoblastic leukemia/lymphoma with
    t(4;11) 297*f*
  B-lymphoblastic leukemia/lymphoma with
    t(12;21) 301*f*
  Burkitt lymphoma 467*f*
  diffuse large B-cell lymphoma 435*f*, 436*f*
  follicular lymphoma 400*f*, 405*f*, 406*f*
  lymphoplasmacytic lymphoma 363*f*
  mantle cell lymphoma 421*f*
  nodal marginal zone lymphoma 386*f*
  overview 25, 26*t*–32*t*, 42–43
  peripheral T-cell lymphoma, not otherwise
    specified 589*f*
CD11, overview 39
CD11b
  acute monoblastic leukemia 270*f*
  acute myelomonocytic leukemia 236*f*
  chronic myelomonocytic leukemia 205*f*
  Down syndrome-associated acute
    megakaryoblastic leukemia 287*f*
CD11c
  hairy cell leukemia 371*f*, 372*f*
  splenic marginal zone lymphoma 379*f*
CD13
  acute monoblastic leukemia 270*f*
  acute myeloid leukemia 228*f*, 264*f*
  acute myelomonocytic leukemia 236*f*
  acute undifferentiated leukemia 318*f*
  B-lymphoblastic leukemia/lymphoma with
    BCR-ABL1 fusion 295*f*
  B-lymphoblastic leukemia/lymphoma with
    t(4;11) 297*f*
  chronic myelogenous leukemia 160*f*
  Down syndrome-associated acute
    megakaryoblastic leukemia 287*f*
  minimal residual disease 727*f*
  myelodysplastic syndrome 139*f*
  overview 36
CD14
  acute myelomonocytic leukemia 236*f*
  acute undifferentiated leukemia 318*f*
  B-lymphoblastic leukemia/lymphoma with
    hyperplody 305*f*
  chronic myelomonocytic leukemia 205*f*,
    206*f*
  overview 36
CD15
  acute monoblastic leukemia 270*f*
  acute myeloid leukemia 264*f*
  acute promyelocytic leukemia 233*f*
  bilineal acute leukemia 321*f*
  B-lymphoblastic leukemia/lymphoma with
    BCR-ABL1 fusion 295*f*
  B-lymphoblastic leukemia/lymphoma with
    t(4;11) 297*f*
  chronic myelomonocytic leukemia 205*f*, 206*f*
  classical Hodgkin lymphoma 602*f*, 603*f*,
    606*f*, 607*f*, 608*f*
  cytomegalovirus lymphadenitis 87*f*
  hepatosplenic T-cell lymphoma 543*f*
  minimal residual disease 727*f*
  myeloid sarcoma 280*f*
  overview 36
  paroxysmal nocturnal hemoglobinuria 107*f*
  primary effusion lymphoma 458*f*
CD16
  chronic lymphoproliferative disorder of
    natural killer cells 514*f*
  mature natural killer cell neoplasm 506*f*
  overview 35
  paroxysmal nocturnal hemoglobinuria
    deficiency 105*t*
  Sézary syndrome 558*f*
  T-cell large granular lymphocytic leukemia
    512*f*, 513*f*
CD19
  acute myeloid leukemia 228*f*
  acute undifferentiated leukemia 318*f*
  angioimmunoblastic T-cell lymphoma 573*f*
  atypical CLL 346*f*
  B-lymphoblastic leukemia/lymphoma with
    BCR-ABL1 fusion 295*f*
    hyperplody 305*f*
    t(4;11) 297*f*
    t(12;21) 301*f*
  B-prolymphocytic leukemia 353*f*
  Burkitt lymphoma 467*f*
  chronic myelogenous leukemia 160*f*
  diffuse large B-cell lymphoma 435*f*
  follicular lymphoma 405*f*
  hairy cell leukemia 371*f*
  lymphoplasmacytic lymphoma 363*f*
  MALT lymphoma 393*f*
  mantle cell lymphoma 421*f*
  mature B-cell neoplasms 329*f*
  minimal residual disease 727*f*
  myelodysplastic syndrome 119*f*
  nodal marginal zone lymphoma 386*f*
  overview 25
  primary effusion lymphoma 458*f*
  small lymphocytic lymphoma 340*f*, 341*f*
  splenic marginal zone lymphoma 379*f*
CD20
  atypical CLL 346*f*
  B-lymphoblastic leukemia/lymphoma with
    BCR-ABL1 fusion 295*f*
  B-lymphoblastic leukemia/lymphoma with
    hyperplody 305*f*
  blastic plasmacytoid dendritic cell neoplasm
    656*f*
  Burkitt lymphoma 467*f*
  Castleman's disease 84*f*
  classical Hodgkin lymphoma 608*f*
  diffuse large B-cell lymphoma 429*f*, 431*f*,
    433*f*, 435*f*, 436*f*
  follicular lymphoma 400*f*, 401*f*, 406*f*
  germinotropic lymphoproliferative disorder
    461*f*
  hairy cell leukemia 369*f*, 372*f*
  infectious mononucleosis 88*f*, 630*f*

intravascular large B-cell lymphoma 452*f*
lymphoplasmacytic lymphoma 363*f*, 364*f*
MALT lymphoma 393*f*
mantle cell lymphoma 413*f*, 414*f*, 415*f*, 416*f*, 418*f*, 420*f*, 421*f*, 423*f*
mature B-cell neoplasms 329*f*
nodal marginal zone lymphoma 386*f*
nodular lymphocyte predominant Hodgkin lymphoma 596*f*
overview 25–32
plasmablastic lymphoma 459*f*
post-transplant lymphoproliferative disorders 623*f*
primary cutaneous follicle center lymphoma 474*f*
primary cutaneous marginal zone lymphoma 476*f*
primary mediastinal (thymic) large B-cell lymphoma 451*f*
Richter syndrome 347*f*
small lymphocytic lymphoma 339*f*, 341*f*
T-cell/histiocyte-rich large B-cell lymphoma 440*f*, 441*f*
CD21
  B-prolymphocytic leukemia 353*f*
  Castleman's disease 84*f*
  follicular dendritic cell sarcoma 659*f*
  follicular lymphoma 405*f*
  nodular lymphocyte predominant Hodgkin lymphoma 597*f*
  overview 32–33
CD22
  atypical CLL 346*f*
  bilineal acute leukemia 321*f*
  B-lymphoblastic leukemia/lymphoma with hyperplody 305*f*
    t(4;11) 297*f*
    t(12;21) 301*f*
  B-prolymphocytic leukemia 353*f*
  chronic myelogenous leukemia 160*f*
  diffuse large B-cell lymphoma 435*f*, 436*f*
  follicular lymphoma 405*f*
  hairy cell leukemia 372*f*
  lymphoplasmacytic lymphoma 363*f*
  MALT lymphoma 393*f*
  mantle cell lymphoma 414*f*, 421*f*
  nodal marginal zone lymphoma 386*f*
  overview 33
  primary effusion lymphoma 458*f*
  small lymphocytic lymphoma 340*f*
  splenic marginal zone lymphoma 379*f*
CD23
  atypical CLL 346*f*
  B-prolymphocytic leukemia 353*f*
  follicular lymphoma 405*f*
  lymphoplasmacytic lymphoma 363*f*
  MALT lymphoma 393*f*
  mantle cell lymphoma 421*f*
  minimal residual disease 727*f*
  nodal marginal zone lymphoma 386*f*
  overview 33
  primary cutaneous follicle center lymphoma 474*f*
  Richter syndrome 347*f*
  small lymphocytic lymphoma 339*f*, 340*f*, 341*f*
  splenic marginal zone lymphoma 379*f*
CD24
  overview 33

paroxysmal nocturnal hemoglobinuria 105*t*, 107*f*
CD25
  adult T-cell leukemia/lymphoma 535*f*
  anaplastic large cell lymphoma 582*f*
  enteropathy-associated T-cell lymphoma 549*f*
  hairy cell leukemia 371*f*
  mastocytosis 195*b*
  mycosis fungoides 556*f*
  nodal marginal zone lymphoma 386*f*
  peripheral T-cell lymphoma, not otherwise specified 589*f*
  Sézary syndrome 558*f*
  splenic marginal zone lymphoma 379*f*
  T-cell polymphocytic leukemia 527*f*
CD26, overview 35
CD30
  anaplastic large cell lymphoma 579*f*, 580*f*, 585*f*
  classical Hodgkin lymphoma 602*f*, 603*f*, 604*f*, 606*f*, 607*f*
  cytomegalovirus lymphadenitis 87*f*
  enteropathy-associated T-cell lymphoma 543*f*
  overview 39
  post-transplant lymphoproliferative disorders 623*f*, 624*f*
  primary cutaneous anaplastic large cell lymphoma 563*f*
  primary effusion lymphoma 458*f*
  primary mediastinal (thymic) large B-cell lymphoma 451*f*
  T-cell/histiocyte-rich large B-cell lymphoma 442*f*
CD31, acute megakaryoblastic leukemia 275*f*
CD33
  acute monoblastic leukemia 270*f*
  acute myeloid leukemia 228*f*, 264*f*
  acute myelomonocytic leukemia 236*f*
  acute promyelocytic leukemia 233*f*
  acute undifferentiated leukemia 318*f*
  chronic myelogenous leukemia 160*f*
  chronic myelomonocytic leukemia 206*f*
  Down syndrome-associated acute megakaryoblastic leukemia 287*f*
  minimal residual disease 727*f*
  myelodysplastic syndrome 139*f*
  overview 36
  paroxysmal nocturnal hemoglobinuria 107*f*
CD34
  acute erythroleukemia 273*f*
  acute megakaryoblastic leukemia 275*f*
  acute myeloid leukemia 228*f*, 252*f*, 254*f*, 264*f*
  acute myelomonocytic leukemia 236*f*
  acute promyelocytic leukemia 233*f*
  acute undifferentiated leukemia 318*f*
  B-lymphoblastic leukemia/lymphoma with *BCR-ABL1* fusion 295*f*
  B-lymphoblastic leukemia/lymphoma with hyperplody 305*f*
  B-lymphoblastic leukemia/lymphoma with t(12;21) 301*f*
  B-lymphoblastic leukemia/lymphoma with t(4;11) 297*f*
  chronic myelogenous leukemia 160*f*
  chronic myelomonocytic leukemia 206*f*
  Down syndrome-associated acute megakaryoblastic leukemia 287*f*
  hypoplastic acute myeloid leukemia 279*f*

minimal residual disease 727*f*
myelodysplastic syndrome 138*f*
overview 38
CD35
  follicular dendritic cell sarcoma 659*f*
  overview 33
CD36
  acute erythroleukemia 273*f*
  acute megakaryoblastic leukemia 275*f*
  Down syndrome-associated acute megakaryoblastic leukemia 287*f*
  overview 38
CD38
  acute megakaryoblastic leukemia 275*f*
  acute monoblastic leukemia 270*f*
  acute myelomonocytic leukemia 236*f*
  B-prolymphocytic leukemia 353*f*
  Burkitt lymphoma 467*f*
  diffuse large B-cell lymphoma 435*f*
  Down syndrome-associated acute megakaryoblastic leukemia 287*f*
  follicular lymphoma 405*f*
  lymphoplasmacytic lymphoma 363*f*
  MALT lymphoma 393*f*
  overview 38
  plasma cell leukemia 492*f*
  plasma cell myeloma 492*f*
  primary effusion lymphoma 458*f*
CD41
  acute megakaryoblastic leukemia 275*f*
  acute undifferentiated leukemia 318*f*
  Down syndrome-associated acute megakaryoblastic leukemia 287*f*
  mantle cell lymphoma 421*f*
  overview 38
CD42, overview 38
CD43
  blastic plasmacytoid dendritic cell neoplasm 655*f*, 658*f*
  overview 39
CD45
  acute erythroleukemia 273*f*
  acute monoblastic leukemia 270*f*
  acute myeloid leukemia 228*f*, 264*f*
  acute myelomonocytic leukemia 236*f*
  acute promyelocytic leukemia 233*f*
  acute undifferentiated leukemia 318*f*
  aggressive natural killer cell leukemia 517*f*
  bilineal acute leukemia 320*f*, 321*f*
  B-lymphoblastic leukemia/lymphoma with *BCR-ABL1* fusion 295*f*
    hyperplody 305*f*
    t(4;11) 297*f*
    t(12;21) 301*f*
  Burkitt lymphoma 467*f*
  chronic lymphoproliferative disorder of natural killer cells 514*f*
  chronic myelogenous leukemia 160*f*
  chronic myelomonocytic leukemia 206*f*
  diffuse large B-cell lymphoma 435*f*
  Down syndrome-associated acute megakaryoblastic leukemia 287*f*
  enteropathy-associated T-cell lymphoma 549*f*
  extranodal natural killer cell/T-cell lymphoma, nasal type 520*f*
  minimal residual disease 727*f*
  myelodysplastic syndrome 139*f*
  overview 35
  plasma cell leukemia 492*f*

CD45 (*Continued*)
  plasma cell myeloma 492*f*
  primary effusion lymphoma 458*f*
  T-cell/histiocyte-rich large B-cell lymphoma 442*f*
  T-lymphoblastic leukemia/lymphoma 311*f*, 312*f*
CD48, paroxysmal nocturnal hemoglobinuria deficiency 105*t*
CD52
  adult T-cell leukemia/lymphoma 535*f*
  paroxysmal nocturnal hemoglobinuria deficiency 105*t*
CD55
  acute megakaryoblastic leukemia 275*f*
  overview 39
  paroxysmal nocturnal hemoglobinuria deficiency 105*t*
CD56
  acute monoblastic leukemia 270*f*
  acute myeloid leukemia 228*f*
  aggressive natural killer cell leukemia 517*f*
  blastic plasmacytoid dendritic cell neoplasm 655*f*, 656*f*, 657*f*
  chronic lymphoproliferative disorder of natural killer cells 514*f*
  chronic myelomonocytic leukemia 205*f*, 206*f*
  Down syndrome-associated acute megakaryoblastic leukemia 287*f*
  enteropathy-associated T-cell lymphoma 549*f*
  extranodal natural killer cell/T-cell lymphoma, nasal type 520*f*
  hepatosplenic T-cell lymphoma 542*f*, 543*f*
  mature natural killer cell neoplasm 506*f*
  myelodysplastic syndrome 119*f*
  overview 35–36
  plasma cell leukemia 492*f*
  plasma cell myeloma 492*f*
  Sézary syndrome 558*f*
  T-cell large granular lymphocytic leukemia 512*f*, 513*f*
CD57
  enteropathy-associated T-cell lymphoma 549*f*
  hepatosplenic T-cell lymphoma 543*f*
  nodular lymphocyte predominant Hodgkin lymphoma 597*f*
  overview 36
  Sézary syndrome 558*f*
  T-cell large granular lymphocytic leukemia 512*f*, 513*f*
CD58, paroxysmal nocturnal hemoglobinuria deficiency 105*t*
CD59
  acute megakaryoblastic leukemia 275*f*
  overview 39
  paroxysmal nocturnal hemoglobinuria 105*t*, 107*f*
CD61
  Down syndrome-associated acute megakaryoblastic leukemia 287*f*
  overview 38
CD64
  acute monoblastic leukemia 270*f*
  acute myeloid leukemia 264*f*
  acute myelomonocytic leukemia 236*f*
  acute undifferentiated leukemia 318*f*
  chronic myelomonocytic leukemia 205*f*, 206*f*
  overview 36–37

CD66, paroxysmal nocturnal hemoglobinuria deficiency 105*t*
CD68
  acute monocytic leukemia 269*f*
  dermatopathic lymphadenitis 93*f*
  hemophagocytosis 645*f*, 647*f*
  myeloid sarcoma 280*f*
  Niemann–Pick disease 639*f*
  overview 37
  sinus histiocytosis 90*f*
CD71
  acute erythroleukemia 273*f*
  acute megakaryoblastic leukemia 275*f*
  Down syndrome-associated acute megakaryoblastic leukemia 287*f*
  overview 37
CD73, paroxysmal nocturnal hemoglobinuria deficiency 105*t*
CD77, overview 33
CD79, overview 33
CD79a
  bilineal acute leukemia 320*f*
  diffuse large B-cell lymphoma 431*f*
CD88, overview 37
CD90
  overview 38
  paroxysmal nocturnal hemoglobinuria deficiency 105*t*
CD99, overview 38
CD103
  hairy cell leukemia 371*f*
  overview 33
CD110, overview 38
CD114, overview 37
CD115
  acute myeloid leukemia 228*f*
  overview 37
CD117
  acute erythroleukemia 273*f*
  acute megakaryoblastic leukemia 275*f*
  acute monoblastic leukemia 270*f*
  acute myeloid leukemia 247*f*, 252*f*, 264*f*
  acute myelomonocytic leukemia 236*f*
  acute promyelocytic leukemia 233*f*
  acute undifferentiated leukemia 318*f*
  B-lymphoblastic leukemia/lymphoma with *BCR-ABL1* fusion 295*f*
    hyperplody 305*f*
  bilineal acute leukemia 321*f*
  chronic myelogenous leukemia 160*f*
  chronic myelomonocytic leukemia 206*f*
  Down syndrome-associated acute megakaryoblastic leukemia 287*f*
  mastocytosis 192*f*, 195*b*, 195*f*, 198*f*
  minimal residual disease 727*f*
  myelodysplastic syndrome 138*f*
  overview 38
  panmyelosis 278*f*
  plasma cell leukemia 492*f*
  plasma cell myeloma 492*f*
CD120
  B-lymphoblastic leukemia/lymphoma with t(4;11) 297*f*
  t(12;21) 301*f*
  overview 37
CD123
  blastic plasmacytoid dendritic cell neoplasm 657*f*

overview 37
CD138
  angioimmunoblastic T-cell lymphoma 572*f*
  Castleman's disease 85*f*, 86*f*
  cutaneous plasmacytoma 478*f*
  lymphoplasmacytic lymphoma 364*f*
  MALT lymphoma 393*f*
  overview 26*t*–32*t*, 33
  plasma cell leukemia 487*f*, 492*f*
  plasma cell myeloma 488*f*, 492*f*
  plasmablastic lymphoma 461*f*
  plasmacytoma 496*f*, 498*f*
  post-transplant lymphoproliferative disorders 622*f*
  primary cutaneous marginal zone lymphoma 476*f*
CD157
  paroxysmal nocturnal hemoglobinuria deficiency 105*t*
  primary effusion lymphoma 457*f*
CD207, overview 37
CD235, overview 37
CD238, overview 37
CD240, overview 37
CD242, overview 37
CD246, overview 35
CDA, *see* Congenital dyserythropoietic anemia
CEBPA, acute myeloid leukemia mutation and features 242–243
Cell proliferation 50–52
CEL, NOS, *see* Chronic eosinophilic leukemia, not otherwise specified
CFU, *see* Colony forming unit
CGD, *see* Chronic granulomatous disease
CGH, *see* Comparative genomic hybridization
CHAD, *see* Cold hemagglutinin disease
Chédiak–Higashi granules 664
Chédiak–Higashi syndrome (CHS) 639–640, 639*f*, 640*f*, 641*f*
Chemotherapy, *see* Acute myeloid leukemia; Bone marrow; Myelodysplastic syndromes
Chromosomal aneuploidy 52, 52*f*, 53*f*
Translocations, *see* Chromosomes; Cytogenetics
Chromosomes, *see also* Cytogenetics, *specific aberrations and diseases*
  anaplastic large cell lymphoma 583*t*
  del 7q 365*f*
  del(13)(q12q22) 545*f*
  dup 15q 322*f*
  i(7)(q10) 101
  inv(2)(p23q35) 578
  inv(16)(p13;q22)
    immunophenotype 235
    molecular/cytogenetic studies 235, 236*f*, 237*f*
    morphology 233–235
  iso(7)(q10) 544*f*
  isolated del(5q) 121–122, 141*f*
  mature B-cell neoplasms 330*t*
  monosomies, *see specific monosomies*
  t(1;2)(q25;p23) [TPM3-ALK] 578
  t(1;19)(q23;p13.3)
    immunophenotype 300–301
    molecular/cytogenetic studies 301, 301*f*
  t(1;22)(p13;q13) 239–240, 239*f*, 240*f*
  t(2;3) (p23;q21) [TFG-ALK] 578
  t(2;9;22) 323*f*

t(2;17)(p23;q23) [CLTC-ALK] 578
t(3;3)(q21;q26.2)
  immunophenotype 239
  morphology 239
t(5;14)(q31;q32)
  immunophenotype 301
  molecular/cytogenetic studies 301, 302f
t(5;17)(q35;q21.1);(NPM1;RARA) 233
t(6;9)(p23;q34)
  immunophenotype 238
  molecular/cytogenetic studies 238
  morphology 238
t(8;16)(p11.2;p13.3) 240, 241f
t(8;21)(q22;q22)
  immunophenotype 228, 228f
  molecular/cytogenetic studies 228, 229f
  morphology 227–228, 228f
t(9;11)(p22;q23)
  immunophenotype 237–238
  molecular/cytogenetic studies 238, 238f
  morphology 237, 237f
t(9;22)(q34;q11.2) 240
t(11;17)(q13;q21.1);(NuMA;RARA) 233
t(11;17)(q23;q21.1);(PLZF;RARA) 233
t(11;19)(q23;p13) 322f
t(12;21)(p13;q22)
  immunophenotype 300, 301f
  molecular/cytogenetic studies 300
t(15;17)(q22;q21.1);(PML;RARA) 233
t(15;17)(q22;q21.1)
  immunophenotype 231–233, 233f
  molecular/cytogenetic studies 233, 234f
  morphology
    hypergranular form 230f, 231
    hypogranular form 231, 232f
t(16;16)(p13;q22)
  immunophenotype 235
  molecular/cytogenetic studies 235
  morphology 233–235
T-lymphoblastic leukemia/lymphoma 312t
trisomies, see specific trisomies
t(v;11q23)
  immunophenotype 296–297, 297f
  molecular/cytogenetic studies 297–300, 297f, 298f, 299f
  morphology 296–300, 296f
t(X;2)(q11-12;p23) [MSN-ALK] 578
Chronic eosinophilic leukemia, not otherwise specified (CEL, NOS)
  differential diagnosis 184
  immunophenotype 186
  molecular/cytogenetic studies 184
  morphology 185–186, 185f
  overview 184–187
Chronic granulomatous disease (CGD) 665–666
Chronic lymphocytic leukemia/small lymphocytic lymphoma (CLL/SLL)
  clinical features 335
  differential diagnosis 349, 349t
  immunophenotype and clinical correlations 338, 342f
  minimal residual disease detection 725
  molecular/cytogenetic studies 338–344, 340f, 342f, 343f, 344f
  morphology 336–338, 336f, 337f, 338f, 339f, 340f, 341f
  prognostic factors 335t

Richter syndrome 346–349, 347f, 348f, 349f
variants
  atypical chronic lymphocytic leukemia 344, 345f, 346f
  mu heavy-chain disease 344–346
Chronic lymphoproliferative disorder of natural killer cells (CLNK)
  clinical features 513–515
  differential diagnosis 519–522, 522t
  immunophenotype 514, 514f
  molecular/cytogenetic studies 514–515
  morphology 514, 514f
Chronic myelogenous leukemia (CML)
  cytogenetics 161–163, 162f, 163f, 164f
  differential diagnosis 163–164, 164t, 211, 211t
  flow cytometry 159–160, 160f
  immunohistochemistry 160
  minimal residual disease detection 725
  molecular studies 160–161, 161f
  morphology and laboratory findings
    accelerated phase 158, 158f
    blast phase 158–159, 159f
    overview 155–159, 156f, 157f
  stages 158t
Chronic myelomonocytic leukemia (CMML)
  cytogenetics 206–207, 206f
  diagnostic criteria 201b
  differential diagnosis 211, 211t
  flow cytometry 204–205, 205f, 206f
  immunohistochemistry 205–206, 206f
  morphology and laboratory findings 202–204, 202f, 203f, 204f, 205f
  overview 201–207
Chronic neutrophilic leukemia (CNL)
  diagnostic criteria 183b
  differential diagnosis 184
  molecular/cytogenetic studies 184, 185f
  morphology 183–184, 184f
  overview 183–184
Chronic renal failure
  anemia 703
  platelet dysfunction 713
CHS, see Chédiak–Higashi syndrome
Classical Hodgkin lymphoma, see Lymphocyte-depleted classical Hodgkin lymphoma; Lymphocyte-rich classical Hodgkin lymphoma; Mixed cellularity classical Hodgkin lymphoma; Nodular sclerosis classical Hodgkin lymphoma
CLL/SLL, see Chronic lymphocytic leukemia/small lymphocytic lymphoma
CLNK, see Chronic lymphoproliferative disorder of natural killer cells
CML, see Chronic myelogenous leukemia
CMML, see Chronic myelomonocytic leukemia
CMV, see Cytomegalovirus
CNL, see Chronic neutrophilic leukemia
Cold hemagglutinin disease (CHAD) 700
Cold-reacting antibodies 700, 700f
Colony forming unit (CFU)
  CFU-Baso 2–3
  CFU-E 2
  CFU-Eo 3–4
  CFU-G 2–3
  CFU-GEMM 2–4
  CFU-GM 2–3, 8

CFU-M 2–3
CFU-Meg 4
Comparative genomic hybridization (CGH) 59–60, 60f
Congenital dyserythropoietic anemia (CDA)
  differential diagnosis 684
  groups 684
  overview 680–684, 682t
  type I 680–682, 683f
  type II 682–683
  type III 683–684
Copper deficiency anemia 703
Cutaneous B-cell lymphoma, see Diffuse large B-cell lymphoma–leg type; Plasmacytoma; Primary cutaneous follicle center lymphoma; Primary cutaneous marginal zone lymphoma
Cutaneous γδ T-cell lymphoma
  clinical features 563–564
  differential diagnosis 565, 567t
  immunophenotype 562
  molecular/cytogenetic studies 562
  morphology 561–562, 564f
Cutaneous mastocytosis, see Mastocytosis
Cutaneous T-cell lymphoprolifative disorders, see Cutaneous γδ T-cell lymphoma; Epstein–Barr virus-associated hydroa vacciforme-like cutaneous lymphoma of childhood; Lymphomatoid papulosis; Mycosis fungoides; Primary cutaneous aggressive epidermotropic CD8+ cytotoxic T-cell lymphoma; Primary cutaneous anaplastic large cell lymphoma; Primary cutaneous CD4+ small/medium T-cell lymphoma; Sézary syndrome; Subcutaneous panniculitis-like T-cell lymphoma
Cytogenetics, see also Chromosomes, specific diseases
  acute myeloid leukemia, see Acute myeloid leukemia
  banding techniques 49
  cell preparation 48–49
  chromosome analysis
    aneuploidy 52, 52f, 53f
    balanced rearrangements 50–52, 51f
    loss of heterozygosity 53
    overview 49–53, 50f
  chronic myelogenous leukemia 161–163, 162f, 163f, 164f
  chronic myelomonocytic leukemia 206–207, 206f
  chronic neutrophilic leukemia 184, 185f
  essential thrombocytocythemia 179–180
  hairy cell leukemia 371–372
  historical perspective 47–48
  juvenile myelomonocytic leukemia 210–211, 211f
  lymphoplasmacytic lymphoma 359–360, 365f
  mastocytosis 194
  mature B-cell neoplasms 330–331, 330t, 331f
  minimal residual disease detection 725–727, 726f
  mucosa-associated lymphoid tissue lymphoma 394–395, 394f
  myelodysplastic neoplasms 151–154, 151t, 152f, 152t, 153f

Cytogenetics, *see also* Chromosomes, *specific diseases* (*Continued*)
  myelodysplastic syndromes 120–127, 120*f*, 121*f*, 122*f*, 123*f*, 124*f*
  nodal marginal zone lymphoma 383–385
  polycythemia vera 170, 170*f*
  splenic marginal zone lymphoma 376–379, 379*f*, 380*f*
Cytokines, *see specific cytokines*
Cytomegalovirus (CMV), lymphadenitis 87*f*

D

Dacrocyte, *see* Teardrop
DBA44 369*f*
DC, *see* Dendritic cell; Dyskeratosis congenita
Dendritic cell (DC)
  bone marrow cell characteristics 8, 9*f*
  CD markers 37, 649*t*
  disorders, *see* Blastic plasmacytoid dendritic cell neoplasm; Follicular dendritic cell sarcoma; Interdigitating dendritic cell sarcoma; Langerhans cell histiocytosis; Langerhans cell sarcoma
  subtypes 649*f*
Dermatopathic lymphadenitis 92, 93*f*
DGS, *see* DiGeorge syndrome
Diamond–Blackfan anemia
  clinical features 101–102, 101*f*
  molecular/cytogenetic studies 102
  morphology 101–102, 102*f*
Diffuse large B-cell lymphoma (DLBCL)
  classification 427, 427*b*
  differential diagnosis 432
  immunophenotype 429–430, 435*f*, 436*f*
  molecular/cytogenetic studies 430–432, 436*f*, 437*f*
  morphological variants 429, 431*f*, 432*f*, 433*f*, 434*f*
  morphology and laboratory findings 428–429, 428*f*, 429*f*, 430*f*
  subtypes, *see* Epstein–Barr virus-associated diffuse large B-cell lymphoma; Primary diffuse large B-cell lymphoma of the central nervous system; T-cell/histiocyte-rich large B-cell lymphoma
Diffuse large B-cell lymphoma associated with chronic inflammation
  clinical features 453–454
  differential diagnosis 454
  immunophenotype 453–454
  molecular/cytogenetic studies 454
  morphology 453, 454*f*
Diffuse large B-cell lymphoma-leg type
  clinical features 476–477
  differential diagnosis 477
  immunophenotype 476
  molecular/cytogenetic studies 476–477, 477*f*
  morphology 476, 477*f*
DiGeorge syndrome (DGS) 614–615, 615*f*, 616*f*
DLBCL, Diffuse large B-cell lymphoma
DNA microarray, principles 64–65, 65*f*
DNA sequencing
  chemical sequencing 65–66
  detection and analysis 66
  limitations 66–67, 66*f*
  next-generation sequencing 67
  overview 65–68

Dohle body, granulocyte 663, 663*f*
Dot blot, principles 64
Down syndrome
  acute myeloid leukemia
    differential diagnosis 288
    forms 285–289
    immunophenotype 287–288, 287*f*
    molecular/cytogenetic studies 288, 288*f*
    morphology 286*f*, 287, 287*f*
  chronic neutrophilic leukemia 185*f*
  transient abnormal myelopoiesis
    immunophenotype 283–285, 285*f*
    molecular/cytogenetic studies 285, 285*f*
    morphology 283, 284*f*
Drug-induced hemolytic anemia 685, 700*f*, 701
Dutcher body 11, 13*f*
Dyskeratosis congenita (DC)
  clinical features 100–101
  molecular/cytogenetic studies 100–101
  morphology 100

E

EBV-T/NK LPD, *see* Epstein–Barr virus-associated T-cell/natural killer cell lymphoproliferative disorder of children and young adults
Echinocyte 675, 677*f*, 696
Elliptocyte 696, 696*f*
EMA 596*f*
Emperipoiesis 644*f*
Endothelial cell, bone marrow cell characteristics 14
Enteropathy-associated T-cell lymphoma
  clinical features 547
  differential diagnosis 549
  immunophenotype 547, 549*f*, 550*f*
  molecular/cytogenetic studies 547–549
  morphology 547, 548*f*, 549*f*
Enumerating probe 57, 59
Eosinophil
  atypical granules 664, 664*f*
  bone marrow cell characteristics 7
Eosinophilia
  associated conditions 213*b*
  causes 672, 672*b*
EPO, *see* Erythropoietin
Epstein–Barr virus, *see* Epstein–Barr virus-associated diffuse large B-cell lymphoma; Germinotropic lymphoproliferative disorder; Infectious mononucleosis
Epstein–Barr virus-associated diffuse large B-cell lymphoma
  clinical features 440–444
  differential diagnosis 443
  immunophenotype 441–443
  molecular/cytogenetic studies 443
  morphology 441, 443*f*, 444*f*, 445*f*
Epstein–Barr virus-associated hydroa vacciforme-like cutaneous lymphoma of childhood 565, 567*f*
Epstein–Barr virus-associated T-cell/natural killer cell lymphoproliferative disorder of children and young adults (EBV-T/NK LPD) 518–519, 521*f*
Erdheim–Chester disease, reactive histiocytic proliferation 642, 644*f*

Erythropoietin (EPO), bone marrow effects 721–722
Essential thrombocytocythemia (ET)
  diagnostic criteria 177*b*
  differential diagnosis 180–182, 180*t*
  molecular/cytogenetic studies 179–180
  morphology and laboratory findings 177–179, 178*f*, 179*f*
  myelofibrosis criteria 178*b*
  overview 176–180
ET, *see* Essential thrombocytocythemia
Extramedullary hematopoiesis
  overview 21, 22*f*, 23*f*
  primary myelofibrosis 172–176, 175*f*, 176*f*
Extranodal natural killer cell/T-cell lymphoma, nasal type
  clinical features 516–518
  immunophenotype 517–518, 520*f*
  molecular/cytogenetic studies 518, 520*f*
  morphology 516–517, 519*f*, 520*f*

F

FA, *see* Fanconi anemia
Factor VIII 38, 156*f*, 171*f*
Faggot cell 231*f*
Fanconi anemia (FA)
  clinical features 99–100, 100*f*
  molecular/cytogenetic studies 100
  morphology 99–100, 100*f*
Fascin 442*f*, 606*f*, 607*f*
FDC, *see* Follicular dendritic cell
FGFR1 rearrangement
  immunophenotype 216
  molecular/cytogenetic studies 216
  morphology and laboratory findings 216
  overview 216–217
Fibrosis, *see also specific diseases*
  bone marrow 76–77, 77*b*, 77*f*, 78*f*
  myelodysplastic syndrome 144, 145*f*
FISH, *see* Fluorescence in situ hybridization
5q-myelodysplastic syndrome
  immunophenotype 137
  morphology 137, 140*f*, 141*f*
  overview 137
FL, *see* Follicular lymphoma
Flow cytometry, *see also* Immunophenotype, *specific diseases*
  acute myeloid leukemia, *see* Acute myeloid leukemia
  chronic myelogenous leukemia 159–160, 160*f*
  chronic myelomonocytic leukemia 204–205, 205*f*, 206*f*
  compensation 41, 41*f*
  data analysis 41, 41*f*, 42*f*
  gating 40–41, 40*f*
  instrumentation 33–34, 40*f*
  mature B-cell neoplasms 328–330, 329*f*, 329*t*
  minimal residual disease detection 727, 727*f*
  myelodysplastic syndromes 115–119, 119*f*
  paroxysmal nocturnal hemoglobinuria findings 106–107, 107*f*
  quality control 41
Fluorescence in situ hybridization (FISH), *see also specific diseases*
  formalin-fixed paraffin-embedded tissues 60, 61*f*

myelodysplastic syndromes 121f, 122f, 123f, 124f, 125–126
principles 55–62
probes 56–59, 56f, 57f, 58f
FMC7, B-cell marker 33
Folate deficiency 684, 685b
Follicular dendritic cell (FDC) 19–20, 20f
Follicular dendritic cell sarcoma
  clinical features 658–661
  differential diagnosis 658t, 660
  immunophenotype 660
  morphology 658–660, 659f
Follicular lymphoma (FL)
  clinical features 397
  differential diagnosis 405–407, 405f, 407t
  grading 403t, 404f
  immunophenotype 398–402, 405f
  intrafollicular neoplasia 403
  molecular/cytogenetic studies 402–403, 408f, 408t
  morphology
    bone marrow 402f
    lymph node 397f, 398f
    overview 397–398
    salivary gland 399f
    small intestine 401f
    stomach 400f
  pediatric lymphoma 403
  primary intestinal lymphoma 403
Fusion probe 57–59

G

Gaucher disease (GD)
  classification 636t
  clinical features 635–638
  differential diagnosis 636–638
  immunophenotype 636
  molecular studies 636
  morphology 636, 637f
G-CSF, see Granulocyte colony-stimulating factor
GD, see Gaucher disease
Gelatinous transformation, bone marrow 69, 69f
Germinotropic lymphoproliferative disorder
  clinical features 460–462
  morphology 454, 462f
Glanzmann thrombasthenia (GT)
  clinical features 711–712
  differential diagnosis 712
  immunophenotype 711
  molecular studies 711–712
  morphology 711
Gleevec, see Imatinib mesylate
Glucose-6-phosphate dehydrogenase deficiency 697–698, 698b, 698f
Granule deficiency
  neutrophil 664, 664f
  platelets
    dense granule deficiency 712
    α-granule deficiency 712
Granulocyte colony-stimulating factor (G-CSF), bone marrow effects 722, 722f
Granuloma, see also specific diseases
  bone marrow 70–74, 71f, 72f, 73b, 73f, 74f
  lymphadenitis
    cat-scratch disease 92, 92f, 93f
    Kikuchi's disease 94, 95f
    Kimura's disease 96, 96f
    sarcoidosis 91–92, 91f
    systemic lupus erythematosus 94, 95f
    toxoplasmosis 94, 94f
Granzyme B 511f, 518f
GT, Glanzmann thrombasthenia

H

Hairy cell 327f
Hairy cell leukemia (HCL)
  clinical features 367t
  differential diagnosis 373, 373t
  immunophenotype 370–371
  molecular/cytogenetic studies 371–372
  morphology 367–370, 368f, 369f, 370f
  variants 372–373, 372f
HCL, see Hairy cell leukemia
HE, see Hereditary elliptocytosis
Heavy chain deposition disease
  alpha heavy chain disease 500–501, 500f
  gamma heavy chain disease 501
  mu heavy chain disease 344–346, 501
Heinz body 641f, 679f, 698f
Hematogone, bone marrow cell characteristics 11, 11f, 12f
Hematopoiesis
  erythropoiesis 2
  extramedullary hematopoiesis, see Extramedullary hematopoiesis
  lymphopoiesis 4
  myelopoiesis 2–4
  overview 1, 2f
  thrombopoiesis 4
Hematopoietic stem cell transplantation
  bone marrow changes after transplant 716–719, 716f, 717f, 718f, 719f, 720f
  complications 684b
  indications 715b
Hemoglobin, see also Thalassemia syndromes
  genes 690f
  unstable hemoglobins 695
Hemoglobin A 691f, 692f
Hemoglobin C 691f, 692f, 694f
Hemoglobin D 691f, 692f
Hemoglobin H 679f
Hemoglobin S 691f
Hemolytic anemia, see also Autoimmune hemolytic anemia
  drug-induced hemolytic anemia 700f, 701
  hypersplenism 702
  infection 702
  thermal hemolysis 701, 702f
  transfusion reaction 701
  traumatic hemolysis 701, 702f
Hemolytic disease of the newborn 701
Hemolytic uremic syndrome (HUS) 708–710
Hemophagocytic histiocytosis, lymphadenopathy 91, 91f
Hemophagocytic lymphohistiocytosis (HLH)
  classification 644
  clinical features 643–647
  diagnostic criteria 647b
  differential diagnosis 646–647
  immunophenotype 646, 647f
  molecular studies 646
  morphology 643–647, 645f, 646f
Hemophagocytosis 646f, 647f

Hepatosplenic T-cell lymphoma
  clinical features 539
  differential diagnosis 545
  immunophenotype 539, 543f
  molecular/cytogenetic studies 539–545, 543f, 544f, 545f
  morphology 539, 540f, 541f, 542f
Hereditary elliptocytosis (HE) 696
Hereditary spherocytosis (HS) 695, 696f
Histiocyte 641f, 644f, 645f, 646f, 647f, see also Reactive histiocytic proliferations
Histiocytic sarcoma 646–647
HIV, see Human immunodeficiency virus
HL, see Hodgkin lymphoma
HLH, see Hemophagocytic lymphohistiocytosis
Hodgkin cell 601f, 602f, 605f
Hodgkin lymphoma (HL), see also Nodular lymphocyte predominant Hodgkin lymphoma
  classical Hodgkin lymphoma, see Lymphocyte-depleted classical Hodgkin lymphoma; Lymphocyte-rich classical Hodgkin lymphoma; Mixed cellularity classical Hodgkin lymphoma; Nodular sclerosis classical Hodgkin lymphoma
  classification 593–594
  epidemiology 593
  Reed–Sternberg cell 594f
  staging 593
  treatment 593
Horse-shoe cell 578f
Howell-Jolly body 676, 682, 686
HS, see Hereditary spherocytosis
HTLV-I, see Adult T-cell leukemia/lymphoma
Human cell differentiation molecules, see also specific CDs and cells
  overview 25–39
  table 26t–32t
Human herpesvirus-8, see Germinotropic lymphoproliferative disorder; Large B-cell lymphoma arising in human herpesvirus-8-associated multicentric Castleman disease
Human immunodeficiency virus (HIV)
  clinical features 617–619
  lymphoma association and classification 619
  molecular/cytogenetic studies 619
  morphology 617–618, 618f, 619f
  myelodysplasia 144
HUS, see Hemolytic uremic syndrome
Hypocellular myelodysplastic syndrome 142–144, 144f
Hypoplastic acute myeloid leukemia 271, 279f

I

IDA, see Iron deficiency anemia
IDC sarcoma, see Interdigitating dendritic cell sarcoma
Idiopathic CD4+ T-lymphocytopenia 619
Idiopathic hypereosinophilic syndrome (IHES) 185f
IHES, see Idiopathic hypereosinophilic syndrome
IM, see Infectious mononucleosis
Imatinib mesylate (Gleevec), bone marrow effects 722–725, 724f

Immunohistochemistry, see also
    Immunophenotype, specific diseases
  chronic myelogenous leukemia 160
  chronic myelomonocytic leukemia 205–206, 206f
  Langerhans cell histiocytosis 650–651, 652f
  myelodysplastic syndromes 119
  principles 42–43, 42f, 43f, 44f, 45f
Immunophenotype, see also Flow cytometry;
    Immunohistochemistry
  acute basophilic leukemia 277
  acute erythroid leukemia 271, 273f
  acute megakaryoblastic leukemia 274–276, 275f
  acute monoblastic leukemia 267–268, 270f
  acute monocytic leukemia
  acute myeloid leukemia
    Down syndrome 287–288, 287f
    myelodysplasia-related changes 246
    overview 223f, 224–225
    t(3;3)(q21;q26.2) 238–239, 239f
    t(6;9)(p23;q34) 238
    t(8;21)(q22;q22) 228, 228f
    t(9;11)(p22;q23) 237–238
    t(16;16)(p13;q22) 235
    therapy-related neoplasms 253
    topoisomerase inhibitor therapy 253
  acute myelomonocytic leukemia 277
  acute panmyelosis with myelofibrosis 277
  acute promyelocytic leukemia 231–233, 233f
  acute undifferentiated leukemia 318, 318f
  adult T-cell leukemia/lymphoma 531–532, 535f
  aggressive natural killer cell leukemia 515–516, 517f
  anaplastic large cell lymphoma
    anaplastic lymphoma kinase-negative 583
    anaplastic lymphoma kinase-positive 578, 579f, 580f, 582f
  anaplastic lymphoma kinase 456
  angioimmunoblastic T-cell lymphoma 570–573, 573f
  B-cell prolymphocytic leukemia 352–353, 353f
  blastic plasmacytoid dendritic cell neoplasm 654, 655f, 656f, 657t
  B-lymphoblastic leukemia/lymphoma
    hyperdiploidy-associated disease 302, 305f
    hypodiploidy-associated disease 304
    not otherwise specified 304, 305f
    recurrent genetic abnormalities 286f, 294
    t(1;19)(q23;p13.3) 300–301
    t(12;21)(p13;q22) 300, 301f
    t(5;14)(q31;q32) 301
    t(v;11q23) 296–297, 297f
  chronic eosinophilic leukemia, not otherwise specified 186
  chronic lymphocytic leukemia/small lymphocytic lymphoma 338, 342f
  chronic lymphoproliferative disorder of natural killer cells 514, 514f
  chronic myelogenous leukemia 159–160, 160f
  chronic myelomonocytic leukemia 204–206, 205f, 206f
  cutaneous γδ T-cell lymphoma 562
  diffuse large B-cell lymphoma associated with chronic inflammation 453–454
  diffuse large B-cell lymphoma overview 429–430, 435f, 436f
  diffuse large B-cell lymphoma–leg type 476
  Down syndrome transient abnormal myelopoiesis 283–285, 285f
  enteropathy-associated T-cell lymphoma 547, 549f, 550f
  Epstein–Barr virus-associated diffuse large B-cell lymphoma 441–443
  extranodal natural killer cell/T-cell lymphoma nasal type 517–518, 520f
  5q-myelodysplastic syndrome 137
  follicular dendritic cell sarcoma 660
  follicular lymphoma 398–402, 405f
  Gaucher disease 636
  Glanzmann thrombasthenia 711
  hairy cell leukemia 370–371
  hemophagocytic lymphohistiocytosis 646, 647f
  immunophenotype 267–268
  infectious mononucleosis 628–630, 630f
  intravascular large B-cell lymphoma 453, 453t
  Langerhans cell histiocytosis 650–651, 652f
  large B-cell lymphoma arising in human herpesvirus-8-associated multicentric Castleman disease 460
  lymphocyte-depleted classical Hodgkin lymphoma 608–609
  lymphocyte-rich classical Hodgkin lymphoma 607–608
  lymphomatoid granulomatosis 455
  lymphomatoid papulosis 562
  lymphoplasmacytic lymphoma 359, 363f, 364f
  mantle cell lymphoma 412, 421f
  mast cell 189t
  mastocytosis 194
  mature B-cell neoplasms 328–330, 329f, 329t
  mixed cellularity classical Hodgkin lymphoma 605, 607f
  mixed phenotype acute leukemia 320–321, 320f, 321f
  monoclonal gammopathy of undetermined significance 482–483
  mucosa-associated lymphoid tissue lymphoma 378f, 392
  mycosis fungoides 554–555, 556f
  myelodysplastic neoplasms 151
  myelodysplastic syndromes 115–119, 119f
  Niemann–Pick disease 639
  nodal marginal zone lymphoma 378f, 383
  nodular lymphocyte predominant Hodgkin lymphoma 597, 597f
  nodular sclerosis classical Hodgkin lymphoma 603, 604f
  paroxysmal nocturnal hemoglobinuria 106–107, 107f
  PDGFRA rearrangement 215
  PDGFRB rearrangement 216
  peripheral T-cell lymphoma, not otherwise specified 589, 589f
  plasma cell myeloma 487f, 488, 492f
  plasmablastic lymphoma 459, 461f
  primary cutaneous anaplastic large cell lymphoma 561, 563f
  primary cutaneous follicle center lymphoma 473
  primary cutaneous marginal zone lymphoma 475, 476f
  primary diffuse large B-cell lymphoma of the central nervous system 445
  primary effusion lymphoma 457, 458f
  primary mediastinal (thymic) large B-cell lymphoma 449–450
  refractory anemia with excess blasts 136–137, 139f
  refractory anemia with ringed sideroblasts 134
  refractory cytopenia with multilineage dysplasia 134–136
  refractory cytopenia with unilineage dysplasia (RCUD) 131–132, 132f
  Sézary syndrome 557, 558f
  splenic marginal zone lymphoma 376, 379f
  T-cell large granular lymphocytic leukemia 511–512, 512f, 513f
  T-cell polymphocytic leukemia 526, 527f
  T-cell/histiocyte-rich large B-cell lymphoma 439
  T-lymphoblastic leukemia/lymphoma 309–310, 311f, 311t
Immunosuppressants, lymphoproliferative disorder induction 625
Infectious mononucleosis (IM)
  clinical features 627–631
  differential diagnosis 631
  immunophenotype 628–630, 630f
  lymphadenopathy 88f
  molecular studies 630
  morphology 627–628, 629f
Interdigitating dendritic cell sarcoma
  clinical features 655–658
  differential diagnosis 657–658, 658t
  immunophenotype 657, 657f, 659f
  morphology 657
Interleukins
  T-helper cell secretion 20
  thrombopoiesis role 4
  types and functions 3t
Intravascular large B-cell lymphoma
  clinical features 451–453
  differential diagnosis 453
  immunophenotype 453, 453t
  molecular/cytogenetic studies 453, 453f
  morphology 452–453, 452f
Iron deficiency anemia (IDA)
  clinical features 688–689
  differential diagnosis 692–693, 693t
  laboratory findings 689t
  morphology 688–689, 689f
Isochromosome 7q 539

J

JAK2 mutation, see Essential thrombocytocythemia; Myelodysplastic neoplasms; Polycythemia vera; Primary myelofibrosis
JMML, see Juvenile myelomonocytic leukemia
Juvenile myelomonocytic leukemia (JMML)
  clinical features 208–211, 209f
  differential diagnosis 211, 211t
  molecular/cytogenetic studies 210–211, 211f
  morphology and laboratory findings 209–210, 210f

K

Kahler's disease, *see* Plasma cell myeloma
Ki-67 39, 443f, 461f, 467f, 470f, 477f, 498f, 550f, 624f
Kikuchi disease
    granulomatous lymphadenitis 94, 95f
    reactive histiocytic proliferation 642, 643f
Kimura's disease, granulomatous lymphadenitis 96, 96f
c-Kit, mutation in mastocytosis 190f

L

LAD, *see* Leukocyte adhesion deficiency
Langerhans cell histiocytosis (LCH)
    classification 650t
    clinical features 649–652
    differential diagnosis 652, 658t
    immunohistochemistry 650–651, 652f
    molecular/cytogenetic studies 651–652
    morphology 650, 650f, 651f, 652f
Langerhans cell sarcoma 652–653, 653f
Large B-cell lymphoma arising in human herpesvirus-8-associated multicentric Castleman disease
    clinical features 460
    differential diagnosis 460
    immunophenotype 460
    molecular/cytogenetic studies 460
    morphology 460
Large B-cell lymphoma, *see* Anaplastic lymphoma kinase-positive large B-cell lymphoma; B-cell lymphoma, unclassifiable; Diffuse large B-cell lymphoma; Diffuse large B-cell lymphoma associated with chronic inflammation; Germinotropic lymphoproliferative disorder; Intravascular large B-cell lymphoma; Large B-cell lymphoma arising in human herpesvirus-8-associated multicentric Castleman disease; Lymphomatoid granulomatosis; Plasmablastic lymphoma; Primary effusion lymphoma; Primary mediastinal (thymic) large B-cell lymphoma
LCH, *see* Langerhans cell histiocytosis
LDCHL, *see* Lymphocyte-depleted classical Hodgkin lymphoma
Leukocyte adhesion deficiency (LAD) 666–667, 666f
Leukoerythroblastosis 174f
Light chain deposition disease 500, 500t
LOH, *see* Loss of heterozygosity
Loss of heterozygosity (LOH) 47, 53
LP cell 597f, 598f
LPL, *see* Lymphoplasmacytic lymphoma
LRCHL, *see* Lymphocyte-rich classical Hodgkin lymphoma
Lymphadenopathies
    follicular pattern 81–87, 83t
    granulomatous lymphadenitis 91–92
    mixed pattern 92–97
    overview of patterns 81t
    paracortical pattern 87
    sinus pattern 88–91

Lymph node
    follicular structures 18–20, 19f, 21f
    medulla 20
    paracortex 20, 22f
    vascular structures 20
Lymphocyte, bone marrow cell characteristics 10
Lymphocyte-depleted classical Hodgkin lymphoma (LDCHL)
    clinical features 608–609
    differential diagnosis 609–610, 610t
    immunophenotype 608–609
    molecular/cytogenetic studies 609
    morphology 608, 609f
Lymphocyte-rich classical Hodgkin lymphoma (LRCHL)
    clinical features 607–608
    differential diagnosis 609–610, 610t
    immunophenotype 607–608
    molecular/cytogenetic studies 608
    morphology 607, 608f
Lymphocytopenia, associated conditions 627, 632t
Lymphocytosis, *see also* Infectious mononucleosis; Persistent polyclonal B-cell lymphocytosis
    associated conditions 627–633, 629f
    benign lymphoid aggregates in bone marrow 631–632, 631b, 632f, 632t
    stress-induced lymphocytosis 631
Lymphomatoid granulomatosis
    clinical features 454–456
    differential diagnosis 456, 456t
    immunophenotype 455
    molecular/cytogenetic studies 455–456
    morphology 455, 455f
Lymphomatoid papulosis (LyP)
    clinical features 561–562
    immunophenotype 562
    molecular/cytogenetic studies 562
    morphology 561–562, 564f
Lymphoplasmacytic lymphoma (LPL)
    clinical features 357
    differential diagnosis 360
    immunophenotype 359, 363f, 364f
    molecular/cytogenetic studies 359–360, 365f
    morphology and laboratory findings 357–358, 357f, 359f, 361f, 362f
    Waldenström's macroglobulinemia 357
LyP, *see* Lymphomatoid papulosis
Lysosomal storage diseases
    classification 636b
    Gaucher disease, *see* Gaucher disease
    Niemann–Pick disease, *see* Niemann–Pick disease

M

Macrocytosis
    anemias, *see specific diseases*
    causes 684b
Macrophage, bone marrow cell characteristics 8, 9f
Mantle cell lymphoma (MCL)
    clinical features 411
    differential diagnosis 419
    immunophenotype 412, 421f
    molecular/cytogenetic studies 412–417, 422f

    morphology 411–412, 412f, 413f, 414f, 415f, 416f, 419f, 420f
    variants 415f, 417t, 418f, 419f, 421f, 423f
Marginal zone, spleen 16–17, 17f, 18f
Mast cell
    bone marrow cell characteristics 8, 8f
    immunophenotype 189t
    mediators 189t
Mast cell leukemia 197
Mast cell sarcoma 197
Mastocytosis
    classification 194–198, 196b
    cutaneous mastocytosis 195f, 196
    diagnostic criteria 195b
    differential diagnosis 198–199, 199t
    immunophenotype 194
    molecular/cytogenetic studies 190f, 194
    morphology and laboratory findings 190f, 191–194, 191f, 192f, 193f, 194f, 198f, 199f
    overview 189
    systemic mastocytosis 196–198, 197f
Mature B-cell neoplasms
    classification 325b
    differential diagnosis 332–333
    frequency by type 326t
    immunophenotype 328–330, 329f, 329t
    molecular/cytogenetic studies 330–331, 330t, 331f
    morphology 326–328, 326f, 327f, 328f
    staging 325–326, 326t
May–Hegglin anomaly, granulocyte 663, 663f
MCCHL, *see* Mixed cellularity classical Hodgkin lymphoma
MCL, *see* Mantle cell lymphoma
MDN, *see* Myelodysplastic neoplasms
MDS, *see* Myelodysplastic syndromes
Mediastinal (thymic) large B-cell lymphoma, *see* Primary mediastinal (thymic) large B-cell lymphoma
Medulla, lymph node 20
MEFV 500f
Megakaryoblast, bone marrow cell characteristics 10, 10f
Megakaryocytic hypoplasia 705, 705f, 706f
Megakaryocytosis 705–706, 706f, 707f
Megaloblastic anemia, *see* Folate deficiency; Vitamin B12 deficiency
Metamyelocyte, bone marrow cell characteristics 7
Metastasis, bone marrow 74–76, 75f, 76f
Methotrexate, lymphoproliferative disorder induction 625
MF, *see* Mycosis fungoides
MGUS, *see* Monoclonal gammopathy of undetermined significance
Microcytic anemia
    differential diagnosis 692–693, 693t
    iron deficiency anemia
        clinical features 688–689
        laboratory findings 689t
        morphology 688–689, 689f
    thalassemia syndromes
        hemoglobin genes 690f
        α-thalassemia 692
        β-thalassemia 689t, 690–692, 690f, 691f
        rare variants 692
Micromegakaryocyte 135f, 143f, 156f, 245f

Minimal residual disease (MRD)
  flow cytometry 727, 727f
  molecular/cytogenetic studies 725–727, 726f
Mixed cellularity classical Hodgkin lymphoma (MCCHL)
  clinical features 604–607
  differential diagnosis 609–610, 610t
  immunophenotype 605, 607f
  molecular/cytogenetic studies 605
  morphology 605, 605f, 606f
Mixed phenotype acute leukemia (MPAL)
  biphenotypic acute leukemia 320f, 321, 321f
  classification 317f
  differential diagnosis 323
  immunophenotype 320–321, 320f, 321f
  morphology 319–320, 319f
  not otherwise specified 322
  recurrent genetic abnormalities 321–322, 322f, 323f
MLBCL, see Primary mediastinal (thymic) large B-cell lymphoma
MLL 256f
Monoclonal antibody therapy, bone marrow effects 722–725, 723t, 724f
Monoclonal gammopathy of undetermined significance (MGUS)
  clinical features 482–483, 482b
  immunophenotype 482–483
  morphology 482, 483f
Monocyte
  bone marrow cell characteristics 8, 8f
  CD markers 36–37
Monocytoid B-cell hyperplasia, lymphadenopathy 89f, 90
Monocytopenia 640
Monocytosis
  associated conditions 642b
  overview 640
Monosomy 7 140f, 211f
Monosomy 13 494f
Monosomy 20 140f
Monosomy Y 422f
Mott cell 11, 13f
MPAL, see Mixed phenotype acute leukemia
MPO, see Myeloperoxidase
MRD, see Minimal residual disease
Mucosa-associated lymphoid tissue (MALT) lymphoma
  clinical features 389
  differential diagnosis 395
  immunophenotype 378f, 392
  molecular/cytogenetic studies 394–395, 394f
  morphology 389–392, 390f, 391f, 392f, 393f
Multiple myeloma, see Plasma cell myeloma
MUM1 461f, 462f, 477f
Mycosis fungoides (MF)
  clinical features 553–557, 559t
  differential diagnosis 558–559, 567t
  immunophenotype 554–555, 556f
  molecular/cytogenetic studies 555–556, 557f
  morphology 553–554, 554f, 555f
  variants 556–557
Myeloblast 7, 640f
Myelocyte, bone marrow cell characteristics 6f, 7
Myelodysplastic neoplasms (MDN), see also specific diseases
  classification 129, 129t, 149
  cytogenetics 151–154, 151t, 152f, 152t, 153f
  immunophenotype 151
  JAK2 mutation diseases 167
  molecular studies 151
  morphology 149–151, 150f
Myelodysplastic syndromes (MDS), see also specific syndromes
  acute myeloid leukemia progression 130t
  chromosomal abnormalities 111t, 112t
  classification
    autoimmune myelodysplasia 144
    fibrosis type 144, 145f
    human immunodeficiency virus-associated myelodysplasia 144
    hypocellular myelodysplastic syndrome 142–144, 144f
    isolated del(5q) 121–122, 141f
    overview 129, 129t
    paraneoplastic myelodysplasia 144
    pediatric myelodysplastic syndrome 142, 143f
    refractory anemia with excess blasts 136–137
    refractory anemia with ringed sideroblasts 132–134
    refractory cytopenia with multilineage dysplasia 134–136
    refractory cytopenia with unilineage dysplasia 129–132
    secondary myelodysplastic syndrome 141–144
    toxicity and exposures 146
    unclassified 113f, 137–141, 142f
  cytogenetics 120–127, 120f, 121f, 122f, 123f, 124f
  differential diagnosis 146–148, 146t
  flow cytometry 115–119, 119f
  fluorescence in situ hybridization 121f, 122f, 123f, 124f, 125–126
  immunohistochemistry 119
  International Prognostic Scoring System 112t
  mastocytosis 198f
  molecular studies 119–120
  morphology
    dyserythropoiesis 114, 114f, 115f
    dysgranulopoiesis 114–115, 116f, 117f
    megakaryocyte and platelet abnormalities 115, 118f
    overview 112–115, 113f
  survival 130t
Myeloid sarcoma 279, 280f
Myelokathexis 669, 669f
Myelolipoma, extramedullary hematopoiesis 21, 23f
Myeloperoxidase (MPO)
  acute myeloid leukemia staining 223, 223f
  deficiency in granulocytes 665
Myelophthistic anemia 703

N

Naphthol AS-D acetate esterase, acute myeloid leukemia staining 223
Naphthol AS-D chloroacetate, acute myeloid leukemia staining 222f, 224
α-Napthyl butyrate esterase, acute myeloid leukemia staining 223, 224f
Natural killer (NK) cell, see also specific neoplasms cell
  aggressive natural killer cell leukemia
    clinical features 515–516
    immunophenotype 515–516, 517f
    molecular/cytogenetic studies 516, 518f
    morphology 515, 515f, 516f, 517f
  CD markers 35–36
  chronic lymphoproliferative disorder
    clinical features 513–515
    immunophenotype 514, 514f
    molecular/cytogenetic studies 514–515
    morphology 514, 514f
  Epstein–Barr virus-associated T-cell/natural killer cell lymphoproliferative disorder of children and young adults 518–519, 521f
  extranodal natural killer cell/T-cell lymphoma nasal type
    clinical features 516–518
    immunophenotype 517–518, 520f
    molecular/cytogenetic studies 518, 520f
    morphology 516–517, 519f, 520f
  morphology 505
  neoplasm classification 506, 507t
  types 482
Necrosis, bone marrow 70, 70f
Neonatal immune-mediated thrombocytopenia 708
Neutropenia
  agranulocytosis 667f
  causes
    bone marrow disorders 668
    drugs 667, 668b
    infection 667
    non-immune chronic idiopathic neutropenia 668
    overview 668b
    primary immune disorders 667–668
  congenital disease 669
Neutrophil
  functional abnormalities 665–667
  morphologic abnormalities
    Alder-Reilly anomaly 663–664, 664f
    Auer rods 664, 664f
    Chédiak–Higashi granules 664
    Dohle bodies 663, 663f
    granule deficiency 664, 664f
    May–Hegglin anomaly 663, 663f
    Pelger–Huet anomaly 664–665, 665f
    toxic granulation 663, 663f
Neutrophilia
  causes 669b
  primary neutrophilia 671
  reactive neutrophilia 669–671, 670f, 671f
  spurious neutrophilia 672
Niemann–Pick disease (NPD)
  classification 638t
  clinical features 638–639
  differential diagnosis 639
  immunophenotype 639
  molecular studies 639
  morphology 638–639, 638f, 639f
NK cell, see Natural killer cell
NLPHL, see Nodular lymphocyte predominant Hodgkin lymphoma
NMZL, see Nodal marginal zone lymphoma
Nodal marginal zone lymphoma (NMZL)
  clinical features 383

differential diagnosis 385–388, 387t
immunophenotype 378f, 383
molecular/cytogenetic studies 383–385
morphology 383, 384f, 385f
pediatric disease 385
Nodular lymphocyte predominant Hodgkin lymphoma (NLPHL)
clinical features 594–599
differential diagnosis 597–598, 598t
immunophenotype 597, 597f
molecular/cytogenetic studies 597, 598f
morphology 594–595, 595f, 596f
Nodular sclerosis classical Hodgkin lymphoma (NSCHL)
clinical features 601–604
differential diagnosis 609–610, 610t
immunophenotype 603, 604f
molecular/cytogenetic studies 603–604, 604f
morphology 601–603, 602f, 603f
Northern blot, principles 64
NPD, see Niemann–Pick disease
NPM1, acute myeloid leukemia mutation and features 242
NSCHL, see Nodular sclerosis classical Hodgkin lymphoma

O

Oct1 39
Oct2 39, 442f, 463f
Open canalicular system 711f
Osteoblast, bone marrow cell characteristics 11–12, 13f
Osteoclast, bone marrow cell characteristics 12–14, 14f
Osteopenia, associated conditions 76
Osteosclerosis, associated conditions 76
Osterosclerotic myeloma 497

P

p63 445f
Panmyelosis, see Acute panmyelosis with myelofibrosis
Pappenheimer body 676, 678f
Paracortex, lymph node 20, 22f
Paraimmunoblast 337f
Paraneoplastic myelodysplasia 144
Paroxysmal cold hemoglobinuria (PCH) 700
Paroxysmal nocturnal hemoglobinuria (PNH)
clinical features 105–108
deficient membrane proteins 105t
flow cytometry findings 106–107, 107f
molecular/cytogenetic studies 107–108
morphology 105–106, 106f
testing indications 106, 106b
Parvovirus B19, pure red cell aplasia 681f, 682f
PAS, see Periodic acid–Schiff reaction
Pax5 34, 441f, 445f, 463f, 585f, 624f
PCFCL, see Primary cutaneous follicle center lymphoma
PCH, see Paroxysmal cold hemoglobinuria
PCM, see Plasma cell myeloma
PCMZL, see Primary cutaneous marginal zone lymphoma
PCR, see Polymerase chain reaction
PDGFRA rearrangement
immunophenotype 215

morphology and laboratory findings 214–215, 214f, 215f
overview 213–215
PDGFRB rearrangement
immunophenotype 216
morphology and laboratory findings 215–216, 215f, 216f
overview 215–216
PEL, see Primary effusion lymphoma
Pelger–Huet anomaly, granulocyte 664–665, 665f
Periodic acid–Schiff reaction (PAS), acute myeloid leukemia staining 223, 223f
Peripheral T-cell lymphoma, not otherwise specified (PTCL, NOS)
differential diagnosis 589–590, 590t
immunophenotype 589, 589f
molecular/cytogenetic studies 589, 589f
morphology 587–589, 588f
prognostic groups 587
Pernicious anemia, see Vitamin B12 deficiency
Persistent polyclonal B-cell lymphocytosis 631
Philadelphia chromosome, see Chronic myelogenous leukemia
Plasma cell, see also specific neoplasms
bone marrow cell characteristics 11, 12f, 13f
lymphoplasmaytic lymphoma/ Waldenström's macroglobulinemia 357f, 358f, 360f
MALT lymphoma 391f, 393f
neoplasm classification 482b
post-transplant lymphoproliferative disorder 622f, 623f
primary cutaneous marginal zone lymphoma 476f
Plasma cell myeloma (PCM)
bone effects 485f
diagnostic criteria 484b
immunophenotype 487f, 488, 492f
molecular/cytogenetic studies 490–497, 493b, 493f, 494f, 494t, 495f
morphology 484f, 485–486, 488f, 489f, 490f
prognostic factors 485b
types
asymptomatic 483–485, 483b
non-secretory 484–485, 486f
plasma cell leukemia 485
symptomatic 483–484
Plasmablastic lymphoma
clinical features 459–460
differential diagnosis 459
immunophenotype 459, 461f
molecular/cytogenetic studies 459
morphology 459, 459f
Plasmacytoma
bone 497, 497b
cutaneous 477–478, 478f
extramedullary plasmacytoma 496f, 497, 498f
Platelet, see also specific disorders
CD markers 38
drug-induced disorders 713
dysfunction in pathological conditions 713–714
morphology 15–16, 16f
PMF, see Primary myelofibrosis
PNH, see Paroxysmal nocturnal hemoglobinuria
Poems syndrome, see Osterosclerotic myeloma

Polycythemia vera (PV)
cytogenetics 170, 170f
molecular studies 169–170, 169f
morphology and laboratory findings
overview 168–169
polycythemic phase 168, 168f
spent phase 168–169, 169f
overview 167–170
Polymerase chain reaction (PCR)
miscellaneous techniques 64
molecular studies, see specific diseases
primer design 62
principles 62–64, 63f
product analysis 63
quality control 62–63
real-time polymerase chain reaction 63
reverse transcriptase polymerase chain reaction 63
Post-transplant lymphoproliferative disorders (PTLD)
classification 621t
clinical features 621–625
molecular/cytogenetic studies 625, 625f
morphology
classical Hodgkin lymphoma type 625
early lesions 622–623, 622f
monomorphic type 623–625, 624f
polymorphic type 623, 623f
PRCA, see Pure red cell aplasia
Previous biopsy site, bone marrow 76, 77f
Primary cutaneous aggressive epidermotropic CD8+ cytotoxic T-cell lymphoma 565, 567t
Primary cutaneous anaplastic large cell lymphoma (C-ALCL)
clinical features 561
differential diagnosis 565, 567t
immunophenotype 561, 563f
molecular/cytogenetic studies 561
morphology 561, 562f
Primary cutaneous CD4+ small/medium T-cell lymphoma 565, 566f, 567t
Primary cutaneous follicle center lymphoma (PCFCL)
clinical features 473–474
differential diagnosis 473–474
immunophenotype 473
molecular/cytogenetic studies 473
morphology 473, 474f
Primary cutaneous marginal zone lymphoma (PCMZL)
clinical features 474–476
differential diagnosis 475
immunophenotype 475, 476f
molecular/cytogenetic studies 475
morphology 474–475, 475f
Primary diffuse large B-cell lymphoma of the central nervous system
clinical features 444–447
immunophenotype 445
molecular/cytogenetic studies 445–447
morphology 444, 445f, 446f
Primary effusion lymphoma (PEL)
clinical features 456–459
differential diagnosis 457, 458t
immunophenotype 457, 458f
molecular/cytogenetic studies 457
morphology 457, 457f, 458f

Primary mediastinal (thymic) large B-cell lymphoma (MLBCL)
  clinical features 449–451
  differential diagnosis 450, 452t
  immunophenotype 449–450
  molecular/cytogenetic studies 450, 451f
  morphology 449, 450f, 451f
Primary myelofibrosis (PMF)
  molecular/cytogenetic studies 176
  morphology and laboratory findings
    extramedullary hematopoiesis 172–176, 175f, 176f
    fibrotic stage 172, 172f, 173f, 174f
    overview 171–176
    prefibrotic stage 171–172, 171f, 172f
  overview 170–176
Prolymphocyte, bone marrow cell characteristics 11, 13f
Promyelocyte, bone marrow cell characteristics 6f, 7
Pseudofollicle 337f
Pseudo-Gaucher cell 157f, 724f
PTCL, NOS, see Peripheral T-cell lymphoma, not otherwise specified
PTLD, see Post-transplant lymphoproliferative disorders
Pure red cell aplasia (PRCA)
  differential diagnosis 680
  morphology 680, 680f
  parvovirus B19 infection 681f, 682f
  types and causes 679–680, 679b
PV, see Polycythemia vera
Pyruvate kinase deficiency 698

R

RA, see Rheumatoid arthritis
Radiation therapy, see Acute myeloid leukemia; Bone marrow; Myelodysplastic syndromes
RAEB, see Refractory anemia with excess blasts
RARA, acute promyelocytic leukemia translocations 233
RARS, see Refractory anemia with ringed sideroblasts
RCUD, see Refractory cytopenia with unilineage dysplasia
Reactive histiocytic proliferations
  autoimmune disease 641–642
  Erdheim–Chester disease 642, 644f
  iatrogenic causes 643
  infection 641
  Kikuchi disease 642, 643f
  Rosai-Dorfman disease 642–643, 644f
  sarcoidosis 642, 643f
  tumor association 642
Red blood cell
  anemia, see specific diseases
  counts in healthy adults 675t
  cytoskeleton structure and gene defects 695f, 697t
  inclusions 640f, 641f
  morphologic variants 637f, 638–639, 638f
  morphology 14–15
Red pulp, spleen 17–18, 17f, 18f
Reed–Sternberg cell 594f, 602f, 605f
Refractory anemia with excess blasts (RAEB)
  immunophenotype 136–137, 139f
  molecular/cytogenetic studies 137, 139f, 140f
  morphology 136, 137f, 138f
  overview 136–137
Refractory anemia with ringed sideroblasts (RARS)
  immunophenotype 134
  molecular/cytogenetic studies 134, 134f
  morphology 133–134, 133f
  overview 132–134
Refractory cytopenia with multilineage dysplasia (RCMD)
  immunophenotype 134–136
  molecular/cytogenetic studies 136, 136f
  morphology 134, 135f
  overview 134–136
Refractory cytopenia with unilineage dysplasia (RCUD)
  immunophenotype 131–132, 132f
  molecular/cytogenetic studies 132, 132f
  morphology 130–131, 130f, 131f
  overview 129–132
Rheumatoid arthritis (RA), lymphadenopathy 82, 82f, 83f
Richter syndrome 346–349, 347f, 348f, 349f
Rosai-Dorfman disease 89, 89f, 90f, 642–643, 644f
Rouleaux formation 484f, 492f, 700f
Rubriblast, bone marrow cell characteristics 9, 9f, 10f
Russell body 11, 13f, 255f, 489f

S

S-100 37, 653f
Sarcoidosis
  granulomatous lymphadenitis 91–92, 91f
  reactive histiocytic proliferation 642, 643f
SCD, see Sickle cell disease
Schistocyte 676, 677f
Schwachman–Diamond syndrome
  clinical features 101
  molecular/cytogenetic studies 101
  morphology 101
SCID, see Severe combined immunodeficiency syndromes
Severe combined immunodeficiency syndromes (SCID)
  gene mutations 614t
  overview 613
Sézary cell 526f, 555f, 558f
Sézary syndrome
  clinical features 557–558
  differential diagnosis 558–559
  immunophenotype 557, 558f
  molecular/cytogenetic studies 557–558, 558f
  morphology 557, 558f
Sickle cell disease (SCD)
  clinical features 693–695, 693f
  differential diagnosis 695
  morphology 694–695, 694f
Sinus histiocytosis, lymphadenopathy 88–89, 89f, 90f
SLE, see Systemic lupus erythematosus
Smudge cell 336f
SMZL, see Splenic marginal zone lymphoma
Southern blot, principles 64, 64f
Spleen
  marginal zone 16–17, 17f, 18f
  red pulp 17–18, 17f, 18f
  white pulp 16, 17f, 18f
Splenic diffuse red pulp small B-cell lymphoma 379–380
Splenic marginal zone lymphoma (SMZL)
  clinical features 375
  differential diagnosis 380–381, 381t
  immunophenotype 376, 379f
  molecular/cytogenetic studies 376–379, 379f, 380f
  morphology 375–376, 376f, 377f, 378f
  variants 372–373, 372f
Stomatocytosis 697
Subcutaneous panniculitis-like T-cell lymphoma 562–563, 567t
Sudan Black B, acute myeloid leukemia staining 223, 223f
Systemic lupus erythematosus (SLE), granulomatous lymphadenitis 94, 95f
Systemic mastocytosis, see Mastocytosis

T

T-ALL/LBL, see T-lymphoblastic leukemia/lymphoma
TAM, see Transient abnormal myelopoiesis
Target cell 676f
T-cell
  CD markers 34–35
  differentiation 503–504, 503f
  neoplasms, see also specific neoplasms
    classification 506, 507t
    cytogenetics 504f, 505f
    flow cytometry 504f
  receptor rearrangement 504f, 513f
T-cell/histiocyte-rich large B-cell lymphoma (THRLBCL)
  differential diagnosis 440, 442t
  immunophenotype 439
  molecular/cytogenetic studies 439–440
  morphology 439, 440f, 441f, 442f
T-cell large granular lymphocytic (T-LGL) leukemia
  clinical features 509–513
  differential diagnosis 519–522, 522t
  immunophenotype 511–512, 512f, 513f
  molecular/cytogenetic studies 512–513, 514f
  morphology 510–511, 510f, 511f
T-cell polymphocytic leukemia (T-PLL)
  clinical features 525
  differential diagnosis 529
  immunophenotype 526, 527f
  molecular/cytogenetic studies 526–529, 528f
  morphology 525–526, 525f, 526f
T-cell receptor-γδ (TCR-γδ) 513f, 539, 545
TCR-γδ, see T-cell receptor-γδ
TdT, see Terminal deoxynucleotidyl transferase
Teardrop 675, 682–683, 703
Telangiectasia macularis eruptiva perstans 196
Terminal deoxynucleotidyl transferase (TdT), precursor-associated marker 38–39
Thalassemia syndromes
  differential diagnosis 692–693, 693t
  hemoglobin genes 690f
  rare variants 692
  α-thalassemia 692
  β-thalassemia 689t, 690–692, 690f, 691f

THRLBCL, see T-cell/histiocyte-rich large B-cell lymphoma
Thrombocytopenia
  associated conditions 710
  autoimmune thrombocytopenic purpura
    clinical features 706–708
    differential diagnosis 708
    morphology 708
  hemolytic uremic syndrome 708–710
  neonatal immune-mediated thrombocytopenia 708
  post-transfusion purpura 708
  thrombotic thrombocytopenic purpura 708–710, 709f
Thrombopoietin (TPO) 4
Thrombotic thrombocytopenic purpura (TTP) 708–710, 709f
TIA-1 550f
T-LGL leukemia, see T-cell large granular lymphocytic leukemia
T-lymphoblastic leukemia/lymphoma (T-ALL/LBL)
  clinical features 309
  differential diagnosis 314
  immunophenotype 309–310, 311f, 311t
  molecular/cytogenetic studies 310–313, 312t, 313f
  morphology 309, 310f
  rare variants 313–314
Touton giant cell 644f
Toxic granulation, granulocyte 663, 663f
Toxoplasmosis, granulomatous lymphadenitis 94, 94f
T-PLL, see T-cell polymphocytic leukemia
TPO, see Thrombopoietin
Transfusion
  post-transfusion purpura 708
  hemolytic anemia 701
Transient abnormal myelopoiesis (TAM), Down syndrome
  morphology 283, 284f
  immunophenotype 283–285, 285f
  molecular/cytogenetic studies 285, 285f
Trisomy 3 380f
Trisomy 8 140f, 142f, 153f, 206f
Trisomy 9 153f
Trisomy 10 322f
Trisomy 15 134f
Trisomy 18 477f
Trisomy 21, see Down syndrome
TTP, see Thrombotic thrombocytopenic purpura

U

Urticaria pigmentosa 196

V

Vitamin B12 deficiency
  causes 685b
  differential diagnosis 687
  morphology 686–687, 686f, 687f
  overview 685
  syndromes 686

W

Waldenström's macroglobulinemia, see Lymphoplasmacytic lymphoma
Warm agglutinins, see Warm-reacting antibodies
Warm-reacting antibodies 699–700
WAS, see Wiscott–Aldrich syndrome
Whipple's disease, lymphadenopathy 90
White pulp, spleen 16, 17f, 18f
Wiscott–Aldrich syndrome (WAS) 613–614, 712

X

X-linked lymphoproliferative syndrome (XLP)
  clinical features 616–617b
  diagnostic criteria 616b
  molecular/cytogenetic studies 617
  morphology 616–617
XLP, see X-linked lymphoproliferative syndrome

Z

ZAP-70
  B-cell marker 34
  small lymphocytic lymphoma 342f